COOPERATION IN WIRELESS NETWORKS:
PRINCIPLES AND APPLICATIONS

Cooperation in Wireless Networks: Principles and Applications

Real Egoistic Behavior is to Cooperate!

Edited by

FRANK H.P. FITZEK
Aalborg University, Denmark

and

MARCOS D. KATZ
Samsung Electronics Co. Ltd., Korea

Springer

A C.I.P. Catalogue record for this book is available from the Library of Congress.

ISBN-10 1-4020-4710-X (HB)
ISBN-13 978-1-4020-4710-7 (HB)
ISBN-10 1-4020-4711-8 (e-book)
ISBN-13 978-1-4020-4711-4 (e-book)

Published by Springer,
P.O. Box 17, 3300 AA Dordrecht, The Netherlands.

www.springer.com

Printed on acid-free paper

Printed in the Netherlands.

To Lilith and Samuel.

ANTS
HUMAN
COOPERATION
TIT FOR TAT

Contents

List of Figures

List of Tables

Contributing Authors

Frank H.P. Fitzek Frank H. P. Fitzek is an Associate Professor in the Department of Communication Technology, University of Aalborg, Denmark heading the Future Vision group. He received his diploma (Dipl.-Ing.) degree in electrical engineering from the University of Technology - Rheinisch-Westfälische Technische Hochschule (RWTH) - Aachen, Germany, in 1997 and his Ph.D. (Dr.-Ing.) in Electrical Engineering from the Technical University Berlin, Germany in 2002 for quality of service support in wireless CDMA networks. As a visiting student at the Arizona State University he conducted research in the field of video services over wireless networks. He co-founded the start-up company acticom GmbH in Berlin in 1999. In 2002 he was Adjunct Professor at the University of Ferrara, Italy giving lectures on wireless communications and conducting research on multi-hop networks. In 2005 he won the YRP award for the work on MIMO MDC and in 2005 he received the Young Elite Researcher Award of Denmark. His current research interests are in the areas of 4G wireless communication networks, cross layer protocol design and cooperative networking.

Marcos D. Katz received the B.S. degree in Electrical Engineering from Universidad Nacional de Tucumán, Argentina in 1987, and the M.S. and Ph.D. degrees in Electrical Engineering from University of Oulu, Finland, in 1995 and 2002, respectively. He worked as a Research Engineer at Nokia Telecommunications from 1987 to 1995, designing analog circuits for high-speed PDH/SDH line interfaces. From 1995 to 2001 he was a Senior Research Engineer at Nokia Networks, Finland, where he developed multiple antenna techniques for several TDMA and CDMA research projects. In 2001-2002 he was a Research Scientist at the Centre for Wireless Communications, University of Oulu, Finland, where he concentrated on synchronization problems of CDMA networks. Since 2003 Dr. Katz has been working as a Principal Engineer at Samsung Electronics, Advanced Research Lab., Telecommunications R&D Center, Suwon, Korea. His current research interests include synchronization, multi-antenna, and cooperative techniques, as well as optical wireless communications for future 4G wireless communication systems. From the beginning of 2006 he is serving as the vice-chair of Working Group 5 (short-range communications) for the Wireless World Research Forum (WWRF).

Anders Brødløs Olsen received his M.Sc.E.E. degree in signal processing with specialization in Applied Signal Processing and Implementation from Aalborg University, Denmark, in 2002. From August 2002 he worked at the Center for Indlejrede Software Systemer (CISS), Aalborg University as research engineer, and from March 2004 he started towards the pursue of a Ph.D. degree at the department for Communication Technology at Aalborg University, His research interests are overall energy conservation, within implementation issues for dynamic resource optimization for mobile wireless systems, with main focus on Dynamic Voltage Scaling (DVS) methods.

Andrea J. Goldsmith is an associate professor of Electrical Engineering at Stanford University, USA, and was previously an assistant professor of Electrical Engineering at Caltech, USA. She has also held industry positions at Maxim Technologies and AT&T Bell Laboratories. Her research includes work on the capacity of wireless channels and networks, wireless communication and information theory, adaptive resource allocation in wireless networks, multiantenna wireless systems, energy-constrained wireless communications, wireless communications for distributed control, and cross-layer design for cellular systems, ad-hoc wireless networks, and sensor networks. She received the B.S., M.S., and Ph.D. degrees in Electrical Engineering from U.C. Berkeley. Dr. Goldsmith is a Fellow of the IEEE and of Stanford, and currently holds Stanford's Bredt Faculty Development Scholar Chair. She has received several awards for her research, including the National Academy of Engineering Gilbreth Lectureship, the Alfred P. Sloan Fellowship, the Stanford Terman Fellowship, the National Science Foundation CAREER Development Award, and the Office of Naval Research Young Investigator Award. She currently serves as editor for the Journal on Foundations and Trends in Communications and Information Theory and in Networks, and was previously an editor for the IEEE Transactions on Communications and for the IEEE Wireless Communications Magazine. Dr. Goldsmith is active in the IEEE Information Theory and Communications Societies and is an elected member of the Board of Governor for both societies. She also serves as Vice President of the Communication Theory Technical Committee of the Communications Society.

Andrej Stefanov received the B.S. degree in electrical engineering from Cyril and Methodius University, Skopje, Macedonia, in 1996. He received the M.S. and Ph.D. degrees in electrical engineering from Arizona State University, Tempe, AZ, in 1998 and 2001, respectively. During the summer 2000, he was with the Advanced Development Group, Hughes Network Systems, Germantown, MD. He joined the Electrical and Computer Engineering Department of the Polytechnic University, Brooklyn, NY, as an assistant professor in October 2001. His current research interests are in communication theory, wireless and mobile communications, wireless networks, channel coding and joint source-channel coding.

David Mazzarese received the Diplome d'Ingenieur in Electrical Engineering from ENSEA (Ecole Nationale Superieure de l'Electronique et de ses Applications), France, in 1998. In 1999, he joined TRLabs and the University of Alberta, Edmonton, Canada, in the Wireless Group directed by Dr. Witold Krzymien. He received the Ph.D. in Computer and Electrical Engineering from the University of Alberta in 2005 for the thesis entitled "High Throughput Downlink Wireless Packet Data Access with Multiple Antennas and Multi-User Diversity". Currently, David Mazzarese is a research engineer with Samsung Electronics in Suwon, South Korea, in the Global Standards and Research Lab, Telecommunications Network R&D Centre. He is currently participating in the IEEE 802.22 Working Group on Wireless Regional Area Networks. His research interests include multiuser MIMO systems, cooperative techniques and cognitive radios.

Desmond S. Lun received the B.Sc. and B.E. (Hons.) degrees from the University of Melbourne, Melbourne, Australia, in 2001, and the S.M. degree in 2002 in electrical engineering and computer science from the Massachusetts Institute of Technology (MIT), Cambridge, where he is currently working toward the Ph.D. degree in the Department of Electrical Engineering and Computer Science. His research interests include network coding and wireless communications.

H. T. Kung received his Ph.D. degree from Carnegie Mellon in 1974 and B.S. degree from National Tsing Hua University in Taiwan in 1968. He is currently William H. Gates Professor of Computer Science and Electrical Engineering at Harvard University. Since 1999, he has co-chaired a joint Ph.D. program with the Harvard Business School on information, technology and management. Prior to joining Harvard, he served on the faculty of Carnegie Mellon University from 1974 to 1992. Dr. Kung has pursued a variety of interests over his career, including complexity theory, database theory, systolic arrays, VLSI design, parallel computing, computer networks, and network security. He led numerous research projects on the design, implementation and experiment of novel computers and networks. In addition to his academic activities, he maintains a strong link with industry by serving as a consultant and board member to numerous companies. Dr. Kung is a member of National Academy of Engineering in US and Academia Sinica in Taiwan.

Halim Yanikomeroglu was born in Giresun, Turkey, in 1968. He received a B.Sc. degree in Electrical and Electronics Engineering from the Middle East Technical University, Ankara, Turkey, in 1990, and a M.A.Sc. degree in Electrical Engineering (now ECE) and a Ph.D. degree in Electrical and Computer Engineering from the University of Toronto, Canada, in 1992 and 1998, respectively. He was with the Research and Development Group of Marconi Kominikasyon A.S., Ankara, Turkey, from January 1993 to July 1994. Since 1998 Dr. Yanikomeroglu has been with the Department of Systems and Computer Engineering at Carleton University, Ottawa, where he is now an Associate Professor with tenure. His research interests include almost all aspects of wireless communications with a special emphasis on cellular multihop networks, radio resource management, and CDMA multi-antenna systems. At Carleton University, he teaches graduate courses on digital, mobile, and wireless communications. Dr. Yanikomeroglu has been involved in the steering committees and technical program committees of numerous international conferences in wireless communications; he has also given several tutorials in such conferences. He was the Technical Program Co-Chair of the IEEE Wireless Communications and Networking Conference 2004 (WCNC'04). He is an editor for IEEE Transactions on Wireless Communications, and a guest editor for Wiley Journal on Wireless Communications & Mobile Computing; he was an editor for IEEE Communications Surveys & Tutorials for 2002-2003. Currently he is serving as the Chair of the IEEE Communication Society's Technical Committee on Personal Communications (2005-06), and he is also a member of the Society's Technical Activities Council (2005-06). Dr. Yanikomeroglu is a member of the Advisory Committee for Broadband Communications and Wireless Systems (BCWS) Centre at Carleton University; he is also a registered Professional Engineer in the province of Ontario, Canada.

Hamid Aghvami is presently the Director of the Centre for Telecommunications Research at King's. He has published over 300 technical papers and given invited talks all over the world on various aspects of Personal and Mobile Radio Communications as well as giving courses on the subject world wide. He was Visiting Professor at NTT Radio Communication Systems Laboratories in 1990 and Senior Research Fellow at BT Laboratories in 1998-1999. He is currently Executive Advisor to Wireless Facilities Inc., USA and Managing Director of Wireless Multimedia Communications LTD. He leads an active research team working on numerous mobile and personal communications projects for third and fourth generation systems, these projects are supported both by the government and industry. He is a distinguished lecturer and a member of the Board of Governors of the IEEE Communications Society. He has been member, Chairman, Vice-Chairman of the technical programme and organising committees of a large number of international conferences. He is also founder of PIMRC & ICT. He is a fellow of the Royal Academy of Engineering, and fellow member of the IEEE and IEE.

Hideki Ochiai received the B.E. degree in communication engineering from Osaka University, Osaka, Japan, in 1996, and the M.E. and Ph.D. degrees in information and communication engineering from The University of Tokyo, Tokyo, Japan, in 1998 and 2001, respectively. From 2001 to 2003, he was with the Department of Information and Communication Engineering, The University of Electro-Communications, Tokyo, Japan. Since April 2003, he has been with the Department of Electrical and Computer Engineering, Yokohama National University, Yokohama, Japan, where he is currently an Associate Professor. From 2003 to 2004, he was a Visiting Scientist at the Division of Engineering and Applied Sciences, Harvard University, Cambridge, MA, USA. Dr. Ochiai was a recipient of a Student Paper Award from the Telecommunications Advancement Foundation in 1999 and the Ericsson Young Scientist Award in 2000.

István Z. Kovács received his B.Sc. from 'Politehnica' Technical University of Timisoara, Romania in 1989, his M.Sc.E.E. from The Franco-Polish School of New Information and Communication Technologies/ Ecole Nationale Supérieure des Télécommunications de Bretagne, Poland/France in 1996, and his Ph.D.E.E. in Wireless Communications from Aalborg University, Denmark in 2002. From 2002 to August 2005 he held the position of assistant research professor with the Center For TeleInFrastruktur, Department of Communication Technology of Aalborg University, in the Antennas and Propagation division. His research interests have been in the field of radio channel propagation measurements and modeling, with major focus on short-range ultra-wideband radio channel and ultra-wideband antenna investigations. He has been actively involved in the European IST PACWOMAN and IST MAGNET projects and participated in several industrial projects with partners such as TeleDanmark, Motorola, IOSpan, and ArrayComm. He has made a number of paper contributions and has contributed to two book chapters on UWB propagation topics. Currently István Z. Kovács holds a position as a Wireless Networks Specialist in Network Systems Research/Nokia Networks, Aalborg, Denmark conducting research in the area of 3G/3.9G wireless networks.

J. Nicholas Laneman received B.S. degrees (summa cum laude) in Electrical Engineering and in Computer Science from Washington University, St. Louis, MO, in 1995. At the Massachusetts Institute of Technology (MIT), Cambridge, MA, he earned the S.M. and Ph.D. degrees in Electrical Engineering in 1997 and 2002, respectively. Since 2002, Dr. Laneman has been on the faculty of the Department of Electrical Engineering, University of Notre Dame, where his current research interest lie in wireless communications and networking, information theory, and detection & estimation theory. From 1995 to 2002, he was affiliated with the Department of Electrical Engineering and Computer Science and the Research Laboratory of Electronics, MIT, where he held a National Science Foundation Graduate Research Fellowship and served as both a Research and Teaching Assistant. During 1998 and 1999 he was also with Lucent Technologies, Bell Laboratories, Murray Hill, NJ, both as a Member of the Technical Staff and as a Consultant, where he developed robust source and channel coding methods for digital audio broadcasting. His industrial interactions have led to five U.S. patents. Dr. Laneman received the MIT EECS Harold L. Hazen Teaching Award in 2001 and the ORAU Ralph E. Powe Junior Faculty Enhancement Award in 2003. He is a member of IEEE, ASEE, and Sigma Xi.

Janne Riihijärvi currently works as a senior project manager (the GOLLUM project) and research assistant at the RWTH Aachen University in the Wireless Networks Research Group (MobNets). Before joining Aachen University he worked in various networking research projects at the University of Oulu, where he received his M.Sc. degree from in 2002, and at VTT Electronics. His current research includes studying networking aspects of small mobile electronic devices and theoretical aspects of large-scale networks.

Jerry C. H. Lin received the B.S. degree in electrical engineering and another B.S. degree (First Class Honors) with double major in pure and applied mathematics from University of Calgary, Calgary, Alberta, Canada in 2002 and 2003, respectively. He received the M.S. degree in electrical engineering from Polytechnic University, Brooklyn, New York, in 2005.

Joseph Mitola III is a consulting scientist with The MITRE Corporation. He published the first paper on software radios back in 1992. Dr. Mitola not only coined the widely used terms “software radio” and its evolutionary development, “cognitive radio”, but he created the fundamental associated technical framework to make these enabling technologies one of the pillars of future wireless communications. He was elected first chair of the global Software-Defined Radio Forum in 1996, and continues to foster dual use military-civilian radio technology. His current research interests center on enhancing the computational intelligence of software radios enabling future cognitive radios. Prior to joining Thc MITRE Corporation in 1993, he was chief scientist of electronic systems, E-Systems Melpar Division (now Raytheon Falls Church), where he was responsible for research and technology development for communications and electronic warfare systems. He holds a Bachelor of Science in electrical engineering from Northeastern University (Highest Honors), a Master's in engineering degree from the Johns Hopkins University, and a licentiate and doctorate in engineering from the Royal Institute of Technology (KTH). Stockholm, Sweden (June 2000). Ever since its introduction almost 15 years ago Dr. Mitola has been a prolific contributor, educator and writer on software radios, and more recently on cognitive radios. He is the author of the books "Software Radio Architecture", Wiley, 2000, and "Cognitive Radio Architecture", Wiley, 2006 and co-editor of Software Radio Technologies: Selected Readings, Wiley-IEEE press, 2001. In addition Dr. Mitola has served as guest editor of several IEEE Communication Magazines, and he has written numerous technical papers on his subject.

Kathiravetpillai Sivanesan received the B.Sc. in Electrical and Electronic Engineering from University of Peradeniya, Sri Lanka in 1996. He received his M.Phil. in Telecommunications from University of Hong Kong in 2000. Then, he joined Icore wireless communication laboratory at University of Alberta, Canada and obtained his Ph.D. in 2005 under the supervision of Prof. N.C. Beaulieu. Currently, he is with Samsung Electronics as a research engineer at Suwon, South-Korea. His research interest lies in the general areas of communication theory and statistical signal processing, more particularly cooperative techniques, multiple antenna systems and interference cancellation.

Keivan Navaie received his B.Sc. degree from Sharif University of Technology, Tehran, Iran, his M.Sc. degree from the University of Tehran, Tehran, Iran, and his Ph.D degree from Tarbiat Modarres University, Tehran, Iran, all in Electrical Engineering in 1995, 1997 and 2004 respectively. From August 2002 to March 2004, he was with Bell University Laboratories, University of Toronto. From March to November 2004, he was with the School of Mathematics and Statistics, Carleton University, Ottawa, Canada, as a Postdoctoral Research Fellow. He is currently with the Broadband Communication and Wireless System Centre, System and Computer Engineering Department, Carleton University, Ottawa, Canada. His research interests are mainly in the radio resource management issues of cellular and multi-hop wireless networks and traffic modelling. Dr. Navaie is a member of the IEEE.

Konrad Wrona is currently a Principal Investigator in Security & Trust at SAP Research Lab in Sophia Antipolis, France. He was born in Warsaw, Poland. He was awarded MSc in Telecommunications from Warsaw University of Technology, Poland in 1997 and PhD in Electrical Engineering from RWTH Aachen University, Germany in 2005. He has over 8 years of work experience in industrial (Ericsson Research and SAP Research) and academic (Media Lab Europe and RWTH Aachen) research and development environment. He is author and co-author of over 20 publications and several patents. His research interests include security in communication networks and distributed systems, secure wireless applications, and electronic commerce.

Marina Petrova is currently working as a senior project manager and research assistant at the RWTH Aachen University in the Wireless Networks Research Group. She has a degree in electronics and telecommunication engineering from the University St. Cyril and Methodius, Skopje, Macedonia. She was also a project manager for Aachen University in the European 6HOP project for heterogeneous multihop wireless networks. Her main research interest are in the areas of ad hoc wireless networks, cognitive communication systems, and service discovery.

Mischa Dohler obtained his MSc degree in Telecommunications from King's College London in 1999, and his Diploma in Electrical Engineering from Dresden University of Technology, Germany, in 2000. He has been lecturer at the Centre for Telecommunications Research, King's College London, until June 2005. He is now in the R&D department of France Télécom working on embedded and future communication systems. Prior to Telecommunications, he studied Physics in Moscow. He has won various competitions in Mathematics and Physics, and participated in the 3rd round of the International Physics Olympics for Germany. He has published numerous research papers and holds several patents. He is a member of the IEEE, has been the Student Representative of the IEEE UKRI Section and a member of the Student Activity Committee of IEEE Region 8. He has also been the London Technology Network Business Fellow for King's College London. He has given five international short-courses, two on UMTS at WPMC02 & ATAMS02 and three on distributed cooperative systems at VTC Spring 2004, COST273 & VTC Spring 2006.

Morten Holm Larsen was born in Denmark, on the 1'st of September 1978. He received the M.Sc. degree in electrical engineering from Aalborg University, Aalborg, Denmark, in 2004. He is currently an PhD student at Aalborg University in the Department of Communication Technology. His main research interests are Multiple Description, data compression and digital communication theory.

Muriel Médard is a Harold E. and Esther Edgerton Associate Professor in the Electrical Engineering and Computer Science at MIT and the Associate Director of the Laboratory for Information and Decision Systems. She was previously an Assistant Professor in the Electrical and Computer Engineering Department and a member of the Coordinated Science Laboratory at the University of Illinois Urbana-Champaign. From 1995 to 1998, she was a Staff Member at MIT Lincoln Laboratory in the Optical Communications and the Advanced Networking Groups. Professor Médard received B.S. degrees in EECS and in Mathematics in 1989, a B.S. degree in Humanities in 1990, a M.S. degree in EE 1991, and a Sc.D. degree in EE in 1995, all from the Massachusetts Institute of Technology (MIT), Cambridge. She serves as an Associate Editor for the Optical Communications and Networking Series of the *IEEE Journal on Selected Areas in Communications*, as an Associate Editor in Communications for the *IEEE Transactions on Information Theory* and as a Guest Editor for the Joint special issue of the *IEEE Transactions on Information Theory* and the *IEEE/ACM Transactions on Networking* on Networking and Information Theory . She has served as a Guest Editor for the *IEEE Journal of Lightwave Technology* and as an Associate Editor for the *OSA Journal of Optical Networking*. Professor Médard's research interests are in the areas of network coding and reliable communications, particularly for optical and wireless networks. She was awarded the IEEE Leon K. Kirchmayer Prize Paper Award 2002 for her paper, "The Effect Upon Channel Capacity in Wireless Communications of Perfect and Imperfect Knowledge of the Channel," *IEEE Transactions on Information Theory*, volume 46, issue 3, May 2000, pages 935–946. She was co-awarded the Best Paper Award for G. Weichenberg, V. Chan, M. Médard, "Reliable Architectures for Networks Under Stress," *Fourth International Workshop on the Design of Reliable Communication Networks (DRCN 2003)*, October 2003, Banff, Alberta, Canada. She received a NSF Career Award in 2001 and was co-winner 2004 Harold E. Edgerton Faculty Achievement Award, established in 1982 to honor junior faculty members "for distinction in research, teaching and service to the MIT community."

Natasha Devroye received her Bachelor's degree in electrical engineering in 2001 from McGill University, Montreal, PQ, Canada. She received her Master's degree in 2003, and is currently working towards her Ph.D. degree both in the Division of Engineering and Applied Sciences at Harvard University, Cambridge, MA, USA.

Niranjan Ratnakar received the B.Tech. degree in electrical engineering from the Indian Institute of Technology, Madras in 2001 and the M.S. degree in electrical and computer engineering from the University of Illinois at Urbana-Champaign in 2003. He is currently working toward the Ph.D. degree at the University of Illinois at Urbana-Champaign. His research interests include network coding, multiterminal information theory, wireless communications, and algebraic coding.

Oh-Soon Shin received the B.S., M.S., and Ph.D. degrees in electrical engineering from Seoul National University, Seoul, Korea, in 1998, 2000, and 2004, respectively. Since March 2004, he has been with the Division of Engineering and Applied Sciences at Harvard University, Cambridge, MA, USA, as a postdoctoral research fellow. His research interests include wireless communications, communication theory, and signal processing for communications. Dr. Shin received the Best Paper Award from CDMA International Conference 2000.

Patrick Claus Friedrich Eggers has since 1992 been project leader of the propagation group of CPK. He is now the research coordinator of the Antennas and Propagation division, Department of Communication Technology (KOM), Aalborg University (AAU). He is on the technical research council of the Center for TeleInFrastruktur (CTIF) at AAU, and has been project and work package manager in several European research projects (TSUNAMI, CELLO etc) and in industrial projects with partners such as Nokia, Ericsson, Motorola, IOSpan, ArrayComm, Avendo Wireless, Samsung etc., as representative of CTIF, KOM (and previously the center for PersonKommunikation-CPK). He is author of over 40 papers, as well as section author and chapter editor in different COST final reports (COST207, 231, 259) and books. He is initiator and coordinator of an internationally targeted M.Sc.E.E. program in Mobile Communications taught in English at Aalborg University, as well as designer and coordinator of the newly starting programme in Software Defined Radio.

Patrick Mitran received the Bachelor's and Master's degrees in electrical engineering, in 2001 and 2002, respectively, from McGill University, Montreal, PQ, Canada. He is currently working toward the Ph.D. degree in the Division of Engineering and Applied Sciences, Harvard University, Cambridge, MA, USA. His research interests include iterative decoding theory, joint source-channel coding, detection and estimation theory as well as information theory.

Persefoni Kyritsi received the B.S. degree in electrical engineering from the National Technical University of Athens, Athens, Greece, in 1996, the M.S. and the Ph.D. degrees in Electrical Engineering from Stanford University, Stanford, CA, in 1998 and 2002 respectively. She has worked in several aspects of wireless communications for Lucent Technologies Bell Labs, in wireline communications for Deutsche Telekom, Frankfurt- Germany, and in circuit design for Intel Corporation and the Nokia Research Center, Helsinki- Finland. She currently holds the position of Assistant Professor at the Department of Communication Technology in Aalborg University, Aalborg, Denmark. Her research interests include radio channel measurements, channel modeling, multiple antenna systems, radio propagation and information theory for multiple input-multiple output systems. Dr. Kyritsi is a member of Sigma Xi.

Petar Popovski received the Dipl.-Ing. in electrical engineering and M.Sc. in communication engineering from the Faculty of Electrical Engineering, Sts. Cyril and Methodius University, Skopje, Macedonia, in 1997 and 2000, respectively. He received a Ph.D. degree from Aalborg University, Denmark, in 2004. From 1998 to 2001 he was a teaching and research assistant at the Institute of Telecommunications, Faculty of Electrical Engineering in Skopje. He is currently Assistant Research Professor at the Department of Communication Technology at the Aalborg University. His research interests are related to wireless ad hoc networks, wireless sensor networks, random access protocols, joint optimization of source coding and network protocols, and error-control coding.

Peter Koch received his MSc and PhD degrees in Electronics Engineering from Aalborg University in 1989 and 1996, respectively. Since 1997 he has been employed as an associate professor at Aalborg University, Institute for Electronic Systems. In 2004 he was one of the funders of Center for TeleInFrastruktur, Aalborg University, and currently he is acting as co-director for this center. His research interests are methodologies for software design for heterogeneous multiprocessor platforms, in particular with energy reduction as the driving cost parameter.

Petri Mähönen is a full professor and Ericsson chair of wireless networks at the RWTH Aachen University. Previously he studied and worked in the United States, the United Kingdom, and Finland. Before accepting his chair at RWTH Aachen in 2002, he was working as a professor and research director of networking at the Center for Wireless Communications, Oulu, Finland. He has been principal investigator in several international research projects, including several large European Union research projects for wireless communications.

He has published over 100 journal and conference articles, and is author of over 20 patents or patent applications. He has been also active on different standardization fora, and is consulting several multinational companies on wireless and 4G research strategies. His current research with his students focuses on wireless Internet, low-power communications including sensors, broadband wireless access, applied mathematical methods for telecommunications, and cognitive networks and radios.

Ralf Koetter received the Diploma degree in electrical engineering from the Technical University Darmstadt, Darmstadt, Germany, in 1990 and the Ph.D. degree from the Department of Electrical Engineering, Linkoping University, Sweden. From 1996 to 1997, he was a Visiting Scientist at the IBM Almaden Research Laboratory, San Jose, CA. He was a Visiting Assistant Professor at the University of Illinois at Urbana-Champaign, and a Visiting Scientist at CNRS, Sophia Antipolis, France, from 1997 to 1998. He joined the faculty of the University of Illinois at Urbana-Champaign in 1999, where he is currently an Associate Professor with the Coordinated Science Laboratory. His research interests include coding and information theory and their application to communication systems. Dr. Koetter received an IBM Invention Achievement Award in 1997, a National Science Foundation CAREER Award in 2000, and an IBM Partnership Award in 2001. He served as Associate Editor for coding theory and techniques for the *IEEE Transactions on Communications* from 1999 to 2001. In 2000, he started a term as Associate Editor for coding theory of the *IEEE Transactions on Information Theory*. He received the 2004 paper award of the Information Theory Society of the IEEE and is a member of the Board of Governors of this society.

Saeed S. Ghassemzadeh received his Ph.D. degree in electrical engineering from the City University of New York in 1994. From 1989-1992, he was a consultant to SCS Mobilecom, a wireless technology development company, where he conducted research in the areas of propagation and CDMA systems. In 1992, SCS Mobilecom merged with IMM (International Mobile Machines) to form InterDigital Communications corp (`http://www.interdigital.com`). From 1992-1995, while pursuing his Ph.D., he worked as a principal research engineer at InterDigital, where he conducted research in the areas of fixed/mobile wireless channels and was involved in system design, development, integration, and testing of B-CDMA technology. During the same time, he was also an adjunct lecturer at City College of New York. In 1995, he joined AT&T Wireless communication center of excellence at AT&T Bell-Labs (`http://www.lucent.com`) as a member of technical staff involved in design and development of the fixed wireless base station development team. He also conducted research in all areas of CDMA access technologies, propagation channel measurement and modeling, satellite communications, wireless local area networks and coding in wireless systems. Currently, he is a senior technical specialist in Communication Technology Research department at AT&T Labs-Research (`http://www.research.att.com`), Florham Park, NJ. His current research interest includes ultra-wideband technologies, wireless channel measurement and modeling, wireless LANs and Cognitive Radio communication. Dr. Ghassemzadeh is a senior member of IEEE, IEEE communication society and IEEE Vehicular technology society.

Shuguang Cui is an assistant professor of Electrical and Computer Engineering at the University of Arizona, USA. He received the B.Eng. degree in Radio Engineering with the highest distinction from Beijing University of Posts and Telecommunications, Beijing, China, in 1997, the M.Eng. degree in Electrical Engineering from McMaster University, Hamilton, Canada, in 2000, and the Ph.D. in Electrical Engineering from Stanford University, California, USA, in 2005. From 1997 to 1998 he worked at Hewlett-Packard, Beijing, P. R. China, as a system engineer. In the summer of 2003, he worked at National Semiconductor, Santa Clara, CA, as a wireless system researcher. His current research interests include cross-layer optimization for energy-constrained wireless networks, hardware and system synergies for high-performance wireless radios, and general communication theories. He is the winner of the NSERC graduate fellowship from the National Science and Engineering Research Council of Canada and the Canadian Wireless Telecommunications Association (CWTA) graduate scholarship.

Søren Vang Andersen received his M.Sc and Ph.D degrees in electrical engineering from Aalborg University, Aalborg, Denmark, in 1995 and 1999, respectively. Between 1999 and 2002 he was with the Department of Speech, Music and Hearing at the Royal Institute of Technology, Stockholm, Sweden, and Global IP Sound AB, Stockholm, Sweden. Since 2002 he is an associate professor with the digital communications (DICOM) group at Aalborg University. Søren Vang Andersen's research interests are within multimedia signal processing: coding, transmission, and enhancement.

Tatiana Kozlova Madsen received the M.Sc and Ph.D degrees in Mathematics from Moscow Lomonosov State University, Moscow, Russia, in 1997 and 2000, respectively. From 2001 she joined the Department of Communication Technology, Aalborg University where she is currently an Assistant Professor. She undertakes research and teaching within wireless communication and networking areas. Her current research interests are in the areas of wireless networking and mathematical modelling of communication systems and protocols. She is particular interested in influence of wireless channel and wireless access technologies on the behavior of the upper layer protocols and protocols optimization for wireless links. Dr. Madsen is an Associate Editor of Springer Journal of Wireless Personal Communications responsible for the networking area.

Tracey Ho received S.B. and M.Eng. degrees in electrical engineering (1999) and a Ph.D. in electrical engineering and computer science (2004) from the Massachusetts Institute of Technology. She has done postdoctoral work at the University of Illinois at Urbana-Champaign and Lucent's Bell Labs. She is currently an Assistant Professor at the California Institute of Technology and an associate editor for the *IEEE Communications Letters*. Her research interests are at the intersection of information theory, networking, wireless communications and machine learning.

Vahid Tarokh received the Ph.D. degree in electrical engineering from the University of Waterloo, Waterloo, ON, Canada, in 1995. He then worked at AT&T Labs-Research, Florham Park, NJ, and AT&T wireless services until August 2000 as a Member, Principal Member of Technical Staff, and finally the Head of the Department of Wireless Communications and Signal Processing. In September 2000, he joined the Department of Electrical Engineering and Computer Science, Massachusetts Institute of Technology (MIT), Cambridge, MA, as an Associate Professor, where he worked until June 2002. Since June 2002, he is with Harvard University, Cambridge, MA, where he is a Professor of Applied Mathematics and a Senior Fellow in Electrical Engineering. Dr. Tarokh has received a number of awards, including the 1987 Gold Tablet of The Iranian Math Society, the 1995 Governor General of Canada's Academic Gold Medal, the 1999 IEEE Information Theory Society Prize Paper Award, and the 2001 Alan T. Waterman Award. In 2002, he was selected as one of the top 100 young inventors of the year by the Technology Review Magazine. He has received honorary degrees from Harvard University and The University of Windsor.

Yasushi Takatori was born in Tokyo, Japan, 1971. He received his B.E. degree in electrical and communication engineering and his M.E. degree in system information engineering from Tohoku University, Sendai, Japan in 1993 and 1995, respectively. He received his Ph.D. degree in wireless communication engineering from Aalborg University, Denmark in 2005. In 1995, he joined the NTT Wireless Systems Laboratories, Nippon Telegraph and Telephone Corporation (NTT), in Japan. He is now working on NTT Network Innovation

Laboratories. He was a visiting researcher at the Center for TeleInFrastructure (CTIF), Aalborg University, Aalborg, Denmark from 2004 to 2005. He received the Young Engineers Award from the IEICE of Japan in 2000, the Excellent Paper Award of WPMC in 2004 and YRP Award in 2005. His current research interest is smart antennas, MIMO systems and spatial signal processing techniques. He is an associate editor of Springer Journal of Wireless Personal Communications. He is a member of the IEEE and the IEICE.

Foreword

by Professor Robert Axelrod

Cooperation has been the subject of intensive study in the social and biological sciences, as well as in mathematics and artificial intelligence. The fundamental finding is that even egoists can sustain cooperation provided the structure of their environment allows for repeated interactions (Axelrod 1984).

Some truly remarkable illustrations of this principle have recently appeared in the realm of information systems. One example is the success of open source (Axelrod and Cohen, 1999; Weber 2005) in which thousands of people cooperate to build a system, such as Linux. Another example is the success of eBay which is based on a feedback system that allows strangers to trust each other based upon the validity of the reputations for cooperating with others in the past. Still another example is Wikipedia, an encyclopedia to which anyone can contribute, and which by means of peer cooperation attains remarkable coverage and quality.

Wireless networks provide yet another realm in which cooperation among large numbers of egoists can be attained, provided that the right institutional structure can designed and implemented. Wireless communications is a rapidly emerging area of technology. Its success will depend in large measure on whether self-interested individuals can be provided a structure in which they are proper incentives to act in a cooperative mode. The editors and contributors to *Cooperation in Wireless Networks* demonstrate that our understanding of how cooperation works in a variety of other realms can illuminate what it will take for wireless technology to realize its full potential.

- Axelrod, Robert, *The Evolution of Cooperation* (NY: Basic Books, 1984).
- Axelrod, Robert and Michael D. Cohen, *Harnessing Complexity* (NY: Free Press, 1999).

- Cohen, Adam. *The Perfect Store: Inside Ebay*, (NY: Little, Brown; 2002).
- Weber, Steven, *The Success of Open Source* (Cambridge: MA: Harvard University Press, 2005).

Robert Axelrod, November, 2005
Gerald Ford School of Public Policy
University of Michigan
Ann Arbor, MI 48109
axe@umich.edu

Foreword

by Professor John G. Proakis

It is with pleasure that I write this Forward to the book on cooperation in wireless communications. This is indeed a unique book that treats an emerging topic in wireless communications. As the reader will observe in reading this book, cooperative techniques can be employed across different layers of a communication system and across different communication networks. Such techniques are based on the premise that through cooperation, all participants engaged in cooperative communication will achieve some benefit.

The editors in their introductory chapter categorize cooperation as either Implicit, or Explicit Macro, or Explicit Micro (Functional) Cooperation. Examples of implicit cooperation are communication protocols such as TCP and ALOHA. In such protocols, participants share a common resource based on fair sharing of that resource but without experiencing any other tangible gain and without the establishment of any particular framework for cooperation.

In contrast, explicit macro cooperation is characterized by a specified framework and established by design. Cooperative entities that fall in this category are wireless terminals and routers, which may cooperate, for example, by employing relaying techniques that extend the range of communication for users beyond their immediate coverage area. Such cooperation potentially provides mutual benefits to all users.

Explicit micro or functional cooperation is also characterized by a specific framework that is established by design. However, the cooperation involves functional parts or components of various entities, such as antennas in wireless terminals, processing units in mobile computing devices, and batteries in mobile devices. Explicit micro cooperation provides the potential for building low complexity wireless terminals with low battery consumption.

In the chapters contained in the book, the reader will find in-depth treatments of various aspects of cooperation in wireless communications. Among topics covered are fundamental limits, enabling technologies and implementation issues for cooperative communications; cooperative coding techniques for wireless networks; security in wireless cooperative networks; relaying and multihop techniques; energy aware allocation techniques in cooperative wireless networks; and perspectives on cooperative techniques in 4th generation networks.

I am confident that this book will stimulate and challenge both the academic and industrial communities in the field of telecommunications, as researchers and groups of engineers investigate, design and implement future wireless communication networks.

John G. Proakis, December, 2005
Northeastern University
516 Dana Research Building
Department of Electrical and Computer Engineering
Boston, MA 02115
proakis@ece.neu.edu

Acknowledgments

The hardest arithmetic to master is that which enables us to count our blessings.

Eric Hoffer

The present book is the result of the coordinated efforts of many people, and it would have not been possible without the key contributions of our invited authors. We are deeply indebted and immensely proud to have each and every one of them contributing in this book. We thank you all for sharing with us your technical expertise, and for the professionalism and endless enthusiasm showed during the writing period.

We would like to thank Prof. Ramjee Prasad from Aalborg University, Denmark and Sr.V.P. Dr. Young Kyun Kim, from Samsung Electronics, Korea. Their support and open minded attitude towards new technical horizons were instrumental to carry out this project. We also thank Prof. Torben Larsen from Aalborg University for his valuable comments and support.

We are greatly beholden to Prof. Robert Axelrod from University of Michigan, USA, and Prof. John Proakis from Northeastern University, USA for their encouragement as well as for their kindness in writing, in their respective fields, both Forewords.

We wish sincerely to thank Dr. Tim Brown and Mr. Robert Sheahan from Aalborg University for their invaluable help in proof reading several chapters of the book.

We are particularly thankful to Mark de Jongh and Cindy Zitter, from Springer for their encouragement, patience and flexibility during the whole edition process. We also thank Rajeshwari Thiagarajan and Werner Hermens for their kind help during the final edition.

We would also like to thank all our colleagues in Aalborg University and Samsung Electronics for their encouragement and interest. Henrik Benner, Ole Benner, Finn Hybjerg Hansen, Per Mejdal Rasmussen, Bo Nygaard Bai, Svend Erik Volsgaard and Torben H. Knudsen, from the KOM Computer WorkShop deserve our sincere thanks and appreciation for keeping this project web site up and running day and night, and for always providing us with technical assistance and handy solutions. We thank Simone Zecchetto for his nice illustration. We are grateful to our secretaries, Hanne Gade and Kirsten Jensen in Aalborg and So-youn Park in Korea, for their helpful and continuous assistance.

Finally, but most importantly, we Editors would like to thank our better halves, Sterica and Paula, and our children, Lilith and Samuel, for their unfailing support and for being so understanding while we worked on this book.

Preface

None is so great that he needs no help, and none is so small that he cannot give it.

King Solomon

It is important for him who wants to discover not to confine himself to one chapter of science, but to keep in touch with various others.

Jacques Hadamard

The idea of a book on cooperation in wireless networks was conceived and essentially implemented through cooperation. While working on a joint academia-industry research project on future wireless communication systems, we realized that a great deal of techniques having the potential to tackle many of the identified technical challenges exploited cooperative principles. Cooperation has been and continues to be a subject of intense research in many disciplines, notably natural, social and economic sciences. More recently, cooperation has started to receive attention in other sciences, particularly in engineering. In the last years, we have witnessed an increasing interest in exploring cooperative techniques in the domain of wireless communications. There are several reasons explaining such an interest. In the first instance, wireless communication systems are increasingly becoming more complex, with local processing influenced or even governed by other entities or global conditions. More and more system parts, protocols and algorithms are designed to support elaborated interactions between entities or functions. Sometimes it is even convenient to see conventional procedures in a wireless system from the cooperation perspective: the simplest communication between two terminals in the ubiquitous cellular architecture can be described by a cooperative model. Also, the steady emergence of distributed access architectures, independent of centralized decisions, is paving the way towards cooperative interactions between nodes. Moreover, opportunities for interaction also arise as new and planned wireless systems offer more fine-grain resource availability than ever, in particular in the time, frequency and spatial domains. Furthermore, cooperation in wireless communications is emerging as a key technology as we are consistently finding technical evidence that cooperating does pay off.

Cooperation can be understood as a joint action for mutual benefit. In wireless networks that definition is still valid but since there is a broad diversity in possible interacting entities, cooperation needs to be approached more widely. Cooperating entities do not necessarily need to be of the same type and the benefits obtained may vary from entity to entity. Cooperative techniques can be applied within and across the OSI layers and they can take place even between heterogeneous networks. Cooperative mechanisms can be embedded in the system and cooperation may then happen as a part of the normal interactions needed to move information across the network, without the knowledge or specific consent of the involved users. However, in some cases users themselves could be in a key position to allow cooperation by sharing their resources, i.e., terminals, with each other. In such a case, clear incentives are needed.

Cooperation has different connotations and meanings in wireless communications. On the *communicational* perspective, cooperation embraces a number of techniques taking advantage of the synergetic interaction of more than one entity as well as the collaborative use of resources, all aiming to enhance performance. Entities in this context can be defined at many levels, including signals, functions, algorithms, processing elements, building blocks and complete units, to name a few. Examples of this type of cooperation abound and it is the main cooperative approach dealt with in this book. From a purely *operational* point of view, cooperation implies the interaction and negotiating procedures between entities needed to establish and maintain interoperation. A typical example in this case would be the setups and procedural actions required to ensure end-to-end operation in an inter-network setup, in particular in environments with heterogeneous networks. In a setup that includes a number of nodes (i.e., wireless terminals, base stations, etc.) the *social* or *collective* aspect of cooperation is also of key importance. Cooperation here can be understood as the process of establishing and maintaining a network of collaborating nodes aiming to a mutual benefit. The process of node engagement has a key role as the node needs to decide on its participation on this ad hoc network. Unlike the previous approaches, where typically cooperation is inherently embedded in the system and hence invisible to the user, in this arrangement each user is in a key position as he or she ultimately decides whether to cooperate or not. These decisions, on an individual or collective scale, are important and both have impact on network performance. Appealing incentives should be offered in order to induce the users to cooperate.

We were initially motivated to compile this book mostly by numerous technical challenges, however, work in other fields had a profound influence in the way we approached cooperation. The simple but thoughtful quote '*Real egoistic behavior is to cooperate!*' by Kurt Edwin always inspired us with its unvarnished truth. We acknowledge the work by Robert Axelrod, which markedly influenced our views. In his work, among others, he identified a number of fundamental answers and behavioral rules from the question *'How can cooperation emerge in a world of egoists without central authority?'* The reader might be intrigued to know how this question is related to our wireless world. A paradigm shift is taking place in wireless and mobile communications, where two network access concepts (and their associated network architectures) are usually considered. In general, the centralized wireless access has been the dominant approach, typically exemplified by cellular communications. However, distributed (i.e., decentralized) access concepts are rapidly gaining ground, making communication links between terminals possible, and eventually giving to the user the power to decide whether his or her terminal can be part of a cooperative network. Indeed, in a cooperation-enabled wireless world, people will ultimately decide to switch their terminals to a *cooperative mode* or not. In a long-term perspective, where virtually everything will be connected and a truly *wireless knowledge society* will emerge, one may wonder how the patterns of cooperation will evolve in such a hyper-connected world, and what will be their technical, social and economic impact? With this book we hope not only to give technical insights on how cooperative techniques can improve performance in wireless communication systems, but also we would like to motivate further exploration in a fertile multidisciplinary research ground.

What are the benefits of cooperation in the wireless world? As the chapters of the book will discuss, cooperative techniques can be used to enhance many fundamental performance figures of a wireless communication system, being perhaps data throughput, quality of service, network coverage as well as spectral and power efficiency being the most relevant ones. In our view, every involved party will benefit from cooperation, namely manufacturers, operators, service providers and ultimately and most importantly, users. While cooperation is the leitmotiv of this book, such collaborative interactions are explored from two different angles. The first part of the book considers the underlying principles of cooperative techniques in wireless communication systems, covering a number of fundamental theoretical concepts. Applications of cooperation are discussed in the final part of the book, where practical examples at different layers are presented. Some chapters deal with both principles and applications.

Cooperation, widely exploited by nature, has been present probably from rather early origins of our time. Complex cooperative interactions take place at many levels, in organisms, within or between groups, or in large communities, to name a few. We people take advantage of cooperation on a daily basis, on small and big scales, thoughtfully and unconsciously. We are starting to apply and even mimic cooperative principles in engineering, and we have a lot to learn from around us. Other sciences have already extensively studied cooperation, developing models and analysis tools. Some of these ideas are now started to be investigated also in the context of wireless communications. We expect that the upcoming wireless and mobile communication systems will make well use of cooperative techniques, though not necessarily on a large scale in their introductory phase. If we look into a more distant future, cooperative techniques have the potential to be widely used in communications, eventually becoming one of the underlying principles to support communications. Our opening quotation, some three thousand years old, contains enough insight and wisdom as to be applied today, not only to people as originally intended, but also to future wireless communication systems.

Even though cooperation in wireless communications has not yet reached its full maturity, its realm is already broad, spreading and growing in depth incessantly. However we felt that the information on cooperative techniques is widely scattered over different, often not well connected, research areas. This book is an attempt to put into one volume a comprehensive and technically rich view of the wireless communications scene from a cooperation point of view. Considering the already vast and diverse existing research output, we thought that the most reasonable way to proceed is to invite leading researchers in the field to share their expertise in this book. Thus, this book owes its existence to the cooperative efforts of many people from all over the globe.

The editors welcome any suggestions, comments or constructive criticism on this book. Such a feedback would be used to improve forthcoming editions. Editors can be contacted at editors@cowinet.com and additional material can be found at `http://cowinet.kom.aau.dk`.

Frank H.P Fitzek, Aalborg, Denmark
Marcos D. Katz, Suwon, Korea
December, 2005

Chapter 1

COOPERATION IN NATURE AND WIRELESS COMMUNICATIONS

Real egoistic behavior is to cooperate!

Frank H. P. Fitzek
Aalborg University
ff@kom.aau.dk

Marcos D. Katz
Samsung Electronics
marcos.katz@ieee.org

Abstract: The intention of this first chapter is two–fold, to provide examples of cooperation in nature and to use these examples to motivate consideration of cooperation in wireless communication systems. Nature has demonstrated that cooperative species out compete selfish species in many ecological niches. We advocate the introduction of cooperative methods into omnipresent wireless communication systems, which can be described as egoistic (or can be described at least as agnostic) so far. Even for highly centralized systems such as the cellular wireless communication networks, cooperation demonstrates its strength in offering increased quality in service with less complex terminals. Later chapters will give a much more detailed view on the potential of cooperation in the wireless domain, but here we would like to derive the first principles and motivate this kind of cross–over.

Keywords: GSM, nash equilibrium, pavlov, prisoner's dilemma, UMTS, WLAN, desmondus rotundus, implicit and explicit cooperation, iterated prisoner's dilemma, multi-hop, power saving, reciprocity, tit for tat, wireless communication, zero games

Frank H. P. Fitzek and Marcos D. Katz (eds.), Cooperation in Wireless Networks: Principles and Applications, 1–27.

1. Basics of Cooperation

The ultimate goal of this chapter is to motivate the use of cooperative techniques in future wireless communication systems. To that end, we will explain some terminology used to discuss cooperation and explore briefly cooperation in other sciences, where this topic has been widely studied before.

The word cooperate derives from the Latin words *co-* and *operare* (to work), thus it connotes the idea of "working together". Cooperation is the strategy of a group of entities working together to achieve a common or individual goal. The main idea behind cooperation is that each cooperating entity gains by means of the unified activity. Cooperation can be seen as the action of obtaining some advantage by giving, sharing or allowing something. Cooperation is extensively applied by human beings and animals, and we would like here to map different cooperation strategies into wireless communication systems. While the term *cooperation* can be used to describe any relationship where all participants contribute, we tend to use it here to describe the more restrictive case in which all participants gain. If we use it in the broader sense of simply working together, it will be apparent from the context or explicitly stated. This restricted definition of cooperation contrasts with altruism, a behavior where one of the participants does not gain from the interaction to support others.

As we will show later, the strategy of cooperation does not mandate contributing in every situation, each entity evaluates each situation and makes a decision based on circumstances. In case cooperative behavior does not lead to a clear benefit, the entity should not help other entities. We describe this refusal to help as being an autarky, being selfish, or acting egotistically.

An analogy between cooperation in natural and human sciences with the world of wireless communications can sometimes be established, though it is not our aim here to identify all such possibilities. Later in this chapter, and as an example, we discuss the correspondence between some typical wireless scenarios and some models developed in human sciences. It is interesting to note that in nature cooperation can take place at a small scale (*i.e.*, few entities collaborate) or large scale (*i.e.*, massive collaboration). The latter includes cooperation between the members of large groups up to the society itself. A similar classification holds in the wireless domain. A few nodes (*e.g.*, terminals, base stations) can cooperate to achieve certain goals. The foreseen wireless knowledge society is expected to be a highly connected (global) network where virtually any entity (man or machine) can be wirelessly connected with each other. Cooperation in such a hyper–connected world will play a key role in shaping the technical and human perspectives of communication.

Desmondus rotundus

Examples for cooperation in nature are numerous including the population of ants, termites, bees, hunting lions, and human beings, corresponding to collaboration between members of the same species. Cross–collaboration between members of different species is not uncommon in nature, including fungi and algae living together as lichens or Egyptian plovers cleaning the teeth of crocodiles. In the case of cooperation among heterogeneous entities gains could be of different nature, that is one might gain food and the other safety. One illustrative example from nature was first reported in [Wilkinson, 1984] and summarized in [Ridley, 1998] about the behavior of vampire bats. Vampire bats live in large groups and spend most of the day in hollow trees. In the night they search for large animals that have some bleeding cuts. Once they find such an animal, they will sit next to the cut and simply sip as much blood as they can. Bats try to feed every day, but are able to survive also some time without feeding. The critical limit for a period without any food is sixty hours, where most of the bats without any blood will starve to death. Fortunately, bats that have found enough blood can help out others that have not enough once they are back at their hollow trees. The donation of blood is not based on altruism, but on a strict scoring table. Bats are able to remember which bats gave them blood previously. Furthermore they can detect if other bats getting asked for blood are declining these requests even though they have enough to donate. Wilkinson reports that bats are paying attention to the stomachs of other bats to detect cheaters. The bats punish cheaters by refusing to help them when they are in need but reward cooperators by helping them when they need it. Bats can live up to eighteen years so cooperation is well established as cheaters are cut out and cooperators will be helped out and starve to death when they fall on hard times but cooperators live to breed over many seasons.

Tour de France

Another example of cooperation can be seen every year at the Tour de France. This is a competition for cyclists touring around France. The competition takes several weeks and every day all cyclists start together to race over hundreds of kilometers. Because they all start at once they form a large group referred to as the *peloton*. The *peloton* provides protection from wind resistance, but also limits the speed of a fast cyclist. To win, a cyclist must escape from the main group at some point. As the way to the finish line may still be far away, a single cyclist would spend too much energy escaping from the *peloton* and fighting the front wind the whole way alone. Therefore alliances are made across the different teams to escape from the *peloton*. Once they escaped, they form a small group that moves faster than the main group but provides protection from the wind resistance. Within the group they change the lead cyclist from time to time so no one cyclist spends all of the energy to fight the front wind. They work together fighting the front wind even though each individual cyclist has the intention to win the race and it is clear to each of them that only one can win. Cooperation among the cyclists is established in this example as any selfish behavior would put the whole group at risk. Even though not cooperating and just joining the group might be beneficial in the beginning (the cyclist saves energy), the small group can not travel as fast as it could if all contributed and the group may even loose enough speed to get caught by the *peloton* again. If the group is not caught by the *peloton*, they stop cooperating shortly before the finish line and make their best speeds individually. This non–cooperating behavior at the end is known and accepted by all cyclists and will not stop them from cooperating on subsequent days.

The Income–Cost Relationship

Generally speaking, human beings (usually subconsciously and often imperfectly) evaluate their own doings by an n–dimensional income–cost relationship optimizing their own profit (P) by maximizing the income (I) related to a given cost (C) in terms of resources. Or the other way round, to achieve a given profit with minimal costs in terms of scarce resources or fixed income. In both cases the income should be larger than the cost as given below

$$I - C = P > 0. \tag{1.1}$$

Historically, the first mass realizations of this n–dimensional income-cost relationship led unfortunately to culturally inadequate solutions such as slavery. Such solutions assumed that every person has a fixed capacity I so P can only be maximized by minimizing C. In the case of slavery the profit per slave was small (as obviously the motivation was low and in turn I was small and the number of overseers and guards was large, which increased C) and had to be compensated for by using a large number of slaves. From a modern working–life perspective based on a well stipulated work remuneration and where slavery should not be an option, cooperation has been identified as optimal regarding the n–dimensional income–cost relationship. A motivated employee will bring large profit by producing more and reducing the costs. C is minimized and I maximized if the goals of the company and the employee are identical where employer and employee mutually support their goals.

A very important boundary condition for cooperation is that each participating entity is gaining more by cooperation than they would by operating alone for the same costs. It is not important that all entities contribute the same effort, gain the same amount, or even have the same gain to cost ratio, but the effect of cooperation should bring advantage or gain to each cooperating entity. This can best be expressed by the simple statement in [Edwin, 1994].

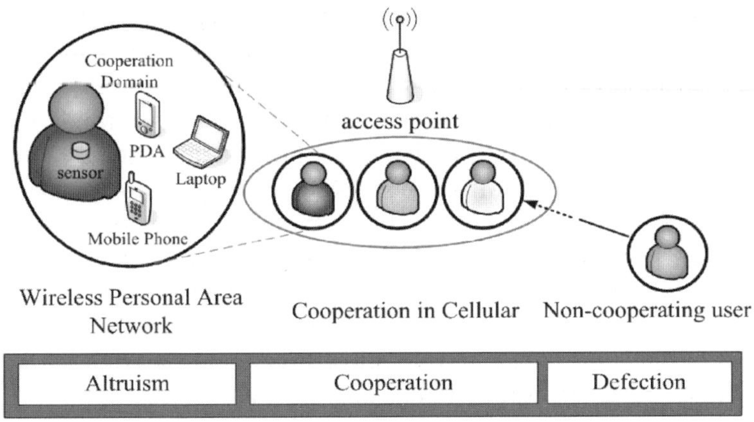

Figure 1.1. Cooperative Horizon.

"Real egoistic behavior is to cooperate!"

Different forms of cooperation exist. As given in Figure 1.1 the most generous form of cooperation is called *altruism*. In this case the contributing entity is not motivated by its own pay–offs, but tries to increase the pay–offs of the others. Altruism is one of the foundations of family. In the wireless domain, this may be the case in the so called wireless personal area networks, where the user has several wireless devices in his immediate proximity. All devices serve the same user and together form a group entity like honey bees collectively form a hive or colony. A device, that sacrifices battery power to assist another device the same user is operating, may seem altruistic when considered alone but selfish when considered as part of the personal area network.

When the cooperating entities are based on their own interests, we distinguish between cooperation and defection. Defection will take place when one entity does not see any gain in cooperation and assumes the highest profit by staying alone, *i.e.*, not cooperating. Cooperation is chosen over defection if the profit of cooperation seems higher that that of defection. In the sequel of this chapter we will see that the evaluation of the profit is not always an easy task and the outcome depends on the prevailing scenario. Figure 1.1 summarizes these concepts by depicting the *cooperative horizon* together with possible examples of cooperation in wireless networks.

2. The Prisoner's Dilemma

One of the most famous examples of the study of cooperation is the prisoner's dilemma. The concept of the prisoner's dilemma was introduced by Merrill Flood and Melvin Dresher in 1950. The name was given by Albert Trucker in the same year. The prisoner's dilemma tries to describe the problem of cooperation of two entities. Each individual entity is trying to maximize its own gain, without concern for the well–being of the other entity. The prisoner's dilemma is a representative of a non–zero sum game regarding the game theory (see [Smith, 1982]). Non–zero and zero sum games refer to the possible outcome of a game. Summing up all gains and losses, the game is referred to as zero if the outcome is null, while if there is a remaining gain or loss we refer to it as a non–zero game. In the zero game case with two entities, the gain of one entity corresponds to loss of another one. This concept can be generalized and holds even for n–entities. In such situations non cooperative and selfish behavior is the best from the point of view of the individual. In non–zero games the most profitable behavior may be different because the non-zero sum allows the gain of a group to be higher than the sum of the gains of individuals operating alone.

The prisoner's dilemma can be described by many examples. One example is found in [Pundstone, 1992]. It describes two thieves who are caught by the police and they are interrogated separately at the same time. Each thief has two options, namely not telling the truth to the police (cooperating with his colleague) or confessing (not cooperating with his colleague). In total there are four possible outcomes. If both confess they will go to jail for a long time. If one confesses and the other is loyal, the latter will go into jail for a very long time, while the former one goes free. But if both cooperate (not telling anything to the police), both will go to jail for a short period only.

Table 1.1. The Prisoner's Dilemma.

		Thief A	
		Stay quiet	Confess (blame it on the other)
Thief B	Stay quiet	Both serve six months	Thief B serves ten years; Thief A goes free
	Confess (blame it on the other)	Thief A serves ten years; Thief B goes free	Both serve two years

The dilemma they are facing is that both decisions are made independently (even though before the police arrested them they might have promised to be loyal) and cannot be sure that the other will remain loyal. Moreover, they may hope the other will cooperate so that they could decline their cooperation to get a larger advantage out of it. The incentive to cheat is the core of the dilemma.

The aforementioned example is referred to as *symmetric* prisoner's dilemma, because the pay–offs are the same even if the role of the players is switched. In other words the pay–offs are reciprocal. The *symmetric* prisoner's dilemma can be considered as a special case of the *asymmetric* prisoner's dilemma. For the later dilemma pay–offs depend on the role of the players. As an example if player one defects and player two cooperates, they will be paid off by $P1$ and

$P2$; if the decisions are exchanged (now player one cooperates and player two defects) the pay–offs are $P3$ and $P4$, respectively. Even though the decision on cooperation or defection is the same, the rewards may be different.

In the cooperative horizon illustrated in Figure 1.1 the prisoner's dilemma lies in the middle of the cooperation and defecting cases. That defection is the better strategy for a single run PD is also proved by the Nash equilibrium [Nash, 1950]. The equilibrium is reached if the strategy of each player is an optimal response to the strategies adopted by other players, and nobody has an incentive to deviate from the chosen strategy as described by [Holt and Roth, 2004]. Or with the words of [Holt and Roth, 2004] defining the Nash equilibrium as a self-enforcing agreement reached by the players without central authority following the self interest of each player. So if defecting pays off more than cooperating, why should cooperation emerge as the better strategy? This can be explained if the prisoner's dilemma is played multiple times instead of only once. This is described in the next section.

3. The Iterated Prisoner's Dilemma

As the prisoner's dilemma describes the problem whether to cooperate or not for a single run, the question is whether the decision of the entities will be different if the number of runs is increased. This is described as the iterated prisoner's dilemma. In case of the thieves, they might be caught a second time and they remember the outcome of their last encounter and realize they will probably be in the same situation again in the future. The possibility to get punished in the next round will force the entities to be nice and cooperate.

The iterated prisoner's dilemma was described in detail by [Axelrod, 1984] in his book about *The Evolution of Cooperation* and some prior work [Axelrod and Hamilton, 1981]. To find out the best strategy of cooperation, in 1979 he invited people around the world to submit computer programs following a given cooperative strategy. All programs entered a tournament against each other to find out the best strategy. There were two tournaments. In the first tournament there were fourteen strategies submitted playing against each other plus one random strategy (so fifteen in total). The smallest program had four lines of code and the largest one 77. The second tournament (knowing the result of the first tournament) had 63 participants with five and 152 lines of code for the smallest and largest program, respectively. There was no limitation on the complexity of the computer program. The programs submitted differed in their complexity and strategy in terms of hostility, forgiveness, etc.

The tournament had more or less the following rules as given by [Axelrod, 1984]. Two strategies are playing against each other represented by one player each. Both players have to choose between two options simultaneously, whether to cooperate or to defect. This setup gives four possible outcomes as given in

Table 1.2. The players are named *Column Player* and *Row Player*. In case both players decide to cooperate they get a reward (R) of three points each ($R = 3$). In the opposite case, where both players defect, they are punished (P), getting only one point each ($P = 1$). The largest values per player is achieved if one player decides to cooperate and the other player defects. In this case the defecting player achieves five points ($T = 5$), while the cooperating player gets no points at all ($S = 0$).

In the single–iteration case with $P > S$ and $T > R$, the best strategy is to defect. However, the dilemma is that the best choice for each entity means that both receive P, the worst possible collective outcome as reported by [Macy, 1996]. The best choice for a given player (if he has no clue what the other one is doing) is to defect. By defecting the possible outcomes are 1 or 5, while cooperating only would give 0 or 3.

Table 1.2. Axelrod's Tournament on the Prisoner's Dilemma.

		Column Player	
		Cooperate	Defect
	Cooperate	R = 3, R = 3 Reward (R) for mutual cooperation	S = 0, T = 5 Sucker's (S) payoff, and temptation (T) to unilaterally defect
Row Player	Defect	T = 5, S = 0 (T) to unilaterally defect and sucker's (S) payoff Temptation	P = 1, P = 1 Punishment (P) for mutual defection

The goal of the project was to find out when an entity should cooperate or when it should be selfish without any overall regulations. In R. Axelrod's words:

> *"Under what conditions will cooperation emerge*
> *in a world of egoists without central authority?"*

The first insight of this tournament was that greedy strategies were outperformed in the long run by cooperative strategies. The best strategy was developed by A. Rapoport referred to as *tit for tat* (TFT). This program was written in BASIC containing only four lines of code (in the second tournament it was resubmitted with five lines). This program outperformed even the more complex submissions. The winning strategy is startlingly simple. It starts to cooperate on the first iteration and mimics (or repeats) the previous move of the opponent in the following iteration. This strategy has two properties. First it starts to cooperate expressing its aim to be nice. Afterwards it will continue to cooperate if the opponent entity is doing the same, but will punish the opponent entity if it declines. An interesting extension of the simple *tit for tat* is the strategy *tit for tat with forgiveness* or *generous tit for tat* as introduced by [Wu and Axelrod, 1995]. The main difference to the original strategy is that after one opponent defects, the other one will still cooperate with a small probability. This behavior is justified to show the opponent that cooperation is still the desired way to interact. The extended version has its application when miscommunication is introduced among entities. A situation that we face often in wireless communication is that the wireless link between entities is often error–prone. Also robust to miscommunication is the Pavlov strategy. The Pavlov strategy embodies an almost reflex–like response to the payoff [Nowak and Sigmund, 1993]. Whenever an entity got rewarded by R or T points, it repeats the last move. But it changes the strategy if it got punished by P or S. Pavlov cooperates with *tit for tat* and itself, exploits unconditional cooperators, but it is more heavily exploited by unconditional defectors than tit for tat [Wedekind and Milinski, 1996]. Comparing the strategies with each other Axelrod listed four simple suggestions to be successful:

1 Do not be envious

2 Do not be the first to defect

3 Reciprocate both cooperation and defection

4 Do not be too clever

The first point was already mentioned before. Each entity should maximize its own gain and not be envious if the other partner entities are gaining more out of it. Each entity should focus on itself and compare the actual situation with that to be isolated and not with others. This is totally in line with Edwin's statement that real egoistic behavior is to cooperate. As reported by [Duncan, 2003] humans and non–humans tend to evaluate their doings by the factor fairness, cooperation may be refused because of the sense of fairness. Even though a lower reward with fairness may be achieved instead of acting purely rationally.

The second suggestion is that a strategy should be nice and not harm the other in the first place. This suggestion is directly derived from the tournament's results. The strategies with the highest scores all behave in a nice way.

Reciprocity is the third suggestion. Reciprocity has two outcomes. First it helps to establish cooperation when both sides are willing to cooperate and secondly it starts punishing the opponent in case it wants to exploit the other by defecting.

The last suggestion is underlined by the fact that the simplest strategy won the tournament and that more complex strategies led to mutual defection. Also permanent retaliation is one possible outcome of being too clever. In case one entity starts to decline (or in case of miscommunication one entity assumes that the other side declines) it may end up in total defection. Therefore forgiveness should be part of the strategy. Axelrod concluded that all entities for their own benefit (whatever selfish or egoistic motivation they have) should cooperate in a nice, non–envious, and forgiving way. For some readers it may seem obvious and logical to cooperate given the tournament settings. If both players are friends and one is asking for help, surely we would suggest being nice and cooperating. In this case we would act as a central entity giving advice, violating the premise of no regulating authority. To understand the dilemma it helps to try it out once with friends, playing a dozen rounds and tracking the points. Or as an alternative, the book by Poundstone has a step–by–step evaluation of the prisoner's dilemma played by two friends. Even though cooperation might be the best solution they often tried to get the higher benefits by defection. For completeness, we would like to mention the results of the 20th anniversary of the Iterated Prisoner's Dilemma in 2004. A group from Southampton University came up with a new strategy being stronger than the known *tit for tat* derivations. The main idea was to submit multiple programs to the tournament, which was fully in line with the regulations of the tournament. The submitted programs were trained to recognize each other in the first iterations of the game. After recognizing each other (to be from Southampton) one entity starts to cooperate while the other one declines. By this method the declining strategies obtained the most points out of it, while the cooperating one obtained nearly nothing. In case the submitted strategies were played against other strategies they used the *tit for tat* behavior. By this strategy they outperformed the *tit for tat* strategy with some of their programs, while the rest of their submissions were ranked very low. One may call this unfair, but it is a valid study of one aspect of cooperation. The situation the Southampton team created is similar to the personal area network situation described previously in this chapter. You can also find this form of altruism among family members, which certainly will not act according to the *tit for tat* strategy. Altruistic behavior can also be found in nature where it usually involves a relative providing assistance without the individual receiving any apparent benefit. In contrast to the Tour de France's *peloton*, the cooperation found in the Southampton example is formed among the same team players. In the Southampton experiment some players take the role of masters while others become slaves (see [Grossman, 2004; Axelrod, 2004]). In the winning strategy, the latter ones sacrifice themselves to help the masters to win.

4. N–person Prisoner's Dilemma

In the previous section we have concentrated on the classical iterated prisoner's dilemma for two competitive entities. In real life there are more entities with different strategies involved. This leads to the so called n–person prisoner's

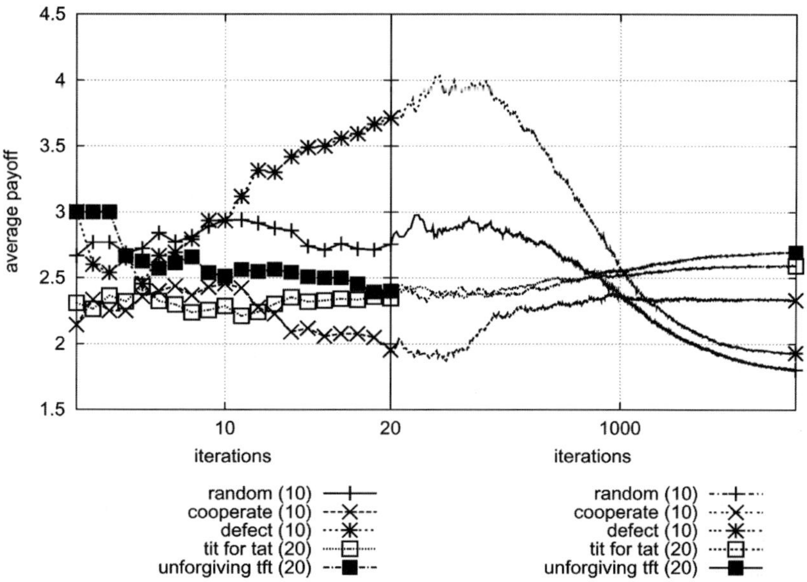

Figure 1.2. Average payoff for seventy entities with five different strategies for 20000 iterations.

dilemma. In [Axelrod, 1997] this field of research is explained more in detail. Here we used a tool from [Wilensky, 2002] to investigate seventy entities with five different strategies. We consider now an illustrative example where seventy terminals are moving around in a virtual two-dimensional world and meet each other from time to time. Every terminal is able to recognize a terminal that has been met before, and is also able to remember every opponent's last moves.

The considered strategies are as follows: Ten terminals use the *defect* always strategy. The total opposite strategy, where terminals are always nice and *cooperate*, is used by another ten entities. A larger group (20) plays the *tit for tat* strategy as explained above. The same amount of entities use *unforgiving tit for tat*. This strategy differs from that of the original version by the fact that any counterpart entity that is defecting once will be defected in the future. The last group with ten terminals just uses a *random* strategy selected out of the prior four strategies.

The average payoff for the five strategies is given in Figure 1.2 using the payoffs given in Table 1.2. Note that the first twenty iterations are given in a linear scale, while the rest up to 20k iterations is shown in a log–scale.

In the initial phase, after the first twenty iterations, it seems that the defecting strategy is the better choice as it pays off over 3.5 points. The worst strategy is the cooperative one with less than 2. In the long run the *tit for tat* strategies are the best, with 2.59 and 2.69 for the original and the unforgiving strategy. Even the pure cooperative strategy with 2.33 outperforms the random (1.80) and the defecting (1.92) one. The random strategy is the worst as it got punished by the unforgiving *tit for tat* and exploited by the defecting strategy.

Here we have investigated the situation when the entities will not change their behavior over the investigation phase. However, an entity may adapt a different strategy when it pays off better than the old strategy. The problem may arise in our example if all terminals judge their situation after iteration 20 and switch to the defecting strategy. This would end up in a final value of 1 for the average pay off. If all terminals would change to the *tit for tat* strategy, the average pay off would become 3 (asking the unforgiving tit for tat to forget about the past).

5. Stimulating Cooperative Behavior

In this section we highlight how cooperative behavior can be stimulated in any given environment. In [Kurzban and Houser, 2005] experiments were conducted to investigate the willingness of human beings to cooperate. The authors state that human being form polymorphic populations formed by individuals varying in their degree of cooperativeness. Without going into details (interested readers are referred to [Kurzban and Houser, 2005]) three main groups were identified. The first group tends to cooperate all the time with the risk to be exploited. A second group, referred to as free–riders, never wanted to cooperate. The third and largest group was evaluating the situation they were in and waited to make the decision to cooperate or not.

As a result and boundary for future application of cooperative strategies in wireless communication system, cheating has to be prevented or kept as small as possible. In line with the results of the *tit for tat* tournament carried out in 2004, cheating may pay off for those doing it at the cost of the others. Moreover it will stop potential cooperators from joining the cooperative group. Therefore, cooperation should provide fairness among entities and prevent cheating. Obviously cooperation would emerge as the only reasonable possibility if the values T, P, R, and S are found to be in favor for cooperation such as P and S are very low and R is as large or larger than T. We talk about a prisoner's dilemma if the payoff inequities hold, that is

$$T > R > P > S \tag{1.2}$$

and

$$R > (T + S)/2. \tag{1.3}$$

Note, this is true for the *symmetric* case. For the asymmetric we refer to [Stanford Encyclopedia of Philosophy, 2003]. Cooperation is a better strategy than alternating defection and cooperation. In case $R > T$, we are not in a dilemma anymore and cooperation would be chosen over any other given strategy by any entity. To stimulate cooperation, the benefit for cooperation should be as high as possible compared to exploiting others. (R close to T).

6. Cooperation in Wireless Communication Systems

Engineering has often been successful in imitating or emulating nature. Initially this took place in the mechanical world such as copying the skin structure of sharks for airplanes. We can also find examples in a large variety of engineering areas, like mimicking ants' behavior for Internet routing, the general concept of splitting and other examples. Here, inspired by the patterns of cooperative behavior in nature, and particularly in human beings, we would like to explore and further apply concepts of cooperation in wireless communication networks. In this chapter, and just as a motivating example, we would consider cooperation from the wireless network architecture standpoint. Certainly, this is just one possible use of cooperation, and as we will see through this book, cooperative techniques can be applied within and across any communication layer. Many well known concepts and techniques in wireless communication can be described by using cooperative principles; though often the cooperative aspects are not necessarily highlighted. As an example, network protocols can be described with a cooperative framework as all entities of a communication group should follow common communication rules. Cooperation and fairness are the underlying principle of many distributed systems and may be found less often in centralized systems. Our focus relies on terminals performing cooperation aiming to improve basic communication capabilities as well as to consuming less power. Whether to cooperate or not should be evaluated by the terminal case by case and realized only when the situation is such that cooperation gives higher gains than being autonomous (autarky case). Being selfish prevents being exploited, such interpretation could result in a mobile relaying scenario, where a terminal is being used by others with no apparent benefit for the relaying unit. Therefore mechanisms are needed that support and motivate the idea of cooperation.

Classification of Cooperation

Cooperation in wireless networks can be approached from different angles, and in fact, it has different meanings and connotations. In this section we will classify and discuss the most relevant approaches to cooperation in the context of wireless communication networks. Probably the most important side of cooperation is its *communicational* aspect, which includes several techniques exploiting the joint collaborative efforts of multiple entities in the system. Entities like signals, functions, algorithms, processing elements, building blocks and complete units interact mutually in order to bring some advantages, including enhancement of performance and better use or resources. This is the most explored area of cooperation, and techniques like cooperative diversity cooperative coding, network coding, cooperative antennas, etc., have and are been extensively studied. Communicational cooperation is inherently embedded in the wireless network and it is generally invisible to the user. This book mostly deals with the communicational aspects of cooperation. In a different approach, and usually in the context of heterogeneous networks, the term cooperation is also used from the *operational* standpoint, referring to the interaction and negotiating procedures between entities required to establish and maintain communication between different networks. The main target here is to ensure end-to-end connectivity, where the main players are (different) terminals operating in different networks. In order to make this possible, the most convenient setups and methodology needs to be developed, including system architecture and procedures. We also distinguish the important *social* aspect of cooperation, pointing out the dynamic process of establishing and maintaining a network of collaborative nodes (*e.g.*, wireless terminals). The process of node engagement is important as each node needs to decide on its participation on this ad hoc network, having each decision an individual and collective impact on performance. Unlike the previous approaches, in this arrangement each user is in a key position as he or she ultimately decides whether to cooperate or not. Appealing incentives need be offered to the users in order to encourage them to cooperate.

In the following we classify different levels of cooperation, mostly referring to the communicational aspect of it. The main difference is the existence of a cooperation framework dividing the classes into implicit and explicit cooperation classes. This classification is motivated by a pragmatic approach aiming for a better explanation of our concepts and ideas.

- The *Implicit Cooperation* is characterized by fairness and respect in a passive way. Network protocols themselves can be seen as a form of cooperation. We refer to this as *implicit (or passive) cooperation*. In this first level of cooperation, the interaction takes place without any pre–established cooperative framework. Furthermore the cooperation focuses on the fair sharing of a given resource without gaining anything else. We could refer to this as a zero–sum game.

- The main characteristic of explicit cooperation is that it is established through a given framework. In other words, cooperative behavior is allowed and supported by design, allowing counterpart entities to actively interact directly with each other. We consider here that the cooperating

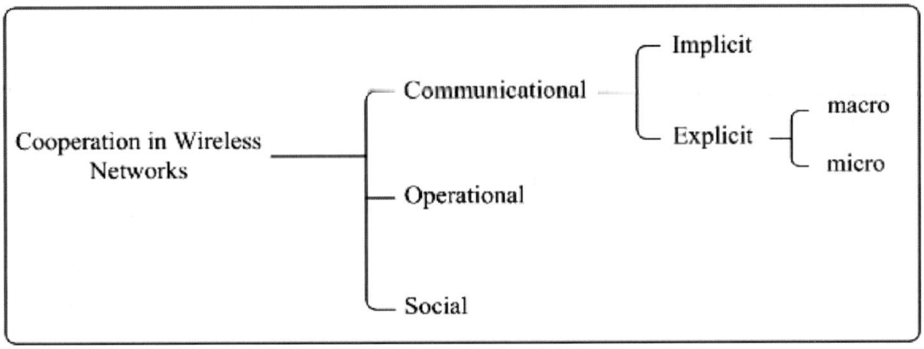

Figure 1.3. A practical classification of cooperation in wireless networks.

entities are wireless terminals, virtual access points, wireless routers and other macroscopic wireless system parts. At this second level of cooperation the aforementioned entities would cooperate with each other and is referred to as *Explicit Macro Cooperation*. Examples of explicit macro cooperation abound, including relaying techniques, coded cooperation as given by [Nosratinia et al., 2004].

- The last level of cooperation goes down into the entities, addressing functional parts or components as candidates for cooperation. Examples of as micro entities include typically processing units, functional parts and algorithms. Important constituent parts of the system, mostly hardware parts like components, batteries, antennas, can be can be involved in the cooperation process, though they cannot be considered as cooperating micro entities per se. This kind of cooperation targeting the individual functions of a entity (*e.g.*, terminal) is referred to as *Explicit Micro (or Functional) Cooperation*. This third level of cooperation will be explained in greater detail shortly.

Figure 1.3 summarizes the classification of cooperation in wireless networks as described previously. Next, we will consider examples of implicit and explicit cooperation.

Implicit cooperation. Two very well known communication protocols such as the Transport Control Protocol and the ALOHA protocol are representative examples of implicit cooperation and they will be briefly described from a cooperative standpoint next.

The Transport Control Protocol (TCP), used as the main transport protocol in the Internet, is a very good example for such a cooperative protocol. The protocol promotes fairness among users by applying a flow control. The flow control prevents a user from taking the maximum link capacity right away as this may overload some routers in the Internet, causing large delays and losses for all traffic streams going through that router. Instead, the flow control rules have a user start with a minimum of capacity and increase it stepwise according to some rules described in the flow control specification. Once too much capacity is used, the flow control detects this situation and reduces the allowed capacity. The same flow control strategy is applied in all Internet-capable terminals assuring fairness among all Internet users. As described by [Akella et al., 2002] it is well known that each user may change his TCP protocol in a way that he can achieve a larger throughput by violating the fairness among users and thus reducing the throughput of the fair users. The more users violate the TCP protocol, the smaller will be the gain of the misbehaving users and simultaneously the overall throughput of the fair users will be reduced further. Even though the overall throughput is decreasing with each unfair user, still users may see a benefit in breaking the fairness rule. Fortunately, the majority of the users is behaving in a fair manner; maybe because they want to be fair or maybe because they do not know how to change the protocol. The main reason is probably that the motivation to *defect* is very low as the rewards for cooperation and defection are nearly equal ($T \simeq R$).

The ALOHA–based medium access scheme is another example of mutual cooperation among wireless end systems. The ALOHA protocol is a simple random access protocol with a minimum of signalling overhead. It was invented in the 70s by the University of Hawaii and became the base for many other protocols. The protocol was designed for distributed wireless end systems without central control. End systems are allowed to access the wireless medium following a set of given rules. Whenever a station has to send a packet it starts to transmit it right away. For the successful reception it is important that only one wireless terminal is transmitting at the time. However, as we deal with a distributed system, we cannot assure that the medium is not being used by more than one entity at the same time. If more than one terminal transmits simultaneously this situation is referred to as *collision*. When a collision occurs all information involved is lost. Let us assume that two wireless end systems transmit packets that collide. As the packets are lost, the stations need to retransmit them. In this case, if both terminals would follow the strategy to retransmit immediately, another collision would follow. Therefore ALOHA foresees this situation and each end system will back–off individually with independently determined periods referred to as back-off time. The back–off time is chosen randomly within a given time window. For each newly occurred collision the time window will increase to reduce the probability of a new collision. After the first successful transmission the time window is set to the minimum value. This protocol could easily be changed by misbehaving nodes to increase their own throughput at the costs of the overall performance by retransmitting right away hoping that the colliding entities will choose a large time window as given by [Cagalj et al., 2005] as an enhancement of ALOHA.

Explicit macro cooperation. As an example of the explicit macro cooperation we use the well known multi–hop or relaying scenario. Here a cooperation framework is given that allows terminals to interact with each other in a cooperative manner. Research efforts on application of cooperative strategies in wireless networks are gaining momentum around the world. Representative work virtually covering all the relevant research areas in cooperation are presented in this book. The term *cooperation* is often used in the field of relaying. Relaying is a technology to extend virtually the communication range by forwarding. In case a source cannot deliver its information to the sink directly, it may ask a relaying node, located within the coverage of both the sink and the source, to relay this information. For such a scenario cooperation may need external motivation to take place.

In multi–hop or relaying networks wireless nodes (including even mobile ones) allow the forwarding of information packets in cases, for instance, that the direct radio path between source and sink is greatly deteriorated by the instantaneous channel conditions. One field of application for multi–hop is coverage extension. In such a case the forwarding (or relaying) node is part of the network (belonging to the network provider) and therefore it is controlled by a central entity. In principle, the forwarding operation in this scenario can not be considered as cooperative as the forwarder is not gaining by it. Formally, we understand by cooperation any (agreed) interaction that brings mutual benefits to the cooperating entities. In case of a mobile and wireless forwarder the question whether it is cooperation or not depends on the scenario. In the following we consider two possible scenarios for a wireless communication system based on WLAN with rate adaptation techniques such as used in the IEEE WLAN standards 802.11a/g.

As illustrated in Figure 1.4 the node A N_A (source) needs to transmit packets to the access point AP (sink). Due to the instantaneous channel conditions a direct link can not be established. By using node B N_B as a relay, the packets may reach their destination. The next question arises: why should node B do the forwarding? While the gain of node A is evident what is the gain in doing it for node B? Certainly some of the charge in the battery of node B will be drained by the relaying process and this may not be the intention of the user of node B.

One of the most highlighted arguments for relaying is the fact that the nodes may switch their roles and the relaying node may benefit the next time. But as we have seen in prior sections, the cooperation is based on reciprocity and for that the nodes need to be identified. Note, that in this context reciprocity can not be taken as a given, but is based on the probability of detecting cheaters

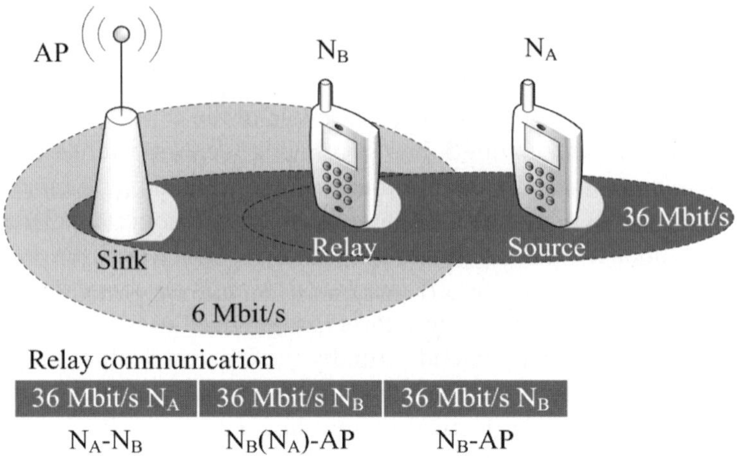

Figure 1.4. Relaying example with source node N_B and relaying node N_A hoping to get paid off later.

and having the capability to punish them. For the relaying example, this means that we in principle would need to remember all terminals that refuse to relay. In the social domain [Ridley, 1998] it is stated that the number of closer human interconnections (number of address in address-book) is around 150. Certainly the mobile handhelds base will be much larger, and therefore it is a problem to store all defecting nodes. It has been shown by [Axelrod, 1984] that if the chance of meeting again is small, defection is the most effective strategy. Furthermore, some terminals may simply be switched off (after taking advantage of the relaying by others) and will not be detected as defectors. The vampire bats had an easier job detecting cheaters, they all have to come back to their roosts at the end of the night and can not hide their state. As the number of terminals that may take part in the relaying is huge, defectors may profit from this and will hardly be recognized as such. If we go back to the example of the vampire bats, if the number of bats would be much larger, defection will pay off. This is especially true if the number of bats is larger compared to the life cycle of a bat.

On the other hand, there are examples of relaying that may be referred to as cooperation. We take the same setup as before except that we assume that the distance between the nodes N_A and N_B is smaller. As given in Figure 1.5 node A has now two possibilities: 1.) send directly to the access point with 6 Mbit/s or, when forwarding is supported, by node B to transmit twice with 36 Mbit/s. In this scenario whether to cooperate or not depends on node B. As node B can only access the medium after node A has finished its transmission, node B may decide to forward the packets to free the channel for its own potential transmissions. This last example underlines once more the principle

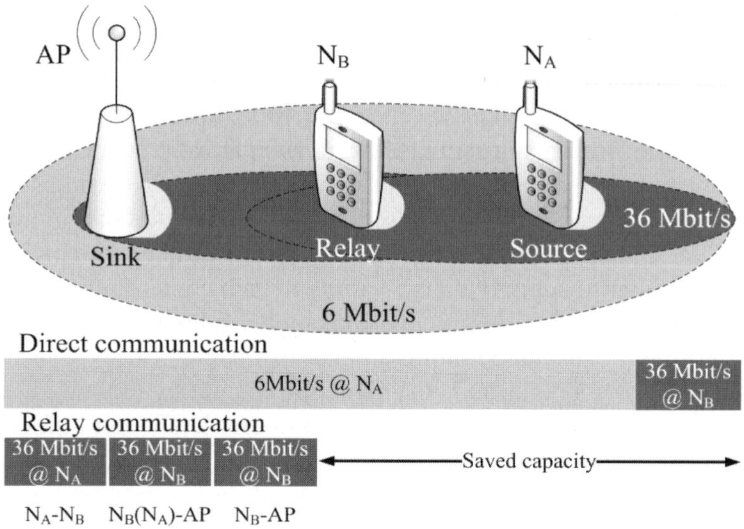

Figure 1.5. Relaying example with source node N_B and relaying node N_A having incentive to cooperate.

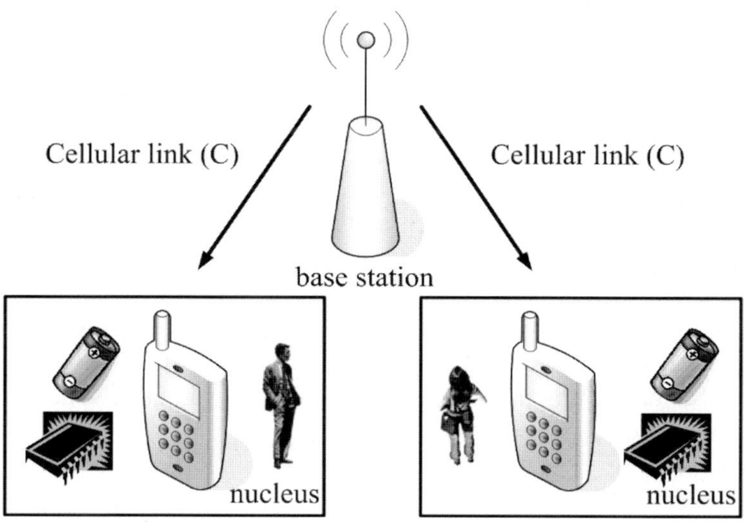

Figure 1.6. State of the art approach for wireless networks with autarky terminals.

of cooperation. Node B will invest some battery to get more wireless capacity. In this example the cooperative attitude pays off and both nodes may gain. This situation can also be referred to as *enlightened self interest* because node B may not have an immediate gain but it looks ahead to the probability that it will soon have traffic to transmit and works to keep the channel free and available.

Explicit micro cooperation. The state of the art architecture in wireless communication is given in Figure 1.6. The base station controls and communicates with the terminals. The terminals need to hold all necessary capabilities to support the services requested by the user. In past and current wireless systems terminals support roughly the same services, such as voice. However, already in the current third generation (3G) mobile communication systems, one can see a trend towards supporting varied services. It is expected that in upcoming wireless communication systems, *e.g.*, 4G, this trend will be more pronounced. Henceforth, in order to support a large variety of rich content services it is natural to expect more complex terminals, with increased power demands. These dependencies will be discussed in more detail in Chapter 14.

An alternative to cooperation is the approach of autarky (or self-sustaining) terminals; that is, terminals that attempt to be completely self sufficient and independent. Thus, by definition such a terminal will never ask for or grant assistance. We argue that an autarky terminal will be more complex in terms of hardware realization and in turn it will likely consume more power than a counterpart exploiting cooperative techniques. Both the complexity and the request for more power deteriorate the market potential of a wireless terminal. Hardware complexity leads to higher prices, while larger power consumption leads to shorter standby times. Both criteria are important for the user's decision buying a mobile terminal as given in [TNS, 2005].

Here we advocate exploiting cooperative behavior in cellular communication systems. As given in Figure 1.7, we assume that each terminal has the capability of communicating with the base station (or access point) and simultaneously with other terminals. In state of the art wireless networks, the former can be realized by technologies such as Global System for Mobility (GSM) or Universal Mobile Telecommunications System (UMTS), while the later can be realized by Bluetooth or Wireless Local Area Network (WLAN). In future wireless communication systems the same air interface may be used for short range as well as cellular communication. Alternatively, multi-mode terminals are expected to have implemented on board several air interfaces being able to operate simultaneously if required.

From the point of view of resources, every terminal has a given battery capacity. Furthermore the user has a certain willingness to pay for a given service and to invest on the mobile terminal in the beginning. Here we provide a rather generic example of cooperation. Later chapters will refine this example. Here we focus on the principle of cooperation. Further assumptions, beside the one that you need for a cellular and a short range communication, are that the short range communication can be done at higher rate and with a lower power budget than the cellular communication. As an example assume the base station conveys multicast information (see Chapter 16 for Multiple Description Coding (MDC)) towards the terminals provided by two different sub-streams. By multicast we

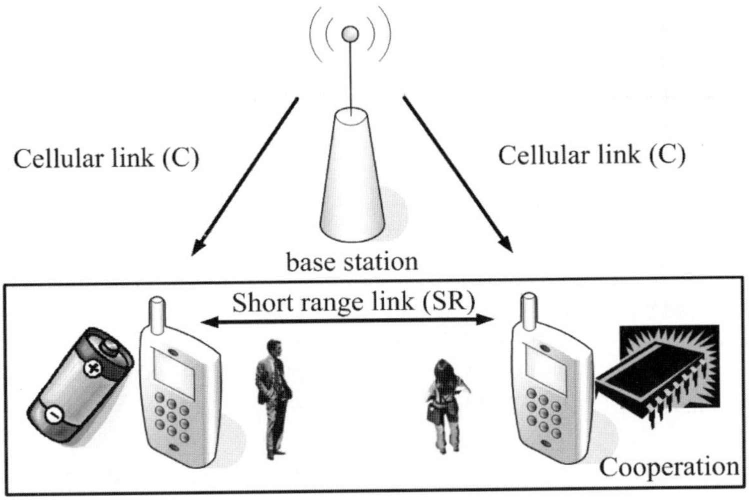

Figure 1.7. Cooperation among terminals exploiting the potential of combined data reception/transmission, battery, and processing unit.

address the scenario where multiple terminals are interested in the same content distributed by the central access point. A stand–alone mobile terminal needs to receive both sub-streams to decode the overall signal and extract the service successfully. But the receiving process of two sub-streams requires twice the power $P_{c,rx}$ needed for the reception of the cellular communication. In a cooperative scenario, the two terminals may receive only one sub-stream each (hopefully disjoint) and forward it by the short range channel to its cooperating counterpart. For the local exchange the terminal needs to invest $P_{sr,tx}$ to send on the short range link and $P_{sr,rx}$ to receive on the same technology. In this case the same service is received at a power budget $P_{c,rx} + P_{sr,tx} + P_{sr,rx}$ as given in Table 1.3.

When the power budget of the cooperative approach does not exceed the corresponding budget for the cellular (non-cooperative) case, that is $P_{c,rx} + P_{sr,tx} + P_{sr,rx} < 2 \cdot P_{c,rx}$, then cooperation is favorable over defecting. In this case both terminals are gaining. But lets consider the following case: One terminal decides to switch off its forwarding streams to save power. Still this terminal would have high quality video services in contrast to the still cooperating one, where the quality drops down. In case the terminal is detecting this situation that it gets exploited, it will do the same and both have to back up with the larger power consuming cellular communication to receive the high quality video. This brings us to one important designing rule, that the defection has to be taken into account.

Table 1.3. The Prisoner's Dilemma Example for Wireless Communication.

		Terminal A	
		Cooperate	Defect
	Cooperate	$P_{c,rx} + P_{sr,tx} + P_{sr,rx}$ High video quality	Terminal A has high quality and spend only $P_{c,rx} + P_{sr,rx}$ Terminal B has low video quality and spends $P_{c,rx} + P_{sr,tx} + P_{sr,rx}$
Terminal B	Defect	Terminal B has high quality and spend only $P_{c,rx} + P_{sr,rx}$; Terminal A has low video quality and spends $P_{c,rx} + P_{sr,tx} + P_{sr,rx}$	$2 \cdot P_{c,rx}$ High video quality

Designing Rules for Cooperative Wireless Networks

In this section we discuss designing rules for cooperative wireless networks at the system level. The rules do not depend on specific protocol layers, but should be addressed by them. To enable cooperation in wireless communication systems the following points need to be fulfilled or be taken into consideration

Access for cellular and short range communication systems. The first point addresses the essential capabilities to communicate among cooperative entities besides the existing cellular communication. In omnipresent wireless networks the cellular link may be represented by UMTS/GSM, while the short range communication may be realized by WLAN or Bluetooth. Future wireless communication systems, as they may identify cooperation as an enabling technology, may come up with a unified air interface that supports both the cellular and the short range communication in a dynamic way.

Cooperation discovery, announcement, and maintenance. When the possibility to communicate with the base station and cooperating terminals is given, the next step is to identify cooperating entities and to announce its own willingness in cooperation on a given task. As cooperation may be applied for different goals, a sophisticated announcement scheme is needed. In particular for mobile networks, where the setup of the nodes may vary, it has to be carefully checked over the run–time if the existing cooperation is still the best option, it may be the case that cooperation has to be dismissed. On the other hand a new cooperative group may be formed achieving better performance than the prior group. We highlight this in Chapter 11 in terms of power consumption.

Timely reciprocity. From the user standpoint, as reciprocity is the main driving force for cooperation, the feedback of a cooperative interaction should come in a timely fashion. In other words, every cooperating party should see his or her benefits in doing so with the shortest possible delay. The incentives should not be hidden and they should be readily available in most of the cases. This can be clearly seen in the reception and distribution of MDC streams, where every new stream increases QoS almost instantaneously. On the other hand, an appealing incentive is hard to see in a typical relaying scenario, therefore external motivation schemes may be needed for this scenario.

Evaluate the n–dimensional cost function for cooperation. Each entity needs to evaluate its own n–dimensional cost function whether to cooperate or not. In the prisoner's dilemma the cost function was easy as only one dimension (points 0, 1, 3, or 5) was given. For mobile phones the n–dimensional cost function is spanned by service costs, battery energy, processing gain, and spectrum. Means are needed to allow nodes to evaluate a certain situation where to dimensions have to be traded against each other. As an example the service costs may decrease at the cost of more power consumption. Whether this is acceptable or not depends on the current situation of the user (maybe he has no possibility within the near future to charge his terminal and therefore he is more picky on the power issue, or just the other way round).

Memory of cooperating and defecting mobile phones. As learned from the prisoner's dilemma, each terminal needs to identify other entities to decide whether to cooperate or not. Each entity may have different strategies and to prevent exploitations by others, the willingness to cooperate depends on the counterpart entity. Also consideration may be given to advanced defecting detection like gossiping (telling others about recently encountered defectors) and associations (User A's PDA will not cooperate but User A's mobile phone will, so forward for the PDA and the mobile phone will forward for you).

Network services supporting cooperative behavior. As we have already mentioned before, the network should try to stimulate the cooperation by services that support cooperative services. MDC, as given in Chapter 16, is one example of such support.

Avoid free riders or cheaters. As given in the second postulate to stimulate cooperation, cheaters and free–riders should have no chance to benefit as this may lead to less willingness in the group to cooperate.

Security. Cooperation involves at least one entity that is intentionally aware of the ongoing communication. For some scenarios that is exactly the way it is

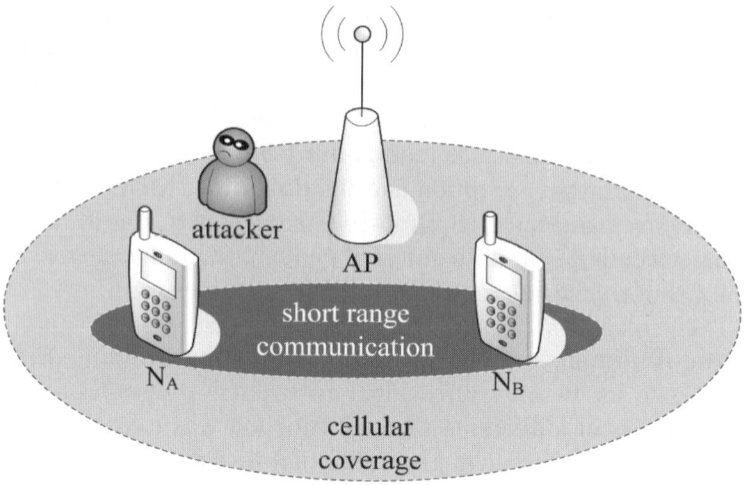

Figure 1.8. Security example for cooperative wireless terminals.

meant to be if we take multicast services into account. On the other side cooperation could also be used in case of private information transmission. In such case some may see the cooperation as a risk. But knowing about the cooperation and the resulting communication paths, the security can even be increased by cooperation. Assume an attacker tries to listen to the ongoing communication between the base station and a wireless node. If partial information is now transmitted over one cooperating entity, the attacker has to be aware of this communication and be physically in range of both communications as given in Figure 1.8. This is even more difficult for the attacker if directed beams are used. Cooperating wireless networks have to be aware of the security issues and exploit the potentials of the cooperation. In case the involved cooperative entity is not fully trustworthy more cooperative entities can be involved creating diversity among the terminals. The concept of impersonation and forgery also comes under the topic of security. If nodes are making decisions about cooperation based on the past behavior of nodes, it is important that nodes cannot assume new identities or make a false report about status (*i.e.*, report that it did forward a packet but not actually forward it).

7. Cooperative Principles in Wireless Communications: The Future

Several examples of cooperation in wireless networks were presented and briefly discussed in this motivating chapter. In most of the cases a few cooperating entities were involved. Even in such simple scenarios there are countless possibilities for cooperation, within and across network protocol layers, as well as across different networks. Along this book many of these ideas and concepts will be presented in detail. In an attempt to extend the concepts of cooperation and motivate further research, we consider approaching cooperation in wireless networks from a wider perspective. It is commonly accepted that the continuous advances in information technology are leading us towards a knowledge society. We particularly emphasize on a *wireless knowledge society*, a hyper–connected world where virtually any entity (now we refer to man and machine) is or can be (wirelessly) connected to any other entity. We can easily imagine complex networks where millions of nodes, local and global, interact. Every man and every object becomes in this vision a potential node. This is by no means a far fetched assumption. In a few years from now half of the world population is expected to be a mobile subscriber. Moreover, advances in short–range communications, low power electronics and sensor networks will make it possible to connect everything wirelessly, resulting, in tens of billions of additional nodes according to some estimates. One could still think of conventional access schemes being used in such a colossal scenario, but is this the best approach? Advanced cooperative techniques can be regarded as promising to transfer information from a given node to another node in such a environment. Cooperation here could have wider connotations than in typically considered cases. First, it could be *massive* as one inherent task of every single node would be to cooperate whenever is possible (positive cooperation attitude). We could assume here that cooperation (or cooperative access) would be the underlying principle by which the communication will takes place. One can understand also that cooperation is implicitly present in such a scenario. An analogy with society would be specifically the tacit cooperation that exists between people. Our society is based not only on formal (and explicit) cooperation but, it mostly relies on invisible massive cooperation (and trust) among countless members. In our private life, we do cooperate with our closest baker, farmers around the country and the engineers who designed our mobile phone half a globe away. Based on needs, experience, personal/group/environmental situation, common interests and other factors, groups are formed and cooperation takes place within them. The level of cooperation between different entities may vary reflecting the upcoming needs and type of relationship. The same may apply in the considered wireless scenario. Intricate interactions between the nodes will occur, targeting individual and/or group benefits. Node cooperation at a given instant will probably be decided based on short– and long–term patterns of behavior (experience), targets (needs), channel conditions (personal/environmental situation) and more. The *division of labor*, through specialization of its members, has had a profound impact on the development of modern society, particularly after the industrial revolution. The beneficiaries of the division of labor are not only the component members but also the society itself. It is commonly understood that the division of labor is what makes human society greater than the sum of its parts. In a hyper-connected wireless society, where entities (*e.g.*, machines) are technically capable of cooperating among each other, evolution may lead to particular situations where we can identify certain patterns of division of labor. In other words, in certain situations or in given scenarios the problem of communication between points A and B can be efficiently solved by a particular arrangement of cooperating entities, where each involved entity takes a given role (specialization) to carry out a specific task. Eventually, in a different situation, entities could take a different role. Note that the referred specialization includes different tasks assigned to a group of similar entities.

8. Conclusion

In this chapter we have explored the basic ideas and concepts of cooperation. The prisoner's dilemma and the iterated version of it were presented and possible strategies were proposed to get the maximal benefit out of those dilemmas. For certain scenario settings such in Axelrod's tournament, cooperation pays off. But cooperation should not be a defacto strategy by itself, but evaluated in each particular case. Cooperation is not pure altruism as this would make those entities vulnerable to selfish and exploiting entities. Some characteristics of how cooperation should be done following the general concept by Axelrod was presented and mapped into the wireless communication systems. As a long term goal, when cooperation is applied to wireless networks, insights of this application may feed back to the social science. We have given the classification of cooperative systems for the wireless domain. We see high potential that cooperative techniques will be applied in future wireless communication systems because of its strength that it has shown on other fields and because it will allow to build less complex terminals with lower power consumption for high quality service provisioning. We motivate the cross over from cooperation in social science towards wireless communication systems.

References

Akella, A., Karp, R., Papadimitrou, C., Seshan, S., and Shenker, S. (2002). Selfish behavior and stability of the internet: A game-theoretic analysis of tcp. In *SIGCOMM 2002*.

Axelrod, R. (1984). *The Evolution of Cooperation*. basic Books.

Axelrod, R. (1997). *The Complexity of Cooperation*. Princeton Paperback.

Axelrod, R. (2004). Tournament on the 20th Anniversary of original Prisoner's Dilemma's competition. *http://www.prisoners-dilemma.com*.

Axelrod, R. and Hamilton, W. D. (1981). The evolution of cooperation. *Science*, 211:1390–1396.

Cagalj, M., Ganeriwal, S., Aad, I., and Hubaux, J. P. (2005). On selfish behavior in csma/ca networks. In *IEEE Infocom 2005*, Miami - FL, USA.

Duncan, K. (2003). Yerkes researchers first to recognize sense of fairness in nonhuman primates – findings shed light on the role of emotion in human economic interaction. *Yerkes National Primate Research Center.*

Edwin, K. (1994). *Mensch und Technik – Methoden systemtechnischer Planung.* trans-aix-press, Aachen.

Grossman, W. M. (2004). New Tack Wins Prisoner's Dilemma. *Wired Magazine.*

Holt, C. A. and Roth, A. E. (2004). The nash equilibrium: A perspective. *Proceedings of the National Academy of Sciences of the United States of America (PNAS)*, 101:3999–4002.

Kurzban, R. and Houser, D. (2005). Experiments investigating cooperative types in humans: A complement to evolutionary theory and simulations. *Proceedings of the National Academy of Sciences of the United States of America (PNAS)*, 102(5):1803–1807.

Macy (1996). Natural selection and social learning in prisoner's dilemma: Co-adaptation with genetic algorithms and artificial neural networks. *Sociological Methods and Research*, 25:103–137.

Nash, J. F. (1950). Equilibrium points in n-person games. *Proceedings of the National Academy of Sciences of the United States of America (PNAS)*, 36:48–49.

Nosratinia, A., Hunter, T. E., and Hedayat, A. (2004). Cooperative communication in wireless networks. *IEEE Communication Magazine*, 42(10):74–80.

Nowak, M. and Sigmund, K. (1993). A strategy of win-stay, lose-shift that outperforms tit-for-tat in the prisoner's dilemma game. *Nature*, 364:56–58.

Pundstone, W. (1992). *Prisoner's Dilemma.* Doubleday.

Ridley, M. (1998). *The Origins of Virtue : Human Instincts and the Evolution of Cooperation.* tbd.

Smith, J. M. (1982). *Evolution of the Theory of Games.* Cambridge University Press.

Stanford Encyclopedia of Philosophy (2003). Prisoner's dilemma. *http://plato.stanford.edu/entries/prisoner-dilemma/.*

TNS (2005). Two-day batter life tops wish list for future all-in-one phone device. Technical report, Taylor Nelson Sofres.

Wedekind, C. and Milinski, M. (1996). Human cooperation in the simultaneous and the alternating prisoner's dilemma: Pavlov versus generous tit-for-tat. *Proceedings of the National Academy of Sciences of the United States of America (PNAS)*, 93(7):2686–2689.

Wilensky, U. (2002). Netlogo pd n-person iterated model. *Center for Connected Learning and Computer-Based Modeling, Northwestern University, Evanston, IL.*

Wilkinson, G. (1984). Reciprocal food sharing in vampire bats. *Nature*, 308:181–184.

Wu, J. and Axelrod, R. (1995). How to Cope with Noise in the Iterated Prisoner's Dilemma. *Journal of Conflict Resolution*, 39:183–189.

Chapter 2

COOPERATIVE COMMUNICATIONS

Fundamental Limits and Practical Implementation

Arnab Chakrabarti
Rice University
arnychak@rice.edu

Ashutosh Sabharwal
Rice University
ashu@rice.edu

Behnaam Aazhang
Rice University
aaz@rice.edu

Abstract: This chapter summarizes theoretically achievable gains and the construction of practical codes for user-cooperation. Most of these results relate to the *relay* channel, which is a three-terminal channel that captures the essence of user-cooperation and serves as one of the primary building blocks for cooperation on a larger scale. In investigating the fundamental limits of relaying, we present information-theoretic results on the achievable throughput of relay channel in mutual-information terms. We also include results on Gaussian channels, and for the practically important case of half-duplex relaying. In the domain of relay coding, we specifically discuss pragmatic code constructions for half as well as full-duplex relaying, using LDPC codes as components.

Keywords: wireless communication, user cooperation, relay, broadcast, multiple access, decode-and-forward, estimate-and-forward, amplify-and-forward, information theory, coding, LDPC, max-flow min-cut

Frank H. P. Fitzek and Marcos D. Katz (eds.), Cooperation in Wireless Networks: Principles and Applications, 29–68.

1. Introduction

Cooperative communication is one of the fastest growing areas of research, and it is likely to be a key enabling technology for efficient spectrum use in future. [1] The key idea in user-cooperation is that of resource-sharing among multiple nodes in a network. The reason behind the exploration of user-cooperation is that willingness to share power and computation with neighboring nodes can lead to savings of overall network resources. Mesh networks provide an enormous application space for user-cooperation strateges to be implemented. In traditional communication networks, the physical layer is only responsible for communicating information from one node to another. In contrast, user-cooperation implies a paradigm shift, where the channel is not just one link but the network itself. The current chapter summarizes the fundamental limits achievable by cooperative communication, and also discusses practical code constructions that carry the potential to reach these limits.

Cooperation is possible whenever the number of communicating terminals exceeds two. Therefore, a three-terminal network is a fundamental unit in user-cooperation. Indeed, a vast portion of the literature, especially in the realm of information theory, has been devoted to a special three-terminal channel, labeled the *relay* channel. The focus of our discussion will be the relay channel, and its various extensions. In contrast, there is also a prominent portion of literature devoted to cooperation as viewed from a network-wide perspective, which we will only briefly allude to.

Our emphasis is on user-cooperation in the domain of wireless communication, and the fundamental limits that we discuss are information theoretic in nature. In this regard, we first bound the achievable rates of relaying using mutual information expressions involving inputs and outputs of the cooperating nodes. We then investigate relaying in the context of Gaussian channels, and summarize known results for well-known relaying protocols. In recent years, half-duplex relaying has been accepted as a practical form of relaying that has potential for implementation in near future. Therefore, we devote a section to the derivation of the fundamental limits of half-duplex relaying. Last, we consider a scenario where the source and the relay exchange roles, which is a departure from the conventional relay channel. This departure, however, captures the essence of user-cooperation where both nodes stand to gain from sharing their resources, which is why this model is a prominent candidate for future implementation.

As regards the coding strategies, we will discuss practical code constructions that emulate random coding strategies used in information theoretic achievability proofs. The component codes of choice are LDPC (Low Density Parity Check) codes, because of their simple factor graph representations, and low-complexity belief propagation decoding. We present code constructions for both half and full-duplex Gaussian relay channels. Many practical challenges encountered in relay coding are exposed in the course of our treatment.

This chapter is organized as follows. First, we present a historical summary of important contributions in the field of relaying. Following that, we include a section to introduce preliminary concepts and terminology for the reader who is unfamiliar with the literature. The next section is devoted to a discussion of information-theoretic limits on the throughput achievable by relaying. In this regard, we pay special attention to Gaussian links; discuss limits of half-duplex relay communication; and finally investigate a scenario where two nodes cooperate with each other without any separate notion of one node being a source and another a relay. The next section is devoted to explicit code constructions for the relay channel, and here we discuss LDPC codes for both full and half-duplex relaying. The final section concludes with a few closing remarks.

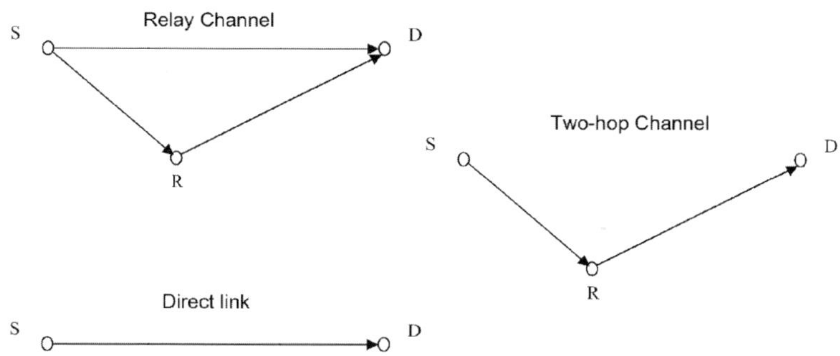

Figure 2.1. Direct, two-hop and relay communications.

2. A Brief History of Relaying

We summarize prominent contributions in the area of user-cooperation. Our list of contributions is by no means exhaustive, but we attempt to touch upon the many approaches to user-cooperation over the years.

Communication from a single source to a single destination without the help of any other communicating terminal is called *direct*, *single-user* or *point-to-point* communication (Figure 2.1). User-cooperation is possible whenever there is at least one additional node willing to aid in communication. The simplest and oldest form of user-cooperation is perhaps multi-hopping, which is nothing but a chain of point-to-point links from the source to the destination (Figure 2.1 shows two-hop communication). No matter what the channel, there is some attenuation of the signal with distance, which makes long-range point-to-point communication impractical. This problem is overcome by replacing a single long-range link with a chain of short-range links, where at each intermediate node there is a *booster* or *repeater* to enhance signal quality. Multi-hopping was conceived about the same time as smoke and drum signals, therefore we do not attempt to put a time stamp on it.

More recently, the three-terminal relay channel (depicted in Figure 2.1) was introduced by [van der Meulen, 1968; van der Meulen, 1971]. In his original work, van der Meulen discovered upper and lower bounds on the capacity of the relay channel, and made several observations that led to improvement of his results in later years. The capacity of the general relay channel is still unknown, but the bounds discovered by van der Meulen were improved significantly by [Cover and El Gamal, 1979]. In the interim, [Sato, 1976] also looked at the relay channel in the context of the Aloha system. Notably, an extensive review of results on several channels that are important to network information theory was published in [van der Meulen, 1977]. The review summarized the state-of-the-art at that time, but our understanding of relaying has improved considerably since then. Other important contributions of the era which contributed to the understanding of user-cooperation include [Slepian and Wolf, 1973; Gaarder and Wolf, 1975; Cover and Leung, 1981; Willems, 1982; Cover, 1972; Cover, 1975; Bergmans and Cover, 1974; Marton, 1979; Gel'fand and Pinsker, 1980; Han, 1981; El Gamal and van der Meulen, 1981; Cover et al., 1980; Wyner, 1978; Wyner and Ziv, 1976].

Undoubtedly, the most prominent work on relaying to date is [Cover and El Gamal, 1979]. Most of the results in this work have still not been superseded. In the years following [Cover and El Gamal, 1979], there was some interest in the relay channel, as is evident from the literature. In [El Gamal and Aref, 1982], the authors discovered the capacity of the semideterministic relay channel, where the received signal at the relay is a deterministic function of the source and relay transmissions. There was an effort to generalize the results of [Cover and El Gamal, 1979] to networks with multiple relays in [Aref, 1980; El Gamal, 1981]. These works also investigated deterministic relay networks with no interference, and deterministic broadcast relay networks.

In parallel with the effort on relaying, there was a prominent body of research on the capacity of the multiple-access channel with generalized feedback (MACGF). This channel was studied in [King, 1978] with a model where two transmitters transmit to a common destination, and these transmitters also receive a common feedback from the destination. In [Carleial, 1982], this model was generalized to include different feedback to the two transmitters. It is easy to see that the relay channel is a special case of Carleial's model. Remarkably, as discussed in [Kramer et al., 2005], Carleial introduced a coding scheme that is different from, and in some respects preferable to the superposition block-Markov encoding introduced by [Cover and El Gamal, 1979].

Perhaps due to the difficulty of finding new and better information-theoretic results, and the technological challenges of implementing user-cooperation, the interest in relaying and user cooperation diminished after the early 80's. Until the turn of the century, there were sporadic contributions on relaying, broadcast, and multiple-access channels as evidenced in [Zhang, 1988; Zeng et al., 1989; Thomas, 1987]. Efforts on relay coding continued until the late 80's and early 90's as evidenced in [Ahlswede and Kaspi, 1987; Kobayashi, 1987; Vanroose and van der Meulen, 1992]. On the other hand, some truly remarkable strides were made during this period in the general area of digital and wireless communications, such as discovering the capacity of multi-antenna systems by [Foschini and Gans, 1998; Telatar, 1999], a great deal of advancement in our understanding of fading channels (summarized in [Biglieri et al., 1998]), and remarkable progress in channel coding including the discovery of Turbo codes in [Berrou et al., 1993], space-time codes in [Tarokh et al., 1998], and the rediscovery of LDPC codes of [Gallager, 1963] in [MacKay, 1999; Luby et al., 2001; Richardson and Urbanke, 2001]. These advances set the stage for a second wave of research on relaying by providing a whole new context and new tools to attack the problem.

One of the prominent works that helped to draw attention to user-cooperation in recent years is [Sendonaris et al., 2003a; Sendonaris et al., 2003b]. In this work, the authors propose user-cooperation as a form of diversity in a mobile uplink scenario, and show its benefits using various metrics. Also noteworthy are the contributions of [Laneman, 2002; Laneman and Wornell, 2003; Laneman et al., 2004] for studying the performance of important relaying protocols in fading environments. Yet another important set of contributions came in the form of novel information theoretic results and new insights into information theoretic coding in [Kramer et al., 2005] (also more recently in [Chong et al., 2005]). In [Schein and Gallager, 2000; Schein, 2001] the authors considered a variation of the relay channel where there is no direct source-destination link, but there are two relays to aid communication. In [Schein and Gallager, 2000; Schein, 2001] the authors considered a variation of the relay channel where there is no direct source-destination link, but there are two relays to aid communication. New information theoretic results and results on power control were also discovered in [Wang et al., 2005; Høst-Madsen and Zhang, 2005]. A variety of contributions to relaying including new bounds, cut-set theorems, power control strategies, LDPC relay code designs, and some of the earliest results on half-duplex relaying were proposed in [Khojastepour, 2004]. Researchers realized that relaying can mimic multiple-antenna systems even when the communicating entities were incapable of supporting multiple antennas. Prominent literature on the use of space-time codes with relays includes [Laneman and Wornell, 2003; Nabar et al., 2004; Mitran et al., 2005]. Other noteworthy recent contributions are by [El Gamal et al., 2004; Reznik et al., 2004; Hasna and

Alouini, 2003; Boyer et al., 2004; Toumpis and Goldsmith, 2003; Liang and Veeravalli, 2005].

In a different direction, [Gupta and Kumar, 2000] proposed a new approach towards finding network information carrying capacity, which led to research on finding scaling laws for wireless networks in a variety of settings. Numerous works were published on studies of networks with large numbers of nodes, as contrasted to the simple few-node channels studied by traditional information theory. In [Gupta and Kumar, 2003], the authors showed that the use of advanced multi-user schemes can improve network transport capacity significantly. Subsequently, [Xie and Kumar, 2004] discovered an achievable rate expression for a degraded Gaussian channel with multiple relays and established bounds on its transport capacity. The results of [Reznik et al., 2004] also treated the case of multiple Gaussian degraded relay stages with a total average power constraint. In another direction, [Gastpar and Vetterli, 2005] showed that the upper-bounds on relay capacity obtained from cut-set theorems coincide with known lower bounds as the number of relays becomes large. Other prominent contributions in this area include [Xue et al., 2005; Xie and Kumar, 2005; Grossglauser and Tse, 2002].

With significant advances in technology over the last two decades, the promise of relaying is very real. A large body of research is currently geared towards developing practical user-cooperation schemes to harvest the gains predicted by information theory. Solutions in this direction include [Sendonaris et al., 2003b; Stefanov and Erkip, 2004; Janani et al., 2004; Hunter et al., 2004; Khojastepour et al., 2004a; Chakrabarti et al., 2005a; Zhang et al., 2004; Zhang and Duman, 2005; Castura and Mao, 2005; Zhao and Valenti, 2003].

Yet another area of user cooperation where recent years have seen an explosion of publications is that of *network coding*. The field grew largely after the publication of [Ahlswede et al., 2000], although an earlier publication [Yeung and Zhang, 1999] also contained seeds of the idea. Subsequently, several important advancements to the field have been made, of which some of the fundamental ones are in [Li et al., 2003; Koetter and Medard, 2003; Chou et al., 2003; Jaggi et al., 2005; Ho et al., 2003; Li and Li, 2004; Yeung et al., 2005].

3. Preliminaries of Relaying

The relay channel is the three-terminal communication channel shown in Figure 2.2. The terminals are labeled the source (S), the relay (R), and the destination (D). All information originates at S, and must travel to D. The relay aids in communicating information from S to D without actually being an information source or sink. The signal being transmitted from the source is labeled X. The signal received by the relay is V. The transmitted signal from the relay is W, and the received signal at the destination is Y. Several notions of

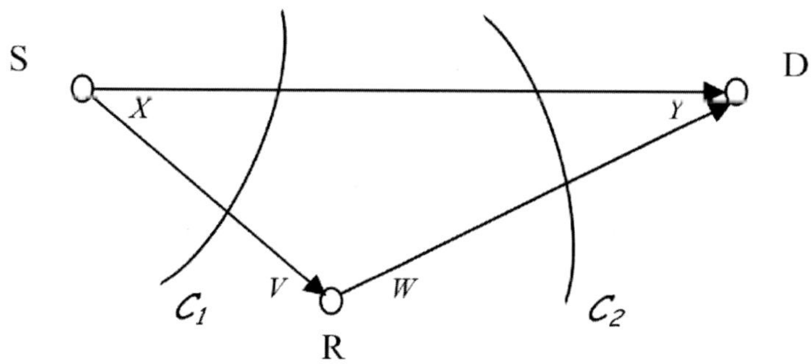

Figure 2.2. The relay channel with three nodes: the source S, the relay R, and the destination D. These three nodes are conceptually divided into two subsets by two cuts of interest: C_1 or the broadcast cut which separates S from $\{R, D\}$, and C_2 or the multiple-access cut, which separates $\{S, R\}$ from D. The channel input at S is given by X, the input at R is W, and the outputs at R and D are V and Y respectively.

relaying exist in the literature. We will list the prominent ones in this section. Conceptually, information is relayed in two phases or modes: first, when S transmits and (R, D) receive, commonly called the broadcast (BC) mode; and second when (S, R) transmit and D receive, also known as the multiple-access (MAC) mode. Note that this differentiation is only conceptual since it is possible for communication in both modes to take place simultaneously. We will elaborate on this a little later, but first we will enumerate four different models of relaying that can be classified based on the above two modes.

1 $S \rightarrow (R, D)$; $(S, R) \rightarrow D$ (most general form of relaying);

2 $S \rightarrow R$; $(S, R) \rightarrow D$ (D ignores signal from S in first mode);

3 $S \rightarrow (R, D)$; $R \rightarrow D$ (S does not transmit in second mode);

4 $S \rightarrow R$; $R \rightarrow D$ (multi-hop communication).

Of these, the first model is the most general, and most early results on relaying were based on the first model. The second and the third are simplified models introduced mainly for analytical tractability. For example, they simplify the analysis of outage probabilities and the design of space-time codes for fading relay channels in [Laneman et al., 2004; Nabar et al., 2004].

The last model of relaying is much older as well as simpler than the other three, and is commonly known as multi-hop communication. Unlike the other three models, multi-hop communication does not yield diversity benefits, and it is primarily used to combat signal attenuation in long-range communication. In wireless communication, usually there is severe attenuation of signal power with distance. This attenuation is characterized by a channel attenuation exponent γ. In other words, if the transmitted power is P, then the received power at a distance d is $\frac{P}{d^\gamma}$. The value of γ lies in the range of 2 to 6 for most wireless channels. This attenuation makes long-range communication virtually impossible. The simplest solution to this problem is to replace a single long-range link with a chain of short-range links by placing a series of nodes in between the source and the destination. A distinguishing feature of multi-hopping is that each node in this chain communicates only with the one before and the one after in the chain, or nodes that are one "hop" away. In a wireless environment, it may be possible for a node to receive or transmit its signal to other nodes that are several hops away, but such capability is ignored in multi-hopping, making it a simple and extremely popular, but suboptimal mode of user-cooperation. Of all the modes of user-cooperation discussed in this chapter, multi-hopping is the only one that is widely implemented today.

Half-duplex versus Full-duplex Relaying

A relay is said to be half-duplex (or 'cheap' as in [Khojastepour et al., 2003]) when it cannot simultaneously transmit and receive in the same band. In other words, the transmission and reception channels must be orthogonal. Orthogonality between transmitted and received signals can be in time-domain, in frequency domain, or using any set of signals that are orthogonal over the time-frequency plane. If a relay tries to transmit and receive simultaneously in the same band, then the transmitted signal interferes with the received signal. In theory, it is possible for the relay to cancel out interference due to the transmitted signal because it knows the transmitted signal. In practice, however, any error in interference cancellation (due to inaccurate knowledge of device characteristics or due to the effects of quantization and finite-precision processing) can be catastrophic because the transmitted signal is typically 100-150dB stronger than the received signal as noted in [Laneman et al., 2004]. Due to the difficulty of accurate interference cancellation, full-duplex radios are not commonly used; however, dvances in analog processing could potentially enable full-duplex relaying.

Although early literature on information theoretic relaying was based almost entirely on full-duplex relaying (eg. [van der Meulen, 1971; Cover and El Gamal, 1979]), in recent years a lot of research, and especially research directed towards practical protocols, has been based on the premise of half-duplex relaying(eg. [Khojastepour et al., 2003; Liang and Veeravalli, 2005; Janani et al., 2004; Laneman et al., 2004; Nabar et al., 2004]).

Relay Protocols

The capacity of the general relay channel of Figure 2.2 is not known even today, over thirty years after the channel was first proposed. Moreover, there is no single cooperation strategy known that works best for the general relay channel. As we will discuss in a subsequent section on fundamental limits, there are at least two fundamental ideas (and a third that is practically less important) based on which the source and relay nodes can share their resources to achieve the highest throughput possible for any known coding scheme. The cooperation strategies based on these different ideas have come to be known as *relay protocols*.

The first idea involves decoding of the source transmission at the relay. The relay then retransmits the decoded signal after possibly compressing or adding redundancy. This strategy is known as the *decode-and-forward* protocol, named after the fact that the relay can and does decode the source transmission. The decode-and-forward protocol is close to optimal when the source-relay channel is excellent, which practically happens when the source are relay are physically near each other. When the source-relay channel becomes perfect, the relay channel becomes a 2×1 multiple-antenna system. Following the naming convention of [Cover and El Gamal, 1979], some authors use the term *cooperation* to strictly mean the decode-and-forward type of cooperation.

The second idea, sometimes called *observation*, is important when the source-relay and the source-destination channels are comparable, and the relay-destination link is good. In this situation, the relay may not be able to decode the source signal, but nonetheless it has an independent observation of the source signal that can aid in decoding at the destination. Therefore, the relay sends an estimate of the source transmission to the destination. This strategy is known as the *estimate-and-forward* (also known as *compress-and-forward* or *quantize-and-forward*) protocol.

The *amplify-and-forward* (also sometimes called *scale-and-forward*) protocol is a special case of the above strategy where the estimate of the source transmission is simply the signal received by the relay, scaled up or down before retransmission. A 1×2 multi-antenna system is a relay channel where amplify-and-forward is the optimal strategy, and the amplification factor is dictated by the relative strengths of the source-relay and source-destination links.

The third idea, known as *facilitation*, is mostly of theoretical interest. When the relay is not able to contribute any new information to the destination, then it simply tries to stay out of the way by transmitting the signal that would be least harmful to source-destination communication.

The names for the protocols that we have described above are generally accepted by the relaying community. However, the reader is cautioned that some authors refer to the aforementioned protocols differently. For example, in [Khojastepour, 2004], *scale-and-forward* and *amplify-and-forward* refer to different schemes. Therefore, it is always a good idea to check the authors' definitions of scientific terms used in a document.

4. Relaying: Fundamental Limits

The following is a brief outline of this section. First, we will summarize well-known information theoretic results on the full-duplex relay channel stated in terms of mutual information expressions. Second, we will present the achievable rates of Gaussian channels for various relay protocols. Following that, we will discuss results on half-duplex relaying. Finally, we will briefly summarize the results of [Sendonaris et al., 2003a; Sendonaris et al., 2003b] for a three-terminal network where each node acts as both source, and (full-duplex) relay for the other node. A discussion on fundamental limits of relaying cannot be complete without results on fading channels; however, they will be treated in a separate chapter by Nicholas Laneman.

The relay channel is shown in Figure 2.2. We will assume that the channel is discrete time and memoryless. The signals X,V,W, and Y are chosen from finite sets $\mathcal{X}$,$\mathcal{V}$,$\mathcal{W}$, and $\mathcal{Y}$ respectively, and the channel is described by the conditional probability densities $p(v, y|X = x, W = w)$ on $(\mathcal{V} \times \mathcal{Y})$ for all $(x \times w) \in (\mathcal{X} \times \mathcal{W})$. In what follows, we briefly summarize important results known for the relay channel in terms of mutual information expressions. Many of these results are due to [Cover and El Gamal, 1979].

For the special case of a degraded relay channel, *i.e.*, when $X \rightarrow (V, W) \rightarrow Y$ is a Markov chain, the following theorem from [Cover and El Gamal, 1979] gives the capacity of the relay channel.

THEOREM 2.1 *[Cover and El Gamal, 1979] The capacity C_d of the degraded relay channel is given by*

$$C_d = \sup_{p(x,w)} \min\left(I(X, W; Y), I(X; V|W)\right). \tag{2.1}$$

The rate of Theorem 2.1 is achievable for any relay channel (not necessarily degraded). The premise of degradedness is used only to prove that this rate cannot be surpassed, and is therefore the capacity when the channel is degraded. The practical utility of Theorem 2.1 stems from the fact that it provides an achievable rate (a lower bound on the capacity) for the general relay channel. This lower bound is fairly tight if the source-relay (SR) channel is better than the relay-destination (RD) channel, which may physically correspond to a scenario where R is closer to S than to D. This can be attributed to the fact that the relay channel resembles a degraded channel more and more closely as the SR link improves relative to the RD link. A 2×1 multi-antenna system can be thought of as a degraded relay channel, where the two transmitter antennas corresponding to the source and the relay have a perfect communication channel in between them.

Achieving the rate of Theorem 2.1 requires a coding scheme where R decodes the signal it receives from S before passing it on to D. Therefore, Theorem 2.1 corresponds to the achievable rate of the *decode-and-forward* relaying protocol. We will discuss practical code designs for this protocol in a subsequent section on relay coding.

For a reversely degraded relay channel *i.e.*, when $X \rightarrow (Y, W) \rightarrow V$ is a Markov chain, the capacity of the relay channel is given by the following theorem in [Cover and El Gamal, 1979].

THEOREM 2.2 *[Cover and El Gamal, 1979] The capacity C_{rd} of the reversely degraded relay channel is given by*

$$C_{rd} = \max_{w \in \mathcal{W}} \max_{p(x)} I(X; Y|w). \tag{2.2}$$

Theorem 2.2 advocates *facilitation*, where the relay transmits a signal that will maximize the capacity of the SD link. Since R receives a signal that is a noisy version of what D receives, R knows nothing that D does not already know - "thus w is set constantly at the symbol that "opens" the channel for the transmission of x, directly to y at rate $I(X; Y|w)$"as quoted from [Cover and El Gamal, 1979]. As in the case of Theorem 2.1, the achievability of the rate in Theorem 2.2 does not require the reverse degradedness assumption, and is true for all relay channels. This theorem is usually not important in practice because practical relay channels that resemble reversely degraded channels have small capacity. A channel where the SR distance is much larger than the SD distance would mimic reverse degradedness. Geometric constraints dictate that R will be far from both S and D in such a scenario, and it is therefore intuitive that such a relay will not be of much use.

For the general relay channel of Figure 2.2, the following upper bound is due to [Cover and El Gamal, 1979].

THEOREM 2.3 *[Cover and El Gamal, 1979] The capacity C of the general relay channel is bounded above as follows*

$$C \leq \sup_{p(x,w)} \min \left(I(X, W; Y), I(X; V, Y|W) \right). \tag{2.3}$$

The above upper bound is a consequence of a general cut-set theorem for information flow in networks. The reader should refer to Page 444 of [Cover and Thomas, 1991] for an extended discussion of this theorem. Here, we will briefly define notation, and state this elegant and powerful theorem without proof.

A network with multiple terminals is shown in Figure 2.3. We closely follow the conventions in [Cover and Thomas, 1991]. The network consists of N

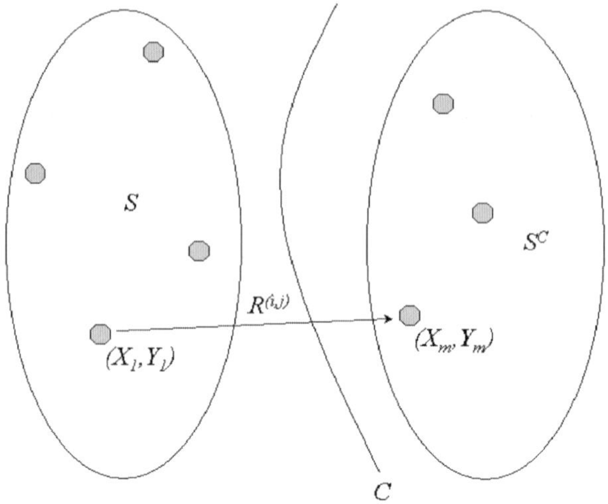

Figure 2.3. A network with multiple nodes divided in two sets S and S^C separated by a cut C.

nodes. Node i is characterized by the input-output pair $(X^{(i)}, Y^{(i)})$, and sends information at a rate $R^{(}ij)$ to node j. The nodes are divided in two sets S and S^C (the complement of S), and a cut C conceptually separates the nodes in these two sets. The channel is represented by the conditional probability mass function $p(y^{(1)}, y^{(2)}, ..., y^{(N)}|x^{(1)}, x^{(2)}, ..., x^{(N)})$ of the outputs given the inputs. The following theorem bounds the rate of information transfer from nodes in S to those in S^C by the conditional mutual information.

THEOREM 2.4 *[Cover and Thomas, 1991] If the information rates $\{R^{(ij)}\}$ are achievable, then there exists a joint input probability distribution $p(x^{(1)}, x^{(2)}, ..., x^{(N)})$ such that*

$$\sum_{i \in S, j \in S^C} R^{(ij)} \leq I\left(X^{(S)}; Y^{(S^C)} | X^{(S^C)}\right) \tag{2.4}$$

for all $S \subset 1, 2, ..., N$.

The above theorem is analogous to the well-known *max-flow min-cut* theorem [Ford and Fulkerson, 1962]. [2] It says that the rate of information flow across any cut (boundary) dividing the set of nodes in a network in two parts (the transmitter and receiver sides) cannot exceed the mutual information between the channel inputs on the transmitter side and the channel outputs on the receiver side conditioned on the knowledge of inputs on the receiver side.

The aforementioned cut-set theorem is a powerful tool for bounding the capacity region from above for several important multi-terminal communication channels. Capacity results in network information theory are usually difficult to prove, and in many important cases such as the multiple-access channel and the degraded relay channel, the capacity of a channel is known because the upper bound given by the max-flow min-cut theorem is achievable. Unfortunately, the upper bound given by the cut-set theorem is not always achievable. There are examples where the capacity has been shown to be smaller than what is indicated by the cut-set bound. One of the prominent cases is that of the Gaussian vector broadcast channel (GVBC) (see [Caire and Shamai, 2003; Vishwanath et al., 2003; Yu and Cioffi, 2004; Viswanath and Tse, 2003]), where the upper bound on sum capacity is derived based on the work of [Sato, 1978], and achievability has been shown using [Costa, 1983; Marton, 1979]. Nevertheless, the cut-set theorem remains one of the most powerful tools in network information theory.

Notably, the above cut-set theorem has recently been extended to networks with multiple states by [Khojastepour, 2004]. We will discuss the extension in the context of half-duplex relaying in a subsequent section.

Returning to our discussion of the relay channel, the reader will observe that Theorem 2.3 follows immediately from Theorem 2.4. For the general relay channel, the upper bound of Theorem 2.3 is not known to be achievable. However, it *is* achievable when feedback is available, as shown in Theorem 3 of [Cover and El Gamal, 1979]. In a relay channel with feedback, S as well as R know the signal received by D (with unit delay, which does not reduce capacity). In fact, a relay channel with feedback is a degraded relay channel where V is replaced with (Y, V), therefore using Theorem 2.1, the rate of (2.3) is achievable. Here, the authors would like to caution the reader that results based on feedback in information theory can be misleading because this feedback does not carry a cost. In practice, feedback, like all other information, must be communicated, and will therefore consume a fraction of the channel capacity.

Apart from decode-and-forward (Theorem 2.1) and facilitation (Theorem 2.2), there is at least one other idea in cooperation that yields useful achievable rates, exceeding those of Theorem 2.1 in some scenarios. The idea corresponds to the estimate-and-forward protocol. Cover and El Gamal realized that in between the two cases of a degraded channel, where the relay can perfectly decode the source signal, and the reversely degraded channel, where the relay can contribute no new information, there exists a regime where the relay can contribute partial information to improve decoding at the destination. This partial information is in the form of an estimate of the source signal received by the relay. As an example, suppose that both SR and SD channels are AWGN links having the same noise variance. The channel is neither degraded nor reversely degraded, but since the relay and the destination have independent views of the noise, the destination would benefit from knowing the received signal at the relay, or an estimate of it. The achievable rate for this strategy is given by the following theorem

THEOREM 2.5 *[Cover and El Gamal, 1979] The rate R_{cf} is achievable for a general relay channel where*

$$R_{cf} = \sup I(X; Y, \widehat{V} | W) \tag{2.5}$$

subject to the constraint

$$I(W; Y) \geq I(V; \widehat{V} | W, Y) \tag{2.6}$$

where the supremum is over all joint distributions of the form $p(x, w, y, v, \widehat{v}) = p(x)p(w)p(y, v|x, w)p(\widehat{v}|v, w)$ *on* $\mathcal{X} \times \mathcal{W} \times \mathcal{Y} \times \mathcal{V} \times \widehat{\mathcal{V}}$ *and* $\widehat{V}$ *has a finite range.*

The above theorem introduces the idea of an estimate $\widehat{V}$ of V that is communicated from R to D when decoding is not possible. The final theorem (Theorem 7) in [Cover and El Gamal, 1979] superimposes the decode-and-forward and compress-and-forward strategies to yield a composite achievable rate that reduces to the achievable rates of the two aforementioned strategies in special cases. We do not present the theorem here, but the interested reader can refer directly to [Cover and El Gamal, 1979].

In addition to the above results, all of which are from [Cover and El Gamal, 1979], there are few other results known for special cases of the relay channel. The capacity of the semideterministic relay channel, where the channel output at the relay is a deterministic function of both source and relay inputs, was derived in [El Gamal and Aref, 1982] using Theorem 7 of [Cover and El Gamal, 1979] and Theorem 2.3 presented above. In addition, two new results were discovered in [Khojastepour, 2004] using a new set of tools developed in the context of half-duplex relay channels. These tools include a new cut-set theorem (presented as Theorem 2.11 in this chapter) as well as a novel coding scheme introduced in [Khojastepour et al., 2002b; Khojastepour, 2004]. We will state these results as follows.

THEOREM 2.6 *[Khojastepour, 2004] If the relay channel transition function can be written in the form*

- $p(y, v|x, w) = p((y_1, y_2), v_1|(x_1, x_2), w_2) = p(y_1|v_1)p(v_1|x_1) p(y_2|x_2, w_2)$ *then the capacity of the relay channel is*

$$C_1 = \sup_{p(x_1)p(x_2,w_2)} \min \Big(I(X_1; V_1) + I(X_2; Y_2|W_2), \\ I(X_1; Y_1) + I(X_2, W_2; Y_2) \Big); \tag{2.7}$$

- $p(y, v|x, w) = p((y_1, y_2), v_1|(x_1, x_2), w_2) = p(y_1, v_1|x_1)p(y_2|x_2, w_2)$ *then the capacity of the channel is*

$$C_2 = \sup_{p(x_1)p(x_2,w_2)} I(X_1; Y_1) + I(X_2, W_2; Y_2). \tag{2.8}$$

subject to the constraint

$$I(X_1;V_1) + I(X_2;Y_2|W_2) \geq I(X_1;Y_1) + I(X_2,W_2;Y_2). \qquad (2.9)$$

The above two results were derived in the context of the half-duplex relay channel, therefore the subscripts 1 and 2 for broadcast and multiple-access modes respectively. However, the results are equally applicable to the full-duplex relay channel.

In addition to the above results, new and better achievable rates have been recently reported in [Chong et al., 2005] based on two new coding strategies that combine ideas of decode-and-forward and compress-and-forward. The encoding and decoding differ from that of [Cover and El Gamal, 1979] in that regular block-Markov sperposition encoding and backward decoding [Willems, 1982] are used.

Capacity Results for the Gaussian Relay Channel

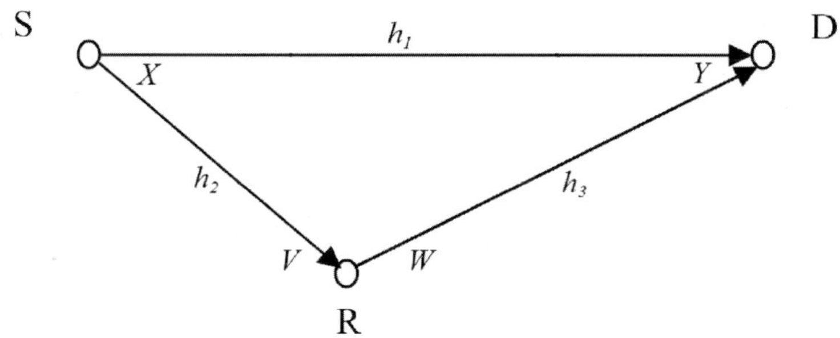

Figure 2.4. Gaussian relay channel.

In this section, we discuss information theoretic results for Gaussian relay channels. We first introduce our notation and state the well-known upper bound that can be derived from Theorem 2.3. Then we discuss several lower bounds including some recently discovered ones. The lower bounds are of practical interest because every lower bound is a consequence of a constructive coding scheme, and carries the potential for implementation in future.

A full-duplex Gaussian relay channel is shown in Figure 2.4. For this discussion, we will use the notation of [Khojastepour, 2004]. The dependence of channel outputs on inputs is as follows: the output at the relay is $v = h_2x + n_1$, and the output at the destination is $y = h_1x + h_3w + n_2$. Here, h_1, h_2 and h_3 are the SD, SR and RD channel gains respectively, and n_1 and n_2 are Gaussian noise with variance N_1 and N_2 respectively. The three variables $\gamma_1 = \frac{|h_1|^2}{N_2}$, $\gamma_2 = \frac{|h_2|^2}{N_1}$, and $\gamma_3 = \frac{|h_3|^2}{N_2}$ denote the SD, SR, and RD channel SNRs respectively as shown in Figure 2.4. The input power constraints are given by $E[X^2] \leq P_1$ and $E[W^2] \leq P_2$.

The following upper bound on the Gaussian relay capacity can be obtained from Theorem 2.3.

THEOREM 2.7 *[Cover and El Gamal, 1979]*

$$C_{AWGN} \leq \max_{\rho:0\leq\rho\leq 1} \min \Big(C((1-\rho^2)(\gamma_1+\gamma_2)P_1), \quad C(\gamma_1 P_1 + \gamma_3 P_2 + 2\rho\sqrt{\gamma_1\gamma_3 P_1 P_2}) \Big) \tag{2.10}$$

$$\textit{where} \quad C(x) = \frac{1}{2}\log(1+x). \tag{2.11}$$

The parameter ρ in the above equation has the physical interpretation of the correlation between source and relay signals. Increasing ρ in equation (2.9) increases the mutual information term corresponding to the multiple-access cut, which is the second argument of the min function above. At the same time, increasing ρ decreases the mutual information term corresponding to the broadcast cut, which is the first argument of the min function above. This interpretation of ρ is also true for several of the lower bounds that we present below.

Perhaps the most prominent lower bound on the capacity of the Gaussian relay channel is a consequence of Theorem 2.1. This lower bound was actually derived in Theorem 5 of [Cover and El Gamal, 1979] as the capacity of the Gaussian degraded relay channel. Using our notation, we present the lower bound in the following theorem

THEOREM 2.8 *[Cover and El Gamal, 1979]*

$$C_{AWGN} \geq R_{DF} = \max_{\rho:0\leq\rho\leq 1} \min \Big(C((1-\rho^2)\gamma_2 P_1), \quad C(\gamma_1 P_1 + \gamma_3 P_2 + 2\rho\sqrt{\gamma_1\gamma_3 P_1 P_2}) \Big) \tag{2.12}$$

where $C(x)$ is defined in (2.11).

Since the above lower bound coincides with the capacity of the degraded AWGN relay channel, we do not expect it to be tight when the channel is far from being degraded, for example when the SD link received SNR exceeds that of the SR link. In this scenario, even the capacity of the direct link $C(\gamma_1 P_1)$ exceeds that of decode-and-forward relaying. This is true even if unbounded power is available at the relay. This is certainly not the best that we can do with the relay. For instance, consider the case where the SR and SD channels have identical SNRs, and the RD channel is noiseless. Here, the relay channel becomes a 1×2 MIMO system, where the best strategy is for the relay to simply forward its analog received signal with appropriate power, so that the destination effectively receives a signal that is the result of maximal-ratio combining. In general, depending on the amount of noise in the RD channel, the relay may spend variable amounts of resources on sending an estimate of its received signal to the destination. Estimate-and-forward relaying, as proposed in [Cover and El Gamal, 1979], does not explicitly state the nature of the estimate to be forwarded. Achievable rates of the estimate-and-forward protocol have been derived for the Gaussian channel in [Khojastepour et al., 2004b; Kramer et al., 2005] using ideas of source coding at the relay and decoding with side information at the destination [Wyner and Ziv, 1976]. The following theorem gives the achievable rate of estimate-and-forward relaying for the Gaussian relay channel of Figure 2.4.

THEOREM 2.9 *[Khojastepour et al., 2004b] The achievable rate of estimate-and-forward relaying is given by*

$$C_{AWGN} \geq R_{EF} = C\left(\gamma_1 P_1 + \frac{\gamma_2 P_1 \gamma_3 P_2}{1 + \gamma_2 P_1 + \gamma_1 P_1 + \gamma_3 P_2}\right) \tag{2.13}$$

where $C(x)$ is defined in (2.11).

The above theorem shows that the rate achievable by estimate-and-forward relaying can always surpass that of the direct link, although not always exceed that of the decode-and-forward protocol.

An achievable rate for the estimate-and-forward protocol has also been derived in [Sabharwal and Mitra, 2005]. Yet another rate is known to be achievable using retransmission of a quantized version of the received signal at the relay. The rate is presented as the achievable rate of the quantize-and-forward protocol in Chapter 3 of [Khojastepour, 2004].

The following theorem of [Khojastepour, 2004] gives the rate of amplify-and-forward relaying under the name of scale-and-forward relaying. Conceptually, amplify-and-forward is a simple (but not always the best) way of implementing the principle of Theorem 2.5. Here, the estimate being forwarded is simply a scaled version of the received signal. We divide the transmission into $L \to \infty$ consecutive sub-blocks of length M. Furthermore, assume that the relay builds its input signal based on all previously received signals in each sub-block. For the simple case of $M = 2$, the following theorem gives the achievable rate.

THEOREM 2.10 *[Khojastepour, 2004] The optimal achievable rate of the scale-and-forward scheme with the sub-block length M = 2 for the Gaussian*

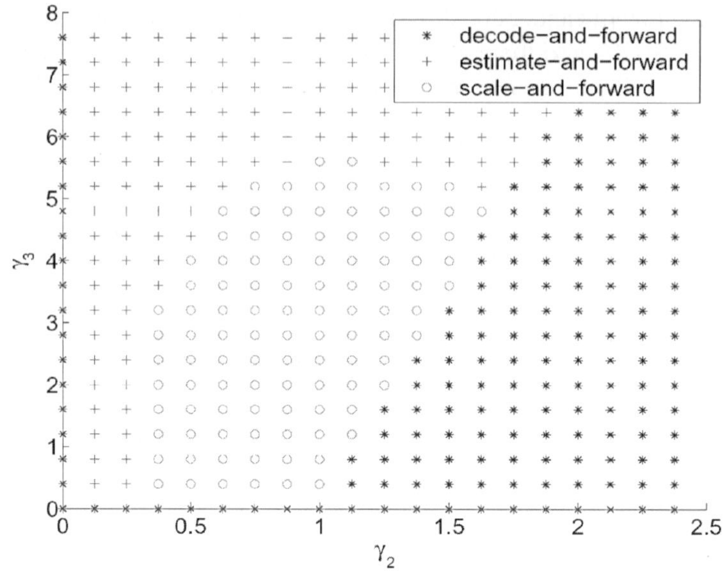

Figure 2.5. Regions where each protocol outperforms all others ($P_1 = P_2 = -10dB$, $\gamma_1 = 1$).

relay channel is given by

$$C_{AWGN} \geq R_{SF(M=2)} = \max_{Q_1, Q_2 : Q_1 + Q_2 \leq 2P_1 ; \beta \leq \frac{2P_2}{h_2^2 Q_1 + N_1}}$$

$$\frac{1}{2} C \left(\frac{h_1^2 Q_1}{N_2} + \frac{h_1^2 Q_2 + \beta h_2^2 h_3^2 Q_1}{N_2 + \beta h_3^2 N_1} + \frac{h_1^2 Q_1 Q_2 + \beta h_2^2 h_3^2 N_2^2}{N_2 (N_2 + \beta h_3^2 N_1)} \right) \tag{2.14}$$

where $C(x)$ is defined in (2.11).

Figure 2.5 shows the region of channel conditions where each protocol performs optimally. In this figure, only the SR and RD links change whereas the SD link is fixed. Therefore, in a way, the plot shows relative performance of these protocols as a function of relay position.

The reader is reminded that several notions coexist in the domain of relaying, and for each notion there may be several achievable rates depending on the assumptions made during derivation. Often, finding the achievable rate corresponding to a given protocol under a given channel model is not analytically feasible. This may be true, for instance, if the throughput maximizing input distribution does not have a familiar form. For this reason, it is important to view each achievable rate expression in the light of its context. We have only summarized some of the best achievable rates known for three prominent protocols.

The interested reader can find additional results on Gaussian channels in [Gupta and Kumar, 2003; Khojastepour et al., 2004b; Laneman et al., 2004; Sabharwal and Mitra, 2005; Gastpar and Vetterli, 2005; Schein and Gallager, 2000].

Fundamental Limits for Half-duplex Relays

In this section, we will first present a new cut-set theorem for networks with multiple states due to [Khojastepour, 2004]. This cut-set theorem yields an upper bound on the half-duplex relay capacity. Following that, we will discuss well-known upper and lower bounds for the half-duplex relay channel.

The following theorem from [Khojastepour, 2004] can be thought of as a generalization of Theorem 2.4. For the convenience of the reader, we define our notation again. The network consists of N nodes. Node i is characterized by the input-output pair $(X^{(i)}, Y^{(i)})$, and sends information at a rate $R^{(ij)}$ to node j. The nodes are divided in two sets S and S^C (the complement of S), and a cut C conceptually separates the nodes in these two sets. The channel is represented by a collection of m conditional probability mass functions $p(y^{(1)}, y^{(2)}, ..., y^{(N)}|x^{(1)}, x^{(2)}, ..., x^{(N)}|m)$ of the outputs given the inputs, where m is the state of the network which takes its values from a set of possible states $\mathcal{M}$, with finite cardinality $M = |\mathcal{M}|$. We denote the state of the channel in the kth network use as m_k. For any state m define $n_m(k)$ as the number of times that the network is used in state m in the first k network uses. Let

$$t_m = \lim_{k\to\infty} \frac{n_m(k)}{k} \tag{2.15}$$

denote the portion of the time that the network has been used in state m as the total number of network uses goes to infinity. The following theorem bounds the rate of information transfer from nodes in S to those in S^C by the conditional mutual information.

THEOREM 2.11 *[Khojastepour, 2004] Consider a general network with M states, where M is finite. If the information rates $\{R^{(ij)}\}$ are achievable, then the sum rate of information transfer from a node set S_1 to a disjoint node set S_2, where $S_1, S_2 \subset \{1, 2, ..., N\}$ and for any choice of network state sequence $(m_k)_{k=1}^{\infty}$, is bounded by:*

$$\sum_{i\in S_1, j\in S_2} R^{(ij)} \le \sup_{t_m} \min_{S} \sum_{m=1}^{M} t_m I\left(X_{(m)}^{(S)}; Y_{(m)}^{(S^C)} | X_{(m)}^{(S^C)}\right) \tag{2.16}$$

for some joint probability distributions $p(x^{(1)}, x^{(2)}, ..., x^{(N)}|m), m = 1, 2, ..., M$ when the minimization is taken over all sets $S \subset \{1, 2, ..., N\}$ subject to $S \cap S_1 = S_1$, $S \cap S_2 = \phi$ and the supremum is over all non-negative t_m subject to $\sum_{i=1}^{M} t_m = 1$.

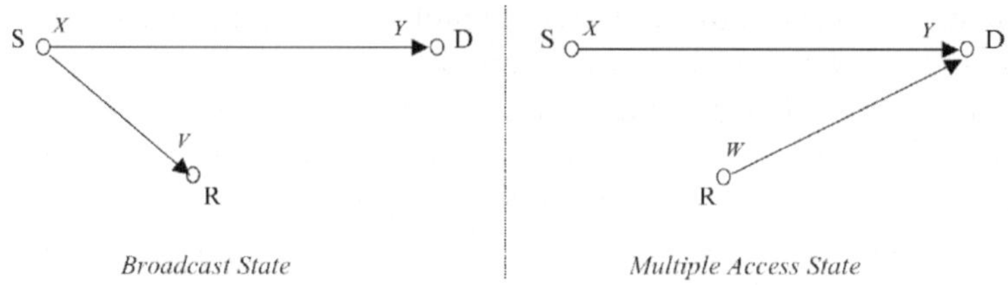

Figure 2.6. Two states of the half-duplex relay channel.

Using the above theorem, an upper bound for the capacity of the half-duplex relay channel can be found. In a half-duplex relay, the network can be in one of two states, based on whether the relay is transmitting or receiving. These states are called the *broadcast* and the *multiple-access* states and are shown in Figure 2.6. The two cuts of interest, which are also called the *broadcast* and *multiple-access* cuts, are shown in Figure 2.2. A simple application of Theorem 2.11 now gives us the following result.

THEOREM 2.12 *[Khojastepour, 2004] The capacity of a general half-duplex relay channel is upper bounded as follows.*

$$C_{hd} \leq \sup_{t:0\leq t\leq 1} \min \Big(tI(X_1;Y_1,V_1) + (1-t)I(X_2;Y_2|W_2), \\ tI(X_1;Y_1) + (1-t)I(X_2,W_2;Y_2)\Big) \quad (2.17)$$

where the subscript 1 *stands for the broadcast state, and* 2 *stands for the multiple-access state.*

The most well-known lower bound for half-duplex decode-and-forward relaying is given by the following theorem.

THEOREM 2.13 *[Khojastepour, 2004] The capacity of a general half-duplex relay channel is upper bounded as follows.*

$$C_{hd} \leq \sup_{t,0\leq t\leq 1} \min \Big(tI(X_1;V_1) + (1-t)I(X_2;Y_2|W_2), \\ tI(X_1;Y_1) + (1-t)I(X_2,W_2;Y_2)\Big) \quad (2.18)$$

where the subscript 1 *stands for the broadcast state, and* 2 *stands for the multiple-access state.*

The similarity between Theorem 2.13 and Theorem 2.12 is easy to see. The relationship between the rates in the half-duplex and full-duplex cases is also visible or closer inspection. Results on half-duplex relaying have been discussed in various papers, including [Liang and Veeravalli, 2005; Khojastepour et al., 2002a; Høst-Madsen and Zhang, 2005; Zahedi et al., 2004]. A comprehensive treatment of this channel can be found in [Khojastepour, 2004], including results on Gaussian half-duplex relay channels, and the development of a comprehensive framework that encompasses the cases of time-division and frequency-division half-duplex relaying as special cases of a channel with multiple states.

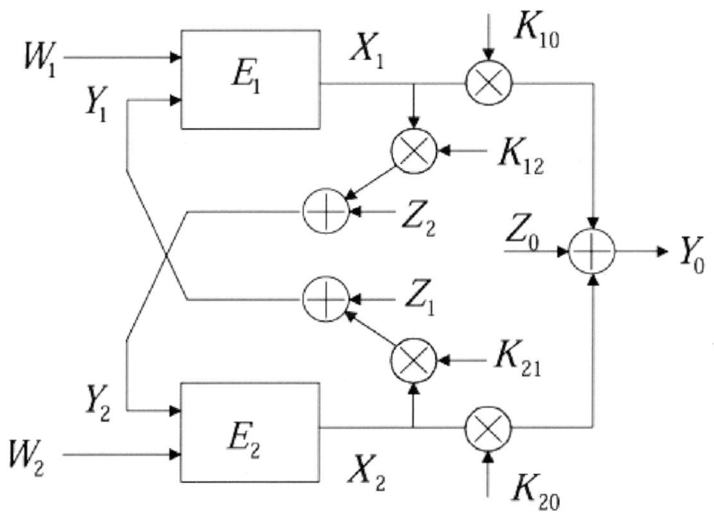

Figure 2.7. Channel model for cooperation when source and relay exchange roles.

Limits of Two-user Cooperation

Here, we consider the channel model of Figure 2.7. There are two sources and a single destination. Both transmitters can overhear each other, and are willing to cooperate by forwarding information from the other. Transmitters are capable of full-duplex communication. The mathematical model for this channel is given by the following equations

$$Y_0 = K_{10}X_1 + K_{20}X_2 + Z_0 \tag{2.19}$$
$$Y_1 = K_{21}X_2 + Z_1 \tag{2.20}$$
$$Y_2 = K_{12}X_1 + Z_2, \tag{2.21}$$

with $Z_0 \sim \mathcal{N}(0, \Xi_0)$, $Z_1 \sim \mathcal{N}(0, \Xi_1)$ and $Z_2 \sim \mathcal{N}(0, \Xi_2)$. In general, we assume that $\Xi_1 = \Xi_2$. The system is causal and transmission is done for B

blocks of length n, therefore the signal of Source 1 at time j, $j = 1, \ldots, n$, can be expressed as $X_1(W_1, Y_1(j-1), Y_1(j-2), \ldots, Y_1(1))$, where W_1 is the message that Source 1 wants to transmit to the destination in that block. Similarly, for Source 2 we have $X_2(W_2, Y_2(j-1), Y_2(j-2), \ldots, Y_2(1))$.

We assume that Source 1 divides its information W_1 into two parts: W_{10}, which is sent directly to the destination, and W_{12}, which is sent to Source 2 and then forwarded by Source 2 to the destination. Source 1 structures its transmit signal so that it is able to send the above information as well as some additional cooperative information to the destination. This is done as follows

$$X_1 = X_{10} + X_{12} + U_1 \tag{2.22}$$

where the power is divided as

$$P_1 = P_{10} + P_{12} + P_{U1}. \tag{2.23}$$

Here, U_1 refers to the part of the signal that carries cooperative information. Thus, X_{10} uses power P_{10} to send W_{10} at rate R_{10} directly to the destination, X_{12} uses power P_{12} to send W_{12} to Source 2 at rate R_{12}, and U_1 uses power P_{U1} to send cooperative information to the destination. Forwarding is based on the principle of decode-and-forward, therefore the transmission rate of W_{12}, i.e. R_{12}, and the power allocated to W_{12}, i.e. P_{12}, should be such that W_{12} can be perfectly decoded by Source 2. Source 2 similarly structures its transmit signal X_2 and divides its total power P_2. In the above setup, the following theorem from [Sendonaris et al., 2003a] gives an achievable rate region.

THEOREM 2.14 *[Sendonaris et al., 2003a] An achievable rate region for the system given in (2.19)-(2.21) is the closure of the convex hull of all rate pairs (R_1, R_2) such that $R_1 = R_{10} + R_{12}$ and $R_2 = R_{20} + R_{21}$ where*

$$R_{12} < E\left\{C\left(\frac{K_{12}^2 P_{12}}{K_{12}^2 P_{10} + \Xi_1}\right)\right\} \tag{2.24}$$

$$R_{21} < E\left\{C\left(\frac{K_{21}^2 P_{21}}{K_{21}^2 P_{20} + \Xi_2}\right)\right\} \tag{2.25}$$

$$R_{10} < E\left\{C\left(\frac{K_{10}^2 P_{10}}{\Xi_0}\right)\right\} \tag{2.26}$$

$$R_{20} < E\left\{C\left(\frac{K_{20}^2 P_{20}}{\Xi_0}\right)\right\} \tag{2.27}$$

$$R_{10} + R_{20} < E\left\{C\left(\frac{K_{10}^2 P_{10} + K_{20}^2 P_{20}}{\Xi_0}\right)\right\} \tag{2.28}$$

$$R_{10} + R_{20} + R_{12} + R_{21} < E\left\{C\left(\frac{K_{10}^2 P_1 + K_{20}^2 P_2 + 2K_{10}K_{20}\sqrt{P_{U1}P_{U2}}}{\Xi_0}\right)\right\} \tag{2.29}$$

for some power assignment satisfying $P_1 = P_{10} + P_{12} + P_{U1}$, and $P_2 = P_{20} + P_{21} + P_{U2}$. The function $C(x)$ is defined in (2.11) and E denotes expectation with respect to the fading parameters K_{ij}.

5. Practical Strategies for Relaying Information

There are several parallel efforts going on to harness the gains of relay coding in practice, of which we mentioned some in an earlier section on the history of relaying. In this section, we will focus only on the results of [Sendonaris et al., 2003a; Sendonaris et al., 2003b; Khojastepour et al., 2004a; Chakrabarti et al., 2005a].

We first describe a CDMA based user-cooperation strategy that was proposed in [Sendonaris et al., 2003a; Sendonaris et al., 2003b]. It was one of the first implementations of user-cooperation to have been proposed, and it was designed keeping in mind the realities of cellular communication. After describing the aforementioned scheme, we present relay code designs using LDPC component codes for both full-duplex ([Khojastepour et al., 2004a]) and half-duplex ([Chakrabarti et al., 2005a]) relays.

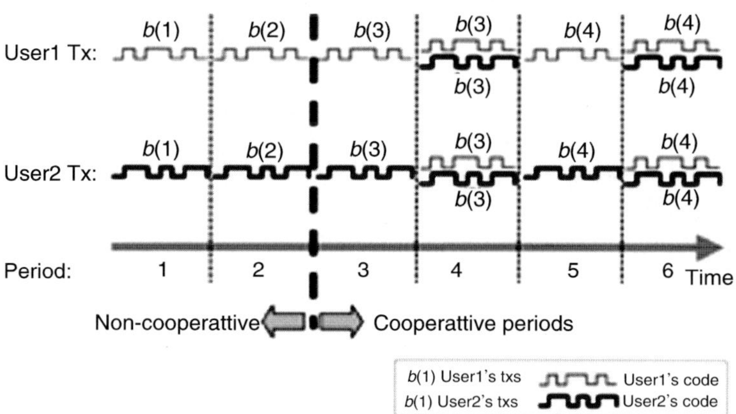

Figure 2.8. User cooperation using spreading codes.

CDMA Implementation for User-cooperation

To begin with, let us assume that each user has a single spreading code, which is orthogonal to the spreading codes of all other users. We further assume that the coherence time of the channel equals L symbol periods, i.e. the channel does not change for L symbol periods. For the simple case of $L = 3$, we describe the transmitted signals. If the sources were not cooperating, they would transmit

$$\begin{array}{rcl} X_1(t) & = & \underbrace{\begin{matrix} a_1\, b_1^{(1)}\, c_1(t)\,, \\ a_2\, b_2^{(1)}\, c_2(t)\,, \end{matrix}}_{\text{Period 1}} \quad \underbrace{\begin{matrix} a_1\, b_1^{(2)}\, c_1(t)\,, \\ a_2\, b_2^{(2)}\, c_2(t)\,, \end{matrix}}_{\text{Period 2}} \quad \underbrace{\begin{matrix} a_1\, b_1^{(3)}\, c_1(t) \\ a_2\, b_2^{(3)}\, c_2(t) \end{matrix}}_{\text{Period 3}} \\ X_2(t) & = & \end{array} \tag{2.30}$$

where $b_j^{(i)}$ is the ith bit from user j, $c_j(t)$ is the spreading code used by user j, and $a_j = \sqrt{\frac{P_j}{T_s}}$ where P_j is the power used by user j and T_s is the symbol period.

For fairness, any cooperative scheme developed in the same framework must satisfy some basic criteria. The total number of spreading codes used by the two users must remain the same as in the non-cooperative scheme, and the cooperative strategy should be of comparable complexity to the non-cooperative scheme. Under the proposed cooperative scheme, the users transmit

$$\begin{array}{rclll} X_1(t) & = & a_{11}\, b_1^{(1)}\, c_1(t)\,, & a_{12}\, b_1^{(2)}\, c_1(t)\,, & a_{13}\, b_1^{(2)}\, c_1(t) \\ & & & & +a_{14}\, \hat{b}_2^{(2)}\, c_2(t) \\ X_2(t) & = & a_{21}\, b_2^{(1)}\, c_2(t)\,, & a_{22}\, b_2^{(2)}\, c_2(t)\,, & a_{23}\, \hat{b}_1^{(2)}\, c_1(t) \\ & & & & +a_{24}\, b_2^{(2)}\, c_2(t) \\ & & \underbrace{\qquad\qquad}_{\text{Period 1}} & \underbrace{\qquad\qquad}_{\text{Period 2}} & \underbrace{\qquad\qquad}_{\text{Period 3}} \end{array} \tag{2.31}$$

where $\hat{b}_j^{(i)}$ is the partner's estimate of user j's i^{th} bit. The parameters $\{a_{ji}\}$ control the amount of power allocated to a user's own bits versus the bits of the partner, while maintaining an average power constraint of P_j for user j, over L periods.

The way to interpret the above is as follows. In Period 1, each user sends data to the base station only. In period 2, users send data to each other in addition to the base station. After this, each user estimates its partner's data and constructs a cooperative signal that is sent to the destination in Period 3. This cooperative signal is a superposition of spreading codes of both users.

To generalize the above scheme to arbitrary number of symbol periods L, we define another parameter L_c. The two users cooperate for $2L_c$ periods, whereas the remaining $L - 2L_c$ periods are used for sending their own information. For example, in (2.31), $L = 3$ and $L_c = 1$, whereas in (2.30), $L = 3$ and $L_c = 0$. By changing the value of L_c over time, it is possible to achieve different points on the capacity region. The $\{a_{ij}\}$ are chosen to satisfy the power constraints

$$\begin{array}{rcl} \frac{1}{L}(L_n a_{11}^2 + L_c(a_{12}^2 + a_{13}^2 + a_{14}^2)) & = & P_1 \\ \frac{1}{L}(L_n a_{21}^2 + L_c(a_{22}^2 + a_{23}^2 + a_{24}^2)) & = & P_2. \end{array} \tag{2.32}$$

This cooperative scheme is depicted in Figure 2.8 for the case of $L = 6$, $L_c = 2$. The performance of the above scheme and the design of optimal receivers for this type of user-cooperation is discussed in [Sendonaris et al., 2003b].

LDPC Codes for Full-duplex Relaying

One of the first attempts on practical full-duplex relay code design was due to [Khojastepour et al., 2004a]. Although the designs of [Khojastepour et al., 2004a] are not optimal in an information-theoretic sense, they perform well, and they incorporate most of the essential components of practical relay LDPC code design, namely - capacity analysis for finite alphabet (eg. BPSK), protocol design, power allocation, factor graph construction, and decoding algorithms. In order to design capacity-approaching relay codes, each of these must be done optimally, and in addition there is often the additional step of code-profile optimization.

Here we will briefly describe two protocols proposed in [Khojastepour et al., 2004a], and the factor graph construction. Decoding is performed using belief propagation, and code profiles are optimized using density evolution. The interested reader can refer to [Khojastepour et al., 2004a] for additional details.

Two protocols are proposed in [Khojastepour et al., 2004a]. The first is called the *simple protocol*, where transmission from the source occurs in B blocks of length N. A pair of consecutive blocks uses a pair of jointly designed *constituent codes*. Odd blocks use one of the constituent codes, and even blocks use the other. The source sends new information in each block. At the end of each block, the relay finds the codeword that is closest to its received signal, and retransmits it without re-encoding.

The second protocol, which is called the *DF protocol* is inspired by the decode-and-forward scheme, and is somewhat similar to the simple protocol. Again, transmission from the source occurs in B blocks of length N. In each block, the source sends a superposition of a new codeword and a repetition of the previous codeword with an appropriate power ratio. In the first and last blocks, only one codeword is sent. At the end of each block, the relay decodes the new codeword from the received signal and retransmits it the same way as in the simple protocol. The constituent codes used in the above protocols are irregular LDPC codes proposed in [Luby et al., 2001; Richardson et al., 2001], chosen for their capacity-approaching performance.

Before proceeding, we present a very brief introduction to factor graphs in the context of LDPC codes. The interested reader will find extensive reading material on factor graphs in [Kschischang et al., 2001; Richardson and Urbanke, 2004].

Any block code (LDPC codes are block codes) can be represented completely by its parity check matrix H. This binary matrix, in turn, can be uniquely represented by a bipartite graph. For a discussion of full-duplex LDPC codes only, we follow the following conventions. A variable node is represented by a circle in the graph and corresponds to a column of the parity check matrix H. A check node is represented by a square and corresponds to a row of the same matrix. Last, there is a connection between a check node and variable node if and only if the parity check matrix H has a 1 in the corresponding row and column (we confine ourselves to binary LDPC codes). For example, the following is the parity check matrix of a rate $\frac{1}{2}$ LDPC code.

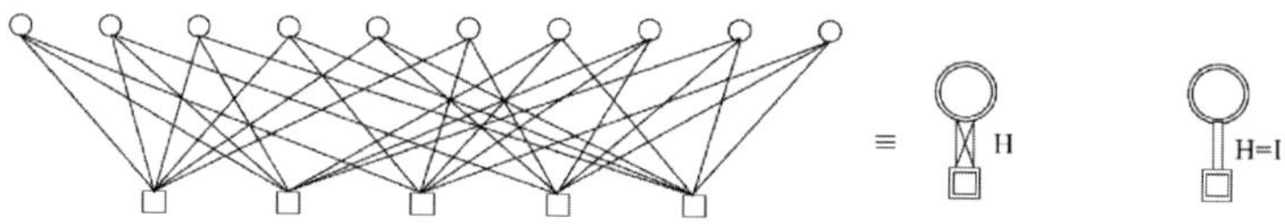

Figure 2.9. Factor graph for parity check matrix H and shorthand noations.

$$H = \begin{bmatrix} 1 & 1 & 1 & 1 & 1 & 1 & 0 & 0 & 0 & 0 \\ 1 & 1 & 1 & 0 & 0 & 0 & 1 & 1 & 1 & 0 \\ 1 & 0 & 0 & 1 & 0 & 1 & 1 & 1 & 0 & 1 \\ 0 & 1 & 0 & 0 & 1 & 1 & 0 & 1 & 1 & 1 \\ 0 & 0 & 1 & 1 & 1 & 0 & 1 & 0 & 1 & 1 \end{bmatrix}.$$

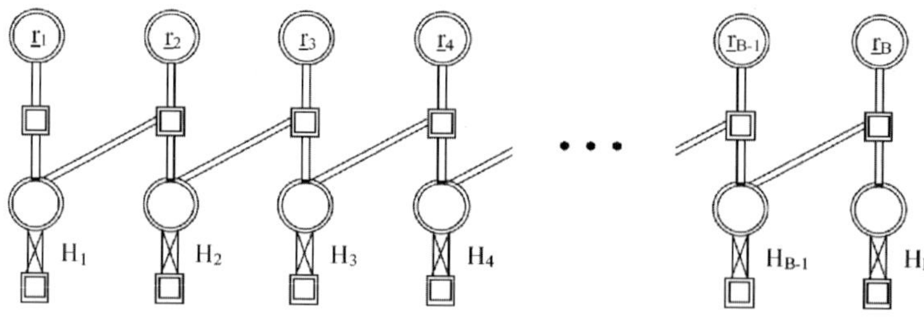

Figure 2.10. Factor graph for optimal decoding at the destination.

Its factor graph is depicted in Figure 2.9. When $H = I$, the factor graph is a series of parallel connections between the check nodes and the variable nodes. Two shorthand notations are introduced in Figure 2.9 for a general parity check matrix and for the special case of $H = I$.

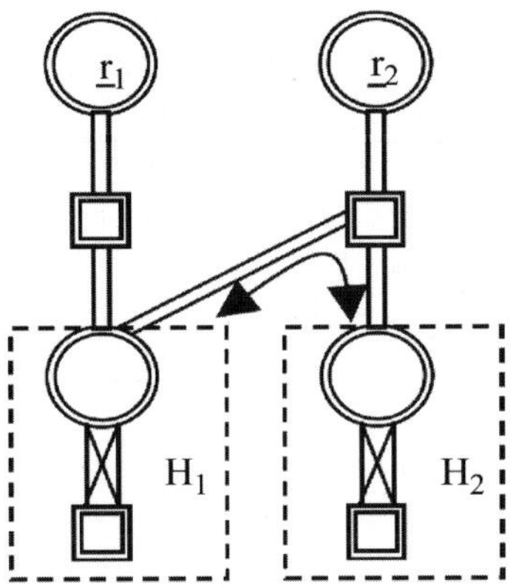

Figure 2.11. Factor graph of a pair of consecutive codes.

The signal received by the destination in each block is a superposition of two codewords. Corresponding to this fact, Figure 2.10 shows the factor graph structure for optimal decoding at the destination based on the entire set of B blocks. It is extremely complex to find optimized LDPC code profiles for the entire factor graph since it requires joint optimization of B matrices. Therefore, as a practical alternative, only pairs of codes, as depicted in Figure 2.11 are optimized at a time. The two codes in a pair are then alternately used over consecutive channel uses. It is to be noted that a set of LDPC code optimized for the entire factor graph of Figure 2.10 would perform optimally only when the decoding is performed jointly over all B blocks, which is infeasible. For a block-by-block successive decoding scheme (see Figure 2.12), the optimization of successive code pairs is actually the optimal strategy.

Two algorithms were proposed in [Khojastepour, 2004] for decoding the received signals at the destination, called the forward and the backward decoding algorithms. Note that the first and the last transmissions in the above coding scheme use only a single code, whereas any intermediate received signal is a superposition of a pair of codes. Therfore, decoding may either commence from the first or the last received codeword, corresponding to forward and backward decoding respectively. Forward decoding has a minimal latency of two blocks, and also performs better when the relay is near the destination. Backward decoding, in contrast, is better when source and relay are close to each other; however, it has a decoding latency of B blocks. The performance of the proposed LDPC codes is ahown in Figure 2.13.

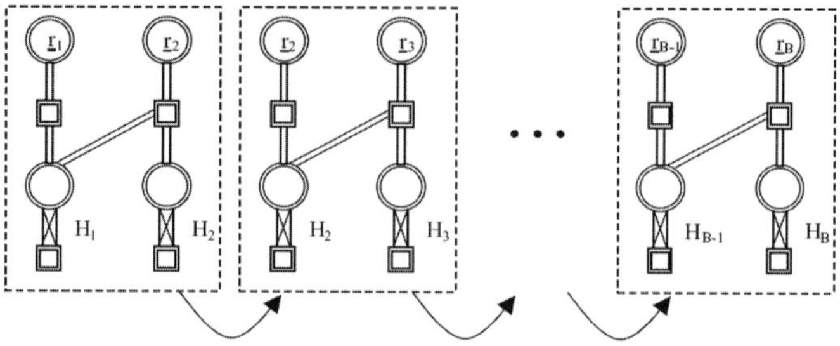

Figure 2.12. Factor graph for block-by-block successive decoding at the destination.

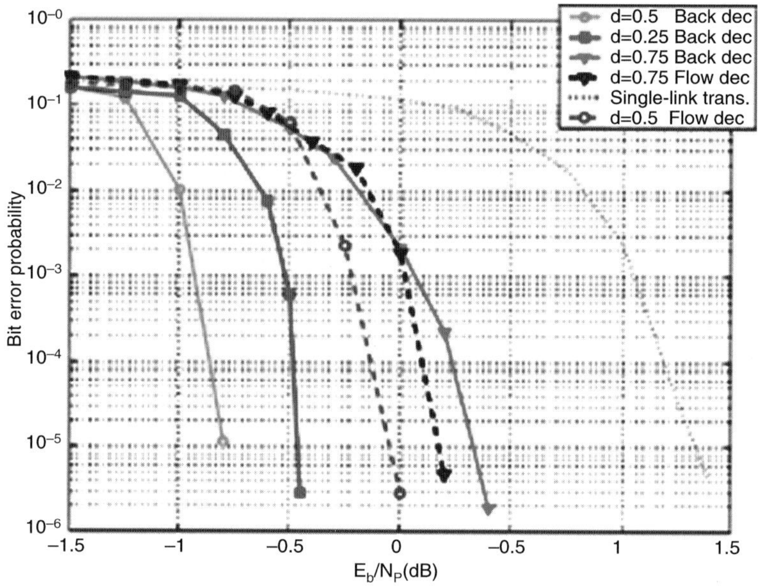

Figure 2.13. Performance of full-duplex relay coding scheme for the simple protocol. Performance of a single-user code using the same constituent parity check matrix is shown for comparison. Here source and relay are unit distance apart and the relay is on the line joining them at a distance d from the source.

LDPC Codes for Half-duplex Relaying

In this section, we discuss LDPC code designs proposed in [Chakrabarti et al., 2005a] for the half-duplex relay channel. The code designs are based on the information theoretic random-coding scheme for half-duplex decode-and-forward relaying. Although the relay channel is commonly visualized as a combination of a broadcast and a multiple-access channel, it is shown that the achievable rate of decode-and-forward relaying can be approached by using single-user codes decoded with single-user receivers. The single-user decodability of these codes supports the practicality of half-duplex relaying.

It is shown in [Høst-Madsen, 2004; Sabharwal, 2004; Chakrabarti et al., 2005b] that the gains of relaying are significant only in low to medium SNRs. At high SNRs, the throughput of relaying is not a significant fraction larger than that of a direct link. And in the low to medium SNR range, binary modulation on each channel dimension (QPSK) approaches the capacity of the AWGN channel. This justifies the use of binary codes. Another challenge in code construction is that the implementation of source-relay correlation in multiple-access mode introduces an added level of complexity. In contrast it is simple to devise coding schemes where this correlation is either 0 or 1. Empirical results in [Chakrabarti et al., 2005b] show that the loss in throughput is negligible when the better of $\rho = 0, 1$ is chosen instead of the optimal correlation.

We assume that the two rate terms in the achievable rate expression (2.17) are equal at the point where the achievable rate is maximized, i.e.

$$\begin{aligned} & tI(X_1;V_1) + (1-t)I(X_2;Y_2|W_2) \\ = \; & tI(X_1;Y_1) + (1-t)I(X_2,W_2;Y_2). \end{aligned} \tag{2.33}$$

The above is easy to prove for Gaussian codebooks when the source and the relay have separate power constraints. Even if this is not true in general, it is easy to see that we can find rates $R_1 \leq I(X_1;V_1)$, $R_2 \leq I(X_2;Y_2|W_2)$, $R_3 \leq I(X_1;Y_1)$, and $R_4 \leq I(X_2,W_2;Y_2)$ satisfying

$$\text{Achievable rate} = tR_1 + (1-t)R_2 = tR_3 + (1-t)R_4. \tag{2.34}$$

The proposed coding scheme then remains the same with the mutual information terms replaced by the corresponding rates.

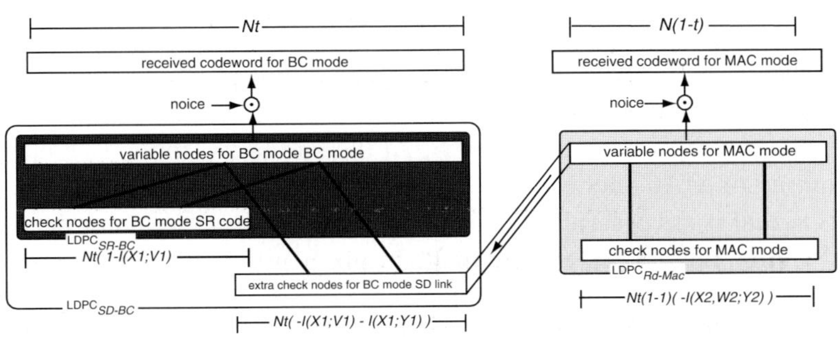

Figure 2.14. LDPC code structure for $\rho = 1$.

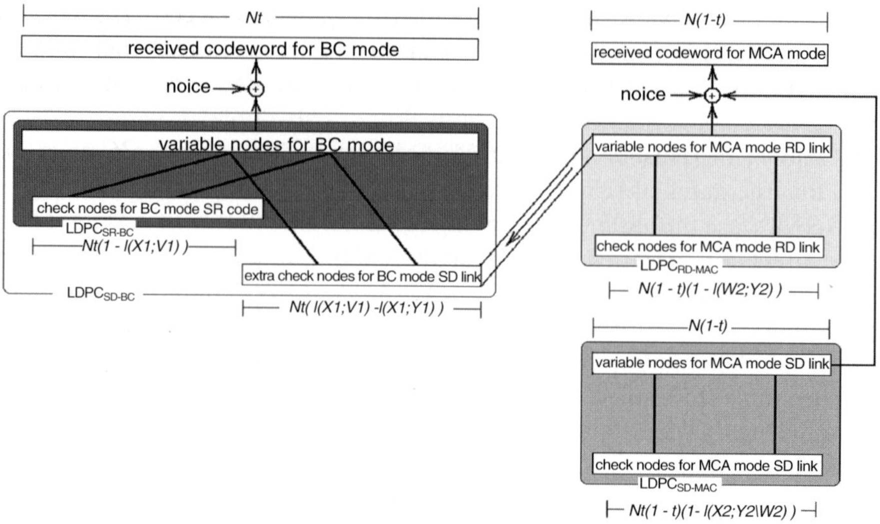

Figure 2.15. LDPC code structure for $\rho = 0$.

The factor-graph structures of the codes for the two cases corresponding to $\rho = 0$ and $\rho = 1$ are shown in Figures 2.14 and 2.15 respectively. The factor graphs in this section are represented as a pair of rectangular blocks representing the variable and check nodes, and connected by a pair of parallel lines denoting the edges of the graph. In contrast to the convention of the previous section, the parallel lines do not indicate that the parity check matrix is identity. Several single-user codes combine to form the factor graphs of Figure 2.14 and Figure 2.15. Any set of variable nodes and associated check nodes that are enclosed in a rectangular box with rounded edges indicates that the enclosed nodes correspond to a single-user LDPC code. We now explain the encoding and decoding for both values of ρ. We assume that the sum length of BC and MAC mode codewords is N bits.

When $\rho = 1$, S and R transmit identical signals in MAC mode. For this case, the following scheme is used. In the beginning of BC mode, S encodes information bits using a code $LDPC_{SR-BC}$ to yield a codeword of length tN bits. This codeword is transmitted by S. At the end of BC mode (which is also the beginning of MAC mode), both R and D receive the BC mode source signal. This signal is successfully decoded by R. However, D cannot decode the received signal, and stores a copy of it. In the beginning of MAC mode, the tN variable bits from BC mode are compressed. Compression is done at both S and R, by multiplying with the same parity matrix. These compressed bits, acting as parity together with the parity bits of $LDPC_{SR-BC}$ form a composite code $LDPC_{SD-BC}$ that can be decoded at D at the end of MAC mode. In order to communicate the compressed bits to D reliably, S, and R treat them as information bits for MAC mode, and re-encode them using a code $LDPC_{MAC}$ to yield a codeword of length $(1-t)N$, which is then transmitted synchronously from S and R with appropriate powers. The structure of the code is shown in Figure 2.14.

For $\rho = 1$, decoding is performed as follows. R decodes $LDPC_{SR-BC}$ at the end of BC mode using belief propagation like any single-user LDPC code. D waits for both BC and MAC mode signals to arrive before it commences decoding. $LDPC_{MAC}$ is decoded like a single-user LDPC code, from which side information in the form of additional parity bits is obtained about the BC mode signal. Using knowledge of the single-user BC mode source-relay code, and with the help of these additional parity bits, $LDPC_{SD-BC}$ is decoded. This final decoding also is performed using belief propagation.

For $\rho = 0$, the BC mode is the same as before. In MAC mode, however, S and R transmit independent (therefore uncorrelated) information using codes $LDPC_{SD-MAC}$ and $LDPC_{RD-MAC}$ respectively. As before, R compresses the information bits received in BC mode to produce additional parity bits, which serve as relay information bits in MAC mode. These bits are re-encoded by R using $LDPC_{RD-MAC}$ to yield $(1-t)N$ coded bits. The source, in MAC mode, sends bits of new information encoded using $LDPC_{SD-MAC}$ to yield another set of $(1-t)N$ coded bits. Thus, $(1-t)N$ coded bits each from S and R are transmitted simultaneously with appropriate power allocation, so that the two (uncorrelated) signals appear superimposed at D. The structure of the code is shown in Figure 2.15.

For $\rho = 0$, decoding proceeds as follows. R decodes the BC mode signal like a single-user LDPC code. D waits for both BC and MAC mode signals. In MAC mode, the rates for SD and RD channels correspond to one of the corner points of the MAC capacity region, for which it is well known [Cover and Thomas, 1991] that capacity can be achieved by successive decoding (onion-peeling). The MAC signal is successively decoded to first reveal the relay codeword, treating both noise and interference from S as noise. Next, the relay codeword is subtracted out to reveal the source codeword in the presence of noise alone, which is then decoded. The MAC mode source information is new information, whereas the relay information provides additional parity bits to aid in decoding the BC mode codeword.

The main challenge is the design of codes $LDPC_{SD-BC}$ and $LDPC_{SR-BC}$, which must be jointly optimized, since the factor graph of the latter is a subgraph of the factor graph of the former. The reader should note that these codes are of different rates, and although the received codeword is same for both R and D, the received SNRs are different. To avoid confusion, we would like to mention that neither S, nor R actually uses $LDPC_{SD-BC}$ to encode information. It is merely a convenience to visualize the side information received by D in MAC mode as extra parity bits in addition to the actual parity bits of $LDPC_{SR-BC}$, and call the composite a code $LDPC_{SD-BC}$. The optimization of code profiles is performed using a modification of the density evolution algorithm. In the implementation of density evolution, the messages have been approximated as Gaussians to speed up the optimization, the cost being usually small inaccuracy in threshold determination. We omit details of code profile optimization using modified density evolution; the interested reader can find these in [Chakrabarti et al.,].

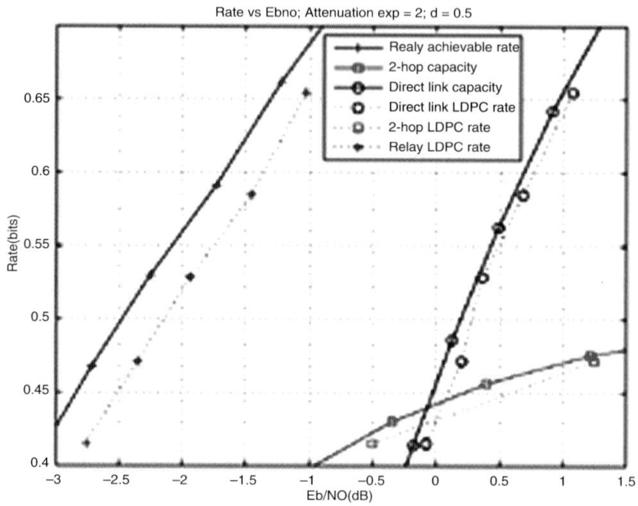

Figure 2.16. Rate vs. E_b/N_0. Theoretical limits and LDPC performance based on thresholds.

Figure 2.16 shows how far the coding schemes for direct, two-hop and relay coding schemes are from their respective theoretical bounds. The thresholds for direct and two-hop channels are calculated using code ensembles with maximum variable node degrees comparable to the relay codes. In this figure, we fix the rates of the codes, and calculate E_b/N_0 from the numerically calculated thresholds of the LDPC codes. For the relay coding scheme, the achievable E_b/N_0 values are less than 0.4 dB away from the theoretical minimum. Note that the thresholds for the relay channel are not thresholds of any single code, but a function of the thresholds of all component codes.

Figure 2.17 plots the BER performance of the overall relay LDPC coding scheme for $\rho = 1$, taking into account the individual BER performances of all three component codes. Some modifications have been incorporated into the coding scheme to improve performance when the number of decoding iterations is small. The code designs correspond to a total power of -1dB ($P = 10^{-0.1}$) and correlation $\rho = 1$ in this case. For this case, the SR code has rate 0.9, the SD code has rate 0.47 and the MAC code has rate 0.8. The results are for a

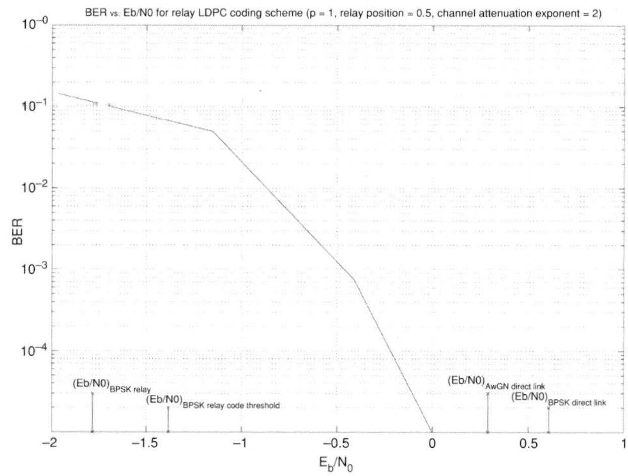

Figure 2.17. BER vs. E_b/N_0 for relay LDPC coding scheme ($\rho = 1$).

blocklength of 100K. The codes are randomly constructed and there is no cycle removal with the exception of removing double edges.

6. Conclusion

Problems in cooperative communication continue to intrigue researchers by their difficulty and the potential for faster and more reliable communication. There is a wide open space for the implementation of principles of user-cooperation in mesh and sensor networks. User-cooperation also has potential for implementation in mobile handheld devices, but here fair sharing of resources must be ensured by a suitable protocol. We anticipate that our current understanding of the principles of user-cooperation, together with advances in technology will enable cooperative communication networks in future.

Notes

1. The authors are grateful to Andrew Sendonaris, Elza Erkip, and Mohammad Ali Khojastepour for permission to use figures and various other content from their publications in this chapter.

2. The max-flow min-cut theorem (Ford-Fulkerson Algorithm) is one of seven major theorems in combinatorics that are all logically equivalent, the other six being Konig's theorem, Konig-Egervary theorem (1931), Menger's theorem (1929), The Birkhoff-Von Neumann theorem (1946), Dilworth's theorem, and Hall's Marriage theorem. The reader may therefore have encountered the max-flow min-cut theorem in a different form.

References

Ahlswede, R., Cai, N., Li, S.-Y. R., and Yeung, R. W. (2000). Network information flow. *IEEE Trans. Inform. Theory*, 46(4):1204–1216.

Ahlswede, R. and Kaspi, A. H. (1987). Optimal coding strategies for certain permuting channels. *IEEE Trans. Inform. Theory*, 33(3):310–314.

Aref, M. R. (1980). *Information flow in relay networks*. PhD thesis, Stanford University.

Bergmans, P. and Cover, T. M. (1974). Cooperative broadcasting. *IEEE Trans. Inform. Theory*, 20:317–324.

Berrou, C., Glavieux, A., and Thitimajshima, P. (1993). Near shannon limit error-correcting coding and decoding: Turbo-codes. In *Proc. of ICC*, volume 2, pages 1064–1070.

Biglieri, E., Proakis, J., and Shamai, S. (1998). Fading channels: Information-theoretic and communications aspects. *IEEE Trans. Inform. Theory*, 44(6): 2619–2692.

Boyer, J., Falconer, D. D., and Yanikomeroglu, H. (2004). Multihop diversity in wireless relaying channels. *IEEE Trans. Commun.*, 52(10):1820–1830.

Caire, G. and Shamai, S. (2003). On the achievable throughput of a multiantenna Gaussian broadcast channel. *IEEE Trans. Inform. Theory*, 49:1691–1706.

Carleial, A. B. (1982). Multiple-access channels with different generalized feedback signals. *IEEE Trans. Inform. Theory*, 28(6):841–850.

Castura, J. and Mao, Yongyi (2005). Rateless coding for wireless relay channels. In *Proc. of ISIT*, pages 810–814.

Chakrabarti, A., de Baynast, A., Sabharwal, A., and Aazhang, B. LDPC codes for decode-and-forward half-duplex relaying. in preparation for IEEE J. Select. Areas Commun.

Chakrabarti, A., de Baynast, A., Sabharwal, A., and Aazhang, B. (2005a). LDPC code design for half-duplex decode-and-forward relaying. In *Proc. of the Allerton Conference*, Monticello, IL.

Chakrabarti, A., Sabharwal, A., and Aazhang, B. (2005b). Sensitivity of achievable rates for half-duplex relay channel. In *Proc. of SPAWC*.

Chong, H. F., Motani, M., and Garg, H. K. (2005). New coding strategies for the relay channel. In *Proc. of ISIT*, pages 1086–1090.

Chou, P. A., Wu, Y., and Jain, K. (2003). Practical network coding. In *Proc. of the Allerton conference*.

Costa, M. H. M. (1983). Writing on dirty paper. *IEEE Trans. Inform. Theory*, 29:439–441.

Cover, T., El Gamal, A., and Salehi, M. (1980). Multiple access channels with arbitrarily correlated sources. *IEEE Trans. Inform. Theory*, 26(6):648–657.

Cover, T. and Thomas, J. (1991). *Elements of Information Theory*. John Wiley and Sons.

Cover, T. M. (1972). Broadcast channels. *IEEE Trans. Inform. Theory*, 18:2–14.

Cover, T. M. (1975). An achievable rate region for the broadcast channel. *IEEE Trans. Inform. Theory*, 21(4):399–404.

Cover, T. M. and El Gamal, A. A. (1979). Capacity theorems for the relay channel. *IEEE Trans. Inform. Theory*, 25:572–584.

Cover, T. M. and Leung, C. S. K. (1981). An achievable rate region for the multiple-access channel with feedback. *IEEE Trans. Inform. Theory*, 27(3): 292–298.

El Gamal, A., Mohseni, M., and Zahedi, S. (2004). On reliable communication over additive white Gaussian noise relay channels. *submitted to IEEE Trans. Inform. Theory*.

El Gamal, A. and van der Meulen, E. C. (1981). A proof of Marton's coding theorem for the discrete memoryless broadcast channel. *IEEE Trans. Inform. Theory*, 27(1):120–122.

El Gamal, A. A. (1981). On information flow in relay networks. In *Proc. of IEEE National Telecommunications Conference*, volume 2, pages D4.1.1–D4.1.4.

El Gamal, A. A. and Aref, M. (1982). The capacity of the semideterministic relay channel. *IEEE Trans. Inform. Theory*, 28(3):536.

Ford, L. and Fulkerson, D. (1962). *Flows in Networks*. Princeton University Press.

Foschini, G. J. and Gans, M. J. (1998). On limits of wireless communications in a fading environment when using multiple antennas. *Wireless Personal Communications*, 6(3):311–335.

Gaarder, N. T. and Wolf, J. K. (1975). The capacity region of a multiple-access discrete memoryless channel can increase with feedback. *IEEE Trans. Inform. Theory*, 21(1):100–102.

Gallager, R. G. (1963). *Low Density Parity Check Codes*. PhD thesis, MIT.

Gastpar, M. and Vetterli, M. (2005). On the capacity of large Gaussian relay networks. *IEEE Trans. Inform. Theory*, 51(3):765–779.

Gel'fand, S. I. and Pinsker, M. S. (1980). Capacity of a broadcast channel with one deterministic component. *Probl. Pered. Inform.*, 16(1):24–34.

Grossglauser, M. and Tse, D. N. C. (2002). Mobility increases the capacity of ad hoc wireless networks. *IEEE/ACM Trans. Networking*, 10(4):477–486.

Gupta, P. and Kumar, P. R. (2000). The capacity of wireless networks. *IEEE Trans. Inform. Theory*, pages 388–404.

Gupta, P. and Kumar, P. R. (2003). Towards an information theory of large networks: An achievable rate region. *IEEE Transactions on Information Theory*, 49(8):1877–1894.

Han, T. S. (1981). The capacity region for the deterministic broadcast channel with a common message. *IEEE Trans. Inform. Theory*, 27(1):122–125.

Hasna, M. O. and Alouini, M.-S. (2003). End-to-end performance of transmission systems with relays over Rayleigh-fading channels. *IEEE Trans. Wireless Commun.*, 2(6):1126–1131.

Ho, T., Koetter, R., Medard, M., Karger, D., and Effros, M. (2003). The benefits of coding over routing in a randomized setting.

Hunter, T. E., Sanayei, S., and Nosratinia, A. (2004). The outage behavior of coded cooperation. In *Proc. of ISIT*, page 270.

Høst-Madsen, A. (2004). Cooperation in the low power regime. In *Proc. of the Allerton Conference*.

Høst-Madsen, A. and Zhang, J. (2005). Capacity bounds and power allocation for the wireless relay channel. *IEEE Trans. Inform. Theory*, 51(6):2020–2040.

Jaggi, S., Sanders, P., Chou, P. A., Effros, M., Egner, S., Jain, K., and Tolhuizen, L. (2005). Polynomial time algorithms for multicast network code construction. *IEEE Trans. Inform. Theory*, 51:1973–1982.

Janani, M., Hedayat, A., Hunter, T. E., and Nosratinia, A. (2004). Coded cooperation in wireless communications: space-time transmission and iterative decoding. *IEEE Trans. Signal Processing*, 52:362–371.

Khojastepour, M. A. (2004). *Distributed Cooperative Communications in Wireless Networks*. PhD thesis, Dept. of Electrical and Computer Engg., Rice University.

Khojastepour, M. A., Ahmed, N., and Aazhang, B. (2004a). Code design for the relay channel and factor graph decoding. In *Proc. of Asilomar Conference*, volume 2, pages 2000–2004.

Khojastepour, M. A., Sabharwal, A., and Aazhang, B. (2002a). Bounds on achievable rates for general multi-terminal networks with practical constraints. In *Proc. of IPSN*.

Khojastepour, M. A., Sabharwal, A., and Aazhang, B. (2002b). On the capacity of 'cheap' relay networks. In *Proc. of CISS*.

Khojastepour, M. A., Sabharwal, A., and Aazhang, B. (2003). On capacity of Gaussian 'cheap' relay channel. *GLOBECOM*, pages 1776–1780.

Khojastepour, M. A., Sabharwal, A., and Aazhang, B. (2004b). Improved achievable rates for user cooperation and relay channels. In *Proc. of ISIT*.

King, R. C. (1978). *Multiple access channels with generalized feedback*. PhD thesis, Stanford University.

Kobayashi, K. (1987). Combinatorial structure and capacity of the permuting relay channel. *IEEE Trans. Inform. Theory*, 33(6):813–826.

Koetter, R. and Medard, M. (2003). An algebraic approach to network coding. *IEEE/ACM Transactions on network coding*, 11:782–795.

Kramer, G., Gastpar, M., and Gupta, P. (2005). Cooperative strategies and capacity theorems for relay networks. *IEEE Trans. Inform. Theory*, 51(9):3037–3063.

Kschischang, F. R., Frey, B. J., and Loeliger, H.-A. (2001). Factor graphs and the sum-product algorithm. *IEEE Trans. Inform. Theory*, 47:498–519.

Laneman, J. N. (2002). *Cooperative diversity in wireless networks: Algorithms and Architectures*. PhD thesis, Massachusetts Institute of Technology, Cambridge, MA.

Laneman, J. N., Tse, D. N. C., and Wornell, G. W. (2004). Cooperative diversity in wireless networks: Efficient protocols and outage behavior. *IEEE Trans. Inform. Theory*, 50:3062–3080.

Laneman, J. N. and Wornell, G. W. (2003). Distributed space-time coded protocols for exploiting cooperative diversity in wireless networks. *IEEE Trans. Inform. Theory*, 49(10):2415–2425.

Li, S.-Y. R., Yeung, R. W., and Cai, N. (2003). Linear network coding. *IEEE Trans. Inform. Theory*, 49:371–381.

Li, Z. and Li, B. (2004). Network coding in undirected networks. *Proc. of CISS*.

Liang, Y. and Veeravalli, V. V. (2005). Gaussian orthogonal relay channels: Optimal resource allocation and capacity. *IEEE Trans. Inform. Theory*, 51:3284–3289.

Luby, M. G., Mitzenmacher, M., Shokrollahi, M. A., and Spielman, D. (2001). Improved low-density parity-check codes using irregular graphs. *IEEE Trans. Inform. Theory*, 47:585–598.

MacKay, D. J. C. (1999). Good error-correcting codes based on very sparse matrices. *IEEE Trans. Inform. Theory*, 45:399–431.

Marton, K. (1979). A coding theorem for the discrete memoryless broadcast channel. *IEEE Trans. Inform. Theory*, 25(3):306–311.

Mitran, P., Ochiai, H., and Tarokh, V. (2005). Space-time diversity enhancements using collaborative communications. *IEEE Trans. Inform. Theory*, 51(6):2041–2057.

Nabar, R. U., Bolcskei, H., and Kneubuhler, F. W. (2004). Fading relay channels: Performance limits and space-time signal design. *IEEE J. Select. Areas Commun.*, 22:1099–1109.

Reznik, A., Kulkarni, S. R., and Verdú, S. (2004). Degraded Gaussian multiple relay channel: capacity and optimal power allocation. *IEEE Trans. Inform. Theory*, 50(12):3037–3046.

Richardson, T. J., Shokrollahi, M. A., and Urbanke, R. L. (2001). Design of capacity-approaching irregular low-density parity-check codes. *IEEE Trans. Inform. Theory*, 47:619–637.

Richardson, T. J. and Urbanke, R. L. (2001). The capacity of low-density parity-check codes under message-passing decoding. *IEEE Trans. Inform. Theory*, 47:599–618.

Richardson, T. J. and Urbanke, R. L. (2004). Modern coding theory (draft of book).

Sabharwal, A. (2004). Impact of half-duplex radios and decoding latencies on mimo relay channels. In *Proc. of the Allerton Conference*, Monticello, IL.

Sabharwal, A. and Mitra, U. (2005). Rate-constrained relaying: A model for cooperation with limited relay resources. submitted to IEEE Trans. Inform. Theory.

Sato, H. (1976). Information transmission through a channel with relay. The Aloha System, University of Hawaii, Honolulu, Tech. Rep.

Sato, H. (1978). An outer bound to the capacity region of broadcast channels. *IEEE Trans. Inform. Theory*, 24(3):374–377.

Schein, B. (2001). *Distributed coordination in network information theory*. PhD thesis, Massachusetts Institute of Technology.

Schein, B. and Gallager, R. (2000). The Gaussian parallel relay network. In *Proc. of ISIT*, page 22, Sorrento, Italy.

Sendonaris, A., Erkip, E., and Aazhang, B. (2003a). User cooperation diversity. Part I. System description. *IEEE Trans. Commun.*, 51:1927–1938.

Sendonaris, A., Erkip, E., and Aazhang, B. (2003b). User cooperation diversity. Part II. Implementation aspects and performance analysis. *IEEE Trans. Commun.*, 51:1939–1948.

Slepian, D. and Wolf, J. (1973). Noiseless coding of correlated information sources. *IEEE Trans. Inform. Theory*, 19:471–480.

Stefanov, A. and Erkip, E. (2004). Cooperative coding for wireless networks. *IEEE Trans. Commun.*, 52:1470–1476.

Tarokh, V., Seshadri, N., and Calderbank, A. R. (1998). Space-time codes for high data rate wireless communication: Performance criterion and code construction. *IEEE Trans. Inform. Theory*, 44:744–765.

Telatar, I. E. (1999). Capacity of multiple-antenna Gaussian channels. *Eur. Trans. Tel.*, 10(6):585–595.

Thomas, J. A. (1987). Feedback can at most double Gaussian multiple access channel capacity. *IEEE Trans. Inform. Theory*, 33(5):711–716.

Toumpis, S. and Goldsmith, A. J. (2003). Capacity regions for wireless ad hoc networks. *IEEE Trans. Wireless Commun.*, 2(4):736–748.

van der Meulen, E. C. (1968). *Transmission of information in a T-terminal discrete memoryless channel*. PhD thesis, Dept. of Statistics, University of California, Berkeley.

van der Meulen, E. C. (1971). Three-terminal communication channels. *Advanced Applied Probability*, 3:120–154.

van der Meulen, E. C. (1977). A survey of multi-way channels in information theory: 1961-1976. *IEEE Trans. Inform. Theory*, 23:1–37.

Vanroose, P. and van der Meulen, E. C. (1992). Uniquely decodable codes for deterministic relay channels. *IEEE Trans. Inform. Theory*, 38(4):1203–1212.

Vishwanath, S., Jindal, N., and Goldsmith, A. J. (2003). Duality, achievable rates, and sum-rate capacity of Gaussian MIMO broadcast channels. *IEEE Trans. Inform. Theory*, 49(10):2658–2668.

Viswanath, P. and Tse, D. N. C. (2003). Sum capacity of the vector Gaussian broadcast channel and uplink-downlink duality. *IEEE Trans. Inform. Theory*, 49:1912–1921.

Wang, Bo, Zhang, Junshan, and Høst-Madsen, A. (2005). On the capacity of MIMO relay channels. *IEEE Trans. Inform. Theory*, 51:29–43.

Willems, F. M. J. (1982). *Informationtheoretical Results for the Discrete Memoryless Multiple Access Channel*. PhD thesis, Katholieke Univ. Leuven, Leuven, Belgium.

Wyner, A. D. (1978). The rate-distortion function for source coding with side information at the receiver—II: General sources. *Inform. Contr.*, 38:60–80.

Wyner, A. D. and Ziv, J. (1976). The rate-distortion function for source coding with side information at the receiver. *IEEE Trans. Inform. Theory*, 22(1): 1–11.

Xie, L. L. and Kumar, P. R. (2005). An achievable rate for the multiple-level relay channel. *IEEE Trans. Inform. Theory*, 51(4):1348–1358.

Xie, Liang-Liang and Kumar, P. R. (2004). A network information theory for wireless communication: Scaling laws and optimal operation. *IEEE Trans. Inform. Theory*, 50(5):748–767.

Xue, F., Xie, Liang-Liang, and Kumar, P. R. (2005). The transport capacity of wireless networks over fading channels. *IEEE Trans. Inform. Theory*, 51(3):834–847.

Yeung, R. W., Li, S.-Y. R., Cai, N., and Zhang, Z. (2005). Theory of network coding.

Yeung, R. W. and Zhang, Z. (1999). Distributed source coding for satellite communications. *IEEE Trans. Inform. Theory*, 45:1111–1120.

Yu, Wei and Cioffi, J. M. (2004). Sum capacity of Gaussian vector broadcast channels. *IEEE Trans. Inform. Theory*, 50:1875–1892.

Zahedi, S., Mohseni, M., and El Gamal, A. (2004). On the capacity of AWGN relay channels with linear relaying functions. In *Proc. of ISIT*, page 399, Chicago, IL.

Zeng, C. M., Kuhlmann, F., and Buzo, A. (1989). Achievability proof of some multiuser channel coding theorems using backward decoding. *IEEE Trans. Inform. Theory*, 35(6):1160–1165.

Zhang, Z. (1988). Partial converse for a relay channel. *IEEE Trans. Inform. Theory*, 34(5):1106–1110.

Zhang, Z., Bahceci, I., and Duman, T. M. (2004). Capacity approaching codes for relay channels. In *Proc. of ISIT*.

Zhang, Zheng and Duman, T. M. (2005). Capacity approaching turbo coding for half duplex relaying. In *Proc. of ISIT*.

Zhao, B. and Valenti, M. C. (2003). Distributed turbo coded diversity for relay channel. 39:786–787.

Chapter 3

COOPERATION, COMPETITION AND COGNITION IN WIRELESS NETWORKS

From Theory to Implementation

Oh-Soon Shin
Division of Engineering and Applied Sciences, Harvard University
osshin@deas.harvard.edu

Natasha Devroye
Division of Engineering and Applied Sciences, Harvard University
ndevroye@deas.harvard.edu

Patrick Mitran
Division of Engineering and Applied Sciences, Harvard University
mitran@deas.harvard.edu

Hideki Ochiai
Department of Electrical and Computer Engineering, Yokohama National University
hideki@ynu.ac.jp

Saeed S. Ghassemzadeh
AT&T Labs-Research
saeedg@research.att.com

H. T. Kung
Division of Engineering and Applied Sciences, Harvard University
kung@harvard.edu

Vahid Tarokh
Division of Engineering and Applied Sciences, Harvard University
vahid@deas.harvard.edu

Frank H. P. Fitzek and Marcos D. Katz (eds.), Cooperation in Wireless Networks: Principles and Applications, 69–100.

Abstract: Nodes and/or clusters of a wireless network operating on the same frequency can operate using three different paradigms: 1) *Competition*: Traditionally, this is information theoretically casted in the framework of interference channels. 2) *Cooperation*: Silent transmitters/receivers can help active transmitters/receivers in the transmission/reception of their messages, but have to extract this message from the underlying transmission or by other methods, and 3) *Cognitive Radio Transmission*: Some devices extract the message(s) of other transmitter(s) from their signals or by other methods, and use it to minimize interference from/to their own transmitted signals.

Competition has been well-studied in the literature. Cooperation has been less studied and *cognitive radio transmission* has not been studied much. For the cooperative case, we demonstrate that most of the multiple-input multiple-output (MIMO) space-time diversity gain can also be achieved through cooperative communications with single-antenna/multiple-antenna nodes when there is one receiving agent. In particular, for the single antenna case, we consider communication to take place between clusters of nearby nodes. We show the existence of cooperative codes for communications for which the intra-cluster negotiation penalty is in principle small and almost all the diversity gain of traditional space-time codes may be realized. For example, for a single transmitter node with two cooperators and one receiver node, if the cooperators have as little as 10 dB path loss advantage over the receiver, the penalty for cooperation over traditional space-time systems is negligible. Furthermore, we demonstrate and discuss the implementation of this idea in an orthogonal frequency division multiplexing (OFDM) system using a software defined ratio (SDR) platform. On the other hand, cooperative beamforming is an alternative way of realizing cooperative gain, particularly for a wireless sensor network. We analyze the statistical average properties and distribution of the beampattern of cooperative beamforming using the theory of random arrays.

For cognitive radio transmissions, which captures a form of asymmetric cooperation, we define a generalized cognitive radio channel as an n-transmitter, m-receiver interference channel in which sender i obtains (causally or non-causally) the messages of senders 1 through $i-1$. For simplicity, only the two sender, two receiver case is considered. In this scenario, one user, a cognitive radio, obtains (genie assisted, or causally) knowledge of the data to be transmitted by the other user. The cognitive radio may then simultaneously transmit over the same channel, as opposed to waiting for an idle channel as in a traditional cognitive radio channel protocol. Gel'fand-Pinsker coding (and the special case of dirty-paper coding) and ideas from achievable region constructions for the interference channel are used, and an achievable region for the cognitive radio channel is computed. In the Gaussian case, the described achievable region is compared to the upper bound provided by the 2×2 Gaussian MIMO broadcast channel, and an interference-free channel. We then extend the results to provide an achievable region for cognitive multiple access networks.

Keywords: cognitive multiple access network, cognitive radio channel, cooperative beamforming, cooperative diversity, dirty-paper coding, interference channel, multiple-input multiple-output, orthogonal frequency division multiplexing, space-time codes, wireless networks.

1. Introduction

All networks can be partitioned into clusters of nodes. When a wireless network is partitioned in this fashion, we can speak of *intra-cluster* as well as *inter-cluster behavior.* For *intra-cluster* behavior, we look at the nodes within each cluster, and notice that nodes can compete for resources (*competitive behavior*), can fully cooperate (*cooperative behavior*), or can partially cooperate in what we will call *cognitive behavior.* Similarly, for *inter-cluster* behavior, entire clusters can behave in a *competitive, cooperative,* or *cognitive* fashion. Figs. 3.1(a), 3.1(b) and 3.1(c) demonstrate examples of inter-cluster competitive, cognitive and cooperative behaviors respectively. The thick solid lines indicate that the messages of one cluster are known to the other. In Fig. 3.1(b) the two competitive clusters are independent and compete for the channel. In Fig. 3.1(b) the lower cluster has knowledge of the two messages to be sent by the upper cluster, but not vice versa. This is an example of inter-cluster cognitive behavior. In Fig. 3.1(c) the two clusters know each others' messages, an example of inter-cluster cooperative behavior. Within one cluster the messages may remain independent when the cluster is operated in an *intra-cluster competitive fashion.* If messages within one cluster were to be shared, it would constitute *intra-cluster cooperation (full) or cognition (partial).*

In a full cooperation paradigm, nodes which would have otherwise remained silent in a traditional competition paradigm, cooperate with the source and destination to increase communication capacity and reliability. Arguably, the initial work on cooperative communications stretches as far back as the pioneering

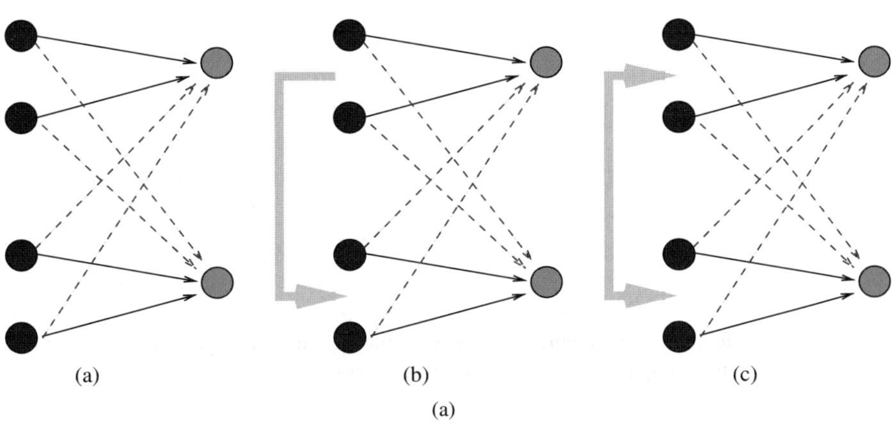

Figure 3.1. Inter-cluster behaviors of wireless networks. (a) Competitive behavior. All messages are independent. (b) Cognitive behavior. The thick solid arrow indicates that the second cluster has knowledge of the messages of the first cluster, but not vice versa. (c) Cooperative behavior. The thick solid two-way arrow indicates that each cluster knows the messages to be sent by the other cluster.

papers by van der Meulen [Van der Meulen, 1971] and Cover et al. [Cover and El Gamal, 1979] on the relay channel. However, the results obtained there do not appear to be directly applicable to inexpensive relays for wireless networks. This is because in realistic wireless models, it is not practically feasible to transmit and receive on the same antenna simultaneously (half-duplex constraint), since the intensity of the near field of the transmitted signal is much higher than that of the far field of the received signal. In wireless systems, the channel fading coefficients are usually not known to the transmit nodes; only the receive nodes have knowledge of the channel, *i.e.,* realistic wireless channels are *compound channels* [Wolfowitz, 1978; Csiszár and Körner, 1981]. Finally, while the degraded relay channel has been completely solved [Cover and El Gamal, 1979; Reznik et al., 2002], in wireless systems most noise is due to thermal noise in the receiver frontend. While it may be reasonable to assume that the relay has a better signal-to-noise ratio (SNR) than the ultimate receiver, it is unrealistic to assume that the receiver is a degraded version of the relay. Various extensions of the non-compound relay channel may be found in [Schein and Gallager, 2000; Gupta and Kumar, 2003; Gastpar et al., 2002; Gastpar and Vetterli, 2002a; Gastpar and Vetterli, 2002b; Khojastepour et al., 2003a; Khojastepour et al., 2003b; Khojastepour et al., 2003c; Kramer et al., 2004; Wang et al., 2005; Xie and Kumar, 2004].

Some more recent work on cooperative communications with emphasis on treating the wireless channel as a compound channel may be found in [Laneman and Wornell, 2003; Laneman et al., 2004; Hunter and Nostratinia, 2002; Hunter et al., 2003; Sendonaris et al., 2003a; Sendonaris et al., 2003b]. In [Laneman and Wornell, 2003], the authors consider a two-stage protocol (where the source transmits for a fixed amount of time followed by a fixed duration cooperation phase) to solve the half-duplex constraint and consider repetition and space-time coding based cooperative diversity algorithms. This is extended in [Laneman et al., 2004] with the consideration of adaptive protocols such as selection relaying and incremental relaying. In [Hunter and Nostratinia, 2002; Hunter et al., 2003] a similar time-division (TD) approach is employed where the relay is permitted to transmit its own information during the second phase if it is unable to cooperate. In [Sendonaris et al., 2003a; Sendonaris et al., 2003b], the authors assume two dedicated orthogonal subchannels between two mobile nodes, derive an achievable region for communication to a base station and consider code division multiple access (CDMA) implementation aspects. These results are derived by employing coding techniques [Willems, 1982; Willems et al., 1983] similar to those used for multiple access channels with generalized feedback [Carleial, 1982].

In Section 2, we present a bandwidth efficient decode and forward approach [Mitran et al., 2005] that does not fix phase durations or orthogonal subchannels to resolve the half-duplex constraint: each relay determines based on its own receive channel when to listen and when to transmit. Furthermore, the transmitters are not aware of the channel and we make no assumption of degradedness. In the case of multiple relays assisting the source, the approach permits one relay to assist another in receiving the message. However, more recent work along this line may be found in [Katz and Shamai, 2004; Azarian et al., 2004]. Finally, we briefly outline a practical cooperative system founded upon orthogonal frequency division multiplexing (OFDM) transmission [Shin et al., 2005].

Cooperative beamforming is an alternative cooperation technique to cooperative diversity especially suited for ad hoc sensor networks [Ochiai et al., 2005]. The advantages and applications of traditional beamforming with antenna arrays are well known; in wireless communications, beamforming is a powerful means for interference suppression which enables space division multiple access. Even though each node is equipped with a single antenna, if the nodes in a cluster share the information *a priori* and synchronously transmit data, it may be possible to beamform when transmitting (and also receiving) the data in a distributed manner. The overhead due to intra-cluster information sharing may be relatively small as this can be done by low-cost short-distance broadcasting-type communication among nodes. Thus, with distributed cooperative beamforming, the nodes can send the collected information to the far-end receiver over long distances with high efficiency. Also, only the cluster in the specified target direction receives the data with high signal power and no significant interference occurs for clusters in other directions. In Section 3, the beampattern aspects of cooperative beamforming using random arrays are analyzed in the framework of wireless networks. The average statistical figure of merits are first developed. In particular, it is shown that with N cooperative nodes, one can achieve a directivity of order N asymptotically. The probability distribution of the far-field beampattern is then analyzed.

While the cooperative schemes described thus far have relied on symmetric cooperation between nodes, asymmetric cooperation is also possible, and has been inspired by an explosion of interest in cognitive and software radios, as is evidenced by FCC proceedings [FCC, 2003; FCC, 2005], talks [Mitola, 1999b], and papers [Mitola, 1999a; Mitola, 2000]. *Software Defined Radios* (SDR) [Mitola, 1995] are devices used to communicate over the wireless medium equipped with either a general purpose processor or programmable silicon as hardware base, and enhanced by a flexible software architecture. They are low-cost, can be rapidly upgraded, and may adapt to the environment in real-time. Such devices are able to operate in many frequency bands under multiple transmission protocols and employ a variety of modulation and coding schemes. Taking this one step further, [Mitola, 2000] coined the term *cognitive radio* for software defined radios capable of sensing their environment and making decisions instantaneously, without any user intervention. This allows them to change their modulation schemes or protocols so as to better communicate with the sensed environment. Apart from their low cost and flexibility, another benefit of software radio technology is spectral efficiency. Currently, FCC measurements [FCC Spectrum Policy Task Force, 2002], indicate that at any time roughly 10% of the unlicensed frequency spectrum is actively in use (leaving 90% unused). If a wireless device such as a cognitive radio is able to sense an idle channel at a particular frequency band (or time), then it can shift to that frequency band (or time slot) to transmit its own information, dramatically increasing spectral (or temporal) efficiency.

Although cognitive radios have spurred great interest and excitement in industry, many of the fundamental theoretical questions on the limits of such technology remain unanswered. In current cognitive radio protocol proposals, the device listens to the wireless channel and determines, either in time or frequency, which part of the spectrum is unused [Horne, 2003]. It then adapts its signal to fill this void in the spectrum space. Thus, a device transmits over a certain time or frequency only when no other node does. In Section 4, the cognitive radio behavior is generalized to allow two users to simultaneously transmit over the same time or frequency [Devroye et al., 2004]. According to this approach, a cognitive radio will listen to the channel and, if sensed idle, proceed with the traditional cognitive radio channel model, that is, transmit during the voids. On the other hand, if another sender is sensed, the radio may decide to proceed with simultaneous transmission. The cognitive radio need not wait for an idle channel to start transmission. Specifically, we will prove achievability, in the information theoretic sense, of a certain set of rates at which two senders (cognitive radios, denoted as $\mathcal{S}_1$ and $\mathcal{S}_2$) can transmit simultaneously over a common channel to two independent receivers $\mathcal{R}_1$ and $\mathcal{R}_2$ when $\mathcal{S}_2$ is aware of the message to be sent by $\mathcal{S}_1$. The methods borrow ideas from Gel'fand and Pinsker's coding for channels with known interference at the transmitter [Gel'fand and Pinsker, 1980], Costa's dirty paper coding [Costa, 1983], the interference channel [Carleial, 1978], the Gaussian MIMO broadcast channel [Weingarten et al., 2004], and the achievable region of the interference channel described by Han and Kobayashi [Han and Kobayashi, 1981]. Finally, we discuss extensions of the results to cognitive multiple access networks [Devroye et al., 2005].

2. Cooperative Diversity

Preliminaries

For simplicity, we consider three nodes denoted as source (s), relay (r) and destination (d) as illustrated in Fig. 3.2, each equipped with N_s, N_r and N_d antennas respectively. The results can readily generalize to multiple relay nodes. We assume that while listening to the channel, the relay does not transmit. Hence, the communications protocol is as follows. The source node wishes

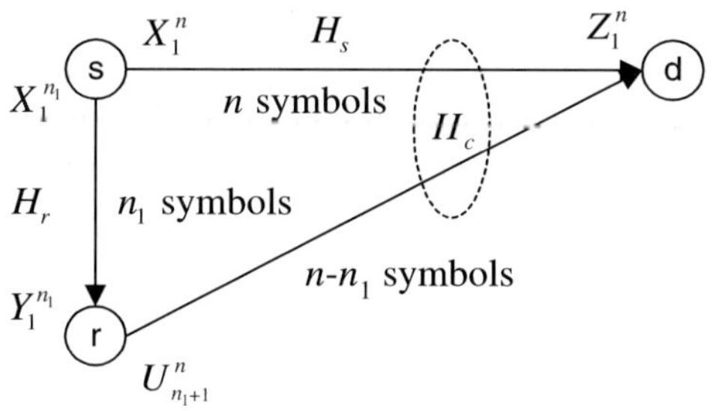

Figure 3.2. The cooperative communications problem for two transmit cooperators and one receiver.

to transmit one of 2^{nR} messages to the destination employing n channel uses. While not transmitting, the relay node listens. Due to the relay node's proximity to the source, after n_1 samples from the channel (a number which the relay determines on its own and for which the source has no knowledge), it may correctly decode the message. After decoding the message, it then proceeds to transmit for the remaining $n - n_1$ transmissions in an effort to improve the reception of the message at the destination. The destination is assumed to be made aware of n_1 before attempting to decode the message. This may be achieved by an explicit low-rate transmission from the relay to the destination. Alternatively, if the value of n_1 is constrained to some integer multiple of a fundamental period n_0 (say $n_0 \sim \sqrt{n}$), then the destination may estimate n_1 accurately using power detection methods. We denote the first phase of the n_1 transmissions as the *listening* phase while the last $n - n_1$ transmissions as the *collaboration* phase.

We assume that all channels are modeled as additive white Gaussian noise (AWGN) with quasi-static fading. In particular, X and U are column vectors representing the transmission from the source and relay nodes respectively and we denote by Y and Z the received messages at the relay and destination respectively. Then during the listening phase we have

$$Z = H_s X + N_Z \tag{3.1}$$
$$Y = H_r X + N_Y, \tag{3.2}$$

where N_Z and N_Y are column vectors of statistically independent complex AWGN with variance 1/2 per row per dimension, H_s is the the fading matrix between the source and destination nodes and likewise, H_r is the fading matrix between the source and relay nodes. During the collaboration phase, we have

$$Z = H_c[X^T, U^T]^T + N_Z, \tag{3.3}$$

where H_c is a channel matrix that contains H_s as a submatrix (see Fig. 3.2).

We further assume that the source has no knowledge of the H_r and H_c matrices (and hence the H_s matrix too). Similarly, the relay has no knowledge of H_c but is assumed to know H_r. Finally, the destination knows H_c. Without loss of generality, we will assume that all transmit antennas have unit average power during their respective transmission phases. Likewise, the receive antennas have unit power Gaussian noise. If this is not the case, the respective H matrices may be appropriately scaled row-wise and column-wise.

Under the above unit transmit power per transmit antenna and unit noise power per receive antenna constraint, it is well known that a MIMO system with Gaussian codebook and with rate R bits/channel use can reliably communicate over any channel with transfer matrix H such that $R < \log_2 \det(I + HH^\dagger) \stackrel{\triangle}{=} C(H)$ [Telatar, 1999; Foschini and Gans, 1998], where I denotes the identity matrix and $H^\dagger$ is the conjugate transpose of H.

Intuition for the above problem then suggests the following. During the listening phase, the relay knowing H_r listens for an amount of time n_1 such that $nR < n_1 C(H_r)$. During this time, the relay receives at least nR bits of information and may reliably decode the message. The destination, on the other hand, receives information at the rate of $C(H_s)$ bits/channel use during the listening phase and at the rate of $C(H_c)$ bits/channel use during the cooperative phase. It may reliably decode the message provided that $nR < n_1 C(H_s) + (n - n_1)C(H_c)$. In the limit as $n \to \infty$, the ratio n_1/n approaches a fraction f and we may conjecture that there exists a "good" code of rate R for the set of channels (H_r, H_c) which satisfies

$$R \leq fC(H_s) + (1 - f)C(H_c) \text{ and } R \leq fC(H_r), \tag{3.4}$$

for some $f \in [0, 1]$. We note that if the channel between the source and the relay is particularly poor, we may fall back on the traditional point-to-point communications paradigm and add the following region to that given in (3.4)

$$R \leq C(H_s). \tag{3.5}$$

Achievable Rate of Cooperative Diversity

In this section, we define and state a theorem on the achievability for compound synchronous relay channels. The codebook for the source will be denoted by $C_s^{(n)}$ and consists of $K2^{nR}$ codewords for some constant $K > 0$. The wth codeword of the source node codebook will be denoted by $x_1^n(w)$. If the source node has N_s transmit antennas, then each codeletter consists of a column vector with dimension N_s and each codeword is in fact an $N_s \times n$ matrix. For the relay, we will denote by $C_r^{(n)}$ a family of n codebooks $C_r^{(n,n_1)}$ indexed by $1 \leq n_1 \leq n$ where $C_r^{(n,n_1)}$ is a codebook with $K2^{nR}$ codewords of length $n - n_1$. The wth codeword of $C_r^{(n,n_1)}$ will be denoted by $u_{n_1+1}^n(w)$. Finally, we will denote by $C^{(n)} = \{C_s^{(n)}, C_r^{(n)}\}$.

Before explaining the encoding procedure, it will help to explain the decoder. With each message $W = w$, pair of channels (H_r, H_c) and value of n_1, we associate some disjoint (over w) subsets of $\mathbb{C}^N$ as follows: $S_{w,H_c,n_1} \subset \mathbb{C}^{N_d \times n}$ and $R_{w,H_r,n_1} \subset \mathbb{C}^{N_r \times n_1}$. We shall refer to $C^{(n)}$ as the encoder or codebook and the sets S_{w,H_c,n_1} and R_{w,H_r,n_1} as the decoder.

Encoding and decoding: The source wishes to transmit message $W = w$ to the destination. To that end, the source looks up the wth codeword in its codebook and proceeds to transmit it to the destination and the relay. The relay, knowing the channel H_r, decides upon the smallest value of n_1 for which $nR + \delta < n_1 C(H_r)$ (for some fixed $\delta > 0$) and for which $\delta < n_1/n < 1 - \delta$. If no such n_1 exists, the relay takes $n_1 = n$, makes no attempt to decode the message and remains silent. If $n_1 < n$, the relay listens to the channel for this duration and lists all the $\hat{w}$ for which $Y_1^{n_1} \in R_{\hat{w},H_r,n_1}$. If $\hat{w}$ exists (and is hence unique), the relay looks up the $\hat{w}$th codeword in the $C_r^{(n,n_1)}$ codebook and proceeds to transmit it for the remaining $n - n_1$ channel uses. Otherwise, the relay declares an error.

After the last transmission, the destination has now received Z_1^n where

$$Z_i = \begin{cases} H_s X_i + N_{Z,i} & i \leq n_1 \\ H_c[X_i^T, U_i^T]^T + N_{Z,i} & i > n_1, \end{cases} \tag{3.6}$$

and is informed of the value of n_1. The destination then proceeds to list all $\hat{w}$ such that $Z_1^n \in S_{\hat{w},H_c,n_1}$. If $\hat{w}$ exists (and is hence unique), the destination declares the transmitted message as $\hat{W} = \hat{w}$. Otherwise an error is declared. We shall abuse notation and denote by the event $\hat{W} \neq W$ the case when either the relay or the destination declares an error or decodes an incorrect message (if the relay makes no attempt at decoding the message, it cannot produce an error).

Since the source, relay and destination nodes each have N_s, N_r and N_d antennas respectively, we note that $H_r \in \mathbb{C}^{N_r \times N_s}$ and $H_c \in \mathbb{C}^{N_d \times (N_s + N_r)}$. We denote by $\mathcal{H}$ a subset of compound relay channels, *i.e.*, $\mathcal{H} \subset \mathbb{C}^{N_r \times N_s} \times \mathbb{C}^{N_d \times (N_s + N_r)}$. Also, for a codebook $C^{(n)}$, we denote by λ_n^s (where the superscript s denotes synchronism)

$$\lambda_n^s = \max_w \sup_{(H_r,H_c) \in \mathcal{H}} P[\hat{W} \neq W | W = w, H_r, H_c]. \tag{3.7}$$

DEFINITION 3.1 *(Achievability for a compound relay channel) A rate R is said to be achievable for a set of pairs $(H_r, H_c) \in \mathcal{H}$ if for any $\epsilon > 0$, there exists a sequence of encoders and decoders $C^{(n)}$, S_{w,H_c,n_1} and R_{w,H_r,n_1} in n*

such that $\lambda_n^s \to 0$ *as* $n \to \infty$ *and each codeword in each sub-codebook of* $C^{(n)}$ *has average power at most* $1 + \epsilon$. □

Before stating a theorem on the existence of good codes, we define a norm on a complex matrix H with entries $H_{i,j}$ as $||H|| \triangleq \max_{i,j}\{|H_{i,j}|\}$.

THEOREM 3.2 *Consider the set* $\mathcal{H}_{\delta,L}(R)$ *of matrices* (H_r, H_c) *such that* $||H_r|| \leq L$ *and* $||H_c|| \leq L$ *and which satisfy either both*

$$R + \delta \leq fC(H_s) + (1 - f)C(H_c) \text{ and } R + \delta \leq fC(H_r) \qquad (3.8)$$

for some $\delta \leq f \leq 1 - \delta$ *(each* f *may depend on* H_r*), or*

$$R + \delta \leq C(H_s). \qquad (3.9)$$

Then, the rate R *is achievable for the compound relay channel* $\mathcal{H}_{\delta,L}(R)$ *for any* $\delta > 0$ *and* $L > 0$. □

This theorem essentially states that the region in equations (3.4) – (3.5) may be arbitrarily approximated by taking $\delta > 0$ sufficiently small and $L > 0$ sufficiently large. The proof is given in [Mitran et al., 2005], where the result was also extended to the case of bounded asynchronism, similar to that in [Cover et al., 1981], between nodes.

Performance Analysis

We analyze the theoretical performance of a code that achieves the compound channels in Theorem 1 for single antenna nodes when the fading is quasi-static Rayleigh distributed. In particular, since L was an arbitrarily large constant, we shall take $L = \infty$. Similarly, since δ was an arbitrarily small positive number, we take $\delta = 0$. Furthermore, we will relax our restrictions on unit power per transmit antenna (as stated earlier, this was allowable since the respective H matrices could be appropriately scaled to compensate). In this section, it will be more convenient to keep the H matrices fixed and show the explicit dependence of the outage probability on the receive signal power at the destination node (during the listening phase) per transmit antenna at the source node, E_s, and the noise power at each receive antenna, σ^2. Under these conditions, we have

$$C(H, \gamma) \triangleq \log_2 \det(I + \gamma HH^{\dagger}), \qquad (3.10)$$

where $\gamma \triangleq \frac{E_s}{\sigma^2}$ and the expression holds regardless of the number of transmit antennas (as E_s is defined as the normalized receive power per transmit antenna). We model the proximity of the relay node to the source node by a reduction $G \in \mathbb{R}$ in path loss, or equivalently, an increase in the achievable rate between cooperator nodes as expressed by $C(H, G\gamma)$. With these conventions, we will assume that the code successfully transmits the message from the source to the destination in a two cooperator scenario provided that either

$$R \leq fC(H_r, G\gamma) \text{ and } R \leq fC(H_s, \gamma) + (1-f)C(H_c, \gamma), \qquad (3.11)$$

for some $0 < f < 1$, or

$$R \leq C(H_s, \gamma) \qquad (3.12)$$

holds.

We note that the fraction f is determined by the relay and depends only on the realization of H_r according to

$$f^* \triangleq \min\{1, R/C(H_r, G\gamma)\}. \qquad (3.13)$$

Since $C(H_c, \gamma) \geq C(H_s, \gamma)$, this is the optimal choice of f to minimize the outage probability of our scheme. Even if the relay knew H_c, it could not do better. Furthermore, given f, the effective receive power at the destination is $(2-f)E_s$ as the relay was only transmitting for a fraction $1-f$ of the total transmission time. The effective receive SNR for the duration of the transmission is then $(2-f)E_s/\sigma^2$. We may thus rewrite the outage probability P_{out} *for our proposed scheme* as

$$P_{out} = P\left[R > f^*C(H_s, \gamma) + (1-f^*)C(H_c, \gamma)\right]. \qquad (3.14)$$

Under the assumption that the relay transmits the same instantaneous energy as the source node and suffers the same amount of path loss, (3.14) can be rewritten using the definition in (3.10) as

$$\begin{aligned} P_{\text{out}} = P[R > & f^* \log_2\left(1+\gamma|H_s|^2\right) \\ & + (1-f^*)\log_2\left(1+\gamma\left(|H_s|^2+|H_{r,d}|^2\right)\right)]. \end{aligned} \qquad (3.15)$$

Evaluation of (3.15) is difficult to carry out even numerically, since exact analysis typically yields double integrals. Using the Jensen's inequality, a tight lower bound for (3.15) can be derived as [Mitran et al., 2005]

$$\begin{aligned} P_{\text{out}} &\geq P\left[R > \log_2\left(1+\gamma\left(|H_s|^2+(1-f^*)|H_{r,d}|^2\right)\right)\right] \\ &= P\left[|H_s|^2+(1-f^*)|H_{r,d}|^2 < \frac{2^R-1}{\gamma}\right]. \end{aligned} \qquad (3.16)$$

Note that the right hand side of (3.16) is the cumulative distribution function (CDF) of the random variable $|H_s|^2 + (1-f^*)|H_{r,d}|^2$, which can be found by standard algebra of random variables. The explicit expression of (3.16) is

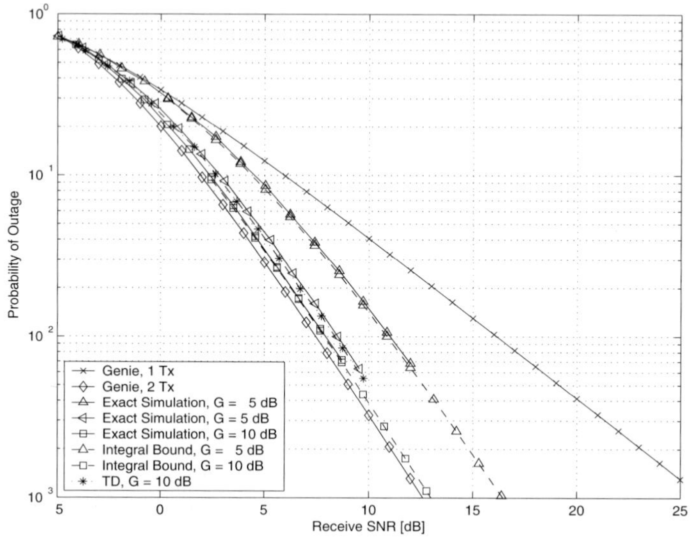

(a)

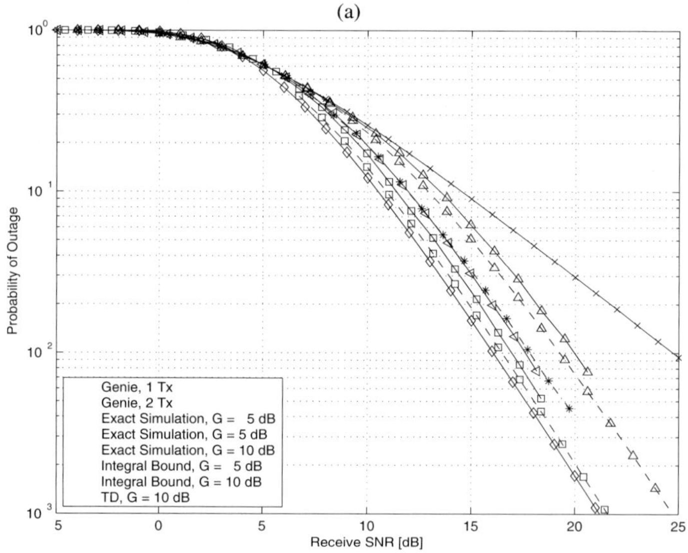

(b)

Figure 3.3. Outage probability of two transmit collaborators and one receiver for various geometric gain factors G. (a) $R = 0.5$. (b) $R = 2$.

given in [Mitran et al., 2005] for the case that all fading channel coefficients $H_r, H_s, H_{r,d}$ are i.i.d. complex Gaussian random variables.

Fig. 3.3 illustrates the simulated outage probability (for $R = 0.5$ and $R = 2$) and the corresponding lower bound (3.16) for various values of G versus the averaged receive SNR for quasi-static Rayleigh fading channels, *i.e.*, channels where each of the H matrices have independent circularly symmetric Gaussian distributed entries with total variance 1. Exact results were obtained by Monte-Carlo simulation of equations (3.13) and (3.14). These simulation results are confirmed by the tightness of the lower bound (3.16). Also illustrated in Fig. 3.3 is the outage probability of a traditional 1×1 and 2×1 space-time system, which is referred to as a genie bound. We see that even with as little gain as $G = -5$ dB, for an outage probability of 10^{-2}, the loss in performance is only 3.5 dB compared to the genie 2×1 bound. With $G = 10$ dB, the genie 2×1 bound is closely approached by the proposed cooperative scheme.

Finally, it is instructive to compare the performance of this scheme (where the relay listens for the smallest fraction of time f that is necessary) to a scheme where f is not allowed such flexibility. In Fig. 3.3, the performance of a scheme where f is constrained to 0.5 or 1.0 is also illustrated. (We use the notation TD for this scheme. Whereas $f = 0.5$ corresponds to a half listening/half cooperation protocol, $f = 1.0$ is equivalent to no cooperation.) For $G = 10$ dB and $R = 0.5$, we see that this TD scheme performs as well as the proposed scheme with $G = 5$ dB at an outage probability of 10^{-2}. Hence, in this case, the penalty for employing a predetermined TD scheme is equivalent to a 5 dB penalty in geometric gain. For higher rates such as $R = 2$, the penalty increases.

The result in Theorem 1 generalizes in a straightforward manner to multiple transmit collaborators. In particular, an extension to the case of three transmit cooperators and one receiver, as illustrated in Fig. 3.4, is presented in [Mitran et al., 2005]. We note a remarkable feature of the scheme in Fig. 3.4. If the relay r_0 has a better channel from the source than the relay r_1, relay r_1 may receive information not only from the source node s, but from the relay node r_0 as soon as r_0 has finished listening. By symmetry, a similar situation is possible

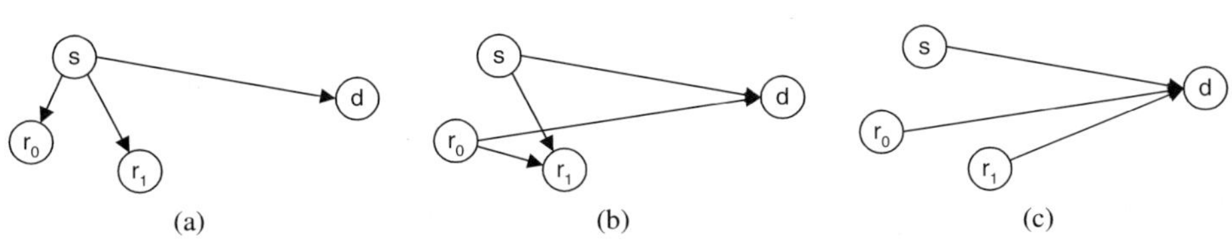

Figure 3.4. The three cooperators problem with one receiver. (a) Nodes r_0, r_1, and d are listening. (b) Node r_0 has stopped listening and started cooperating. It transmits to nodes d *and* r_1. (c) Node r_1 has stopped listening and started cooperating.

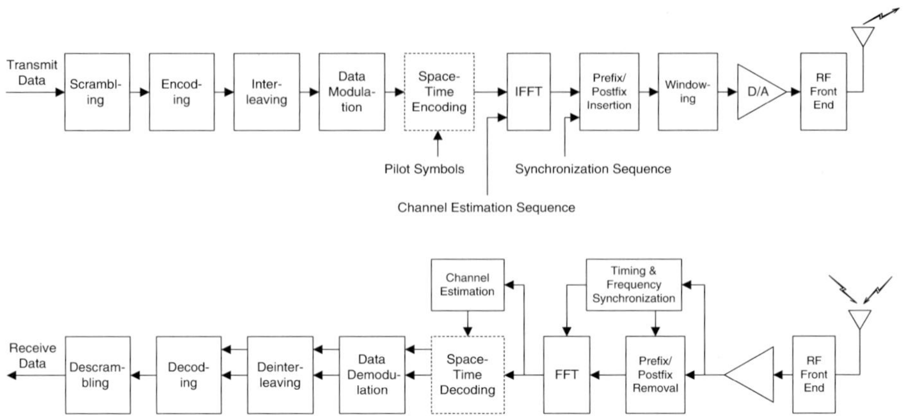

Figure 3.5. Block diagram of the CO-OFDM transmitter and receiver (dotted blocks are used only in the cooperation phase).

if relay r_1 has a better channel than r_0. If the number of cooperative nodes were further increased to m, we would see a cascade effect by which the relays would quickly share among themselves the message by way of $m(m-1)/2$ possible paths. More recent literature that allows for this sort of information sharing strategy may be found in [Katz and Shamai, 2004; Azarian et al., 2004].

Implementation: Cooperative OFDM System

A space-time cooperative system based on orthogonal frequency division multiplexing (OFDM), which is referred to as a cooperative (CO)-OFDM system, has been designed in [Shin et al., 2005]. The system will be implemented on a software radio platform. We briefly outline the main features of the CO-OFDM system and some performance results. Details can be found in [Shin et al., 2005].

Fig. 3.5 illustrates a block diagram of the CO-OFDM transmitter and receiver. The structure is similar to that of the IEEE 802.11a standard [IEEE 802.11a-1999, 1999] except for the use of space-time cooperation. Note that transmit symbols are encoded according to a form of time-division cooperative diversity protocol discussed in Section 2.0. The transmission of each frame involves two subsequent phases with fixed duration: the *listening phase* and the *cooperation phase*. In the *listening phase*, the source broadcasts a listening subframe to the relays and destination. Space-time coding is not employed in this phase, since the source is equipped with only one transmit antenna. If the destination succeeds in decoding the listening subframe, the following cooperation phase is ignored at the destination. Otherwise, the destination attempts to decode the succeeding cooperation subframe. Note that the relays and destination can

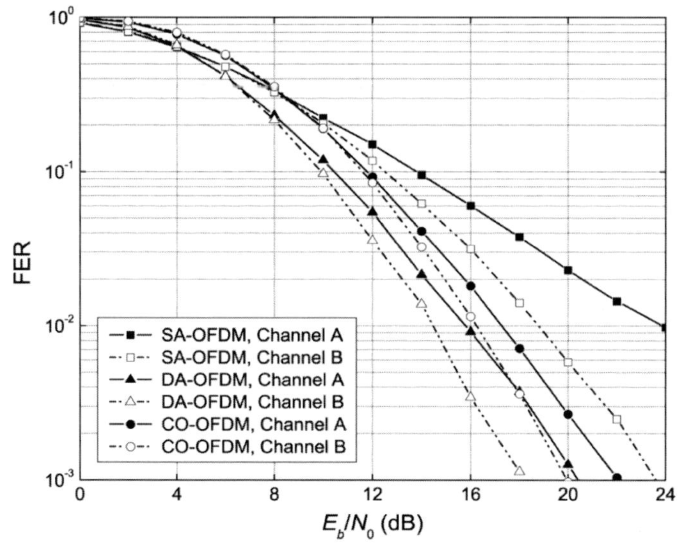

Figure 3.6. The overall FER performance of the CO-OFDM system.

realize whether decoding of each subframe is successful or not by computing the checksum of the frame check sequence.

In the *cooperation phase*, the source constructs and transmits a cooperation subframe, which corresponds to a portion of the space-time coded version of the listening subframe. The behavior of the relay depends on whether it has succeeded or not in decoding the preceding listening subframe. If a relay has succeeded in decoding, the relay also constructs and transmits a cooperation subframe, which corresponds to another portion of space-time coded signal. Then the destination may receive the complete space-time coded signal from the source and relay, enabling the reliable decoding of the cooperation subframe. Otherwise, if the relay has failed to decode the listening subframe, it is silent in the cooperation phase. The listening and cooperation subframes are allowed to be transmitted at different transmission rates. For the case of a single relay node, [Shin et al., 2005] has also devised a frame structure including preamble sequences, and provided simple and effective timing and frequency synchronization algorithms and a channel estimation algorithm.

Fig. 3.6 shows the overall frame error rate (FER) performance of the CO-OFDM system, when the synchronization and channel estimation algorithms proposed in [Shin et al., 2005] are adopted. The performance of a single-antenna OFDM (SA-OFDM) system and a double-antenna OFDM (DA-OFDM) system without cooperation is also presented for comparison. The geometric gain G is assumed to be 10 dB. Other details of simulation conditions are given in [Shin et al., 2005]. From the figure, we can observe significant performance

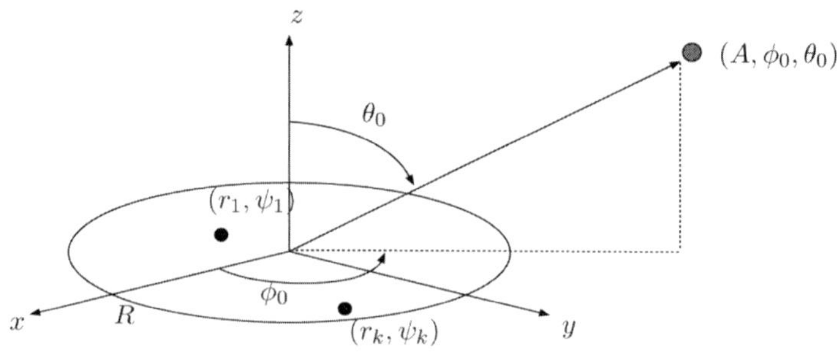

Figure 3.7. Definitions of notation for cooperative beamforming.

improvement of the CO-OFDM system over the SA-OFDM system. At a FER of 10^{-2}, for example, the energy gain of the CO-OFDM system over the SA-OFDM system is found to be as much as 6.7 dB for channel A, and 2.5 dB for channel B, where the channel models are given in [European Telecommunications Standards Institute, 1998; Shin et al., 2005]. From the slopes of the FER curves, we notice that the CO-OFDM system achieves a diversity order comparable to that of the DA-OFDM system, as predicted by the theory.

3. Cooperative Beamforming

System Model and Beampattern

The geometrical configuration of the distributed nodes and destination (or target) is illustrated in Fig. 3.7 where, without loss of generality, all the cooperative nodes are assumed to be located on the x-y plane. The kth node location is thus denoted in polar coordinates by (r_k, ψ_k). The location of the destination is given in spherical coordinates by (A, ϕ_0, θ_0). Following the standard notation in antenna theory [Balanis, 1997], the angle $\theta \in [0, \pi]$ denotes the elevation direction, whereas the angle $\phi \in [-\pi, \pi]$ represents the azimuth direction. In order to simplify the analysis, the following assumptions are made:

- The location of each node is chosen randomly, following a uniform distribution within a disk of radius R.
- Each node is equipped with a single ideal isotropic antenna.
- All nodes transmit identical energies, and the path losses of all nodes are also identical. Thus the underlying model falls within the framework of phased arrays.
- There is no reflection or scattering of the signal. Thus, there is no multipath fading or shadowing.
- The nodes are sufficiently separated that any mutual coupling effects [Balanis, 1997] among the antennas of different sensor nodes are negligible.

- All the nodes are perfectly synchronized so that no frequency offset or phase jitter occurs.

Let a realization of node locations $r = [r_1, r_2, \ldots, r_N] \in [0, R]^N$ and $\boldsymbol{\psi} = [\psi_1, \psi_2, \ldots, \psi_N] \in [-\pi, \pi]^N$ be given, where N denotes the number of nodes. We are interested in the radiation pattern in the far-field region, and we assume that the far-field condition $A \gg r_k$ holds. Then under the above assumptions, the array factor in the far field can be approximated as [Ochiai et al., 2005]

$$F(\phi, \theta | \boldsymbol{r}, \boldsymbol{\psi}) \approx \frac{1}{N} \sum_{k=1}^{N} e^{j\frac{2\pi}{\lambda} r_k [\sin\theta_0 \cos(\phi_0 - \psi_k) - \sin\theta\cos(\phi - \psi_k)]} \tag{3.17}$$

Of particular interest in practice is the case where $\theta_0 = \frac{\pi}{2}$, *i.e.*, the destination node is in the same plane as the cooperative transmit nodes. Without loss of generality, we also assume that $\phi_0 = 0$, since the choice of ϕ_0 does not change the results. Under this assumption and the assumption on the distribution of node location, the array factor in (3.17) is simplified to [Ochiai et al., 2005]

$$F(\phi | \boldsymbol{z}) \triangleq F(\phi, \theta = \pi/2 | \boldsymbol{r}, \boldsymbol{\psi}) = \frac{1}{N} \sum_{k=1}^{N} e^{-j 4\pi \tilde{R} \sin\left(\frac{\phi}{2}\right) z_k}, \tag{3.18}$$

where each element z_k of $\boldsymbol{z}$ is given as $z_k \triangleq \frac{r_k}{R} \sin(\psi_k - \phi/2)$, and $\tilde{R} \triangleq \frac{R}{\lambda}$ is the radius of the disk normalized by the wavelength λ. Finally, the far-field beampattern can be defined as

$$P(\phi | \boldsymbol{z}) \triangleq |F(\phi | \boldsymbol{z})|^2 = \frac{1}{N} + \frac{1}{N^2} \sum_{k=1}^{N} e^{-j\alpha(\phi) z_k} \sum_{\substack{l=1 \\ l \neq k}}^{N} e^{j\alpha(\phi) z_l}, \tag{3.19}$$

where $\alpha(\phi) \triangleq 4\pi\tilde{R}\sin(\phi/2)$.

Average Far-Field Beampattern

By taking the average of (3.19) over all realizations of $\boldsymbol{z}$, we obtain the average beampattern as

$$P_{\text{av}}(\phi) \triangleq E_{\boldsymbol{z}}\{P(\phi | \boldsymbol{z})\} = \frac{1}{N} + \left(1 - \frac{1}{N}\right) \left| 2 \cdot \frac{J_1(\alpha(\phi))}{\alpha(\phi)} \right|^2, \tag{3.20}$$

where $J_n(x)$ is the nth order Bessel function of the first kind. Using this formula, several statistical properties have been investigated in [Ochiai et al., 2005], including positions of peaks and zeros, 3 dB beamwidth, 3 dB sidelobe region, and average directivity. The most important amongst them is the average directivity, which characterizes how much radiated energy is concentrated on in the desired direction relative to a single isotropic antenna. The average directivity

D_{av} is defined as

$$D_{\text{av}} \triangleq E_{\boldsymbol{z}}\{D(\boldsymbol{z})\} = E_{\boldsymbol{z}}\left\{\frac{2\pi}{\int_{-\pi}^{\pi} P(\phi|\boldsymbol{z})d\phi}\right\}. \tag{3.21}$$

In [Ochiai et al., 2005], the following Theorem on a lower bound of the normalized directivity was proved.

THEOREM 3.3 (NORMALIZED DIRECTIVITY LOWER BOUND) *For large $\tilde{R}$ and N, D_{av}/N is lower bounded by $\frac{1}{1+\mu N/\tilde{R}}$, where μ is a positive constant independent of N and $\tilde{R}$ ($\mu \approx 0.09332$).*

Note that the factor $N/\tilde{R}$ in the lower bound can be seen as a *one-dimensional node density*. Theorem 3.3 indicates that the node density almost uniquely determines the normalized directivity D_{av}/N. It is important to note that in order to achieve a certain normalized directivity with a large number of nodes N, the node density should be maintained to the desired value by spreading the nodes as sparsely as possible. The above theorem also indicates that in order to achieve high directivity (i.e. directivity close to N), the distribution of nodes should be as sparse as possible.

Distribution of Far-Field Beampattern of Cooperative Beamforming

For random arrays, the above average behavior does not necessarily approximate a beampattern of any given realization unless $N \to \infty$. In fact, even though the average beampattern has a sharp mainbeam and sidelobes always close to $1/N$, there is a large dynamic range of sidelobes among randomly generated beampatterns. Therefore, in practice, the statistical *distribution* of beampatterns and sidelobes in particular, is of interest. By approximating the beampattern sidelobes as a complex Gaussian process, Lo [Lo, 1964] has derived the distribution of the beampattern in the case of linear random arrays. In the context of cooperative beamforming, we briefly summarize the exact complementary CDF (CCDF) of the beampattern and a Gaussian approximation of it similar to [Lo, 1964].

- Exact distribution:

$$\Pr[P(\phi) > P_0] = \iint_{x^2+y^2>N^2P_0} f_{\tilde{X},\tilde{Y}}(x,y)\,dx\,dy, \tag{3.22}$$

  where kth entries of $\tilde{X}$ and $\tilde{Y}$ are, respectively, given as $\tilde{x}_k \triangleq \cos(z_k\alpha(\phi))$ and $\tilde{y}_k \triangleq \sin(z_k\alpha(\phi))$ ($k = 1, 2, \cdots, N$).

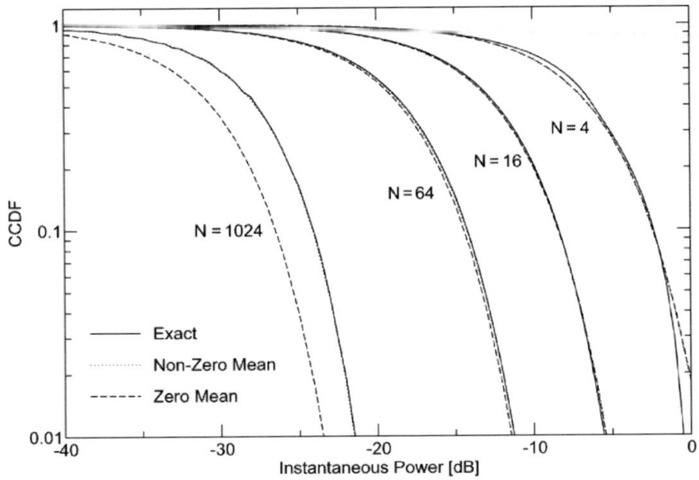

Figure 3.8. CCDF of beampattern with $\tilde{R} = 2$ and $\phi = \pi/4$.

- Gaussian approximation:

$$\Pr\left[P(\phi) > P_0\right] \approx Q\left(2\sqrt{2N}\frac{J_1\left(\alpha(\phi)\right)}{\alpha(\phi)}, \sqrt{2NP_0}\right), \tag{3.23}$$

where $Q(\cdot, \cdot)$ denotes the Marcum-Q function.

- Zero-mean Gaussian approximation:

$$\Pr\left[P(\phi) > P_0\right] \approx e^{-NP_0}. \tag{3.24}$$

In Fig. 3.8, the CCDFs computed with various formulae are shown with $\tilde{R} = 2$ and $\phi = \pi/4$, which corresponds to the sidelobe region. The exact formula of (3.22), the Gaussian approximation of (3.23), and the zero-mean Gaussian approximation of (3.24) are shown in the figure. As observed from Fig. 3.8, even the zero-mean Gaussian approximation may be valid for this sidelobe region, but for $N = 1024$ the Gaussian approximation will have some noticeable discrepancy with the exact value. This is due to the fact that the zero-mean approximation does not hold for this case. In [Ochiai et al., 2005], the distribution of the beampattern was discussed in more detail, and an approximate upper bound on the distribution of the peak sidelobes was derived. Furthermore, [Ochiai et al., 2005] also considered both open-loop and closed-loop scenarios and investigated the effects of phase ambiguities and location estimation errors on the resultant average beampatterns.

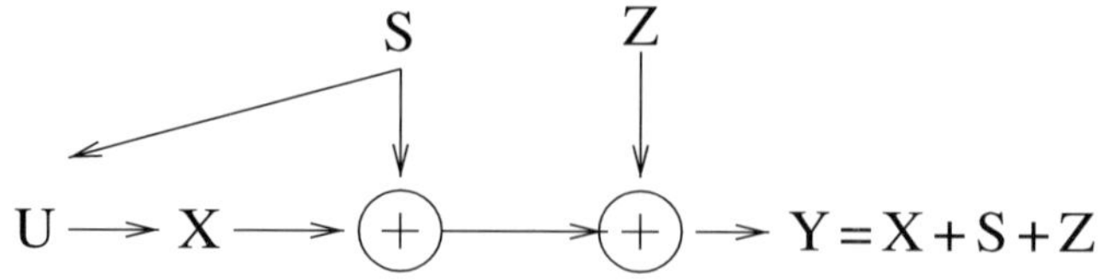

Figure 3.9. Dirty paper coding channel with input X, auxiliary random variable U, interference S known non-causally to the transmitter, additive noise Z and output Y.

4. Cognitive Radio

Preliminaries

A key idea behind achieving high data rates in an environment where two senders share a common channel is interference cancellation or mitigation. When side-information is known at the transmitter only, the channel capacity is given by the well-known formula obtained by Gel'fand and Pinsker [Gel'fand and Pinsker, 1980] as

$$C = \max_{p(u,x|s)} \left[I(U;Y) - I(U;S) \right], \tag{3.25}$$

where X is the input to the channel, Y is the output, S is the interference, and U is an auxiliary random variable chosen to make the channel $U \to Y$ appear causal. The channel model and variables are shown in Fig. 3.9 for additive interference and noise. We refer to the coding technique used in [Gel'fand and Pinsker, 1980] as Gel'fand-Pinkser coding or binning. In the Gaussian noise and interference case, Costa achieves the capacity of an interference-free channel by assuming the input X to the channel is Gaussian, and then considering an auxiliary variable U of the form $U = X + \alpha S$ for some parameter α whose optimal value is equal to the ratio of the signal power to the signal plus noise power. Since the rate thus obtained is equal to the capacity of an interference-free channel, which provides an upper bound, optimality is achieved by the assumed Gaussian input X. *Dirty paper coding* is the term first used by Costa [Costa, 1983] to describe a technique which completely mitigates *a-priori* known interference over an input power constrained additive white Gaussian noise channel. We will make use of the coding techniques of Costa [Costa, 1983], Gel'fand and Pinsker [Gel'fand and Pinsker, 1980], as well as Cover and Chiang [Cover and Chiang, 2002] in Section 4.0.

The cognitive radio channel is also closely related to the interference channel, which is briefly described next. Consider a discrete memoryless *interference channel* [Carleial, 1978], with random variables $X_1 \in \mathcal{X}_1$, $X_2 \in \mathcal{X}_2$ as inputs to the channel characterized by the conditional probabilities $p(y_1|x_1, x_2)$, $p(y_2|x_1, x_2)$ with resulting channel output random variables $Y_1 \in \mathcal{Y}_1$, $Y_2 \in \mathcal{Y}_2$. The interference channel corresponds to two independent senders $\mathcal{S}_1$, $\mathcal{S}_2$, with

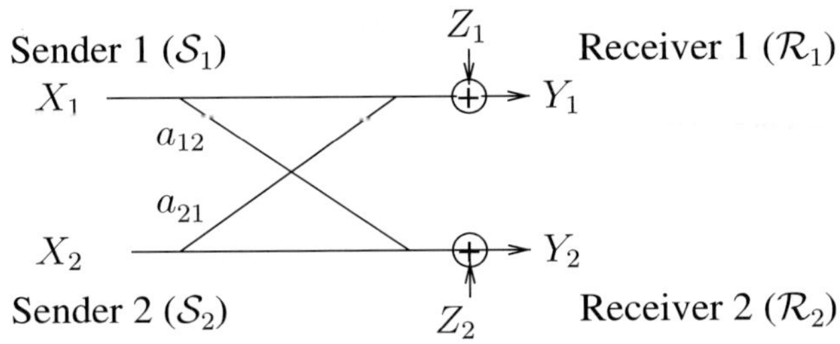

Figure 3.10. The additive interference channel with inputs X_1, X_2, outputs Y_1, Y_2, additive noise Z_1, Z_2 and interference coefficients a_{12}, a_{21}.

independent non-cooperating receivers $\mathcal{R}_1, \mathcal{R}_2$, transmitting over the same channel. The additive interference channel is shown in Fig. 3.10, where the parameters a_{12}, a_{21} capture the effects of the interference. In addition to the additive interference from the other sender, each output is affected by independent additive noise Z_1, Z_2.

The interference channel capacity, in the most general case, is still an open problem. In the case of strong interference, as defined in [Han and Kobayashi, 1981; Sato, 1981], and very strong interference, as defined in [Carleial, 1978], the capacity is known. Achievable regions of the interference channel have been calculated in [Han and Kobayashi, 1981], and recently in [Sason, 2004]. We will make use of techniques as in [Han and Kobayashi, 1981], merged with Gel'fand-Pinsker coding [Gel'fand and Pinsker, 1980] to provide an achievable region for the cognitive radio channel.

Genie-Aided Cognitive Radio Channel

We define a *cognitive radio channel* to be an interference channel in which $\mathcal{S}_2$ has knowledge of the message to be transmitted by $\mathcal{S}_1$. This is either obtained causally, or could possibly be given to the sender non-causally by a *genie*. We first focus on the non-causal scenario, *i.e., genie-aided cognitive radio channel* C_{COG}. $\mathcal{S}_2$ can exploit the knowledge of $\mathcal{S}_1$'s message, and potentially improve the transmission rate. It can do so using a dirty paper coding technique [Costa, 1983] and an achievable region construction for the interference channel [Han and Kobayashi, 1981]. Intuitively, the achievable region in [Han and Kobayashi, 1981] should lie entirely within the achievable region of C_{COG}, since senders are permitted to at least partially cooperate. An upper bound for our region in the Gaussian case is provided by the 2×2 MIMO broadcast channel whose capacity has recently been calculated in [Weingarten et al., 2004]. In [Weingarten et al., 2004], dirty paper coding techniques are shown to be optimal for non-degraded vector broadcast channels. The cognitive radio channel model resembles that of [Weingarten et al., 2004], with one important difference: the relation between

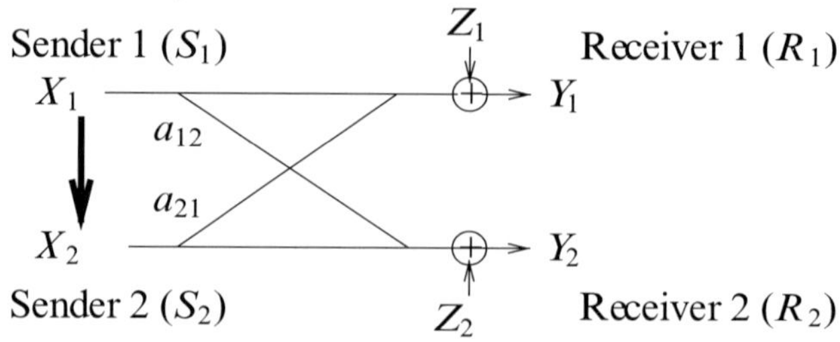

Figure 3.11. The additive interference genie-aided cognitive radio channel with inputs X_1, X_2, outputs Y_1, Y_2, additive noise Z_1, Z_2 and interference coefficients a_{12}, a_{21}. S_1's input X_1 is given to S_2 (indicated by the arrow), but not the other way around.

the two senders is asymmetric. The rate of $\mathcal{S}_2$ is also bounded by the rate achievable in an interference-free channel, with $a_{12} = 0$.

An (n, K_1, K_2, ϵ) *code* for the *genie-aided cognitive radio channel* consists of K_1 codewords $x_1^n(i) \in \mathcal{X}_1^n$ for $\mathcal{S}_1$, and $K_1 \cdot K_2$ codewords $x_2^n(i,j) \in \mathcal{X}_2^n$ for $\mathcal{S}_2, i \in \{1, 2, \ldots, K_1\}, j \in \{1, 2, \ldots, K_2\}$, which together form the *codebook*, revealed to both senders and receivers such that the average error probabilities under some decoding scheme are less than λ.

DEFINITION 3.4 *(Achievable Rate and Region) A rate pair (R_1, R_2) is said to be* achievable *for the genie-aided cognitive radio channel if there exists a sequence of $(n, 2^{\lceil nR_1 \rceil}, 2^{\lceil nR_2 \rceil}, \epsilon_n)$ codes such that $\epsilon_n \to 0$ as $n \to \infty$. An* achievable region *is a closed subset of the positive quadrant of $\mathbb{R}^2$ of achievable rate pairs.*

The Modified Genie-Aided Cognitive Channel C_{COG}^m

As in [Han and Kobayashi, 1981], we introduce a modified genie-aided cognitive radio channel, C_{COG}^m, (m for modified) and demonstrate an achievable region for C_{COG}^m. Then, a relation between achievable rates for C_{COG}^m and C_{COG} is used to establish an achievable region for the latter. The modified genie-aided cognitive radio channel C_{COG}^m is defined in Fig. 3.12, and we reuse the notation of the interference channel.

The modified genie-aided cognitive radio channel introduces two pairs of auxiliary random variables: (M_1, N_1) and (M_2, N_2). The random variables $M_1 \in \mathcal{M}_1$ and $M_2 \in \mathcal{M}_2$ represent, as in [Han and Kobayashi, 1981], the private information to be sent from $\mathcal{S}_1 \to \mathcal{R}_1$ and $\mathcal{S}_2 \to \mathcal{R}_2$ respectively. In contrast, the random variables $N_1 \in \mathcal{N}_1$ and $N_2 \in \mathcal{N}_2$ represent the public information to be sent from $\mathcal{S}_1 \to (\mathcal{R}_1, \mathcal{R}_2)$ and $\mathcal{S}_2 \to (\mathcal{R}_1, \mathcal{R}_2)$ respectively. The function of these M_1, N_1, M_2, N_2 is as in [Han and Kobayashi, 1981]: to decompose or define *explicitly* the information to be transmitted between various input and output pairs.

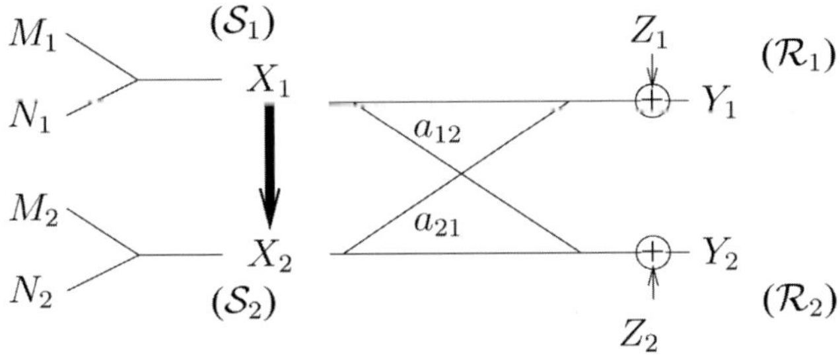

Figure 3.12. The modified cognitive radio channel with auxiliary random variables M_1, M_2, N_1, N_2, inputs X_1, X_2, additive noise Z_1, Z_2, outputs Y_1, Y_2 and interference coefficients a_{12}, a_{21}.

In C_{COG}^m, M_2 and N_2 also serve a dual purpose: these auxiliary random variables are analogous to the auxiliary random variables of Gel'fand and Pinsker [Gel'fand and Pinsker, 1980] or Cover and Chiang [Cover and Chiang, 2002]. They serve as fictitious inputs to the channel, so that after $\mathcal{S}_2$ is informed of the encoded message of $\mathcal{S}_1$ non-causally, the channel still behaves like a Discrete Memoryless Channel (DMC) from $(M_1, N_1, M_2, N_2) \rightarrow (Y_1, Y_2)$. As in [Cover and Chiang, 2002; Gel'fand and Pinsker, 1980], there is a penalty in using this approach which will be reflected by a reduction in achievable rates (compared to the fictitious DMC from (M_1, N_1, M_2, N_2) to (Y_1, Y_2)) for the links which use the non-causal information.

A code and an achievable region for C_{COG}^m can be defined similarly to those in C_{COG}. As mentioned in [Han and Kobayashi, 1981], the introduction of a time-sharing random variable W is thought to strictly extend the achievable region obtained using a convex hull operation. Thus, let $W \in \mathcal{W}$ be a time-sharing random variable whose n-sequences $w^n \triangleq (w^{(1)}, w^{(2)}, \ldots, w^{(n)})$ are generated independently of the messages, according to $\prod_{t=1}^n p(w^{(t)})$. The n-sequence w^n is given to both senders and both receivers. The following theorem and corollary on achievable rates for C_{COG}^m were proved in [Devroye et al., 2004].

THEOREM 3.5 *Let $Z \triangleq (Y_1, Y_2, X_1, X_2, M_1, N_1, M_2, N_2, W)$, and let $\mathcal{P}$ be the set of distributions on Z that can be decomposed into the form*

$$p(w)p(m_1|w)p(n_1|w)p(x_1|m_1, n_1, w)p(m_2|x_1, w)p(n_2|x_1, w)$$
$$\times\, p(x_2|m_2, n_2, w)p(y_1|x_1, x_2)p(y_2|x_1, x_2).$$

For any $Z \in \mathcal{P}$, let $S(Z)$ be the set of all quadruples $(R_{11}, R_{12}, R_{21}, R_{22})$ of non-negative real numbers such that there exist non-negative real (L_{21}, L_{22})

satisfying:

$$\begin{aligned} R_{11} &\le I(M_1; X_1|N_1, W) \\ R_{12} &\le I(N_1; X_1|M_1, W) \\ R_{11} + R_{12} &\le I(M_1, N_1; X_1|W) \end{aligned}$$

$$\begin{aligned} R_{21} &\le L_{21} - I(N_2; M_1, N_1|W) \\ R_{22} &\le L_{22} - I(M_2; M_1, N_1|W) \end{aligned}$$

$$\begin{aligned} R_{11} &\le I(Y_1, N_1, N_2; M_1|W) \\ R_{12} &\le I(Y_1, M_1, N_2; N_1|W) \\ L_{21} &\le I(Y_1, M_1, N_1; N_2|W) \\ R_{11} + R_{12} &\le I(Y_1, N_2; M_1, N_1|W) \\ R_{11} + L_{21} &\le I(Y_1, N_1; M_1, N_2|W) \\ R_{12} + L_{21} &\le I(Y_1, M_1; N_1, N_2|W) \\ R_{11} + R_{12} + L_{21} &\le I(Y_1; M_1, N_1, N_2|W) \end{aligned}$$

$$\begin{aligned} L_{22} &\le I(Y_2, N_1, N_2; M_2|W) \\ R_{12} &\le I(Y_2, N_2, M_2; N_1|W) \\ L_{21} &\le I(Y_2, N_1, M_2; N_2|W) \\ L_{22} + L_{21} &\le I(Y_2, N_1; M_2, N_2|W) \\ L_{22} + R_{12} &\le I(Y_2, N_2; M_2, N_1|W) \\ R_{12} + L_{21} &\le I(Y_2, M_2; N_1, N_2|W) \\ L_{22} + R_{21} + L_{12} &\le I(Y_2; M_2, N_1, N_2|W). \end{aligned}$$

Let S be the closure of $\cup_{Z \in \mathcal{P}} S(Z)$. Then any element of S is achievable for the modified genie-aided cognitive radio channel C^m_{COG}. □

Another important rate pair for C^m_{COG} is achievable: that in which $\mathcal{S}_2$ transmits no information of its own to $\mathcal{R}_2$, and simply aids $\mathcal{S}_1$ in sending its message to $\mathcal{R}_1$. When this is the case, the rate pair $(R_1^*, 0)$ is achievable, where R_1^* is the capacity of the vector channel $(\mathcal{S}_1, \mathcal{S}_2) \to \mathcal{R}_1$. Note however, that the analogous rate pair $(0, R_2^*)$ is not achievable, since that would involve $\mathcal{S}_1$ aiding $\mathcal{S}_2$ in sending its message, which cannot happen under our assumptions: $\mathcal{S}_2$ knows $\mathcal{S}_1$'s message, but not vice versa. The overall achievable region is given by the following Corollary.

COROLLARY 3.6 *Let $\mathcal{C}_0$ be the set of all points $(R_{11}+R_{12}, R_{21}+R_{22})$ where $(R_{11}, R_{12}, R_{21}, R_{22})$ is an achievable rate tuple of Theorem 3.5. Consider*

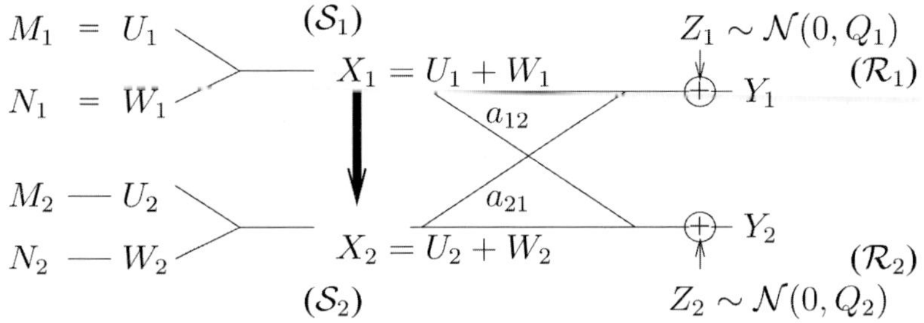

Figure 3.13. The modified Gaussian genie-aided cognitive radio channel with interference coefficients a_{12}, a_{21}.

the vector channel $(\mathcal{S}_1, \mathcal{S}_2) \to \mathcal{R}_1$ *described by the conditional probability density* $p(y_1|x_1, x_2)$ *for all* $y_1 \in \mathcal{Y}_1$, $x_1 \in \mathcal{X}_1$, $x_2 \in \mathcal{X}_2$, *and define* $R_1^* \triangleq \max_{p(x_1,x_2)} I(X_1, X_2; Y_1)$. *Then the convex hull of the region* $\mathcal{C}_0$ *with the point* $(R_1^*, 0)$ *is achievable for the genie-aided cognitive radio channel* C_{COG}^m. □

The Gaussian Cognitive Radio Channel

Consider the genie-aided cognitive radio channel, depicted in Fig. 3.13 with independent additive noise $Z_1 \sim \mathcal{N}(0, Q_1)$ and $Z_2 \sim \mathcal{N}(0, Q_2)$. We assume the two transmitters are power limited to P_1 and P_2 respectively. In order to determine an achievable region for the modified Gaussian genie-aided cognitive radio channel, specific forms of the random variables described in Theorem 3.5 are assumed. As in [Costa, 1983; Gallager, 1968; Han and Kobayashi, 1981], Theorem 3.5 and its Corollary can readily be extended to memoryless channels with discrete time and continuous alphabets by finely quantizing the input, output, and interference variables (Gaussian in this case). Let the time-sharing random variable W be constant. Consider the case where, for certain $\alpha, \beta \in \mathbb{R}$ and $\lambda, \overline{\lambda}, \gamma, \overline{\gamma} \in [0, 1]$, with $\lambda + \overline{\lambda} = 1$, $\gamma + \overline{\gamma} = 1$, and additional independent auxiliary random variables U_1, W_1, U_2, W_2 as in Fig. 3.13, the following hold:

$$
\begin{aligned}
U_1 &= M_1 \sim \mathcal{N}(0, \lambda P_1) \\
W_1 &= N_1 \sim \mathcal{N}(0, \overline{\lambda} P_1) \\
X_1 &= U_1 + W_1 = M_1 + N_1 \sim \mathcal{N}(0, P_1) \\
M_2 &= U_2 + \alpha X_1 \text{ where } U_2 \sim \mathcal{N}(0, \gamma P_2) \\
N_2 &= W_2 + \beta X_1 \text{ where } W_2 \sim \mathcal{N}(0, \overline{\gamma} P_2) \\
X_2 &= U_2 + W_2 \sim \mathcal{N}(0, P_2).
\end{aligned}
$$

The achievable regions thus obtained for the Gaussian genie-aided cognitive radio channel are plotted in Fig. 3.14. The innermost region (black) corresponds

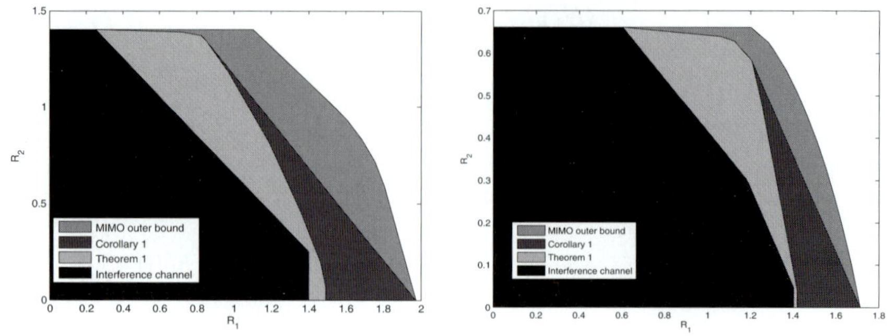

(a) (b)

Figure 3.14. Achievable region of [Han and Kobayashi, 1981] (innermost polyhedron), Theorem 3.5 (the next to smallest), and Corollary 3.6 (the second to largest), and the intersection of the capacity region of the 2×2 MIMO broadcast channel with the outer bound on R_2 of an interference-free Gaussian channel of capacity $1/2 \log(1 + P_2/Q_2)$ (the largest region). (a) $Q_1 = Q_2 = 1$, $a_{12} = a_{21} = 0.55$, $P_1 = P_2 = 6$. (b) $Q_1 = Q_2 = 1$, $a_{12} = a_{21} = 0.55$, $P_1 = 6$, $P_2 = 1.5$.

to the achievable region of [Han and Kobayashi, 1981], and is obtained by setting $\alpha = \beta = 0$. As expected, because of the extra information at the encoder and the partial use of a dirty-paper coding technique, the achievable region in Theorem 3.5, the second to smallest region (cyan) in Fig. 3.14, extends that of [Han and Kobayashi, 1981]. The overall achievable region, that of Corollary 3.6, further extends that of Theorem 3.5, as seen by the second largest (red) region in Fig. 3.14. An upper bound on our achievable rate region is provided by the 2×2 Gaussian MIMO broadcast channel, whose capacity was computed in [Weingarten et al., 2004]. Here, the two senders can fully cooperate (fully symmetric system). We calculate this region for input covariance constraint matrix of the form $S = \begin{pmatrix} P_1 & c \\ c & P_2 \end{pmatrix}$, for some $-\sqrt{P_1 P_2} \leq c \leq \sqrt{P_1 P_2}$ (which ensures S is positive semi-definite), and which mimics the power constraints P_1 and P_2 on each individual sender (asymmetric problem). The largest region in Fig. 3.14 is the intersection of the 2×2 Gaussian MIMO broadcast channel capacity region with the bound on $\mathcal{S}_2$'s rate $R_2 \leq \frac{1}{2}\log(1 + P_2/Q_2)$ provided by the interference-free channel in which $a_{12} = 0$.

Cognitive Radio Channel: the Causal Case

In practice, the message x_1^n that $\mathcal{S}_1$ wants to transmit cannot be non-causally given to $\mathcal{S}_2$. The transmitter $\mathcal{S}_2$ must obtain the message in real time, and one possible way to do so is by exploiting proximity to $\mathcal{S}_1$. As in Section 2, this proximity can be modeled by a reduction G in path loss, or equivalently, an

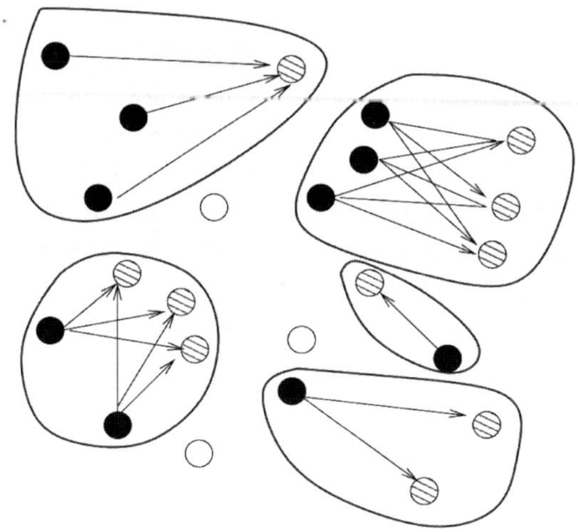

Figure 3.15. A wireless network consisting of cognitive and possibly non-cognitive devices. Black nodes are senders ($\mathbf{S_i}$), striped nodes are receivers ($\mathbf{R_i}$), and white nodes are neither (*i.e.,* single node clusters). A directed edge is placed between each desired sender-receiver pair at each point/period in time. The graph has been partitioned into subsets of *generalized MIMO channels.*

increase in capacity between $\mathcal{S}_1$ and $\mathcal{S}_2$, relative to the channels between the senders and the receivers. If, for example, the channel between $\mathcal{S}_1$ and $\mathcal{S}_2$ is an AWGN channel, then the capacity would increase, for a factor $G \geq 1$, to $C = \frac{1}{2}\log\left(1 + G \cdot \frac{P_1}{Q}\right)$, where Q is the additive Gaussian noise power. Alternatively, if $\mathcal{S}_1$ and $\mathcal{S}_2$ are base-stations, then it may be possible for S_2 to obtain S_1's message through a high bandwidth wired connection (if one exists) in real time. In the Gaussian cognitive radio channel model, all receivers know the channel between themselves and the relevant sender(s). In addition, both senders and receivers know the interference channel parameters a_{12} and a_{21}. In [Devroye et al., 2004], several protocols that allow $\mathcal{S}_2$ to causally obtain $\mathcal{S}_1$'s message were proposed, and corresponding achievable regions were derived. Three of them use a two-phase approach as in Section 2. Details of the protocols and achievable regions can be found in [Devroye et al., 2004]. We should note that the genie-aided cognitive radio channel achievable region provides an outer bound on a causal achievable region which uses the same coding strategy.

Cognitive Multiple Access Networks

In the previous section, we provided an achievable region for a two sender, two receiver *cognitive radio channel.* The achievable region was extended to *cognitive multiple access networks* in [Devroye et al., 2005]. We consider an arbitrary wireless network consisting of cognitive and possibly non-cognitive radio devices, as illustrated in Fig. 3.15. At each point /period in time, certain devices in sending mode wish to transmit to other devices in receiving mode. At each point/period in time, the wireless network can be represented as a directed graph by drawing a directed edge between every sender-receiver pair, as in Fig. 3.15. We define a *generalized MIMO channel* $(\mathbf{S}, \mathbf{R})$ as a connected bipartite directed graph where each sender node in $\mathbf{S}$ transmits to a subset of the receiver nodes in $\mathbf{R}$, and the channel is fully described by the conditional probability $P(\mathbf{r}|\mathbf{s})$. The *generalized MIMO channel* reduces to well-studied channels in certain cases. When a cluster consists of a single sender, it becomes a broadcast channel. When a cluster consists of a single receiver, it becomes a multiple access channel (MAC). When all receivers in a cluster are connected to all senders in a cluster, we have a vector MAC. The following lemma can readily be proved [Devroye et al., 2005].

LEMMA 3.7 *Any cognitive network can be partitioned into a set of* generalized MIMO channels $(\mathbf{S_i}, \mathbf{R_i})$ *where each sender node in* $\mathbf{S_i}$ *only transmits to a subset of the receiver nodes* $\mathbf{R_i}$. □

As described in the Introduction, after partitioning a wireless network into clusters, one can consider both *inter* and *intra* cluster competitive, cognitive, and cooperative behavior. In a cognitive multiple access network, clusters consist of classical information theoretic multiple access channels. The capacity region of the *intra-cluster cognitive MAC* has previously been considered in the classic information theoretic context of [Van der Meulen, 1971; Van der Meulen, 1977]. In [Devroye et al., 2005], an achievable region for two MAC channel clusters that simultaneously transmit and interfere has been computed in the case that one MAC cluster knows the messages to be sent by the other MAC cluster. In the Gaussian case, [Devroye et al., 2005] also numerically evaluated an achievable region for cognitive behavior and compared it to the achievable regions under competitive behavior as well as cooperative behavior.

5. Summary and Remarks

In this chapter, we have developed a general framework of wireless networks in the context of competition, cooperation and cognition. For the cooperation paradigm, we have provided a bandwidth efficient approach to compound Gaussian relay channels, and shown the existence of a cooperative code which is good over wide range of channels. We have also presented the design of a cooperative diversity system on an OFDM platform to demonstrate cooperative diversity gains in practical wireless systems. As an alternative to cooperative diversity, we have also introduced cooperative beamforming concepts along with their respective performance analyses. For cognitive radio channels, we have defined a more flexible and potentially more efficient transmission model, and constructed an achievable region. Finally, we have discussed extensions of the idea to cognitive multiple access networks.

We conclude by discussing some research opportunities in this emerging field. One possible extension of the cooperative diversity scheme is to investigate the effect of full asynchronism between nodes. Another extension is to investigate more refined cooperation on part of the relays. A number of open issues remain for cooperative beamforming as well, such as applicability of beamforming when the destination or nodes in the cluster are in rapid motion or the channel suffers severe multipath fading. There is still a lot of work to do in the context of practical implementation, such as development of more efficient protocols, and synchronization and channel estimation algorithms. From the viewpoint of higher layer protocols, development of clustering protocols and link layer error control protocols is an important topic for both approaches.

As the cognitive radio channel, which captures the essence of asymmetric cooperation, has only recently been introduced, numerous promising research directions exist. From a theoretical perspective, the capacity of this channel, as well as its causal version are still open problems. Some achievable regions have already been calculated, and the development of tight upper bounds on the cognitive and causal cognitive radio channels will advance the field towards this final goal. Extensions of the cognitive radio channel to fading and compound channels is another research area to be explored. Practically, coding protocols and schemes that enable cognitive transmission must be devised, and issues similar to those encountered in full cooperation, such as synchronization and channel estimation, will naturally arise here as well.

References

Azarian, K., El Gamal, H., and Schniter, P. (2004). On the achievable diversity-multiplexing tradeoff in half-duplex cooperative channels. Submitted to *IEEE Trans. Inf. Theory*.

Balanis, C. A. (1997). *Antenna Theory: Analysis and Design*. Wiley, New York.

Carleial, A. B. (1978). Interference channels. *IEEE Trans. Inf. Theory*, IT-24(1):60–70.

Carleial, A. B. (1982). Multiple-access channels with different generalized feedback signals. *IEEE Trans. Inf. Theory*, 28(6):841–850.

Costa, M. H. M. (1983). Writing on dirty paper. *IEEE Trans. Inf. Theory*, IT-29:439–441.

Cover, T. M. and Chiang, M. (2002). Duality between channel capacity and rate distortion. *IEEE Trans. Inf. Theory*, 48(6).

Cover, T. M. and El Gamal, A. A. (1979). Capacity theorems for the relay channel. *IEEE Trans. Inf. Theory*, 25(5):572–584.

Cover, T. M., McEliece, R. J., and Posner, E. C. (1981). Asynchronous multiple-access channel capacity. *IEEE Trans. Inf. Theory*, 27(4):409–413.

Csiszár, I. and Körner, J. (1981). *Information Theory: Coding Theorems for Discrete Memoryless Systems*. Academic Press, New York.

Devroye, N., Mitran, P., and Tarokh, V. (2004). Achievable rates in cognitive radio channels. Submitted to *IEEE Trans. Inf. Theory*.

Devroye, N., Mitran, P., and Tarokh, V. (2005). Cognitive multiple access networks. Submitted to *IEEE Trans. Inf. Theory*.

European Telecommunications Standards Institute (1998). Universal mobile telecommunications system (UMTS): Selection procedures for the choice of radio transmission technologies for the UMTS. Technical report, ETSI.

FCC (2003). FCC ET docket no. 03-108: Facilitating opportunities for flexible, efficient, and reliable spectrum use employing cognitive radio technologies. Technical report, FCC.

FCC (2005). http://www.fcc.gov/oet/cognitiveradio/.

FCC Spectrum Policy Task Force (2002). FCC report of the spectrum efficiency working group. Technical report, FCC.

Foschini, G. J. and Gans, M. J. (1998). On limits of wireless communications in a fading environment when using multiple antennas. *Wireless Personal Communications*, 6:311–335.

Gallager, R. G. (1968). *Information Theory and Reliable Communication*, chapter 7. Wiley, New York.

Gastpar, M., Kramer, G., and Gupta, P. (2002). The multiple relay channel: Coding and antenna-clustering capacity. In *Proc. IEEE Int. Symp. Inf. Theory*, page 136, Lausanne, Switzerland.

Gastpar, M. and Vetterli, M. (2002a). On the asymptotic capacity of Gaussian relay networks. In *Proc. IEEE Int. Symp. Inf. Theory*, page 195, Lausanne, Switzerland.

Gastpar, M. and Vetterli, M. (2002b). On the capacity of wireless networks: The relay case. In *Proc. IEEE INFOCOM*, pages 1577–1586, New York, NY.

Gel'fand, S. I. and Pinsker, M. S. (1980). Coding for channels with random parameters. *Probl. Contr. and Inf. Theory*, 9(1):19–31.

Gupta, P. and Kumar, P. R. (2003). Towards an information theory of large networks: An achievable rate region. *IEEE Trans. Inf. Theory*, 49:1877–1894.

Han, T. S. and Kobayashi, K. (1981). A new achievable rate region for the interference channel. *IEEE Trans. Inf. Theory*, IT-27(1):49–60.

Horne, W. D. (2003). Adaptive spectrum access: Using the full spectrum space.

Hunter, T., Hedayat, A., Janani, M., and Nosratinia, A. (2003). Coded cooperation with space-time transmission and iterative decoding. In *Proc. WNCG Wireless Networking Symposium*.

Hunter, T. and Nostratinia, A. (2002). Coded cooperation under slow fading, fast fading, and power control. In *Proc. Asilomar Conference on Signals, Systems, and Computers*.

IEEE 802.11a-1999 (1999). Part 11: Wireless LAN medium access control (MAC) and physical layer (PHY) specifications: high-speed physical layer in the 5 GHz band. Technical report, IEEE.

Katz, M. and Shamai, S. (2004). Communicating to co-located ad-hoc receiving nodes in a fading environment. In *Proc. IEEE Int. Symp. Inf. Theory*, page 115, Chicago, IL.

Khojastepour, M., Aazhang, B., and Sabharwal, A. (2003a). Bounds on achievable rates for general multi-terminal networks with practical constraints. In *Proc. Information Processing in Sensor Networks*.

Khojastepour, M., Sabharwal, A., and Aazhang, B. (2003b). On capacity of Gaussian 'cheap' relay channel. In *Proc. IEEE Global Telecommun. Conf.*, pages 1776–1780.

Khojastepour, M., Sabharwal, A., and Aazhang, B. (2003c). On the capacity of 'cheap' relay networks. In *Proc. Conference on Information Sciences and Systems*.

Kramer, G., Gastpar, M., and Gupta, P. (2004). Cooperative strategies and capacity theorems for relay networks. Submitted to *IEEE Trans. Inf. Theory*.

Laneman, J. N., Tse, D. N. C., and Wornell, G. W. (2004). Cooperative diversity in wireless networks: Efficient protocols and outage behavior. *IEEE Trans. Inf. Theory*, 50(12):3062–3080.

Laneman, J. N. and Wornell, G. W. (2003). Distributed space-time-coded protocols for exploiting cooperative diversity in wireless networks. *IEEE Trans. Inf. Theory*, 49(10):2415–2425.

Lo, Y. T. (1964). A mathematical theory of antenna arrays with randomly spaced elements. *IRE Trans. Antennas Propagat.*, 12:257–268.

Mitola, J. (1995). The software radio architecture. *IEEE Commun. Mag.*, 33(5): 26–38.

Mitola, J. (1999a). Cognitive radio for flexible mobile multimedia communications. In *Proc. IEEE Mobile Multimedia Conference*.

Mitola, J. (1999b). Future of signal processing–Cognitive radio. In *Proc. IEEE ICASSP*. Keynote address.

Mitola, J. (2000). *Cognitive Radio*. PhD thesis, Royal Institute of Technology (KTH), Sweden.

Mitran, P., Ochiai, H., and Tarokh, V. (2005). Space-time diversity enhancements using collaborative communications. *IEEE Trans. Inf. Theory*, 51(6): 2041–2057.

Ochiai, H., Mitran, P., Poor, H. V., and Tarokh, V. (2005). Collaborative beamforming for distributed wireless ad hoc sensor networks. To appear in *IEEE Trans. Signal Processing*.

Reznik, A., Kulkarni, S., and Verdú, S. (2002). Capacity and optimal resource allocation in the degraded Gaussian relay channel with multiple relays. In *Proc. Allerton Conf. Commun., Control and Comp.*, Monticello, IL.

Sason, I. (2004). On achievable rate regions for the Gaussian interference channel. *IEEE Trans. Inf. Theory*.

Sato, H. (1981). The capacity of Gaussian interference channel under strong interference. *IEEE Trans. Inf. Theory*, IT-27(6).

Schein, B. and Gallager, R. G. (2000). The Gaussian parallel relay network. In *Proc. IEEE Int. Symp. Inf. Theory*, page 22, Sorrento, Italy.

Sendonaris, A., Erkip, E., and Aazhang, B. (2003a). User cooperation diversity–Part I: System description. *IEEE Trans. Commun.*, 51(11):1927–1938.

Sendonaris, A., Erkip, E., and Aazhang, B. (2003b). User cooperation diversity–Part II: Implementation aspects and performance analysis. *IEEE Trans. Commun.*, 51(11):1939–1948.

Shin, O.-S., Chan, A., Kung, H. T., and Tarokh, V. (2005). Design of an OFDM cooperative space-time diversity system. Submitted to *IEEE Trans. Signal Processing*.

Telatar, I. E. (1999). Capacity of multi-antenna Gaussian channels. *European Trans. on Telecomm.*, 10(6):585–595.

Van der Meulen, E. C. (1971). Three-terminal communication channels. *Adv. Appl. Prob.*, 3:120–154.

Van der Meulen, E. C. (1977). A survey of multi-way channels in information theory. *IEEE Trans. Inf. Theory*, IT-23(1):1–37.

Wang, B., Zhang, J., and Høst-Madsen, A. (2005). On the capacity of MIMO relay channels. *IEEE Trans. Inf. Theory*, 51(1):29–43.

Weingarten, H., Steinberg, Y., and Shamai, S. (2004). The capacity region of the Gaussian MIMO broadcast channel. Submitted to *IEEE Trans. Inf. Theory*.

Willems, F. M. J. (1982). *Information theoretic results for the discrete memoryless multiple access channel*. PhD thesis, Katholieke Universiteit Leuven, Belgium.

Willems, F. M. J., Van der Meulen, E. C., and Schalkwijk, J. P. M. (1983). An achievable rate region for the multiple access channel with generalized feedback. In *Proc. Allerton Conf. Commun., Control and Comp.*, pages 284–293.

Wolfowitz, J. (1978). *Coding Theorems of Information Theory*. Springer-Verlag, New York.

Xie, L.-L. and Kumar, P. R. (2004). A network information theory for wireless communication: Scaling laws and optimal operation. *IEEE Trans. Inf. Theory*, 50(5):748–767.

Chapter 4

COOPERATION TECHNIQUES IN CROSS-LAYER DESIGN

Shuguang Cui
University of Arizona
cui@ece.arizona.edu

Andrea J. Goldsmith
Stanford University
andrea@wsl.stanford.edu

Abstract: We explore node cooperation in a cross-layer design framework to optimize wireless network performance. We focus on the design methodology for networks with hard energy constraints, where each network node is equipped with a finite energy source that cannot be recharged. We consider network topologies where transmission distances may be small, so that transmission energy does not necessarily dominate total energy consumption. We assume that network nodes can cooperate across multiple layers of the protocol stack: in signal transmissions at the link layer, spectrum sharing at the multiple access layer, information relaying at the network layer, and information processing at the application layer. We investigate how to incorporate node cooperation at individual protocol layers into an overall cross-layer design framework using several design examples. In the first example we consider joint design of the routing, MAC, and link layer protocols to minimize total energy consumption. The joint design is posed within an optimization framework subject to given system constraints, which can be easily solved using existing convex optimization techniques. In the second example we propose a cooperative MIMO technique where multiple nodes within a cluster cooperate in signal transmission and/or reception to reduce energy and delay. By treating each cooperating cluster as a super node, the design optimization falls within the same framework as our first example. The third example considers cooperating nodes trying to jointly estimate an unknown target. In this setting we optimize link layer power control and the number of cooperating nodes to meet a

Frank H. P. Fitzek and Marcos D. Katz (eds.), Cooperation in Wireless Networks: Principles and Applications, 101–126.

desired distortion while minimizing the total power consumption. These design examples illustrate several of the key features and tradeoffs in cross-layer design of wireless networks with cooperating nodes, but they only scratch the surface of possible techniques and applications. The goal is for these design examples to motivate further exploration of cooperative techniques in cross-layer design.

Keywords: node cooperation, cooperative MIMO, cross-layer design.

1. Introduction

Wireless networks are a pervasive technology that is changing the way people work and play. However, wireless networks have unique characteristics that make it difficult for them to support demanding applications, especially applications with high data rate requirements, hard delay constraints, and hard energy constraints. The layered approach to wireless (and wired) network design, where each layer of the protocol stack is oblivious to the design and operation of other layers, has not worked well in general, especially under stringent performance requirements. Layering precludes the benefits of joint optimization across protocol layers, which can significantly improve performance. Moreover, good protocol designs for isolated layers often interact in negative ways across layers, which can significantly degrade end-to-end performance. Thus, stringent performance requirements for wireless networks can only be met through a cross-layer design.

In this chapter we focus on cross-layer design and node cooperation under hard energy constraints, motivated by sensor network applications. Wireless sensor networks consist of small nodes with sensing, computation, and wireless networking capabilities, representing a convergence of three important technologies. Sensor networks can provide the enabling technology for smart homes and buildings, automated assisted living for the elderly and disabled, environmental monitoring, automated emergency response and homeland security, and automated highways and battlefields. Sensor nodes typically have hard energy constraints, since each node is powered by a small battery that may not be rechargeable or renewable. Therefore, reducing energy consumption is the most important design consideration for such networks. Since all the layers in the network protocol stack affect the overall system performance, synergies between the design of different layers must be exploited to optimize the system performance while satisfying given resource constraints (see [Goldsmith and Wicker, 2002]).

In addition to cross-layer design, node cooperation is a powerful technique to improve wireless network performance (see [Laneman and Wornell, 2003; Laneman et al., 2004; Cui et al., 2004]). Node cooperation is especially appealing in sensor networks since each sensor node has only limited functionality and the whole network is usually deployed to cooperate on the same mission. However, there are many problems associated with optimizing node cooperation, especially within a cross-layer framework. This chapter aims to articulate some of these design issues and to illustrate the potential benefits of node cooperation via cross-layer design.

This chapter is organized as follows. We first introduce the concept of cross-layer design. We then discuss possible node cooperation strategies at individual layers, as well as how to combine cross-layer design with node cooperation. We next illustrate these ideas through several design examples. The first example jointly optimizes the routing, MAC, and link layer protocols to minimize energy consumption. In the second example we investigate cooperative MIMO techniques at the link layer to reduce energy and delay. The last example considers power minimization in distributed estimation to meet a given distortion constraint. We conclude with some observations and a discussion of open research questions.

2. Cross-layer Design

Protocol layering is a common abstraction in network design. Layering provides design modularity for network protocols that facilitates standardization and implementation. Unfortunately, the layering paradigm does not work well in wireless networks, where many protocol design issues are intertwined. In this section we describe protocol layering as it applies to wireless networks, as well as the interactions between protocol layers, which motivate the need for cross-layer design.

A simplified network protocol stack is shown in Fig. 4.1 and the main functions of each individual layer are described as follows:

1 The hardware layer is composed of the fundamental hardware blocks where the upper layer algorithms are implemented. Since all the power is consumed by the hardware, the hardware layer must be considered in a cross-layer energy minimization framework.

2 The link layer, also referred to as the physical layer, deals primarily with transmitting bits reliably over a point-to-point wireless link. The design tradeoffs associated with the link layer include modulation, coding, diversity, adaptive techniques, MIMO, equalization, multi-carrier modulation, and spread spectrum.

3 The MAC layer controls how different users share the given spectrum and ensures reliable packet transmissions. Allocation of signaling dimensions to different users is done through either deterministic access or random access. For deterministic access, the signaling dimensions are divided into dedicated channels, where the most common methods are Time Division Multiple Access (TDMA), Frequency Division Multiple Access (FDMA), and Code Division Multiple Access (CDMA). For random access, the channels are assigned to active users dynamically, and

the most common methods are different forms of ALOHA, Carrier Sense Multiple Access (CSMA), and scheduling.

4 The network layer establishes and maintains end-to-end connections in the network. The main functions of the network layer are neighbor discovery, routing, and dynamic resource allocation. Routing is often done based on the Internet Protocol (IP).

5 The transport layer provides the end-to-end functions of retransmission, error recovery, reordering, and flow control. The most common protocol used at this layer is the Transport Protocol (TCP). The TCP protocol does not typically work well in wireless networks since it assumes that all packet losses are due to congestion and thus implements flow control in response to lost packets. In wireless networks packets are typically lost due to channel noise, fading, interference, and distortion, and flow control is not an appropriate mechanism to deal with these impairments.

6 The application layer generates the data to be sent over the network and processes the data received over the network. The main functions of the application layer are source coding/decoding and error concealment.

A more detailed description of the functions supported by each layer can be found in [Goldsmith, 2005; Bertsekas and Gallager, 1992]. In a purely layered structure each layer is individually designed with predefined interfaces between neighboring layers. Typically no layers talk directly to non-neighbor layers. To account for performance dynamics of neighboring layers, each layer needs to have a certain degree of adaptability. However, since the inter-layer interface is predefined and the dynamics of non-neighbor layers are shielded by neighboring layers, the overall system adaptability is limited.

Protocol design based on a layered structure provides reasonable performance in wired networks, where individual layer dynamics are limited and the system bandwidth and power are relatively unconstrained. Thus, over-allocation of system resources can make up for inefficiencies associated with separate protocol layer designs. However, this is not the case for networks built on top of wireless channels, since wireless links exhibit large random dynamics due to multipath fading, Doppler, and interference. In addition, wireless systems have more stringent resource constraints: bandwidth is expensive as companies must spend billions of dollars to acquire licensed spectrum; and power is limited, due to finite battery life. Moreover, increasing power on a link causes more interference to other links, so this is not a useful mechanism to improve overall system performance. Due to the dynamics of wireless channels along with their bandwidth and power constraints, wireless networks based on a layered protocol design approach can perform poorly in wireless environments (see Chapter 16 in [Goldsmith and Wicker, 2002]). For example, in a typical wireless network, packets can be lost due to bursty interferences at the MAC layer or congestion at the network layer. However, due to the shielding effect of the network layer, the transport layer assumes that all the loss is due to network congestion. As a result, the TCP protocol will adjust window size to reduce congestion even when no links are congested. Consequently, the overall network throughput will be dramatically degraded, as shown in [Tian et al., 2005; Fu and Liew, 2003].

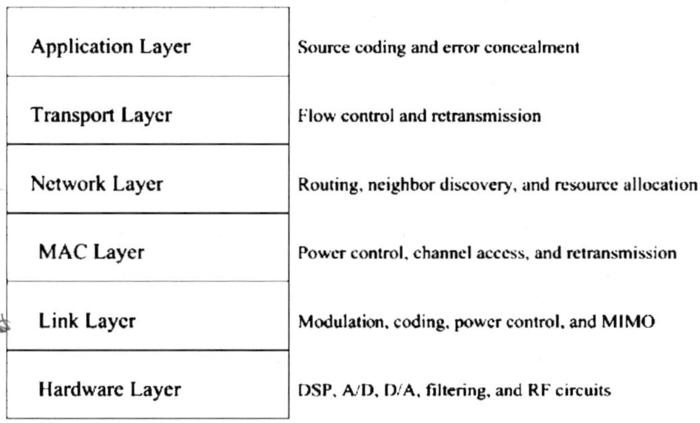

Figure 4.1. Simplified layered structure.

Cross-layer design is a joint design optimization across all or several layers in the protocol stack under given resource constraints. Joint optimization across protocol layers exploits synergies between protocol layers and avoids negative interactions to improve the network performance if designed properly. Cross-layer design can include information exchange between different layers (not necessarily neighboring layers), adaptivity at each layer to this information, and diversity built into each layer to insure robustness. As an example, a joint design of the TCP window control and the MAC layer power control can reduce the impact of packet loss on the overall network throughput by responding with either congestion or power control, whichever is more effective. Generally speaking, the link layer can employ adaptive modulation and coding to exploit or compensate for the time-varying wireless channel. Adaptivity at the link layer can then be used by higher layer protocols to achieve better performance. For example, the MAC layer can assign a longer channel usage time to links with low-rate modulation schemes to meet a throughput constraint; the routing layer can reroute traffic to links supporting high-rate modulation schemes to minimize congestion; and the application layer can use multiple-description codes to leverage the diversity of different routes. Significant performance gain can be achieved by these interactions between different layers (see Chapter 16 in [Goldsmith, 2005]).

Cross-layer design for throughput maximization has received a lot of attention over the past few years. An achievable rate region for wireless networks with joint optimization of the routing, MAC, and link layer designs was computed in [Toumpis and Goldsmith, 2003]. Throughput maximization by joint design of power control, link scheduling, and routing was considered in, for example, [O'Neill et al., 2004; Johansson and Xiao, 2004; Kodialam and Nandagopal, 2003; Jain et al., 2003; Radunovic and Boudec, 2003]. The design challenges and the importance of cross-layer design for meeting application requirements in energy-constrained networks were described in [Goldsmith and Wicker, 2002]. An overview of the synergy between the various layers was given in [Kozat et al., 2004].

In the previous work discussed above, it has been widely demonstrated that cross-layer design can lead to performance improvement over a layered approach, especially in networks with hard resource constraints. However, as pointed out in [Kawadia and Kumar, 2005], cross-layer design is not a panacea and requires some caution, as it may conflict with long-term architectural principles or cause unintended protocol interactions that lead to negative consequences. The motivation for the layered structure is the desire for a good architectural design that leads to proliferation and longevity, as has been experienced in the Internet. To inherit these properties, cross-layer optimization for wireless networks should strive to maintain clearly-defined interfaces between different layers, while allowing each layer to interface with both neighboring and non-neighboring layers, as shown in Fig. 4.2. While the modularity of layering is preserved in this way, it may create oscillating control loops at a given layer due to contradictory control commands from other layers. To minimize such oscillations and maintain the simplicity of a layered structure as much as possible, it is important to determine the key layers for cross-layer design, and the layers that can be omitted from cross-layering with minimal performance penalty. To do so, cross-layer optimization across all protocol layers must first be determined, and then different layers can be omitted in the optimization to find how much these omissions degrade performance. It should be noted that most work on cross-layer design entails numerical optimization and does not lead to closed-form results. This trend indicates the need for interdisciplinary expertise and tools in communication theory, network theory, and optimization to address these challenging problems.

Although cross-layer design has been a focus of much recent research, a general design framework has yet to be developed. Moreover, it is not clear how to maintain the flexibility to support new applications while tailoring a protocol design to specific applications. These are but a few of the challenges associated with cross-layer design.

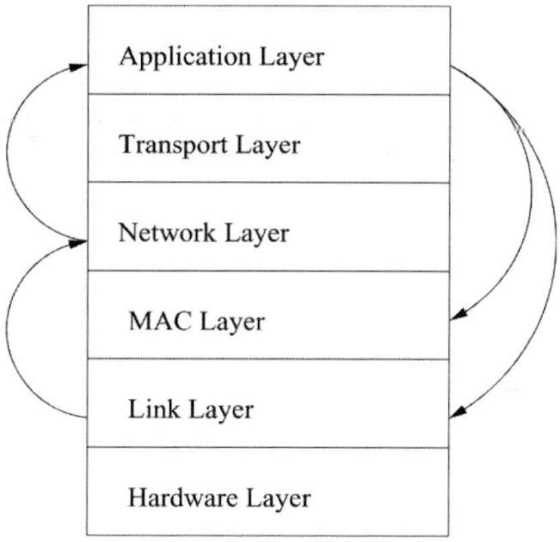

Figure 4.2. Cross-layer Interaction.

3. Node Cooperation in Wireless Networks

Node cooperation to improve wireless network performance has been the focus of much recent research (see [Laneman and Wornell, 2003; Laneman et al., 2004; Cui et al., 2004]). Node cooperation takes place between peer nodes, where nodes can join a cooperative effort or behave independently. For example, two terminal nodes could help each other relay information to other nodes based on some mutually beneficial agreement.

Node cooperation is not a new concept, but work in this area has mainly focused on the network and transport layers of the protocol stack. In particular, multihop routing or relaying is a form of node cooperation. However, the concept of node cooperation can be extended to other layers in the protocol stack, and such extensions are compelling given the limited functionality of nodes in applications such as sensor networks. For example, in sensor networks node cooperation at the application layer can provide significant performance gains. In particular, assume that sensors are deployed to observe or track a target of interest. Due to the size and cost limitation of each sensor node, the observation quality of individual sensors is usually somewhat limited. However, if multiple sensor nodes cooperate to jointly observe the target, the resulting observations can be intelligently combined to generate an estimate with high fidelity.

In addition to cooperation at the application layer, nodes can also cooperate at the MAC and link layers. In the MAC layer, multiple nodes can cooperate in sharing the spectrum to achieve a desired throughput with minimum energy consumption. If there is no cooperation at all, the multiuser environment can be modeled as an interference channel (see [Carleial, 1978]), where each transmission is treated as interference to all other transmissions. If nodes associated with the transmitting side fully cooperate then the channel can be viewed as a multiple antenna transmitter, while if nodes associated with the receiving side fully cooperate then the channel can be viewed as a multiple antenna receiver. If nodes on both the transmitting and receiving ends fully cooperate then the channel can be viewed as having multiple antennas on both sides (a MIMO channel). However, the optimal form of cooperation depends on the network topology: if the cost of full information exchange is too high, then cooperating nodes may just act as relays. The tradeoffs associated with the best use of cooperating nodes is analyzed in [Ng and Goldsmith, 2005].

In the link layer, node cooperation is different from that in the MAC layer in the sense that nodes help a given source node transfer its information bits to its destination node (see [Laneman and Wornell, 2003]). Although these information bits are transferred by multiple nodes, they may be coded in different ways, which can be further unified into a MIMO diversity coding framework (see [Cui et al., 2004]). In other words, node cooperation at the link layer provides a form of link diversity.

4. Node Cooperation with Cross-layer Design

Since node cooperation can take place at multiple levels of the protocol stack, it naturally couples with cross-layer design. However, as with any cross-layer design, care must be taken to avoid negative interactions between layers. For example, consider node cooperation in the link layer to obtain link diversity. While diversity will improve performance over a given link, node cooperation requires some power and bandwidth to achieve this diversity. If some bandwidth must be allocated to node cooperation, then less bandwidth will be available on the channel between the transmit and receive clusters, which may force these clusters to use higher level modulation or higher rate codes. Such higher rate techniques may have a higher probability of error than the original system even with the diversity gain achieved through cooperation. Thus, optimizing node cooperation to improve performance requires a joint link and MAC layer design. More generally, optimizing node cooperation must generally be done within a cross-layer framework to avoid performance losses associated with the cooperation at different protocol layers. We now describe some of the recent research in this area.

Cross-layer optimization of multihop routing is inherently a joint optimization between cross-layer design and node cooperation. The first results in this area investigate throughput maximization by joint design of power control, link scheduling, and routing [O'Neill et al., 2004; Johansson and Xiao, 2004; Kodialam and Nandagopal, 2003; Jain et al., 2003; Radunovic and Boudec, 2003]. For a network with multiple source nodes and a single destination node (common in sensor networks), node cooperation or relaying is usually unidirectional since information transmitted by one node is relayed only by the nodes following it along the route. For a network with multiple source and multiple destination nodes, the concept of node cooperation is more obvious, since the relaying could be mutual between two nodes: node A may relay information for node B on flow 1 and node B may relay information for node A on flow 2. In addition, joint design between routing and coding can be viewed under the general framework of network coding [Koetter and Medard, 2003].

In [Liu and Ge, 2004], the authors investigate the joint design between routing and the link layer via an orthogonal space-time coding framework. In this work the authors demonstrate that the joint optimization significantly enhances the end-to-end diversity. A similar cross-layer design problem is discussed in [Hares et al., 2003], where the authors emphasize the joint optimization between cooperative diversity routing and adaptive modulation in the link layer based on a TDMA MAC layer. In [Cui and Goldsmith, 2005a], joint optimization among routing, MAC, and link adaptation is modeled with cooperative signal transmission at the link layer, where node cooperation is implemented with distributed diversity MIMO codes. This results in significant reduction of both energy and end-to-end transmission delay.

Another cluster of research results in this area focus on joint optimization between the application layer and other layers, where node cooperation is deployed at the application layer. In [Ganesan et al., 2004], joint power minimization at the link layer and node placement at the application layer are investigated, where multiple nodes cooperate on data gathering at the application layer. The joint optimization leads to remarkable energy savings. In [Xiao et al., 2004], joint estimation problems in sensor networks are presented, where transmission rate and power at the link layer are jointly optimized with node cooperation in the application layer. The optimal number of cooperating nodes can be found by minimizing the total transmit power under certain end-to-end distortion requirement. In [Begen et al., 2003] application layer cooperating schemes, such as multi-description coding, are jointly designed with the routing layer to improve the quality of video/image transmission over lossy wireless networks.

Although these results shed much light on the tradeoffs and design processes of cross-layer design in networks with cooperating nodes, there remains much research to be done in finding a comprehensive framework for this cross-layer design and analysis. In the following section, we present several cross-layer design examples for wireless networks with cooperating nodes under a hard energy constraint. We hope that these design examples will provide insight and motivation for additional investigation of cross-layer design in networks with cooperating nodes.

5. Design Examples

In the first example we consider joint design optimization between the routing, MAC, and link layers. We show that the joint design can be modeled or approximated as a convex optimization problem when Time Division Multiple Access (TDMA) is used as the multiple access technique. In the second example we allow multiple nodes to cooperate on signal transmission and reception by forming a virtual antenna array, and then incorporate this link layer cooperation into the joint optimization model of our first design example. The last example investigates joint design of the application layer and power control at the link layer, with node cooperation at the application layer to minimize distortion in a distributed estimation problem. More details on these design examples can be found in [Cui et al., 2005a; Cui et al., 2005b; Cui et al., 2005c; Cui and Goldsmith, 2005a; Cui and Goldsmith, 2005b].

Joint Routing, MAC, and Link Layer Optimization

In a typical sensor network, sensors are powered by small batteries that cannot be replaced. Under this hard energy constraint, sensor nodes can only transmit a finite number of bits in their lifetime. Consequently, reducing the energy consumption per bit for end-to-end transmissions becomes an important design consideration for such networks. Since all layers of the protocol stack contribute to the energy consumed in a bit's end-to-end transmission, energy minimization requires a joint design across all these layers as well as the underlying hardware where the energy is actually expended [Goldsmith and Wicker, 2002]. We consider the simplified case where interference is eliminated by using Time Division Multiple Access (TDMA) schemes, which is appropriate for small-scale sensor networks where synchronization between nodes is feasible. We consider a joint design across the link, MAC, and routing protocol layers, and we allow multiple nodes to cooperate in multihop relaying to minimize transmission energy.

We assume that information collected by multiple sensors needs to be transmitted to a remote central processor that we call a hub node. If the hub node is far away, the information may first be transmitted to a relay node that is willing to cooperate, then multihop routing will be used to forward the data to its final destination. The corresponding scenario is illustrated in Fig. 4.3.

We now discuss the system model in detail. For node i, we use $\mathcal{N}_i$ to denote the set of nodes that send data to node i, and use $\mathcal{M}_i$ to denote the set of nodes that receive data from node i. We denote the normalized time slot length for the transmission over link $i \to j$ (from node i to node j) as $\delta_{ij} = \frac{t_{ij}}{T}$, where $\sum_{i=1}^{N-1} \sum_{j \in \mathcal{M}_i} \delta_{ij} \leq 1$ and T is the TDMA frame length. We use W_{ij} to denote the number of packets transmitted over link $i \to j$ during each period T and use ν to denote the number of bits per packet. As discussed in [Cui et al., 2005c] we

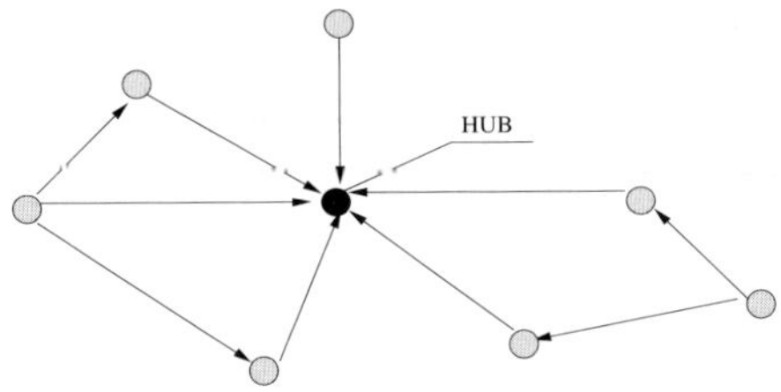

Figure 4.3. Data collection in a sensor network.

assume three modes of operation for each node: active mode, sleep mode, and transient mode. To simplify the formulation we neglect the effect of the transient mode. Thus, nodes i and j will be in active mode when link $i \to j$ is active, and will otherwise be in sleep mode where all the circuits are turned off to save energy. For node i we use P_{ct}^i and P_{cr}^i to denote the circuit power consumption values for the transmitting circuits and the receiving circuits, respectively. The transmit power needed for satisfying a target probability of bit error P_b from node i to node j is denoted as P_0^{ij}. Therefore, the total average power spent on link $i \to j$ is given as

$$P_{ij} = \delta_{ij}(P_{cr}^j + P_{ct}^i + (1+\alpha)P_0^{ij})$$

where α is determined by the power amplifier efficiency such that $(1+\alpha)P_0^{ij}$ is the total power consumed in the transmitter power amplifier. The total energy consumed over link $i \to j$ per cycle is thus given as $\epsilon_{ij} = TP_{ij}$.

As discussed in [Cui et al., 2005c], to increase the network lifetime we can choose to minimize the total energy consumption via the following optimization:

$$\begin{aligned} \min \quad & \textstyle\sum_{i=1}^{N-1} \sum_{j \in \mathcal{M}_i} \epsilon_{ij} \\ \text{s. t.} \quad & \textstyle\sum_{i=1}^{N-1} \sum_{j \in \mathcal{M}_i} t_{ij} \leq T, \\ & \textstyle\sum_{j \in \mathcal{M}_i} W_{ij} - \sum_{j \in \mathcal{N}_i} W_{ji} = L_i, \quad i = 1, \cdots, N, \\ & \tfrac{\nu W_{ij}}{C_{ij}B} \leq t_{ij} \leq \tfrac{\nu W_{ij}}{2B}, \quad j \in \mathcal{M}_i, \quad i = 1, \cdots, N-1, \end{aligned} \tag{4.1}$$

where the first constraint is the TDMA constraint, the second constraint is the flow conservation constraint, which guarantees that at each node the difference between the total outgoing traffic and the total incoming traffic is equal to the traffic generated by the node itself (denoted as $L_i = R_i T$ with R_i the packet rate), and C_{ij} in the third constraint is the maximum constellation size each link can use without violating given peak power constraints, with B the symbol rate. For general modulation schemes such as uncoded MQAM, the problem can be approximated as a convex one over t_{ij} and W_{ij} if we relax these parameters to take on real values. To reduce the relative error caused by the relaxation, we can use integer programming techniques such as the Branch and Bound algorithm (see [Garfinkel and Nemhauser, 1972]).

We now provide a numerical example based on the string topology shown in Fig. 4.4. In this figure the three source nodes 1, 2, and 3 are sending data to the destination node 4 at rates $R_1 = 60$ pps (packets per second), $R_2 = 80$ pps, and $R_3 = 20$ pps, respectively. All other system parameters as defined the same as in [Cui et al., 2005c]. As a result, Fig. 4.4 shows the optimal routing, scheduling, and modulation constellation size b for uncoded MQAM based on this optimization framework, assuming that both processing and transmission circuit energy are taken into account in the total energy consumption. Specifically, the number above each link in Fig. 4.4 is the time slot length assigned to that link, and the number below each link is the optimal constellation size used for that link. The model for circuit energy consumption is discussed in [Cui et al., 2005c]. It is well known that when only the transmission energy is considered, multihop routing saves energy. However, as shown in [Cui et al., 2005c], when the circuit processing energy is included, single-hop transmissions may be more efficient than multihop routing schemes due to the processing energy consumed at the relays. However, if rate adaptation is allowed, multihop transmission may regain the performance advantage since a higher constellation size reduces transmission time, and thus circuit energy consumption in the relaying nodes is reduced.

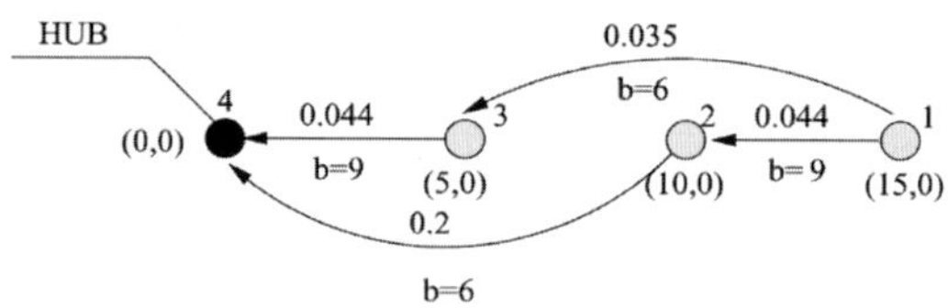

Figure 4.4. Minimizing total energy consumption, $R_1 = 60$ pps, $R_2 = 80$ pps, $R_3 = 20$ pps.

Different scheduling (ordering) of the optimal time slot assignments, the t_{ij}'s, will lead to different delay performance, although they all have the same energy efficiency. It is shown in [Cui et al., 2005b] that the minimum packet delay among all possible schedules is equal to the frame length T, and a simple algorithm exists to find such a minimum-delay schedule for any loop-free network with one sink node. Thus, by solving the problem in Eq. (4.1), we can find the minimum possible energy required to transfer a given number of packets within a delay deadline T. Alternatively, instead of minimizing energy under a delay constraint, we can also consider the dual problem of minimizing delay under an energy constraint. Specifically, given a total energy budget E_M per period, we can find the minimum possible value for $T = \sum_{i=1}^{N-1} \sum_{j \in \mathcal{M}_i} t_{ij}$ that is required to finish the transfer of a given number of packets. The dual problem is characterized as

$$\begin{aligned} \min. \quad & \textstyle\sum_{i-1}^{N-1} \sum_{j \subset \mathcal{M}_i} t_{ij} \\ \text{s. t.} \quad & \textstyle\sum_{j \in \mathcal{M}_i} W_{ij} - \sum_{j \in \mathcal{N}_i} W_{ji} = L_i, \quad i = 1, \cdots, N, \\ & \textstyle\sum_{i=1}^{N-1} \sum_{j \in \mathcal{M}_i} \epsilon_{ij} \leq E_M, \\ & t_{ij} \geq 0, \quad j \in \mathcal{M}_i, \quad i = 1, \cdots, N-1, \end{aligned}$$

For a given network topology, the achievable energy-delay region consists of all the achievable energy-delay pairs. The energy-delay region is a convex set. This is because if energy-delay points (ϵ_1, T_1) and (ϵ_2, T_2) are contained in the energy-delay region, then any convex combination of these points can be achieved by time-sharing between the transmission schemes corresponding to the two end points. Hence, any convex combination of these points are contained in the achievable energy-delay region. Here, we calculate the Pareto-optimal energy-delay tradeoff which characterizes the minimum possible delay for a given energy consumption (or vice versa), and the optimal tradeoff curve defines the boundary of the achievable energy-delay region.

The optimal tradeoff curve can be found by varying the value of β in the following optimization problem.

$$\begin{aligned} \min. \quad & \textstyle\sum_{i=1}^{N-1} \sum_{j \in \mathcal{M}_i} t_{ij} + \beta \sum_{i=1}^{N-1} \sum_{j \in \mathcal{M}_i} P_{ij} t_{ij} \\ \text{s. t.} \quad & \textstyle\sum_{j \in \mathcal{M}_i} W_{ij} - \sum_{j \in \mathcal{N}_i} W_{ji} = L_i, \quad i = 1, \cdots, N, \\ & t_{ij} \geq 0, \quad j \in \mathcal{M}_i, \quad i = 1, \cdots, N-1, \end{aligned}$$

where the first term in the objective function is the delay and the second term is the total energy consumption weighted by a scanning parameter β.

The models we introduced in this example are quite general and flexible for solving various cross-layer optimization problems. In addition, they have clear structure related to layers: energy cost is defined by the link layer; delay performance is defined by MAC and scheduling; and the flow conservation constraint is defined by routing. Such structure helps us unify node collaboration effects at individual layers into the cross-layer optimization model. In the next example, we show how to deploy node cooperation at the link layer and solve the overall problem in the same cross-layer optimization framework.

Cooperative MIMO with Cross-layer Optimization

In this example, node collaboration is deployed in both the routing layer (multihop relaying) and the link layer through cooperative Multiple Input Multiple Output (MIMO) transmission. To incorporate node cooperation at the link layer into the cross-layer design framework we apply the following strategy: We first abstract cooperative MIMO links into virtual Single Input Single

Output (SISO) links with equivalent link layer performance; We then apply the cross-layer optimization methods that are proposed for systems with SISO links in the first design example to find the optimal solution.

System model. We consider a sensor network composed of multiple clusters of nodes, as shown in Fig. 4.5. This figure shows clusters of nodes where the nodes within the same cluster are closely spaced and cooperate in signal transmission and/or reception. Here we extend the cooperative strategy proposed in [Cui et al., 2004] to a multihop networking scenario, where we find the routing and scheduling that optimize energy and/or delay performance based on cooperative MIMO transmissions at each hop.

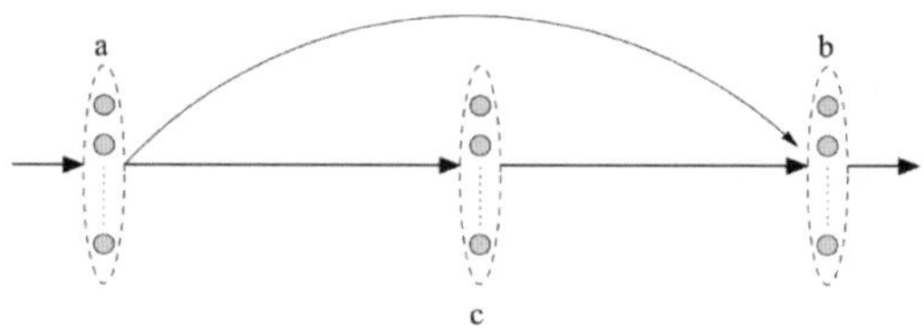

Figure 4.5. A clustered network.

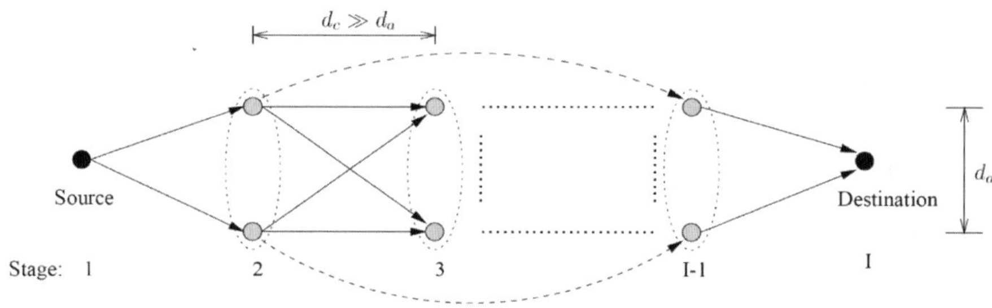

Figure 4.6. A double-string network.

We restrict our attention to the double-string network topology shown in Fig. 4.6, which represents regularly spaced sensors for data collection. In this topology there are clusters of two nodes, where within a cluster the nodes are separated by distance d_a while the distance between clusters is d_c with $d_c \gg d_a$. While Fig. 4.6 shows clusters of size $M = 2$, our design methodology applies to any cluster size. The highly regular topology of the double string network facilitates analysis, and also demonstrates potential performance gains for more general topologies. For the network in Fig. 4.6, there are $I - 2$ stages of node clusters between the source and the destination. Thus, if the distance between the source and the destination is d, then the distance between the neighboring stages is $d_c = \frac{d}{I-1}$. We also assume that transmissions from stage m to stage

n is allowed for any m and n with $1 \leq m \leq n \leq I$, where the source node is at stage 1 and the destination node is at stage I.

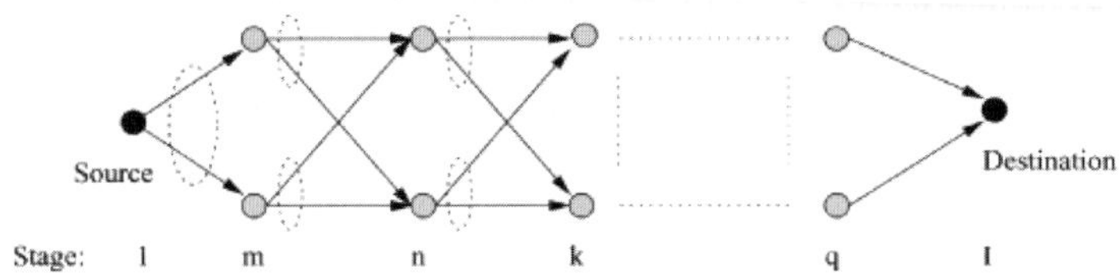

Figure 4.7. Cooperative transmission.

We assume that the source node generates data at L_1 packets per collection period T with a fixed packet size $v = 100$ bits. Therefore, the network needs to support a throughput of $S_0 = \frac{L_1}{T}$ packets per second (pps) between the source node and the destination node. We also assume a TDMA-based transmission scheme where the frame length is equal to T. Therefore, the network needs to convey L_1 packets from the source to the destination within each frame. We want to find a variable-length TDMA scheme where each transmission is assigned an optimal transmission time with the total sum bounded by T to minimize the energy consumed across the network within each frame. Due to the nature of TDMA, there is only one transmission in the network at any given time.

The nodes cooperate in the following manner. As shown in Fig. 4.7, within the first slot in each frame, the source node broadcasts a certain number of packets to the two nodes of the cluster at stage m, $2 \leq m \leq I$. If $m < I$, then the upper node at stage m acts as antenna 1 and the lower node acts as antenna 2. These antennas transmit two streams of codewords that are encoded according to a 2×1 Alamouti code (see [Paulraj et al., 2003]). Note that for a given time slot, the pair of nodes at stage m is allowed to transmit to any pair of nodes at stage n, $n > m$. The two nodes at stage n will decode the information independently and repeat the cooperative coding and transmission process. In addition, a pair of nodes may be assigned more than one time slot within each frame to transmit packets to different stages. Note that it is possible that the source node transmits all the packets directly to the destination node, if that is more efficient. The optimization of which clusters participate in the multihop routing and the corresponding transmission scheduling is performed off-line using the model introduced later: this information is then communicated to all nodes prior to transmission. We assume that the receivers can correctly decode the packets if the raw bit error rate (before error correction) is below a certain threshold $\bar{P}_b$. We also assume that the network is synchronized, which may be enabled by utilizing beacon signals in a separate control channel. Although the scheme just proposed is for node clusters of size $M = 2$, similar ideas can be applied to larger clusters.

For each link, we assume a flat Rayleigh fading channel, *i.e.*, the channel gain between each transmitter and each receiver is a scalar. In addition, the mean path loss is modeled by a power falloff proportional to the distance squared. As derived in [Cui and Goldsmith, 2005a], the total power consumed during the transmission from stage m to stage n is given by

$$P_{mn} = P_{ct}^{m} + P_{cr}^{n} + (1+\alpha)P_0^{mn}, \tag{4.2}$$

where P_{ct}^{m} is the total transmitter circuit power consumption across stage m, P_{cr}^{n} is the total receiver circuit power consumption across stage n, and P_0^{mn} is the total transmit power across stage m that is required to achieve certain probability of bit error target $\bar{P}_b$. As defined previously, α is determined by the power amplifier efficiency such that $(1+\alpha)P_0^{mn}$ is the total power consumed in the transmitter power amplifiers across stage m. Note that when $m = 1$, *i.e.*, for SISO transmission from the source node to other stages, P_0^{mn} is defined differently from the cases when $m > 1$, *i.e.*, for cooperative 2×1 MISO transmissions. The formula for P_0^{mn} in different cases is given in [Cui and Goldsmith, 2005a].

After calculating P_{mn}, it is still difficult to incorporate the cooperative MIMO structure into the optimization models proposed in the first design example, which originally addresses systems with SISO links. Fortunately, we can apply a simple trick to make the problem manageable. Since all the transmissions occur between different pairs of cooperating nodes and the pairing relationship is fixed, we can treat each pair of nodes in the same stage as one super node. Then the double-string network is simplified to a single-string network as shown in Figure 4.8, which can be treated as a virtual SISO system with the total number of nodes given by $N = I$. The total power required for transmission between two super nodes is given by Eq. (4.2). The corresponding energy or delay minimization problem can thus be modeled in the same way as in the SISO case, which was discussed in the previous design example. For networks with an arbitrary cluster size M, similar equivalent SISO systems can be obtained with P_{mn} modified according to an $M \times 1$ MISO system. After such equivalent SISO systems are obtained, they can be optimized based on the models proposed in the first design example.

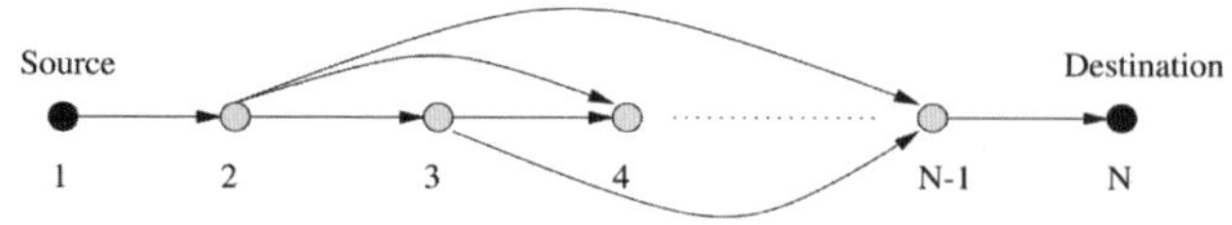

Figure 4.8. Equivalent SISO system.

We now give numerical results for the cooperative MIMO system for both the fixed-rate and adaptive-rate cases (for a fixed rate we assume QPSK transmissions with a $B = 10$ KHz symbol rate). We consider a double-string network with ten stages ($I = 10$), $d = 270$ m, $S_a = 200$ pps, and $L_1 = 60$ packets. The other system parameters are defined the same as in [Cui and Goldsmith, 2005a]. For both the non-cooperative and cooperative MIMO systems, if the frame length $T \leq \frac{L_1}{S_a} = 0.3$ s, single-hop transmission is the only option since the frame length T is not large enough for multiple hops to take place. When $T > 0.3$ s, we have the option to use multihop routing to save transmission energy.

For the case where we only consider the transmission energy, the optimal energy-delay tradeoff curve is shown in Fig. 4.9, where we see that node cooperation at the link layer reduces both energy and delay, but the benefit of rate adaptation is not obvious except that the delay can be further reduced at the expense of energy. On the left side of point A, the two curves for the cooperative MIMO systems have almost merged due to the fact that QPSK is used in both systems to minimize the transmission energy. The slight difference between the two curves on the left side of point A is just due to some numerical rounding errors.

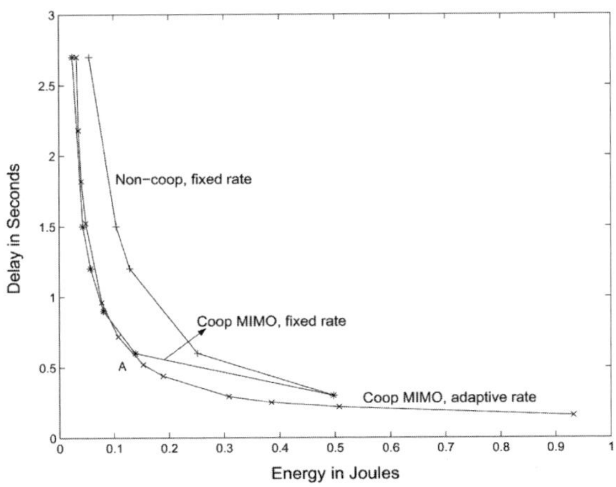

Figure 4.9. Transmission Energy only.

For the case where we consider both the transmission energy and the circuit processing energy, the optimal energy-delay tradeoff curve is shown in Fig. 4.10. We see in this curve dramatic performance improvement achieved by the cooperative MIMO system with rate adaptation, since this adaptation minimizes the sum of the transmission energy and the circuit processing energy and gives the upper layers more freedom to choose optimal multihop routes. The circuit related parameters used in this optimization are described in [Cui and Goldsmith, 2005a].

From this design example we see that cooperative MIMO coupled with cross-layer optimization can significantly improve the energy-delay tradeoff in wire-

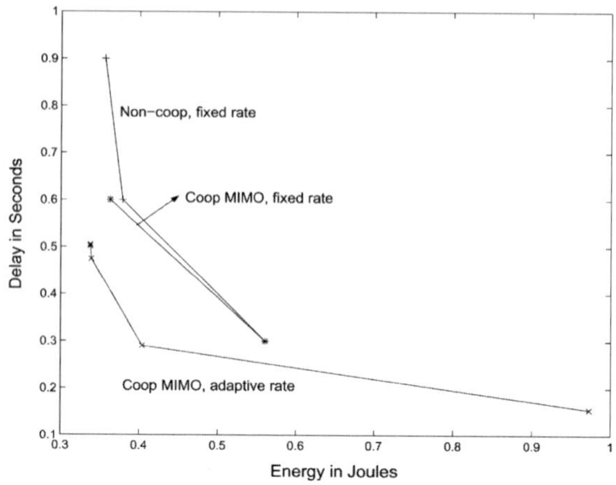

Figure 4.10. Circuit energy included.

less networks. If the cooperation is properly executed and jointly designed with upper layers, no local information exchange between cooperating nodes is needed, and the optimal routing and transmission schemes can be found using convex optimization techniques. Numerical examples demonstrate the performance improvement of cooperative MIMO over non-cooperative methods. The performance difference is especially dramatic when node cooperation is jointly optimized with rate adaptation at the link layer.

Cooperative Estimation with Optimal Power Scheduling

In this example we consider an energy-efficient method for joint estimation of an unknown analog source under a given distortion constraint. This can be viewed as node cooperation at the application layer to minimize distortion. Our approach is purely analog: each sensor amplifies and forwards its noise-corrupted analog observation to the fusion center for joint estimation, as shown in Fig. 4.11. The link layer optimization entails minimizing the total transmission power across all the sensor nodes while satisfying a distortion requirement on the joint estimate. We will see that the optimal solution entails turning off sensors with bad transmission quality to the fusion center, which in turn defines the optimal number of nodes for the cooperation.

System model. We assume that there are K sensors and the observation $x_k(t)$ at sensor k is represented as a random signal $\theta(t)$ corrupted with the observation noise $n_k(t)$: $x_k(t) = \theta(t) + n_k(t)$. Each sensor transmits the signal $x_k(t)$ to the fusion center where $\theta(t)$ is estimated from the $x_k(t)$'s, $k = 1, \cdots, K$. We further assume that $n_k(t)$ is of unknown statistics and the amplitude of $x(t)$ is bounded within $[-W, W]$, which is defined by the sensing

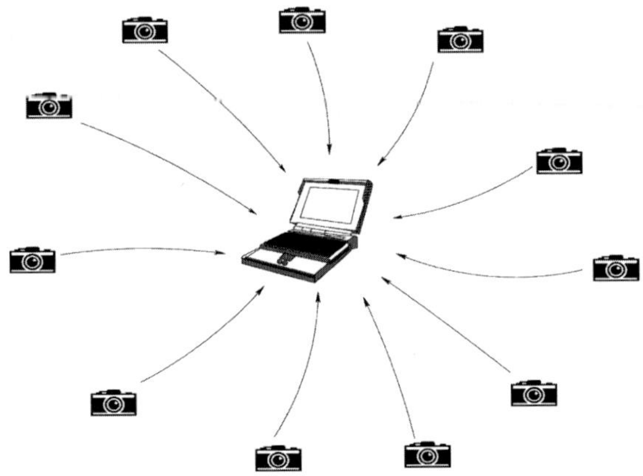

Figure 4.11. Sensor network with a fusion center.

range of each sensor. For simplicity we assume $W = 1$, but our analysis can be easily extended to any values.

We assume that $x_k(t)$ is also band-limited and the information is contained within the frequency range $[-B/2, B/2]$. We consider an analog Single Side-Band (SSB) system with a coherent receiver (see [Haykin, 1994]). The transmitted signal is given by

$$y_t(t) = 2\sqrt{\alpha}\cos(\omega_c t)x(t) + 2\sqrt{\alpha}\sin(\omega_c t)\hat{x}(t),$$

for which the average transmission power is

$$P = 4\alpha P_x \leq 4\alpha W^2, \tag{4.3}$$

where $\hat{x}(t)$ is the Hilbert transform of $x(t)$, ω_c is the carrier frequency, 4α is the transmitter power gain, and P_x is the peak power of $x(t)$.

The received signal at the fusion center is given by

$$y_r(t) = 2\sqrt{\alpha}\sqrt{g}\cos(\omega_c t)x(t) + 2\sqrt{\alpha}\sqrt{g}\sin(\omega_c t)\hat{x}(t) + n_c(t),$$

where g is the channel power gain and $n_c(t)$ is the channel AWGN. Hence, at the output of the coherent detector the signal is

$$y(t) = \sqrt{\alpha}\sqrt{g}x(t) + \frac{1}{2}n_c^I(t)\cos(\pi\frac{B}{2}t) + \frac{1}{2}n_c^Q(t)\sin(\pi\frac{B}{2}t),$$

where $n_c^I(t) + jn_c^Q(t)$ is the complex envelope of $n_c(t)$. After passing $y(t)$ through a low-pass filter and sampling the baseband signal at a sampling rate B, we can obtain an equivalent discrete-time system. Since we have K such sensors, the overall system is shown in Fig. 4.12. The K transmitters share the channel via Frequency Division Multiple Access (FDMA).

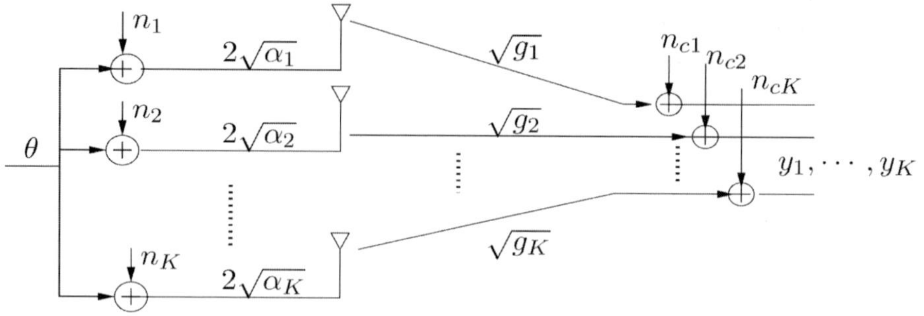

Figure 4.12. Amplify and Forward.

The received signal vector at each snapshot is given by

$$\mathbf{y} = \mathbf{h}\theta + \mathbf{v},$$

where

$$\begin{aligned} \mathbf{y} &= [y_1, y_2, \cdots, y_K]^\dagger, \\ \mathbf{h} &= [\sqrt{\alpha_1 g_1}, \sqrt{\alpha_2 g_1}, \cdots, \sqrt{\alpha_K g_K}]^\dagger, \\ \mathbf{v} &= [\sqrt{\alpha_1 g_1} n_1 + n_{c1}, \cdots, \sqrt{\alpha_K g_K} n_k + n_{cK}]^\dagger, \end{aligned}$$

and † indicates transpose.

According to [Mendel, 1995], the best linear unbiased estimator (BLUE) for θ is given by

$$\begin{aligned} \hat{\theta} &= [\mathbf{h}^\dagger \mathbf{R}^{-1} \mathbf{h}]^{-1} \mathbf{h}^\dagger \mathbf{R}^{-1} \mathbf{y} \\ &= \left(\sum_{k=1}^{K} \frac{\alpha_k g_k}{\sigma_k^2 \alpha_k g_k + \xi_k^2} \right)^{-1} \sum_{k=1}^{K} \frac{\sqrt{\alpha_k g_k} y_k}{\sigma_k^2 \alpha_k g_k + \xi_k^2}, \end{aligned}$$

where the noise variance matrix $\mathbf{R}$ is a diagonal matrix with $R_{kk} = \sigma_k^2 \alpha_k g_k + \xi_k^2$ with σ_k^2 the variance of the sensor observation noise $n_k(t)$, $k = 1, \cdots, K$. The channel noise variance ξ_k^2 is defined by the noise power spectral density and the bandwidth B.

The mean squared error of this estimator is given as (see [Mendel, 1995])

$$\begin{aligned} Var[\hat{\theta}] &= [\mathbf{h}^\dagger \mathbf{R}^{-1} \mathbf{h}]^{-1} \\ &= \left(\sum_{k=1}^{K} \frac{\alpha_k g_k}{\sigma_k^2 \alpha_k g_k + \xi_k^2} \right)^{-1}. \end{aligned}$$

According to Eq. (4.3), the transmit power for node k is bounded by $4W^2\alpha_k$. Therefore, the minimum power analog information collection problem can be cast as

$$\begin{aligned} \min \quad & \sum_{k=1}^{K} W^2 \alpha_k \\ \text{s. t.} \quad & \left(\sum_{k=1}^{K} \frac{\alpha_k g_k}{\sigma_k^2 \alpha_k g_k + \xi_k^2} \right)^{-1} \le D_0, \\ & \alpha_k \ge 0, \quad k = 1, \cdots, K, \end{aligned}$$

where D_0 is the distortion target. Although this problem is not convex over the α_k's, it can be transformed into a convex problem by introducing an intermediate variable $r_k = \frac{\alpha_k g_k}{\sigma_k^2 \alpha_k g_k + \xi_k^2} = \frac{1}{\sigma_k^2 + \frac{\xi_k^2}{g_k \alpha_k}}$ [Cui et al., 2005a].

If we we rank the channel quality according to the inverse of the channel signal-to-noise power ratio (SNR), *i.e.*, $\frac{\xi_1^2}{g_1} \le \frac{\xi_2^2}{g_2} \le \ldots \le \frac{\xi_K^2}{g_K}$, then the optimal solution is given as [Cui et al., 2005a]

$$\begin{aligned} \alpha_k^{opt} &= \frac{\xi_k^2}{g_k} \frac{r_k^{opt}}{1 - \sigma_k^2 r_k^{opt}} \qquad k = 1, \cdots, K_1 \\ &= \frac{\xi_k^2}{g_k \sigma_k^2} \left(\sqrt{\frac{g_k}{\xi_k^2}} \eta_0 - 1 \right) \qquad k = 1, \cdots, K_1, \end{aligned} \tag{4.4}$$

and $\alpha_k^{opt} = 0$ otherwise, where K_1 is an index threshold that can be uniquely defined and η_0 is a system constant (see [Cui et al., 2005a]).

Therefore, the optimal power allocation strategy is divided into two steps. In the first step, a threshold K_1 for k is obtained. For channels with index higher than K_1 (with low channel SNR), the corresponding sensors are turned off and no power is wasted. In other words, there is no need for these sensors to participate in the cooperation. For the remaining active sensors, power should be assigned according to Eq. (4.4). From Eq. (4.4) we see that when the channel is fairly good, *i.e.*, $\sqrt{\frac{g_k}{\xi_k^2}}\eta_0 \gg 1$, we have $\alpha_k^{opt} \propto \sqrt{\frac{\xi_k^2}{g_k}} \frac{\eta_0}{\sigma_k^2}$, which means the optimal solution is inversely proportional to the square root of the channel SNR. When $\sqrt{\frac{g_k}{\xi_k^2}}\eta_0$ is close to one, the optimal solution may no longer have such properties. For all channel conditions, the power is scaled by the factor $\frac{1}{\sigma_k^2}$, which means that more power is used to transmit the signals from the sensors with better observation quality, *i.e.*, the importance is weighted among all the sensors participating in the cooperation.

Numerical results. We now solve the optimization problem for some specific examples to show how much gain we can obtain with the joint node cooperation and cross-layer design process. We assume that the channel power gain $g_k = \frac{G_0}{d_k^{3.5}}$ where d_k is the transmission distance from sensor k to the fusion center and $G_0 = -30$ dB is the gain at $d = 1$ m. As in [Xiao et al., 2004], we generate σ_k^2 uniformly within the range $[0.01, 0.08]$. We take $B = 10$ KHz and $\xi_k^2 = -90$ dBm, $k = 1, \cdots, K$. For an example with 100 sensors, Fig. 4.13 (a) shows the relative power savings compared with the uniform transmission strategy where all the sensors use the same transmission power to achieve the given distortion target. The relative power savings is plotted as a function of $R = \frac{\sqrt{\mathbf{Var}(d)}}{\mathbf{E}(d)}$, the distance deviation normalized by the mean distance. For each value of R, we average the relative power savings over 100 random runs where in each run the d_k's are randomly generated according to the given R. As expected, a larger variation of distance and corresponding channel quality leads to a higher power savings when this variation is exploited with optimal node cooperation.

In Fig. 4.13 (b), the number of active sensors over R is shown for an example with 10 sensors. We see that when the transmission distances for different sensors span a wide range of values (*i.e.*, R is large), more sensors can be eliminated from cooperation to save energy, since the remaining sensors have very good channels.

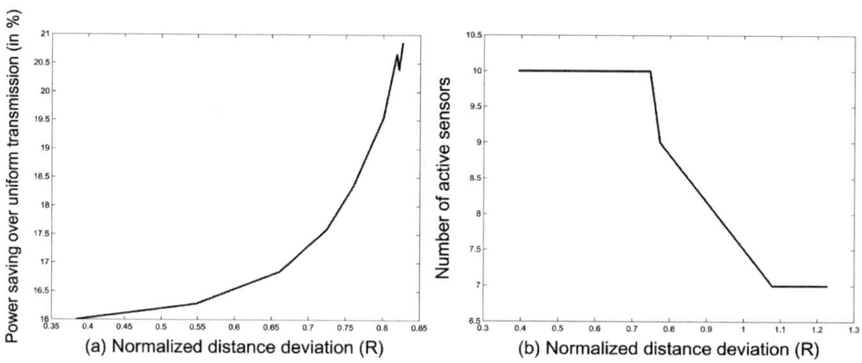

Figure 4.13. (a) Power savings of optimal power allocation vs. uniform power; (b) Number of active sensors versus distance deviation.

From this design example we see that by combining node cooperation at the application layer and power scheduling at the link layer, we obtain the optimal number of cooperating sensors to minimize total power consumption while satisfying the distortion constraint. Although finding the optimal number of cooperative nodes is a nonconvex integer programming problem, by using the transmission power as an intermediate variable, we can indirectly determine which nodes participate in the cooperation. In other words, if the optimal transmission power for a particular node is zero, we know that this node should not participate in the cooperation. Thus, the node cooperation problem is integrated into the cross-layer design framework, which leads to a convex formulation that can be efficiently solved.

Observations and Concluding Remarks

We have provided an overview of a general framework and process for cross-layer design in networks with node cooperation. The framework requires multi-disciplinary expertise across network protocol layers, and draws from the fields of communication theory, network theory, and optimization. While cross-layer design introduces many challenges, it is critical to achieve good performance in wireless networks with hard constraints. However, cross-layer design must be undertaken with caution to avoid eliminating the benefits of a layered design, including flexibility, modularity, simplicity, and network scalability. Finding the right balance between preserving the advantages of a layered design while tailoring protocols to specific network and application performance requirements via cross-layer design is a broad and deep open question in wireless network research. This balance is particularly challenging in light of the different requirements associated with different applications: an application such as video, with hard delay constraints and high data rates, would perhaps require a different cross-layer design than an application such as sensor networks, where data rates are low and energy is the hard constraint. It indeed seems unlikely that a single cross-layer design framework could be applicable to all applications and networks, unless it is made so generic that most of the cross-layer benefits are lost.

Both cross-layer design and node cooperation are research areas in their infancy, so we still have little insight or experience to draw from in developing a cross-layer design framework. At this stage, much can be learned by investigating cross-layer design for specific applications, and then trying to extrapolate the knowledge gained to more general settings. In this vein we have provided several examples of cross-layer design under energy constraints. These examples illustrate that cross-layer design with node cooperation can provide significant performance gains, and also leads to designs that differ significantly from existing protocols optimized for individual layers. While performance is best when all protocol layers are included in the cross-layer design, often a judicious choice of jointly optimizing just a few layers yields near-optimal results. In particular, significant performance improvement can come from co-design of non-neighboring layers, such as joint optimization of the application layer and power control at the link layer. This observation is particularly important in networks where some of the protocol layer designs are hard to change, such as the IP at the network layer in the Internet. However, performance gains based on jointly optimizing a few layers critically depends on choosing the right layers in the cross-layer design: design across the wrong subset of layers can lead to minimal performance improvement or degraded performance relative to individual layer design due to unintended interactions between layers. Much work on general cross-layer design principles for networks with cooperating nodes as well as designs for specific systems remains to be done, and this field will likely remain a rich source of open research problems for the foreseeable future.

References

Begen, A. C., Altunbasak, Y., Ergun, O., and Begen, M. A. (2003). Real-time multiple description and layered encoded video streaming with optimal diverse routing. In *Eighth IEEE International Symposium on Computers and Communication.*

Bertsekas, D. and Gallager, R. (1992). *Data Networks*. Prentice Hall, 2nd edition.

Carleial, A. B. (1978). Inteference channels. *IEEE Transactions on Information Theory*, 24(1):60–70.

Cui, S. and Goldsmith, A. J. (2005a). Cross-layer optimization of sensor networks based on cooperative mimo techniques with rate adaptation. In *Proc. IEEE SPAWC*.

Cui, S. and Goldsmith, A. J. (2005b). Energy efficient routing using cooperative mimo techniques. In *Proc. IEEE ICASSP*.

Cui, S., Goldsmith, A. J., and Bahai, A. (2004). Energy efficiency of mimo and cooperative mimo in sensor networks. *IEEE J. Selected Areas of Commun.*, 22(6):1089–1098.

Cui, S., Goldsmith, A. J., Xiao, J., Luo, Z. Q., and Poor, H. V. (2005a). Energy-efficient joint estimaiton in sensor networks: Anlog vs. digital. In *Proc. IEEE ICASSP*.

Cui, S., Madan, R., Goldsmith, A. J., and Lall, S. (2005b). Energy-delay tradeoff for data collection in sensor networks. In *Proc. of IEEE ICC*.

Cui, S., Madan, R., Goldsmith, A. J., and Lall, S. (2005c). Joint routing, mac, and link layer optimization in sensor networks with energy constraints. In *Proc. of IEEE ICC*.

Fu, C. P. and Liew, S. C. (2003). Tcp veno: Tcp enhancement for transmission over wireless access networks. *IEEE J. Selet. Areas Commun.*, 21(2):216–228.

Ganesan, D., Cristescu, R., and Beferull-Lozano, B. (2004). Power-efficient sensor placement and transmission structure for data gathering under distortion constraints. In *Third International Symposium on Information Processing in Sensor Networks*.

Garfinkel, R. S. and Nemhauser, G. L. (1972). *Integer Programming*. John Wiley & Sons.

Goldsmith, A. J. (2005). *Wireless Communications*. Cambridge University Press.

Goldsmith, A. J. and Wicker, S. W. (2002). Design challenges for energy-constrained ad hoc wireless networks. *IEEE Wireless Communications Magazine*, 9(4):8–27.

Hares, S., Yanikomeroglu, H., and Hashem, B. (2003). Diversity-and amc (adaptive modulation and coding)-aware routing in tdma multihop networks. In *Proc. Globecom*.

Haykin, S. (1994). *Communication Systems*. John Wiley & Sons, 3rd edition.

Jain, K., Padhye, J., Padmanabhan, V. N., and Qiu, L. (2003). Impact of interference on multi-hop wireless network performance. In *Proc. MobiCom*.

Johansson, M. and Xiao, L. (2004). Scheduling, routing and power allocation for fairness in wireless networks. In *Proc. IEEE Vehicular Technology Conference, Spring*.

Kawadia, V. and Kumar, P. R. (2005). A cautionary perspective on cross-layer design. *IEEE Wireless Communications*, 12(1):3–11.

Kodialam, M. and Nandagopal, T. (2003). Characterizing achievable rates in multi-hop wireless networks: The joint routing and scheduling problem. In *Proc. MobiCom*.

Koetter, R. and Medard, M. (2003). An algebraic approach to network coding. *IEEE Transactions on Information Theory*, 11(5):782–795.

Kozat, U. C., Koutsopoulos, I., and Tassiulas, L. (2004). A framework for cross-layer design of energy-efficient communication with QoS provisioning in multi-hop wireless networks. In *Proc. IEEE INFOCOM*.

Laneman, J. N., Tse, D. N. C., and Wornell, G. W. (2004). Cooperative diversity in wireless networks: Efficient protocols and outage behavior. *IEEE Transactions on Information Theory*, 50(12):3062–3080.

Laneman, J. N. and Wornell, G. W. (2003). Distributed space-time-coded protocols for exploiting cooperative diversity in wireless networks. *IEEE Transactions on Information Theory*, 49(10):2415–2425.

Liu, L. and Ge, H. (2004). Space-time coding for wireless sensor networks with cooperative routing diversity. In *The Thirty-Eighth Asilomar Conference on Signals, Systems, and Computers*.

Mendel, J. M. (1995). *Lessons in Estimation Theory for Signal Processing, Communications, and Control*. Prentice Hall.

Ng, C. T. K. and Goldsmith, A. J. (2005). Capacity gain from transmitter and receiver cooperation. In *Proc. International Symposium on Information Theory (ISIT)*.

O'Neill, D., Julian, D., and Boyd, S. (2004). Seeking Foschini's genie: optimal rates and powers in wireless networks. *IEEE Transactions on Vehicular Technology*. To Appear.

Paulraj, A., Nabar, R., and Gore, D. (2003). *Introduction to Space-Time Wireless Communications*. Cambridge University Press.

Radunovic, B. and Boudec, J. Y. Le (2003). Joint scheduling, power control and routing in symmetric, one-dimensional, multi-hop wireless networks. In *Proc. Modeling and Optimization in Mobile, Ad Hoc and Wireless Networks (WiOpt)*.

Tian, Y., Xu, K., and Ansari, N. (2005). Tcp in wireless environments: problems and solutions. *IEEE Communications Magazine*, 43(3):27–32.

Toumpis, S. and Goldsmith, A. J. (2003). Capacity regions for wireless ad hoc networks. *IEEE Transactions on Wireless Communications*, 2(4):736–748.

Xiao, J., Cui, S., Luo, Z. Q., and Goldsmith, A. J. (2004). Joint estimation in sensor networks under energy constraint. In *Proc. IEEE 1st Conference on Sensor and Ad Hoc Communications and Networks*.

Chapter 5

NETWORK CODING IN WIRELESS NETWORKS

A survey of techniques for efficient operation of coded wireless packet networks

Desmond S. Lun
Massachusetts Institute of Technology
dslun@mit.edu

Tracey Ho
California Institute of Technology
tho@caltech.edu

Niranjan Ratnakar
University of Illinois at Urbana-Champaign
ratnakar@uiuc.edu

Muriel Médard
Massachusetts Institute of Technology
medard@mit.edu

Ralf Koetter
University of Illinois at Urbana-Champaign
koetter@uiuc.edu

Abstract: The advent of network coding promises to change many aspects of networking. Network coding moves away from the conventional approach to networking, where packets are treated as inviolable, atomic units to be transported through

Frank H. P. Fitzek and Marcos D. Katz (eds.), Cooperation in Wireless Networks: Principles and Applications, 127–161.

the network, and instead allows packets to be mixed and combined, so packets outgoing from a node are allowed to be arbitrary, causal functions of packets received at that node. This approach has shown much promise in multi-hop wireless networks, affording gains including more efficient use of resources and the ability for decentralized operation. In this chapter, we overview some of the issues and technical approaches associated with network coding in the wireless domain. We hope, thereby, to provide the reader with a firm theoretical basis from which practical implementations and theoretical extensions can be developed.

Keywords: ad hoc networks, forward error correction, multicast communication, multi-hop wireless networks, network coding.

1. Introduction

The notion of coding at the packet level—commonly called network coding—has attracted much recent interest. In this survey, we provide an overview of some of the issues and technical approaches associated with network coding in the wireless domain. Considering the wireless applications is not merely an extension or simple modification of the wireline case. One could argue that most wireless routing schemes have indeed sought to replicate, in the wireless domain, topologies that resemble wireline networks. This approach has tended to neglect characteristics of wireless transmissions, such as inherent broadcast, interference, fading and mobility, in order to re-use the vast algorithmic and protocol knowledge established for routing in wireline networks. However, limited acknowledgment, in protocol design, of wireless transmission peculiarities has generally led to inefficient use of limited resources, such as spectrum and battery life, as well as to considerable complications in deployment. It is therefore our goal for this paper to afford some insights, in these early stages of the development of network coding, for the application of network coding to wireless environments, in such a way that identifies techniques that are common to both wireless and wireline networks, but embraces the peculiarities of wireless media.

To illustrate the differences and similarities between network coding for wireless and wireline applications, we commence with the simple canonical example from the initial work on the topic of network coding by [Ahlswede et al., 2000]. Figure 5.1 shows this example. Each link is assumed to have unit capacity, be error-free, and provide a unidirectional link which does not interfere with other links emerging from or incident upon a common node. The simple coding shown in the figure affords a multicast connection conveying two bits, b_1 and b_2, from the sender at node 1 to the receivers at nodes 6 and 7. We consider bits rather than packets because, from bits, it is clear how packets should be coded.

Figure 5.2 considers the same topology but seeks to represent the behavior of wireless links sharing bandwidth and enabled by a single omnidirectional

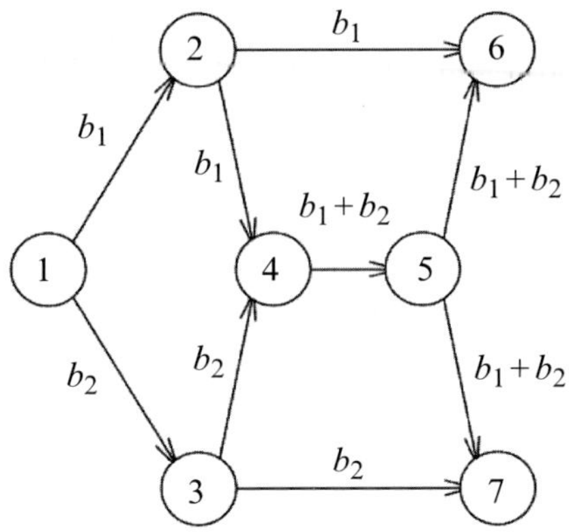

Figure 5.1. Canonical example of network coding in wireline networks given by [Ahlswede et al., 2000]. We denote by $b_1 + b_2$ the binary sum of bits b_1 and b_2.

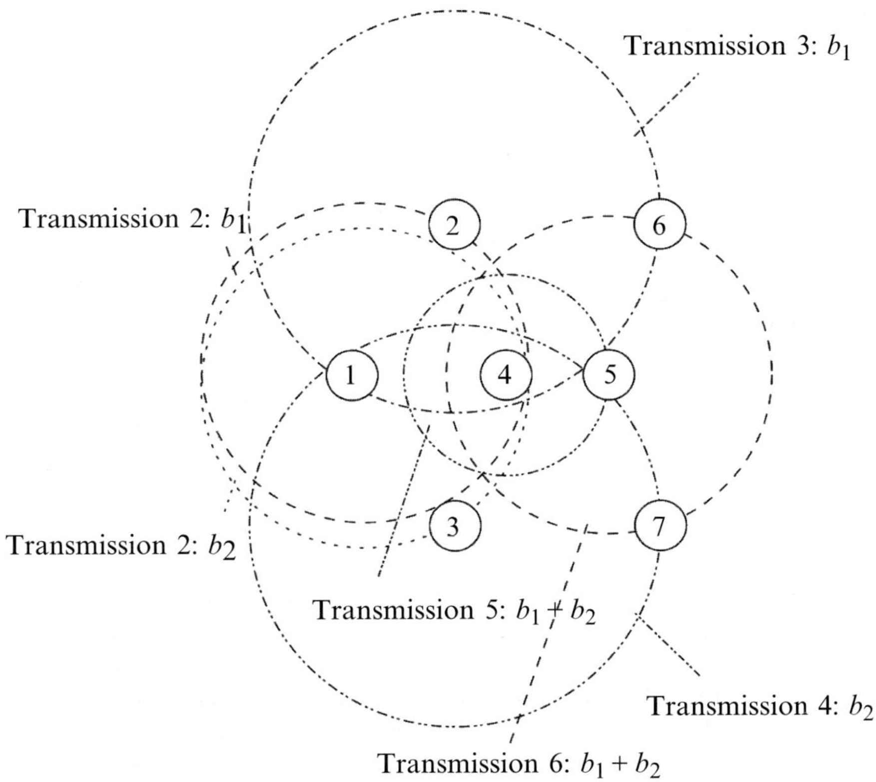

Figure 5.2. Figure 5.1 redrawn for wireless links.

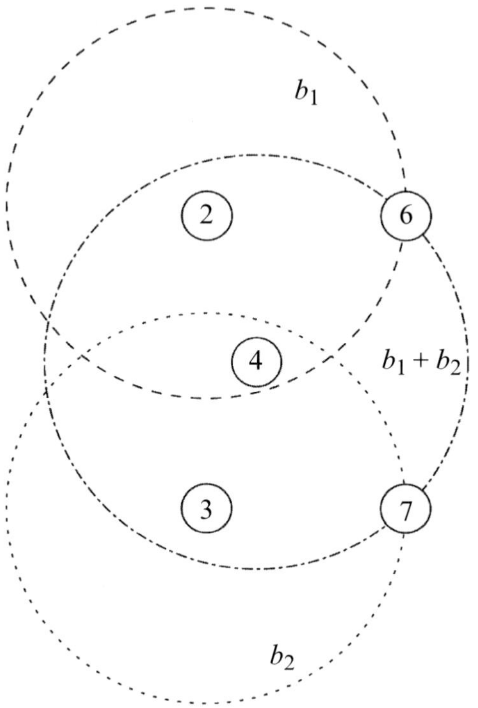

Figure 5.3. A simple five-node wireless network employing network coding.

transmit antenna and a single omnidirectional receive antenna at each node. Rather than using arcs to represent links, we draw circles whose radii indicate the range of a transmission. Moreover, the constraints of shared spectrum preclude a node transmitting and receiving simultaneously, as well as simultaneous reception of more than one transmission at a node. The sequential transmissions represented in Figure 5.2 indicate a possible instantiation of a schedule of transmissions and their associated ranges, which satisfy the simple wireless transmission modalities we detailed above. Note that network coding is still useful at transmission 5, in which node 4 communicates directly to node 5. In the absence of network coding, nodes 4 and 5 would need to perform two transmissions, one for b_1 and one for b_2. Note, however, that other instances of radii and schedule besides that represented in Figure 5.2 are possible. For example, at transmission 5, node 4 could have selected a wider range of transmission, including the receivers, nodes 6 and 7. In such a scheme, network coding would still be advantageous. Some different choices of transmission ranges would obviate the usefulness of network coding altogether, for example, if the source node, node 1, transmitted b_1 and b_2 successively over wide enough regions to include both receivers.

Network size	Approach	Average multicast energy			
		2 sinks	4 sinks	8 sinks	16 sinks
20 nodes	MIP algorithm	30.6	33.8	41.6	47.4
	Network coding	15.5	23.3	29.9	38.1
30 nodes	MIP algorithm	26.8	31.9	37.7	43.3
	Network coding	15.4	21.7	28.3	37.8
40 nodes	MIP algorithm	24.4	29.3	35.1	42.3
	Network coding	14.5	20.6	25.6	30.5
50 nodes	MIP algorithm	22.6	27.3	32.8	37.3
	Network coding	12.8	17.7	25.3	30.3

Table 5.1. Average energy of random multicast connections of unit rate for various approaches in random wireless networks of varying size. Nodes were placed randomly within a 10×10 square with a radius of connectivity of 3. The energy required to transmit at unit rate to a distance d was taken to be d^2. Source and sink nodes were selected according to an uniform distribution over all possible selections.

The dependence of connectivity on choice of transmission radii renders operation highly dependent on physical repartition of nodes. The simple example of Figure 5.3 illustrates this dependence. The topology is similar to that of Figures 5.1 and 5.2, except that nodes 1 and 5 have been removed and, so, b_1 originates at node 2 and b_2 at node 3. The representation of the transmission radii and schedules follow naturally from Figure 5.2. In this case, node 6 can be seen as "overhearing" the transmission of b_1 to the center of the network and, similarly, node 7 receives b_2 as part of the requisite transmission of b_2 to a distance sufficient to establish connectivity. Network coding in this case is a natural relaying with combining. The coding establishes a multicast of two sources originating at nodes 2 and 3 to two receivers at nodes 6 and 7. It can also be used to unicast a single source (b_1) from node 2 to node 7 and a single source (b_2) from node 3 to node 6. Thus, the coding establishes a natural multicast scenario from two unicast scenarios. The cause for this natural multicast is that nodes, because of the intrinsic properties of wireless transmission, overhear communications which may not be of direct interest to them but which will allow them to infer the transmission that are.

The examples that we have discussed clearly show that there is some potential for improving the performance of wireless networks by using network coding. We therefore wish to generalize the technique and make it broadly applicable. The remainder of this chapter is largely devoted to discussing such a generalization. We give a general prescription for the operation of coded wireless networks, *i.e.* we give a method for determining which node should send what when. The techniques that we discuss are very recent and have not yet been throughly tested, but preliminary results are promising. For example, in the problem of minimum-energy multicast, these techniques have been found to produce reductions in average energy consumption ranging from 13%

to 49% when compared to the Multicast Incremental Power (MIP) algorithm by [Wieselthier et al., 2002]—one of the few good approaches to minimum-energy multicast in non-coded, routed wireless networks (see Table 5.1). The results in Table 5.1 are for lossless networks, where links are, owing presumably to some underlying retransmission scheme, effectively lossless. We believe that performance gains for lossy networks, where network coding can be used as a means of ensuring reliable transmission and retransmission is obviated, may be even more significant. Before proceeding any further with the prescription for operation, however, we first present a model for wireless packet networks.

2. Model

We model the network with a directed hypergraph $\mathcal{H} = (\mathcal{N}, \mathcal{A})$, where $\mathcal{N}$ is the set of nodes and $\mathcal{A}$ is the set of hyperarcs. A hypergraph is a generalization of a graph, where, rather than arcs, we have hyperarcs. A hyperarc is a pair (i, J), where i, the start node, is an element of $\mathcal{N}$ and J, the set of end nodes, is a non-empty subset of $\mathcal{N}$.

Each hyperarc (i, J) represents a wireless broadcast link from node i to nodes in the non-empty set J. This link may be lossless or lossy, *i.e.* it may or may not be subject to packet erasures. Let A_{iJK} be the counting process describing the arrival of packets that are injected on hyperarc (i, J) and received by exactly the set of nodes $K \subset J$, *i.e.* for $\tau \geq 0$, $A_{iJK}(\tau)$ is the total number of packets that are injected on hyperarc (i, J) and received by all nodes in K (and no nodes in $\mathcal{N} \setminus K$) between time 0 and time τ. For example, suppose that three packets are injected on hyperarc $(1, \{2, 3\})$ between time 0 and time 1 and that, of these three packets, one is received by node 2 only, one is lost entirely, and one is received by both nodes 2 and 3; then we have $A_{1(23)\emptyset}(1) = 1$, $A_{1(23)2}(1) = 1$, $A_{1(23)3}(1) = 0$, and $A_{1(23)(23)}(1) = 1$.

We assume that A_{iJK} has an average rate z_{iJK}; more precisely, we assume that

$$\lim_{\tau \to \infty} \frac{A_{iJK}(\tau)}{\tau} = z_{iJK} \tag{5.1}$$

almost surely. When links are lossless, we have $z_{iJK} = 0$ for all $K \subsetneq J$.

Let $z_{iJ} := \sum_{K \subset J} z_{iJK}$ be the average rate at which packets are injected into hyperarc (i, J). The rate vector z, consisting of z_{iJ}, $(i, J) \in \mathcal{A}$, is called the coding subgraph and can be varied within a constraint set Z dictated to us by lower layers (for examples of such constraint sets, see [Cruz and Santhanam, 2003; Jain et al., 2003; Johansson et al., 2003; Xiao et al., 2004; Kodialam and Nandagopal, 2005; Wu et al., 2005]). We reasonably assume that Z is a convex subset of the positive orthant containing the origin. We associate with the network a cost function f that maps feasible coding subgraphs to real numbers and that we seek to minimize. For wireless networks, it is common for the cost function to reflect energy consumption, but it could also represent, for example, average latency, monetary cost, or a combination of these considerations.

We focus on network coding within a multicast session consisting of one or more sources multicasting to the same set of receiver nodes, or sinks. Coding within individual multicast sessions is a reasonable practical approach. There are cases where capacity can be increased by coding across sessions, such as the one mentioned in Section 1, but much less is known at present about such codes, which are discussed briefly in Section 5. We consider multicast sessions as they are the most general type of session, including unicast and broadcast as special cases. We denote the source processes by $X_1, X_2, \ldots, X_r$. Source X_k is generated at node $a(k)$, where $a : \{1, \ldots, r\} \to \mathcal{N}$ is an arbitrary mapping. The source processes are multicast to a set $T \subset \mathcal{N}$ of sinks. For simplicity, we assume subsequently that $a(k) \notin T$ for all $k \in \{1, \ldots, r\}$. Processes $X_1, \ldots, X_r$, mapping a, and set T specify a set of *multicast connection requirements*.

3. Distributed random network coding

In this section, we discuss how distributed random network coding can achieve any feasible set of multicast connection requirements in a given coding subgraph z. As a consequence, in setting up a single multicast session in a network, there is no loss of optimality in separating the problems of subgraph selection and coding, *i.e.* separating the optimization for a minimum-cost subgraph, which we discuss in Section 4, and the construction of a code for a given subgraph, which we now discuss. We ultimately aim for practicable distributed solutions for both problems, which can be simultaneously and, to a large degree, independently implemented.

We first consider idealized, lossless, "static" networks before considering the "dynamic" packet networks of our model in Section 2.

Static networks

To illustrate the basics of algebraic network coding, let us first consider the simple five-node network of Figure 5.2. Following our model, we represent the network using the hypergraph shown in Figure . We have $a(1) = 2$, $a(2) = 3$, and $T = \{6, 7\}$. We suppose that the source processes X_1 and X_2 are bits, so node 2 transmits X_1, node 3 transmits X_2, and node 4 transmits $X_1 + X_2$, the binary sum of X_1 and X_2. Recall that this coding permits both node 6 and node 7 to recover X_1 and X_2, establishing a multicast session consisting of two sources and two receivers.

We can generalize these basic ideas to more complex networks, with coding functions that can be arbitrary functions instead of just binary sums of bits. More specifically, we can consider the random process transmitted on hyperarc

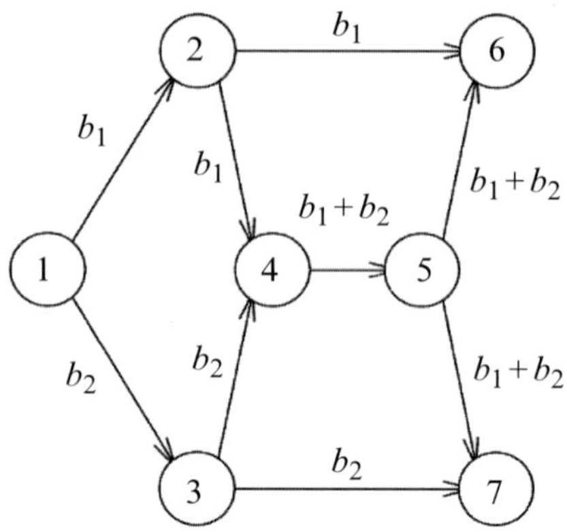

Figure 5.4. Figure 5.3 redrawn in its hypergraph representation.

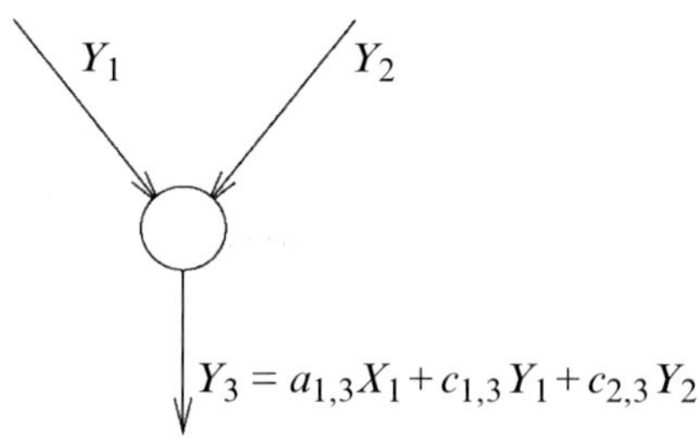

Figure 5.5. Illustration of linear coding at a node.

(i, J), denoted Y_{iJ}, to be a binary sequence obtained as a function θ_{iJ} of hyperarc (i, J)'s *inputs*, *i.e.* source processes X_k for which $a(k) = i$ and random processes $Y_{i'J'}$ for which $i \in J'$, if any. Functions $\{\theta_{iJ} | (i, J) \in \mathcal{A}\}$ specify a network code. The kth output process $Z_{t,k}$ at a receiver node t is the binary sequence obtained as a function $\phi_{t,k}$ of the random processes $Y_{i'J'}$ for which $t \in J'$. If, for a given network code, functions $\{\phi_{t,k}\}$ exist for which the output processes $\{Z_{t,k}\}$ are equal to the corresponding source processes $\{X_k\}$ (possibly with some delay), the network code is called *permissible* for t. A network code that is permissible for all receivers simultaneously is called *valid*, and the network code together with the corresponding decoding functions $\{\phi_{t,k}\}$ form a *solution* to the multicast connection problem. If a solution exists for a multicast connection problem, it is called *feasible*, and the connection requirements are said to be feasible on the network.

It turns out that for any feasible multicast connection problem, there exists a linear network coding solution, in which coding operations are done on length-u vectors of bits in a finite field $\mathbb{F}_q$, $q = 2^u$. We can express the coding functions in compact mathematical form by considering unit rate sources and unit capacity hyperarcs, allowing for multiple sources at a node and multiple hyperarcs with the same origin and destination nodes. In an acyclic network where each node waits for inputs on all its incoming hyperarcs before sending an output, the formation of the process Y_{iJ} transmitted on a hyperarc (i, J) is represented by the equation

$$Y_{iJ} = \sum_{\{k\,:\,a(k)=i\}} a_{k,iJ} X_k + \sum_{\{(i',J')\,:\,i \in J'\}} c_{i'J',iJ} Y_{i'J'}.$$

This is illustrated in Figure 5.5. Each sink receives as many linearly independent input processes as the number of source processes, and is able to decode by taking a linear combination of its input processes:

$$Z_{t,k} = \sum_{\{(i',J')\,:\,t \in J'\}} b_{t,k,i'J'} Y_{i'J'}.$$

For general networks with cycles, we need to explicitly consider transmission delays to ensure stability. For instance, if each link has the same delay, the linear coding equations are

$$\begin{aligned} Y_{iJ}(t+1) &= \sum_{\{k\,:\,a(k)=i\}} a_{k,iJ} X_k(t) \\ &\quad + \sum_{\{(i',J')\,:\,i \in J'\}} c_{i'J',iJ} Y_{i'J'}(t), \\ Z_{t,k}(t+1) &= \sum_{u=0}^{\mu} b'_{t,k}(u) Z_{t,k}(t-u) \\ &\quad + \sum_{\{(i',J')\,:\,t \in J'\}} \sum_{u=0}^{\mu} b''_{t,k,i'J'}(u) Y_{i'J'}(t-u), \end{aligned}$$

where $X_k(t), Y_{iJ}(t), Z_{t,k}(t), b'_{t,k}(t)$ and $b''_{t,k,i'J'}(t)$ are the values of the corresponding variables at time t respectively. The variable μ represents the memory required at receiver t for decoding when link delays are considered. These equations and random processes can be represented algebraically in terms of a delay variable D:

$$\begin{aligned} Y_{iJ}(D) &= \sum_{\{k\,:\,a(k)=i\}} D a_{k,iJ} X_k(D) \\ &\quad + \sum_{\{(i',J')\,:\,i \in J'\}} D c_{i'J',iJ} Y_{i'J'}(D), \\ Z_{t,k}(D) &= \sum_{\{(i',J')\,:\,t \in J'\}} b_{t,k,i'J'}(D) Y_{i'J'}(D), \end{aligned}$$

where

$$b_{t,\,k,\,i'J'}(D) = \frac{\sum_{u=0}^{\mu} D^{u+1} b''_{t,k,i'J'}(u)}{1 - \sum_{u=0}^{\mu} D^{u+1} b'_{t,k}(u)}$$

and

$$X_k(D) = \sum_{t=0}^{\infty} X_k(t) D^t,$$

$$Y_{iJ}(D) = \sum_{t=0}^{\infty} Y_{iJ}(t) D^t, \qquad Y_{iJ}(0) = 0,$$

$$Z_{t,k}(D) = \sum_{t=0}^{\infty} Z_{t,k}(t) D^t, \qquad Z_{t,k}(0) = 0.$$

Furthermore, for a given feasible coding subgraph, choosing the code coefficients $\{a_{k,iJ}, c_{i'J',iJ}\}$ uniformly at random from a sufficiently large field $\mathbb{F}_q$ gives, with high probability, a solution to any multicast connection problem that is feasible on the subgraph. The field size q must be at least greater than the number of receivers d. Some of the coefficients can be fixed rather than randomly chosen, as long as there exists a solution to the network connection problem with the same values for these fixed coefficients. For instance, if a node i receives linearly dependent processes on two incoming links $(i_1, J_1), (i_2, J_2)$, it can fix $c_{i_1 J_1, iJ} = 0$ for all its outgoing links (i, J).

While not the only way to find a valid network code, this random linear coding approach offers a particularly convenient distributed way to set up a network code, in which each node makes independent random choices of coding functions. The coefficient vectors needed for decoding can be sent through the network with each data block over which the code remains constant. The coefficient vector sent with each block of data from source X_k, $k = 1, \ldots, r$, is the length-r unit vector with a single nonzero entry, 1, in the kth position. Each coding node applies the same linear mappings to the coefficient vectors as to their corresponding data. In this way, the inputs received at a sink are accompanied by coefficient vectors specifying their composition as a linear combination of the original source processes.

To see why random linear network coding is sufficient to solve any feasible multicast connection problem with high probability, consider the hypergraph $\mathcal{H}$ used by some (possibly nonlinear) solution to a feasible multicast problem. Note that the multicast rate can be no larger than the minimum of the rates that can be sent to each sink separately on $\mathcal{H}$. We show that, with high probability, random linear network coding achieves this rate, which implies that it is sufficient.

Suppose we do random linear network coding over $\mathcal{H}$. Consider a subgraph $\mathcal{H}_t$ of $\mathcal{H}$ that transmits the desired rate to sink t separately. The network coding solution can be reduced to a flow solution to sink t over $\mathcal{H}_t$ if the code coefficients

$\{a_{k,iJ}, c_{i'J',iJ}\}$ associated with $\mathcal{H}_t$ take the value 1 and the rest of the code coefficients take the value 0. In this case, sink t receives a set of inputs whose coefficient vectors together form an identity matrix. Now consider the matrix of coefficient vectors of the same set of inputs, but with the code coefficients $\{a_{k,iJ}, c_{i'J',iJ}\}$ as indeterminate variables. The determinant of this matrix is a polynomial in $\{a_{k,iJ}, c_{i'J',iJ}\}$ (and in D, if we consider link delays) which we know, from considering the flow solution to t, is not identically zero. Each sink similarly has a set of inputs whose associated coefficient vectors form a matrix with a nonzero determinant polynomial. Multiplying these matrices together, we obtain a polynomial in $\{a_{k,iJ}, c_{i'J',iJ}\}$ that is not identically zero. By the Schwartz-Zippel theorem (see, for example, [Motwani and Raghavan, 1995]), if we choose the code coefficients uniformly at random from a finite field $\mathbb{F}_q$ where q is greater than the degree of the polynomial, then the polynomial takes a zero value with probability inversely proportional to q.

Owing to the particular structure of these polynomials, the probability of obtaining a valid random code is actually higher than that given by the Schwartz-Zippel theorem. We can bound this probability more tightly as follows. We denote by η the number of hyperarcs (i, J) with associated random coefficients $\{a_{k,iJ}, c_{i'J',iJ}\}$.

THEOREM 5.1 *Consider a multicast connection problem on an arbitrary static network and a network code in which some or all network code coefficients* $\{a_{k,iJ}, c_{i'J',iJ}\}$ *are chosen uniformly at random from a finite field* $\mathbb{F}_q$ *where* $q > d$, *and the remaining code coefficients, if any, are fixed. If there exists a solution to the network connection problem with the same values for the fixed code coefficients, then the probability that the random network code is valid is at least* $(1 - d/q)^\eta$.

For a proof of Theorem 5.1, see [Ho et al., 2003]. Recall that $q = 2^u$, so the error probability decreases exponentially with u. Thus, random linear coding achieves maximum multicast capacity with probability exponentially approaching 1 with the number of bits in the coding field.

We can extend the basic network coding model and results for static networks described above to dynamic packet networks with bursty sources and varying link delays and capacities. In the static case, the code is the same for all vectors of bits originating at the same source or traversing the same hyperarc. In the dynamic case, the code may change from packet to packet, but is the same for all vectors of bits in the same packet. Thus, we may draw an analogy between sources and hyperarcs in the static case, and source packets and coded/forwarded packets respectively in the dynamic case. In the dynamic case, each packet would contain a coefficient vector of length equal to the number of source packets that may be coded together in the network.

One way to operate a dynamic network is for each coding node to wait until it receives a packet corresponding to each of its inputs, before coding them together and transmitting packets on its outgoing hyperarcs. But such waiting seems unnecessary. An alternative approach divides packets formed around the same time into batches. Each packet is labeled with a batch number and some information specifying its intended sink node(s), and coding is done, without waiting, only across packets of the same batch. We consider applying such a batched approach to dynamic networks in the following section.

Dynamic networks

The specific coding scheme we consider, which we hereafter refer to as distributed random network coding, is as follows. We suppose that, at the beginning of each batch, all nodes flush their memories of packets associated with previous batches. Now, at node $a(k)$, source X_k generates a batch of K' message packets $X_{k,1}, X_{k,2}, \ldots, X_{k,K'}$, which are vectors of length u over the finite field $\mathbb{F}_q$. (If the packet length is b bits, then we take $u = \lceil b/\log_2 q \rceil$.) These message packets are placed in node $a(k)$'s memory.

The coding operation performed by each node is simple to describe and is the same for every node: Received packets are stored into the node's memory, and packets are formed for injection with random linear combinations of its memory contents whenever a packet injection occurs on an outgoing link. The coefficients of the combination are drawn uniformly from $\mathbb{F}_q$.

Since all coding is linear, we can write any packet x in the network as a linear combination of the $K := rK'$ message packets; namely, we have $x = \sum_{k=1}^{r} \sum_{l=1}^{K'} \gamma_{kl} X_{k,l}$. We call γ the *global encoding vector* of x, and we assume that it is sent along with x as side information in its header. The overhead this incurs (namely, $K \log_2 q$ bits) is negligible if packets are sufficiently large.

Nodes are assumed to have unlimited memory. The scheme can be modified so that received packets are stored into memory only if their global encoding vectors are linearly-independent of those already stored. This modification keeps our conclusions regarding the scheme unchanged while ensuring that nodes never need to store more than K packets.

A sink node collects packets and, if it has K packets with linearly-independent global encoding vectors, it is able to recover the message packets. Decoding can be done by Gaussian elimination, and the scheme can be operated ratelessly, *i.e.* it can be run indefinitely until successful reception (at which stage that fact is signaled to other nodes).

We suppose that sink $t \in T$ wishes to achieve rate arbitrarily close to R_t, *i.e.* to recover the K message packets, sink t wishes to wait for a time Δ_t that is only marginally greater than K/R_t. The main result that we have in relation to distributed random network coding is as follows.

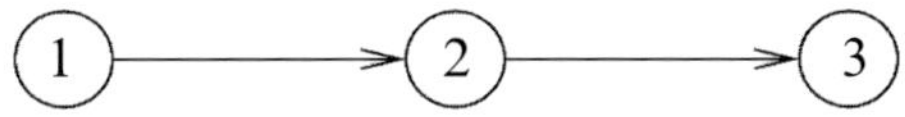

Figure 5.6. A network consisting of two links in tandem.

THEOREM 5.2 *Distributed random network coding achieves capacity for multicast sessions; more precisely, for K sufficiently large, it can satisfy, with arbitrarily small error probability, the set of multicast connection requirements $(X_1, \ldots, X_r; a; T)$ at rate arbitrarily close to R_t packets per unit time for each $t \in T$ if there exists, for all $t \in T$, a non-negative flow vector $x^{(t)}$ satisfying*

$$\sum_{\{J|(i,J)\in\mathcal{A}\}} \sum_{j\in J} x_{iJj}^{(t)} - \sum_{\{j|(j,I)\in\mathcal{A}, i\in I\}} x_{jIi}^{(t)} = \sigma_i^{(t)}, \tag{5.2}$$

for all $i \in \mathcal{N}$, and

$$\sum_{j\in K} x_{iJj}^{(t)} \leq \sum_{\{L\subset J|L\cap K\neq\emptyset\}} z_{iJL}$$

for all $(i, J) \in \mathcal{A}$ and $K \subset J$, where

$$\sigma_i^{(t)} := \begin{cases} \frac{|\{k|a(k)=i\}|}{r} R_t & \text{if } i \neq t, \\ -R_t & \text{if } i = t. \end{cases}$$

We see from Theorem 5.2 that distributed random network coding is remarkably robust: If run over sufficiently large batch sizes K, it achieves the maximum feasible rate of a given coding subgraph, with only assumption (5.1) on the arrival of received packets on a link. Assumption (5.1) makes no claims on loss correlation or lack thereof—all we require is that a long-run average exists. This fact is particularly important in wireless packet networks, where slow fading and collisions often cause packets not to be received in a steady stream.

To see why Theorem 5.2 is true, consider first the simplest non-trivial case: that of a single unicast connection over two links in tandem (see Figure 5.6).

Suppose we wish to establish a connection of rate arbitrarily close to R packets per unit time from node 1 to node 3. Suppose further that the coding scheme is run for a total time Δ, from time 0 until time Δ, and that, in this time, a total of N packets is received by node 2. We call these packets $v_1, v_2, \ldots, v_N$.

Any received packet y in the network is a linear combination of $v_1, v_2, \ldots, v_N$, so we can write

$$y = \sum_{n=1}^{N} \beta_n v_n.$$

Since v_n is formed by a random linear combination of the message packets $w_1, w_2, \ldots, w_K$, we have

$$v_n = \sum_{k=1}^{K} \alpha_{nk} w_k$$

for $n = 1, 2, \ldots, N$. Hence

$$y = \sum_{k=1}^{K} \left(\sum_{n=1}^{N} \beta_n \alpha_{nk} \right) w_k,$$

and it follows that the kth component of the global encoding vector of y is given by

$$\gamma_k = \sum_{n=1}^{N} \beta_n \alpha_{nk}.$$

We call the vector β associated with y the *auxiliary encoding vector* of y, and we see that any node that receives $\lfloor K(1+\varepsilon) \rfloor$ or more packets with linearly-independent auxiliary encoding vectors has $\lfloor K(1+\varepsilon) \rfloor$ packets whose global encoding vectors collectively form a random $\lfloor K(1+\varepsilon) \rfloor \times K$ matrix over $\mathbb{F}_q$, with all entries chosen uniformly. If this matrix has rank K, then node 3 is able to recover the message packets. The probability that a random $\lfloor K(1+\varepsilon) \rfloor \times K$ matrix has rank K is, by a simple counting argument, $\prod_{k=1+\lfloor K(1+\varepsilon) \rfloor - K}^{\lfloor K(1+\varepsilon) \rfloor} (1 - 1/q^k)$, which can be made arbitrarily close to 1 by taking K arbitrarily large. Therefore, to determine whether node 3 can recover the message packets, we essentially need only to determine whether it receives $\lfloor K(1+\varepsilon) \rfloor$ or more packets with linearly-independent auxiliary encoding vectors.

Our proof is based on tracking the propagation of what we call *innovative* packets. Such packets are innovative in the sense that they carry new, as yet unknown, information about $v_1, v_2, \ldots, v_N$ to a node. It turns out that the propagation of innovative packets through a network follows the propagation of jobs through a queueing network, for which fluid flow models give good approximations. We present the following argument in terms of this fluid analogy.

Since the packets being received by node 2 are the packets $v_1, v_2, \ldots, v_N$ themselves, it is clear that every packet being received by node 2 is innovative. Thus, innovative packets arrive at node 2 at a rate of z_{122}, and this can be approximated by fluid flowing in at rate z_{122}. These innovative packets are stored in node 2's memory, so the fluid that flows in is stored in a reservoir.

Packets, now, are being received by node 3 at a rate of z_{233}, but whether these packets are innovative depends on the contents of node 2's memory. If node 2 has more information about $v_1, v_2, \ldots, v_N$ than node 3 does, then it is likely that new information will be described to node 3 in the next packet that it receives. Otherwise, if node 2 and node 3 have the same degree of information about

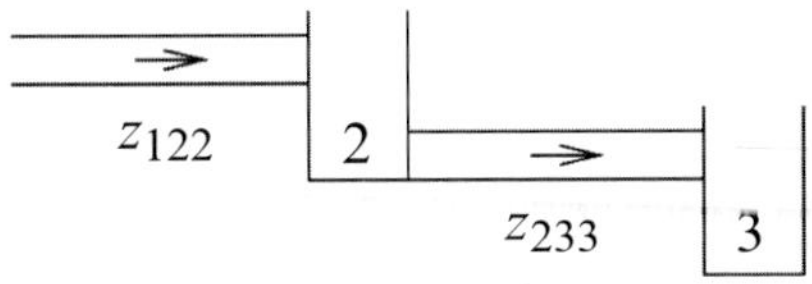

Figure 5.7. Fluid flow system corresponding to two-link tandem network.

$v_1, v_2, \ldots, v_N$, then packets received by node 3 cannot possibly be innovative. Thus, the situation is as though fluid flows into node 3's reservoir at a rate of z_{233}, but the level of node 3's reservoir is restricted from ever exceeding that of node 2's reservoir. The level of node 3's reservoir, which is ultimately what we are concerned with, can equivalently be determined by fluid flowing out of node 2's reservoir at rate z_{233}.

We therefore see that the two-link tandem network in Figure 5.6 maps to the fluid flow system shown in Figure 5.7. It is clear that, in this system, fluid flows into node 3's reservoir at rate $\min(z_{122}, z_{233})$. This rate determines the rate at which packets with new information about $v_1, v_2, \ldots, v_N$—and, therefore, linearly-independent auxiliary encoding vectors—arrive at node 3. Hence the time required for node 3 to receive $\lfloor K(1+\varepsilon) \rfloor$ packets with linearly-independent auxiliary encoding vectors is, for large K, approximately $K(1+\varepsilon)/\min(z_{122}, z_{233})$, which implies that a connection of rate arbitrarily close to R packets per unit time can be established provided that

$$R \leq \min(z_{122}, z_{233}). \tag{5.3}$$

If, as in Theorem 5.2, there exists a flow vector $x^{(3)}$ such that $x^{(3)}_{122} = x^{(3)}_{233} = R$, $x^{(3)}_{122} \leq z_{122}$, and $x^{(3)}_{233} \leq z_{233}$, then it is clear that condition (5.3) is satisfied, and the connection can be established.

From this simple case of a single unicast connection over two links in tandem, it is in fact not difficult to extend to the general case described by Theorem 5.2. We first extend from two links in tandem to arbitrarily many links in tandem, which is quite straightforward. From arbitrarily many links in tandem, we then consider any unicast connection by decomposing the hypergraph flow into a set of paths, each of which can be considered as a unicast connection over a number of links in tandem. Extending to multiple sources is straightforward, as is extending to multiple sinks, where, because the coding scheme is quite oblivious to the flow vectors $x^{(t)}$ for each $t \in T$, each flow behaves more or less independently of the others. For a formal proof of Theorem 5.2, see [Lun et al., 2005b].

4. Cost minimization

We now turn to the subgraph selection problem, which we see is the problem of finding a coding subgraph z of minimum cost satisfying (5.2). Thus, the subgraph selection problem, for a single session, equates to the following optimization problem.

$$\begin{aligned}
&\text{minimize } f(z) \\
&\text{subject to } z \in Z, \\
&\sum_{j \in K} x_{iJj}^{(t)} \leq \sum_{\{L \subset J | L \cap K \neq \emptyset\}} z_{iJL}, \qquad \forall\, (i, J) \in \mathcal{A},\, K \subset J,\, t \in T, \\
&x^{(t)} \in F^{(t)}, \qquad \forall\, t \in T,
\end{aligned} \tag{5.4}$$

where $x^{(t)}$ is the vector consisting of $x_{iJj}^{(t)}$, $(i, J) \in \mathcal{A}$, $j \in J$, and $F^{(t)}$ is the bounded polyhedron of points $x^{(t)}$ satisfying the conservation of flow constraints

$$\sum_{\{J | (i,J) \in \mathcal{A}\}} \sum_{j \in J} x_{iJj}^{(t)} - \sum_{\{j | (j,I) \in \mathcal{A}, i \in I\}} x_{jIi}^{(t)} = \sigma_i^{(t)}, \qquad \forall\, i \in \mathcal{N},$$

and non-negativity constraints

$$x_{iJj}^{(t)} \geq 0, \qquad \forall\, (i, J) \in \mathcal{A},\, j \in J,$$

In the lossless case, problem (5.4) simplifies to the following optimization problem.

$$\begin{aligned}
&\text{minimize } f(z) \\
&\text{subject to } z \in Z, \\
&\sum_{j \in J} x_{iJj}^{(t)} \leq z_{iJ}, \qquad \forall\, (i, J) \in \mathcal{A},\, t \in T, \\
&x^{(t)} \in F^{(t)}, \qquad \forall\, t \in T.
\end{aligned} \tag{5.5}$$

As an example, let us return to the wireless network shown in Figure 5.4. The network is lossless, and we have $a(1) = 2$, $a(2) = 3$, and $T = \{6, 7\}$. We wish to achieve unit rate to both sinks, so $R_6 = R_7 = 1$. We suppose that $Z = [0, 1]^{|\mathcal{A}|}$ and $f(z) = \sum_{(i,J) \in \mathcal{A}} z_{iJ}$. An optimal solution to problem (5.5) is shown in Figure 5.8. We have flows $x^{(6)}$ and $x^{(7)}$, each with half a unit of flow originating from each of the sources and going to their respective sinks. For each hyperarc (i, J), $z_{iJ} = \max(\sum_{j \in J} x_{iJj}^{(6)}, \sum_{j \in J} x_{iJj}^{(7)})$, as we expect from the optimization. To achieve the optimal cost, we can apply distributed random network coding to the subgraph defined by z.

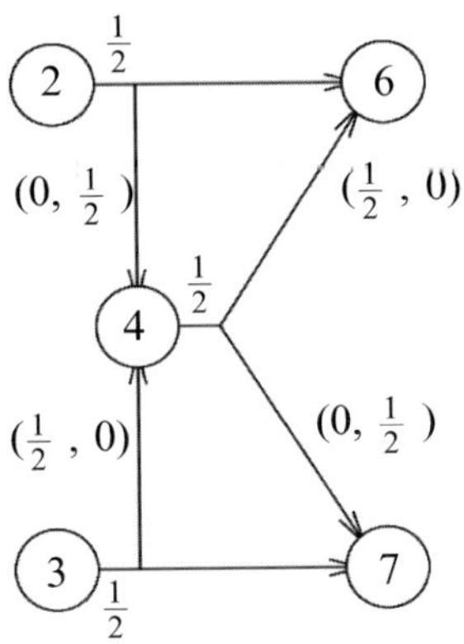

Figure 5.8. Cost optimization of a wireless network with multicast. Each hyperarc is marked with z_{iJ} at the start and with the pair $(x^{(6)}_{iJj}, x^{(7)}_{iJj})$ at the ends.

Neither problem (5.4) nor (5.5) as they stand are easy to solve. But they are very general. Their complexities improve if we assume that the cost function is separable and convex, or even linear; *i.e.* if we suppose $f(z) = \sum_{(i,J)\in A} f_{iJ}(z_{iJ})$, where f_{iJ} is a convex or linear function, which is a very reasonable assumption in many situations. Their complexities also improve if we make some assumptions on the form of the constraint set Z.

A simplification can also be made if we assume that, when nodes transmit in a lossless network, they reach all nodes in a certain area, with cost increasing as this area is increased. More precisely, suppose that we have separable cost, so $f(z) = \sum_{(i,J)\in \mathcal{A}} f_{iJ}(z_{iJ})$. Suppose further that each node i has M_i outgoing hyperarcs $(i, J_1^{(i)}), (i, J_2^{(i)}), \ldots, (i, J_{M_i}^{(i)})$ with $J_1^{(i)} \subsetneq J_2^{(i)} \subsetneq \cdots \subsetneq J_{M_i}^{(i)}$. (We assume that there are no identical links, as duplicate links can effectively be treated as a single link.) Then, we assume that $f_{iJ_1^{(i)}}(\zeta) < f_{iJ_2^{(i)}}(\zeta) < \cdots < f_{iJ_{M_i}^{(i)}}(\zeta)$ for all $\zeta \geq 0$ and nodes i. For $(i, j) \in \mathcal{A}' := \{(i, j)|(i, J) \in A, J \ni j\}$, we introduce the variables

$$\hat{x}_{ij}^{(t)} := \sum_{m=m(i,j)}^{M_i} x^{(t)}_{iJ_m^{(i)}j},$$

where $m(i, j)$ is the unique m such that $j \in J_m^{(i)} \setminus J_{m-1}^{(i)}$ (we define $J_0^{(i)} := \emptyset$ for all $i \in \mathcal{N}$ for convenience). Then, provided that $a_{iJ_1^{(i)}} < a_{iJ_2^{(i)}} < \cdots < a_{iJ_{M_i}^{(i)}}$ for all nodes i, problem (5.5) can be reformulated as the following optimization problem, which has substantially fewer variables.

$$\begin{aligned} &\text{minimize} \sum_{(i,J)\in\mathcal{A}} f_{iJ}(z_{iJ}) \\ &\text{subject to } z \in Z, \\ &\sum_{k\in J^{(i)}_{M_i}\setminus J^{(i)}_{m-1}} \hat{x}^{(t)}_{ik} \le \sum_{n=m}^{M_i} z_{iJ^{(i)}_n}, \qquad \forall\, i \in \mathcal{N}, m = 1,\ldots,M_i, t \in T, \\ &\hat{x}^{(t)} \in \hat{F}^{(t)}, \qquad \forall\, t \in T, \end{aligned} \tag{5.6}$$

where $\hat{F}^{(t)}$ is the bounded polyhedron of points $\hat{x}^{(t)}$ satisfying the conservation of flow constraints

$$\sum_{\{j|(i,j)\in\mathcal{A}'\}} \hat{x}^{(t)}_{ij} - \sum_{\{j|(j,i)\in\mathcal{A}'\}} \hat{x}^{(t)}_{ji} = \sigma^{(t)}_i, \qquad \forall\, i \in N,$$

and non-negativity constraints

$$0 \le \hat{x}^{(t)}_{ij}, \qquad \forall\, (i,j) \in \mathcal{A}'.$$

PROPOSITION 5.1 *Suppose that* $f(z) = \sum_{(i,J)\in\mathcal{A}} f_{iJ}(z_{iJ})$ *and that* $f_{iJ^{(i)}_1}(\zeta) < f_{iJ^{(i)}_2}(\zeta) < \cdots < f_{iJ^{(i)}_{M_i}}(\zeta)$ *for all* $\zeta \ge 0$ *and* $i \in \mathcal{N}$*. Then problem (5.5) and problem (5.6) are equivalent in the sense that they have the same optimal cost and* z *is part of an optimal solution for (5.5) if and only if it is part of an optimal solution for (5.6).*

Proof: Suppose (x, z) is a feasible solution to problem (5.5). Then, for all $(i, j) \in \mathcal{A}'$ and $t \in T$,

$$\begin{aligned} \sum_{m=m(i,j)}^{M_i} z_{iJ^{(i)}_m} &\ge \sum_{m=m(i,j)}^{M_i} \sum_{k\in J^{(i)}_m} x^{(t)}_{iJ^{(i)}_m k} \\ &= \sum_{k\in J^{(i)}_{M_i}} \sum_{m=\max(m(i,j),m(i,k))}^{M_i} x^{(t)}_{iJ^{(i)}_m k} \\ &\ge \sum_{k\in J^{(i)}_{M_i}\setminus J^{(i)}_{m(i,j)-1}} \sum_{m=\max(m(i,j),m(i,k))}^{M_i} x^{(t)}_{iJ^{(i)}_m k} \\ &= \sum_{k\in J^{(i)}_{M_i}\setminus J^{(i)}_{m(i,j)-1}} \sum_{m=m(i,k)}^{M_i} x^{(t)}_{iJ^{(i)}_m k} \\ &= \sum_{k\in J^{(i)}_{M_i}\setminus J^{(i)}_{m(i,j)-1}} \hat{x}^{(t)}_{ik}. \end{aligned}$$

Hence $(\hat{x}, z)$ is a feasible solution of problem (5.6) with the same cost.

Now suppose $(\hat{x}, z)$ is an optimal solution of problem (5.6). Since $f_{iJ_1^{(i)}}(\zeta) < f_{iJ_2^{(i)}}(\zeta) < \cdots < f_{iJ_{M_i}^{(i)}}(\zeta)$ for all $\zeta \geq 0$ and $i \in \mathcal{N}$ by assumption, it follows that, for all $i \in \mathcal{N}$, the sequence $z_{iJ_1^{(i)}}, z_{iJ_2^{(i)}}, \ldots, z_{iJ_{M_i}^{(i)}}$ is given recursively, starting from $m = M_i$, by

$$z_{iJ_m^{(i)}} = \max_{t \in T} \left\{ \sum_{k \in J_{M_i}^{(i)} \setminus J_{m-1}^{(i)}} \hat{x}_{ik}^{(t)} \right\} - \sum_{m'=m+1}^{M_i} z_{iJ_{m'}^{(i)}}.$$

Hence $z_{iJ_m^{(i)}} \geq 0$ for all $i \in \mathcal{N}$ and $m = 1, 2, \ldots, M_i$. We then set, starting from $m = M_i$ and $j \in J_{M_i}^{(i)}$,

$$x_{iJ_m^{(i)}j}^{(t)} := \min \left(\hat{x}_{ij}^{(t)} - \sum_{l=m+1}^{M_i} x_{iJ_l^{(i)}j}, z_{iJ_m^{(i)}} - \sum_{k \in J_{M_i}^{(i)} \setminus J_{m(i,j)}^{(i)}} x_{iJ_m^{(i)}k}^{(t)} \right).$$

It is now not difficult to see that (x, z) is a feasible solution of problem (5.5) with the same cost.

Therefore, the optimal costs of problems (5.5) and (5.6) are the same and, since the objective functions for the two problems are the same, z is part of an optimal solution for problem (5.5) if and only if it is part of an optimal solution for problem (5.6). ■

One specific problem of interest is that of minimum-energy multicast (see, for example, [Wieselthier et al., 2002; Liang, 2002]). In this problem, we wish to achieve minimum-energy multicast in a lossless wireless network without explicit regard for throughput or bandwidth, so the constraint set Z can be dropped altogether. Moreover, the cost function is separable and linear, *i.e.* $f(z) = \sum_{(i,J) \in \mathcal{A}} a_{iJ} z_{iJ}$, where a_{iJ} represents the energy required to transmit a packet to nodes in J from node i. Hence problem (5.6) becomes a linear optimization problem with a polynomial number of constraints, which can therefore be solved in polynomial time. By contrast, the same problem using traditional routing-based approaches is NP-complete—in fact, the special case of broadcast in itself is NP-complete, a result shown by [Ahluwalia et al., 2002; Liang, 2002]. The problem must therefore be addressed using polynomial-time heuristics such as the MIP algorithm by [Wieselthier et al., 2002]. Even if an optimal routing solution is found, it is in general worse than an optimal coding solution because coding subsumes routing. Thus coding promises to significantly outperform routing for practical multicast, and, indeed, simulation results reported by [Lun et al., 2005a] show significant reductions in the average total energy of random multicast connections in random wireless networks of varying size as a result of coding as opposed to routing with the MIP algorithm.

It is, however, not sufficient to have polynomial-time algorithms. For practical applications, it is usually important that solutions can be computed in a distributed manner, with each node making computations based only on local knowledge and knowledge acquired from message exchanges. Thus, we seek distributed algorithms to solve optimization problems (5.4), (5.5), and (5.6), which, when paired with distributed random network coding gives us a fully distributed approach for establishing minimum-cost connections in wireless networks. To this end, we simplify the problem by assuming that the objective function is of the form

$$f(z) = \sum_{(i,J)\in\mathcal{A}} f_{iJ}(z_{iJ}),$$

where f_{iJ} is a monotonically increasing, convex function, and by assuming that, as z_{iJ} is varied, z_{iJK}/z_{iJ} is constant for all $K \subset J$. Therefore,

$$b_{iJK} := \frac{\sum_{\{L\subset J | L\cap K \neq \emptyset\}} z_{iJL}}{z_{iJ}}$$

is a constant. We also drop the constraint set Z, noting that separable constraints, at least, can be handled by making f_{iJ} approach infinity as z_{iJ} approaches its upper constraint. Moreover, in energy-limited scenarios where energy is the principal concern, the rate of the multicast connection can always be dropped so that the constraint set Z is not restrictive; we discuss bandwidth-limited scenarios in a later section.

Hence problem (5.4) becomes

$$\begin{aligned} &\text{minimize} \sum_{(i,J)\in\mathcal{A}} f_{iJ}(z_{iJ}) \\ &\text{subject to} \sum_{j\in K} x_{iJj}^{(t)} \leq z_{iJ} b_{iJK}, \qquad \forall\, (i,J)\in\mathcal{A},\, K\subset J,\, t\in T, \\ &\qquad\qquad x^{(t)} \in F^{(t)}, \qquad \forall\, t\in T. \end{aligned} \tag{5.7}$$

Since the f_{iJ} are monotonically increasing, the constraint

$$\sum_{j\in K} x_{iJj}^{(t)} \leq z_{iJ} b_{iJK}, \qquad \forall\, (i,J)\in\mathcal{A},\, K\subset J,\, t\in T \tag{5.8}$$

gives

$$z_{iJ} = \max_{K\subset J, t\in T} \left\{ \frac{\sum_{j\in K} x_{iJj}^{(t)}}{b_{iJK}} \right\}. \tag{5.9}$$

Expression (5.9) is, unfortunately, not very useful for algorithm design because the max function is difficult to deal with, largely as a result of it not being everywhere differentiable. One way to overcome this difficulty is to approximate z_{iJ} by replacing the max in (1.9) with an l^n-norm (see [Deb and Srikant, 2004]), *i.e.* to approximate z_{iJ} with z'_{iJ}, where

$$z'_{iJ} := \left(\sum_{K \subset J, t \in T} \left(\frac{\sum_{j \in K} x_{iJj}^{(t)}}{b_{iJK}} \right)^n \right)^{1/n}.$$

The approximation becomes exact as $n \to \infty$.

Now the relevant optimization problem is

$$\begin{aligned} &\text{minimize } \sum_{(i,J) \in \mathcal{A}} f_{iJ}(z'_{iJ}) \\ &\text{subject to } x^{(t)} \in F^{(t)}, \qquad \forall\, t \in T, \end{aligned}$$

which is no more than a convex multicommodity flow problem. There are many algorithms for convex multicommodity flow problems (see [Ouorou et al., 2000] for a survey), some of which (*e.g.* the algorithms by [Bertsekas, 1980; Bertsekas et al., 1984]) are well-suited for distributed implementation. Thus, there exists a significant number of distributed algorithms for the subgraph selection problem. We present two: The first, which we call the subgradient method, does not reformulate the problem as a convex multicommodity flow problem and attempts to deal with the constraint (5.8) directly, while the second, which we call the primal-dual method, applies a particular method for solving convex multicommodity flow problems to our problem.

Subgradient method

We present the subgradient method for linear cost functions, though, with some modifications, it may be made to apply to convex ones. Thus, we assume that the objective function f is of the form

$$f(z) := \sum_{(i,J) \in \mathcal{A}} a_{iJ} z_{iJ},$$

where $a_{iJ} > 0$.

Consider the Lagrangian dual of problem (5.7):

$$\begin{aligned} &\text{maximize } \sum_{t \in T} q^{(t)}(p^{(t)}) \\ &\text{subject to } \sum_{t \in T} \sum_{K \subset J} p_{iJK}^{(t)} = a_{iJ} \qquad \forall\, (i,J) \in \mathcal{A}, \qquad (5.10) \\ &\qquad p_{iJK}^{(t)} \geq 0, \qquad \forall\, (i,J) \in \mathcal{A},\, K \subset J,\, t \in T, \end{aligned}$$

where

$$q^{(t)}(p^{(t)}) := \min_{x^{(t)} \in F^{(t)}} \sum_{(i,J) \in \mathcal{A}} \sum_{j \in J} \left(\sum_{\{K \subset J | K \ni j\}} \frac{p_{iJK}^{(t)}}{b_{iJK}} \right) x_{iJj}. \tag{5.11}$$

In the lossless case (optimization problem (5.5)), the dual problem defined by equations (5.10) and (5.11) simplifies somewhat, and we require only a single dual variable $p_{iJJ}^{(t)}$ for each hyperarc (i, J). In the case of optimization problem (5.6), the dual problem simplifies more still, as there are fewer primal variables associated with it. Specifically, we obtain, for the Lagrangian dual,

$$\begin{aligned} &\text{maximize} \sum_{t \in T} \hat{q}^{(t)}(p^{(t)}) \\ &\text{subject to} \sum_{t \in T} p_{iJ_m^{(i)}}^{(t)} = s_{iJ_m^{(i)}}, \qquad \forall\, i \in \mathcal{N}, m = 1, \dots, M_i, \\ &\qquad p_{iJ}^{(t)} \geq 0, \qquad \forall\, (i, J) \in \mathcal{A}, t \in T, \end{aligned} \tag{5.12}$$

where

$$s_{iJ_m^{(i)}} := a_{iJ_m^{(i)}} - a_{iJ_{m-1}^{(i)}},$$

and

$$\hat{q}^{(t)}(p^{(t)}) := \min_{\hat{x}^{(t)} \in \hat{F}^{(t)}} \sum_{(i,j) \in \mathcal{A}'} \left(\sum_{m=1}^{m(i,j)} p_{iJ_m^{(i)}}^{(t)} \right) \hat{x}_{ij}^{(t)}. \tag{5.13}$$

In all three cases, the dual problems are very similar, and essentially the same algorithm can be used to solve them. We present the subgradient method for the case of optimization problem (5.6) and its associated dual (5.12) with the understanding that straightforward modifications can be made for the other cases.

We outline the subgradient method below. [Lun et al., 2005c] give a proof that the algorithm does indeed converge to an optimal solution for appropriate choices of the parameters $\{\theta[n]\}$ and $\{\mu_l[n]\}$. We later explain how to choose these parameters.

1 Each node $i \in N$ computes s_{iJ} for its outgoing hyperarcs and initializes $p_{iJ}[0]$ to a point in the feasible set of (5.12). We take, for our purposes,

$$p_{iJ}^{(t)}[0] := \frac{s_{iJ}}{|T|}.$$

The values of s_{iJ} and $p_{iJ}[0]$ are then sent over hyperarc (i, J).

2 In the nth iteration, use $p^{(t)}[n]$ as the hyperarc costs, and run a distributed shortest path algorithm (*e.g.* distributed Bellman-Ford) to determine

$\hat{x}^{(t)}[n]$ for all $t \in T$. This can be done because subproblem (5.13) is in fact a standard shortest path problem.

3 Based on the $\hat{x}[n]$ obtained, we calculate a subgradient of the dual function with respect to $p^{(t)}_{iJ^{(i)}_m}[n]$, namely $g^{(t)}_{iJ^{(i)}_m}[n] := \sum_{k \in J^{(i)}_{M_i} \setminus J^{(i)}_{m-1}} \hat{x}^{(t)}_{ik}[n]$. We then compute, at node i,

$$p_{iJ}[n+1] := \arg\min_{v \in P_{iJ}} \sum_{t \in T} (v^{(t)} - (p^{(t)}_{iJ}[n] + \theta[n] g^{(t)}_{iJ}[n]))^2$$

for each $(i, J) \in \mathcal{A}$, where P_{iJ} is the $|T|$-dimensional simplex

$$P_{ij} = \left\{ v \,\middle|\, \sum_{t \in T} v^{(t)} = s_{iJ}, v \geq 0 \right\},$$

and $\theta[n] > 0$ is an appropriate step size. In other words, $p_{iJ}[n+1]$ is the Euclidean projection of $p_{iJ}[n] + \theta[n] g_{iJ}[n]$ onto the feasible set P_{iJ}. This projection can be done using the method described by [Lun et al., 2005c]. The value of $p_{iJ}[n+1]$ is sent over hyperarc (i, J).

4 At the end of each subgradient iteration, nodes recover a primal solution $\{\tilde{x}[n]\}$ by setting

$$\tilde{x}[n] := \sum_{l=1}^{n} \mu_l[n] \hat{x}[l], \tag{5.14}$$

where $\{\mu_l[n]\}_{l=1,\ldots,n}$ is an appropriate sequence of convex combination weights.

5 Finally, we need to compute the current coding subgraph $z[n]$ based on the recovered primal solution $\tilde{x}[n]$. Therefore, for each $i \in \mathcal{N}$, we set

$$z_{iJ^{(i)}_m}[n] := \max_{t \in T} \left\{ \sum_{k \in J^{(i)}_{M_i} \setminus J^{(i)}_{m-1}} \tilde{x}^{(t)}_{ik}[n] \right\} - \sum_{m'=m+1}^{M_i} z_{iJ^{(i)}_{m'}}[n]$$

recursively, starting from $m = M_i$ and proceeding through to $m = 1$.

6 Steps (2) to (5) are repeated until the primal solution has converged.

We see that the subgradient method is indeed a distributed algorithm, though it does, unfortunately, operate in synchronous rounds. We expect that, in practice, this synchronicity can be slightly relaxed.

For the choice of $\{\theta[n]\}$ and $\{\mu_l[n]\}$, we first define

$$\gamma_{ln} := \frac{\mu_l[n]}{\theta[n]}, \quad l = 1, \ldots, n, \; n = 0, 1, \ldots,$$

and

$$\Delta\gamma_n^{\max} := \max_{l=2,\ldots,n}\{\gamma_{ln} - \gamma_{(l-1)n}\}.$$

Now, if the step sizes $\{\theta[n]\}$ and convex combination weights $\{\mu_l[n]\}$ are chosen such that

1. $\gamma_{ln} \geq \gamma_{(l-1)n}$ for all $l = 2, \ldots, n$ and $n = 0, 1, \ldots$,
2. $\Delta\gamma_n^{\max} \to 0$ as $n \to \infty$, and
3. $\gamma_{1n} \to 0$ as $n \to \infty$ and $\gamma_{nn} \leq \delta$ for all $n = 0, 1, \ldots$ for some $\delta > 0$,

then the subgradient method converges to an optimal solution.

The required conditions on the step sizes and convex combination weights are satisfied by the following choices ([Sherali and Choi, 1996, Corollaries 2–4]):

1. step sizes $\{\theta[n]\}$ such that $\theta[n] > 0$, $\lim_{n\to 0}\theta[n] = 0$, $\sum_{n=1}^{\infty}\theta_n = \infty$, and convex combination weights $\{\mu_l[n]\}$ given by $\mu_l[n] = \theta[l]/\sum_{k=1}^{n}\theta[k]$ for all $l = 1, \ldots, n$, $n = 0, 1, \ldots$;
2. step sizes $\{\theta[n]\}$ given by $\theta[n] = a/(b + cn)$ for all $n = 0, 1, \ldots$, where $a > 0$, $b \geq 0$ and $c > 0$, and convex combination weights $\{\mu_l[n]\}$ given by $\mu_l[n] = 1/n$ for all $l = 1, \ldots, n$, $n = 0, 1, \ldots$; and
3. step sizes $\{\theta[n]\}$ given by $\theta[n] = n^{-\alpha}$ for all $n = 0, 1, \ldots$, where $0 < \alpha < 1$, and convex combination weights $\{\mu_l[n]\}$ given by $\mu_l[n] = 1/n$ for all $l = 1, \ldots, n$, $n = 0, 1, \ldots$.

Moreover, for all three choices, we have $\mu_l[n+1]/\mu_l[n]$ independent of l for all n, so primal iterates can be computed iteratively using

$$\begin{aligned}\tilde{x}[n] &= \sum_{l=1}^{n}\mu_l[n]\hat{x}[l] \\ &= \sum_{l=1}^{n-1}\mu_l[n]\hat{x}[l] + \mu_n[n]\hat{x}[n] \\ &= \phi[n-1]\tilde{x}[n-1] + \mu_n[n]\hat{x}[n],\end{aligned}$$

where $\phi[n] := \mu_l[n+1]/\mu_l[n]$.

Primal-dual method

For the primal-dual method, we assume that the cost functions f_{iJ} are strictly convex, as this guarantees a unique optimal solution to problem (5.7). We present the algorithm for the lossless case, with the understanding that it can be straightforwardly extended to the lossy case. Thus, the optimization problem we address is

$$\begin{aligned} &\text{minimize } \sum_{(i,J)\in\mathcal{A}} f_{iJ}(z'_{iJ}) \\ &\text{subject to } x^{(t)} \in F^{(t)}, \qquad \forall\, t \in T, \end{aligned} \tag{5.15}$$

where

$$z'_{iJ} := \left(\sum_{t\in T} \left(\sum_{j\in J} x^{(t)}_{iJj} \right)^n \right)^{1/n}.$$

Let $(y)^+_a$ denote the following function of y:

$$(y)^+_a = \begin{cases} y & \text{if } a > 0, \\ \max\{y, 0\} & \text{if } a \leq 0. \end{cases}$$

To solve problem (5.15) in a distributed fashion, consider the following primal-dual algorithm:

$$\dot{x}^{(t)}_{iJj} = -k^{(t)}_{iJj}(x^{(t)}_{iJj}) \left(\frac{\partial f_{iJ}(z'_{iJ})}{\partial x^{(t)}_{iJj}} + q^{(t)}_{ij} - \lambda^{(t)}_{iJj} \right), \tag{5.16}$$

$$\dot{p}^{(t)}_i = h^{(t)}_i(p^{(t)}_i)(y^{(t)}_i - \sigma^{(t)}_i), \tag{5.17}$$

$$\dot{\lambda}^{(t)}_{iJj} = m^{(t)}_{iJj}(\lambda^{(t)}_{iJj}) \left(-x^{(t)}_{iJj} \right)^+_{\lambda^{(t)}_{iJj}}, \tag{5.18}$$

where

$$q^{(t)}_{ij} := p^{(t)}_i - p^{(t)}_j,$$

$$y^{(t)}_i := \sum_{\{J|(i,J)\in\mathcal{A}\}} \sum_{j\in J} x^{(t)}_{iJj} - \sum_{\{j|(j,I)\in\mathcal{A}, i\in I\}} x^{(t)}_{jIi},$$

and $k^{(t)}_{iJj}(x^{(t)}_{iJj}) > 0$, $h^{(t)}_i(p^{(t)}_i) > 0$, and $m^{(t)}_{iJj}(\lambda^{(t)}_{iJj}) > 0$ are non-decreasing continuous functions of $x^{(t)}_{iJj}$, $p^{(t)}_i$, and $\lambda^{(t)}_{iJj}$ respectively.

It can be shown that the above primal-dual algorithm is globally, asymptotically stable (see [Lun et al., 2005c]). Such stability of the algorithm implies that no matter what the initial choice of (x, p) is, the primal-dual algorithm will converge to the unique solution of problem (5.15). We have to choose λ, however, with non-negative entries as the initial choice.

We associate a processor with each node. We assume that the processor for node i keeps track of the variables $\{p^{(t)}_i\}_{t\in T}$, $\{\lambda^{(t)}_{iJj}\}_{t\in T}$, and $\{x^{(t)}_{iJj}\}_{t\in T}$. With such an assignment of variables to processors, the algorithm can be shown to be distributed in the sense that a node exchanges information only with its neighbors at every iteration of the primal-dual algorithm.

In implementing the primal-dual algorithm, we must bear the following points in mind.

- The primal-dual algorithm in (5.16)–(5.18) is a continuous time algorithm. To discretize the algorithm, we consider time steps $m = 1, 2, \ldots$ and replace the derivatives by differences:

$$\begin{aligned} x_{iJj}^{(t)}[m+1] =& x_{iJj}^{(t)}[m] \\ & - \alpha_{iJj}^{(t)}[m] \left(\frac{\partial f_{iJ}(z'_{iJ}[m])}{\partial x_{iJj}^{(t)}[m]} + q_{ij}^{(t)}[m] - \lambda_{iJj}^{(t)}[m] \right), \\ p_i^{(t)}[m+1] =& p_i^{(t)}[m] + \beta_i^{(t)}[m](y_i^{(t)}[m] - \sigma_i^{(t)}), \\ \lambda_{iJj}^{(t)}[m+1] =& \lambda_{iJj}^{(t)}[m] + \gamma_{iJj}^{(t)}[m] \left(-x_{iJj}^{(t)}[m]\right)^+_{\lambda_{iJj}^{(t)}[m]}, \end{aligned}$$

  where

$$q_{ij}^{(t)}[m] := p_i^{(t)}[m] - p_j^{(t)}[m],$$
$$y_i^{(t)}[m] := \sum_{\{J|(i,J)\in\mathcal{A}\}} \sum_{j\in J} x_{iJj}^{(t)}[m] - \sum_{\{j|(j,I)\in\mathcal{A}, i\in I\}} x_{jIi}^{(t)}[m],$$

  and $\alpha_{iJj}^{(t)}[m] > 0$, $\beta_i^{(t)}[m] > 0$, and $\gamma_{iJj}^{(t)}[m] > 0$ can be thought of as step sizes.

- While the algorithm is guaranteed to converge to the optimal solution, the value of the variables at any time instant m is not necessarily a feasible solution. A start-up time is required before a feasible solution is computed.

Bandwidth-limited scenarios and medium access control

While constraints on feasible coding subgraphs are not needed in the energy-limited case, they are central to operation in bandwidth-limited scenarios, *e.g.* optimizing throughput or congestion. The set of feasible vectors of instantaneous and average rates of hyperarcs in $\mathcal{A}$ is determined by physical layer modulation and coding, channel characteristics, and constraints on transmit power. Given the modulation, coding and channel characteristics, medium access control allocates channels or power among interfering transmitters, determining the realizable coding subgraphs.

The number of possible coding subgraphs grows exponentially with the number of nodes and power levels. Although obviously inefficient subgraphs can be eliminated from consideration, optimal medium access in the bandwidth-limited case is generally very complex, even for centralized methods. The only known way to guarantee an optimal solution in general is to specify all feasible subgraphs and optimize over this set. Since this is computationally prohibitive, heuristics are used to reduce the number of subgraphs considered.

To illustrate two existing approaches, we consider the problem of supporting a set $\mathcal{C}$ of multicast sessions on a network, where each session $c \in \mathcal{C}$ consists of a source node $a_c \in \mathcal{N}$ at which exogenous data arrives with average rate R_c and is required to be multicast to a set $T_c \subset \mathcal{N}$ of sinks.

Iterative optimization. One approach, due to [Jain et al., 2003; Wu et al., 2005], starts with a set $\mathcal{Z}$ of K feasible coding subgraphs $z^{(1)}, \ldots, z^{(K)}$, called *elementary capacity graphs*, or ECGs for short. Each ECG is formed by adding unit rate hyperarcs using the following random greedy procedure. A typical communication range R and interference range $R' > R$ are chosen. At each step, a node i is chosen randomly from among those not within distance R' of any end nodes of hyperarcs previously added to the ECG. Let J be the set of nodes within distance R of i but not within distance R' of previously added transmitters. The hyperarc (i, J) is added if power can feasibly be allocated to support unit rate on (i, J) and each previously added hyperarc. This is repeated until no further hyperarcs can be added.

Given $\mathcal{Z} = \{z^{(k)}\}$, we solve the following linear optimization problem.

$$\begin{aligned}
\text{minimize } & \sum_{k=1}^{K} \lambda_k \\
\text{subject to } & \lambda_k \geq 0, \qquad \forall\, k = 1, \ldots, K, \\
& \sum_{j \in J} x_{iJj}^{(tc)} \leq y_{iJ}^{(c)}, \qquad \forall\, (i, J) \in \mathcal{A}, c \in \mathcal{C}\,, t \in T_c, \\
& \sum_{c \in \mathcal{C}} y_{iJ}^{(c)} \leq \sum_{k=1}^{K} \lambda_k z_{iJ}^{(k)}, \qquad \forall\, (i, J) \in \mathcal{A}, \\
& x^{(tc)} \in F^{(tc)}, \qquad \forall\, c \in \mathcal{C}, t \in T_c.
\end{aligned} \tag{5.19}$$

where $F^{(tc)}$ is the bounded polyhedron of points $x^{(tc)}$ forming a flow solution of rate R_c from source node a_c to sink t of session c.

If the solution satisfies $\sum_k \lambda_k \leq 1$, then variables λ_k specify a valid time-sharing among the ECGs in $\mathcal{Z}$ that supports the connection requirements. Otherwise, we carry out an iterative algorithm that alternates between heuristically modifying the set $\mathcal{Z}$ and solving the linear optimization problem (5.19). The set

$\mathcal{Z}$ is modified as follows. For some large integer Q, say 200, the new set $\mathcal{Z}$ is initialized with $\lceil \lambda_k Q \rceil$ copies of ECG $z^{(k)}$, for $k = 1, \ldots, K$. Based on the calculated flows $\{x^{(tc)}\}$, the algorithm removes from these ECGs any hyperarcs that are not needed, and shrinks the destination sets of remaining hyperarcs where possible. It then tries to remove one ECG at a time from $\mathcal{Z}$ by shifting all of its remaining hyperarcs into other ECGs. Each ECG removed reduces the objective function $\sum_k \lambda_k$ by $1/Q$. The iterative algorithm produces a monotonically non-increasing sequence of objective function values, and is carried out until the objective function cannot be further reduced. For more details, see [Jain et al., 2003; Wu et al., 2005].

Dynamic operation. Another approach, due to [Ho and Viswanathan, 2005], dynamically controls medium access, packet scheduling, routing and network coding, without using knowledge of long-term average rates. This algorithm extends to multicast a *back pressure* approach for multi-commodity flow in which routing and flow prioritization are locally determined by gradients in packet queue length. One difference is that the multicast network coding algorithm uses virtual queues to keep track of information intended for each sink. All control decisions are locally made, except for medium access control among interfering transmitters, which is guided by the virtual queue lengths. If all feasible subgraphs are considered, the algorithm stably supports any set of source rates stabilizable with intra-session network coding.

More precisely, we consider a dynamic network model with time slots of length τ_0. We assume that channel state is described by a vector $\underline{S}(\tau)$ that is constant over each time slot $[\tau, \tau + \tau_0)$, takes values from a finite set and is ergodic. Control decisions are made at most once a slot. For simplicity, we assume fixed length packets and link transmission rates that are restricted to integer multiples of the packet-length/time-slot quotient, *i.e.* an integer number of packets can be transmitted in each slot. In each time slot $[\tau, \tau + \tau_0)$, the coding subgraph z takes a value $z(\tau)$, determined by the medium access control policy, from a set $Z(\underline{S}(\tau))$ of feasible subgraphs that depends on the channel state vector $\underline{S}(\tau)$.

Each node i maintains a virtual queue Q_i^{ct} for each sink t of each session c, whose length is denoted U_i^{ct}. The queues are called virtual as the same actual data may be associated with more than one virtual queue. Exogenous session c data arriving at source node a_c undergoes random linear coding to produce coded packets at $(1 + \epsilon)$ times the exogenous arrival rate, which are then associated with queues $Q_{a_c}^{ct}, t \in T_c$. In each time slot $[\tau, \tau + \tau_0)$, the following are carried out:

- Scheduling: For each hyperarc (i, J), one session

$$c_{iJ}^* = \arg\max_{c \in \mathcal{C}} \left\{ \sum_{t \in T_c} \max \left(\max_{b \in J} \left(U_i^{ct} - U_b^{ct} \right), 0 \right) \right\}$$

  is chosen, and link weights

$$w_{iJ}^* = \sum_{t \in T_{c_{iJ}^*}} \max \left(\max_{b \in J} \left(U_i^{c_{iJ}^* t} - U_b^{c_{iJ}^* t} \right), 0 \right)$$

  are defined.

- Medium access control: The state $\underline{S}(\tau)$ is observed, and a coding subgraph

$$z(\tau) = \arg\max_{z \in Z(\underline{S}(\tau))} \sum_{(i,J) \in \mathcal{A}} w_{iJ}^* z_{iJ} \tag{5.20}$$

  is chosen.

- Network coding: For each hyperarc (i, J), a random linear combination of data corresponding to each (session, sink) pair $(c_{iJ}^*, t \in T_{c_{iJ}^*})$ for which $\max_{b \in J} \left(U_i^{c_{iJ}^* t} - U_b^{c_{iJ}^* t} \right) > 0$ is sent at the rate $z_{iJ}(\tau)$ determined by the medium access control, decreasing $U_i^{c_{iJ}^* t}$ by an amount $\min \left\{ U_i^{c_{iJ}^* t}(\tau) , \tau_0 z_{iJ}(\tau) \right\}$ and increasing $U_d^{c_{iJ}^* t}$ by the same amount, where $d = \arg\max_{b \in J} \left(U_i^{c_{iJ}^* t} - U_b^{c_{iJ}^* t} \right)$.

If the optimizations in the algorithm are done exactly, we have the following theorem.

THEOREM 5.3 *Let Λ be the set of all exogenous arrival rate vectors (R_c) such that the connection requirements can be stably supported by some control algorithm with full knowledge of future events. If $((1 + \epsilon)R_c)$ is strictly interior to Λ, the probability that not all sinks are able to decode their respective information decreases exponentially in the length of the code.*

In practice, the medium access control optimization (5.20) can be done heuristically using a greedy approach similar to that in the static case, but with the added guidance of weights w_{iJ}^* for prioritization among candidate hyperarcs (i, J).

5. Further directions and results

Network coding has come a long way in the last few years and it has been extended in many different ways. While we have focused in the previous sections on random coding and optimization problems for network coding in the wireless multicast scenario, it seems appropriate to shine a light also on other applications and research directions that have been inspired by this new and exciting techniques. Some of these advances reside most naturally in wireline networks and we will describe them in this context when so indicated. Nevertheless, almost all results can be without much effort be translated into a wireless context.

Network code construction

The multicast case. Random network coding is an extremely useful tool for multicast operation. However, any randomly chosen network solution does not come with *guarantees* or a certificate indicating that a "good" solution was found. Moreover, it may be possible to significantly reduce the complexity of the network coding itself by finding solution that operate in finite fields of minimal allowable size. Clearly, a network code that can operate over a binary field, *i.e.* utilizing only XOR operations on bits would be far easier to implement than finite field arithmetic over, say, $GF(2^{12})$. An efficient, *i.e.* running in time polynomial in the description of the network, algorithm to find network coding solutions with guaranteed performance was given by [Jaggi et al., 2005b]. The algorithm operates in a centrally controlled manner by first identifying a set of disjoint paths between the source and each of the individual receivers. In subsequent steps a network coding solution is iteratively grown starting from the source node to each of the intended receivers by considering one edge at the time. The encoding function for each edge is here chosen so as to guarantee that the information flowing on any cut in the network is as rich and different as possible.[1] Since all we have to guarantee in order to find a valid network coding solution is that random processes of maximal entropy have to reach the receivers in order to satisfy the multicast requirements this approach will be successful. This realization also led to the notion of a network code as a "linear dispersion" by [Li et al., 2005].[2] A distinguishing feature of the iterative process is that it is possible to keep the field size at its minimum (which often means binary operations). The construction algorithm, originally formulated for acyclic graphs, was later extended by [Erez and Feder, 2005] to the case of graphs with cycles so that we can now avail of efficient algorithms for practically all interesting scenarios. In particular, the extension to constructing wireless network coding solutions with guaranteed performance is straight forward once a set of source/sink flows is known for each receiver. For this purpose the algorithm of Section 4 can be effectively combined with the iterative encoding function choice of [Jaggi et al., 2005b]. Nevertheless, we want to emphasize that the added complexity of constructing a highly efficient network code is only justified for very stable and well known network topologies.[3] Wireless networks would typically operate very efficiently with random network coding.

The construction of network codes has spawned further research on low complexity solutions. A natural question is the effect of limiting the number of nodes or edges in a network that are capable of performing network coding operations (see [Bhattad et al., 2005]). However, it appears that these considerations are far more important in a wireline network where one might be confronted with, *e.g.* converting optical signals into electrical signals in order to perform network coding. The added cost of network coding in a wireless network are likely to be relatively minor when compared to the already necessary modulation and demodulation of RF signals.

The non-multicast case. Constructing codes for the non-multicast case is considerably more difficult than the corresponding task in the multicast setup. The main problem we are now facing is that network coding solutions have to make sure that the right information is delivered to each node. In particular, this renders random network coding as an almost useless tool. Indeed, at present we do not know of an efficient, structured approach to solve the network coding problem optimally. For linear network coding the algebraic characterization of valid solutions given by [Koetter and Médard, 2003] leads to an algorithm that, in principle, can decide if a given network problem has a solution or not. However, the running time of this algorithm is not bounded by a polynomial function in the size of the network.[4] The situation is exacerbated by the fact that linear network coding does not achieve the capacity of networks with arbitrary demands (see [Dougherty et al., 2005]). We find ourselves in a situation where, in practice, we have to resort to various ad-hoc solutions that can be used to demonstrate network coding advantages.

Ad hoc approaches to network coding

Finding optimal solutions to network coding problems involving more than one multicast session is a difficult problem and, at present, there is no practical approach known to achieve the *optimal* throughput in a network. Nevertheless, it is clear that this problem is of prime importance and a number of suboptimal solutions have been investigated. In many of these ad hoc schemes network coding operations are restricted to binary addition, which seems to realize a large portion of the possible gains. Opportunistic network coding in a wireless context has been described by [Katti et al., 2005] in a wireless 802.11 environment. This approach to network coding in a wireless network is motivated by the idea to start from a given wireless system and to opportunistically identify and exploit opportunities for improvement, such as the one mentioned in Section 1. The scheme proposed by [Katti et al., 2005] is characterized by an assumed knowledge of a node of a list of packets that its direct neighbors have. Moreover, network coding is used only if combining packets can give some immediate advantage for the reception of the direct neighbors of a node. The scheme was implemented in an 802.11 network and shows surprisingly large throughput advantages. For further details and for a quantification of the gains, see [Katti et al., 2005]. A fairly similar thinking has been applied by [Hausl et al., 2005] in the context of relay aided up/downlink improvement in a cellular network. Here a relay provides a sum of an uplink and a downlink package to both the mobile and the base station. Owing to the ability of mobile and base station to subtract its own packet from the relay signal, both can improve their reception link quality in a turbo coded system setup; see [Hausl et al., 2005] for further details.

Security aspects

While the main thrust of network coding research aims at increasing bandwidth and/or energy efficiency it also can greatly increase security of transmission. The main idea here is that in random network coded systems transmissions are typically mixtures of many portions of a set of data. Thus no individual packet reveals any information about the individual packets contributing to its makeup. Moreover, a wire-tapper or attacker cannot undo the random linear combination of the observed packets unless he/she has opportunity to observe enough information to retrieve the entire data. In other words, there exists a threshold behavior to the secrecy of the packets which is similar to more traditional wire-tapper or secret sharing schemes. The interested reader is referred to [Ho et al., 2004] for an application of network coding to Byzantine security and to [Jaggi et al., 2005a] for the situation where an attacker's resources are limited. Further references include [Cai and Yeung, 2002; Feldman et al., 2004; Bhattad and Narayanan, 2005].

Notes

1. Loosely speaking we try to maximize the degrees of freedom on each cut
2. Indeed, Jaggi et al.'s algorithm readily solves the problem of constructing a "linear dispersion" as well.
3. Network coding has, *e.g.* been proposed in a VLSI context to distribute multicast signals on a chip. Such a extremely stable environment seems predestined for a very careful construction of the encoding function.
4. In fact it has been shown by [Rasala Lehman and Lehman, 2003] that this question is NP-hard.

References

Ahlswede, Rudolf, Cai, Ning, Li, Shuo-Yen Robert, and Yeung, Raymond W. (2000). Network information flow. *IEEE Trans. Inform. Theory*, 46(4): 1204–1216.

Ahluwalia, Ashwinder, Modiano, Eytan, and Shu, Li (2002). On the complexity and distributed construction of energy-efficient broadcast trees in static ad hoc wireless networks. In *Proc. 2002 Conference on Information Sciences and Systems (CISS 2002)*.

Bertsekas, Dimitri P. (1980). A class of optimal routing algorithms for communication networks. In *Proc. 5th International Conference on Computers and Communication (ICCC '80)*, pages 71–76.

Bertsekas, Dimitri P., Gafni, Eli M., and Gallager, Robert G. (1984). Second derivative algorithms for minimum delay distributed routing in networks. *IEEE Trans. Commun.*, 32(8):911–919.

Bhattad, Kapil and Narayanan, Krishna R. (2005). Weakly secure network coding. In *Proc. WINMEE, RAWNET and NETCOD 2005 Workshops*.

Bhattad, Kapil, Ratnakar, Niranjan, Koetter, Ralf, and Narayanan, Krishna R. (2005). Minimal network coding for multicast. In *Proc. 2005 IEEE International Symposium on Information Theory (ISIT 2005)*, pages 1730–1734.

Cai, Ning and Yeung, Raymond W. (2002). Secure network coding. In *Proc. 2002 IEEE International Symposium on Information Theory (ISIT 2002)*, page 323.

Cruz, R. L. and Santhanam, Arvind V. (2003). Optimal routing, link scheduling and power control in multi-hop wireless networks. In *Proc. IEEE Infocom 2003*, volume 1, pages 702–711.

Deb, Supratim and Srikant, R. (2004). Congestion control for fair resource allocation in networks with multicast flows. *IEEE/ACM Trans. Networking*, 12(2):274–285.

Dougherty, Randall, Freiling, Christopher, and Zeger, Kenneth (2005). Insufficiency of linear coding in network information flow. In *Proc. 2005 IEEE International Symposium on Information Theory (ISIT 2005)*, pages 264–267.

Erez, Elona and Feder, Meir (2005). Convolutional network codes for cyclic networks. In *Proc. WINMEE, RAWNET and NETCOD 2005 Workshops*.

Feldman, Jon, Malkin, Tal, Servedio, Rocco A., and Stein, Cliff (2004). On the capacity of secure network coding. In *Proc. 42nd Annual Allerton Conference on Communication, Control, and Computing*.

Hausl, Christoph, Schreckenbach, Frank, Oikonomidis, Ioannis, and Bauch, Gerhard (2005). Iterative network and channel decoding on a tanner graph. In *Proc. 43rd Annual Allerton Conference on Communication, Control, and Computing*.

Ho, Tracey, Leong, Ben, Koetter, Ralf, Médard, Muriel, Effros, Michelle, and Karger, David R. (2004). Byzantine modification detection in multicast networks using randomized network coding. In *Proc. 2004 IEEE International Symposium on Information Theory (ISIT 2004)*, page 144, Chicago, IL.

Ho, Tracey, Médard, Muriel, Shi, Jun, Effros, Michelle, and Karger, David R. (2003). On randomized network coding. In *Proc. 41st Annual Allerton Conference on Communication, Control, and Computing*.

Ho, Tracey and Viswanathan, Harish (2005). Dynamic algorithms for multicast with intra-session network coding. In *Proc. 43rd Annual Allerton Conference on Communication, Control, and Computing*.

Jaggi, Sidharth, Langberg, Michael, Ho, Tracey, and Effros, Michelle (2005a). Correction of adversarial errors in networks. In *Proc. 2005 IEEE International Symposium on Information Theory (ISIT 2005)*, pages 1455–1459.

Jaggi, Sidharth, Sanders, Peter, Chou, Philip A., Effros, Michelle, Egner, Sebastian, Jain, Kamal, and Tolhuizen, Ludo M. G. M. (2005b). Polynomial time algorithms for multicast network code construction. *IEEE Trans. Inform. Theory*, 51(6):1973–1982.

Jain, Kamal, Padhye, Jitendra, Padmanabhan, Venkata N., and Qiu, Lili (2003). Impact of interference on multi-hop wireless network performance. In *MobiCom '03: Proc. 9th Annual International Conference on Mobile Computing and Networking*, pages 66–80.

Johansson, Mikael, Xiao, Lin, and Boyd, Stephen (2003). Simultaneous routing and power allocation in CDMA wireless data networks. In *Proc. 2003 IEEE International Conference on Communications (ICC 2003)*, volume 1, pages 51–55.

Katti, Sachin, Katabi, Dina, Hu, Wenjun, Rahul, Hariharan, and Médard, Muriel (2005). The importance of being opportunistic: Practical network coding for wireless environments. In *Proc. 43rd Annual Allerton Conference on Communication, Control, and Computing*.

Kodialam, Murali and Nandagopal, Thyaga (2005). Characterizing achievable rates in multi-hop wireless mesh networks with orthogonal channels. *IEEE/ACM Trans. Networking*, 13(4):868–880.

Koetter, Ralf and Médard, Muriel (2003). An algebraic approach to network coding. *IEEE/ACM Trans. Networking*, 11(5):782–795.

Li, Shuo-Yen Robert, Cai, Ning, and Yeung, Raymond W. (2005). On theory of linear network coding. In *Proc. 2005 IEEE International Symposium on Information Theory (ISIT 2005)*, pages 273–277.

Liang, Weifa (2002). Constructing minimum-energy broadcast trees in wireless ad hoc networks. In *Proc. 3rd ACM International Symposium on Mobile Ad Hoc Networking & Computing (MOBIHOC '02)*, pages 112–122.

Lun, Desmond S., Médard, Muriel, and Koetter, Ralf (2005a). Efficient operation of wireless packet networks using network coding. In *Proc. International Workshop on Convergent Technologies (IWCT) 2005*. Invited paper.

Lun, Desmond S., Médard, Muriel, Koetter, Ralf, and Effros, Michelle (2005b). Further results on coding for reliable communication over packet networks. In *Proc. 2005 IEEE International Symposium on Information Theory (ISIT 2005)*, pages 1848–1852.

Lun, Desmond S., Ratnakar, Niranjan, Koetter, Ralf, Médard, Muriel, Ahmed, Ebad, and Lee, Hyunjoo (2005c). Achieving minimum-cost multicast:

A decentralized approach based on network coding. In *Proc. IEEE Infocom 2005*, volume 3, pages 1608–1617, Miami, FL.

Motwani, Rajeev and Raghavan, Prabhakar (1995). *Randomized Algorithms*. Cambridge University Press, Cambridge.

Ouorou, A., Mahey, P., and Vial, J.-Ph. (2000). A survey of algorithms for convex multicommodity flow problems. *Manage. Sci.*, 46(1):126–147.

Rasala Lehman, April and Lehman, Eric (2003). Complexity classification of network information flow problems. In *Proc. 41st Annual Allerton Conference on Communication, Control, and Computing*.

Sherali, Hanif D. and Choi, Gyunghyun (1996). Recovery of primal solutions when using subgradient optimization methods to solve Lagrangian duals of linear programs. *Oper. Res. Lett.*, 19:105–113.

Wieselthier, Jeffrey E., Nguyen, Gam D., and Ephremides, Anthony (2002). Energy-efficient broadcast and multicast trees in wireless networks. *Mobile Networks and Applications*, 7:481–492.

Wu, Yunnan, Chou, Philip A., Zhang, Qian, Jain, Kamal, Zhu, Wenwu, and Kung, Sun-Yuan (2005). Network planning in wireless ad hoc networks: A cross-layer approach. *IEEE J. Select. Areas Commun.*, 23(1):136–150.

Xiao, Lin, Johansson, Mikael, and Boyd, Stephen (2004). Simultaneous routing and resource allocation via dual decomposition. *IEEE Trans. Commun.*, 52(7):1136–1144.

Chapter 6

COOPERATIVE DIVERSITY

Models, Algorithms, and Architectures

J. Nicholas Laneman
Department of Electrical Engineering
University of Notre Dame
jnl@ieee.org

Abstract: Cooperative diveristy allows a collection of radio terminals that relay signals for each other to emulate an antenna array and exploit spatial diversity in wireless fading channels. For a variety of processing algorithms and transmission protocols, performance improvements in terms of transmission rate and reliability have been demonstrated. This chapter summarizes some low-complexity processing algorithms and transmission protocols for cooperative diversity, summarizes performance predictions in terms of information-theoretic outage probability, and how such constructs can be integrated into wireless network architectures, both existing and new.

Keywords: relay channel, outage probability, spatial diversity, layered architectures, cross-layer design.

1. Introduction

In wireless communication systems, the physics of electromagnetic waves lead to multipath propagation of wireless signals and, in turn, variations in received signal strength as a function of transciever location and frequency. Combined with transciever motion, these effects produce wireless channel variations, generally called *fading*, in space, frequency, and time. Diversity techniques for mitigating, and even exploiting, multipath fading are central to improving the performance of wireless communication systems and networks.

Frank H. P. Fitzek and Marcos D. Katz (eds.), Cooperation in Wireless Networks: Principles and Applications, 163–187.

As such they are prevalent, in one form or another, in all modern wireless systems. Indeed, numerous forms of time diversity, frequency diversity, and spatial diversity are leveraged in modern systems [Proakis, 2001; Rappaport, 2002].

Cooperative diversity [Laneman et al., 2004; Laneman and Wornell, 2003] is a relatively new class of spatial diversity techniques that is enabled by relaying [van der Meulen, 1971; Cover and El Gamal, 1979] and cooperative communications [Sendonaris et al., 2003a; Sendonaris et al., 2003b] more generally. As illustrated by other chapters of this book, cooperative communications refers to scenarios in which distributed radios interact to jointly transmit information in wireless environments. Cooperative diversity results when cooperative communications is used primarily to leverage the spatial diversity available among distributed radios. The main motivation here is to improve the reliability of communications in terms of, for example, outage probability, or symbol- or bit-error probability, for a given transmission rate. By contrast, as discussed in other chapters, cooperative communications can also be used primarily to increase the transmission rate. In both cases, cooperation allows for tradeoffs between target performance and required transmitted power, and thus provides additional design options for energy-efficient wireless networks.

The remainder of this chapter is organized as follows. Section 2 summarizes the simplest model within which cooperative diversity is well motivated, namely, wireless network environments with limited time and frequency diversity. Some low-complexity algorithms are defined and evaluated in terms of information-theoretic outage probability. Section 3 describes, at a high level, how such algorithms can be integrated into existing wireless network architectures such as infrastructure (cellular and wireless local area network (WLAN)) and ad hoc networks with cluters. Finally, Section 4 concludes the chapter with some discussion and future directions for research and development.

2. Elements of Cooperative Diversity

In this section, we summarize some of the main elements of cooperative diversity protocols and illustrate some of their performance advantages. To illustrate the issues associated with cooperative communications, we consider a single source, two relays, and a single destination as shown in Figure 6.1. Generalizations to multi-source, and multi-stage cooperation have also been considered by a variety of authors; see, for example, [Gupta and Kumar, 2003; Xie and Kumar, 2004; Kramer et al., 2005; Bölcskei et al., 2004].

Cooperative communications exploits the broadcast nature of the wireless medium and allows radios to jointly transmit information through relaying. As illustrated in Figure 6.1(a), the two relays can receive signals resulting from the source transmission, suitably process those received signals, and transmit signals of their own so as to increase the capacity and/or improve reliability of end-to-end transmissions between the source and destination radios. Figure 6.1(b)

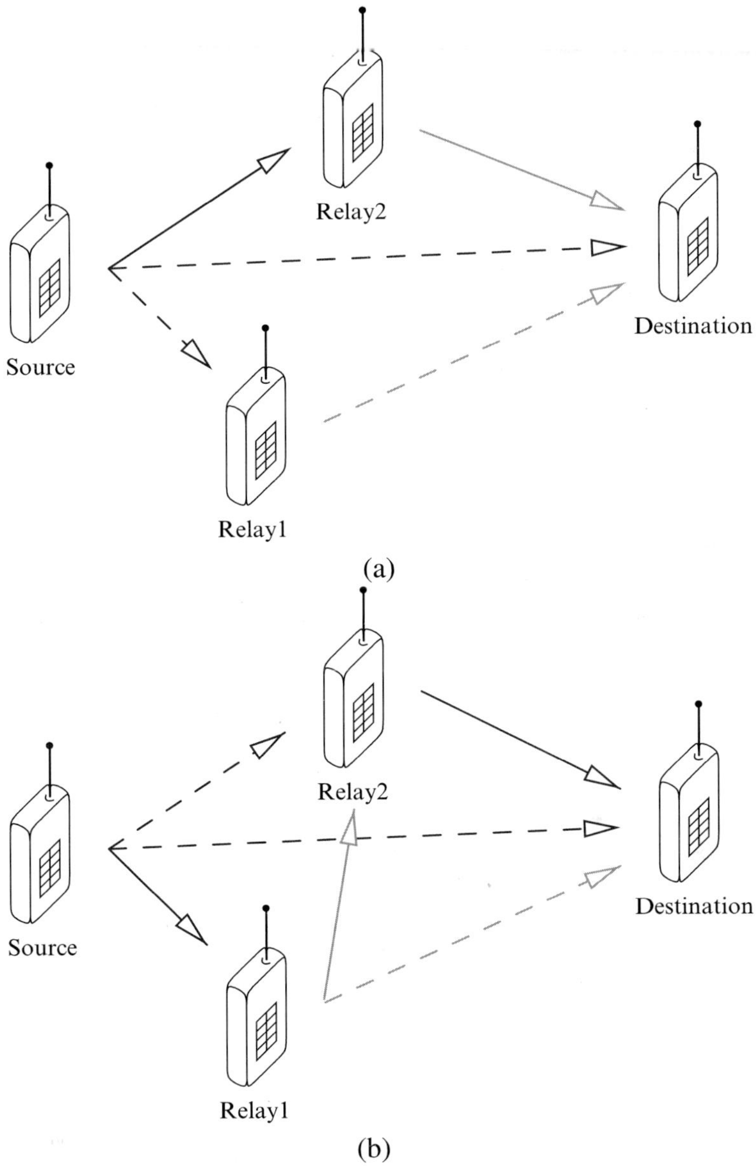

Figure 6.1. Illustration of cooperative wireless transmission with (a) parallel relays and (b) serial relays. Colors indicate transmissions that occur in different time slots or frequency bands. Solid arrows indicate transmissions that are utilized in traditional multihop transmission. Cooperative communications utilizes transmissions corresponding to both solid and dashed arrows by having the appropriate receivers perform some form of combining of their respective incoming signals.

illustrates that relaying can be performed in multiple stages so that relays as well as the destination benefit from spatial diversity. As we will see, among other potential benefits, cooperative communications leverages the spatial diversity available when multiple transmissions experience fading and/or shadowing that is essentially independent. For example, if the source signal experiences a deep fade at the destination, there remains a significant chance that it can be effectively communicated to the destination via one of the relays.

Because cooperative communications is inherently a network problem, issues of protocol layering and cross-layer architectures naturally arise. Starting as low as the physical layer, encoding and signal processing algorithms are required in at the source(s) and relay(s), and signal processing and decoding algorithms are required at the destination(s). However, such issues can be addressed as part of link layer coding and retransmissions, such as automatic repeat request (ARQ). Organizing the schedule for transmissions in time and frequency must be addressed by protocols in the link layer and medium-access control sublayer in coordination with the physical layer. Synchronization of signals in terms of carrier, symbol, and frame synchronization is also particularly important at the physical and link layers. Finally, collecting sets of radios into cooperating groups is inherently a cross-layer issue that can involve the physical, medium-access control, link, and even network layers. Designing an effective cooperative communication system requires insights about all of these issues. As we will further elaborate, the right combination of architecture (what logical components are identified and how they can interact) and algorithms (specific signal encoding, processing, and decoding techniques) can depend upon the application context, radio hardware, and complexity of the system.

Background

Early formulations of general relaying problems appeared in the information theory community [van der Meulen, 1968; van der Meulen, 1971; Cover and El Gamal, 1979] and were inspired by the concurrent development of the ALOHA system at the University of Hawaii. The relay channel model is comprised of three terminals: a source that transmits information, a destination that receives information, and a relay that both receives and transmits information in order to enhance communication between the source and destination. More recently, models with multiple relays have been examined [Schein and Gallager, 2000; Schein, 2001; Gupta and Kumar, 2003; Xie and Kumar, 2004; Khojastepour et al., 2004; Kramer et al., 2005]. Cooperative communications [Sendonaris et al., 2003a; Sendonaris et al., 2003b] is a generalization of the relay channel to multiple sources with information to transmit that also serve as relays for each other. Combinations of relaying and cooperation are also possible, and are often referred to generically as "cooperative communications". Less well know is the fact that all of these models fall within the broader class of channels with "generalized feedback" [King, 1978; Carleial, 1982; Willems, 1982; Willems et al., 1983].

Although problems of relaying and cooperation have been examined in the information theory community for years, the fundamental performance limits, in terms of the Shannon capacity or capacity region, are not known in general. Nevertheless, useful bounds on capacity have been obtained for various approaches. When applied to wireless channel models in particular, relaying and cooperation can be shown to offer significant performance enhancements in terms of various performance metrics, including: increased capacity (or larger capacity region) [Sendonaris et al., 2003a; Kramer et al., 2005; Host-Madsen and Zhang, 2005; Host-Madsen, 2004]; improved reliability in terms of diversity gain [Laneman et al., 2004; Laneman and Wornell, 2003; Nabar et al., 2003; Mitran et al., 2005], diversity-multiplexing tradeoff [Laneman et al., 2004; Laneman and Wornell, 2003; Azarian et al., 2005], and bit- or symbol-error probabilities [Laneman and Wornell, 2000; Sendonaris et al., 2003b; Ribeiro et al., 2005; Chen and Laneman, 2005]. These modern perspectives on and applications of relaying and cooperation have generated considerable research activity on relaying and cooperation within the communications, signal processing, and networking communities, and renewed interest within the information theory community.

System Model

This section summarizes a simple model for cooperative diversity within which later sections describe algorithms, characterize performance, and suggest network architectures. More details on system modeling are available in, for example, [Laneman, 2002; Laneman et al., 2004; Laneman and Wornell, 2003; Kramer, 2004].

Cooperative diversity is motivated by a need to mitigate wireless channel effects resulting from slowly-time varying, frequency non-selective multipath fading, large-scale shadowing, and path-loss. One cost of employing relays in practical systems is that current radios cannot transmit and receive simultaneously in the same frequency band, *i.e.* they must operate in *half-duplex* mode. In addition to power constraints, half-duplex constraints must be an integral part of the model.

More specifically, consider a network with $t \geq 2$ radios. Each radio has a baseband-equivalent, discrete-time transmit signal $X_i[k]$, with average power constraint $\sum_{k=1}^{n} |X_i[k]|^2 \leq nP_i$, and receive signal $Y_i[k]$, $i = 1, 2, \ldots, t$. Incorporating the half-duplex constraints, we model the receive signal at radio i and time sample k as [Kramer, 2004]

$$Y_i[k] = \begin{cases} \sum_{j \neq i}^{t} A_{i,j} X_j[k] + Z_i[k] \,, & \text{if radio } i \text{ receives at time } k \\ 0 \,, & \text{if radio } i \text{ transmits at time } k \end{cases}, \tag{6.1}$$

where $A_{i,j}$ captures the combined effects of frequency-nonselective, quasi-static multipath fading, shadowing, and path-loss between radios i and j, and $Z_i[k]$ captures the thermal noise and other interference received at radio i. Note that $A_{i,j}$ is assumed to be fixed throughout the transmission blocklength. These effects are captured in the simplest settings possible to isolate the benefits of spatial diversity.

Motivated by quasi-static conditions, we consider the scenario in which the coefficients $A_{i,j}$ are known to, *i.e.* accurately measured by, the appropriate receivers, but not fully known to, or not exploited by, the transmitters. That is, radio i knows the realized $A_{i,j}$ but not $A_{i',j}$, for $i' \neq i$, and $j = 1, 2, \ldots, t$. Statistically, we model $A_{i,j}$ as independent complex-valued random variables, which is reasonable for scenarios in which the radios are separated by a number of carrier wavelengths. Furthermore, we model $Z_i[n]$ as zero-mean mutually independent, white circularly-symmetric, complex Gaussian random sequences with common variance N_0.

We expect that some level of synchronization between the terminals is required for cooperative diversity to be effective. For purposes of exposition, we consider the scenario in which the terminals are block, carrier, and symbol synchronous. Given some form of network block synchronization, carrier and symbol synchronization for the network can build upon the same between the individual transmitters and receivers. Although a discussion of how synchronization is achieved is beyond our scope, we note that much of the performance benefits of cooperative diversity appear to be robust to small symbol synchronization errors [Wei et al., 2005] and lack of carrier phase synchronizatin [Chen and Laneman, 2005].

Example Relaying Algorithms

General relaying allows sophisticated joint encoding in the transmit signals of the source(s) and relay(s) as well as intricate processing and decoding of the received signals at the relay(s) and destination(s). Since a growing number of relaying algorithms are appearing in the literature, we summarize only a few simple algorithms for illustration.

Amplify-and-Forward. For amplify-and-forward, relays simply amplify what they receive subject to their power constraint. Amplifying corresponds to a linear transformation at the relay.

Consider first the case of a single relay. The simplest algorithm described below divides transmissions into two blocks of equal duration, one block for the source transmission and one block for the relay transmission; more elaborate amplify-and-forward algorithms, as well as more general linear relaying schemes, have been considered in [Nabar et al., 2003; Jing and Hassibi, 2004].

For the simplest algorithm, the source transmits $X_s[k]$ for $k = 1, 2, \ldots, n$. The relay processes its corresponding received signal $Y_r[k]$ for $k = 1, 2, \ldots, n$, and relays the information by transmitting

$$X_r[k] = \beta_r Y_r[k-n], \quad k = n+1, n+2, \ldots, 2n. \tag{6.2}$$

To remain within its power constraint, an amplifying relay must use gain

$$\beta_r \leq \sqrt{\frac{P_s}{|A_{r,s}|^2 P_r + N_0}}, \tag{6.3}$$

where the gain is allowed to depend upon the fading coefficient $A_{r,s}$ between the source and relay. The destination processes its received signal $Y_d[k]$ for $k = 1, 2, \ldots, 2n$ by some form of diversity combining of the two subblocks of length n.

When multiple relays are active, they can each relay in their own block of channel uses so that their transmissions do not interfere at the destination, or they can relay simultaneously so that their transmissions interfere at the destination. The former approach offers better diversity benefits, but decreases bandwidth efficiency.

Decode-and-Forward. For decode-and-forward, relays apply some form of detection and/or decoding algorithms to their received signals and re-encode the information into their transmit signals. This decoding and re-encoding process often corresponds to a non-linear transformation of the received signals. Although decoding at the relays has the advantages of reducing the impact of receiver noise, as we will see, it can limit performance because of the incoming fading effects.

Again, consider first the case of a single relay. The simplest algorithm described below again divides transmissions into two blocks of equal duration, one block for the source transmission and one block for the relay transmission; more elaborate decode-and-forward algorithms have been considered in [Azarian et al., 2005; Mitran et al., 2005].

For the simplest algorithm, the source transmits $X_s[k]$ for $k = 1, 2, \ldots, n$. The relay forms an estimate $\hat{X}_s[k]$ by decoding its corresponding received signal $Y_r[k]$ for $k = 1, 2, \ldots, n$, and relays a re-encoded version of $\hat{X}_s[k]$. For example, the relay can implement repetition coding by transmitting the signal

$$X_r[k] = \sqrt{\frac{P_r}{P_s}} \hat{X}_s[k-n], \quad k = n+1, n+2, \ldots, 2n. \tag{6.4}$$

Again, the destination processes its received signal $Y_d[k]$ for $k = 1, 2, \ldots, 2n$ by some form of diversity combining of the two subblocks of length n.

Instead of repetition coding, the relay can encode the source message using a codeword that is generally correlated with, by not necessarily identical to, the source codeword. Within the context of the simple algorithms, this corresponds to a form of parallel channel coding. When multiple relays are involved, they can all employ repetition coding or a more general space-time code to transmit information jointly with the source to the destination. Like amplify-and-forward, repetition coding in separate blocks has the advantage of low complexity, but the disadvantages of scheduling and low spectral efficiency.

Selection and dynamic relaying. As we might expect, fixed decode-and-forward is limited by direct transmission between the source and relay. However, since the fading coefficients are known to the appropriate receivers, $A_{r,s}$ can be measured to high accuracy by the cooperating terminals; thus, they can adapt their transmission format according to the realized value of $A_{r,s}$.

This observation suggests the following class of selection relaying algorithms. If the measured $|A_{r,s}|^2$ falls below a certain threshold, the source simply continues its transmission to the destination, in the form of repetition or more powerful codes. If the measured $|A_{r,s}|^2$ lies above the threshold, the relay forwards what it received from the source, using either amplify-and-forward or decode-and-forward, in an attempt to achieve diversity gain.

Informally speaking, selection relaying of this form should offer diversity because, in either case, two of the fading coefficients must be small in order for the information to be lost. Specifically, if $|A_{r,s}|^2$ is small, then $|A_{d,s}|^2$ must also be small for the information to be lost when the source continues its transmission. Similarly, if $|A_{r,s}|^2$ is large, then both $|A_{d,s}|^2$ and $|A_{d,r}|^2$ must be small for the information to be lost when the relay employs amplify-and-forward or decode-and-forward. We formalize this notion when we consider outage performance of selection relaying in Section 2.0.

A further improvement of decode-and-forward is dynamic decode-and-forward [Azarian et al., 2005; Mitran et al., 2005]. In dynamic decode-and-forward, the relay starts by receiving from the source and does not begin transmitting until it is sure it has correctly received the source transmission. Because of quasi-static conditions, the reception time at the relay can be modeled as a random variable, and the coding scheme must take this into account.

Incremental relaying. Fixed and selection relaying can make inefficient use of the degrees of freedom of the channel, especially for high rates, because the relays repeat all the time. In this section, we describe incremental relaying protocols that exploit limited feedback from the destination terminal, *e.g.* a single bit indicating the success or failure of the direct transmission. These incremental relaying protocols can be viewed as extensions of incremental redundancy, or hybrid automatic-repeat-request (ARQ), to the relay context. In ARQ, the source retransmits if the destination provides a negative acknowledgment via feedback; in incremental relaying, the relay retransmits in an attempt to exploit spatial diversity.

As one example, consider the following protocol utilizing feedback and amplify-and-forward transmission. First, the source transmits its information to the destination. The destination indicates success or failure by broadcasting a single bit of feedback to the source and relay, which we assume is detected reliably by at least the relay. If the source-destination signal-to-noise ratio (SNR) is sufficiently high, the feedback indicates success of the direct transmission, and the relay does nothing. If the source-destination SNR is not sufficiently high for successful direct transmission, the feedback requests that the relay amplify-and-forward what it received from the source. In the latter case, the destination tries to combine the two transmissions. Protocols of this form make more efficient use of the degrees of freedom of the channel, because they repeat rarely, and only when necessary.

Performance Benefits

Having described some basic relaying algorithms, we now turn to illustrating their performance. To evaluate algorithms, we utilize outage probability [Ozarow et al., 1994] as a metric throughout. Because the channels are quasi-static, channel mutual informations become random variables as functions of the fading coefficients. The outage probability is then the probability that a mutual information random variable falls below some fixed rate chosen *a priori*. There are several advantages to such an information-theoretic treatment of the problem, including abstracting away many of the details of the channel coding and decoding algorithms as well as accounting for the decreased spectral efficiency required by half-duplex operation in the relays. We note that results similar to those obtained for outage probability can also be obtained for symbol- and bit-error rates of uncoded cooperative modulation and demodulation; see, *e.g.* [Laneman and Wornell, 2000; Ribeiro et al., 2005; Chen and Laneman, 2005].

Non-Cooperative transmission. To be more precise, and for comparison with the results to follow, let us compute the outage probability of a system without cooperative diversity in the model (6.1).

Consider non-cooperative transmission from radio s to radio d. In this case, the mutual information random variable, in bits per channel use, viewed as a function of the fading coefficient $A_{d,s}$, satisfies [Cover and Thomas, 1991; Telatar, 1999]

$$I_{\mathrm{NC}} \leq \log\left(1 + \frac{|A_{d,s}|^2 P_s}{N_0}\right), \tag{6.5}$$

with equality achieved for independent, identically distributed complex circular Gaussian inputs with zero-mean and variance P_s. The outage probability for rate R, in bits per channel use, is then given by [Ozarow et al., 1994]

$$
\begin{aligned}
p_{\text{out}}^{\text{NC}} :=& \Pr[I_{\text{NC}} \le R] \\
=& \Pr\left[|A_{d,s}|^2 \le \frac{2^R - 1}{(P_s/N_0)} \right].
\end{aligned}
$$

Note that if radios s and r transmit and receive, respectively, in only k out of the n channel uses, the mutual information random variable becomes

$$I_{\text{NC}} = \frac{k}{n} \log\left(1 + \frac{n}{k}\frac{|A_{d,s}|^2 P_s}{N_0}\right).$$

Because of the reduced number of channel uses, radio s can increase its transmitted power per channel use and remain within its average power constraint for the entire block.

Cooperative transmission. Outage results for cooperative transmission can be obtained by extending similar results for multiple-input, multiple-output (MIMO) systems [Telatar, 1999].

The simplest amplify-and-forward algorithm for a single source and relay produces an equivalent one-input, two-output conditionall complex Gaussian noise channel with different noise levels in the outputs. As [Laneman et al., 2004] details, the mutual information random variable between the input and the two outputs is

$$I_{\text{AF}} = \frac{1}{2}\log\left(1 + 2\frac{|A_{d,s}|^2 P_s}{N_0} + f\left(2\frac{|A_{r,s}|^2 P_s}{N_0}, 2\frac{|A_{d,r}|^2 P_r}{N_0}\right)\right) \tag{6.6}$$

as a function of the fading coefficients, where

$$f(x,y) := \frac{xy}{x+y+1}. \tag{6.7}$$

We note that (6.6) is achieved by i.i.d. complex Gaussian inputs, and that the amplifier gain β_r does not appear in (6.6), because the constraint (6.3) is met with equality.

For the simplest decode-and-forward algorithm with repetition coding, the mutual information random variable is

$$I_{\text{RDF}} = \frac{1}{2}\min\left\{\log\left(1+2\frac{|A_{r,s}|^2 P_s}{N_0}\right), \log\left(1+2\frac{|A_{d,s}|^2 P_s}{N_0}+2\frac{|A_{d,r}|^2 P_r}{N_0}\right)\right\}, \tag{6.8}$$

and is achieved by i.i.d. zero-mean complex Gaussian inputs. If parallel channel coding is used instead of repetition coding, then the mutual information random variable for independent source and relay codebooks is

$$I_{\text{PDF}} = \frac{1}{2}\min\left\{\log\left(1+2\frac{|A_{r,s}|^2 P_s}{N_0}\right), \log\left(1+2\frac{|A_{d,s}|^2 P_s}{N_0}\right) + \log\left(1+2\frac{|A_{d,r}|^2 P_r}{N_0}\right)\right\}. \tag{6.9}$$

It is the sum of the signal-to-noise ratio random variables $|A_{i,j}|^2 P_j/N_0$ in (6.6) and (6.8) that can lead to diversity gains when compared to (6.5). Since parallel channel coding is superior to repetition coding, (6.9) can also offer diversity gains. However, in either case, the source-relay link can limit performance.

By contrast, selection decode-and-forward can be shown to offer *full diversity* regardless of the quality of the source-relay link. For the simplest selection decode-and-forward algorithm with repetition coding, the mutual information random variable is

$$I_{\text{SRDF}} = \begin{cases} \frac{1}{2}\log\left(1+2\frac{|A_{d,s}|^2 P_s}{N_0}\right), & \text{if } \frac{1}{2}\log\left(1+2\frac{|A_{r,s}|^2 P_s}{N_0}\right) \leq R \\ \frac{1}{2}\log\left(1+2\frac{|A_{d,s}|^2 P_s}{N_0} + 2\frac{|A_{d,r}|^2 P_r}{N_0}\right), & \text{if } \frac{1}{2}\log\left(1+2\frac{|A_{r,s}|^2 P_s}{N_0}\right) > R \end{cases} \tag{6.10}$$

If parallel channel coding is used instead of repetition coding, then the mutual information random variable for independent source and relay codebooks is

$$I_{\text{SPDF}} = \begin{cases} \frac{1}{2}\log\left(1+2\frac{|A_{d,s}|^2 P_s}{N_0}\right), & \text{if } \frac{1}{2}\log\left(1+2\frac{|A_{r,s}|^2 P_s}{N_0}\right) \leq R \\ \frac{1}{2}\log\left(1+2\frac{|A_{d,s}|^2 P_s}{N_0}\right) + \log(\left(1+2\frac{|A_{d,r}|^2 P_r}{N_0}\right), & \text{if } \frac{1}{2}\log\left(1+2\frac{|A_{r,s}|^2 P_s}{N_0}\right) > R \end{cases} \tag{6.11}$$

Figure 6.2 illustrates example outage performance for non-cooperative transmission and cooperative transmission with up to two relays. The outage probabilities corresponding to the mutual informations for no cooperation (6.5), amplify-and-forward (6.6), repetition decode-and-forward (6.10), and parallel/space-time decode-and-forward (6.11) are shown. We observe from Figure 6.2 that cooperation increases the diversity order, *i.e.* the negative slope of a plot of log-outage vs. SNR in dB, and provides full spatial diversity in the number of cooperating nodes, *i.e.* the source plus the number of relays. Although the two forms of decode-and-forward have similar performance for the case of one relay for the particular network geometry, path-loss exponent, and spectral efficiency considered, for two relays the advantages of parallel/space-time decode-and-forward are apparent in Figure 6.2. More general analytical results in this direction are available in, *e.g.* [Laneman et al., 2004; Laneman and Wornell, 2003].

3. Cooperative Diversity in Existing Network Architectures

In this section, we suggest some simple ways for integrating cooperative diversity into existing network architectures such as infrastructure networks and clustered ad hoc networks.

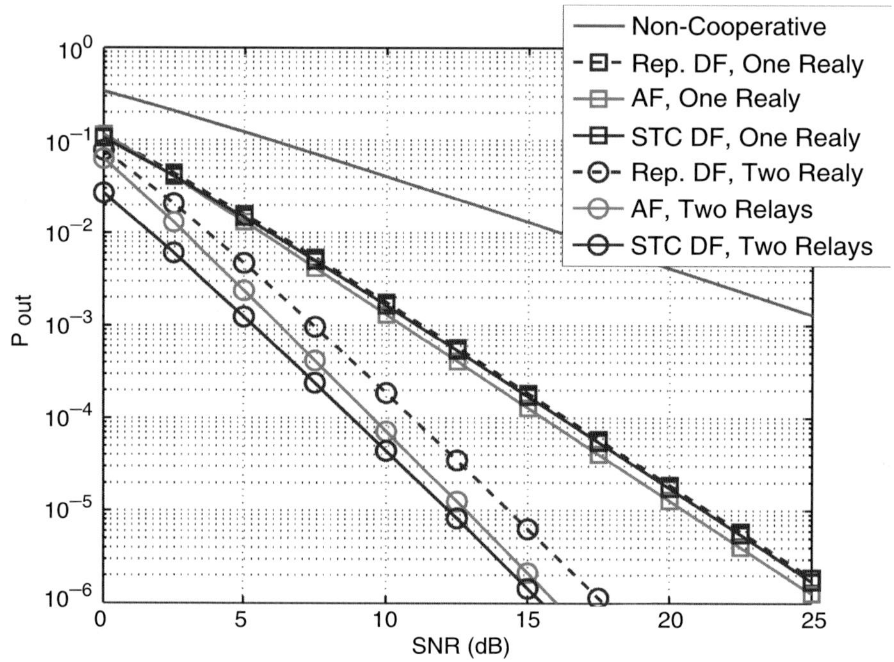

Figure 6.2. Example outage performance of non-cooperative and cooperative transmission computed via Monte Carlo simulations. The model has: path-loss with exponent $\alpha = 3$, independent Rayleigh fading, network geometry with relays located at the midpoint between the source and destination, spectral efficiency $R = 1/2$, and uniform power allocation.

Centralized Partitioning for Infrastructure Networks

Our focus in this section is on infrastructure networks, in which all terminals communicate through an access point (AP). In such scenarios, the AP can gather information about the state of the network, *e.g.* the path-losses among terminals, select a cooperative mode based upon some network performance criterion, and feed back its decision on the appropriate control channels. Here cooperative diversity lives across the medium-access control, and physical layers; routing is not considered.

Matching Algorithms. In this section, we consider grouping terminals into cooperating *pairs*. Additional studies of grouping algorithms appear in [Hunter and Nosratinia, 2006; Lin et al., 2006]. As we will see, choosing pairs of co-operating terminals is an instance of a more general set of problems known as *matching* problems on graphs [Rosen, 2000]. To outline the general matching framework, let $\mathcal{G} = (\mathcal{V}, \mathcal{E})$ be a graph, with $\mathcal{V}$ a set of vertices and $\mathcal{E} \subseteq \mathcal{V} \times \mathcal{V}$ a set of edges between vertices. A subset $\mathcal{M}$ of $\mathcal{E}$ is called a *matching* if edges in $\mathcal{M}$ are pairwise disjoint, *i.e.* no two edges in $\mathcal{M}$ are incident on the same vertex. Note that

$$|\mathcal{M}| \leq \lfloor |\mathcal{V}|/2 \rfloor, \tag{6.12}$$

where $|\mathcal{M}|$ is the cardinality of the set $\mathcal{M}$ and $\lfloor x \rfloor$ denotes the usual floor function. When the bound (6.12) is achieved with equality, the matching is called a *perfect matching*. Since we will be working with *complete* graphs, *i.e.* there is an edge between each pair of vertices, there will always be a perfect matching for $|\mathcal{V}|$ even. As a result, we will not be concerned with so-called *maximal matching* problems.

Instead, we focus on *weighted matching* problems. Given an edge e in $\mathcal{E}$, the *weight* of the edge is some real number $w(e)$. Given a subset $\mathcal{S}$ of $\mathcal{E}$, we denote its sum weight by

$$w(\mathcal{S}) = \sum_{e \in \mathcal{S}} w(e). \tag{6.13}$$

The *minimal weighted matching* problem is to find a matching $\mathcal{M}$ of minimal weight [Rosen, 2000]. We also consider two other matching algorithms, both based upon randomization, that approximate minimal weighted matching and offer lower complexity.

Specifically, we consider the following algorithms:

- **Minimal Weighted Matching:** Since algorithms for implementing minimal weighted matching are well-studied and readily available [Ahuja et al., 1993; Rosen, 2000], we do not go into their details. We note, however, that more recent algorithms for minimal weighted matching have complexity $O(|\mathcal{V}|^3)$ [Rosen, 2000].

- **Greedy Matching:** To reduce complexity and approximate minimal weighted matching, we consider a greedy algorithm in which we randomly select a free vertex v and match it with another free vertex v' such that the edge $e = (v, v')$ has minimal weight. The process continues until all of the vertices have been matched. Since each step of the algorithm takes at most $|\mathcal{V}|$ comparisons, and there are $|\mathcal{V}|/2$ steps, the complexity of this algorithm is $O(|\mathcal{V}|^2)$. We note that this greedy algorithm need not be optimal for this order of complexity.

- **Random Matching:** To reduce complexity still further, we consider a random matching algorithm where we pair vertices randomly. The complexity of this algorithm is $O(|\mathcal{V}|)$.

In addition to the algorithms outlined above for matching cooperating terminals, there are a variety of other possibilities. Instead of the general weighted matching approach, we can randomly partition the terminals into two sets and utilize *bipartite weighted matching* algorithms, which have slightly lower complexity (in terms of their coefficients, not order) and are conceptually simpler to implement than general weighted matching algorithms [Rosen, 2000]. Another possibility is to again randomly partition the terminals into two sets and utilize *stable marriage* algorithms with still lower complexity $O(|\mathcal{V}|^2)$. Such algorithms may be suitable for decentralized implementation [Ahuja et al., 1993].

Example performance. Figure 6.3 shows a set of example results from the various matching algorithms described above. Terminals are independently and uniformly distributed in a square of side 2000 m, with the basestation/access point located in the center of the square. Variances for Rayliegh fading are computed using a $d^{-\alpha}$ path-loss model, with $\alpha = 3$. The weight of an edge $e = (v, v')$ is the average of the outage probabilities for terminal v using v' as a relay, and vice versa. In particular, we utilize the amplify-and-forward result (6.6) for this example; more generally, we can employ any of the outage probability expressions for a pair of cooperating terminals. Each set of results is averaged over 100 trial networks with the various matching algorithms applied. The results are normalized so that the performance of non-cooperative transmission is the same in each trial, *i.e.* the received SNR for direct transmission averaged over all the terminals in the network is normalized to be the horizontal axis in Figure 6.3.

We note several features of the results in Figure 6.3. First, all the matching algorithms exhibit full diversity gain of order two with respect to non-cooperative transmission. As we would expect, random, greedy, and minimal matching perform increasingly better, but only in terms of SNR gain. Although diversity gain remains constant because we only group terminals into cooperating pairs, the relative SNR gain does improve slightly with increasing network size. This

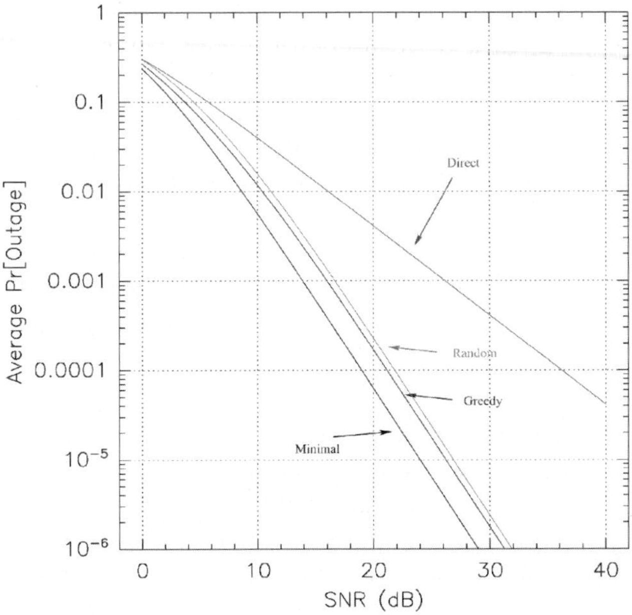

(a) $n = 10$

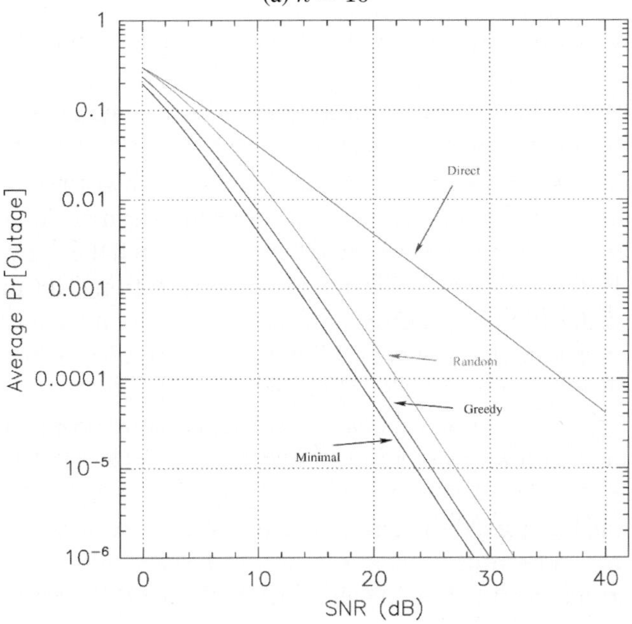

(b) $n = 50$

Figure 6.3. Matching algorithm performance in terms of average outage probability vs. received SNR (normalized for direct transmission).

effect appears most pronounced in the case of greedy matching. This observation suggests that optimal matching is more crucial to good performance in small networks, because there are fewer choices among a small number of terminals. In general, the SNR gains of the more computationally demanding matching algorithms are most beneficial in low to moderate SNR regimes where the benefits of the diversity gains are smallest. As the diversity gains increase for higher SNR, its becomes less crucial to utilize complex matching algorithms.

Figure 6.4 compares the results of minimal and greedy matching for a sample network with 50 terminals. We see that the minimal matching tends to have pairs such that one of the terminals is almost on the line connecting the basestation and the other terminal. By comparison, the greedy matching algorithm exhibits much more randomness.

Clustering in Ad-Hoc Networks

Our focus in this section is on clustering in ad-hoc networks, and how cooperative diversity can be integrated into such architectures. Clustering algorithms partition a large ad-hoc network into a set of clusters, each centered around a clusterhead. Terminals communicate directly to their associated clusterhead, and routing is usually performed between clusterheads. In this sense, clustering mimics some of the features of infrastructure networks: clusters correspond to cells and clusterheads correspond to basestations. However, in ad-hoc settings the clusters and clusterheads may be varying as the network operates, the clusterheads themselves can have information to transmit, and the clusterhead network must share the wireless bandwidth.

There are many tradeoffs in the design of clustering algorithms, too many to fully address here. For example, clustering algorithms can be designed in order to reduce the complexity and overhead of routing through the network [Das and Bharghavan, 1997]; they can be designed in coordination with turning radios on and off in order to reduce power consumption in the network [Chen et al., 2002]; and they can be designed to facilitate fusion of measurements in sensor networks [Heinzelman et al., 2000; Heinzelman et al., 2002].

Instead our objective in this section is to suggest how cooperative diversity can be integrated into an existing clustered ad-hoc network. To this end, we consider three clusters as in Figure 6.5. Figure 6.5(a) illustrates how direct transmission can be utilized to communicate information between terminals in different clusters. A terminal transmits to its clusterhead, clusterheads route the transmission to the destination cluster, and finally the destination clusterhead transmits to the destination terminal. Figure 6.5(b) illustrates how each of the inter-cluster direct transmissions can be converted into cooperative diversity transmissions. If the average inter-terminal spacing is $\overline{r}$, we see that inter-cluster transmissions occur roughly over a distance $2\overline{r}$ on average. There are likely to be

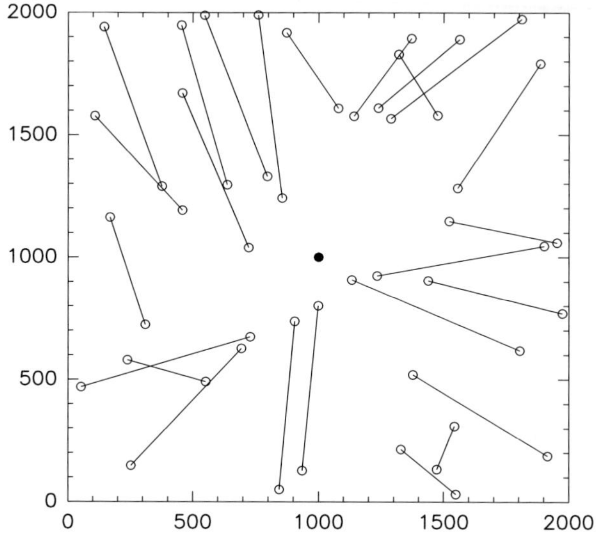

(a) Minimal Matching

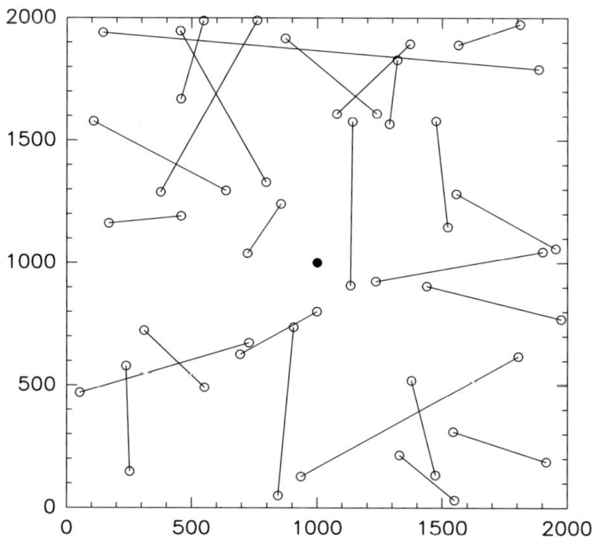

(b) Greedy Matching

Figure 6.4. Matching algorithm results for an example network: (a) minimal matching, (b) greedy matching. Terminals are indicated by circles, and matched terminals are connected with lines.

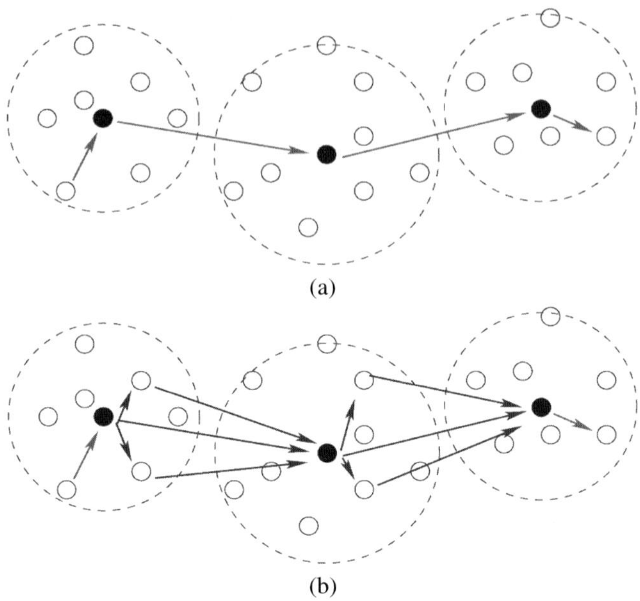

Figure 6.5. Clustering with (a) direct transmission and (b) cooperative diversity transmission.

useful relays between clusterheads, and in order to coordinate the transmissions we utilize relays in the originating cluster.

As we increase the cluster size, the number of clusters in the network decreases. Thus the complexity of routing across the entire network decreases as well; however, the utility and complexity of routing within a cluster increases. From a diversity standpoint, the inter-cluster direct transmissions are over longer and longer distances, but there are more and more potential relays to exploit. The complexity and benefits of cooperative diversity between clusters thus increases as well. Again, at least from a diversity standpoint because of reduced bandwidth efficiency and diminishing returns of diversity gains, there is no reason to grow the cluster size too large. On the other hand, growing the cluster size allows more and more terminals to be asleep when they are not transmitting and still conserve the transport of the network [Chen et al., 2002]. The point is, there are a variety of issues to explore here, and cooperative diversity can be one of them.

4. Discussion and Future Directions

Although it seems safe to suggest that some form of cooperative communications will emerge in future wireless networks, the final form of the network architectures and the resulting impact on performance for various applications

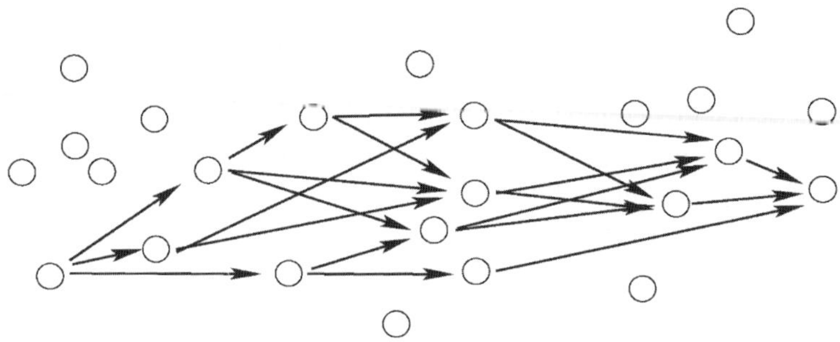

Figure 6.6. Illustration of multi-stage cooperative transmission. Downstream receivers can combine signals from all upstream transmitters. Only one complicated "link" is presented to the network layer.

remain unclear. In this section, we discuss some important directions for clarifying these issues.

Rethinking the Link Abstraction

Cooperative communications raises interesting questions about the right "link" abstraction for wireless networks. Much of the work on cooperation focuses on a relatively small number of cooperating terminals relaying in parallel, or in a single stage. As we saw in Section 2 and Section 3, such approaches can lead to performance improvements and can be readily integrated into existing network architectures by replacing individual links with single-stage cooperative links.

For multi-hop networks, one rather provacative idea [Scaglione and Hong, 2003; Scaglione et al., 2006] is to replace a multi-hop route consisting of many links by a single, multi-stage cooperative link such as the one shown in Figure 6.6. Typical multi-hop architectures for ad hoc networks present the network layer with a large number of relatively simple links, and routing over all those links becomes non-trivial. By contrast, multi-stage cooperation presents the network layer with a smaller number of more complicated links. Between these two extremes, one can imagine shifting complexity among the network, link and multi-access, and physical layers depending upon the network size and terminal processing capabilities. Although some analysis of individual multi-stage cooperative links has appeared [Boyer et al., 2004; Ribeiro et al., 2005], more work should focus on how such complicated links interact in the context of a network architecture.

Hardware Testbeds

Because cooperative communications involves more than a single transmitter and a single receiver, accurate modeling and performance prediction can become challenging without many simplifying assumptions such as those considered in Section 2. Verifying that such assumptions are reasonable, or developing better models, requires hardware testbeds for cooperative communications.

The first, and perhaps only, public demonstration of cooperative diversity in radio hardware appears in [Bletsas, 2005] as part of work performed at the MIT Media Laboratory. Custom terminals were built on a printed circuit board consisting of a 916 MHz on-off keying radio from RF Monolithics and a low-cost Cygnal 8051 microcontroller running at 22 MHz. Software for transmission, synchronization, and reception was written from scratch.

The protocols developed within this testbed reflect the design objective of simple radio hardware. Among multiple potential relays between a source and destination, the "best" relay is selected in a distributed fashion using a request-to-send (RTS) followed by clear-to-send (CTS). The destination performs simple selection combining of the direct and relayed transmissions. Tags in the packet headers are used to display the selected relay at the destination visually with different colors. During a demonstration indoors, one can often see the colors change, indicating selection of different relays, as people and other objects move about the room, making the benefits of relaying much more tangible than a plot of outage probability against signal-to-noise ratio.

Dumb Cooperation, Smart Routing, and Distributed Beamforming

Varying amounts of channel state information (CSI) at the source and relay terminals for their channels to the destination leads to a spectrum of transmission strategies. We briefly summarize some classes of strategies here.

One class of strategies involves no transmit CSI at the source or relay terminals. That is, the source and relays do not obtain CSI on their outgoing channels, but they can obtain receive CSI on their incoming channels. This scenario is the one considered in the model in Section 2, which is based upon the original work on cooperative diversity in [Laneman et al., 2004; Laneman and Wornell, 2003]. Relative to more elaborate schemes to be discussed next, the transmission strategies can be viewed as "dumb" cooperation since they spend little overhead to obtain CSI. On the other hand, these strategies effectively waste transmit energy by having multiple relays transmit in order to achieve diversity gain at the destination.

A more informed class of strategies involves some transmit CSI at the relays, and perhaps the source. However, the key distinction between this class and the one to follow is that the CSI is used only to select a single relay. Such "opportunistic relaying" protocols, as developed within the context of the hardware testbed described above, are in some sense simpler than dumb cooperation protocols, but are more power and bandwidth efficient than dumb cooperation due to the additional CSI [Bletsas et al., 2006]. From a higher perspective, opportunistic relaying can be viewed as a form of "smart routing" in a local area with frequent routing table updates.

The most informed class of strategies involves complete transmit CSI at the source and relays, including both amplitude and phase information, and allows for "distributed beamforming" [Sendonaris et al., 2003a; Host-Madsen and Zhang, 2005; Barriac et al., 2004; Ochiai et al., 2005]. Beamforming exploits coherent combination of signals at the receiver, and can be very power efficient and bandwidth. Effective distributed beamforming requires accurate carrier synchronization among the source and relays [Brown et al., 2005], but much simpler codes than those designed for dumb cooperation can be employed.

Having quickly summarized a broader spectrum of cooperative strategies beyond those of Section 2, we conclude by saying that it is unclear which of these approaches will have the largest impact in practical networks. For simplicity, analysis suppresses many interactions among signal processing for channel estimation, beamforming, and synchronization; channel coding for error control; and network protocols for data transfer. Overhead, particularly for dissemination of CSI and network protocols, is often not taken into account. It seems unlikely that a model general enough to evaluate all these interactions and tradeoffs will be tractable. Instead, extensive simulations and implementations within wireless testbeds will have to be pursued in order to complete the story.

References

Ahuja, Ravindra K., Magnanti, Thomas L., and Orlin, James B. (1993). *Network Flows: Theory, Algorithms, and Applications*. Prentice-Hall, Englewood Cliffs, NJ.

Azarian, Kambiz, El Gamal, Hesham, and Schniter, Philip (2005). On the Achievable Diversity-Multiplexing Tradeoff in Half-Duplex Cooperative Channels. *IEEE Trans. Inform. Theory*, 51(12):4152–4172.

Barriac, G., Mudumbai, R., and Madhow, U. (2004). Distributed Beamforming for Information Transfer in Sensor Networks. In *Proc. International Symposium on Information Processing in Sensor Networks (IPSN)*, Berkeley, CA.

Bletsas, Aggelos (2005). *Intelligent Antenna Sharing in Cooperative Diversity Wireless Networks*. PhD thesis, Massachusetts Institute of Technology, Cambridge, MA.

Bletsas, Aggelos, Khisti, Ashish, Reed, D. P., and Lippman, A. (2006). A Simple Cooperative Diversity Method Based on Network Path Selection. *IEEE J. Select. Areas Commun.* To appear.

Bölcskei, Helmut, Nabar, Rohit U., Oyman, Özgür, and Paulraj, Arogyaswami J. (2004). Capacity Scaling Laws in MIMO Relay Networks. *IEEE Trans. Wireless Commun.* Submitted for publication.

Boyer, John, Falconer, David, and Yanikomeroglu, Halim (2004). Multihop Diversity in Wireless Relaying Channels. *IEEE Trans. Commun.*, 52(10): 1820–1830.

Brown, III, D. Richard, Prince, Gregary B., and McNeill, John A. (2005). A Method for Carrier Frequency and Phase Synchronization of Two Autonomous Cooperative Transmitters. In *Proc. IEEE Workshop on Sig. Proc. Adv. in Wireless Comm. (SPAWC)*, New York, NY.

Carleial, Aydano B. (1982). Multiple-Access Channels with Different Generalized Feedback Signals. *IEEE Trans. Inform. Theory*, 28(6):841–850.

Chen, Benjie, Jamieson, Kyle, Balakrishnan, Hari, and Morris, Robert (2002). Span: An Energy-Efficient Coordination Algorithm for Topology Maintenance in Ad Hoc Wireless Networks. *ACM Wireless Networks Journal*, 8(5).

Chen, Deqiang and Laneman, J. Nicholas (2005). Modulation and Demodulation for Cooperative Diversity in Wireless Systems. *IEEE Trans. Wireless Commun.* To appear.

Cover, Thomas M. and El Gamal, Abbas A. (1979). Capacity Theorems for the Relay Channel. *IEEE Trans. Inform. Theory*, 25(5):572–584.

Cover, Thomas M. and Thomas, Joy A. (1991). *Elements of Information Theory*. John Wiley & Sons, Inc., New York.

Das, Bevan and Bharghavan, Vaduver (1997). Routing in Ad-Hoc Networks Using Minimum Connected Dominating Sets. In *Proc. IEEE Int. Conf. Communications (ICC)*, volume 1, pages 376–380, Montreal, Canada.

Gupta, Piyush and Kumar, P. R. (2003). Towards an Information Theory of Large Networks: An Achievable Rate Region. *IEEE Trans. Inform. Theory*, 49(8):1877–1894.

Heinzelman, Wendi, Chandrakasan, Anantha, and Balakrishnan, Hari (2000). Energy-Efficient Communication Protocols for Wireless Microsensor Networks. In *Proc. of the Hawaii Int. Conf. on System Sciences*, pages 3005–3014, Maui, HW.

Heinzelman, Wendi B., Chandrakasan, Anantha P., and Balakrishnan, Hari (2002). An Application-Specific Protocol Architecture for Wireless Microsensor Networks. *IEEE Trans. Wireless Commun.*, 1(4):660–670.

Host-Madsen, Anders (2004). Capacity Bounds for Cooperative Diversity. *IEEE Trans. Inform. Theory*. Submitted for publication.

Host-Madsen, Anders and Zhang, Junshan (2005). Capacity Bounds and Power Allocation in Wireless Relay Channel. *IEEE Trans. Inform. Theory*, 51(6): 2020–2040.

Hunter, Todd E. and Nosratinia, Aria (2006). Coded Cooperation in Multi-User Wireless Networks. *IEEE Trans. Wireless Commun.* To appear.

Jing, Yindi and Hassibi, Babak (2004). Wireless Networks, Diversity, and Space-Time Codes. In *Proc. IEEE Information Theory Workshop (ITW)*, San Antonio, TX.

Khojastepour, M. A., Sabharwal, A., and Aazhang, B. (2004). Improved Achievable Rates for User Cooperation and Relay Channels. In *Proc. IEEE Int. Symp. Information Theory (ISIT)*, Chicago, IL.

King, Roger C. (1978). *Multiple Access Channels with Generalized Feedback.* PhD thesis, Stanford University, Palo Alto, CA.

Kramer, Gerhard (2004). Models and Theory for Relay Channels with Receive Constraints. In *Proc. Allerton Conf. Communications, Control, and Computing*, Monticello, IL.

Kramer, Gerhard, Gastpar, Michael, and Gupta, Piyush (2005). Cooperative Strategies and Capacity Theorems for Relay Networks. *IEEE Trans. Inform. Theory*, 51(9):3037–3063.

Laneman, J. Nicholas (2002). *Cooperative Diversity in Wireless Networks: Algorithms and Architectures.* PhD thesis, Massachusetts Institute of Technology, Cambridge, MA.

Laneman, J. Nicholas, Tse, David N. C., and Wornell, Gregory W. (2004). Cooperative Diversity in Wireless Networks: Efficient Protocols and Outage Behavior. *IEEE Trans. Inform. Theory*, 50(12):3062–3080.

Laneman, J. Nicholas and Wornell, Gregory W. (2000). Energy-Efficient Antenna Sharing and Relaying for Wireless Networks. In *Proc. IEEE Wireless Comm. and Networking Conf. (WCNC)*, Chicago, IL.

Laneman, J. Nicholas and Wornell, Gregory W. (2003). Distributed Space-Time Coded Protocols for Exploiting Cooperative Diversity in Wireless Networks. *IEEE Trans. Inform. Theory*, 49(10):2415–2525.

Lin, Zinan, Erkip, Elza, and Stefanov, Andrej (2006). Cooperative Regions and Partner Choice in Coded Cooperative Systems. *IEEE Trans. Commun.* To appear.

Mitran, P., Ochia, H., and Tarokh, V. (2005). Space-Time Diversity Enhancements using Cooperative Communications. *IEEE Trans. Inform. Theory*, 51(6):2041–2057.

Nabar, R. U., Bölcskei, Helmut, and Kneubuhler, F. W. (2003). Fading Relay Channels: Performance Limits and Space-Time Signal Design. *IEEE J. Select. Areas Commun.*, 22(6):1099–1109.

Ochiai, H., Mitran, P., Poor, H. V., and Tarokh, V. (2005). Collaborative Beamforming for Distributed Wireless Ad Hoc Sensor Networks. *IEEE Trans. Signal Processing*, 53(11):4110–4124.

Ozarow, Lawrence H., Shamai (Shitz), Shlomo, and Wyner, Aaron D. (1994). Information Theoretic Considerations for Cellular Mobile Radio. *IEEE Trans. Veh. Technol.*, 43(5):359–378.

Proakis, John G. (2001). *Digital Communications*. McGraw-Hill, Inc., New York, fourth edition.

Rappaport, Theodore S. (2002). *Wireless Communications: Principles and Practice*. Prentice-Hall, Inc., Upper Saddle River, New Jersey, second edition.

Ribeiro, Alejandro, Cai, Xiaodong, and Giannakis, Georgios B. (2005). Symbol Error Probabilities for General Cooperative Links. *IEEE Trans. Wireless Commun.*, 4(3):1264–1273.

Rosen, Kenneth H., editor (2000). *Handbook of Discrete and Combinatorial Mathematics*. CRC Press, Boca Raton, FL.

Scaglione, Anna, Goeckel, Dennis, and Laneman, J. Nicholas (2006). Cooperative Communications in Mobile Ad-Hoc Networks: Rethinking the Link Abstraction. *IEEE Signal Processing Mag.* Submitted for publication.

Scaglione, Anna and Hong, Yao-Win (2003). Opportunistic Large Arrays: Cooperative Transmission in Wireless Multihop Ad Hoc Networks to Reach Far Distances. *IEEE Trans. Signal Processing*, 51(8):2082–2092.

Schein, Brett (2001). *Distributed Coordination in Network Information Theory*. PhD thesis, Massachusetts Institute of Technology, Cambridge, MA.

Schein, Brett and Gallager, Robert G. (2000). The Gaussian Parallel Relay Network. In *Proc. IEEE Int. Symp. Information Theory (ISIT)*, page 22, Sorrento, Italy.

Sendonaris, Andrew, Erkip, Elza, and Aazhang, Behnaam (2003a). User Cooperation Diversity, Part I: System Description. *IEEE Trans. Commun.*, 15(11): 1927–1938.

Sendonaris, Andrew, Erkip, Elza, and Aazhang, Behnaam (2003b). User Cooperation Diversity, Part II: Implementation Aspects and Performance Analysis. *IEEE Trans. Commun.*, 51(11):1939–1948.

Telatar, I. Emre (1999). Capacity of Multi-Antenna Gaussian Channels. *European Trans. on Telecomm.*, 10(6):585–596.

van der Meulen, Edward C. (1968). *Transmission of Information in a T-Terminal Discrete Memoryless Channel*. Department of Statistics, University of California, Berkeley, CA.

van der Meulen, Edward C. (1971). Three-Terminal Communication Channels. *Adv. Appl. Prob.*, 3:120–154.

Wei, Shuangqing, Goeckel, Dennis, and Valenti, Matthew (2005). Asynchronous Cooperative Diversity. *IEEE Trans. Wireless Commun.* Submitted for publication.

Willems, Frans M. J. (1982). *Informationtheoretical Results for the Discrete Memoryless Multiple Access Channel.* PhD thesis, Katholieke Universiteit Leuven, Leuven, Belgium.

Willems, Frans M. J., van der Meulen, Edward C., and Schalkwijk, J. Pieter M. (1983). An Achievable Rate Region for the Multiple Access Channel with Generalized Feedback. In *Proc. Allerton Conf. Communications, Control, and Computing*, pages 284–292, Monticello, IL.

Xie, Liang-Liang and Kumar, P. R. (2004). An Achievable Rate for the Multiple Level Relay Channel. In *Proc. IEEE Int. Symp. Information Theory (ISIT)*, Chicago, IL.

Chapter 7

COOPERATION IN AD-HOC NETWORKS

Wireless Multihop Networks from Theory to Practice

Petri Mähönen
RWTH Aachen University
Wireless Networks Group, Kackertstrasse 9,
Aachen, Germany
pma@mobnets.rwth-aachen.de

Marina Petrova
RWTH Aachen University
Wireless Networks Group, Kackertstrasse 9,
Aachen, Germany
mpe@mobnets.rwth-aachen.de

Janne Riihijärvi
RWTH Aachen University
Wireless Networks Group, Kackertstrasse 9,
Aachen, Germany
jar@mobnets.rwth-aachen.de

Abstract: In this chapter we are outlining the major challenges encountered when one is trying to deploy realistic ad hoc networks. We are also emphasizing that the current trend towards less mobile mesh networks and sensor networks are actually most probably enabling the emergence of ad hoc type of networks for the civilian markets. Our main focus is to give enough starting points for interested reader to read more specific literature on the existing research in this fast moving field. One of the main conclusions of the chapter is that providing cooperative ad hoc networks requires much more than deploying ad hoc routing capability into network. Especially if one is limited to use IEEE 802.11 technologies, one has to be very careful with the performance limitations. Apart of advocating the

Frank H. P. Fitzek and Marcos D. Katz (eds.), Cooperation in Wireless Networks: Principles and Applications, 189–222.

less-mobile mesh networks, we point out that the recent suggestion to use "multi-radio" approach is a very sensible one. We also speculate on more advanced research possibilities, namely we point out that cognitive radio and networking principles especially, if combined with efficient topology awareness might be an effective way to ensure optimal and cooperative ad hoc networks in the future.

Keywords: cooperation; wireless ad hoc and mesh networks; coloring algorithm; multihop communications

1. Introduction

More than a decade the research community has been quite intensively studying the mobile ad hoc networks, popularly known as MANETs. The great vision since the beginning of their development has been to create autonomous and self-organizing network without any pre-established infrastructure or centralized administration. This enables the randomly distributed nodes to form a temporary functional network and support seamless leaving or joining of nodes.

A tremendous amount of work has been done towards solving research problems related to wireless ad hoc networks (see *e.g.*, [Toh, 2003; Toh, 2001a; Perkins, 2001] and references therein). Although a considerable amount of successful research is done, especially when considering military ad hoc networks, the deployment of *large-scale (massive)* ad hoc networks in the civilian context has been limited to very few cases. There are certainly many reasons for this lack of commercial success, one of those being that the time has not been ripe for ad hoc networking, and certainly many practical engineering problems have been underestimated during the first phase of enthusiasm.

In certain sense all wireless digital communication requires *cooperation* as the systems are required to share resources, and at the very least the end-to-end hosts need to have transmission systems and protocols that are compatible, and somehow standardized. The requirement for cooperation in the case of ad hoc systems, however, is a very stringent one and is present at all system levels. These challenges have not always been foreseen in the right context.

We do not claim to give a comprehensive review on ad hoc networking in this chapter due to fact that it would be out of the scope, and most importantly due to space limitations. There are many excellent treatments available today, and we refer the reader for example to the excellent book by Siva Ram Murthy & Manoj ([Murthy and Manoj, 2004] and reference therein), one should look also other recent reviews (*e.g.*, [Toh, 2001a; Perkins, 2001; Basagni, 2004; Sheu and Jie, W. (Editors), 2005]). This chapter is a quick glance to main issues that one must take in account, when one is designing *multihop, ad hoc* networks.

Our subtitle "from theory to practice" is emphasizing the goal to understand the realistic limitations that we encounter with the present day multihop MANET approaches. One of the key issues is to stress the fact (that was not entirely new insight for the first generation packet network researchers, but has sometimes been not valued enough) that routing itself is not the only problem for multihop wireless ad hoc networks. In fact, for guaranteeing reasonable quality of service, one needs to consider many other aspects than routing, and depending on the chosen transmission and network layer technologies, there are always limits on how many hops and what level of mobility can be supported. This statement is very much a practical engineering based, *i.e.*, regardless of some asymptotic, theoretical limits that are derived on ad hoc capacity, in practice, when one is deploying real systems (at least with the foreseeable technology) there are limits for ad hoc network practicality even in the capacity domain. Due the chosen "practicality" -theme, we are mostly considering only IEEE 802.11 -type of deployment scenarios in our examples. This is done on purpose as most of the practical experimentation is done by using WiFi-systems. However, this is also a limitation of this chapter, which we are wholeheartedly admitting.

In the following, we start with the quick review on some useful historical facts, and also give some framework suggestions for our work. This framework part includes also some definitions and scoping for the "cooperation" itself. We are also commenting some possible interesting research domains that may become more important in the future. After the introduction part, we progress on analyzing the case of multihop wireless ad hoc networks, especially in IEEE 802.11 context. This work is mostly based on the wireless mesh (low mobility) case, and we have chosen to use mostly experimental treatment from our and others' previous research work. Finally, we progress from the multihop capacity work towards some more recent possibilities, such as showing how distributed coloring algorithms and topology control can be used in the wireless ad hoc and mesh context. In the very end, we are drawing some conclusions and dare also to present some longer-term research visions. We are also specifically commenting as requested by editors on possibility to have ad hoc networks as a part of "4G infrastructure" in the future.

Ad Hoc and Sensor Networks Drivers

There are indications that ad hoc networking is finally finding its place, and has also good possibilities to be adopted for commercial purposes, perhaps not as an alternative, but as an extension to existing paradigms. There is on-going interest to apply ad hoc networking principles towards a range of possibilities such as (community) *mesh networking*, *range-extension* of cellular and mesh networks, and small-scale special purpose ad hoc -networks such as Personal Area Networks for games and entertainment. This is in part reflecting the enhanced technological capabilities, but also the fact that real applications cases have been found. This is good, as only slightly over simplifying, a lot of research in the case of ad hoc networking has been almost purely technology push driven.

More recently wireless sensor networks (WSN) have emerged as equally strong research topic, and many of the fundamental problems are shared between "traditional" ad hoc network and WSN research. However, we are emphasizing that WSN research is not a recent spin-off from ad hoc research, as it has a long history in the industrial automation and military domain. In fact, the Distributed Sensor Networks (DSN) program of DARPA (Defense Advanced Research Projects Agency) was launched already around 1980 in the U.S.A.

As mentioned ad hoc networks and wireless sensor networks share many problems (see recent treatments like [Akyildiz et al., 2002; Karl and Willig, 2005]). Specifically problems related to self-configuration, ad hoc routing, and power consumption are shared between these two domains. However, from the drivers of research point of view, there are some differences. Let us oversimplify and somewhat exaggerate. Ad hoc networking research has been more strongly technology push related, and apart from few special cases (such as military networks) there are only a limited number of well-recognized and accepted application cases available to draw system requirements for *commercial systems*. This is also a challenge, when one tries to compare different research proposals and solutions, because without systems level requirements the comparison might become arbitrary at least from the industrial point of view. Overall, it is quite remarkable how little we have real civilian, mass-market wireless ad hoc products available, taking in to account the massive amount of research done.

Wireless Sensor Networks are somewhat different in their status. Although, there is equally strong technology push, especially if one is looking for design on low-power radio technologies and microelectronics, there has been from the beginning a strong emphasis on prototyping. This is probably due to the fact that WSN–research is also closely related to embedded systems development in general that has always been very much application driven. However, many of the current uses of WSNs are very much "on the spot" applications or simple technology-demonstrators, *i.e.*, narrowly chosen to fulfill some specific project and partnership requirements.

Multihop Packet Radio Networks

The requirement to have a cooperative behavior to enable efficient multihop ad hoc networks has been known at least for 30 years. In fact, it might be useful to emphasise that packet radio networking research started around the mid-1970s is a clear precursor of the work done in the ad hoc and multihop context today. The seminal work done by the first generation of packet radio research is still very valuable today. We refer the reader to such contributions as the series of Kleinrock & Tobagi authored articles on packet switching and packet radio networks ([Kleinrock and Tobagi, 1975], [Tobagi and Kleinrock, 1975], [Kleinrock, 1978]), the early packet radio *network* article by Kahn ([Kahn, 1977], see also [Kahn et al., 1978]), spatial reuse paper by Kleinrock & Silvester ([Kleinrock and Silvester, 1987]), and many others (see *e.g.*, [Jubin and Tornow, 1987; Shacham and Westcott, 1987; Tobagi, 1987; Kahn, R. E. (ed.), 1978] just to mention a few).

Cooperation and Challenges

The challenges in the case of ad hoc networking are broadly related to issue to ensure enough cooperation between distinct nodes, and at the same time using the scarce wireless resources efficiently. The *cooperation* is defined as "the action of co-operating, *i.e.* of working together towards the same end, purpose, or effect; joint operation" ([OED, 2001]). In the case of ad hoc networks one should be careful to understand that there are two distinct co-operation domains;

1 *"Communications Cooperation"*, in the strict communications stack domain, means that we need to provide a common set of communications protocols and transmission methods for all the corresponding hosts so that the network can be established. This problem is shared with all communication systems, but the dynamical nature of the ad hoc networks makes this quite difficult. In the case of the ad hoc networks the challenges are rising from the need to support *distributed* algorithms and protocols, and dynamic topology without sacrificing too much of efficiency.

2 *"Social Cooperation"* of the forwarding nodes for a common good is another aspect. There the challenge is the question how to guarantee that nodes between the source and the destination are cooperating on packet forwarding. In the case of the closed ad hoc network applications (such as military or emergency networks) this is easier to ensure than if one is considering highly dynamic privately owned network hosts. This sort of "social cooperation" is beyond the scope of this chapter, but other chapters in this book are addressing at least in part this problem domain. We also note that the recent trends towards community mesh networks, and range-extending, commercial ad hoc applications are also making this problem easier to tackle in this limited domain, that are not as dynamic as full MANETs.

A very large amount of research has been invested towards ad hoc routing. Although there are still some problems to be solved, we mostly comment here certain mature technologies that have emerged. We believe that the major part of the future research work will be directed to new problem domains. In fact, some engineering problems need to be solved even before intelligent link-aware routing solutions can be implemented easily.

From the cooperative behavior point of view clearly more work with MAC (Media Access Control) layer algorithms is required. Most of the current testbeds are using IEEE 802.11 MAC (or slightly modified versions of it). The popularity of 802.11 makes it difficult to envision quick departure from it, but regardless more efficient MAC protocols are required (cf. [Chandra et al., 2000; Murthy and Manoj, 2004]). In the case of WSNs there has been increased activity on designing low-power, low bit-rate MAC solutions for ad hoc networks. The idea of building smart-antenna based MAC-protocols for ad hoc networks has recently gained popularity and can be a promising solution under some certain conditions (see [Ko et al., 2000; Fahmy et al., 2002; Choudhury et al., 2002; Ramanathan et al., 2005; Vilzmann et al., 2005], see also survey by Vilzmann & Bettstetter, [Vilzmann and Bettstetter, 2005]). Power control, especially when related to *topology control*, is another important research challenge that has been gaining a merited interest (see [Santi, 2005] and references therein).

Finally, we mention in this introductional part that energy efficiency is still a challenge to be met, and it is very demanding problems as it requires cross-layer optimization approach, including also careful design of underlying electronics itself.

Cooperation Domains and Metrics

The challenge with the cooperative networks is that even in the case of communication cooperation there are different domains of cooperation. The domains, in fact, rise quite naturally from the fact that there are disjoint resources that need to be shared between hosts. These include most notably need to share *frequency, time* and quite often *space*[1].

Apart from the need to share resources in cooperative manner, there is also an issue of relevant **metrics**. Some of the metrics are related to physical resources, most notably to available energy. Other metrics are typically related to communication domain itself (*e.g.* bit-rate, latency, ...). Designing and operating an efficient ad hoc network is fundamentally a dynamical *optimization problem*. As the ad hoc network itself can be relatively dynamic, the system itself must be able to adapt to changes (and this certainly goes beyond "simple" routing). However, in order to make optimization decisions one needs to have a performance metric. This is a difficulty, as in the end any reasonable ad hoc metric would be a multivariate and multiparameter function. Moreover, as the decisions should be done in the distributed fashion it is not always clear how to guarantee global convergence or fairness. In fact, one has to remember that although we talk about cooperative system, different users ("players" in a game theoretical sense) can have highly different goals, hence different performance metrics. Although the performance of computer networks have been studied for decades, the issue of highly distributed performance optimization in the case of ad hoc networks still needs more fundamental research before we really understand all the limitations.

One of our own recent contributions on the discussion is to point out that ad hoc cooperation is not only optimization and game modelling problem, but it is also policy optimization issue. Policy optimization here means that as the system will inevitably encounter situations, where mutually exclusive optimization issues and race-conditions occur, there needs to be a way to describe policies or preferences on how to solve such situations. Apart from some recent work done in the case of spectrum agility (cognitive radios), and some relevant analysis with BGP and software radio work, as far as we are aware of the ad hoc "policy languages" have not really been considered in depth.

2. Limits of Multihop

MANETs find their applications mostly in multihop scenarios where there is no wired infrastructure available. The envisioned applications of ad hoc communication include commercial and educational use, emergency cases, on road vehicle networks, military communication, sensor networks, etc. However, many analytical and practical studies have already shown various drawbacks of multihop ad hoc communications (both technological and human limitations) in terms of throughput, fairness, energy and bandwidth limitations, which make it difficult to envisage commercial deployment of very large ad hoc networks.

Although stand-alone ad hoc networks might provide support for interesting applications, they have not really taken up outside military domain. While the presently deployed hotspots offer only single hop connection to the infrastructure these wireless multihop technologies can be leveraged to increase the reach of such networks. This can be accomplished without wired infrastructure in several ways. On one hand ad hoc routing among clients can increase the coverage area of an access point and on the other hand a wireless mesh network can be established to interconnect APs. The combination of multihop wireless networks with fixed/cellular networks seems very attractive, because it allows usage of even wider range of services. However, it is *not* without its practical limitations.

In the next sections we will walk through several issues that characterize the ad hoc multihop networks and discuss their performance taking into account a large number of analytical and practical studies carried out both in the industry and academia in the past decade.

Routing Metrics Challenge

Since the ad hoc network is a cooperative set of mobile nodes, each node plays a role of a logical router and forwards packets from other nodes. Due to the dynamic nature of the ad hoc networks, highly adaptive routing protocols are required to cope with the frequent topology changes. There has been a substantial work done in the ad hoc routing resulting in design of number of different MANET routing protocols such as DSR ([Johanson et al., 2001]), AODV ([Perkins and Royer, 1999]), DSDV ([Perkins and Watson, 1994]), OLSR ([Clausen et al., 2001]) etc. Depending on the technique of acquiring the route to the destination, the existing ad hoc routing protocols can be divided into three groups (see also [Feeney, 1999] for further discussion on the taxonomy of routing protocols). *Reactive* protocols acquire and maintain the routes in an on-demand fashion and/or the route discovery is initiated only when needed. Examples of reactive protocols include DSR and AODV. *Proactive* routing protocols, on the other hand, maintain the routes to all destinations in the network constantly. OLSR is a typical example of a proactive routing protocol. *Hybrid* routing protocols are both reactive and proactive in nature. The protocols allow the nearby nodes, grouped into zones, clusters or trees, to maintain the routes pro-actively and discover the routes to the far away nodes in reactive manner. ZRP ([Haas and Pearlman, 2001]), is one of the most known belonging to this group of protocols. Recently there has been also interest on how to use extended OSPF in the wireless, ad hoc context (see [Ahrenholz et al., 2005]).

The routing in MANETs has traditionally focused on finding out solutions that minimize hop-count and provide fast adaptation in the case of highly dynamic (mobile) networks. One of the problems with the most minimal hop-count approaches is that it does not take the link-quality into account . Especially in the case of IEEE 802.11 based networks that are deployed into large area, the difference between link qualities can be very large indeed. As a result, it is not rare case that the minimum hop-count based routing schemes chose routes with significantly less capacity than the high-quality paths available in the network. This issue has been pointed out in details, *e.g.* by [De Couto et al., 2002].

A number of different performance metrics, such as the ETX in [Couto et al., 2003] (expected transmission count metric), per-hop RTT ([Adya et al., 2004]), link-quality dual (SNR, BER), and per-hop packet-pair ([Draves et al., 2004]), that characterize the quality of the wireless link have emerged in the recent years. For example, ETX finds high-throughput paths using per-link measurements of the packet loss in both directions of the wireless links. In the per-hop RTT approach, the nodes probe periodically their neigbours measuring the RTT. The RTT samples are averaged using TCP-like low-pass filter and the path with the least sum of RTT is selected. The per-hop packet-pair technique, on the other hand, uses two two-back-to-back periodic probings to the each neigbour. The receiving node measures the arrival delay between the two probes and reports it back to the sender. The sender averages the delay samples and the finally the route with the least delay is chosen. Both the per-hop RTT and the PckPair metric implicitly take into account the load, the bandwidth and the loss rate of the wireless link. One problem related to link-quality aware routing is the practical issue, how to actually measure some of the lower-layer (MAC and PHY) parameters and use them at the upper layers (e.g network layer). We discuss this problem in more detail below.

Energy Consumption in Multihop

The nodes in a mobile ad hoc network rely on batteries for proper operation. Since they need to relay their messages through other nodes toward their intended destinations, depletion of the batteries will have a great impact on the overall network performance. Especially if the power consumption rate is not evenly distributed across all nodes, some nodes may expire sooner than others leading to partitioning of the network ([Toh, 2001b]).

Increasing the lifetime of each node is a rather complex process and can be done at different layers. The so-called non-communication power consumption is very dependent upon the actual hardware implementation. Further on an adaptive power control at the physical layer can help to conserve the battery life of the hosts. On the other hand, data link and routing protocol design can also significantly impact the processing and the transceiver power dissipated in wireless communication. At the data link layer, energy conservation can be achieved by using effective retransmission schemes. To maximize the lifetime of an ad hoc network, the routing protocols could introduce sleep periods so that the hosts can stop transmitting and/or receiving for arbitrary periods of time without causing any serious consequences in the network operation. Moreover, transmission power can he used as a routing metric.

When talking about conserving the life of the battery in the ad hoc networks it is maybe necessary to mention the well known myth saying that the multihop communication *always* saves energy. Seen from the prospective of pure radio propagation theory, the power necessary to transmit a bit of information over a radio is proportional to the distance. If we introduce a multihop communication between two nodes, there should be less power needed to transmit over shorter distances. However, in order to avoid misleading results, particular care should be taken that the energy efficient communication protocols are designed around accurate energy models of the used hardware. In such case, multihop can save energy only if the path attenuation dominates the energy consumption of the hardware which is far less probable then believed ([Min and Chandrakasan, 2003]).

TCP/UDP over Multihop 802.11

In a large number of recent studies on ad hoc networks and specially WLANs, the authors have studied the performance of TCP over IEEE 802.11. The "misbehaviour" of TCP over wireless is a consequence of several issues, and is well recognized problem (see, for example, [Fu et al., 2003; Gurtov and Floyd, 2004; Xylomenos et al., 2001]). The main reason for the unsatisfactory performance of the protocol is the fact that TCP has been primarily designed for wireline networks, where the channel error rates are very low and the congestion is the main cause for packet loss. As a result, there have been several approaches how to optimize TCP for wireless networks. For more details the reader is referred to [DeSimone et al., 1993; Balakrishnan et al., 1995; Sinha et al., 2002]. There has been also number of studies that aim to final optimal parameter values for TCP over wireless systems; the parameter set typically includes, *e.g.*, packet size, congestion window size and buffer size.

In the following we highlight the most common anomalies of TCP in the mobile ad hoc networks. First, due to the dynamic nature of the topology of the ad hoc network some of the wireless links can break. As a result, TCP may experience timeouts that will seriously impact the performance. Moreover the fact that in the wireless network a packet can be lost not only due to congestion but also because of the errors in the wireless channel, leads to undesired TCP behaviour. Additional losses and transmission errors can be caused also by a hidden terminal in the network. Anyhow, regardless of the loss nature, TCP will incorrectly interpret it as a sign of congestion, which will cause adaptation of its window size and reduction of the data flow. Several efficient mechanisms have been proposed in the literature for improving the TCP performance in wireless ad hoc networks ([Bakshi et al., 1996; Xylomenos et al., 2001]). Second, it is shown that TCP performance in an ad hoc multihop environment is sensitive to different parameters such as packet size and TCP window size ([Fu et al., 2003]). Several measurement studies has verified that for a specific network topology and traffic flow, there is a TCP window size at which the throughput reaches the highest value. Further increase of the window size does not lead to a better result. Finally, degradation of the network performance caused due to the interaction with the IEEE 802.11 MAC protocol. More specifically, the present IEEE 802.11 MAC protocol may cause unfairness between competing TCP traffic flows, and a capture of the whole wireless channel by a single node can occur relatively easily.

Performance of Ad Hoc Networks Based on Measurements

Most of the research studies tackling mobile ad hoc networks are based on simulations. However, the simulation results not always reflect the real scenario and can only give a good approximation of the simulated environment. That is why measurements obtained from real hardware testbeds are always recommended and welcome. In order to illustrate some of the challenges and the performance limitations a simple multihop network has, we present here only a small part of results from a comprehensive set of measurements that we carried out in the past couple of years.

A common way to quickly estimate the performance of a specific wireless network is to measure the most common parameters such as throughput and delay. By throughput we mean the actual transport layer payload without any headers successfully received per second. In this section we shall give a quick look into some performance issues of a simple ad hoc network based on measurements and simulations. Here we shall analyze what is the TCP and UDP throughput to be expected in a homogeneous 802.11a/b/g multihop setup. We also give an overview of the outcome from the throughput measurements in a heterogeneous 802.11b, 802.11g and Bluetooth environment.

The measurement setup in the multihop case is a simple string topology in office environment. All measurements were performed using laptops in a Linux environment and TCP NewReno with selective acknowledgement and enabled timestamps. Although we are sure that the tested system could benefit from different protocol boosters or TCP modifications, we are leaving them strictly out from our study. We were limiting our measurement campaign to unmodified, off-the-shelf solutions, since these are building blocks that are mostly used in testbeds, community networks, and simulation studies.The simulation results given in some of the figures are performed using the network simulator ns-2.

On figure 7.1 both measurement and simulation results of a TCP/UDP throughput as a function of a number of hops are depicted. Having in mind the shortcomings of ns-2, we improved the 802.11 MAC module and included enhanced error modelling. The reader is referred to [Wellens et al., 2005] for further details.

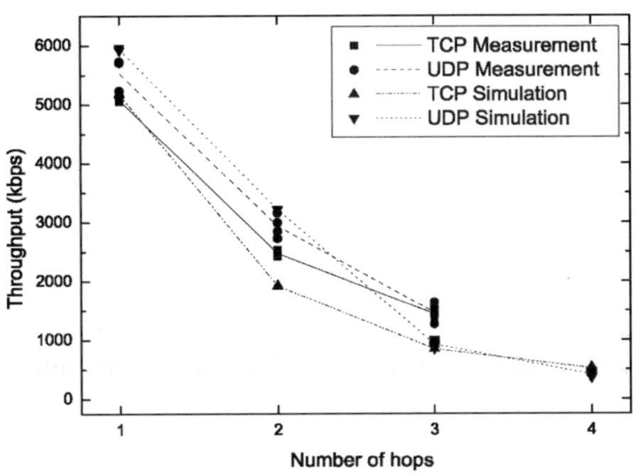

Figure 7.1. TCP/UDP multihop throughput measurements and simulations.

The figure clearly indicates that large number of wireless hops, in a single-radio per node case, is very inefficient, as throughput is lost rapidly. This is unavoidable even in the perfect environment without transmission errors, delays, etc., as it is inherent for single radio repeaters. It is even more serious in the realistic Wi-Fi multihop environment. One can notice that already three hops is quite suboptimal for many purposes. Further increasing of the number of hops will result in unacceptably low throughput. In our tests there were no external nodes contending for the channel. At a public hot spot other users will also produce interference, so the end throughput would fluctuate more.

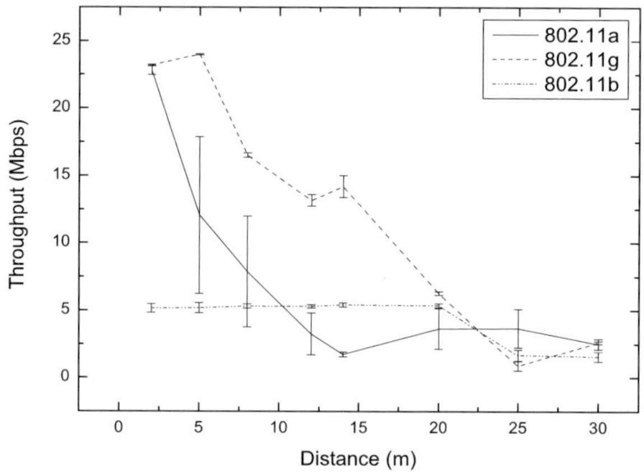

Figure 7.2. TCP throughput measured with 802.11a, 802.11b and 802.11g.

In order to answer the uncertainty on how 802.11a compares to 802.11g, we measured the throughput of 802.11a and 802.11g technologies in a real indoor environment. Both technologies are quite similar on PHY layer but use different frequencies. Figure 7.2 shows the TCP throughput of 802.11a, 802.11b, and 802.11g as a function of distance. We can identify three segments in the graph: at short distances with LoS (line of sight), as expected, both technologies reach the same maximum throughput of about 23 Mbps. When the nodes are further away from each other (>20 m), with obstacles in between, both technologies adapt their bit rate to the lowest possible to maintain the connection. In the range in between, the throughput of 802.11g clearly outperforms 802.11a. Due to the higher path loss of 802.11a, the physical layer mode switches to more robust modulation and coding which leads to lower bit rate at the distance of 5 meters LoS.

Recently the number of different wireless and radio technologies has increased dramatically. These diversity will require efficient interworking of technologies for the deployment of the future wireless heterogeneous systems. In this occasion we address the impact of heterogeneity on the performance of a network, comprised of radios which operate both in the 2.4 and 5 GHz ISM bands. We opted for a three-hop connection consisting of BT, 802.11b, and 802.11g links. It is obvious that the BT link a the bottleneck in the network. Figure 7.3 shows the TCP throughput both over the described three-hop configuration and over a single BT-link. We see that the performance degradation due to multihop is rather acceptable. In general, the heterogeneous multihop connections are more or less limited to the performance of the slowest link involved. This behaviour limits the usability and the applications of such networks. However, having heterogeneous connections in the network can be used to, *e.g.*, extend the range of BT or interconnect technologies.

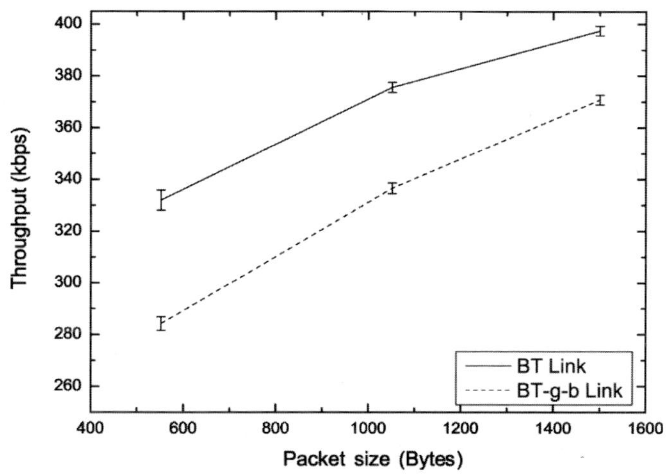

Figure 7.3. Comparison of a single BT link and a heterogeneous 3-hop connection consisting of BT, 802.11g and 802.11b.

Multiradio Approach

We already discussed, in the previous sections, some of the key limitations in the ad hoc multihop wireless networks. Number of them are rising from the MAC-layer or the IEEE 802.11 physical layers themselves. However, not all of those limitations require design of new radios or modifications to MAC-layer. Our practical experimentation and relatively trivial theoretical considerations indicate that the key challenges are mainly related to heterogeneity, interference and collision avoidance. Moreover, due to usage of a single radio for forwarding the traffic the existing bit rate will be halved even in the ideal condition without *e.g.* buffering and scheduling overheads. Anyhow, the above challenges are partially interrelated, and we believe that they could be tackled by enabling "multi-radio" concepts. Emphasizing the need of *multiradio* approach is, surprisingly, quite rare in the ad hoc research literature. The main proponent with very interesting and high quality results on the benefits of using multiple radios has been the Microsoft Research Networking Research Group (see, for example, [Bahl et al., 2004]).

The multi-radio concept means that the wireless nodes could have more than one radio NIC (Network Interface Card) available. In the case of heterogeneous networks this is natural, but we point out that even in the single technology there will be benefits if the nodes have, for example, one radio for receiving and one for sending. The simultaneous use of the radio interfaces, operating on different channels, will boost the performance of the multihop network by minimizing the delay in the data transmission. Multiple interfaces could be also useful for minimizing the handoff latency in the WLANs. However, this is easier to stipulate than to do due to a number of technological problems. One issue is to provide robust software to ensure cooperation and bridging between radio cards. The state of the art in this field is still far from perfect. One major problem is to provide auto-configurability in order to manage co-channel interference between radios. For example in the case of 802.11b cards, two cards would be virtually useless if both were to use same frequency band for their operation.

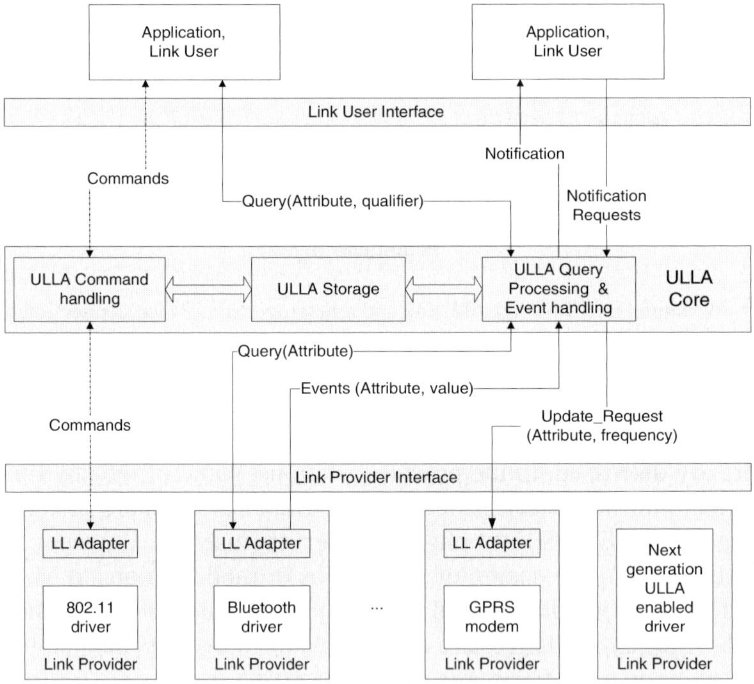

Figure 7.4. The architecture of the unified link-layer API.

Another major problem in realizing any of these multiradio approaches is the difficulty of programming in relation to several wireless technologies. In present-day operating systems the interfaces used to access wireless LAN network interfaces are completely different compared to Bluetooth ones, for example. Additionally, existing programming interfaces offer no useful abstraction of the different attributes and parameters characterising wireless technologies, thus requiring rather intimate understanding of the technology in question from the programmer. One solution to these problems is the *Universal Link-Layer API*, or ULLA for short, that is being developed in the European *GOLLUM*-project ([Farnham et al., 2005]). In addition to offering a unified interface towards different link technologies, the ULLA supports setting up different types of asynchronous notifications on changes in link conditions, and collection of statistical information on the conditions of wireless channels. Figure 7.4 illustrates the overall ULLA architecture.

This kind of API will obviously make implementation of multi-radio, link-aware routing protocols considerably easier than it is today. The work in the GOLLUM project is also very practise-oriented, and early reference implementations of the API have already been developed. Several exiting possibilities for applying this API are being explored at present, especially in relation to combining cellular network connections with more traditional technologies, such as IEEE 802.11, that are often used in the ad hoc community. We definitely see ULLA as very powerful enabling technology for link-aware ad hoc routing protocols and multiradio approaches. More details on the GOLLUM architecture design can be found from [Farnham et al., 2005].

3. Spectrum Cooperation

With the exception of multiradio issues, in the previous section we discussed classical Ad-Hoc networks in which all nodes utilize the same channel to communicate with each other. It is intuitively clear that this leads to highly inefficient use of the radio spectrum, and thus yields suboptimal capacity for end-to-end connections, especially in dense networks with considerable amount of offered traffic. In this section we shall have a look at mechanisms using which nodes in Ad-Hoc networks can also cooperate in the frequency domain[2].

We shall begin by considering a graph-theoretic approach we originally suggested in [Riihijärvi et al., 2005] as a solution to frequency allocation problems in infrastructure-mode wireless LANs. The scheme is based on the solving of the graph coloring problem on an approximation of an *interference graph*, a concept well known in frequency assignment problems. We discuss two variations of the approach. We focus on the one suitable for assigning frequencies for clusters of nodes, and discuss briefly modifications and extensions for per-connection assignments. Naturally, these approaches can be combined in case of traditional clustering algorithms are used, as first scheme can be used amongst clusters, and second one to establish communications between cluster heads.

Let us begin by considering a collection V of nodes amongst which set of frequencies F has to be assigned by some function $f : V \rightarrow F$. We shall for the moment assume that the frequencies corresponding to elements of the set F are non-overlapping, a restriction we shall remove later. With this assumption, it suffices to assign frequencies to nodes with the constraint that two nodes are assigned different frequencies if they would interfere with each others' transmissions if this was not done. We can formalise the interference relation as the *interference graph* $G = (V, E)$, where $\{v, w\} \in E$ if and only if $v \in V$ and $w \in V$ would interfere if $f(v) = f(w)$. Formulated in this manner, the frequency assignment problem becomes the classical graph colouring problem with colour set F and colouring f (see, for example, [Diestel, 2000] for references and more detailed discussion).

Solving the graph colouring problem exactly is well-known to be NP-hard, and thus takes exponentially increasing time as the number of nodes is increased. Due to this, the colouring approach has mainly been applied in frequency planning of cellular systems, to arrive at static or rarely changing frequency allocations. For a review of this work, see [Eisenblätter et al., 2002] and references therein. Nevertheless, effective heuristics make it possible to apply these techniques dynamically, even on nodes with limited processing power. Particularly appropriate is the "degree of saturation" heuristic proposed in [Brélaz, 1979], as it has attractive scaling properties, running in $O(|E| \log |V|)$ time, and is still among the best known heuristics for colouring *geometric graphs*[3].

The DSATUR heuristic is a greedy algorithm based on *degree of saturation*. The degree of saturation for a vertex v is defined as the number of different colours already used to colour vertices in its ("one-hop") neighbourhood $\gamma(v)$. The vertex degree calculated from the uncoloured vertices can be used to break the ties. More formally, the DSATUR algorithm can be described in terms of the following pseudocode, following [Buckley and Lewinter, 2002]. The algorithm takes as an input the set of uncoloured vertices U, the neighbourhood structure of the graph, and the total number of vertices.

```
DSATUR (U, γ(v) ∀v ∈ U, N) {
  Sort U from largest to smallest degree
  Colour first vertex v of U by 1
  i := 1
  Delete v from U
  while (i < N) {
   j := 1
   found := "no"
   Select first w from U with maximum degree of saturation
   while (found = "no") {
```

```
        if (Some x ∈ γ(w) has colour j)
          j := j + 1
        else
          found := "yes"
          Colour w by j
          i := i + 1
          Remove w from U
      }
    }
    All done; Output the colouring
  }
```

For more comprehensive discussion on DSATUR performance, and for some proposed variations to the basic algorithm, we refer reader to [Turner, 1988] and [Battiti et al., 2001].

To give an example of this scheme, consider the left panel of figure 7.5, illustrating the interference graph of a small wireless network in a typical office environment. The interference graph is shared amongst the nodes (we discuss the practical problems and corresponding solutions related to this process below), which then all apply the DSATUR algorithm. Initially, the degree of saturation of all nodes is zero, so the node with the highest degree is coloured in greedy manner, and is assigned the first frequency from F. Now all the uncoloured nodes have degree of saturation of one, so degree is again used to break the tie for assigning the second frequency to the node on lower-right corner of the map. As the two of the nodes have degree of two, additional tie-break rule is required on the third round of the algorithm. Simplest solution is to use the MAC-address of the nodes interpreted as 48-bit integer for this. These two nodes are then coloured on successive rounds, both assigned the third frequency, and finally the remaining node of degree one is coloured. The right panel of figure 7.5 illustrates the resulting "cell structure".

In the clustered and infrastructure cases, this simple scheme turns out to result in good channel assignments. We expect more refined applications of similar techniques to surface, where more information about the wireless network is encoded into the model graph. This would enable more refined optimizations, including consideration of the propagation environments in different channels, and inclusion of dynamic characteristics of the wireless environment into consideration. For frequency assignments between individual nodes variations of this scheme must be considered. In the graph-theoretical framework *edge colourings* and *matching problems* are two of the appropriate tools in this context. For an example application of edge colouring into channel allocation, see [Gandham et al., 2005].

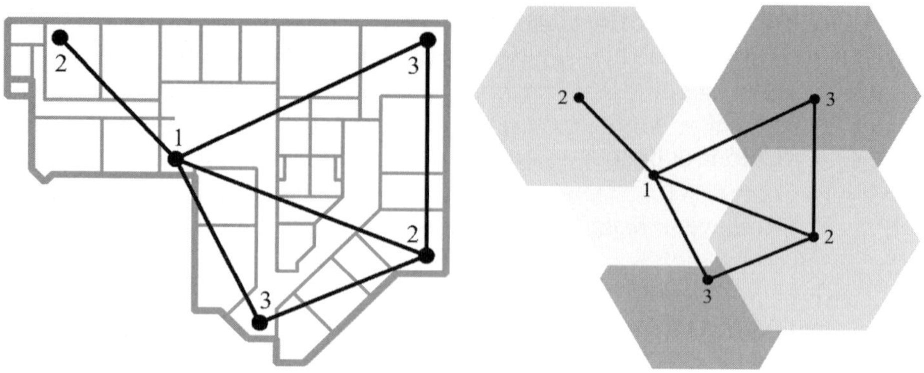

Figure 7.5. An example of a four-node wireless network, with the interference graph together with the values assigned by the colouring algorithm on the left, and illustration of the resulting "cell structure" on the right.

Two variations of the classical colouring problem often considered in the frequency allocation context are the on-line colouring problem, and T-colouring. As the name suggests, in on-line colouring the vertices are presented to the algorithm sequentially, and it must irrevocably colour those vertices without knowing the future inputs. The main application of these types of on-line colouring algorithms is to assign frequencies to nodes or clusters while retaining the existing allocations. The theoretical performance bounds of on-line algorithms tend to be poor, forcing the use of regular algorithms in occasion to optimize the frequency allocations network-wide. For more references on performance of on-line algorithms and example applications to wireless networks see [Halldórsson and Szegedy, 1992] and [Tsai et al., 2002], respectively.

The T-colouring algorithms can be used if the assumption of non-interfering adjacent channels is dropped. More precisely, the definition of a proper colouring is changed to include the condition $|f(v) - f(w)| \notin T$, where v and w are adjacent vertices in the interference graph, and T is a set of integers. If $T = \{0\}$, T-colouring reduces to classical graph colouring. As suggested in the seminal paper of [Hale, 1980], the structure of the set T can be used to put constraints barring use of "too adjacent" channels in nearby nodes. For a review of basic variants of the T-colouring problem, see also [Roberts, 1991].

Another practical problem surfaces if the colouring algorithm returns a colouring using too many frequencies. This indicates that a frequency allocation completely without interference cannot be achieved, at least using the particular heuristic. The most straightforward solution is to apply a graph transform trimming away some of the edges of the interference graph, thus reducing the chromatic index. If the edges of the interference graph carry weights, this trimming can simply be done in the order of least severe interference.

Construction of the interference graph in collaborative manner is also not entirely straightforward. Typical approximation is not to use the connectivity graph of the network instead, even though this disregards any hidden terminal-type of problems that might result. Thus the connectivity graph should be supplemented with additional interference information, if possible. Using fully distributed colouring algorithms, such as one presented in [Hedetniemi et al., 2003], avoids the communication overhead required to explicitly obtain the connectivity information at every node, but the price to be paid is the long convergence time. We do not expect algorithms of this type to be usable in mobile Ad Hoc networks, but they might be practical in some fairly static mesh networks.

Spectrum Agile "Cognitive Radios"

A small note is warranted also on the spectrum agile radios, which are also often called as *cognitive radios*, although Mitola's original cognitive radio definition goes beyond a simple dynamic spectrum allocation (see [Mitola, 2000]). In principle, the general idea of the dynamic spectrum management is to see a large part of the spectrum domain available for the cooperative use within some predefined policies (see, for example, [Buddhikot et al., 2005] and references therein). This has lead to some highly interesting recent R&D activities, where the issue has been to study, if the primary licensee spectrum domains (*e.g.*, TV-bands) could be used by secondary users in opportunistic manners. One of the best known approaches has been DARPA funded spectrum policy language project XG ([DARPA XG Working Group, 2003]) and IEEE 802.22 ([IEEE 802.22 WRAN WG, 2005]) that is developing a standard for cognitive radio - based air-interface for utilizing unused spectrum in TV broadcasting bands. The work done in the domain of dynamic spectrum management may have a large impact for ad hoc networking, *if* efficient spectrum management mechanisms can be defined the actual deployed systems would benefit from technologies developed by ad hoc research community. In fact, it is highly possible that dynamic spectrum management could be a crucial enabling technology to make ad hoc and mesh networks commercially more interesting and viable.

4. Topology Aware Ad Hoc Networks

As we have already discussed above, topology control has surfaced as a very active research area in ad hoc and sensor networks. We distinguish here two highly prominent subfields, namely *clustering research* and traditional *topology control*, which we understand to mean the tuning of the transmit power of nodes (possibly in combination with smart antennae) to optimize the network structure with respect to some metric of interest. Typical objective is to minimize the energy consumption of the network as a whole.

In clustering protocols nodes are organized into groups, and the "leader" of each group, the *cluster head*, is responsible for management of the cluster. Immediate power savings are possible by, for example, using a simple time division scheme, with the cluster head assigning activity schedules. This way other nodes can remain in sleep mode, turning off unnecessary parts of their circuitry, large portion of the time. Since pre-assigning the cluster heads as part of the network configuration process is obviously unfeasible, some form of automation must be applied. Several proposals have appeared in the literature on algorithms for automatically selecting the cluster heads, see, for example, [Karl and Willig, 2005] and references therein. For evening out the energy consumption, the responsibility for being a cluster head should be rotated amongst the nodes as time passes. Further important requirements on the clustering process are uniform distribution of cluster heads amongst the node population, and uniform distribution of energy consumed.

Classical topology control, on the other hand, deals with configuration of the radio coverages of the network nodes. Perhaps the most well-known classical problem in this domain is the *range assignment problem* and its variants. In these problems a simplified radio propagation model is assumed (such as circular radio coverage of tunable radius), and the coverages of the nodes is tuned to make the network connected with minimal overall coverage areas. If power consumption is taken to be dominated by the power-law attenuation, these kinds of problems would be equivalent to finding the radio configuration that minimizes overall power consumption. However, as we pointed out above, this assumption does not always hold. Thus, care should be taken when applying classical topology control research results on real networks. For a thorough review on these matters, and also on discussion on the problems involved, we refer the reader to [Santi, 2005] and references therein.

We shall now turn from describing and analyzing the state-of-the-art toward discussing likely future developments. Main theme in this section is enabling of network self-organization and optimization using more advanced topology control techniques. We will argue that development of new network abstractions becomes necessary, especially for including the effects of geometric relations of nodes in wireless networks.

A very likely short-term trend is that of improved understanding of the effects of topological dynamics (such as preferential attachment processes) on various network types. Also the spectrum of useful probabilistic graph abstractions is likely to grow. Although there exists models that exhibit the small world, scale-free and rich-club properties familiar from fixed networks, it is still too early to claim that these models capture all the intricacies of especially wireless communication networks at the necessary level of detail. An example of a recent model with these properties is given in [Li et al., 2004], where the appropriateness and precise definitions of these abstractions is also discussed in depth. An important piece of the puzzle that must be put into place before these findings can be used in building self-organizing and self-optimizing networks is the mapping of the efficiency of different protocols into topological characteristics. Some of these mappings are already established, or are trivial to derive, like the signaling and computational overhead of different routing protocols as the function of network topology. This research on network topologies will most probably lead to several surprising insights, like issues related to self-similarity transformed our view on how traffic behaves in networks.

When we have solved the research problems outlined above, autonomic optimization of network operation via topological tuning beyond "simple" power control becomes possible. Network nodes can gather information about the network topology and the protocols being used in the network, and map that information into optimization decisions that can be carried out by changing the network topology. At present, network layer topology can be changed by adapting the routing tables, while different overlay networks already have tunable elements in their topology formation. Another interesting aspect that has only been studied a little is the effect of cross-layer correlations in network topology on the performance of various network overlays and hierarchies.

To enable these kinds of optimization mechanisms based on topology control, we need to develop ways to exchange topology information. Routing protocols already do this on the detailed level. However, for scalability and overhead reasons it might be better to apply suitable network abstractions here as well, and exchange information like average path length, or the exponent related to scale-free degree distribution.

We expect topological optimization to become highly active research area especially in the peer-to-peer networking context but also in wireless ad hoc and mesh networks. Instead of a simple shortest path routing the topology formation takes place under numerous constraints (such as collaboration levels, reputation, reliability and energy considerations, and topology of the underlay), reminiscent of policy based routing. In some peer-to-peer environments economical considerations should also be included into considerations. This is also important observation in the fixed domain, where population densities and economic forces can drive the evolution of the network. However, due to limitations in scope, and especially due to lack of space, we shall not discuss these highly interesting issues in detail. At higher abstraction level, there has already been interesting work (see, for example, [Felegyhazi et al., 2003]) related to ad hoc network self-organization and stability in presence of different user behaviors (from cooperating to free-riding), and we expect further exiting research to emerge from this area.

Inclusion of Geometry

In wireless networks topology at any given layer becomes a secondary quantity. It is mainly constrained by the geometry of the network, that is, the spatial relations of the nodes and their environment and, of course, the dynamics of those. No equivalent of the topological abstractions discussed earlier has arisen for describing wireless networks, save for simple lattice or cellular models, and uniformly distributed point fields often used in simulations.

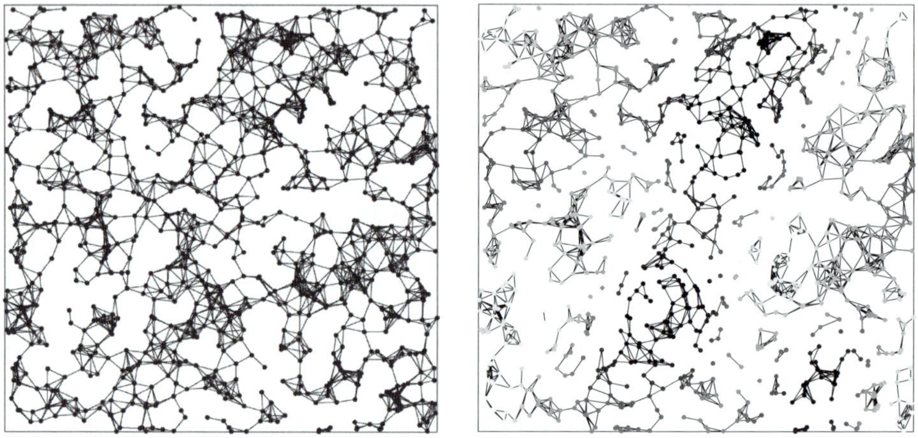

Figure 7.6. A geometric graph modeling an Ad Hoc network with uniformly distributed nodes in flat and curved terrain.

Generic mathematical frameworks such as the random geometric graphs that are suitable for basic analysis of connectivity graphs have been of course developed, see, for example, [Penrose, 2003]. Random geometric graphs are formed by placing nodes on a suitable chosen space, and connected if their distance is small enough. In Figure 7.6 simple model of this type is illustrated, with nodes placed on both flat plane, and on a surface of varying curvature, modeling terrain shapes in hilly or mountainous environments. The difference in the connectivity graphs in these two scenarios is obviously dramatic. Similar differences can easily arise from different mobility patterns of nodes, and also in the fixed network domain from various constraints placed on the network topologies (based again on, for example, trust relationships). Random geometric graphs are, of course, a very simple abstraction, and more refined models will emerge. These will be based at least on correlations in node locations, and on tools of stochastic geometry, see, *e.g.*, [Baccelli et al., 1997].

We expect the models and abstractions of wireless networks to be significantly more complicated than the purely topological models of fixed networks. This is a direct consequence of the complex phenomena wireless communication is associated with. However, significant optimization and self-organization

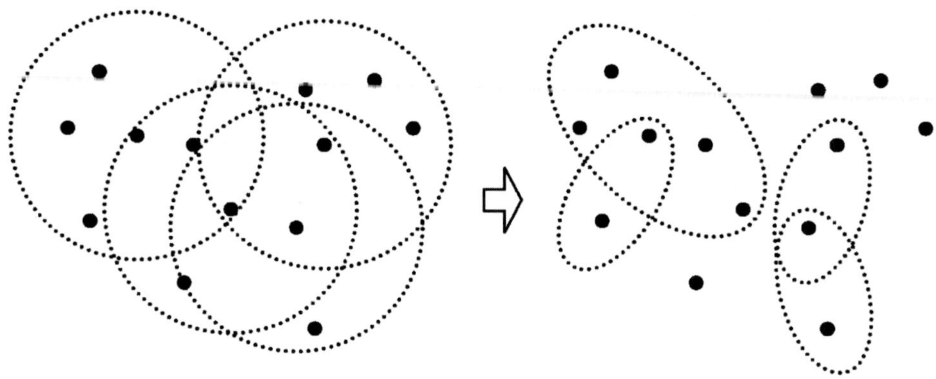

Figure 7.7. Using directional antennae in Ad Hoc networks for interference minimization and topology control.

work can already be done with simple models, such as the geometric graphs discussed above.

As an example, one can consider ad hoc and mesh networks, where adaptive and directional antennae can be used to enhance reliability and minimize interference, see Figure 7.7 for an illustration. The geometrical and topological information plays here a very crucial role. If nodes have information on, at least, their local topology the beam formation and sectorization can be planned much more efficiently and in self-organizing fashion. Especially in the case of unmanaged mesh networks this provides a tempting possibility. There remains also some practical issues, *e.g.* topology and geolocation sharing protocols and "markup" languages need to be standardized.

In the peer-to-peer domain there exists already a number of algorithms and protocols that use information about the underlying topology for optimizing the structure of the overlays they employ, and search processes conducted in the overlay. Similar kind of topology-aware approach can be used to enhance the performance of other types of discovery processes as well, such as capability and resource discoveries in wireless networks. This approach suits particularly well for protocols based on probabilistic and epidemic communications. The key idea here, especially in the resource limited wireless networks is to save resources by trying to send service discovery queries towards "information rich" pointers. Moreover, the abstraction allows to try to send certain queries towards more powerful nodes with fixed communication capability. The issue of self-organization becomes important here since we would like to ensure through self-organization that "richly connected" information nodes are available in suitable places. Hence, self-organization is not only a question of simply organizing communication capability, but to also make some topology control and service location optimization.

Another rich research field is that of dynamical properties related to geometry. It has been well established that many of the mobility models available in modern network simulators are not really satisfactory, and that the choice of mobility model can have a large impact on simulation results. For a well-known, but effective illustration on the differences between mobility models see Figure 7.8. However, the validation of the models is a difficult problem, due to scarcity of user mobility data available. In the practical self-organization design the awareness of mobility abstraction is a useful concept. If different components in the network, *e.g.* ad hoc network nodes, can be aware what sort of mobility is in average occurring around them, they can adapt their self-organization and protocol parameters accordingly. This would lead to self-organizing network, which will adapt based on conceived mobility.

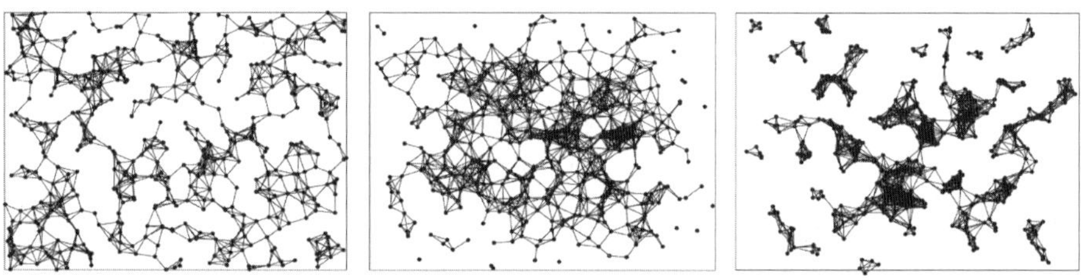

Figure 7.8. Geometric graphs corresponding to stationary node location distributions for random walk, random waypoint, and nomadic group mobility models. Differences, especially in terms of clustering, are clearly visible.

To summarize, the main short-term research problems we have seen are related to inclusion of geometric characteristics to network abstractions. This is a necessity for enabling effective network self-organization and optimization via topology control. Enabling reaction and optimization based on dynamic network characteristics also requires deep research work for developing abstractions to describe network dynamics. Further development of dynamic graph theory is certainly one of the major issues we see.

5. Hybrid Networks and 4G

Although there will be some commercial ad hoc systems that can be made successful as a standalone solutions, it is highly probable that different *hybrid network* approaches might be a real commercial route towards large-scale use of ad hoc networking principles. As the user mobility is increasing and many different local area applications become possible, there will be new opportunities to deploy limited local area ad hoc networks – these applications can include different games, community services, shopping-mall guides, etc. The need to provide very high bit-rates and cope with variety of network loads will also mean that the use of hybrid network architecture may become very attractive also for operators and manufactures.

The research community has increased its attention towards different hybrid wireless network architectures. The considered hybrid architectures are including full integration of MANETs with cellular and WLAN infrastructures, combination of multihop radio relying with cellular systems, and infrastructure support to provide high-capacity wireless networks on demand. There are different possibilities to classify hybrid architectures. Siva Ram Murthy & Manoj ([Murthy and Manoj, 2004]) divide hybrid architectures to (a) *Systems with Host-cum-Relay Stations* and (b) *Systems with Dedicated Relay* stations, which is a quite good basic differentiation. The former class has got somewhat more attention, but there has been also recent work in the latter domain.

Approximately the systems with dedicated relay stations are less flexible, and in certain sense are trying to use multihop and ad hoc principles to *enhance* existing network architectures. The systems with host-cum-relay stations offer more radical opportunities also in the field of business models, billing, and deployment. We are referring the interested reader to specific suggestions such as *multihop Cellular Network* (MCN) ([Lin and Hsu, 2000; Ananthapadmanabha et al., 2001]), iCAR (Integrated Cellular and ad hoc relaying system) by Wu et al. (in [Wu et al., 2001]), Hybrid Wireless Network Architecture (presented in [Hsieh and Sivakumar, 2001]), and highly interesting SOPRANO-architecture (Self-Organizing Packet Radio Networks with Overlay) proposed by Zadeh et al. ([Zadeh et al., 2002]). Other interesting proposals include, *e.g.*, MuPAC (Multi-power Architecture for Cellular Networks), TWiLL (Throughput Enhanced Wireless in Local Loop), A-GSM (Ad Hoc-GSM) and even 3GPP discussion on Opportunity Driven Multiple Access (ODMA) can be in part seen as a step towards hybrid architectures ([Aggelou and Tafazolli, 2001; 3GPP TSG RAN WG2, 1999; Manoj et al., 2004; Kumar et al., 2002]). Recently amount of submissions and discussion on hybrid and relay based systems has been increasing also in Wireless World Research Forum (WWRF) especially by the industry members showing that there might be momentum building up towards industrial development.

Mesh and Hybrid Deployments

Our own recent theoretical work has been focusing on understanding the opportunities and limitations of N-layer hybrid architectures. The N-layer architecture refers here to the possibility to build hierarchical architectures or overlays, where for example 802.16 network can be attached to provide broadband backbone support for 802.11 wireless hot spots. It seems that mesh networks as 1-layer or 2-layer architectures can be built quite efficiently based on ad hoc networking principles, but a lot of work is still required to understand business models, provide adequate billing mechanisms, optimize network performance (especially if multiradio approach is used).

Hybrid architecture deployments with different relying options require even more research and development work, but in our opinion they are very promising. They should not be seen as a threat by incumbent operators and manufacturers, in fact, the hybrid architectures open up many possibilities for better optimization and innovations on billing and applications domains. However, the pure spectral efficiency should not be over emphasized, especially if one is considering voice-communications (or limited video-streaming) and large-coverage areas. It is very difficult to build up commercially viable alternatives for the single-hop cellular networks. Hence, one has to be careful on the objectives of the hybrid ad hoc network approach, in our opinion they should not be "marketed" as research alternatives for cellular systems, but as systems that increase flexibility and business opportunities (and in the case of data communications can also increase a *local* spectral efficiency and network capacity).

Ad Hoc Boundaries and Applications

We expect that if the hybrid ad hoc network architectures become ubiquitous there will be clear application based boundaries for the network use. The localized applications, such as applications using Personal Area Networks (PANs) for gaming and file-sharing will be based to pure ad hoc networking, a number of application are based to hybrid architecture and can opportunistically use either ad hoc, or longer-range single-hop radio capability for communications, this application domains may well include for example vehicular applications. Even many local area applications will exhibit hybrid performance, as some parts of the system may require infrastructure support, *e.g.* due to need for billing or authentication.

6. Discussion and Conclusions

Although ad hoc networking has not become rapidly as ubiquitous as some of the proponents had been estimated, it seems that many principles that have been developed are finally finding their place towards real deployed systems. Many of the mobile ad hoc network principles are directly usable in the case of mesh networks and range extending wireless relying systems. Moreover, the sensor network applications and hybrid architectures might mean that ad hoc networks might become also a direct part of the future systems. One of the major advantages that commercial, hybrid architectures are exhibiting is the fact that cooperation between nodes can be measured and ensured much more efficiently than in the case of completely free ad hoc networks.

Complexity and Cognition

In this last section, we dare to speculate with the longer-term research focus. We have already seen that building efficient wireless ad hoc networks (hybrid or standalone) requires integration of several different techniques, and adaptivity. This leads fundamentally to issue that network must be *aware of the changes in its environment* and it must be capable of *adapting to changes in optimized and cooperative manner*. This means the systems will become rather complex. Over the last decade or so, the adaptivity boundary has been pushed from Physical Layer towards the whole system. In the case of relatively well defined systems, it is possible to use "classical" algorithmic adaptivity. However, if we need to have distributed decision making and support for optimization that takes in account many parameters and different, even mutually orthogonal, goals, it is quite possible that we need to consider the use of *machine learning* based methods. This is, in fact, the initial suggestion that Mitola was making, when he was introducing the concept of *cognitive radio* (see [Mitola, 2000]).

Mitola presented a model-based competence for software radios, and defined also an early prototype of RKRL (Radio Knowledge Presentation Language). We argue that this approach should be scaled further to include the use of cognitive decision making (optimization) also at the network layers ([Clark et al., 2003; Mähönen, 2004]). One of the work items emerging from our group has been the idea of Network Knowledge and Policy Representation Language - NKPRL). The main idea is that the high level goals and policies for ad hoc network should be presented in the machine readable form, and these representations could be then used by intelligent network elements on deciding what is the best interoperability and cooperation mode between nodes. In some sense, one could see NKPRL as a superset of the spectrum policy language XG in the networking domain.

An interesting question related to the NKPRL development is the choice of abstractions in network descriptions. As an example, graphs form perhaps the most fundamental network abstraction, used by practically all networking protocols suggested until today. Typically nodes that are handling the packets of the protocol under discussion are identified with vertices of the graph, and connections between these nodes are represented as edges. Naturally, depending on the layer of the protocol in question, each edge may consist of a number of actual, physical links. However, we believe there is a need also in network research for developing higher level abstractions for describing and, perhaps much more importantly, reasoning about large classes of networking phenomena.

Further, we would like to see the abstractions as groundwork for developing *network archetypes*, common abstractions with substantial reasoning power that could be applied into cognitive networks. Essentially, we see the network archetypes as a way cognitive network could describe its "self-image" in mathematically precise sense. Reasoning mechanisms could then be applied to decide on which actions to take if this self-image is unsatisfactory, that is, the network requires modifying. In fixed networks the observations of "small world", "scale-free" and "rich club" phenomena could be seen as first steps towards this, but inclusion of geometric relations and network dynamics inherent in ad hoc environments requires great deal of further research.

Conclusions

Wireless Multihop and Ad Hoc network paradigms have certainly many attractive features that can be used to enhance the future 4G network architectures. We have also argued that if the cognitive radio and network capabilities with the topology awareness is included to network architecture, then MANET -type of networking can become quite viable and efficient way to deploy wireless services. One of the challenges for ad hoc networking has been actually on getting enough *standardization* momentum behind the technology. Although, the basic paradigm of MANET is specifically to build ad hoc systems, we nevertheless need to agree on transmission and protocol issues. The only major effort in more formal standardization domain has been MANET working group of IETF (Internet Engineering Task Force). However, the charter of MANET is focused to routing issues, and even in the routing domain the work has progressed relatively slowly towards the consensus. The standardization efforts (*de facto* or *de jure*) definitely are needed to be increased, before ad hoc networking could become more commonplace as pointed out by Toh et al. ([Toh et al., 2005]).

The rapid expansion of IEEE 802.11 based ("WiFi") networks combined with a high potential of IEEE 802.16 ("WiMAX") is leading us to believe that at least low-mobility, mesh-type of ad hoc networks may become quite ubiquitous in the near future. We are arguing, based on experimental results done by ourselves and others, that if one is keeping the hop-count relatively low a quite good quality of service with low complexity of network can be provided. This kind of network architecture with low-mobility can provide an excellent platform for a lot of new data services, but can also provide a tempting alternative for VoIP service provision that include *wireless roaming* support without (costly) support for high mobility. Our scenario to use low-mobility ad hoc networks for data and VoIP services as a *low-cost networking model*, and perhaps some special applications (such as vehicular networks), may be a right economical incentive to make ad hoc networking reality. Hence, we conclude by pointing out that it is now time to start to also consider economical and business models for ad hoc networking.

Acknowledgments

We acknowledge a partial financial support from the European Commission (through projects GOLLUM, RUNES, 6HOP, and MAGNET), DFG (Deutsche Forschungsgemeinschaft) and Ericsson Research.

Notes

1. Spatial domain cooperation might be an important issue in the case of interference limited systems that are based, *e.g.* to spectrum agile cognitive radio technology or for topology controlled systems.

2. We shall speak of only frequency domain to simplify the discussion. Most of the techniques presented are as valid in other multiple access and channelization mechanisms, including TDMA and CDMA based networks. Nevertheless frequency domain cooperation seems to be at present most topical, as it is the method of choice for, for example, IEEE 802.11 based Ad-Hoc networks.

3. Geometric graphs (see [Penrose, 2003]) are graphs where vertices are located on a planar region for example, and edges connect vertices that are close enough in a given norm. Thus their structure can be expected to be similar to those of interference graphs of ad hoc networks.

References

3GPP TSG RAN WG2 (1999). 3GPP TR25.924 V.1.0.0: Opportunity Driven Multiple Access (ODMA).

Adya, A., Bahl, P., Padhye, J., Wolman, A., and Zhou, L. (2004). Protocol for IEEE 802.11 Wireless Networks. In *Proc. of BROADNETS 2004*, San José.

Aggelou, G. N. and Tafazolli, R. (2001). On the Relaying Capacity of Next-Generation GSM Cellular Networks. *IEEE Pers. Comm. Mag.*, 8(1):40–47.

Ahrenholz, J., Henderson, T., Spagnolo, P., Baccelli, E., Clausen, T., and Jacquet, P. (2005). Ospfv2 wireless interface type. IETF draft.

Akyildiz, I. F., Su, W., Sankarasubramaniam, Y., and Cayirci, E. (2002). A Survey on Sensor Networks. *IEEE Communications Magazine*, 40(8):102–114.

Ananthapadmanabha, R., Manoj, B. S., and Murthy, C. Siva Ram (2001). Multihop Cellular Networks: The Architecture and Routing Protocol. In *Proc. of PIMRC 2001*, pages 78–82.

Baccelli, F., Klein, M., Lebourges, M., and Zuyev, S. (1997). Stochastic geometry and architecture of communication networks. *Journal of Telecommunication Systems*, 7:209–227.

Bahl, P., Adya, A., Padhye, J., and Wolman, A. (2004). Reconsidering Wireless Systems with Multiple Radios. *ACM SIGCOMM Computer Communications Review (CCR)*, 34(5).

Bakshi, B., Krishna, P., Vaidya, N. H., and Pradhan, D. K. (1996). Improving Performance of TCP over Wireless Networks. *Technical Report 96-014, Texas A&M University*.

Balakrishnan, H., Seshan, S., and Katz, R. (1995). Improving reliable transport and handover performance in cellular wireless networks. *ACM Wireless Networks*, 1(4):469–481.

Basagni, S. (2004). *Ad Hoc Networking*. John Wiley & Sons.

Battiti, R., Bertossi, A., and Cavallaro, D. (2001). A randomized saturation degree heuristic for channel assignment in cellular radio networks. *IEEE Transactions on Vehicular Technology*, 50(2):364–374.

Brélaz, D. (1979). New methods to color the vertices of a graph. *Communications of the ACM*, 22:251–256.

Buckley, Fred and Lewinter, Marty (2002). *A Friendly Introduction to Graph Theory*. Prentice Hall.

Buddhikot, M. M., Kolodzy, P., Miller, S., Ryan, K., and Evans, J. (2005). DIMSUMNet: New Directions in Wireless Networking Using Coordinated Dynamic Spectrum Access. In *Proc. of IEEE WoWMoM'05*.

Chandra, A., Gummala, V., and Limb, J. O. (2000). Wireless Medium Access Control Protocols. *IEEE Communications Surveys, Second Quarter*.

Choudhury, R. R., Yang, X., Ramanathan, R., and Vaidya, N. H. (2002). Using directional antennas for medium access control in ad hoc networks. In *Proc. ACM MobiCom*.

Clark, D. D., Partridge, C., Ramming, J. C., and Wroclawski, T. (2003). A Knowledge Plane for the Internet. In *Proc. ACM SIGCOMM 2003*.

Clausen, T., Jacquet, P., Laouiti, A., Muhlethaler, P., Qayyum, A., and Viennot, L. (2001). Optimized Link State Routing Protocol. In *Proc. of INMIC '01*.

Couto, D. S. J. De, Aguayo, D., Bicket, J., and Morris, R. (2003). A high-throughput path metric for multi-hop wireless routing. In *Proc. of MobiCom '03*.

DARPA XG Working Group (2003). The XG Vision. Request For Comments.

De Couto, D. S. J., Aguayo, D., Chambers, B. A., and Morris, R. (2002). Performance of Multihop Wireless Networks: Shortest Path is Not Enough. In *Proc. of HotNets-I*.

DeSimone, A., Chuah, M., and Yue, O. (1993). Throughput performance of transport-layer protocols over wireless LANs. In *Proc. of IEEE GLOBECOM'93*.

Diestel, R. (2000). *Graph Theory*. Springer-Verlag.

Draves, Richard, Padhye, Jitendra, and Zill, Brian (2004). Comparison of routing metrics for static multi-hop wireless networks. In *SIGCOMM '04: Proceedings of the 2004 conference on Applications, technologies, architectures, and protocols for computer communications*, pages 133–144, New York, NY, USA. ACM Press.

Eisenblätter, Andreas, Grötschel, Martin, and Koster, Arie M. C. A. (2002). Frequency planning and ramifications of coloring. *Discussiones Mathematicae, Graph Theory*, (22):51–88.

Fahmy, N. S., Todd, T. D., and Kezys, V. (2002). Ad hoc networks with smart antennas using IEEE 802.11-based protocols. In *Proc. IEEE Intern. Conf. Commun. (ICC)*.

Farnham, T., Gefflaut, A., Ibing, A., Mähönen, P., Melpignano, D., Riihijärvi, J., and Sooriyabandara, M. (2005). Toward Open and Unified Link-Layer API. In *Proceedings of the IST Mobile and Wireless Summit*.

Feeney, Laura Marie (1999). A Taxonomy for Routing Protocols in Mobile Ad Hoc Networks. Technical Report T1999:07, Swedish Institute of Computer Science.

Felegyhazi, M., Buttyan, L., and Hubaux, J. P. (2003). Equilibrium analysis of packet forwarding strategies in wireless ad hoc networks – the static case. In *Proceedings of Personal Wireless Communications (PWC '03)*.

Fu, Z., Zerfos, P., Luo, H., Lu, S., Zhang, L., and Gerla, M. (2003). The Impact of Multihop Wireless Channel on TCP Throughput and Loss. In *Proc. of INFOCOM '03*.

Gandham, S., Dawande, M., and Prakash, R. (2005). Link Scheduling in Sensor Networks: Distributed Edge Coloring Revisited. In *Proc. of IEEE INFOCOM'05*.

Gurtov, A. and Floyd, S. (2004). Modeling wireless links for transport protocols. *ACM SIGCOMM Computer Communication Review*, 34(2):85–96.

Haas, Z. J. and Pearlman, M. R. (2001). *ZRP: A Hybrid Framework for Routing in Ad Hoc Networks*, chapter Chapter 7, pages 221–253. Addison-Wesley.

Hale, W. K. (1980). Frequency assignment: theory and applications. *Proceedings of the IEEE*, 68:1497–1514.

Halldórsson, M. M. and Szegedy, M. (1992). Lower bounds for on-line graph coloring. In *Proceedings of the 3rd annual ACM-SIAM symposium on Discrete algorithms*.

Hedetniemi, S. T., Jacobs, D. P., and Srimani, P. K. (2003). Linear time self-stabilizing colorings. *Inf. Process. Lett.*, 87(5):251–255.

Hsieh, H. Y. and Sivakumar, R. (2001). Performance comparison of cellular and multi-hop wireless networks: A quantitative study. In *Proc. of ACM SIGMETRICS*, pages 113–122.

IEEE 802.22 WRAN WG (2005). The IEEE 802.22 WRAN Working Group website. http://www.ieee802.org/22/.

Johanson, D. B., Maltz, D. A., and Broch, J. (2001). *DSR: The Dinamic Source Routing Protocol*, chapter Chapter 5, pages 139–172. Addison-Wesley.

Jubin, J. and Tornow, J. D. (1987). The DARPA Packet Radio Network Protocols. *Proceedings of the IEEE*, 75(1):21–32.

Kahn, R. E. (1977). The Organization of Computer Resources into a Packet Radio Network. *IEEE Transactions on Communications*, 25(1):169–178.

Kahn, R. E., Gronemeyer, S. A., Burchfiel, J., and Kunzelman, R. C. (1978). Advances in Packet Radio Technology. *Proceedings of the IEEE*, 66(11): 1468–1496.

Kahn, R. E. (ed.) (1978). Special Issue on Packet Communication Networks. *Proceedings of the IEEE*, 66(11).

Karl, H. and Willig, A. (2005). *Protocols and Architectures for Wireless Sensor Networks*. John Wiley & Sons.

Kleinrock, L. (1978). Principles and Lessons in Packet Communications. *Proceedings of the IEEE*, 66(11):1320–1329.

Kleinrock, L. and Silvester, J. (1987). Spatial Reuse in Multihop Packet Radio Networks. *Proceedings of the IEEE*, 75(1):156–166.

Kleinrock, L. and Tobagi, F. A. (1975). Packet Switching in Radio Channels: Part I – Carrier Sense Multiple-Access Modes and Their Throughput-Delay Characteristics. *IEEE Transactions on Communications*, 23(12):1400–1416.

Ko, Y.-B., Shankarkumar, V., and Vaidya, N. H. (2000). Medium access control protocols using directional antennas in ad hoc networks. In *Proc. IEEE Infocom*.

Kumar, K. J., Manoj, B. S., and Murthy, C. Siva Ram (2002). MuPAC: Multi-Power Architecture for Packet Data Cellular Networks. In *Proc. of PIMRC 2002*, volume 4, pages 1670–1674.

Li, L., Alderson, D., Willinger, W., and Doyle, J. (2004). A first-principles approach to understanding the internet's router-level topology. In *SIGCOMM '04: Proceedings of the 2004 conference on Applications, technologies, architectures, and protocols for computer communications*, pages 3–14, New York, NY, USA. ACM Press.

Lin, Y. D. and Hsu, Y. C. (2000). Multihop-Cellular: A New Architecture for Wireless Communications. In *Proc. of IEEE INFOCOM 2000*, pages 1273–1282.

Mähönen, P. (2004). Cognitive trends in making: future of networks. In *Proc of PIMRC 2004*, pages 1449–1454.

Manoj, B. S., Frank, C., and Murthy, C. Siva Ram (2004). Throughput Enhanced Wireless in Local Loop. *ACM Mobile Comp. and Comm. Review*, 7(1):95–116.

Min, R. and Chandrakasan, A. (2003). Top five myths about the energy consumption of wireless communication. *Mobile Computing and Communications Review*, 7(1):65–67.

Mitola, J. (2000). *Cognitive Radio*. PhD thesis, KTH.

Murthy, C. Siva Ram and Manoj, B. S. (2004). *Ad Hoc Wireless Networks: Architectures and Protocols*. Prentice Hall.

OED (2001). *Oxford English Dictionary*. Oxford University Press.

Penrose, M. (2003). *Random geometric graphs*. Oxford University Press.

Perkins, C. and Royer, E. M. (1999). Ad-hoc On-Demand Distance Vector Routing. In *Proc. of WMCSA '99*.

Perkins, C. E. (2001). *Ad Hoc Networking*. Addison Wesley.

Perkins, C. E. and Watson, T. J. (1994). Highly Dynamic Destination Sequence Vector Routing (DSDV) for Mobile Computers. In *ACM SIGCOMM'94 Conference of Communications Architecture*, London, UK.

Ramanathan, R., Redi, J., Santivanez, C., Wiggins, D., and Polit, S. (2005). Ad hoc networking with directional antennas: a complete system solution. *IEEE J. Select. Areas Commun.*, 23(3):496–506.

Riihijärvi, J., Petrova, M., and Mähönen, P. (2005). Frequency allocation for WLANs using graph colouring techniques. In *Proceedings of WONS'05.*

Roberts, F. S. (1991). T-colorings of graphs: recent results and open problems. *Discrete mathematics*, 93(2–3):229–245.

Santi, P. (2005). *Topology Control in Wireless Ad Hoc and Sensor Networks.* John Wiley & Sons.

Shacham, N. and Westcott, J. (1987). Future Directions in Packet Radio Architectures and Protocols. *Proceedings of the IEEE*, 75(1):83–98.

Sheu, J.-P. and Jie, W. (Editors) (2005). *Handbook on Theoretical and Algorithmic Aspects of Sensor, Ad Hoc Wireless, and Peer-to-Peer Networks.* Auerbach Publishers.

Sinha, P., Nandagopal, T., Venkitaraman, N., Sivakumar, R., and Bharghavan, V. (2002). WTCP: A reliable transport protocol for wireless wide-area networks. *Wireless Networks*, 8(2-3):301–316.

Tobagi, F. A. (1987). Modeling and Performance Analysis of Multihop Packet Radio Networks. *Proceedings of the IEEE*, 75(1):135–155.

Tobagi, F. A. and Kleinrock, L. (1975). Packet Switching in Radio Channels: Part II – The Hidden Terminal Problem in Carrier Sense Multiple-Access and the Busy-Tone Solution. *IEEE Transactions on Communications*, 23(12):1417–1433.

Toh, C. K. (2001a). *Ad Hoc Mobile Wireless Networks: Protocols and Systems.* Prentice Hall.

Toh, C. K. (2001b). Maximum Battery Life Routing to Support Ubiquitous Mobile Computing in Wireless Ad Hoc Networks. *IEEE Communication Magazine*, (6):138–147.

Toh, C. K. (2003). *Mobile Wireless Internet*, chapter in Future Research Challenges for Mobile Ad Hoc Networks. John Wiley & Sons Publishers.

Toh, C.-K., Mähönen, P., and Uusitalo, M. (2005). Standardization efforts & future research issues for wireless sensors & mobile ad hoc networks. *IEICE Trans. on Comm.*, E88B(9):3500–3507.

Tsai, Y.-T., Lin, Y.-L., and Hsu, F. R. (2002). The on-line first-fit algorithm for radio frequency assignment problems. *Information processing letters*, 84:195–199.

Turner, J. S. (1988). Almost all k-colorable graphs are easy to color. *Journal of Algorithms*, 9:63–82.

Vilzmann, R. and Bettstetter, C. (2005). A Survey on MAC Protocols for Ad Hoc Networks with Directional Antennas. In *Proc. EUNICE Open European Summer School.*

Vilzmann, R., Bettstetter, C., and Hartmann, C. (2005). On the Impact of Beamforming on Mutal Interference in Wireless Mesh Networks. In *Proc. IEEE Workshop on Wireless Mesh Networks (WiMesh)*.

Wellens, M., Petrova, M., Riihijärvi, J., and Mähönen, P. (2005). Building a Better Wireless Mousetrap: Need for More Realism in Simulations. In *Proc. of WONS '05*.

Wu, H., Qiao, C., De, S., and Tonguz, O. (2001). Integrated Cellular and Ad Hoc relaying Systems: iCar. *IEEE J. Select. Areas Commun.*, 19(10):2105–2115.

Xylomenos, X., Polyzos, G. C., Mähönen, P., and Saaranen, M. (2001). TCP Performance Issues over Wireless Links. *IEEE Communication Magazine*.

Zadeh, A. N., Jabbari, B., Pickholtz, R., and Vojcic, B. (2002). Self-Organizing Packet Radio Ad Hoc Networks with Overlay. *IEEE Comm. Mag.*, 40(6):140–157.

Chapter 8

MULTI-ROUTE AND MULTI-USER DIVERSITY IN INFRASTRUCTURE-BASED MULTI-HOP NETWORKS

Keivan Navaie
Broadband Communications and Wireless Systems (BCWS) Centre
System and Computer Engineering Department, Carleton University
keivan@sce.carleton.ca

Halim Yanikomeroglu
Broadband Communications and Wireless Systems (BCWS) Centre
System and Computer Engineering Department, Carleton University
halim@sce.carleton.ca

Abstract: In this chapter multi-route and multi-user diversity in multi-hop infrastructure-based wireless networks are studied. We also propose a network coordinated relaying method, Cooperative Induced Multi-user Diversity Relaying (CIMDR), to overcome the fundamental limitations on the average achieved throughput per-user. In the proposed method, multi-user diversity is induced in a 2-hop forwarding scheme and then exploited in order to improve per-user achieved throughput. Simulation results show that by using the proposed method, the net throughput per-user and the packet-drop-ratio are significantly improved.

Keywords: multi-hop wireless networks, infrastructure-based wireless network, multi-user diversity, multi-route diversity, multiple access protocol.

1. Introduction

Relay-based deployment concepts will play an important role in the cost-effective provision of very high data rates in an almost-ubiquitous manner. Cost-effectiveness is a crucial point for the success of 4G cellular networks.

Frank H. P. Fitzek and Marcos D. Katz (eds.), Cooperation in Wireless Networks: Principles and Applications, 223–241.

There has been an increasing interest in the infrastructure-based wireless multi-hop networks in academia, industry, and standardization bodies.

In the IEEE 802 Wireless World framework, a number of working groups are focusing on developing multi-hop standards:

- IEEE 802.11s - WLAN mesh networking: The goal is to develop a standard for auto-configuring multi-hop paths between access-points (APs) in a wireless distribution system. The standard is targeted to be approved by 2008.
- IEEE 802.15.5 - Wireless Personal Area Network (WPAN) mesh networking: This Task Group aims at determining the necessary mechanisms that must be present in the physical and medium access control (MAC) layers of WPANs to enable multi-hop networking. The standard is targeted to be approved by 2007.
- IEEE 802.16 - Wireless Metropolitan Area Network (WMAN): IEEE 802.16-2004 standard entitled "Air Interface for Fixed Broadband Wireless Access Systems" is approved in July 2004. The MAC layer supports a primarily point-to-multipoint architecture, with an optional multi-hop topology. The 802.16e standard amends the currently approved 802.16 standard in order to support mobility for the devices operating in the 2-6 GHz licensed bands. An optional multi-hop mode is also being considered in 802.16e. IEEE ratification of the 802.16e standard is expected in late 2005.
- IEEE 802.20 - Mobile Broadband Wireless Access (MBWA): The scope of this Task Group is to develop the specification of PHY and MAC layer of an air interface for inter-operable mobile broadband wireless access systems, operating in licensed bands below 3.5 GHz, optimized for IP-data transport, with peak data rates per user in excess of 1 Mbps. IEEE 802.20 standard is also expected to support the multi-hop architecture.

For the next generation of cellular networks, relay-based multi-hop cellular deployment concept has been considered as a potential air interface technology by Wireless World Research Forum (WWRF) as well as the Wireless world INitiative NEw Radio (WINNER) project supported by European Commission.

In addition to the above highlighted on-going standardization efforts, various proprietary multi-hop networks solutions in the unlicensed bands are being developed by the industrial players.

With the emergence of the relay-enabled standards in the IEEE 802 family, much higher interest and activity can be predicted in relay-based communications towards the end of this decade.

It is worth noting that the main goal in using the multi-hop architecture in the current proprietary solutions, as well as in the upcoming first generation relay-enabled standards, is to provide cost-effective high data rate coverage. However, once there is a relay-enabled standard it may be possible to achieve further benefits through the cooperation of the nodes in the network.

In this chapter, we study multi-user and multi-route diversity in multi-hop infrastructure-based wireless networks. We investigate the fundamental limitations on the throughput of single hop infrastructure-based networks. We then propose a simple two-hop relaying method, Cooperative Induced Multi-user Diversity Relaying (CIMDR), to overcome the fundamental limitations on the average achieved throughput per-user. We then present the CIMDR protocol details and investigate its performance. The presented simulation results also show that using CIMDR the throughput and packet drop ratio are significantly improved.

The organization of this chapter is as follows: In Section 2 we study route diversity and multi-user diversity. We also investigate the fundamental limitations of single-hop transmission. In Section 3 we present the CIMDR protocol. Simulation results are presented in Section 4. The chapter in concluded in Section 5.

2. Multi-route Diversity and Multi-user Diversity

A fundamental characteristic of wireless networks is the time variations in wireless channels. An important means to mitigate the destructive effects of the channel time variations is *diversity*, where the basic idea is to improve the system performance by creating several independent paths or, not significantly correlated, between the transmitter and the receiver.

In infrastructure-based wireless networks, the data packets are transmitted to the destination through intermediate relays. Such networks can be considered as a very rich *multi-route diversity* environment. Multi-route diversity is a potential form of diversity, which is achieved as a result of having independent wireless routes between the access-point and each user (see Fig. 8.1).

In a wireless network with multiple users, *multi-user diversity* is achieved as a result of having independent time-varying wireless channels between the access-point and different users in the coverage area (sec Fig. 8.2).

In order to improve system throughput and connectivity performance, appropriate mechanisms should be adopted in appropriate time-scales to exploit multi-user diversity and multi-route diversity.

Multi-route Diversity

In an infrastructure-based multi-hop wireless network "source" and "destination" are defined according to the radio access network. Therefore, for the uplink (downlink) "destination" ("source") means the access-point, and "source" ("destination") means the mobile user.

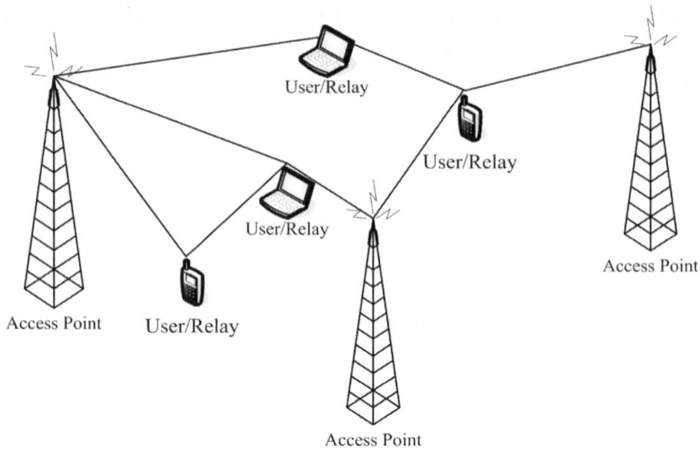

Figure 8.1. Multi-route diversity in infrastructure-based multi-hop networks.

Routing in an infrastructure-based wireless network is a functionality that recognizes, allocates, maintains, and manages wireless routes between the sources and the destinations.

Routing in multi-hop ad-hoc networks. In a wireless ad-hoc network in which there is no infrastructure, the network topology frequently changes. The time-scale of topology change is in the order of nodes' mobility. Therefore, routing in such networks is a challenging task. There are two general approaches for routing in ad-hoc networks: *topology-based* and *position-based routing* (see *e.g.*, [Mauve et al., 2001], and [Hong et al., 2002]).

In *topology-based routing* protocols the connectivity information that exists in the network is utilized for routing. Topology-based routing protocols can be further divided into *proactive*, *reactive*, and *hybrid* approaches. Proactive algorithms employ classical routing methods and maintain routing information of the available routes in the network even if these routes are not currently in use. Obviously, the main drawback of such approaches is the computational complexity and signaling overhead due to the maintenance of the routes that are not actually in use. In contrast, reactive routing protocols only maintain the routes that are currently in use. Reactive routing protocols also have some inherent limitations; these protocols need to perform a route discovery between communication peers before transmission. Obviously, performing route discovery may create a large delay in the transmission of the first packet. Moreover, even though route maintenance for reactive algorithms is restricted to the routes currently in use, it may still generate a significant amount of signalling overhead in cases that the network topology changes frequently.

In order to achieve a higher level of efficiency and scalability, a combination of a local proactive routing mechanisms and a global reactive one are considered as a hybrid routing protocol. However, even a combination of both methods still needs to maintain at least those routes which are currently in use.

The above mentioned limitations of topology-based routing are eliminated by using *position-based* routing protocols, which utilize the physical position information of the participating nodes. In these methods, each node determines its own position through the use of Global Positioning System (GPS) or some other type of positioning service. This position information is then included in the packet's destination address. The routing decision at each node is then being made based on the destination's position contained in the packet header and the position of the forwarding node's neighbors in such way that a performance metric is maximized. This performance metric indicates the efficiency of the routing algorithm in terms of the length of the route between the source and the destination and/or the transmission delay.

Note that due to the lack of network infrastructure, the main challenge for ad-hoc routing is to establish and maintain the connectivity between the source and the destination. This is not the case for infrastructure based wireless networks. In multi-hop infrastructure-based networks, selecting a particular route and transmission on it can be envisaged as a part of the resource management mechanism. Therefore, routing might be implemented jointly with or as a part of other radio resource control mechanisms ([Qiang and Acampora, 1999], [Tsirios and Haas, 2001], [Zhenzhen and Hua, 2004]).

Routing in multi-hop infrastructure-based network. For infrastructure-based multi-hop wireless networks, the stationarity (or low mobility) of the infrastructure nodes motivates the utilization of topology-based proactive methods. In this case, the routing information corresponding to the users within the coverage area of an access-point can be stored in and maintained by that access-point. Reactive routing methods can also be considered as a part of a hybrid method especially for providing ubiquitous network coverage for inter-system interconnection.

Routing techniques for multi-hop infrastructure-based networks should exploit the inherent characteristics of this network architecture:

- Network-oriented processing: Part of the routing in an infrastructure-based multi-hop network can be implemented in the infrastructure entities as these entities have more processing power. Having a network-centric routing technique not only simplifies the routing process but also provides the opportunity of performing routing jointly with other layers' functionalities.

- Position information and data flow direction: The position information and flow direction in both uplink and downlink are available. This information can be utilized for developing efficient position-based routing mechanisms.

- Cooperation incentive: Referring to the fact that the infrastructure deals with the charging issues, there could be a network coordinated framework, which promotes users' participation in cooperative communication schemes. Users' cooperation can also be very helpful in the process of routing particularly in the case of mobile relays.

Multi-route diversity can be exploited in different radio resource management mechanisms including, admission control, hand-over, load balancing, congestion control, and failure recovery.

In admission control, network resources should be allocated to a call/session to support its Quality-of-Service (QoS) during its service time. In multi-hop infrastructure-based networks there should be a close cooperation between admission control and routing mechanism. As a part of admission control, there should be a mechanism to assign a certain network access entity (*e.g.*, an access-point or a fixed relay) to the corresponding user. For a user in the network coverage area, there are likely to have more than one route to a network access entity. Multi-route diversity can be exploited in admission control. Once a certain access-point does not have any available radio resources to accept a new call/session, call admission control mechanism may consider other available routes (even if they are not optimal), and a suitable access-point may then be assigned accordingly. The multi-route diversity also makes the hand-over process easier. Having multiple routes can also be utilized in load balancing and congestion control in which users' traffic is re-routed away from the congested area.

An appropriate routing method may consider "routes" as actual network resources that should be managed and utilized opportunistically to improve the system efficiency through utilizing the most available knowledge. Accordingly, there are a number of challenges for designing a routing mechanism which includes the followings:

- Complexity: Computational complexity and signalling overhead are the fundamental challenges for any radio resource management mechanism. Usually, computational complexity is a function of the number of parameters involved in making a decision or performing an action. However, because of the availability of high processing power in the access-points, the signalling overhead is more critical. Note, that in some circumstances the signaling overhead can be replaced by computational complexity through employing more complex decision making procedures.

- Measurements: Most of the routing methods are based on the assumption of the availability of perfect measurements in appropriate times (*e.g.*, channel state, upstream queue length, etc.). This may not be precisely the case in practice which should be taken into account in designing practical routing mechanisms.

- Supporting advanced communication techniques: Using advanced communication techniques such as multi-antennas, beam-forming and cooperative relaying, are very promising in designing and developing future communication systems. In multi-hop infrastructure-based networks using such techniques may be considered for improving transmission rates. Therefore, a routing mechanism should be flexible enough to be extendable such that one or more of the previously mentioned techniques can be incorporated in the physical layer. An appropriate extension of a routing technique is the one that can efficiently exploit advanced communication techniques to improve the system performance.

- Integration into other radio resource management functionalities: Basically, routing is a functionality located in the networking layer (layer 3). However, in multi-hop infrastructure-based networks, due to time variations in channel characteristics and network topology, routing may be considered as an important entity in a cross-layer design framework which has interaction particularly with resource scheduling, admission control, and handover. Examples of such routing methods are joint routing and scheduling ([Cruz and Santhanam, 2003]), joint routing and load balancing ([Pabst et al., 2005]), inter-system routing for ubiquitous coverage ([Ai-Chun et al., 2004]), and integrated power control and routing ([Yun and Ephremides, 2005]).

Multi-user Diversity in Multi-hop Infrastructure-based Networks

The delay tolerance of data services, alongside with wireless channel fluctuations in the physical layer have been opportunistically utilized to provide efficient resource allocation in data services (see *e.g.*, [Knopp and Humblet, 1995], [Tse, 1997], [Viswanath et al., 2002]). In such techniques, the packet transmission is scheduled when time varying channel capacity happens to be at (or near) its peak. The resulting throughput improvement is referred to as *multi-user diversity gain*. This approach has been employed in high-speed downlink standards for the third generation (3G) cellular wireless communications standards, HSDPA and 1xEV-DV.

In multi-hop infrastructure-based networks, the packets are transmitted to the destination through intermediate relays. An immediate potential advantage

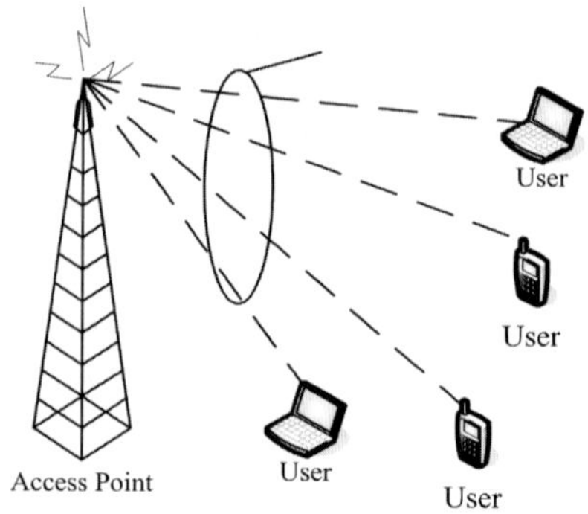

Figure 8.2. Multi-user diversity in an infrastructure-based network with multiple users.

of this architecture is the opportunity of exploiting multi-user diversity in each hop.

Here we consider an infrastructure-based wireless network in which access points with a maximum transmit power level of P_{max} are located at the center of their coverage area. An access-point transmits a signalling channel that can be received by all users in the coverage area.

In our modelling, there are n mobile users, indexed by i, distributed uniformly in the coverage area. Each packet has a large delay tolerance and includes the identity (*e.g.*, physical address) of the destination user. The wireless channel gain between the access-point and ith user at time t is given by the process $\{g_i(t)\}$ which is assumed to be stationary and ergodic. Moreover, for different users in the coverage area, the corresponding channel processes are assumed to be independent and identically distributed (i.i.d.).

At any time t, a resource allocation policy, Π, coordinates the data transmissions from the access-point to relays, or relays to destination users. For a resource allocation policy, $\Gamma_i^{\Pi}(t)$ is defined as the *achieved downlink throughput of user i at time t*, that is the number of bits received by user i at time t. For a resource allocation policy, we define *the feasible long-term achieved downlink throughput per-user*, $\Gamma^{\Pi}(n)$, such that

$$\lim_{T\to\infty} \inf_i \frac{1}{T} \sum_{t=1}^{T} \Gamma_i^{\Pi}(t) \to \Gamma^{\Pi}(n). \tag{8.1}$$

$\Gamma^{\Pi}(n)$ depends on various factors including the maximum supported bit-rate, number of users in the coverage area, and the wireless channel temporal characteristics. The definition in (8.1) is similar to that presented in ([Gupta and Kumar, 2000]), for ad-hoc networks.

To exploit multi-user diversity, a resource allocation policy, $\Pi_{\mathcal{D}}$, is employed. This policy, in its simplest form, allocates the maximum access-point transmit bit-rate to a user i^* at each time t, where

$$i^*(t) = \arg\max\{g_i(t)\}. \tag{8.2}$$

Selecting $i^*(t)$ based on the channel condition may result in an unfair resource allocation. To resolve the fairness issue, some corrective scheduling methods are often used (see *e.g.*, [Viswanath et al., 2002], [Shakkottai and Stolyar, 2002], [Navaie et al., 2005]). Since our focus is on the multi-user diversity gain, we simply consider a long-term fairness requirement in which

$$\lim_{T\to\infty} \inf_{i,j} \frac{1}{T} \sum_{t=1}^{T} \left| \Gamma_i^{\Pi}(t) - \Gamma_j^{\Pi}(t) \right| \to 0;$$

that is a direct consequence of the i.i.d. wireless channels across different users in the coverage area.

Note that, in order to exploit multi-user diversity, according to $\Pi_{\mathcal{D}}$, a user's packets have to be delayed until the channel becomes the best relative to other users. Therefore, the time-scale of channel variations that can be exploited by $\Pi_{\mathcal{D}}$ is limited by the delay tolerance of the corresponding application.

It is shown that for the described resource allocation policy, $\Pi_{\mathcal{D}}$, the overall system throughput performance is significantly higher than that of simultaneous transmission ([Knopp and Humblet, 1995]). The greater the number of users in the coverage area, the higher is the probability of occurrence of a good channel, which results in a greater improvement in the access-point throughput. However, the achieved downlink throughput per-user is still limited by the maximum transmission bit-rate and coverage area, thus limited by fundamental architectural constraints.

Consider a CDMA-based radio interface; for transmission with a rate $R_i(t)$ bits/s to a user i, the basic bit-energy to the interference-plus-noise spectral density constraint should be satisfied. Thus

$$\frac{W}{R_i(t)} \frac{P_{max} g_i(t)}{I_0} \geq \rho_i(t), \tag{8.3}$$

where I_0 is the background interference plus noise power, and $\rho_i(t)$ is the minimum required bit-energy to the interference-plus-noise spectral density for the data transmission with bit-rate $R_i(t)$. For a user i selected for transmission, using (8.3) we write,

$$R_i(t) \leq \xi_0 g_i(t) \tag{8.4}$$

where $\xi_0 = \left(\rho_i(t) I_0\right)^{-1} W P_{max}$. Therefore, for user i,

$$\lim_{T\to\infty} \inf_{i} \frac{1}{T} \sum_{t=1}^{T} \Gamma_i^{\Pi_{\mathcal{D}}}(t) = \lim_{T\to\infty} \inf_{i} \frac{1}{T} \sum_{t=1}^{T} a_i(t) R_i(t) \tag{8.5}$$

where $a_i(t)$ is the selection indicator; *i.e.*, $a_i(t) = 1$, if user i is selected for transmission at time t, and 0 otherwise. Summing (8.5) over all users, we have

$$\Gamma^{\Pi_{\mathcal{D}}}(n) \leq \frac{\xi_0}{n} \lim_{T\to\infty} \inf_i \frac{1}{T} \sum_{i=1}^{n} \sum_{t=1}^{T} a_i(t) g_i(t) \tag{8.6}$$

$$= \frac{\xi_0}{n} \lim_{T\to\infty} \inf_i \frac{1}{T} \sum_{t=1}^{T} g_{i^*}(t). \tag{8.7}$$

Eq. (8.7) shows that the downlink throughput per-user is upper-bounded by $g_{i^*}(t)$.

To increase multi-user diversity gain, in ([Viswanath et al., 2002]), multiple transmit antennas are used to induce large and fast channel fluctuations, *i.e.*, greater $g_{i^*}(t)$. Also in a multiple-cell scenario, the independent time variations of the wireless channels between a user and the neighboring access-points is introduced in ([Navaie and Yanikomeroglu, 2005]), as a new dimension in multi-user diversity. This form of diversity is exploited by joint access-point assignment and packet scheduling, which results in greater $g_{i^*}(t)$ and thus greater multi-user diversity gain per-user.

To exploit the multi-user diversity in a multi-hop network, a relaying method is proposed in ([Larsson and Johansson, 2005]). In this method, using a sequential optimization approach, multi-user diversity is exploited in each hop by selecting the next relay based on the instantaneous channel quality. However, selecting only one relay reduces the opportunity of capturing a good channel in the next hop. In the following section, we propose an access-point coordinated cooperative relaying method, Cooperative Induced Multi-user Diversity Relaying (CIMDR).

3. Cooperative Induced Multi-user Diversity Routing for Multi-hop Infrastructure-based Networks with Mobile Relays

CIMDR (Fig. 8.3) exploits the broadcast nature of wireless channel to induce multi-user diversity through a two-phase process. The basic idea is as follows. In the first phase, access-point broadcasts data packets with its maximum bit-rate. Some users in the coverage area are likely to receive the transmitted data packets. These users, act as potential relays in the second phase; each potential relay wait until the occurrence of a "good channel" to the destination user and then transmit the data packets. As soon as the transmission is carried out by one of the potential relays, the access-point manages to release the packets buffered in other potential relays.

We consider a 2-hop infrastructure-based network. Access-point is located at the center of the coverage area and its maximum transmit power is P_{max}.

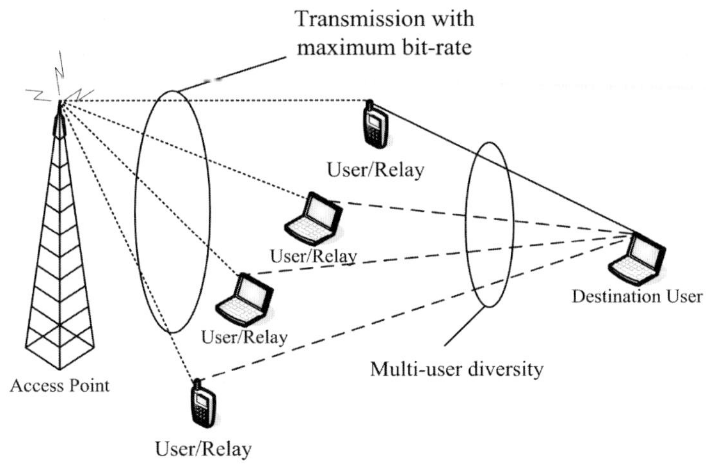

Figure 8.3. CIMDR for 2-hop transmission.

Packets can be transmitted directly from the access-point to the users, or they can go through another mobile user serving as a relay. Access-point transmits signalling on dedicated control channel(s) that can be received by all users in the coverage area.

There are n mobile users, indexed by i, distributed uniformly in the coverage area. Mobile users are able to receive, temporarily save and relay packets in the same frequency band of access-point transmission. They also transmit signaling information on an uplink signaling channel. Mobile terminals have a large enough buffer to store relay packets. Each packet has a large delay tolerance and includes the identity (*e.g.*, physical address) of the destination user. Each user in the coverage area broadcasts a pilot signal to indicate its identity. This pilot signal is also utilized by the relays for channel estimation. To decrease power consumption, broadcasting of users' pilot channel can be activated upon receiving a signal (from the access-point) indicating the existence of a data packet destined to that mobile user.

Since by this scenario we *induce* multi-user diversity through generating independent paths between the destination user and m relays, we name it Co-operative Induced Multi-user Diversity Relaying (CIMDR).

CIMDR Protocol

The proposed scenario, $\Pi_{\mathcal{I}}$, has two phases: the *feeding phase* and the *delivery phase*. These two phases occur sequentially in time (Fig. 8.4). The time-span of each phase (*i.e.*, τ_F and τ_D) is assigned based on the network

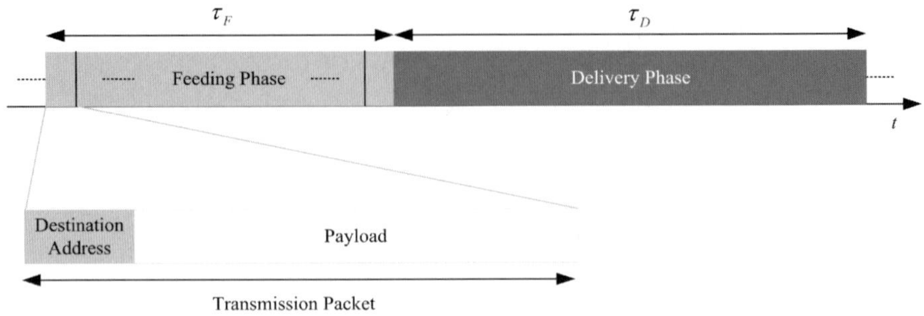

Figure 8.4. CIMDR two-phase protocol and packet structure.

traffic and the communication environment characteristics. Fig. 8.5 shows the signalling procedure of CIMDR protocol.

Feeding Phase: In the feeding phase, packets are broadcasted by the access-point with its maximum bit-rate; the total number of transmitted bits in the feeding phase would be $\tau_F R_{max}$, where τ_F is the time duration of the feeding phase. During the feeding phase multiple packets are transmitted using time domain scheduling. Any user which receives a data packet in the feeding phase acts as potential relays in the delivery phase.

The transmission order of the queued packets in the access-point is managed by a higher-layer functionality. If the destination user is among those who receive packets in the feeding phase, it sends a received acknowledge signal, R-ACK, to the access-point. Consequently, the access-point broadcasts a data release signal, D-REL, and all other relays release that data packet.

Here we assume that the number of users in the coverage area is high enough that in each time instant there is, at least, one user that can receive the transmitted data in the feeding phase. In cases where no mobile user in the coverage area can receive the transmitted packet in the feeding phase, the access-point should reduce its bit rate.

In the feeding phase, multi-user diversity gain comes from the fact that the access-point radio resource is only allocated for transmission with its maximum bit-rate. Note that for a large number of users in the coverage area, it is likely that some users will have a channel state that supports the access-point's highest bit-rate.

Delivery Phase: In the delivery phase, the access-point is kept inactive and only transmissions from relays to the final destinations are allowed. Each relay continuously tracks the quality of the wireless links to the neighboring users as well as their identity. If a relay is able to achieve a transmission bit-rate greater than or equal to a system parameter R_0 bits/s, then that relay transmits to the

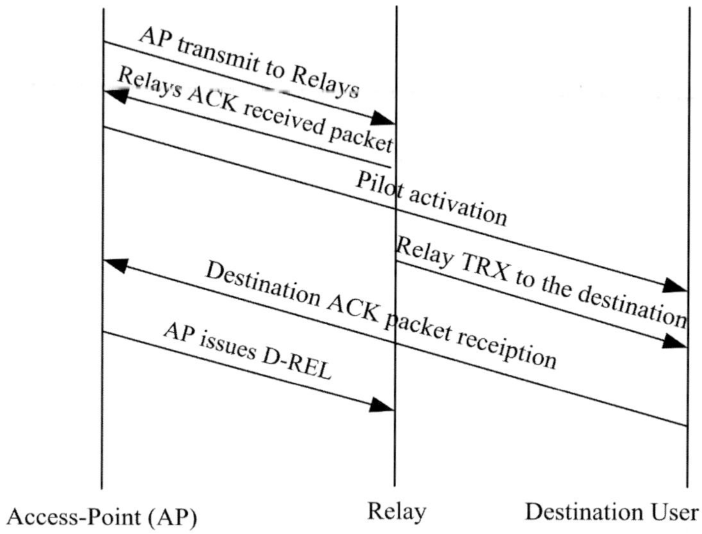

Figure 8.5. CIMDR signaling: Normal transmission.

destination user. The selection of R_0 critically affects the system performance and is elaborated in the Proposition given below.

Medium access control can be either a contention-based method or an access-point coordinated non-contention based method. Upon successful transmission, destination user sends an R-ACK signal to the access-point. Consequently, the access-point broadcasts a D-REL signal and other relays release that packet. If the access-point does not receive an R-ACK corresponding to a packet in a predefined time interval, τ_{max} seconds, that packet is considered lost and a D-REL signal is broadcasted by the access-point (see Fig. 8.6). That packet may be considered for retransmission in a later time.

Multi-user diversity in the delivery phase is exploited by transmission on channels with the achieved bit-rate greater than or equal to R_0. Note that in practice the transmit bit-rate may be adjusted based on the channel status which is fed back into the access-point by the users.

For a given medium access control technique, $0 < \gamma \leq 1$ is defined as the medium access control gain, which shows the average portion of the radio resource (*e.g.*, transmission time) that can be allocated to the competitors for a shared media. For non-contention based medium access control mechanisms $\gamma = 1$. Let R be the average access-point transmission bit-rate for single hop transmission with multi-user scheduling. The following proposition provides the condition on the system parameters for CIMDR.

Proposition 1: For a large number of users in the coverage area, by using CIMDR the access-point throughput is increased compared to the single-hop

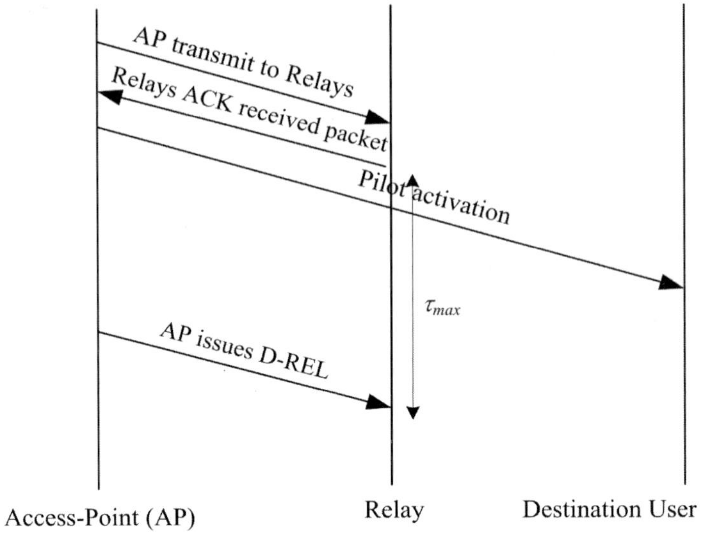

Figure 8.6. CIMDR signaling: Lost packet.

transmission if

$$\frac{1}{R_0} < \gamma\Big(\frac{1}{R} - \frac{1}{R_{max}}\Big). \tag{8.8}$$

Proof: Let $\Phi_D(t)$ and $\Phi_F(t)$ be the indicator functions for delivery and feeding phases, respectively. Therefore, $\Phi_D(t) = 1$ ($\Phi_F(t) = 1$) when the system is in delivery (feeding) phases and $\Phi_D(t) = 0$ ($\Phi_F(t) = 0$), otherwise. The total data bits in the system at time t, $\Theta(t)$, can be calculated as

$$\Theta(t) = \int_0^t \Big(\gamma R_0 \Phi_D(\alpha) - R_{max}\Phi_F(\alpha)\Big)d\alpha.$$

For system stability

$$\lim_{T\to\infty} \Theta(T) \to 0, \tag{8.9}$$

thus

$$\tau_D \gamma R_0 = \tau_F R_{max}. \tag{8.10}$$

On the other hand, compared to the single-hop transmission, the total access-point throughput will be increased if

$$\tau_F R_{max} > (\tau_F + \tau_D)R. \tag{8.11}$$

Hence, using (8.10) and (8.11),

$$\frac{1}{R_0} < \gamma\Big(\frac{1}{R} - \frac{1}{R_{max}}\Big). \tag{8.12}$$

This proves the proposition ■.

On one hand, if R_{max} is very large, then a smaller number of users in the coverage area will receive the data packets in the feeding phase. On the other hand, decreasing R_{max} will increase the number of potential relays but will decrease the overall rate. The transmission rate R_{max} may also be adjusted based on the number of potential relays; if data packets are not received by a reasonable number of mobile users, then R_{max} may be decreased.

Given the condition in (8.8) holds, within the interval $[0,\ T]$ for $T \to \infty$ all packets transmitted to the relays will be delivered to the users. Therefore, for CIMDR, it is simple to show that (8.7) is modified as

$$\Gamma^{\Pi_{\mathcal{I}}}(n) \leq \frac{\xi_1}{n}\breve{g}, \tag{8.13}$$

where ξ_1 is defined similar to ξ_0 in (8.7) and $\breve{g}$ is the minimum time-average value of the channel gain between the access-point and the relay with the maximum transmission bit-rate. Note that

$$\lim_{T\to\infty}\ \inf_{i}\ \frac{1}{T}\sum_{t=1}^{T} g_{i^*}(t) < \breve{g}, \tag{8.14}$$

which is a direct consequence of smaller path-losses because of multi-hop transmission. This directly results in $\Gamma^{\Pi_{\mathcal{I}}}(n) > \Gamma^{\Pi_{\mathcal{D}}}(n)$. In other words, using CIMDR, the achieved average throughput per user is increased.

Note that τ_{max} has an important role in the performance of CIMDR. If $\tau_{max} \to \infty$, then a packet can be kept waiting in a potential relay until the occurrence of a very high rate channel (*i.e.*, very large R_0). For moderate values of τ_{max}, the mobility is very important. The higher the users' mobility the higher is the probability of the occurrence of a high bit-rate channel in the second hop. For a given mobility profile, a larger value of τ_{max} results in the exploitation of the mobility in a more efficient way.

Incentive system for users' cooperation. In CIMDR users which act as relays, participate in the transmission process. However, there should be an incentive system to provide reasonable motivation to the users for cooperation in the relaying process.

This problem is heavily studied in the context of mobile ad-hoc networks (see *e.g.*, [Buttyan and Hubaux, 2003; Zhong et al., 2003]). In an infrastructure-based multi-hop system it is possible to have a network-based incentive system which makes this problem a lot easier.

Here, we propose a simple credit based incentive system. In this system a user i is granted a *participation credit* of $\mu(t)$ upon participating in relaying process at time t. This is due to the fact that the mobile user allocates a part of its processing power for tracking the neighboring mobile stations and for involving

Table 8.1. Simulation Parameters.

Parameter	Value
Physical layer	Based on UMTS
Cell radius	100 m
Access-points transmit power	10 W
Standard dev. of log-normal fading	8 dB
Background noise density	-174.0 dBm/Hz
Propagation loss exponent	4
Time-slot length	10 ms
R_{max}	2 Mbps
R	384 Kbps
Medium access control gain (γ)	≈ 1
Minimum required E_b/I_0	2 dB

in the corresponding signaling processes. As soon as finding the destination user and detecting a channel with available bit-rate greater than or equal to R_0, the relay transmits the data packet thus allocates a portion of its transmission power to relaying. In this case, the network grants a *relaying credit* of $\nu(t)$ to that mobile user. The values of $\mu(t)$ and $\nu(t)$ are related to the network charging strategy and can be varied in different times of the day based on the network traffic. In such a scenario with m mobile users participating in CIMDR, the total granted credit per packet transmission is $m.\mu(t) + \nu(t)$ which would be considered as part of the transmission cost for each data packet.

4. Simulation Results

We simulate a single-cell DS-CDMA system with n active users based on UMTS standard ([Holma and Toskala, 2000]). Users are uniformly distributed in the coverage area. The simulation parameters are presented in Table 8.1. A simple mobility model has been implemented, in which at each time instant, a user randomly located within a circle with its previous location in the center and a diameter of 2.5 meters.

To show the effect of multi-user diversity, we consider three different systems: in System I, for each user the access-point transmits packets in first-come-first-serve fashion using a time domain scheduling scheme. In System II, packets are scheduled based on $\Pi_{\mathcal{D}}$. Transmission in System III is based on $\Pi_{\mathcal{I}}$, with a non-contention based medium access control technique in the delivery phase.

System I is considered as the benchmark, and the average achieved net throughput of Systems II and III are normalized by the average achieved net throughput of System I. Fig. 8.7 illustrates the normalized average achieved net throughput versus the number of users in the coverage area for $\tau_{max} = 2$ s.

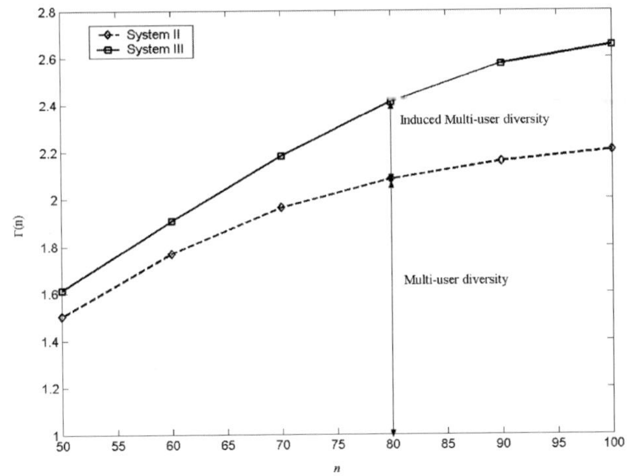

Figure 8.7. Normalized average achieved net throughput versus the number of users.

The difference between the throughput gains of Systems II and III indicates the achieved multi-user diversity gain resulting from exploiting the induced multi-user diversity by CIMDR. As it is expected, this gain is increased by the number of users. Note that normalized throughput curve will saturate because of the access-point total throughput constraint.

We also compare the packet-drop-ratio for System II and System III. Packets are considered lost when they cannot be transmitted within a delay threshold of $\tau_{max} = 2$ s. As it can be seen in Fig. 8.8, using CIMDR improves the packet-drop-ratio performance. The greater improvement in the packet drop ratio is archived by a larger number of users in the coverage area and a larger delay tolerance of 10 s.

5. Conclusion

In this chapter we study the multi-user diversity gain in the downlink of single-hop and multi-hop infrastructure-based networks. We propose an network coordinated cooperative relaying method, Cooperative Induced Multi-user Diversity Relaying (CIMDR), to overcome the fundamental limitations on the average achieved net throughput per-user. In the proposed method, multi-user diversity is induced in a 2-hop forwarding scheme and then exploited to improve per user achieved data throughput. We show that by using the proposed method, the throughput per-user and the packet-drop-ratio are significantly improved.

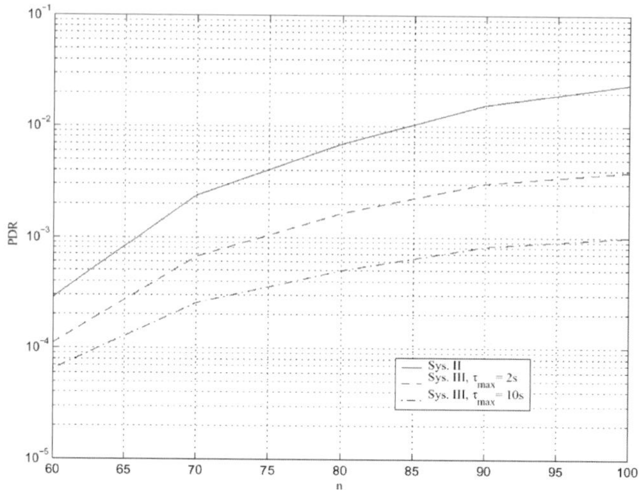

Figure 8.8. Packet-drop-ratio (PDR) of CIMDR and single-hop multi-user diversity scheduling for $\tau_{max} = 2$ and 10 seconds.

References

Ai-Chun, P., Jyh-Cheng, C., Yuan-Kai, C., and Agrawal, P. (2004). Mobility and session management: UMTS vs. cdma2000. *IEEE Wireless Communications*, 11(4):30–43.

Buttyan, L. and Hubaux, J. P. (2003). Stimulating cooperation in self-organizing mobile ad hoc networks. *ACM Mobile Networks and Applications*, 8(5):579–592.

Cruz, R. L. and Santhanam, A. V. (2003). Optimal routing, link scheduling and power control in multihop wireless networks. *in Proc. INFOCOM 2003*, 1:702–711.

Gupta, P. and Kumar, P. R. (2000). The capacity of wireless networks. *IEEE Trans. on Info. Theory*, pages 388–404.

Holma, H. and Toskala, A. (2000). *WCDMA for UMTS: Radio Access for Third Generation Mobile Communications*. John Willy and Sons.

Hong, X., Xu, K., and Gerla, M. (2002). Scalable routing protocols for mobile ad hoc networks. *IEEE Network*, 16(4):11–21.

Knopp, R. and Humblet, P. A. (1995). Information capacity and power control in single-cell multiuser communications. *in Proc. IEEE ICC'95*, pages 331–335.

Larsson, P. and Johansson, Niklas (2005). Multi-user diversity forwarding in multi-hop packet radio networks. *in Proc. IEEE WCNC'05*.

Mauve, M., Widmer, A., and Hartenstein, H. (2001). A survey on position-based routing in mobile ad hoc networks. *IEEE Network*, 15(6):30–39.

Navaie, K., Montuno, D., and Zhao, Y. Q. (2005). Fairness of resource allocation in cellular networks: A survey. In Li, W. and Pan, Y., editors, *Resource Allocation in Next Generation Wireless Networks*. Nova Science Publishers.

Navaie, K. and Yanikomeroglu, H. (2005). Optimal downlink resource allocation for elastic traffic in cellular CDMA/TDMA networks. *Preprint*.

Pabst, R., Walke, B. H., Schultz, D. C., Herhold, P., Yanikomeroglu, H., Mukherjee, S., Viswanathan, H., Lott, M., Zirwas, W., Dohler, M., Aghvami, H., Falconer, D. D., and Fettweis, G. P. (2005). Relay-based deployment concepts for wireless and mobile broadband cellular radio. *IEEE Communications Magazine*, 42(9):80–89.

Qiang, G. and Acampora, A. (1999). Routing and handoff support for a novel wireless broadband access network (UniNet). *in Proc. IEEE GLOBCOM'99*, 5:2788–2793.

Shakkottai, S. and Stolyar, A. (2002). Scheduling for multiple flows sharing a time-varying channel: The exponential rule. In Suhov, Yu. M., editor, *AMS Translations*, 2.

Tse, D. (1997). Optimal power allocation over parallel Gaussian channels. *in Proc. IEEE ISIT'97*, page 27.

Tsirios, A. and Haas, Z. J. (2001). Multipath routing in the presence of frequent topological changes. *IEEE Communications Magazine*, 39(11):132–138.

Viswanath, P., Tse, D. N. C., and Laroia, R. (2002). Opportunistic beamforming using dumb antennas. *IEEE Trans. on Info. Theory*, pages 1277–1294.

Yun, L. and Ephremides, A. (2005). Joint scheduling, power control, and routing algorithm for ad-hoc wireless networks. *HICSS '05*, page 322b.

Zhenzhen, Y. and Hua, Y. (2004). Networking by parallel relays: diversity, lifetime and routing overhead. *in Proc. of Thirty-Eighth Asilomar Conference on Signals, Systems and Computers*, 2:1302–1306.

Zhong, S., Chen, J., and Yang, Y. (2003). Sprite: A simple, cheat-proof, credit-based system for mobile ad-hoc networks. *in Proceedings of IEEE INFOCOM'03*, 3:1987–1997.

Chapter 9

COGNITIVE RADIO ARCHITECTURE

Organizing Computational Intelligence for Peer and Network Collaboration

Joseph Mitola III
*The Mitre Corporation**
jmitola@mitre.org

Abstract: This chapter explores in detail the concept of cognitive radio architecture in centralized and ad hoc networks. Such architecture is paramount for establishing a cooperating platform of environment-aware nodes. The chief constituent elements of the cognitive radio architecture include functions implemented through components using a set of design rules. The synergy between software defined radio and cognitive principles is fully exploited by assuming that the cognitive radio architecture is based on a software defined radio endowed with a perceptive system capable of sensing the radio environment, in particular the use of the radio spectrum. This capability is complemented with an intelligent system able to understand the radio environment and, based on the user's needs or expectations, dynamically reconfigure the system taking into account spectrum usage. The so called cognition cycle is an architecture subsystem including an inference hierarchy, the temporal organization and flow of inferences and control states. The cycle continually observes the environment, orients itself, creates plans, decides, and then acts. The phases of inference from observation to action show the flow of inference, a top-down view of how cognition is implemented algorithmically. The inference hierarchy is the part of the algorithm architecture that organizes the data structures. The cognition functions are implemented via cognition elements consisting of data structures, processes and flows. These include data structures and related processing elements may be modeled as topological maps over abstract domains. The processing elements of the architecture, modeled topological maps like input, best-match and other maps, are described in detail in

*The author's affiliation with the MITRE Corporation is provided for identification purposes only, and is not intended to convey or imply MITRE's concurrence with, or support for, the positions, opinions or viewpoints expressed by the author.

Frank H. P. Fitzek and Marcos D. Katz (eds.), Cooperation in Wireless Networks: Principles and Applications, 243–311.

this chapter. As software defined radio is the fundamental platform for cognitive radio, the former is reviewed also in this chapter, including overviews of software radio architecture (SRA) and software communications architecture (SCA), the transition and evolution towards cognitive radio architecture, and other relevant technical aspects.

1. Introduction

The Cognitive Radio Architecture (CRA) frames the Cognitive Wireless Networks (CWNs) for cooperation among computationally intelligent nodes whether orchestrated by a host CWN or forming dynamic peer CWN on the fly. This chapter contributes to cooperative networks by examining the functions, components, and design rules -the architecture - that enhances software-defined radio SDR to cognition. One critical differentiator between SDR and cognitive radio is the ability to learn from the RF environment and the user. For this learning to occur without burdening the user with "dumb questions" the cognitive radio needs perception systems by which to perceive just those aspects of the environment from RF and user viewpoints that enables the radio to better support the quality of information (QoI) needed by the user.

Cognitive Radios Know Radio Like TellMe® Knows 800 Numbers

When you dial 1-800-555-1212, an algorithm says "Toll Free Directory Assistance powered by TellMe® [TellMe Networks, 2005]. Please say the name of the listing you want." If you mumble it says, "OK, United Airlines. If that is not what you wanted press 9, otherwise wait while I look up the number." Ninety-nine point nine, nine percent of the time TellMe® gets it right, replacing thousands of directory assistance operators of yore. TellMe®, a speech-understanding system, achieves a high degree of success by its focus on just one task: finding a toll-free telephone number. Narrow task focus is the key to algorithm successes.

The Cognitive Radio Architecture (CRA) frames the Cognitive Wireless Networks (CWNs), offspring of TellMe®. CWNs are emerging as practical, real-time, highly focused applications of computationally intelligent technology. CWNs differ from the more general Artificial Intelligence (AI) based services like intelligent agents, computer speech, and computer vision in degree of focus. Like TellMe®, CWNs focus on a very narrow tasks. Broader than TellMe®, the task is to create adapt radio-enabled information services for to the specific needs of a specific user. TellMe®, a network service, requires substantial network computing resources, teraflops of computing capability and terabytes of memory to serve thousands of users at once. CWNs, on the other hand, may start with a Cognitive Radio (CR) in your purse or on your belt, a cell phone on steroids, focused on the narrow task of creating from the myriad available wireless information networks and resources just what is needed by just one user, you. TellMe® interacts with anybody, but each CR is self-aware and owner-aware via sensory perception and Autonomous Machine Learning (AML) technologies, earning the term "cognitive." Each CR fanatically serves the needs and protects the personal information of just one user, you, its owner. Thus, via the CRA incorporates with perception and AML.

TellMe® is here and now, while CWNs are emerging in global wireless research centers and industry forums like the SDR Forum and Wireless World Research Forum (WWRF). This chapter provides an overview of CRA systems architecture challenges and approaches, emphasizing CR as a technology enabler for rapidly emerging commercial CWN services and generation-after-next military communications, based on the foundation technologies computer vision, computer speech, AML, and SDR.

CRs See What You See to Discover RF, Needs, and Preferences

In 2002, GRACE (Graduate Robot Attending ConferencE) [Anne Watzman, 2002], Roombah's big sister with a CRT for a face entered the International Joint Conference on Artificial Intelligence (IJCAI); found the registration desk; registered by talking to the receptionist; followed the signs that said "ROBOTS" this way and "HUMANS" the other way; when called on gave a five-minute talk about herself; and then answered questions. She was the first to complete the mobile robot challenge first articulated in the 1980's. There were no joysticks and no man behind the curtain: just a robot that can autonomously see, hear, and interact with the people and the environment to accomplish those specific tasks.

Compared to GRACE, the standard cell phone is not too bright. Although the common cell phone has a camera, it lacks GRACE's vision algorithms so it does not know what it is seeing. It can send a video clip, but it has no perception of the visual scene in the clip. If it had GRACE-like vision algorithms, it could perceive the visual scene. It could tell if it were at home, in the car, at work, shopping, or driving up the driveway on the way home. If GRACE-like vision algorithms show it that you are entering your driveway in your car, a Cognitive SDR could learn to open the garage door for you wirelessly. Thus, you would not need to fish for the garage door opener, yet another wireless gadget. In fact, you do not need a garage door opener anymore, once CRs enter the market. To open the car door, you will not need a key fob either. As you approach your car, your personal CR perceives the common scene and, as trained, synthesizes the fob RF transmission and opens the car door for you.

Your CR perceives visual scenes continuously searching visual - RF correlations, cues to your needs for wireless services. A CR radio learns to open your garage door when you arrive home from your use of the garage door opener. When first you open the garage door with the wireless garage-door opener, your

CR correlates the visual and RF scenes: owner's hand on device, then RF signal in the ISM band, and then the garage door opens. The next time, your CR verifies through reinforcement learning that your hand on the button, the RF signal, and the opening of the garage door form a sequential script, a use-case. The third time, your cognitive radio detects the approach to the garage door and offers to complete the RF use case for you, saying, "I see we are approaching the garage. Would you like me to open the door for us?" Thereafter, it will open the garage door when you drive up the driveway unless you tell it not to. It has transformed one of your patterns of RF usage, opening the garage door; into a cognitive (self-user perceptive) service, offloading one of your daily tasks.

Since the CR has learned to open the garage door, you may un-clutter your car by just one widget, that door opener. Since your CR learned to open the garage door by observing your use of the radio via AML, you did not pay the cell phone company, and you did not endure pop-up advertising to get this personalized wireless service. As you enter the house with arms full of packages, your CR closes the garage door and locks it for you, having learned that from you as well. For the CR vision system to see what you see, today's Bluetooth earpieces evolve to CR Bluetooth glasses, complete with GRACE-like vision.

CRs do not attempt everything. They learn about your radio use patterns because they know a LOT about radio and about generic users and legitimate uses of radio. CRs have the a-priori knowledge needed to detect opportunities to assist you with your use of the radio spectrum accurately, delivering that assistance with minimum intrusion. TellMe® is not a generic speech understanding system and CR is not a generic AI service in a radio. Products realizing the visual perception of this vignette are realizable on laptop computers today. Reinforcement learning (RL) and Case-based Reasoning (CBR) are mature AML technologies with radio network applications now being demonstrated in academic and industrial research settings as technology pathfinders for CR [Joseph Mitola III, 2000a] and CWN [Petri Mahonen, 2004]. Two or three Moore's law cycles or three to five years from now, these vision and learning algorithms will fit in your cell phone. In the interim, CWNs will begin to offer such services, offering consumers new tradeoffs between privacy and ultra-personalized convenience.

Cognitive Radios Hear What You Hear, Augmenting Personal Skills

Compared to GRACE, the cell phone on your waist is deaf. Although your cell phone has a microphone, it lacks GRACE's speech understanding technology, so it does not perceive what it hears. It can let you talk to your daughter, but it has no perception of your daughter, nor of the content of your conversation. If it had GRACE's speech understanding technology, it could perceive your speech dialog. It could detect that you and your daughter are talking about common subjects like homework, or your favorite song. With CR, GRACE-like speech algorithms would detect your daughter saying that your favorite song is now playing on WDUV. As an SDR, not just a cell phone, your CR tunes to FM 105.5 so that you can hear "The Rose." With your CR, you no longer need a transistor radio. Your CR eliminates from your pocket, purse or backpack yet another RF gadget. In fact, you may not need iPOD®, GameBoy® and similar products as high-end CR's enter the market. Your CR will learn your radio listening and information use patterns, accessing the songs, downloading games, snipping broadcast news, sports, stock quotes as you like as the CR re-programs its internal SDR to better serve your needs and preferences. Combining vision and speech perception, as you approach your car your CR perceives this common scene and, as you had the morning before, tunes your car radio to WTOP to your favorite "Traffic and weather together on the eights." With GRACE's speech understanding algorithms, your CR recognizes such regularly repeated catch phrases, turning up the volume for the traffic report and then turning it down or off after the weather report, avoiding annoying commercials and selecting relevant ones. If you actually need a tax deduction, it will record ***those*** radio commercials for your listening pleasure at tax time when you need them.

For AML, CRs need to save speech, RF, and visual cues, all of which may be recalled by the user, expanding the user's ability to remember details of conversations and snapshots of scenes, augmenting the skills of the <Owner/>[1]. Because of the brittleness of speech and vision technologies, CRs try to "remember everything" like a continuously running camcorder. Since CRs detect content such as speakers' names, and keywords like "radio" and "song," they can retrieve some content asked for by the user, expanding the user's memory in a sense. CRs thus could enhance the personal skills of their users such as memory for detail.

High performance dialog and audio-video retrieval technologies are cutting-edge but not out of reach for suitably narrow domains like TellMe® and customization of wireless services. Casual dialog typically contains anaphora and ellipsis, using words like "this" and "that" to refer to anonymous events like playing a favorite song. Although innovative, speech research systems already achieve similar dialogs in limited domains [Victor Zue, 2005]. When the user says, "How did you do that?" the domain of discourse is limited to the <Self/> and its contemporaneous actions. Since CR can do only one or two things at once, the question, "How did you do that?" has only one primary semantic referent, playing the song. Reasoning using analogy, also cutting edge, is no longer beyond the pale for tightly limited domains like CR and thus is envisioned in the CRA.

CRs Learn to Differentiate Speakers to Reduce Confusion

To further limit combinatorial explosion in speech, CR may form speaker-models, statistical summaries of the speech patterns of speakers, particularly of the <Owner/>. Speaker modeling is particularly reliable when the <Owner/> uses the CR as a cell phone to place a phone call. Contemporary speaker recognition algorithms differentiate male from female speakers with high (>95%) probability. With a few different speakers to be recognized (*e.g.*, fewer than 10 in a family) and with reliable side information like the speaker's telephone number, today's algorithms recognize individual speakers with 80 to 90% probability. Speaker models can become contaminated, such as erroneously including both <Owner/> and <Daughter/> speech in the <Owner/> model. Insightful product engineering could circumvent such problems, rendering <Owner/> interactions as reliable as TellMe® over the next few years.

Over time, each CR learns the speech patterns of its <Owner/> in order to learn from the <Owner/> and not be confused by other speakers. CR thus leverages experience incrementally to achieve increasingly sophisticated dialog. Directional microphones are rapidly improving to service video teleconference (VTC) markets. Embedding these VTC microphones into CR *glasses* would enable CR to differentiate user speech from backgrounds like radio and TV.

Today, a 3 GHz laptop supports this level of speech understanding and dialog synthesis in real-time, making it likely available in a cell phone in three to five years. Today, few consumers train the speech recognition systems embedded in most laptop computers. Its too much work and the algorithms do not take dictation well enough. Thus, although speech recognition technology exists, it is not as effective at the general task of converting speech to text as TellMe® is in finding an 800 number. The CR value proposition, overcomes this limit by embedding machine learning so your CR continually learns about you by analyzing your voice, speech patterns, visual scene, and related use of the RF spectrum from garage door openers to NOAA weather, from cell phone and walkie-talkie to wireless home computer network. Do you want to know if your child's plane is in the air? Ask your CR and it could find "NiftyAir 122 Heavy cleared for takeoff by Dulles Tower." Again, in order to customize services for you, the <Owner/>, the CR must both know a lot about radio and learn a lot about you, the <Owner/>, recording and analyzing personal information and thus placing a premium on trustable privacy technologies. Increased autonomous (and thus free) customization of wireless service include secondary use of broadcast spectrum. The CRA therefore incorporates speech recognition.

More Flexible Secondary Use of Radio Spectrum

Consider a vignette with Lynne' the <Owner/> and Barb, the <Daughter/>. Barb drives to Lynne's house in her car. Coincidentally, Lynne' asks Genie, the CR <Self/> "Can you call Barb for me?"

Genie: "Sure. She is nearby so I can use the TV band for a free video call if you like."

Lynne': "Is that why your phone icon has a blue TV behind it?"

Genie: "Yes. I can connect you to her using unused TV channel 43 instead of spending your cell phone minutes. The TV icon shows that you are using free airtime as a secondary user of TV spectrum. I sent a probe to her cognitive radio to be sure it could do this."

Lynne': "OK, thanks for saving cell time for me. Let me talk to her." Barb's face appears on the screen.

Barb: "Wow, where did you come from?" Barb had never seen her cell phone display her Mom in a small TV picture in real time before, only in video clips.

Lynne': "Isn't this groovy. Genie, my new cognitive radio, hooked us up on a TV channel. It says you are nearby. Oh, I see you are out front and need help with the groceries. Here I come."

In 2004, the US Federal Communications Commission (FCC) issued a Report and Order that radio spectrum allocated to TV, but unused in a particular broadcast market could be used by CR as secondary users under Part 15 rules for low power devices, *e.g.*, to create ad-hoc networks. SDR Forum member companies have demonstrated CR products with these elementary spectrum-perception and use capabilities. Wireless products - military and commercial - realizing the FCC vignettes exist already. Complete visual and speech perception capabilities are not many years distant. Productization is underway. Thus, the CRA emphasizes CR spectrum agility, but in a context of enhanced perception technologies, a long term growth path.

SDR Technology Underlies Cognitive Radio

To conclude the overview, take a closer look at the enabling radio technology, SDR. Samuel F B Morse's code revolutionized telegraphy in the late 1830's, becoming the standard for "telegraph" by the late 1800's. Thus when Marconi and Tesla brought forward wireless technology in 1902, Morse code was already a standard language for HF communication. Today as then, a radio includes an antenna, a RF power amplifier to transmit, and RF conversion to receive; along with a modulator/demodulator (modem) to impart the code to and from the RF channel; and a coder -decoder (codec) to translate information from human-readable form to a form coded for efficient radio transmission. Today as then, RF conversion depended on capacitors and inductors to set the radio frequency, but then some devices were the size of a refrigerator, while today they can be chip-scale devices. Then, the modulator consisted of the proverbial telegraph key, a switch to open and close the transmission circuit for on-off- keyed (OOK) data encoding. Morse code, a series of short (dits) or long marks (dahs) and spaces - sounds and silence - is still the simplest, cheapest way to communicate across a continent, and Morse code over HF radio still is used today in remote regions from the Australian outback to Africa and Siberia. Then and now, the "coder" was the person who had memorized the Morse code, manually converting dit-dit-dah-dit from and to the letters of the alphabet. Radio engineers almost never abandon an RF band (HF) or mode (Morse code). Instead, the use morphs from mainstream to a niche market like sports, amateur radio, remote regions, or developing economies. Today there are nice user interfaces and digital networking, but radio engineering has not taken anything away. At the relatively low data rates of mobile radio (<1 Mbps), networking (routing and switching) is readily accomplished in software, unlike wired networks where data rates reach gigabits per second and dedicated hardware is needed for high speed switching.

The essential functional blocks of radio have not changed for a century and are not likely to change either because the laws of physics define them: antenna, RF conversion, power amplification, modem, and codec. Today, however, microelectronics technologies enable one to pack low power RF, modem, and codecs into single-chip packages while antennas fit neatly into the palm of your hand. Today, there are a myriad of modems evolved from the single RF of Morse to the sharing of RF bands in frequency, time, and code-space. The manual codec has evolved to include communications security (COMSEC) coding, authentication, and multi-layered digital protocol stacks. Cognitive radio embraces all the broad classes of modulation, each with unique modems, codecs and most importantly content, the reason people use the radio, after all.

The SDR Forum and Object Management Group (OMG) have standardized software architecture for wireless plug and play of the myriad band-mode combinations: the Software Communications Architecture (SCA) and Software Radio Architecture (SRA) respectively. But, the real enabler for SDR is the increasingly programmable analog RF of SDR: antennas, RF conversion, and amplifiers. Historically, the analog RF had fixed frequency and bandwidth, optimized for a small RF band such as 88 to 108 MHz for FM broadcast, 850-950 MHz for cell phones and 1.7 to 1.8 GHz for personal communications systems (PCS), a third generation cellular band. Today's cellular radios typically include three chip-sets, one optimized for first-generation "roaming" where infrastructure is not well built out, one for second generation digital service such as GSM, and one for PCS or NexTelr.

Each of these chip-sets accesses only the narrow band needed for the service, so today's cell phones can't open the garage door, not without another (expensive) chip set. In 1990-95, DARPA demonstrated SPEAKeasy II, the first SDR with continuous RF from 2 MHz to 2 GHz in just three analog RF bands: HF (2-30 MHz), mid- band (30-500 MHz) and high band (0.5 to 2 GHz).

Micro Electro-Mechanical Systems (MEMS) technology makes it possible to reprogram analog RF components digitally, so a cell phone could in fact synthesize the garage door opener and key fob as the new digitally controlled analog RF MEMS technology emerges. RF MEMS digitally controls analog RF devices. (See the figure of the RF fingers from UCLA). A computer commands a microscale motor to move the interdigitated fingers of a capacitor, changing its analog value and hence changing the RF center frequency of the analog radio circuit. As the fingers move in and out by a few microns, the RF changes up and down by MHz. As this technology matures and enters service, RF chipsets will be reconfigurable across radio bands and modes, realizing affordable nearly ideal SDRs. FM, and TV/Broadcasts inform large markets with news, sports, weather, music, and the like. From boom box to weather radio, people around the world still depend on AM, FM and TV broadcasts for such information. In the past, you had to buy a specialized radio receiver and tune it manually to the station you like. With RF MEMS SDR, you tell the CR what you want to hear and it finds it for you. Your approval or disapproval constitutes training of the AML algorithms that tuned the MEMS SDR for your user-specific content preferences. RF MEMS have been demonstrated to reduce the size, weight, and power of analog RF subsystems by two to three orders of magnitude, and by over 1000:1 in some cases, but they have been slow to enter markets because of lower than necessary reliability, a focus of both academic and commercial RF MEMS research and development. To facilitate the insertion of RF MEMS and other enabling technologies, the CRA embraces hardware abstraction.

Privacy Is Paramount

A CR that remembers all your conversations for several years needs only a few hundred gigabytes of data memory, readily achieved in wearable CR-PDA even today. Many such conversations will be private, and some will include credit card numbers, social security numbers, bank account information, and the like. When my laptop was stolen with five years of tax returns, the process of dealing with identity theft was daunting and not foolproof. How can one trust a CR with all that personal information? Why would it need to remember all that stuff anyway?

One value proposition of CWNs is the reduction of tedium. Thus, asking the new owner to program the CR or to train it for an hour in the way that one is supposed to trains the speech recognition system in a new laptop would be to increase tedium, not to decrease it. CR therefore aggregates experience, reprocessing the raw speech, vision, and RF data during sleep cycles so that it learns from experience with minimum tedious training interactions with the user. Although based solidly on contemporary RL and CBR technology, task-focused introspective learning for nearly unsupervised dialog acquisition is on the cutting edge of autonomous product development while the more general problem of minimally supervised dialog acquisition in general is at the cutting edge of language research. Thus, CR products will always "cheat" the way TellMe® cheats; CR products pick a small, workable set of tasks that consumers will pay for and use, mini-killer apps. The resulting revenue streams build technology for increasingly capable tasks, evolving towards the vision-RF-dialog skills of the previous vignettes. However, to learn this way, CR really must remember all the raw data - all your keystrokes, emails, and conversations, to learn your use patterns and preferences autonomously, thus capturing private personal data.

If CR must remember your private personal data, then it must protect that information. Finger print readers are not perfect, as is any single information assurance (IA) measure, so CR may use a mix of IA measures. Candidate IA measures include soft biometrics like face and voice recognition along with more obtrusive measures like iris recognition. Layers of public key infrastructure (PKI), GSM-like randomized challenges and signed responses with network validation of identity, and battery backup of IA protection skills, *e.g.*, that erase all user data when the CR detects that it is being physically compromised, *e.g.*, by the unexplained removal of screws of its case. Privacy is paramount, and practical products must protect personal information, identity, medical information, and the like with high reliability. Thus, a mix of soft biometrics likes face and voice recognition coupled with selective hard biometrics like a fingerprint reader, PKI, and other encryption methods. Given the limits of speech and visual perception technologies, CRs employ a large fraction of their sensory perception resources recognizing the face, voice, and daily habits of the <Owner/>. Some robots accumulate stimuli in a way that simulate human emotion, *e.g.*, happiness or distress. If the robot detects its <Owner's> voice and face, then it knows what to expect based on having learned the owners' patterns. If the voice and face are not recognized, then the CR might become defensive, protecting the owner's data and potentially erasing it rather than divulge personal data to someone the <Owner/> has not previously authorized. Embedding a backup battery deep within the motherboard and embedding sensors in screws in the motherboard might dissuade all but the most sophisticated criminals from stealing such CRs. Therefore, CRA explicitly includes hardware and software facilities to implement trustable protection of privacy.

Military Applications Abound

Military applications of CR in CWNs abound. It is easy to imagine realistic vignettes where radios relay the commander's change to an operations order in his own words, "Coalition partners are now located at grid square 76-11, so hold your fire. Rendezvous at Checkpoint Charlie at 1700." Little doubt about the authenticity of an order if it can be recalled and distributed digitally, authenticated and suitably protected to military standards, of course. Tactical military radio communications are notoriously noisy. Thus, a radio that conveys such critical information error-free and in the voice of the commander reduces the fog of war, potentially saving lives.

With autonomous machine learning skills, military CRs (mCRs) would learn coalition RF use patterns. Autonomously re- programming of their SDR transceivers, coalition mCRs could learn to connect commanders directly with each other, avoiding the need for dedicated military radio operators per se and either reducing the size of a squad from ten to nine or enhancing the squad's capabilities by the 10% no longer needed to just operate the radio.

Although one can never completely replace the flexibility and insight of skilled people, as mCRs offload mundane radio operation tasks from the radio operator, the team's effectiveness will increase, beneficial in the short run even if it takes decades to realize "Radar O'Reilly" in software. Although Phraselator experiment showed the promise of real-time language translation in a handheld device for coalition operations, a Phraselator is yet another widget like the garage door opener. Envisioned mCR offer a flexible hardware platform in which to embed Phraselator algorithms invoked by language identification algorithms that detect non-native language and hence the need for real-time translation. Since mCR is about enhancing the effectiveness of communications, language translation embedded in CR to translate when and where needed certainly has the potential to enhance communications among coalition partners who speak different languages, again reducing the fog of war and improving the likelihood of success.

The CRA is not specifically designed for military applications, but its open and evolutionary nature enable a wide range of commercial and military applications.

2. Architecture

Architecture is a comprehensive, consistent set of ***design rules*** by which a specified set of ***components*** achieves a specified set of ***functions*** in products and services that evolve through multiple design points over time [Joseph Mitola III, 2000b]. This section introduces the fundamental design rules by which SDR, sensors, perception, and AML may be integrated to create Aware, Adaptive, and Cognitive Radios (AACR's) with better Quality of Information (QoI) through capabilities to Observe (sense, perceive), Orient, Plan, Decide, Act and Learn (the OOPDAL loop) in RF and user domains, transitioning from merely adaptive to demonstrably cognitive radio, CR.

This chapter develops five complementary perspectives of architecture called CRA I through CRA V. CRA I defines six functional components, black boxes to which are ascribed a first level decomposition of AACR functions and among which important interfaces are defined. One of these boxes is SDR, a proper subset of AACR. One of these boxes performs cognition via the <Self/>, a self-referential subsystem that strictly embodies finite computing (*e.g.*, no while or until loops) avoiding the G'odel-Turing paradox.

CRA II examines the flow of inference through a cognition cycle that arranges the core capabilities of ***ideal*** CR (iCR) in temporal sequence for a logical flow and circadian rhythm for the CRA. CRA III examines the related levels of abstraction for AACR to sense elementary sensory stimuli and to perceive Quality of Service (QoS)-relevant aspects of a <Scene/> consisting of the <User/> in an <Environment/> that includes <RF/>. CRA IV examines the mathematical structure of this architecture, identifying mappings among topological spaces represented and manipulated to preserve set-theoretic properties. Finally, CRA V reviews SDR architecture, sketching an evolutionary path from the SCA/SRA to the CRA. The CRA <Self/> provided in CRA Self .xml expresses in RXML the CRA introduced in this chapter along with a-priori knowledge for AML.

3. CRA I: Functions, Components and Design Rules

The ***functions*** of AACR exceed those of SDR. Reformulating the AACR <Self/> as a *peer* of its own <User/> establishes the need for added functions by which the <Self/> accurately perceives the local scene including the <User/> and autonomously learns to tailor the information services to the specific <User/> in the current RF and physical <Scene/>.

AACR Functional Component Architecture

The SDR components and the related cognitive components of iCR appear in Figure 9.1. The cognition components describe the SDR in Radio XML so that the <Self/> can know that it is a radio and that its goal is to achieve high QoI tailored to its own users. RXML intelligence includes a priori radio background and user stereotypes as well as knowledge of RF and space-time <Scenes/> perceived and experienced. This includes both structured reasoning with iCR peers and CWNs, and ad-hoc reasoning with users, all the while learning from experience.

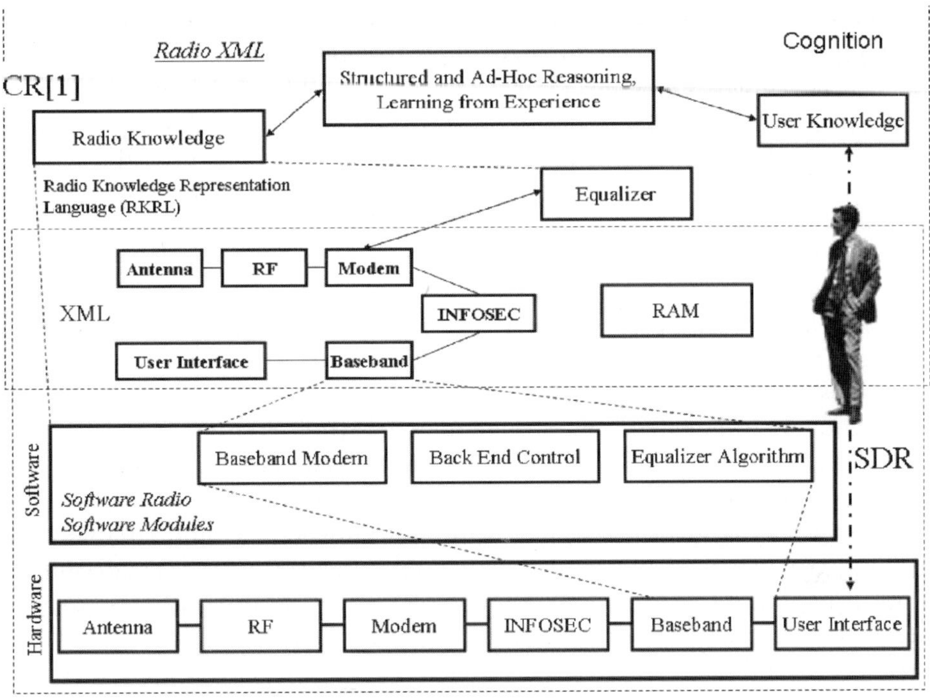

Figure 9.1. The CR Systems Engineer Augments SDR with Computational Intelligence.

The detailed allocation of functions to components with interfaces among the components requires closer consideration of the SDR component as the foundation of CRA.

SDR Components

SDRs include a hardware platform with RF access and computational resources, plus at least one software-defined personality. The SDR Forum has defined its Software Communications Architecture (SCA) and the Object Management Group (OMG) has defined its Software Radio Architecture (SRA), similar fine-grain architecture constructs enabling reduced cost wireless connectivity with next-generation plug and play. These SDR architectures are defined in Unified Modeling Language (UML) object models [Eriksson and Penker, 1998], CORBA Interface Design Language (IDL) [T. Mowbray and R. Malveau, 1997], and XML descriptions of the UML models. The SDR Forum's SCA [SDR Forum, 2006] and OMG SRA [OMG, 2006] standards describe the technical details of SDR both for radio engineering and for an initial level of wireless air interface ("waveform") plug and play. The SCA/SRA was sketched in 1996 at the first DoD-inspired MMITS Forum, developed by the US DoD in the 1990's and is the commercial version of the architecture of US military radios [JTRS, 2006]. This architecture emphasizes plug-and-play wireless personalities on computationally capable mobile nodes where network connectivity is often intermittent at best. The commercial wireless community [WWRF, 2004], on the other hand, led by cell phone giants Ericsson and Nokia envisions a much simpler architecture for mobile wireless devices, consisting of two APIs, one for the service provider and another for the network operator. They define a knowledge plane in the future intelligent wireless networks that is not dissimilar from a distributed CWN. That forum promotes the business model of the user -> service provider -> network operator -> large manufacturer -> device, where the user buys mobile devices consistent with services from a service provider, and the technical emphasis is on ***intelligence in the network***. This perspective no doubt will yield computationally intelligent networks in the near- to mid-term. The CRA developed in this text, however, envisions the computational intelligence to create ad-hoc and flexible networks with the ***intelligence in the mobile device***. This technical perspective enables the business model of user -> device -> heterogeneous networks, typical of the Internet model where the user buys a device (*e.g.*, a wireless laptop) that can connect to the Internet via any available Internet Service Provider (ISP). The CRA builds on both the SCA/SRA and the commercial API model but integrates Semantic Web intelligence in Radio XML for mobile devices to enable more of an Internet business model to advance. This chapter describes how SDR, AACR, and iCR form a continuum facilitated by RXML.

AACR Node Functional Components

The simplest CRA is the minimalist set of functional components of Figure 9.2. A functional component is a black box to which functions have been allocated, but for which implementing components do not exist. Thus, while the Applications component is likely to be primarily software, the nature of those software components is yet to be determined. User Interface functions, on the other hand, may include optimized hardware, *e.g.*, for computing video flow vectors in real time to assist scene perception. At the level of abstraction of the figure, the components are functional, not physical.

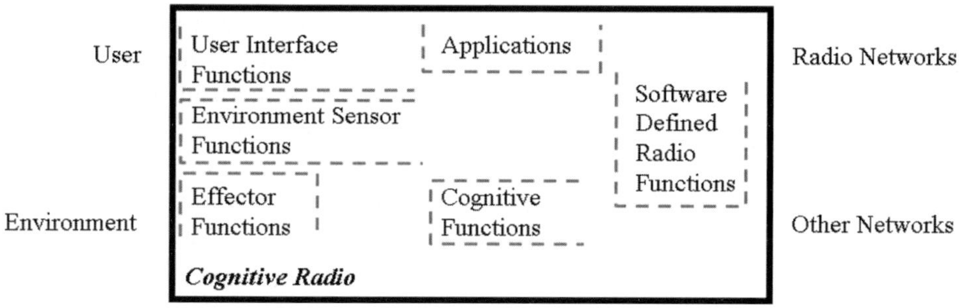

Figure 9.2. Minimal AACR Node Architecture .

These functional components are

1 **The user sensory perception (User SP)** interface includes haptic, acoustic, and video sensing and perception functions,

2 The local **environment** sensors (location, temperature, accelerometer, compass, etc.),

3 The **system applications** (media independent services like playing a network game),

4 The **SDR** functions (which include RF sensing and SDR radio applications),

5 The **cognition** functions (symbol grounding for system control, planning, learning) and

6 The **local effector** functions (speech synthesis, text, graphics, and multimedia displays).

These functional components are embodied on an iCR-Platform, a hardware realization of the six functions. In order to support the capabilities described in the prior chapters, these components go beyond SDR in critical ways. First, the user interface goes well beyond buttons and displays. The traditional user interface has been partitioned into a substantial user sensory subsystem and a set of local effectors. The user sensory interface includes buttons (the haptic interface) and microphones (the audio interface) to include acoustic sensing that is directional, capable of handling multiple speakers simultaneously and to include full motion video with visual scene perception. In addition, the audio subsystem does not just encode audio for (possible) transmission; it also parses and interprets the audio from designated speakers such as the <User/> for a high performance spoken natural language interface. Similarly, the text subsystem parses and interprets the language to track the user's information states, detecting plans and potential communications and information needs unobtrusively as the user conducts normal activities. The local effectors synthesize speech along with traditional text, graphics, and multimedia displays. Systems applications are those *information services* that define value for the user. Historically, voice communications with a phone book, text messaging, and the exchange of images or video clips comprised the core value proposition of systems applications for SDR. These applications were generally integral to the SDR application, such as data services via GPRS, which is really a wireless SDR personality more than an information service. AACR systems applications break the service out of the SDR waveform so that the user need not deal with details of wireless connectivity unless that is of particular interest. Should the user care whether he plays the distributed video game via 802.11 or Bluetooth over the last 3 meters? Probably not. The typical user might care if the AACR wants to switch to 3G at $5 per minute, but a particularly affluent user might not care and would leave all that up to the AACR. The Cognition component provides all the cognition functions from the semantic grounding of entities from the perception system to controlling the overall system through planning and initiating actions, learning user preferences and RF situations in the process. Each of these subsystems contains its own processing, local memory, integral power conversion, built-in-test (BIT) and related technical features.

The Ontological <Self/>

AACR may consist of the six functional components User SP, Environment, Effectors, SDR, Sys Apps, and Cognition. Describing those components to the <Self/> enables external communications and internal reasoning about the <Self/> via ontology : RXML to the rescue.

```
<Self/>
     <iCR-Platform/>
     <Functional-Components>
          <User SP/><Environment/><Effectors/><SDR/><Sys Apps/>
          <Cognition/>
     </Functional-Components>
</Self>
```

Equation 1 The AACR Has Six Functional Components

Given the top-level outline of these functional components along with the requirement that they be embodied in physical hardware and software (the

"Platform"), the six functional components are defined ontologically in Equation 1. In part, this equation states that the hardware-software platform and the functional components of the AACR are independent. Platform-independent computer languages like Java are well understood. This ontological perspective envisions platform-independence as an architecture design principle for AACR. In other words, the burden is on the (software) functional components to adapt to whatever RF-hardware-OS platform might be available.

Design Rules Include Functional Component Interfaces

These functional components imply associated functional interfaces. In architecture, design rules may include a list of the quantities and types of components as well as the interfaces among those components. This section addresses the interfaces among the six functional components.

The AACR N-Squared Diagram of Table 9.1 characterizes AACR interfaces. These constitute and initial set of AACR Applications Programmer Interfaces - AACR API's. In some ways these API's augment the established SDR APIs. For example, the Cognition API brings a planning capability to SDR. This is entirely new and much needed in order for basic ACAR's to accommodate even the basic ideas of XG.

In other ways, these API's supersede the existing SDR APIs. In particular, the SDR user interface becomes the User Sensory and Effector API. User Sensory API's include acoustics, voice, and video, while the effectors include speech synthesis to give the AACR <Self/> its own voice. In addition, wireless applications are growing rapidly. Voice and short message service become an ability to exchange images and video clips with ontological tags among wireless users. The distinctions between cell phone, PDA, and game box continue to disappear.

These interface changes enable the AACR to sense the situation represented in the environment, to interact with the user and to access radio networks on behalf of the user in a situation-aware way.

Interface Notes

User SP - User SP Cross-media correlation interfaces (video- acoustic, haptic-speech, etc) to limit search and reduce uncertainty (*e.g.*, if video indicates user is not talking, acoustics may be ignored or processed less aggressively for command inputs than if user is speaking)

Environment - User SP Environment sensors parameterize user sensor-perception. Temperature below freezing may limit video;

Sys Apps - User SP Systems Applications may focus scene perception by identifying entities, range, expected sounds for video, audio, and spatial perception processing.

Table 9.1. AACR N-Squared Diagram Characterizes AACR Node Internal Interfaces.

From To	User SP	Environment	Sys Apps	SDR	Cognition	Effectors
User SP	1	7	13 PA ◇	19	25 PA ♣	31
Environment	2	8	14 SA ◇	20	26 PA ♣	32
Sys Apps	3	9	15 SCM ◇	21 SD ◇	27 PDC ♠	33 PEM ◇
SDR	4	10	16 PD ◇	22 SD	28 PC ♣	34 SD
Cognition	5 PEC ♣	11 PEC ♣	17 PC ♠	23 PAE ♣	29 SC ♣	35 PE ♣
Effectors	6 SC	12	18 ◇	24	30 PCD ♣	36

Legend: P - Primary; A - Afferent; E- Efferent; C- Control; M - Multimedia; D - Data; S - Secondary; Others not designated P or S are ancillary
The Information Services API (◇ and ♠) consists of interfaces 13-18, 21, 27, and 33
The Cognition API (♣ and ♠) consists of interfaces 25-30, 5, 11, 23, and 35
Interface Notes follow the numbers of the table:

SDR - User SP SDR applications may provide expectations of user input to the perception system to improve probability of detection and correct classification of perceived inputs.

Cognition - User SP This is the ***primary control efferent*** path from cognition to the control of the user sensory perception subsystem, controlling speech recognition, acoustic signal processing, video processing, and related sensory perception. Plans from Cognition may set expectations for user scene perception, improving perception.

Effectors - User SP Effectors may supply a replica of the effect to user perception so that self-generated effects (*e.g.*, synthesized speech) may be accurately attributed to the <Self/>, validated as having been expressed, and/or cancelled from the scene perception to limit search.

User SP - Environment Perception of rain, buildings, indoor/outdoor can set GPS integration parameters.

Environment - Environment Environment sensors would consist of location sensing such as GPS or Glonast; temperature of the ambient; light level to detect inside versus outside locations; possibly smell sensors to detect spoiled food; and others that may surprise one even more. There seems to be little benefit to enabling interfaces among these elements directly.

Sys Apps - Environment Data from the systems applications to environment sensors would also be minimal.

SDR - Environment Data from the SDR personalities to the environment sensors would be minimal.

Cognition - Environment (Primary Control Path) Data from the cognition system to the environment sensors controls those sensors, turning them on and off, setting control parameters, and establishing internal paths from the environment sensors.

Effectors - Environment Data from effectors directly to environment sensors would be minimal.

User SP - Sys Apps Data from the user sensory perception system to systems applications is a ***primary afferent path*** for multimedia streams and entity states that effect information services implemented as systems applications. Speech, images, and video to be transmitted move along this path for delivery by the relevant systems application or information service to the relevant wired or SDR communications path. Sys Apps overcomes the limitations of individual paths by maintaining continuity of conversations, data integrity, and application coherence, *e.g.*, for multimedia games. While the cognition function sets up, tears down, and orchestrates the systems applications, the primary API between the user scene and the information service consist of this interface and its companions, the environment afferent path; the effector efferent path; and the SDR afferent and efferent paths.

Environment - Sys Apps Data on this path assists systems applications in providing location-awareness to services.

Sys Apps - Sys Apps Different information services interoperate by passing control information through the cognition interfaces and by passing domain multimedia flows through this interface. The cognition system sets up and tears down these interfaces.

SDR - Sys Apps This is the primary afferent path from external communications to the AACR. It includes control and multimedia information flows for all the information services. Following the SDR Forum's SCA, this path embraces wired as well as wireless interfaces.

Cognition - Sys Apps Through this path the AACR <Self/> exerts control over the information services provided to the <User/>.

Effectors - Sys Apps Effectors may provide incidental feedback to information services through this afferent path, but the use of this path is deprecated. Information services are supposed to control and obtain feedback through the mediation of the cognition subsystem.

User SP - SDR Although the sensory perception system may send data directly to the SDR subsystem, *e.g.*, in order to satisfy security rules that user biometrics must be provided directly to the wireless security subsystem, the use of this path is deprecated. Perception subsystem information is supposed to be interpreted by the cognition system so that accurate information can be conveyed to other subsystems, not raw data.

Environment - SDR Environment sensors like GPS historically have accessed SDR waveforms directly, such as providing timing data for air interface signal generation. The cognition system may establish such paths in cases where cognition provides little or no value added, such as providing a precise timing reference from GPS to an SDR waveform. The use of this path is deprecated because all of the environment sensors including GPS are unreliable.

Cognition has the capability to de-glitch GPS, *e.g.*, recognizing from video that the <Self/> is in an urban canyon and therefore not allowing GPS to report directly, but reporting on behalf of GPS to the GPS subscribers location estimates based perhaps on landmark correlation, dead reckoning, etc.

Sys Apps - SDR This is the primary efferent path from information services to SDR through the services API.

SDR - SDR The linking of different wireless services directly to each other is deprecated. If an incoming voice service needs to be connected to an outgoing voice service, then there should be a bridging service in Sys Apps through which the SDR waveforms communicate with each other. That service should be set up and taken down by the Cognition system.

Cognition - SDR This is the primary control interface, replacing the control interface of the SDR SCA and the OMG SRA.

Effectors - SDR Effectors such as speech synthesis and displays should not need to provide state information directly to SDR waveforms, but if needed, the cognition function should set up and tear down these interfaces.

User SP - Cognition This is the primary afferent flow for the results from acoustics, speech, images, video, video flow and other sensor-perception subsystems. The primary results passed across this interface should be the specific states of <Entities/> in the scene, which would include scene characteristics such as the recognition of landmarks, known vehicles, furniture and the like. In other words, this is the interface by which the presence of <Entities/> in the local scene is established and their characteristics are made known to the Cognition system.

Environment - Cognition This is the primary afferent flow for environment sensors.

Sys Apps - Cognition This is the interface through which information services request services and receive support from the AACR platform. This is also the control interface by which Cognition sets up, monitors, and tears down information services.

SDR - Cognition This is the primary afferent interface by which the state of waveforms, including a distinguished RF-sensor waveform is made known to the Cognition system. The cognition system can establish primary and backup waveforms for information services enabling the services to select paths in real time for low latency services. Those paths are set up, monitored for quality and validity (*e.g.*, obeying XG rules) by the cognition system, however.

Cognition - Cognition The cognition system as defined in this six component architecture entails (1) orienting to information from <RF/> sensors in the SDR subsystem and from scene sensors in the user sensory perception and environment sensors, (2) planning, (3) making decisions, and (4) initiating actions, including the control over all of the resources of the <Self/>. The <User/> may directly control any of the elements of the systems via paths through the cognition system that enable it to monitor what the user is doing in order to learn from a user's direct actions, such as manually tuning in the user's favorite radio station when the <Self/> either failed to do so properly or was not asked.

Effectors - Cognition This is the primary afferent flow for status information from the effector subsystem, including speech synthesis, displays, and the like.

User SP - Effectors In general, the user sensory-perception system should not interface directly to the effectors, but should be routed through the cognition system for observation. Environment - Effectors The environment system should not interface directly to the effectors. This path is deprecated.

Sys Apps - Effectors Systems applications may display streams, generate speech, and otherwise directly control any effectors once the paths and constraints have been established by the cognition subsystem.

SDR - Effectors This path may be used if the cognition system establishes a path, such as from an SDR's voice track to a speaker. Generally, however, the SDR should provide streams to the information services of the Sys Apps. This path may be necessary for legacy compatibility during the migration from SDR through AACR to iCR but is deprecated.

Cognition - Effectors This is the primary efferent path for the control of effectors. Information services provide the streams to the effectors, but cognition sets them up, establishes paths, and monitors the information flows for support to the user's <Need/> or intent.

Effectors - Effectors These paths are deprecated, but may be needed for legacy compatibility.

The above information flows aggregated into an initial set of AACR APIs define an Information Services API by which an information service accesses the other five components (ISAPI consisting of interfaces 13-18, 21, 27, and 33). They would also define a Cognition API by which the cognition system obtains status and exerts control over the rest of the system (CAPI consisting of interfaces 25-30, 5, 11, 23, and 35). Although the constituent interfaces of these APIs are suggested in the table, it would be premature to define these APIs without first developing detailed information flows and interdependencies, which are defined in this chapter and analyzed in the next. It would also be premature to develop such APIs without a clearer idea of the kinds of RF and User domain knowledge and performance that are expected of the AACR architecture over time. These aspects are developed in the balance of the text, enabling one to draw some conclusions about these API's in the final chapters.

A fully defined set of interfaces and APIs would be circumscribed in RXML. For the moment, any of the interfaces of the N-squared diagram may be used as needed.

Near Term Implementations

One way to implement this set of functions is to embed into an SDR a reasoning engine such as a rule base with an associated inference engine as the Cognition Function. If the Effector Functions control parts of the radio, then we have the simplest AACR based on the simple six component architecture of Figure 9.2- 2. Such an approach may be sufficient to expand the control paradigm from today's state machines with limited flexibility to tomorrow's AACR control based on reasoning over more complex RF states and user situations. Such simple approaches may well be the next practical steps in AACR evolution from SDR towards iCR.

This incremental step doesn't suggest how to mediate the interfaces between multi-sensory perception and situation- sensitive prior experience and a-priori knowledge to achieve situation-dependent radio control that enables the more sophisticated information services of the use cases. In addition, such a simple architecture does not pro-actively allocate machine learning functions to fully understood components. For example, will AML require an embedded radio propagation modeling tool? If so, then what is the division of function between a rule base that knows about radio propagation and a propagation tool that can predict values like RSSI? Similarly, in the user domain, some aspects of user behavior may be modeled in detail based on physics, such as movement by foot and in vehicles. Will movement modeling be a separate subsystem based on physics and GPS? How will that work inside of buildings? How is the knowledge and skill in tracking user movements divided between physics-based computational modeling and the symbolic inference of a rule base or set of Horn clauses with a Prolog engine? For that matter, how will the learning architecture accommodate a variety of learning methods like neural networks, PROLOG, forward chaining, SVM if learning occurs entirely in a cognition subsystem?

While hiding such details may be a good thing for AACR in the near term, it may severely limit the mass customization needed for AACRs to learn user patterns and thus to deliver RF services dramatically better than mere SDRs. Thus, we need to go "inside" the cognition and perception subsystems further to establish more of a fine-grained architecture. This enables one to structure the data sets and functions that mediate multi-sensory domain perception of complex scenes and related learning technologies that can autonomously adapt to user needs and preferences. The sequel thus pro-actively addresses the embedding of Machine Learning (ML) technology into the radio architecture.

Next, consider the networks. Network-independent SDRs retain multiple personalities in local storage, while network-dependent SDRs receive alternate personalities from a supporting network infrastructure - CWNs. High-end SDRs both retain alternate personalities locally and have the ability to validate and accept personalities by download from trusted sources. Whatever architecture emerges must be consistent with the distribution of RXML knowledge aggregated in a variety of networks from a tightly- coupled CWN to the Internet, with a degree of <Authority/> and trust reflecting the pragmatics of such different repositories.

The first two sections of this chapter therefore set the stage for the development of CRA. The next three sections address the cognition cycle, the inference hierarchies, and the SDR architecture embedded both into the CRA with the knowledge structures of the CRA.

The Cognition Components

Figure 9.1 above shows three computational-intelligence aspects of CR:

- Radio Knowledge - RXML:RF
- User Knowledge - RXML:User
- The Capacity to Learn

The minimalist architecture of Figure 9.2 and the functional interfaces of the subsequent table do not assist the radio engineer in structuring knowledge, nor does it assist much in integrating machine learning into the system. The fine grain architecture developed in this chapter, on the other hand, is derived from the functional requirements to fully develop these three core capabilities.

Radio Knowledge in the Architecture

Radio knowledge has to be translated from the classroom and engineering teams into a body of computationally accessible, structured technical knowledge about radio. Radio XML is the primary enabler and product of this foray into formalization of radio knowledge. This text starts a process of RXML definition and development that can only be brought to fruition by industry over time. This process is similar to the evolution of the Software Communications Architecture (SCA) of the SDR Forum [SDR Forum, 2006]. The SCA structures the technical knowledge of the radio components into UML and XML. RXML will enable the structuring of sufficient RF and user world knowledge to build advanced wireless- enabled or enhanced information services. Thus while the SRA and SCA focus on building radios, RXML focuses on using radios.

The World Wide Web is now sprouting with computational ontologies some of which are non-technical but include radio, such as the open CYC ontology. They bring the radio domain into the Semantic Web, which helps people know about radio. This informal knowledge lacks the technical scope, precision and accuracy of authoritative radio references such as the ETSI documents defining GSM and ITU definitions, *e.g.*, of 3GPP.

Not only must radio knowledge be precise, it must be stated at a useful level of abstraction, yet with the level of detail appropriate to the use-case. Thus, ETSI GSM in most cases would over-kill the level of detail without providing sufficient knowledge of the user-centric functionality of GSM. In addition, AACR is multi-band, multi-mode radio (MBMMR), so the knowledge must be comprehensive, addressing the majority of radio bands and modes available to a MBMMR. Therefore, in the development of CR technology below, this text captures radio knowledge needed for competent CR in the MBMMR bands from HF through millimeter wave. This knowledge is formalized with precision that should be acceptable to ETSI, the ITU and Regulatory Authorities (RAs) yet at a level of abstraction appropriate to internal reasoning, formal dialog with a CWN or informal dialog with users.

This kind of knowledge is to be captured in RXML:RF.

The capabilities required for an AACR node to be a cognitive entity are to sense, perceive, orient, plan, decide, act, and learn. To relate ITU standards to these required capabilities is a process of extracting content from highly formalized knowledge bases that exist in a unique place and that bear substantial authority, encapsulating that knowledge in less complete and therefore somewhat approximate form that can be reasoned with on the AACR node and in real time to support RF-related use cases. Table 9.2 illustrates this process.

The table is illustrative and not comprehensive, but it characterizes the technical issues that drive an information- oriented AACR node architecture. Where ITU, ETSI, (meaning other regional and local standards bodies) and CWN supply source knowledge, the CWN is the repository for authoritative knowledge derived from the standards bodies and Regulatory Authorities (RAs), the <Authorities/>. A user-oriented AACR may note differences in the interpretation of source knowledge from <Authorities/> between alternate CWNs, precipitating further knowledge exchanges.

User Knowledge in the Architecture

Next, user knowledge is formalized at the level of abstraction and degree of detail necessary to give the CR the ability to acquire from its Owner and other designated users, the user - knowledge relevant to information services incrementally. Incremental knowledge acquisition was motivated in the introduction to AML by describing how frequent occurrences with similar activity sequences identifies learning opportunities. AML machines may recognize these opportunities for learning through joint probability statistics <Histogram/>. Effective use-cases clearly identify the classes of user and the specific knowledge learned to customize envisioned services. Use cases may also supply sufficient initial knowledge to render incremental AML not only effective, but also - if possible - enjoyable to the user.

Table 9.2. Radio Knowledge in the Node Architecture.

Need	Source Knowledge	AACR Internalization
Sense RF	RF Platform	Calibration of RF, noise floor, antennas, direction
PerceiveRF	ITU, ETSI, ARIB,RA's	Location-based table of radio
Observe RF (Sense&Perceive)	Unknown RF	RF sensor measurements & knowledge of basic types (AM, FM, simple digital channel symbols, typical TDMA, FDMA, CDMA signal structures)
Orient	XG-like policy	Receive, parse, and interpret policy language
	Known Waveform	Measure parameters in RF, Space and Time
Plan	Known Waveform	Enable SDR for which licensing is current
	Restrictive Policy Optimize transmitted waveform, space-time plan	
Decide	Legacy waveform,policy Defer spectrum use to legacy users per policy	
Act	Applications layer	Query for available services (White/yellow pages)
	ITU, ETSI, CWN	Obtain new skills encapsulated as download
Learn	Unknown RF	Remember space-time-RF signatures; discover spectrum use norms and exceptions
	ITU, ETSI, CWN	Extract relevant aspects such as new feature

This knowledge is defined in RXML:User. As with RF knowledge, the capabilities required for an AACR node to be a cognitive entity are to observe (sense, perceive), orient, plan, decide, act, and learn. To relate a use case to these capabilities, one extracts specific and easily recognizable <Anchors/> for stereotypical situations observable in diverse times, places, and situations. One expresses the anchor knowledge in using RXML for use on the AACR node.

Cross-domain Grounding for Flexible Information Services

The knowledge about radio and about user needs for wireless services must be expressed internally in a consistent form so that information services relationships may be autonomously discovered and maintained by the <Self/> on behalf of the <User/>. Relationships among user and RF domains are shown in Figure 9.3.

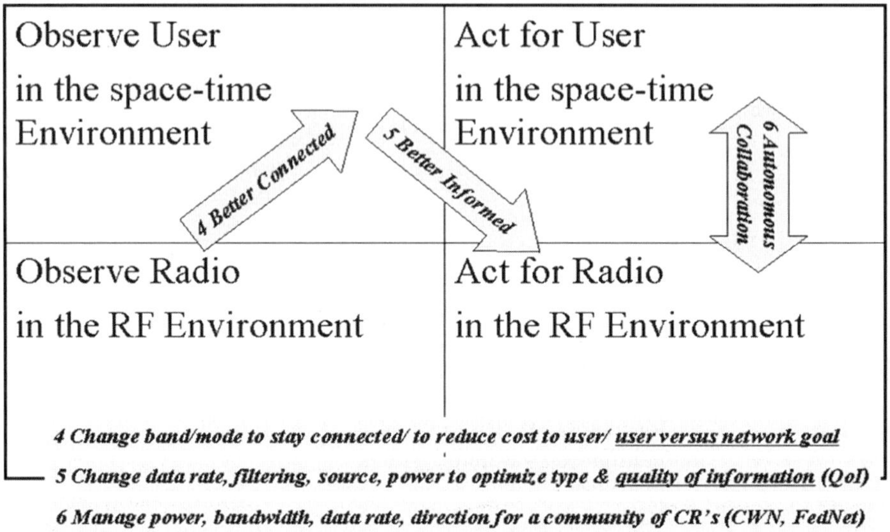

Figure 9.3. Discovering and Maintaining Services.

Staying better connected requires the normalization of knowledge between <User> and <RF> domains. If, for example, the <User/> says "What's on one oh seven - seven," near the Washington, DC area, then the dynamic <User/> ontology should enable the CR to infer that the user is talking about the current FM radio broadcast, the units are in MHz and the user wants to know what is on WTOP. If it can't infer this, then it should ask the user or discover by first dialing a reasonable default, such as 107.7 FM, a broadcast radio station and asking "Is this the radio station you want?" Steps 4, 5, and 6 in the figure all benefit from agreement across domains on how to refer to radio services. Optimizing behavior to best support the user requires continually adapting the <User> ontology with repeated re-grounding of terms in the <User/> domain to conceptual primitives and actions in the <RF/> domain.

The CRA facilitates this by seeding the speech recognition subsystem with the most likely expressions a particular <User/> employs when referring to information services. These would be acquired from the specific users via text and speech recognition, with dialogs oriented towards continual grounding by posing yes/no questions, either verbally or in displays or both, obtaining reinforcement verbally or via haptic interaction or both. The required degree of mutual grounding would benefit from specific grounding-oriented features in the AACR information architecture, developed below.

The process of linking user expressions of interest to the appropriate radio technical operations sometimes may be extremely difficult. Military radios, for example, have many technical parameters. For example, a "channel" in SINCGARS consists of de- hopped digital voice in one context (voice communications) or a 25 kHz band of spectrum in another context (and that may be either an FM channel into which its frequency hop waveform has hopped or an FDMA channel when in single channel mode). If the user says "Give me the Commander's Channel" the SINCGARS user is talking about a "de-hopped CVSD voice stream". If the same user a few seconds later says "This sounds awful. Who else is in this channel?" the user is referring to interference with a collection of hop sets. If the CR observes "There is strong interference in almost half of your assigned channels," then the CR is referring to a related set of 25 kHz channels. If the user then says "OK, notch the strongest 3 interference channels" he is talking about a different subset of the channels. If in the next breath the user says "Is anything on our emergency channel?" then the user has switched from SINCGARS context to <Self/> context, asking about one of the cognitive military radio's physical RF access channels. The complexity of such exchanges demands cross- domain grounding; and the necessity of communicating accurately under stress motivates a structured Natural Language (NL) and rich radio ontology aspects of the architecture developed further below.

Thus, both commercial and military information services entail cross-domain grounding with ontology oriented to NL in the <User> domain and oriented to RXML formalized a-priori knowledge in the <RF> domain. Specific methods of cross-domain grounding with associated architectural features include:

<RF> to <User> Shaping dialog to express precise <RF> concepts to non-expert users in an intuitive way, such as

Grounding: "If you move the speaker box a little bit, it can make a big difference in how well the remote speaker is connected to the wireless transmitter on the TV".

AACR Information Architecture: Include facility for *rich set of synonyms* to mediate cognition-NL-synthesis interface (<Antenna> <Wireless-remote-speaker> "Speaker box").

<RF> to <User> Learning jargon to express <RF> connectivity opportunities in <User> terms.

Grounding: "tee oh pee" for "WTOP", "Hot ninety two" for FM 97.7, "Guppy" for "E2C Echo Grand on 422.1 MHz".

AACR Information Architecture: *NL-visual facility for single- instance update of user jargon.*

<User> to <RF> Relating values to actions: Relate <User> expression of values ("low cost") to features of situations ("normal") that are computable

(<NOT> (<CONTAINS> <Situation> <Unusual/>)) and that relate directly to <RF> domain decisions.

Grounding: Normally wait for free WLAN for big attachment; if situation is <unusual>, ask if user wants to pay for 3G.

AACR Information Architecture: *Associative inference hierarchy* that relates observable features of a <Scene> to user sensitivities, such as <Late-for-work> => <Unusual>; "The President of the company needs this" => <Unusual> because "President" => <VIP> and <VIP> is not in most scenes.

Self-Referential Components

The Cognition component must to assess, manage, and control all of its own resources, including validating downloads. Thus, in addition to <RF> and <User> domains, RXML must describe the <Self/>, defining the AACR architecture to the AACR itself in RXML.

Self-referential Inconsistency

This class of self-referential reasoning is well known in the theory of computing to be a potential black hole for computational resources. Specifically, any Turing-Capable (TC) computational entity that reasons about itself can encounter unexpected Godel-Turing situations from which it cannot recover. Thus TC systems are known to be "partial" - only partially defined because the result obtained when attempting to execute certain classes of procedure are not definable because the computing procedure will never terminate.

To avoid this paradox, CR architecture mandates the use of only "total" functions, typically restricted to bounded-minimalization [J. Mitola III, 1998b]. Watchdog "step-counting" functions [R. Hennie, 1997] or timers must be in place in all its self-referential reasoning and radio functions. The timer and related computationally indivisible control construct is equivalent to the computer-theoretic construct of a step-counting function over "finite minimalization." It has been proven that computations that are limited with certain classes of reliable watchdog timers on finite computing resources can avoid the Godel-Turing paradox or at least reduce it to the reliability of the timer. This proof is the fundamental theorem for practical self-modifying systems.

Briefly: If a system can compute in advance the amount of time or the number of instructions that any given computation should take, then if that time or step-count is exceeded, the procedure returns a fixed result such as "Unreachable in Time T." As long as the algorithm does not explicitly or implicitly re-start itself on the same problem, then with the associated invocation of a tightly time- and computationally-constrained alternative tantamount to giving up, it

- (a) is not Turing capable, but
- (b) is sufficiently computationally capable to perform real-time communications tasks such as transmitting and receiving data as well as bounded user interface functions, and
- (c) is not susceptible to the Turing-G"del incompleteness dilemma and thus
- (d) will not crash because of consuming unbounded or unpredictable resources in unpredictable self-referential loops. This is not a general result.

This is a highly radio domain- specific result that has been established only for isochronous communications domains in which processes are defined in terms of a-priori tightly bounded time epochs such as CDMA frames and SS7 time-outs and for every situation, there is a default action that has been identified in advance that consumes O(1) resources, and the watchdog timer or step-counting function is reliable.

Since radio air interfaces transmit and receive data, there are always defaults such as "repeat the last packed" or "clear the buffer" that may degrade the performance of the overall communications system. A default has O (1) complexity and the layers of the protocol stack can implement the default without using unbounded computing resources.

Watchdog Timer

Without the reliable watchdog timer in the architecture and without this proof to establish the rules for acceptable computing constructs on cognitive radios, engineers and computer programmers would build CRs that would crash in extremely unpredictable ways as their adaptation algorithms get trapped in unpredictable unbounded self-referential loops. Since there are planning problems that can't be solved with algorithms so constrained, either an unbounded community of CR's must cooperatively work on the more general problems or the CN must employ a Turing capable algorithm to solve the harder problems (*e.g.*, NP-hard with large N) off line. There is also the interesting possibility of trading off space and time by remembering partial solutions and re-starting NP-hard problems with these sub-problems already solved. While it doesn't actually avoid any necessary calculations, with O(N) pattern matching for solved subproblems, it may reduce the total computational burden, somewhat like the FFT which converts O(N2) steps to O(N log N) by avoiding the re-computation of already computed partial products. This class of approach to parallel problem solving is similar to the use of pheromones by ants to solve the traveling salesman problem in less than (2N)/M time with M ants. Since this is an engineering text, not a text on the theory of computing, these aspects are not developed further here, but it suffices to show the predictable finiteness and proof that the approach is boundable and hence compatible with the real-time performance needs of cognitive radio.

This timer-based finite computing regime also works for user interfaces since users will not wait forever before changing the situation *e.g.*, by shutting the radio off or hitting another key; and the CR can always kind of throw up its hands and ask the user to take over.

Thus, with a proof of stability based on the theory of computing, the CRA structures systems that not only can modify themselves, but can do it in such a way that they are not likely to induce non-recoverable crashes from the "partial" property of self- referential computing.

Flexible Functions of the Component Architecture

Although this chapter develops the six-element component architecture of a particular information architecture and one reference implementation, there are many possible architectures. The purpose is not to try to sell a particular architecture, but to illustrate the architecture principles. The CRA and research implementation, CR1 [Joseph Mitola III, 2006], therefore, offer open-source licensing for non-commercial educational purposes. Table 9.3 further differentiates architecture features.

Table 9.3. Features of AACR to be Organized via Architecture.

Feature	Function	Examples (RF; vision; speech; location; motion)
Cognition	Monitor & Learn	Get to know user's daily patterns & model the local RF scene over space, time, and situations
Adaptation	Respond to changing environment	Use unused RF, protect owner's data
Awareness Extract information from Sense or perceive sensor domain Perception	Continuously identify knowns, unknowns and backgrounds in the sensor domain	TV channel; Depth of visual scene, identity of objects; location of user, movement and speed of <Self/>
Sensing	Continuously sense & preprocess single sensor-field in single sensory domain	RF FFT; Binary vision; binaural acoustics; GPS; accelerometer; etc.

These functions of the architecture are not different from those of the six-component architecture, but represent varying degrees of instantiation of the six components. Consider the following degrees architecture instantiations:

Cognition functions of radio entail the monitoring and structuring knowledge of the behavior patterns of the <Self/>, the <User>, and the environment (physical, user situation and radio) in order to provide information services, learning from experience to tailor services to user preferences and differing radio environments.

Adaptation functions of radio respond to a changing environment, but can be achieved without learning if the adaptation is pre- programmed.

Awareness functions of radio extract usable information from a sensor domain. Awareness stops short of perception. Awareness is required for adaptation, but awareness does not guarantee adaptation. For example, embedding a GPS receiver into a cell phone makes the phone more location-aware, but unless the value of the current location is actually used by the phone to do something that is location-dependent, the phone is not location-adaptive, only location aware. These functions are a subset of the CRA that enable adaptation.

Perception functions of radio continuously identify and track knowns, unknowns and backgrounds in a given sensor domain. Backgrounds are subsets of a sensory domain that share common features that entail no particular relevance to the functions of the radio. For a CR that learns initially to be a single-Owner-radio, in a crowd, the Owner is the object that the radio continuously tracks in order to interact when needed. Worn from a belt as a Cognitive Wireless Personal Digital Assistant (CWPDA), the iCR perception functions may track the entities in the scene. The non-Owner entities comprise mostly irrelevant background because no matter what interactions may be offered by these entities, the CR will not obey them, just the perceived owner. These functions are a subset of the CRA that enable cognition.

The sensory functions of radio entail those hardware and/or software capabilities that enable a radio to measure features of a sensory domain. Sensory domains include anything that can be sensed, such as audio, video, vibration, temperature, time, power, fuel level, ambient light level, sun angle (*e.g.*, through polarization), barometric pressure, smell, and anything else you might imagine. Sensory domains for vehicular radios may be much richer if less personal than those of wearable radios. Sensory domains for fixed infrastructure could include weather features such as ultra-violet sunlight, wind direction and speed, humidity, or rain rate. These functions are a subset of the CRA that enable perception.

The Platform Independent Model (PIM) in the Unified Modeling Language (UML) of SDR [OMGUML, 2006] provides a convenient, industry-standard computational model that an AACR can use to describe the SDR and computational resource aspects of its own internal structure, as well as describing facilities that enable radio functions. The general structure of hardware and software by which a CR reasons about the <Self/> in its world is also part of its architecture defined in the SDR SCA/SRA as resources.

4. CRA II: The Cognition Cycle

The Cognitive Radio Architecture (CRA) consists of a set of design rules by which the cognitive level of information services may be achieved by a specified set of components in a way that supports the cost-effective evolution of increasingly capable implementations over time [Joseph Mitola III, 2000b]. The cognition subsystem of the architecture includes an inference hierarchy and the temporal organization and flow of inferences and control states, the cognition cycle.

The Cognition Cycle

The cognition cycle developed for CR1 [Joseph Mitola III, 2000a] is illustrated in Figure 9.4. This cycle implements the capabilities required of iCR in a reactive sequence. Stimuli enter the cognitive radio as sensory interrupts, dispatched to the cognition cycle for a response. Such an iCR continually observes (senses and perceives) the environment, orients itself, creates plans, decides, and then acts. In a single-processor inference system, the CR's flow of control may also move in the cycle from observation to action. In a multi-processor system, temporal structures of sensing, preprocessing, reasoning, and acting may be parallel and complex. Special features synchronize the inferences of each phase. The tutorial code all works on a single processor in a rigid inference sequence defined in the figure. This process is called the Wake Epoch because the primary reasoning activities during this large epoch of time are reactive to the environment. There may also be "sleep epochs" for introspective reasoning or "prayer epochs" for asking for help from a higher authority.

During the wake epoch, the receipt of a new stimulus on any of a CR's sensors or the completion of a prior cognition cycle initiates a new primary cognition cycle. The cognitive radio observes its environment by parsing incoming information streams. These can include the monitoring speech-to-text conversion of radio broadcasts, *e.g.*, the weather channel, stock ticker tapes, etc. Any RF-LAN or other short-range wireless broadcasts that provide environment awareness information may be also parsed. In the observation phase, a CR also reads location, temperature, and light level sensors, etc. to infer the user's communications context.

Observe (Sense and Perceive). The iCR senses and perceives the environment (via "Observation Phase" code) by accepting multiple stimuli in many dimensions simultaneously and by binding these stimuli - all together or more typically in subsets - to prior experience so that it can subsequently detect time-sensitive stimuli and ultimately generate plans for action.

Thus, iCR continuously aggregates experience and compares prior aggregates to the current situation. A CR may aggregate experience by remembering

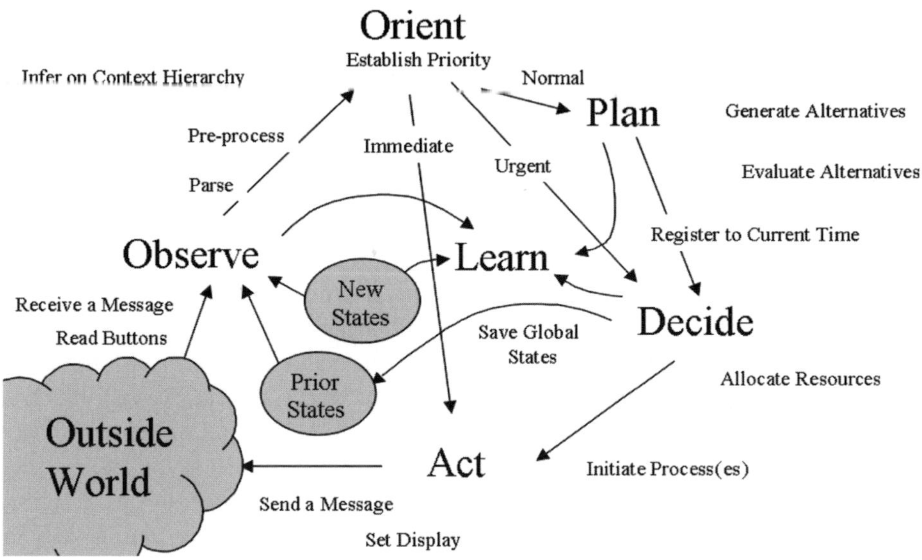

Figure 9.4. Simplified Cognition Cycle.

everything. This may not seem like a very smart thing to do until you calculate that all the audio, unique images, and emails and the radio might experience in a year only takes up a few hundred gigabytes of memory, depending on image detail. So the computational architecture for remembering and rapidly correlating current experience against everything known previously is a core capability of the CRA. A ***novelty*** detector identifies new stimuli, using the new aspects of partially familiar stimuli for incremental learning.

In the six-component (User SP, Environment, Effectors, SDR, Sys Apps, and Cognition) functional view of the architecture defined above, the Observe phase comprises both the User Sensory and Perception (User SP) and the Environment (RF and physical) sensor subsystems. The subsequent Orient phase is part of the Cognition component in this model of architecture.

Orient. The "Orient Phase" determines the significance of an observation by binding the observation to a previously known set of stimuli or "scene."

The Orient phase contains the internal data structures that constitute the equivalent of the Short Term Memory (STM) that people use to engage in a dialog without necessarily remembering everything with the same degree of long term memory. Typically people need repetition to retain information over the long term. The natural environment supplies the information redundancy needed to instigate transfer from STM to Long Term Memory (LTM). In the

CRA, the transfer from STM to LTM is mediated by the sleep cycle in which the contents of STM since the last sleep cycle are analyzed both internally and with respect to existing LTM. How to do this robustly remains an important CR research topic, but the overall framework is defined in CRA.

Matching of current stimuli to stored experience may be achieved by stimulus recognition or by "binding". The orient phase is the first collection of activity in the cognition component.

Stimulus Recognition

Stimulus recognition occurs when there is an exact match between a current stimulus and a prior experience. CR1 is continually recognizing exact matches and recording the number of exact matches that occurred along with the time in number of cognition cycles between the last exact match. By default, the response to a given stimulus is to merely repeat that stimulus to the next layer up the inference hierarchy for aggregation of the raw stimuli. But if the system has been trained to respond to a location, a word, an RF condition, a signal on the power bus, etc, then it may either react immediately or plan a task in reaction to the detected stimulus. If that reaction were in error, then it may be trained to ignore the stimulus given the larger context which consists of all the stimuli and relevant internal states, including time.

Sometimes, the Orient Phase causes an action to be initiated immediately as a "reactive" stimulus-response behavior. A power failure, for example, might directly invoke an act that saves the data (the "Immediate" path to the Act Phase in the figure). A non-recoverable loss of signal on a network might invoke reallocation of resources, *e.g.*, from parsing input to searching for alternative RF channels. This may be accomplished via the path labeled "Urgent" in the figure.

Binding

The binding occurs when there is a nearly exact match between a current stimulus and a prior experience and very general criteria for applying the prior experience to the current situation are met. One such criterion is the number of unmatched features of the current scene. If only one feature is unmatched and the scene occurs at a high level such as the phrase or dialog level of the inference hierarchy, then binding is the first step in generating a plan for behaving similarly in the given state as in the last occurrence of the stimuli. In addition to numbers of features that match exactly, which is a kind of Hamming code, Instance-Based Learning (IBL) supports inexact matching and binding. Binding also determines the priority associated with the stimuli. Better binding yields higher priority for autonomous learning, while less effective binding yields lower priority for the incipient plan.

Plan. Most stimuli are dealt with "deliberatively" rather than "reactively." An incoming network message would normally be dealt with by generating a plan (in the Plan Phase, the "Normal" path). Planning includes plan generation. In research-quality or industrial-strength CR's, formal models of causality must be embedded into planning tools. The Plan phase should also include reasoning about time. Typically, reactive responses are pre- programmed or defined by a network (the CR is "told" what to do), while other behaviors might be planned. A stimulus may be associated with a simple plan as a function of planning parameters with a simple planning system. Open source planning tools enable the embedding of planning subsystems into the CRA, enhancing the Plan component. Such tools enable the synthesis of RF and information access behaviors in a goal-oriented way based on perceptions from the visual, audio, text, and RF domains as well as RA rules and previously learned user preferences.

Decide. The "Decide" phase selects among the candidate plans. The radio might have the choice to alert the user to an incoming message (*e.g.*, behaving like a pager) or to defer the interruption until later (*e.g.*, behaving like a secretary who is screening calls during an important meeting).

Act. "Acting" initiates the selected processes using effector modules. Effectors may access the external world or the CR's internal states.

Externally Oriented Actions

Access to the external world consists primarily of composing messages to be spoken into the local environment or expressed in text form locally or to another CR or CN using KQML, RKRL, OWL, RXML, or some other appropriate knowledge interchange standard.

Internally Oriented Actions

Actions on internal states include controlling machine- controllable resources such as radio channels. The CR can also affect the contents of existing internal models, *e.g.*, by adding an serModel to an existing internal model structure. The new concept itself may assert related concepts into the scene. Multiple independent sources of the same concept in a scene reinforce that concept for that scene. These models may be asserted by the <Self/> to encapsulate experience. The experience may be reactively integrated into RXML knowledge structures as well, provided the reactive response encodes them properly.

Learning. Learning is a function of perception, observations, decisions and actions. Initial learning is mediated by the Observe-phase perception hierarchy in which all sensory perceptions are continuously matched against all prior stimuli to continually count occurrences and to remember time since last occurrence of the stimuli from primitives to aggregates.

Learning also occurs through the introduction of new internal models in response to existing models and CBR bindings. In general, there are many opportunities to integrate ML into AACR. Each of the phases of the cognition cycle offers multiple opportunities for discovery processes like <Histogram> above, as well as many other ML approaches to be developed below. Since the architecture includes internal reinforcement via counting occurrences and via serModels, ML with uncertainty is also supported in the architecture.

Finally, there is a learning mechanism that occurs when a new type of serModel is created in response to an Action to instantiate an internally generated serModel. For example, prior and current internal states may be compared with expectations to learn about the effectiveness of a communications mode, instantiating a new mode-specific serModel.

Self monitoring timing. Each of the prior phases must consist of computational structures for which the execution time may be computed in advance. In addition, each phase must restrict its computations to consume not more resources (time x allocated processing capacity) than the pre-computed upper bound. Therefore, the architecture has some prohibitions and some data set requirements needed to obtain an acceptable degree of stability of behavior for CR as self- referential self-modifying systems.

Since first order predicate calculus (FOPC) used in some reasoning systems is not decidable, one cannot in general compute in advance how much time an FOPC expression will take to run to completion. There may be loops that will preclude this, and even with loop detection, the time to resolve an expression may be only loosely approximated as an exponential function of some parameters (such as the number of statements in the FOPC data base of assertions and rules). Therefore unrestricted FOPC is not allowed.

Similarly, unrestricted For, Until and While loops are prohibited. In place of such loops are bounded iterations in which the time required for the loop to execute is computed or supplied independent of the computations that determine the iteration control of the loop. This seemingly unnatural act can be facilitated by next-generation compilers and CASE tools. Since self-referential self-modifying code is prohibited by structured design and programming practices, there are no such tools on the market today. But since CR is inherently self- referential and self-modifying, such tools most likely will emerge, perhaps assisted by the needs of CR and the architecture framework of the cognition cycle.

Finally, the cognition cycle itself can't contain internal loops. Each iteration of the cycle must take a defined amount of time, just as each frame of a 3G air interface takes 10 milliseconds. As CR computational platforms continue to progress, the amount of computational work done within the cycle will increase, but under no conditions should explicit or implicit loops be introduced into the cognition cycle that would extend it beyond a given cycle time.

Retrospection. Since the assimilation of knowledge by machine learning can be computationally intensive, cognitive radio has "sleep" and "prayer" epochs that support machine learning. A sleep epoch is a relatively long period of time (*e.g.*, minutes to hours) during which the radio will not be in use, but has sufficient electrical power for processing. During the sleep epoch, the radio can run machine learning algorithms without detracting from its ability to support its user's needs. Machine learning algorithms may integrate experience by aggregating statistical parameters. The sleep epoch may re-run stimulus-response sequences with new learning parameters in the way that people dream. The sleep cycle could be less anthropomorphic, employing a genetic algorithm to explore a rugged fitness landscape, potentially improving the decision parameters from recent experience.

Reaching. Out Learning opportunities not resolved in the sleep epoch can be brought to the attention of the user, the host network, or a designer during a prayer epoch. The sleep and prayer epochs are possibilities.

5. CRA III: The Inference Hierarchy

The phases of inference from observation to action show the flow of inference, a top-down view of how cognition is implemented algorithmically. The inference hierarchy is the part of the algorithm architecture that organizes the data structures. Inference hierarchies have been in use since Hearsay II in the 1970s, but the CR hierarchy is unique in its method of integrating machine learning with real-time performance during the Wake Epochs. An illustrative inference hierarchy includes layers from atomic stimuli at the bottom to information clusters that define action contexts as in Table 9.4.

The pattern of accumulating elements into sequences begins at the bottom of the hierarchy. Atomic stimuli originate in the external environment including RF, acoustic, image, and location domains among others. The atomic symbols extracted from them are the most primitive symbolic units in the domain. In speech, the most primitive elements are the phonemes. In the exchange of textual data (*e.g.*, in email), the symbols are the typed characters. In images, the atomic symbols may be the individual picture elements (pixels) or they may be small groups of pixels with similar hue, intensity, texture, etc.

A related set of atomic symbols forms a primitive sequence. Words in text, tokens from a speech tokenizer, and objects in images (or individual image regions in a video flow) are the primitive sequences. Primitive sequences have spatial and/or temporal coincidence, standing out against the background (or noise), but there may be no particular meaning in that pattern of coincidence. Basic sequences, on the other hand, are space-time- spectrum sequences that entail the communication of discrete messages.

Table 9.4. Standard Inference Hierarchy.

Sequence	Level of Abstraction
Context Cluster	***Scenes*** in a play, Session
Sequence Clusters	***Dialogs***, Paragraphs, Protocol
Basic Sequences	***Phrases***, video clip, messages
Primitive Sequences	***Words***, token, immage
Atomic Symbols	***Raw Data***, Phoneme, pixel
Atomic Stimuli	External Phenomena

These discrete messages (*e.g.*, phrases) are typically defined with respect to an ontology of the primitive sequences (*e.g.*, definitions of words). Sequences cluster together because of shared properties. For examples, phrases that include words like "hit," "pitch," "ball," and "out" may be associated with a discussion of a baseball game. Knowledge Discovery and Data Mining (KDD) and the Semantic Web offer approaches for defining, or inferring the presence of such clusters from primitive and basic sequences.

A scene is a context cluster, a multi-dimensional space-time- frequency association, such as a discussion of a baseball game in the living room on a Sunday afternoon. Such clusters may be inferred from unsupervised machine learning, *e.g.*, using statistical methods or nonlinear approaches such as Support Vector Machines (SVM).

Although presented above in a bottom-up fashion, there is no reason to limit multi-dimensional inference to the top layers of the inference hierarchy. The lower levels of the inference hierarchy may include correlated multi-sensor data. For example, a word may be characterized as a primitive acoustic sequence coupled to a primitive sequence of images of a person speaking that word. In fact, since infants seem to thrive on multi-sensory stimulation, the key to reliable machine learning may be the use of multiple sensors with multi-sensor correlation at the lowest levels of abstraction.

Each of these levels of the inference hierarchy is now discussed further.

Atomic Stimuli

Atomic stimuli originate in the external environment and are sensed and pre-processed by the sensory subsystems which include sensors of the RF environment (*e.g.*, radio receiver and related data and information processing) and of the local physical environment including acoustic, video, and location sensors. Atomic symbols are the elementary stimuli extracted from the atomic stimuli. Atomic symbols may result from a simple noise- riding threshold algorithm, such as the squelch circuit in RF that differentiates signal from noise. Acoustic signals may be differentiated from simple background noise this way, but generally the result is the detection of a relatively large speech epoch which contains various kinds of speech energy. Thus, further signal processing is typically required in a preprocessing subsystem to isolate atomic symbols.

The transformation from atomic stimuli to atomic symbols is the job of the sensory preprocessing system. Thus, for example, acoustic signals may be transformed into phoneme hypotheses by an acoustic signal pre-processor. Some speech-to-text software tools may not enable this level of interface via an API, however. To develop industrial strength CR, contemporary speech- to-text and video processing software tools are needed. Speech to text tools yield an errorful transcript in response to a set of atomic stimuli. Thus, the speech to text tool is an example of a mapping from atomic stimuli to basic sequences. One of the important contributions of architecture is to identify such maps and to define the role of the level mapping tools.

Image processing software available for the Wintel-Java development environment JBuilder has the ability to extract objects from images and video clips. In addition, research such as that of Goodman et al defines algorithms for what the AAAI calls cognitive vision [AAAI, 2004].

But there is nothing about the inference hierarchy that forces data from a pre-processing system to be entered at the lowest level. In order for the more primitive abstractions such as atomic symbols to be related to more aggregate abstractions, one may either build up the aggregates from the primitive abstractions or derive the primitive abstractions from the aggregates. Since people are exposed to "the whole thing" by immersion in the full experience of life - touch, sight, sound, taste, and balance - all at once, it seems possible - even likely - that the more primitive abstractions are somehow derived through the analysis of aggregates, perhaps by cross-correlation. This can be accomplished in a CRA sleep cycle. The idea is that the wake cycle is optimized for immediate reaction to stimuli, such as our ancestors needed to avoid predation, while the sleep cycle is optimized for introspection, for analyzing the day's stimuli to derive those objects that should be recognized and acted upon in the next cycle.

Stimuli are each counted. When an iCR that conforms to this architecture encounters a stimulus, it both counts how many such stimuli have been encountered and resets a timer to zero that keeps track of the time since the last occurrence of the stimulus.

Primitive Sequences: Words and Dead Time

The accumulation of sequences of atomic symbols forms primitive sequences. The key question at this level of the data structure hierarchy is the sequence boundary. The simplest situation is one in which a distinguished atomic symbol separates primitive sequences, which is exactly the case with white space between words in typed text. In general, one would like a machine- learning system to determine on its own that the white space separates the keyboard input stream into primitive sequences.

Basic Sequences

The pattern of aggregation is repeated vertically at the levels corresponding to words, phrases, dialogs, and scenes. The data structures generated by PDA Nodes create the concept hierarchy of Table 9.4. These are the reinforced hierarchical sequences. They are reinforced by the inherent counting of the number of times each atomic or aggregated stimulus occurs. The phrase level typically contains or implies a verb (the verb to-be is implied if no other verb is implicit).

Unless digested (*e.g.*, by a sleep process), the observation phase hierarchy accumulates all the sensor data, parsed and distributed among PDA Nodes for fast parallel retrieval. Since the hierarchy saves everything and compares new data to memories, it is a kind of memory-base learning technique. This is a memory-intensive approach, taking a lot of space. When the stimuli retained are limited to atomic symbols and their aggregates, the total amount of data that needs to be stored is relatively modest. In addition, recent research shows the negative effects of discarding cases in word pronunciation. In word pronunciation, no example can be discarded even if "disruptive" to an intentional model. Each exception has to be followed. Thus in CR1, when multiple memories match partially, the most nearly exact match informs the orientation, planning, and action.

Basic sequences are each counted. When an iCR that conforms to this architecture encounters a basic sequence, it both counts how many such sequences have been encountered and resets a timer to zero that keeps track of the time since the last occurrence.

Natural Language in the CRA Inference Hierarchy

In speech, words spoken in a phrase may be co-articulated with no distinct boundary between the primitive sequences in a basic sequence. Therefore, speech detection algorithms may reliably extract a basic sequence while the parsing of that sequence into its constituent primitive sequences may be much less reliable. Typically, the correct parse is within the top ten candidates for contemporary speech-to-text software tools. But the flow of speech signal processing may be something like:

- Isolate a basic sequence (phrase) from background noise using an acoustic squelch algorithm

- Analyze the basic sequence to identify candidate primitive sequence boundaries (words)

- Analyze the primitive sequences for atomic symbols
- Evaluate primitive and basic sequence hypotheses based on a statistical model of language to rank-order alternative interpretations of the basic sequence.

So a practical speech processing algorithm may yield alternative strings of phonemes and candidate parses "all at once.". NLP tool sets may be embedded into the CRA inference hierarchy as illustrated in Figure 9.5. Speech and/or text channels may be processed via natural language facilities with substantial a-priori models of language and discourse. The use of those models should entail the use of mappings among the word, phrase, dialog, and scene levels of the observation phase hierarchy and the encapsulated component(s).

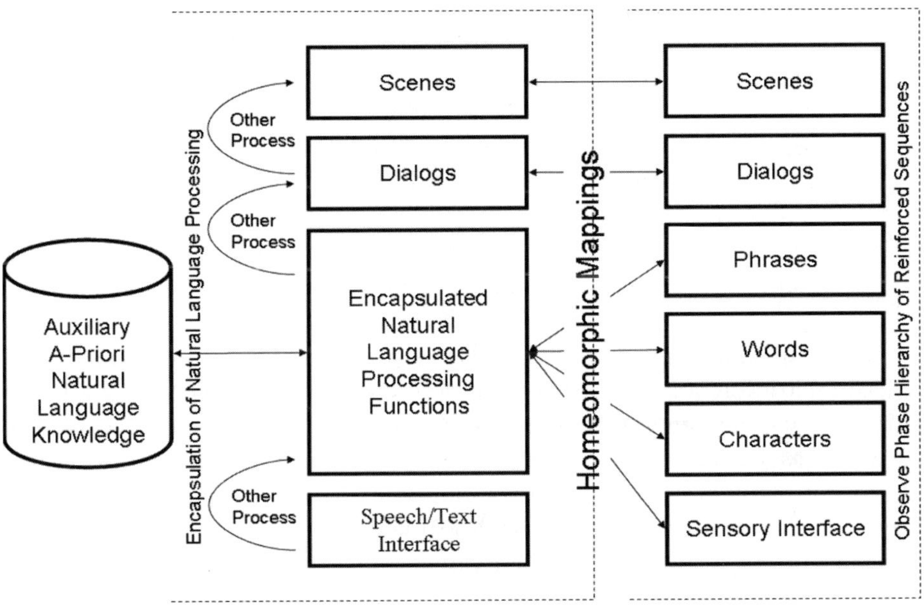

Figure 9.5. Natural Language Encapsulation in the Observation Hierarchy.

It is tempting to expect cognitive radio to integrate a commercial natural language processing system such as IBM's ViaVoice or a derivative of an NLP research system such as SNePS [SNePS Web, 1998], AGFL [Koser, 1999], or XTAG [The XTAG Research Group, 1999] perhaps using a morphological analyzer like PCKimmo [IBM, 2006]. These tools both go too far and not far enough in the direction needed for CRA. One cannot just express a radio ontology in a semantic web Interlingua and plug it neatly into XTAG to get a working cognitive radio. The internal data structures of radio mediate the performance of radio tasks (*e.g.*, "transmit a waveform"). The data structures of XTAG, AGFL, etc mediate the conversion of language from one form to another. Thus, XTAG wants to know that "transmit" is a verb and "waveform" is a noun. The CR needs to know that if the user says "transmit" and a message has been defined, then the CR should call the SDR function *transmit*(). NLP systems also need scoping rules for transformations on the linguistic data structures. The way in which domain knowledge is integrated in linguistic structures of these tools tends to obscure the radio engineering aspects. ViaVoice ans similar tools thus require substantial domain engineering for cognitive radio applications.

Natural language processing systems work well on well-structured speech and text, such as the prepared text of a news anchor. But they do not work well yet on noisy, non-grammatical data structures encountered when a user is trying to order a cab in a crowded bar. Thus, less-linguistic or meta-linguistic data structures may be needed to integrate core cognitive radio reasoning with speech and/or text-processing front ends. The CRA has the flexibility illustrated in the figure above for the subsequent integration of evolved NLP tools. The emphasis of this version of the CRA is a structure of sets and maps required to create a viable cognitive radio architecture. Although introducing the issues required to integrate existing natural language processing tools, the text does not pretend to present a complete solution to this problem.

Observe-Orient Links for Scene Interpretation

CR may use an algorithm-generating language with which one may define self-similar inference processes. In one example, the first process (Proc1) partitions characters into words, detecting novel characters and phrase boundaries as well. Proc2 detects novel words and aggregates known words into phrases. Proc3 detects novel phrases, aggregating known phrases into dialogs. Proc4 aggregates dialogs into scenes, and Proc5 detects known scenes. In each case, a novel entity at level N will be bound in the context of the surrounding known entities at that level to the closest match at the next highest level, N + 1. For example at the word-phrase intersection of Proc2, would map the following phrases:

Equation 2: "Let me introduce Joe"

Equation 3: "Let me introduce *Chip*"

Since "Chip" is unknown while "Joe" is known from a prior dialog, integrated CBR matches the phrases, binding <Chip> = <Joe>. In other words, it will try to act with respect to Chip in the way it was previously trained (at the dialog level) to interact with Joe. In response to the introduction, the system may say "Hello, Chip, How are you?" mimicking the behavior it had been trained with respect to Joe previously. Not too bright, but not all that bad either for a relatively simple machine learning algorithm.

There is a particular kind of dialog that is characterized by reactive world knowledge in which there is some standard way of reacting to given speech-act inputs. For example, when someone says "Hello", you may typically reply with "Hello" or some other greeting. The capability to generate such rote responses is pre-programmed into a lateral component of the Hearsay knowledge source (KS). The responses are not pre-programmed, but the general tendency to imitate phrase level dialogs is a pre-programmed tendency that can be over-ruled by plan generation, but that is present in the orient-phase, which is Proc6.

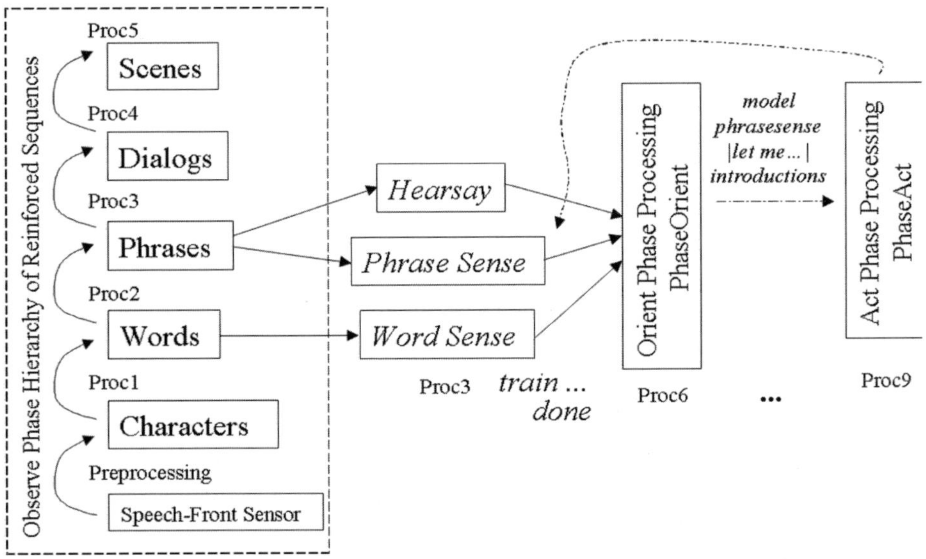

Figure 9.6. The Inference Hierarchy Supports Lateral Knowledge Sources.

Words may evoke a similar tendency towards immediate action. What do you do when you hear the words "Help!!" or "Fire, fire, get out, get out!!" You, the CR programmer, can capture reactive tendencies in your CR by pre-programming an ability to detect these kinds of situations in the Word-sense knowledge source (Figure 9.6). When confronted with them (which is preferred), CR should react appropriately if properly trained, which is one of the key aspects of this text. To cheat, you can pre-program a wider array of stimulus-response pairs so that your CR has more a- priori knowledge, but some of it may not be appropriate. Some responses are culturally conditioned. Will your CR be too rigid? If it has too much a-priori knowledge, it will be perceived by its users as too rigid. If it doesn't have enough, it will be perceived as too stupid.

Observe-Orient Links for Radio Skill Sets

Radio knowledge may be embodied in components called radio skills. Radio knowledge is static, requiring interpretation by an algorithm such as an inference engine in order to accomplish anything. Radio skills, on the other hand, are knowledge embedded in serModels through the process of training or sleeping/dreaming. This knowledge is continually pattern-matched against all stimuli in parallel. That is, there are no logical dependencies among knowledge components that mediate the application of the knowledge. With FOPC, the theorem-prover must reach a defined state in the resolution of multiple axioms in order to initiate action. In contrast, serModels are continually compared to the level of the hierarchy to which they are attached, so their immediate responses are always cascading towards action. Organized as maps primarily among the wake-cycle phases "observe" and "orient," the radio procedure skill sets (SS's) control radio personalities as illustrated in Figure 9.7.

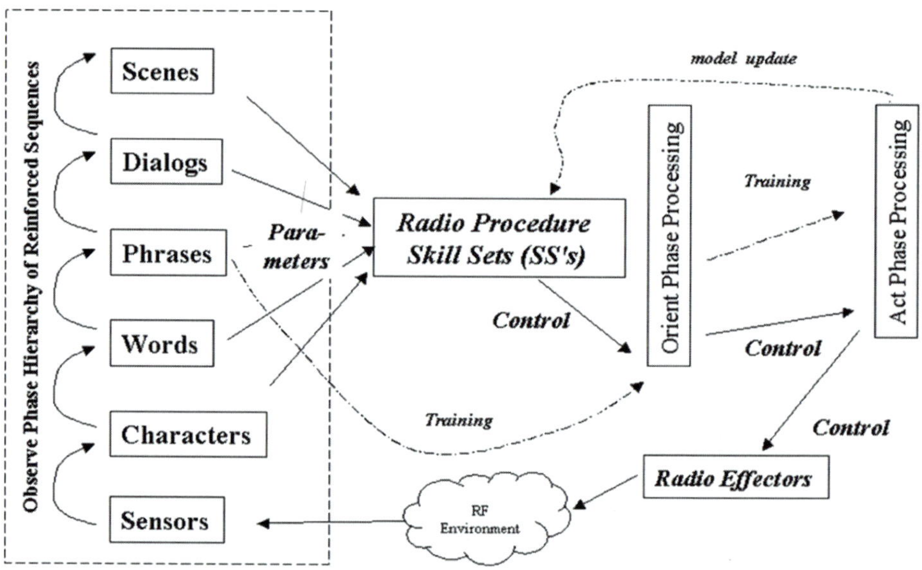

Figure 9.7. Radio Skills Respond to Observations.

These skill sets may either be reformatted into serModels directly from the a-priori knowledge of an RKRL frame, or they may be acquired from training or sleep/dreaming. Each skill set may also save the knowledge it learns into an RKRL frame.

General World Knowledge

An AACR needs substantial knowledge embedded in the inference hierarchies. It needs both external RF knowledge and internal radio knowledge. Internal knowledge enables it to reason about itself as a radio. External radio knowledge enables it to reason about the role of the <Self/> in the world, such as respecting rights of other cognitive and not-so-cognitive radios.

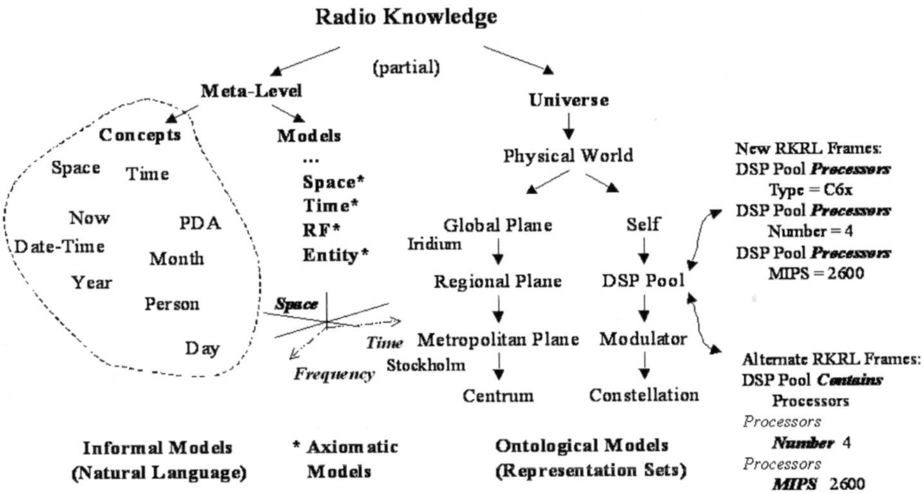

Figure 9.8. External Radio Knowledge Includes Concrete and Abstract Knowledge.

Figure 9.8 illustrates the classes of knowledge an AACR needs to employ in the inference hierarchies and cognition cycle. It is one thing to write down that the Universe includes a Physical World (there could also be a spiritual world, and that might be very important in some cultures). It is quite another thing to express that knowledge in a way that the AACR will be able to use that knowledge effectively. Symbols like "Universe" take on meaning by their relationships to other symbols and to external stimuli. In this ontology, meta-level knowledge consists of ***abstractions***, distinct from existential knowledge of the physical Universe. In RXML, this ontological perspective includes all in a universe of discourse, <Universe> expressed as follows.

```
<Universe>
    <Abstractions> <Time> <Now/> </Time> <Space> <Here/></Space>
    .<RF/>.
     <Intelligent-Entities/> . </Abstractions>
     <Physical-universe>. <Instances/> of Abstractions.
     </Physical-universe>
</Universe>
```

Equation 4 The Universe of Discourse Consists of Abstractions plus the Physical Universe

Abstractions include informal and formal meta-level knowledge from unstructured knowledge of concepts to the more mathematically structured models of space, time, RF, and entities that exist in space-time. To differentiate "now" as a temporal concept from "Now" as the Chinese name of a plant, the CRA includes both the a-priori knowledge of "now" as a space-time locus, <Now/> as well as functions that access and manipulate instances of the concept <Now/>. <Now/> is axiomatic in the CRA, so code refers to "now" (as n.o.w) in planning actions. The architecture allows an algorithm to return the date-time code from Windows to define instances of <Now/>. Definition-by-algorithm permits an inference system like the cognition subsystem to reason about whether a given event is in the past, present, or future. What is the present? The present is some region of time between "now" and the immediate past and future. If you are a paleontologist, "now" may consist of the million year epoch in which we all are thought to have evolved from apes. If you are a rock star, "now" is probably a lot shorter than that to you. How will your CR learn the user's concept of now? The CRA design offers an axiomatic treatment of time, but the axioms were not programmed into the Java explicitly. THE CRA aggregates knowledge of time by a temporal CBR that illustrates the key principles. The CRA does not fix the definition of <Now><Now/> but enables the <Self/> to define the details in an <Instance> in the physical world about which it can learn from the user, paleontologist or rock star.

Given the complexity of a system that includes both a multi- tiered inference hierarchy and the cognition cycle's observe- orient-plan-decide-act sequence with AML throughout, it is helpful to consider the mathematical structure of these information elements, processes, and flows. The mathematical treatment is the subject of the next section.

6. CRA IV: Architecture Maps

Cognition functions are implemented via cognition elements consisting of data structures, processes and flows. These include data structures and related processing elements may be modeled as topological maps over the abstract domains identified in Figure 9.9.

The <Self/> is an entity in the world, while the internal organization of the <Self/> (annotated PDA in the figure) is an abstraction that models the <Self/>. The hierarchy of words, phrases, and dialogs from sensory data to scenes is not inconsistent with visual perception. Words correspond to visual entities, phrases to detectable movement and juxtaposition of entities in a scene. Dialogs correspond to a coherent sequence of movement within the scope of a scene, such as walking across the room. Occlusion may be thought of as a dialog in

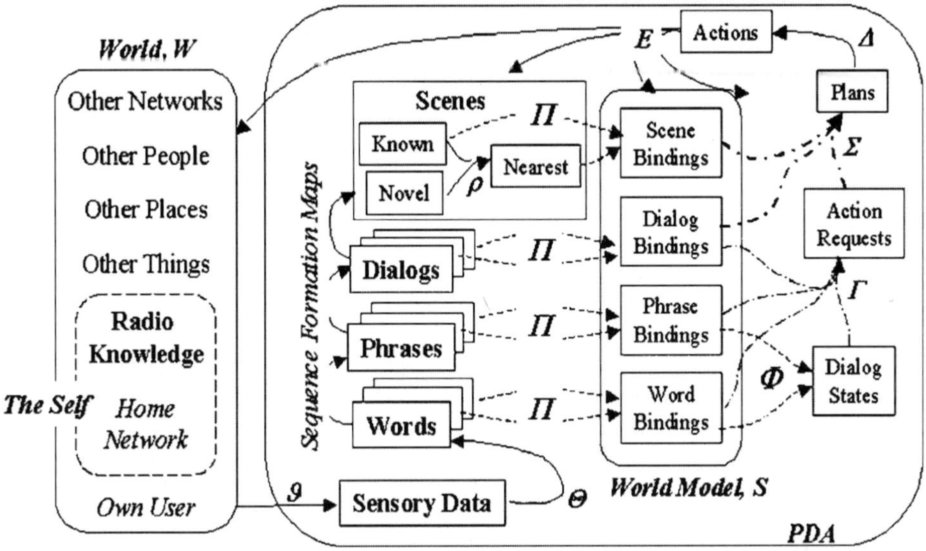

Figure 9.9. Architecture Based on The Cognition Cycle.

which the room asserts itself in part of the scene while observations correspond to assertion of the object. The model data structures may be read as generalized words, phrases, dialogs, and scenes that may be acoustic, visual, or perceived in other sensory domains (*e.g.*, infrared). These structures refer to set-theoretic spaces consisting of a set X and a family of subsets Ox that contain X and the null set and that are closed under union and countable intersection. In other words, each is a topological space induced over the domain. Proceeding up the hierarchy, the scope of the space (X, Ox) increases. A <Scene> is a subset of space-time that is circumscribed by the entity by sensory limits.

The cognition functions modeled in these spaces are topology- preserving maps. Data and knowledge storage spaces are shown as rectangles (*e.g.*, Dialog States, Plans), while processing elements that transform sets are modeled as homeomorphisms, topology- preserving maps, shown as directed graphs (*e.g.*, Π).

CRA Topological Maps

The processing elements of the architecture are modeled topological maps, as shown in Figure 9.9. The input map ϑ consists of components that transform external stimuli to the internal data structure Sensory Data. The transformation Θ consists of entity recognition (via acoustic, optical, and other sensors), lower-level software radio waveform interface components, etc., that create streams of primitive reinforced sequences. The model includes maps that form successively higher level sequences from the data on the immediately lower level.

Reasoning components include the map ρ that identifies the best match of known sequences to novel sequences. These are bound to scene variables by projection components, Π. The maps ϑ, Θ, ρ and Π , constitute Observe Phase processing. Generalized word and phrase level bindings are interpreted by the components Φ to form dialog states. Train, for example, is the dialog-state of a training experience in THE CRA. The components Γ create action requests from bindings and dialog states. The maps Φ and Γ constitute Orient Phase processing. Scene bindings include user communications context. Context-sensitive plans are created by the component Σ that evaluates action requests in the Plan Phase. The Decision Phase processing consists of map Δ that maps plans and scene context to actions. Finally, the map E consists of the effector components that change the PDA's internal states, change displays, synthesize speech, and transmit information on wireless networks using the software radio personalities.

CRA Identifies Self, Owner, and Home Network

The sets of entities in the world that are known to the CR are modeled graphically as rounded rectangles. These include the self grounded in the outside world ("Self"), as well as its knowledge of the self as self (*e.g.*, as "PDA"). The critical entities are world, W, the PDA, and the PDA's World Model, S. [In THE CRA, S are the Orient-phase data structures and processes.] Entities in the world include the differentiated entities "Own User" or Owner, and "Home Network." The architecture requires that the PDA be able to identify these entities so that it may treat them differentially. Other networks, people, places, and things may be identified in support of the primary cognition functions, but the architecture does not depend on such a capability.

CRA Reinforced Hierarchical Sequences

The data structures for perception include the reinforced hierarchical sequences Words, Phrases, Dialogs, and Scenes of the Observe Phase. Within each of these sequences, the novel sequences represent the current stimulus-response cases of the cognitive behavior model. The known sequences represent the integrated knowledge of the cognitive behavior model. Known sequences may consist of RXML statements embedded in the PDA or of knowledge acquired through independent machine learning. The Nearest sequence is the known sequence that is closest in some sense to the novel sequence. The World Model, S, consists primarily of bindings between a-priori data structures and the current scene. These associative structures are also associated with the Observe Phase. Dialog states, action requests, plans, and actions are additional data structures needed for the Observe, Orient, Plan, and Act Phases respectively. Each internal data structure maps to an RXML frame consisting of element (*e.g.*, set, or stimulus), model (*e.g.*, embedded procedure, parameter values), content - typically a structure of elements terminating in either primitive concepts <concept/> (*e.g.*, subset, or response) or instance data, and associated resources. Context is defined as the RXML URL or root from <Universe>, to include source, time, and place of the <Scene>.

Behaviors in the CR Architecture

CRA entails three modes of behavior: waking, sleeping, and praying. Behavior that lasts for a specific time interval is called a behavioral epoch. The axiomatic relationships among these behaviors are expressed in the topological maps of Figure 9.10.

Waking behavior. The waking behavior is optimized for real-time interaction with the user, isochronous control of software radio assets, and real-time sensing of the environment. The conduct of the waking behavior is informally referred to as the awake-state, although it is not a specific system state, but a set of behaviors. Thus, the awake-state cognition-actions (α) map the environment interactions to the current stimulus-response cases. These cases are the dynamic subset of the embedded serModels. Incremental machine learning (δ) maps these interactions to integrated knowledge, the persistent subset of the serModels.

Sleeping and dreaming behaviors. Cognitive PDAs detect conditions that permit or require sleep and dreaming. For example, if the PDA predicts or becomes aware of a long epoch of disuse (*e.g.*, overnight), then the CPDA may autonomously initiate sleeping behavior. Sleep is intentional inactivity, *e.g.*, to recharge batteries quickly. Dreaming behavior employs energy to retrospectively examine experience since the last period of sleep. In THE CRA,

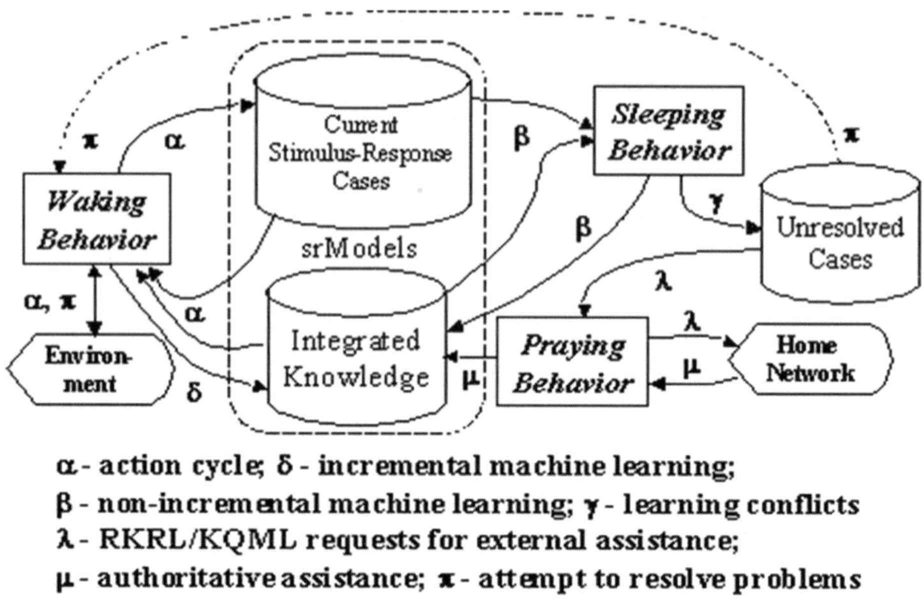

Figure 9.10. Cognitive Behavior Model Consists of Domains and Topological Maps.

all sleep is dreaming. In some situations, the CPDA may request permission to enter sleeping/ dreaming behavior from the user, *e.g.*, if pre- defined limits of aggregate experience are reached. Regular sleeping/ dreaming limits the combinatorial explosion of the process of assimilating aggregated experience into the serModels needed for real-time behavior during the waking behaviors. During the dreaming epochs, the CPDA process experience from the waking behavior using non-incremental machine-learning algorithms. These algorithms map current cases and new knowledge into integrated knowledge (β). A conflict is a context where the user overrode a CPDA decision about which the CPDA had no little or no uncertainty. Map β may resolve the conflict. If not, then it will place the conflict on a list of unresolved conflicts (map γ).

Prayer behavior. Attempts to resolve unresolved conflicts via the mediation of the PDA's home network may be called prayer behavior, reference of the issue to a completely trusted source with substantially superior capabilities. The unresolved-conflicts-list γ is mapped (λ) to RXML XML queries to the PDA's home cognitive network expressed in XML, OWL, KQML, RKRL, RXML or a mix of declared knowledge types. Successful resolution maps network responses to integrated knowledge (μ). There are many research issues surrounding the successful download of such knowledge including the set of support for referents in the unresolved-conflicts lists and the updating of knowledge in the CPDA needed for full assimilation of the new knowledge or procedural fix to the unresolved conflict. The prayer behavior may not be reducible to finite-resource introspection and thus may be susceptible to the partial-ness of TC even though the CPDA and CWN enforce watchdog timers and the like. Alternatively, the PDA may present the conflict sequence to the user, requesting the user's advice during the wake cycle (map π).

From Maps to API's

Each of these maps has a domain and a range. Axiomatically, the domain is the set of subsets of internal data structures over which the map is defined, while the range is the set of subsets onto which the map projects its effects. Thus for each map

M:D R, there is an associated API or API component.

API-M: m ϵ M: d ϵ D -> r ϵ R

In other words, the API for the map M specifies methods or attached procedures defined over subsets d of the domain D that map onto subsets r of the range R. So each map can be interpreted as a generalized API. Some API's may entail more than one map. A planning API for example, might include the maps that generate the plans and the maps that select among plan components and schedule plans for actions. In fact, APIs for many CR functions from perception to planning and action include more functionality than is needed for embedding into a CR, such as visualization tools and user interfaces. Therefore, the representation of API components as maps establishes the foundations of the API without over-constraining the definition of API's for a given CR design. The evolution of CRA from this set of maps to a set of API's with broad industry support may be facilitated by the framework of the maps.

Industrial Strength Inference Hierarchy

Although the CRA provides a framework for API's, it doesn't specify the details of the data structures nor of the maps. The CR1THE CRA research prototype emphasizes ubiquitous learning via serModels and Case Based Reasoning, but it doesn't implement critical features that would be required in deployable CR's. Other critical aspects of such industrial-strength architectures include more capable scene perception and situation interpretation specifically addressing:

Noise, in utterances, images, objects, location estimates and the like. Noise sources include thermal noise, conversion error introduced by the process of converting analog signals (audio, video, accelerometers, temperature, etc) to digital form, error in converting from digital to analog form, preprocessing algorithm biases and random errors, such as the accumulation of error in a digital filter, or the truncation of a low energy signal by threshold logic. Dealing effectively with noise differentiates a tutorial demonstration from an industrially useful product.

Hypothesis management, keeping track of more than one possible binding of stimuli to response, dialog sense, scene, etc. Hypotheses may be managed by keeping the N-best hypotheses (with an associated degree of belief), by estimating the prior probability or other degree of belief in a hypothesis, and keeping a sufficient number of hypotheses to exceed a threshold (*e.g.*, 90 or 99% of all the possibilities), or keeping multiple hypotheses until the probability for the next most likely (2nd) hypothesis is less than some threshold. The estimation of probability requires a measurable space, a sigma-algebra that defines how to accumulate probability on that space, proof that the space obeys the axioms of probability and a certainty calculus that defines how to combine degrees of belief in events as a function of the measures assigned to the probability of the event.

Training Interfaces, the reverse flow of knowledge from the inference hierarchy back to the perception subsystems. The recognition of the user by a combination of face and voice could be more reliable than single-domain recognition either by voice or by vision. In addition, the location, temperature, and other aspects of the scene may influence object identification. Visual recognition of the Owner outdoors in a snow storm, for example, is more difficult than indoors in an office. While the CR might learn to recognize the user based on weaker cues outdoors, access to private data might be constrained until the quality of the recognition exceeds some learned threshold.

Non-Linear Flows: Although the cognition cycle emphasizes the forward flow of perception enabling action, it is crucial to realize that actions may be internal, such as advising the vision subsystem that its recognition of the user is in *e.g.*, error because the voice does not match and the location is wrong. Because of the way the cognition cycle operates on the self, these reverse flows from perception to training are implemented as forward flows from the perception system to the self, directed towards a specific subsystem such as vision or audition. There may also be direct interfaces from the CWN to the CR to upload data structures representing a-priori knowledge integrated into the UCBR learning framework.

7. CRA V: Building the CRA on SDR Architectures

A Cognitive Radio is a Software Radio (SWR) or SDR with flexible formal semantics based entity to entity formal messaging via RXML and integrated machine learning of the self, the user, the environment, and the "situation." This section reviews SWR, SDR, and the Software Communications Architecture (SCA) or Software Radio Architecture (SRA) for those who may be unfamiliar with these concepts. While it is not necessary for an AACR to use the SCA/SRA as its internal model of itself, it certainly must have some model, or it will be incapable of reasoning about its own internal structure and adapting or modifying its radio functionality.

Review of SWR and SDR Principles

Hardware-defined radios such as the typical AM/FM broadcast receiver convert radio to audio using radio hardware, such as antennas, filters, analog demodulators, and the like. SWR is the ideal radio in which the Analog to Digital Converter (ADC) and Digital to Analog Converter (DAC) convert digital signals to and from radio frequencies (RF) directly, and all RF channel modulation, demodulation, frequency translation and filtering are accomplished digitally. For example, modulation may be accomplished digitally by multiplying sine and cosine components of a digitally sampled audio signal (called the "baseband" signal, *e.g.*, to be transmitted) by the sampled digital values of a higher frequency sine wave to up-convert it, ultimately to RF.

Figure 9.11 shows how SDR principles apply to a cellular radio base station. In the ideal SWR, there would be essentially no RF conversion, just ADC/DAC blocks accessing the full RF spectrum available to the (wideband) antenna elements. Today's SDR base- stations approach this ideal by digital access (DAC and ADC) to a band of spectrum allocations, such as 75 MHz allocated to uplink and downlink frequencies for third-generation services. In this architecture, RF conversion can be a substantial system component, sometimes 6-amenable to cost improvements through Moore's Law. The ideal SDR would access more like 2.5 GHz from, say 30 MHz to around 2.5 GHz, supporting all kinds of services in television (TV) bands, police bands, air traffic control bands - you name it. Although considered radical when introduced in 1991 [J. Mitola III, 1992] and popularized in 1995 [J. Mitola III, 1995], recent regulatory rulings are encouraging the deployment of such "flexible spectrum" use architectures.

This SWR ideal typically may not be practical or affordable, so it is important for the radio engineer to understand the tradeoffs (again, see [Joseph Mitola III, 2000b] for SDR architecture tradeoffs). In particular, the physics of RF devices (*e.g.*, antennas, inductors, filters) makes it easier to synthesize narrowband RF and intervening analog RF conversion and Intermediate Frequency (IF) conversion. Given narrowband RF, the hardware-defined radio might employ baseband (*e.g.*, voice frequency) ADC, DAC, and digital signal processing. The Programmable Digital Radios (PDR) of the 1980's and 90;s used this approach.

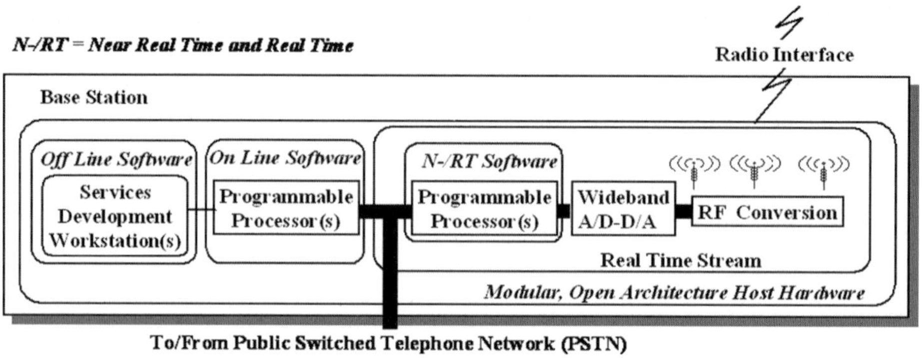

Figure 9.11. SWR Principle Applied to Cellular Base Station.

Historically, this approach has not been as expensive as wideband RF (antennas, conversion), ADCs and DACs. Handsets are less amenable to SWR principles than the base station (Figure 9.12). Base stations access the power grid. Thus, the fact that wideband ADCs, DACs, and DSP consume many watts of power is not a major design driver. Conservation of battery life, however, is a major design driver in the handset.

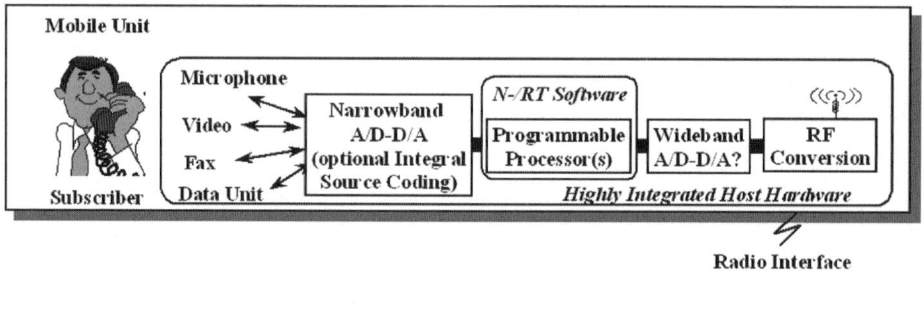

Figure 9.12. Software Radio Principle - "ADC and DAC At the Antenna" May Not Apply.

Thus, insertion of SWR technology into handsets has been relatively slow. Instead, the major handset manufacturers include multiple single-band RF chip sets into a given handset. This has been called the Velcro radio [BellSouth, 1995].

Since the ideal SWR is not readily approached in many cases, the SDR has comprised a sequence of practical steps from the baseband DSP of the 1990's towards the ideal SWR. As the economics of Moore's Law and of increasingly wideband RF and IF devices allow, implementations move upward and to the right in the SDR design space (Figure 9.13).

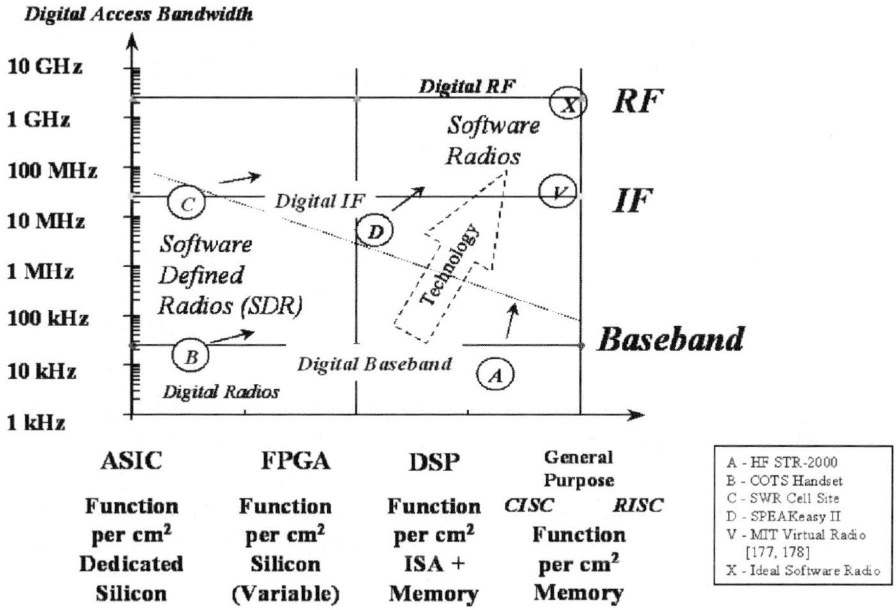

Figure 9.13. SDR Design Space Shows How Designs Approach the Ideal SWR.

This space consists of the combination of digital access bandwidth and programmability. Access bandwidth consists of ADC/DAC sampling rates converted by the Nyquist criterion or practice into effective bandwidth. Programmability of the digital subsystems is defined by the ease with which logic and interconnect may be changed after deployment. Application-Specific Integrated Circuits (ASICs) cannot be changed at all, so the functions are "dedicated" in silicon. Field Programmable Gate Arrays (FPGAs) can be changed in the field, but if the new function exceeds some parameter of the chip, which is not uncommon, then one must upgrade the hardware to change the function, just like ASICs. Digital Signal Processors (DSPs) are typically easier or less expensive to program and are more efficient in power use than FPGAs. Memory limits and instruction set architecture (ISA) complexity can drive up costs of reprogramming the DSP. Finally, general purpose processors, particularly with Reduced Instruction Set Architectures (RISC) are most cost-effective to change in the field. To assess a multi-processor, such as a cell phone with a

CDMA-ASIC, DSP speech codec, and RISC micro-controller, weight the point by equivalent processing capacity.

Where should one place an SDR design within this space? The quick answer is so that you can understand the migration path of radio technology from the lower left towards the upper right, benefiting from lessons learned in the early migration projects captured in *Software Radio Architecture* [Joseph Mitola III, 2000b].

This section contains a very brief synopsis of the key SDR knowledge you will need in order to follow the AACR examples of this text.

Radio Architecture

The discussion of the software radio design space contains the first elements of radio architecture. It tells you what mix of critical components are present in the radio. For SDR, the critical hardware components are the ADC, DAC, and processor suite. The critical software components are the user interface, the networking software, the information security (INFOSEC) capability (hardware and/or software), the RF media access software, including the physical layer modulator and demodulator (modem) and media access control (MAC), and any antenna-related software such as antenna selection, beamforming, pointing and the like. INFOSEC consists of Transmission Security, such as the frequency hopping spreading code selection, plus Communications Security encryption.

The SDR Forum defined a very simple, helpful model of radio in 1997, shown in Figure 9.14. This model highlights the relationships among radio functions at a tutorial level. The CR has to "know" about these functions, so every CR must have an internal model of a radio of some type. This one is a good start because it shows both the relationships among the functions and the typical flow of signal transformations from analog RF to analog or with SDR, digital modems, and on to other digital processing including system control of which the user interface is a part.

This model and the techniques for implementing a SWR and the various degrees of SDR capability are addressed in depth in the various texts on SDR, *e.g.*, see [Walter Tuttlebee, 2002].

The SCA

The US DoD developed the Software Communications Architecture (SCA) for its Joint Tactical Radio System (JTRS) family of radios [JTRS, 2006; SDR Forum, 2006].

The architecture identifies the components and interfaces shown in Figure 9.15. The API's define access to the physical layer, to the media access control (MAC) layer, to the logical link layer (LLC), to security features, and to the input/output of the physical radio device. The physical components consist

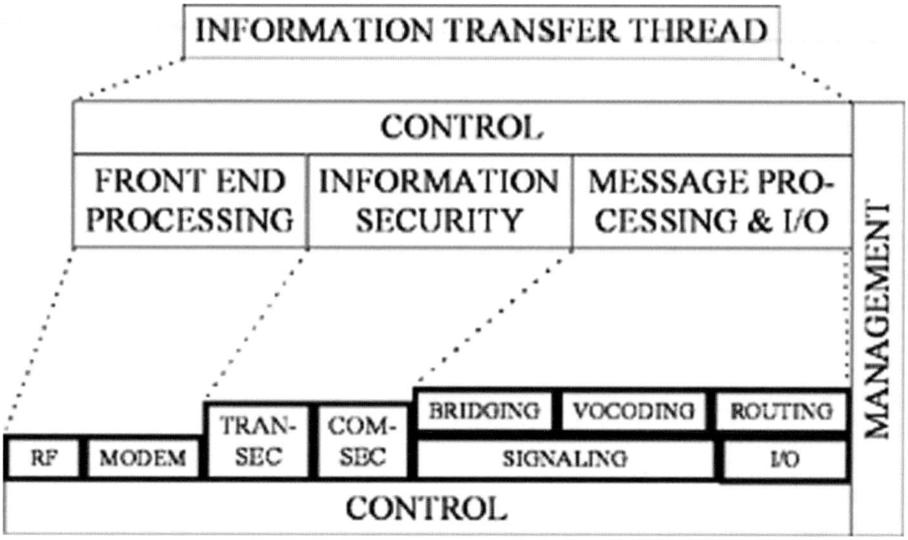

Figure 9.14. SDR Forum (MMITS) Information Transfer Thread Architecture.

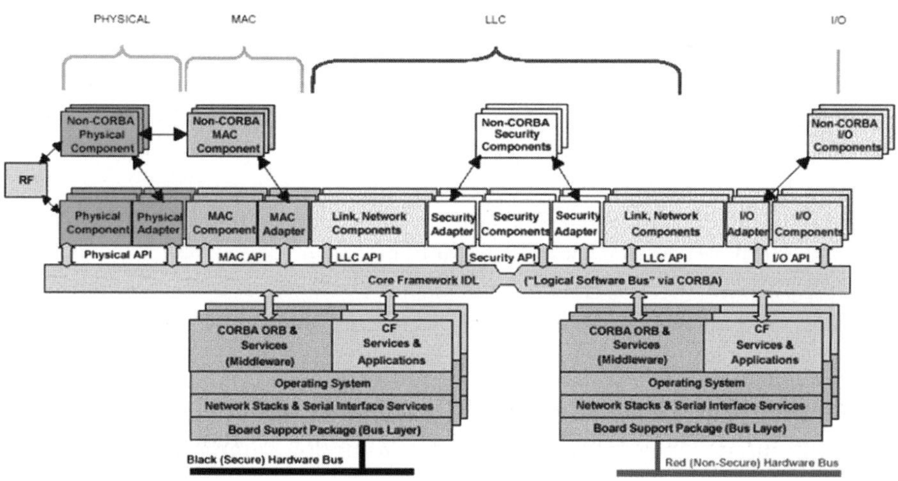

Figure 9.15. JTRS SCA Version 1.0;© SDR Forum, Reprinted with Permission.

of antennas and RF conversion hardware that are mostly analog and that therefore typically lack the ability to declare or describe themselves to the system. Most other SCA-compliant components are capable of describing themselves to the system to enable and facilitate plug and play among hardware and software components. In addition, the SCA embraces POSIX and CORBA.

The model evolved through several stages of work in the SDR Forum and Object Management Group (OMG) into a UML-based object- oriented model of SDR (Figure 9.16). Waveforms are collections of load modules that provide wireless services, so from a radio designer's perspective, the waveform is the key application in a radio. From a user's perspective of a wireless PDA, the radio waveform is just a means to an end, and the user doesn't want to know or have to care about waveforms. Today, the cellular service providers hide this detail to some degree, but consumers sometimes know the difference between CDMA and GSM, for example, because CDMA works in the US, but not in Europe. With the deployment of the third generation of cellular technology (3G), the amount of technical jargon consumers will need to know is increasing. So the CR designer is going to write code (Java code in this book) that insulates the user from those details, unless the user really wants to know.

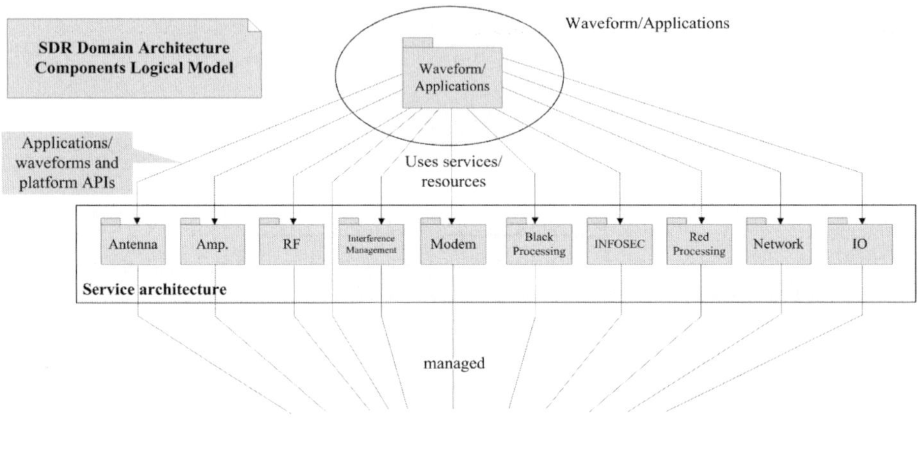

Figure 9.16. SDR Forum UML Model of Radio Services c SDR Forum, Used with Permission.

In the UML model, Amp refers to amplification services, RF refers to RF conversion, interference-management refers to both avoiding interference and filtering it out of one's band of operation. In addition, the jargon for US military radios is that the "red" side contains the user's secret information, but when it is encrypted it becomes "black" or protected, so it can be transmitted. Black processing occurs between the antenna and the decryption process. Notice also in the figure that there is no user interface. The UML model contains a sophisticated set of management facilities, illustrated further in Figure 9.17, to which Human Machine Interface (HMI) or user interface is closely related.

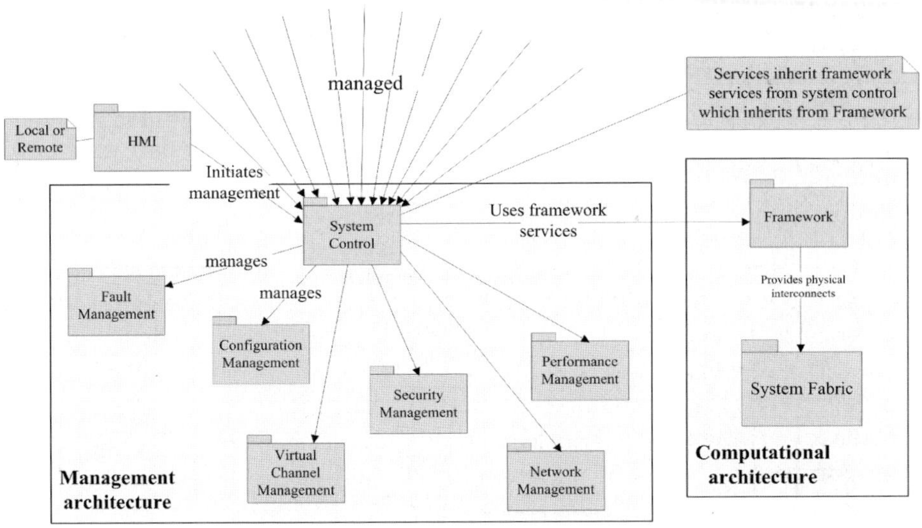

Figure 9.17. SDR Forum UML Management and Computational Architectures; © SDR Forum, Used with Permission.

Systems control is based on a framework that includes very generic functions like event logging, organized into a computational architecture, heavily influenced by CORBA. The management features are needed to control radios of the complexity of 3G and of the current generation of military radios. Although civil sector radios for police, fire, and aircraft lag these two sectors in complexity and are more cost- sensitive, baseband SDR's are beginning to insert themselves even into these historically less technology-driven markets.

Fault management features are needed to deal with loss of a radios processors, memory, or antenna channels. CR therefore interacts with fault management to determine what facilities may be available to the radio given recovery from hardware and/or software faults (*e.g.*, error in a download). Security management is increasingly important in the protection of the user's data by the CR, balancing convenience and security which can be very tedious and time consuming. The CR will direct virtual channel management and (VCM) will learn from the VCM function what radio resources are available, such as what bands the radio can listen to and transmit on and how many it can do at once. Network management does for the digital paths what VCM does for the radio paths. Finally, SDR performance depends on the availability of analog and digital resources, such as linearity in the antenna, millions of instructions per second (MIPS) in a processor, and the like.

Functions-Transforms Model of Radio

The self-referential model of a wireless device used by the CRA and used to define the RKRL and to train THE CRA is the function-transforms model illustrated in Figure 9.18. In this model, the radio knows about sources, source coding, networks, INFOSEC, and the collection of front-end services needed to access RF channels. Its knowledge also extends to the idea of multiple channels and their characteristics (the channel set), and that the radio part may have both many alternative personalities at a given point in time, and that through evolution support, those alternatives may change over time.

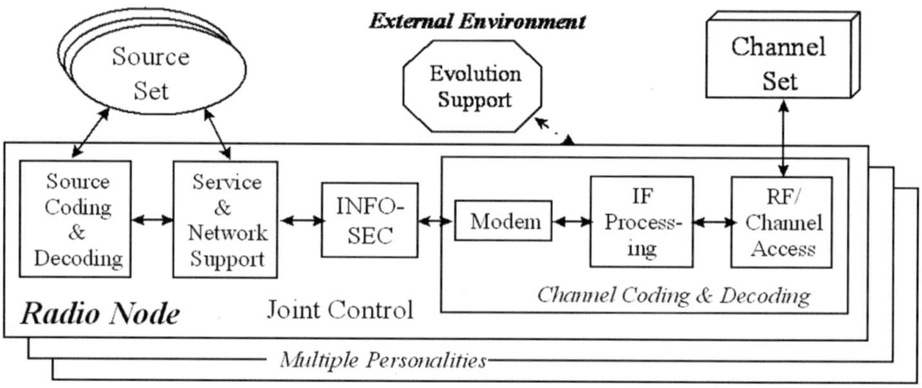

Figure 9.18. Functions-Transforms Model of a Wireless Nod.

Since CR reasons about all of its internal resources, it also must have some kind of computational model of analog and digital performance parameters and how they are related to features it can measure or control. MIPS, for example, may be controlled by setting the clock speed. A high clock speed generally uses more total power than a lower clock speed, and this tends to reduce battery life. Same is true for the brightness of a display. The CR only "knows" this to the degree that it has a data structure that captures this information and some kind of algorithms, pre- programmed and/or learned, that deal with these relationships to the benefit of the user. Constraint languages may be used to express interdependencies, such as how many channels of a given personality are supported by a given hardware suite, particularly in failure modes. CR algorithms may employ this kind of structured reasoning as a specialized knowledge source (KS) when using case-based learning to extend its ability to cope with internal changes.

The ontological structure of the above may be formalized as follows:

```
<SDR>
    <Sources/> <Channels/> <Personality>
```

```
        <Source-Coding-Decoding/><Networking/> <INFOSEC/>
<Channel-Codec><Modem/><IF-Processing/><RF-Access/></Channel-Codec>
</Personality>
      <SDR-Platform/> <Evolution-Support/>
</SDR>
```

Equation 5 SDR Subsystem Components

While this text does not spend a lot of time on the computational ontology of SDR, semantically based dialogs among AACRs about internal issues like downloads may be mediated by developing the RXML above to more fully develop the necessary ontological structures.

Architecture Migration: From SDR to AACR

Given the CRA and contemporary SDR architecture, one must address the transition of SDR, possibly through a phase of AACRs toward the ideal CR. As the complexity of hand-held, wearable, and vehicular wireless systems increase, the likelihood that the user will have the skill necessary to do the optimal thing in any given circumstance goes down. Today's cellular networks manage the complexity of individual wireless protocols for the user, but the emergence of multiband multimode AACR moves the burden for complexity management towards the PDA. The optimization of the choice of wireless service between the "free" home WLAN and the for-sale cellular equivalent moves the burden of radio resource management from the network to the WPDA.

Cognitive Electronics

The increasing complexity of the PDA-user interface also accelerates the trend towards increasing the computational intelligence of personal electronics. AACR is in some sense just an example of computationally intelligent personal electronics system. For example, using a laptop computer in the bright display mode uses up the battery power faster than when the display is set to minimum brightness. A cognitive laptop could offer to set the brightness to low level when it was turned on in battery powered mode. It would be even nicer if it would recognize operation aboard a commercial aircraft and therefore I prefer the brightness down. It should learn that my preference is to set the brightness low on an aircraft to conserve the battery. A cognitive laptop shouldn't make a big deal over that, and it should let me turn up the brightness without complaining. If it had an ambient light sensor or ambient light algorithm for an embedded camera, it could tell that a window shade is open, so I have to deal with the brightness. By sensing the brightness of the *on-board aircraft* scene and associating my control of the brightness of my display with the brightness of the environment a hypothetical cognitive laptop could learn do the right thing in the right situation.

How does this relate to the CRA? For one thing, the CRA could be used as-is to increase the computational intelligence of the laptop. In this case, the self is the laptop and the PDA knows about itself as a laptop, not as a WPDA. It knows about its sensors suite, which includes at least a light level sensor if not a camera through the data structures that define the Self. It knows about the user by observing keystrokes and mouse action as well as by interpreting the images on the camera, *e.g.*, to verify that the Owner is still the user since that is important to building user-specific models. It might build a space-time behavior model of any user or it might be a one-user laptop. Its actions then must include the setting of the display intensity level. In short, the CRA accommodates the cognitive laptop with suitable knowledge in the knowledge structures and functions implemented in the map sets.

When Should a Radio Transition towards Cognition?

If a wireless device accesses only a single RF band and mode, then it is not a very good starting point for cognitive radio. It's just too simple. Even as complexity increases, as long as the user's needs are met by wireless devices managed by the network(s), then embedding computational intelligence in the device has limited benefits. In 1999, Mitsubishi and AT&T announced the first "four-mode handset." The T250 operated in TDMA mode on 850 or 1900 MHz, in first generation Analog Mobile Phone System (AMPS) mode on 850 MHz, and in Cellular Digital Packet Data (CDPD) mode on 1900 MHz. This illustrates early development of multiband, multimode, multimedia (M3) wireless. These radios enhanced the service provider's ability to offer national roaming, but the complexity was not apparent to the user since the network managed the radio resources in the handset.

Even as device complexity increases in ways that the network does not manage, there may be no need for cognition. There are several examples of capabilities embedded in electronics that typically are not heavily used. Do you use your laptop's speech recognition system? What about its IRDA port? If you were the typical user circa 2004, you didn't use either capability of your Windows XP laptop all that much. So complexity can increase without putting a burden on the user to manage that complexity since if the capability isn't central to the way in which the user employs the system.

For radio, as the number of bands and modes increases, the SDR becomes a better candidate for the insertion of cognition technology. But it is not until the radio or the wireless part of the PDA has the capacity to access multiple RF bands that cognition technology begins to pay off. With the liberalization of RF spectrum use rules, the early evolution of AACR may be driven by RF spectrum use etiquette for ad-hoc bands such as the FCC use case. In the not-too-distant future, SDR PDAs could access a satellite mobile services, cordless

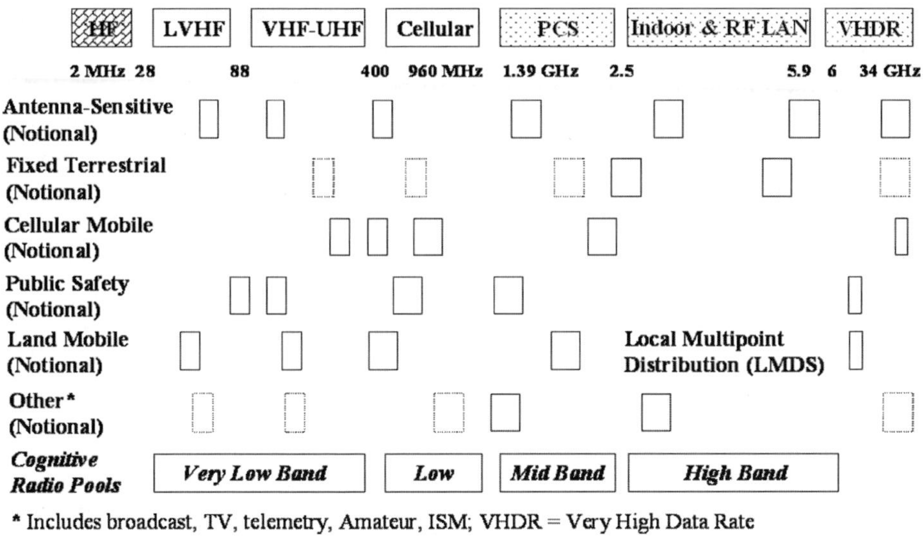

Figure 9.19. Fixed Spectrum Allocations versus Pooling with Cognitive Radio.

telephone, WLAN, GSM, and 3G bands. An ideal SDR device with these capabilities might affordably access octave bands from 0.4 to 0.96 GHz, (skip the air navigation and GPS band from .96 to 1.2 GHz),1.3 to 2.5 GHz, and from 2.5 to 5.9 GHz. Not counting satellite mobile and radio navigation bands, such radios would have access to over 30 mobile sub-bands in 1463 MHz of potentially sharable outdoor mobile spectrum. The upper band provides another 1.07 GHz of sharable indoor and RF LAN spectrum. This wideband radio technology will be affordable first for military applications, next for base station infrastructure, then for mobile vehicular radios and later for handsets and PDAs. When a radio device accesses more RF bands than the host network controls, it is time for CR technology to mediate the dynamic sharing of spectrum. It is the well-heeled conformance to the radio etiquettes afforded by cognitive radio that makes such sharing practical.

Radio Evolution towards the CRA

Various protocols have been proposed by which radio devices may share the radio spectrum. The US FCC Part 15 rules permit low power devices to operate in some bands. In 2003, a Rule and Order (R&O) made unused television (TV) spectrum available for low power RF LAN applications, making the manufacturer responsible for ensuring that the radios obey this simple protocol. DARPA's NeXt Generation (XG) program developed a language for expressing spectrum use policy [Preston Marshall, 2003]. Other more general protocols based on peek-through to legacy users have also been proposed [J. Mitola III, 1999].

Does this mean that a radio must transition instantaneously from the SCA to the CRA? Probably not. The simple six-component AACR architecture may be implemented with minimal sensory perception, minimal learning, and no autonomous ability to modify itself. Regulators want to hold manufacturers responsible for the behaviors of such radios. The simpler the architecture, the simpler the problem of explaining it to regulators and of getting concurrence among manufacturers regarding open architecture interfaces that facilitate technology insertion through teaming. Manufacturers who fully understand the level to which a highly autonomous CR might unintentionally reprogram itself to violate regulatory constraints may decide they want to field aware- adaptive (AA) radios, but may not want to take the risks associated with a self-modifying CR's just yet.

Thus, one can envision a gradual evolution towards the CRA beginning initially with a minimal set of functions mutually agreeable among the growing community of AACR stakeholders. Subsequently, the introduction of new services will drive the introduction of new capabilities and additional API's, perhaps informed by the CRA developed in this text.

Cognition Architecture Research Topics

The cognition cycle and related inference hierarchy imply a large scope of hard research problems for cognitive radio. Parsing incoming messages requires natural language text processing. Scanning the user's voice channels for content that further defines the communications context requires speech processing. Planning technology offers a wide range of alternatives in temporal calculus [C. Phillips, 1997], constraint-based scheduling [C. Phillips, 1997], task planning [S.K. Das, 1997], causality modeling [J. Pearl, 2000], and the like. Resource allocation includes algebraic methods for wait-free scheduling protocols, Open Distributed Processing (ODP), and Parallel Virtual Machines (PVM). Finally, machine learning remains one of the core challenges in artificial intelligence research [R. Michalski and I. Bratko and M. Kubat, 1998]. The focus of this cognitive radio research, then, is not on the development of any one of these technologies per se. Rather, it is on the organization of cognition tasks and on the development of cognition data structures needed to integrate contributions from these diverse disciplines for the context-sensitive delivery of wireless services by software radio.

Learning the difference between situations in which a reactive response is needed versus those in which deliberate planning is more appropriate is a key challenge in machine learning for CR. THE CRA framed the issues. THE CRA goes further, providing useful KS's and related ML so that the CR designer can start there in developing good engineering solutions to this problem for a given CR applications domain.

8. Commercial CRA

The CRA allocates functions to components based on design rules. Typically design rules are captured in various interface specifications including Applications Programmers Interfaces (APIs), and Object Interfaces, such as Java's JINI/JADE structure of intelligent agents. While the previous section introduced the CRA, this section suggests additional design rules by which user domains, sensory domains and radio knowledge of RF Band knowledge may be integrated into industrial-strength AACR products and systems.

Industrial Strength AACR Design Rules

The following design rules circumscribe the way cognitive radio functions are mapped to the components of a wireless PDA within the envisioned architecture

1 The cognition functions shall maintain an explicit topological model of space-time

- Of the user,
- Of the physical environment,
- Of the radio networks, and
- Of the internal states of the radio.

2 The CRA internalizes knowledge as skills, *e.g.*, serModels with no cycles.

- The CRA requires each CR to maintain a trusted count of the number of serModels it contains, the number of associations stored per serModel, and
- The CRA requires each CR to maintain a directed graph of the connections among its serModels. Cycles are precluded from the serModels skills graph. A CR conforming to the CRA must have a reliable way of detecting cycles formed in error (*e.g.*, during a sleep epoch or via a download) and of breaking detected cycles.

3 The CRA requires each CR to predict in advance, the amount of time required for each cognition cycle. The CR is required to set a trusted (hardware) watchdog timer before entering a cognition cycle. If the timer is violated, the system must detect that event, log that event, and mark the serModels invoked in that event as entailing non-determinism.

4 Context shall be formally represented using a topologically sound internal model of space-time-context.

5 Each CR conforming to the CRA shall include an explicit grounding map, M that maps its internal data structures onto elements sensed in the real world represented in its sensory domains, including itself. If the CR cannot map a sensed entity to a space-time-context entity with specified time allocated to attempt that map, then the entity shall be mapped to the unique unmappable entity "UNGROUNDABLE".

6 The model of the world shall follow a formal treatment of time, space, radio frequency, radio propagation, and the identity of entities.

7 Models shall be represented in an open architecture radio knowledge representation language suited to the representation of radio knowledge (*e.g.*, RKRL 0.3). That language shall support topological properties and formal if not axiomatic models.

8 The cognition functions shall maintain location awareness, including

- the sensing of location from global positioning satellites,
- sensing position from local wireless sensors and networks
- and sensing precise position visually.
- Location shall be an element of all contexts.

9 The cognition functions shall maintain awareness of time to the accuracy necessary to support the user and radio functions.

- Time shall be an element of all contexts.

10 The cognition functions shall maintain an awareness of the identity of the PDA, of its Owner, of its primary user, and of other legitimate users designated by the Owner or primary user.

- Current user shall be an element of all contexts.

11 The cognition functions shall reliably infer the user's communications context and apply that knowledge to the provisioning of wireless access by the SDR function.

12 The cognition functions shall model the propagation of its own radio signals with sufficient fidelity to estimate interference to other spectrum users.

- The cognition function shall also assure that interference is within limits specified by the spectrum use protocols in effect in its location (*e.g.*, in spectrum rental protocols).
- It shall defer to the wireless network in contexts where the network manages interference.

13 The cognition functions shall model the domain of applications running on the host platform, sufficient to infer the parameters needed to support the application. Parameters modeled include QoS, data rate, probability of link closure (Grade of Service), and space - time - context domain within which wireless support is needed.

14 The cognition functions shall configure and manage the SDR assets to include hardware resources, software personalities, and functional capabilities as a function of network constraints and use context.

15 The cognition functions shall administer the computational resources of the platform. The management of software radio resources may be delegated to an appropriate SDR function (*e.g.*, the SDR Forum domain manager). Constraints and parameters of those SDR assets shall be modeled by the cognition functions. The cognition functions shall assure that the computational resources allocated to applications, interfaces, cognition and SDR functions are consistent with the user communications context.

16 The cognition functions shall represent the degree of certainty of understanding in external stimuli and in inferences. A certainty calculus shall be employed consistently in reasoning about uncertain information.

17 The cognition functions shall recognize preemptive actions taken by the network and/or the user. In case of conflict, the cognition functions shall defer the control of applications, interfaces, and/or SDR assets to the Owner, to the network or to the primary user, according to appropriate priority and operations assurance protocol.

9. Future Direction

AACR seems headed for the Semantic Web, but the evolution of practical radio devices will shape that evolution. Although many information processing technologies relevant to AACR exist in eBusiness Solutions and are emerging for the Semantic Web, the integration of sensory-perception into SDR and the creation of suitable cognition architectures remain both a research challenge for academic pursuits and a series of increasingly interesting challenges for radio systems engineers. A well formulated CRA that is broadly supported by industry can be a great enabler for such an evolution.

Notes

1. Semantic Web: Researchers formulate CRs as sufficiently speech- capable to answer questions about <Self/> and the <Self/> use of <Radio/> in support of its <Owner/>. When an ordinary concept like "owner" has been translated into a comprehensive ontological structure of computational primitives, *e.g.*, via Semantic Web technology [J. Mitola III, 1998a], the concept becomes a computational primitive for autonomous reasoning and information exchange. Radio XML, an emerging CR derivative of the eXtensible Markup Language, XML, offers to standardize such radio-scene perception primitives. They are highlighted in this

brief treatment by <Angle- brackets/>. All CR have a <Self/>, a <Name/>, and an <Owner/>. The <Self/> has capabilities like <GSM/> and <SDR/>, a self- referential computing architecture, which is guaranteed to crash unless its computing ability is limited to real-time response tasks [J. Mitola III, 1998a]; this is fine for CR but not workable for general purpose computing.

References

AAAI (2004). Cognitive Vision. Palo Alto, CA: AAAI.

Anne Watzman (2002). Robotic Achievements: GRACE Successfully Completes Mobile Robot Challenge at Artificial Intelligence Conference. Pittsburg, PA: Carnegie Melon Views.

BellSouth (1995). The software defined radio request for information. Atlanta, CA: BellSouth.

Phillips C. (1997). Optimal Time-Critical Scheduling. STOC 97.

Eriksson and Penker (1998). UML Toolkit. NY: John Wiley & Sons Inc.

IBM (2006). PC-KIMMO Version 1.0.8 for IBM PC, 18-Feb-92.

Mitola III J. (1992). Software Radios: Survey, Critical Evaluation and Future Directions. Proc. IEEE National Telesystems Conference (NY: IEEE Press.

Mitola III J. (1995). Software Radio Architecture. Communications Magazine (NY: IEEE Press).

Mitola III J. (1998a). Software Radio Architecture: A Mathematical Perspective.

Mitola III J. (1998b). Appendix B: Software Radio Architecture: A Mathematical Perspective. IEEE JSAC.

Mitola III J. (1999). Cognitive Radio. MoMuC.

Pearl J. (2000). Clausity: Models, Reasoning, and Interference. Morgan-Kaufmann.

Joseph Mitola III (2000a). Cognitive Radio: An Integrated Agent Architecture for Software Defined Radio.

Joseph Mitola III (2000b). Software Radio Architecture. Wiley Interscience.

Joseph Mitola III (2006). Aware, Adaptive, and Cognitive Radio. Wiley & Sons.

JTRS (2006). www.jtrs.mil.

Koser (1999). read.me. www.cs.kun.nl:(The Netherlands:University of Nijmegen).

OMG (2006). www.omg.org.

OMGUML (2006). www.omg.org/UML.

Petri Mahonen (2004). Cognitive Wireless Networks. RWTH Aachen.

Preston Marshall (2003). Remarks to the SDR Forum.

Hennie R. (1997). Introduction to Computability. Addison-Wesley.

Michalski R. and Bratko I. and Kubat M. (1998). Machine Learning and Data Mining. John Wiley & Sons, LTD.

SDR Forum (2006). www.sdrforum.org.

Das S.K. (1997). Decision making and plan management by intellegent agents: theory, implementation, and applications. ACM Autonomous Agents 97.

SNePS Web (1998). Sneps. ftp.cs.buffalo.edu:/pub/sneps/.
Mowbray T. and Malveau R. (1997). CORBA Design Patterns. John Wiley & Sons.
TellMe Networks (2005). www.tellme.com.
The XTAG Research Group (1999). A Lexicalized Tree Adjoining Grammar for English Institute for Research in Cognitive Science. Philadelphia, PA: University of Pennsylvania.
Victor Zue (2005). Speech Understanding System. Boston, MIT.
Walter Tuttlebee (2002). Software defined radio:enabling technologies. Wiley.
WWRF (2004). Wireless world research forum. www.wireless-world-research.org.

Chapter 10

STABILITY AND SECURITY IN WIRELESS COOPERATIVE NETWORKS

Providing incentives for cooperation

Konrad Wrona
SAP Research, France
konrad.wrona@sap.com

Petri Mähönen
Department of Wireless Networks, RWTH Aachen University, Germany
pma@mobnets.rwth-aachen.de

Abstract: In this chapter we review and analyse various ways of encouraging cooperation and of mitigating misbehaviour in *cooperative communication systems*. Our results are, in principle, applicable to both wireless and wired networks. We start with discussing possible approaches to accountability in cooperative communication systems and incentives, which can to be use in order to foster cooperation in such systems. We present both an analytical and a simulation model of cooperation in wireless ad hoc networks. The analytical model is based on *evolutionary game theory*. The nodes are adaptive and can dynamically adjust their strategies in order to maximise their own utility. We show that in the case of agents learning by imitation, a cooperative behaviour is an asymptotically stable equilibrium of our model. This promising result suggests that correctly designed reputation and trust mechanisms can facilitate the emergence of sustainable ad hoc communication networks. We also introduce a generic simulation model, which utilises a multi-agent simulation platform and can be easily adapted in order to investigate other kinds of cooperative communication systems, such as file-sharing networks.

Keywords: cooperation, evolutionary game theory, dynamic systems, security, stability.

Frank H. P. Fitzek and Marcos D. Katz (eds.), Cooperation in Wireless Networks: Principles and Applications, 313–363.

1. Introduction

Cooperative communication systems, such as wireless ad hoc and peer-to-peer networks, comprise of cooperating autonomous nodes. While cooperation between nodes is critical for the existence of such distributed systems, it is not obvious why cooperative behaviour should be a dominant strategy for the nodes. In this chapter we investigate various mechanisms for preventing both selfishly-motivated misbehaviour and malicious activities, such as various forms of denial-of-service attacks, in this kind of systems.

We review in this chapter some key issues and findings related to security and stability of cooperative networks. Our review is based in large part of our own previous work, this bias is in part explained by the space limits, since the field is becoming very active and large. We refer the interested reader to more in depth review and bibliography in [Wrona, 2005]. We also point out that the security is an important enabling technology aspect itself for cooperative networks. The security is often treated as an separate aspect from the general cooperative communications. However, as will be emphasised in the following, and we are certainly not the first ones to point this out, the security mechanisms can be used as one way to provide incentives and guarantees for the cooperative communications. The another issue of this chapter is stability. In our context the stability is related to dynamic behaviour of network nodes. Instead, of expecting *a priori* that all nodes are cooperating, we are proposing that we should study from the first principles the situations, where part of the populations is misbehaving. In this context the stability analysis is aimed to answer questions like: *What fraction of population can misbehave so that the network is still robust and stable for most of the users?*

All cooperative communication systems share some common characteristics:

- They are highly distributed, with no or only rudimentary central control.
- Their users voluntarily commit bandwidth, data storage, CPU cycles, battery power, etc., forming a common pool of resources which can be used either by all of them or in order to achieve a common goal.
- The utility which users can obtain from the pooled resources is much higher than they can obtain on their own. For example, they can have access to a better variety of music, build a communication network, find solutions to complex computational problems in a shorter time, or achieve faster transfer of data to mobile terminals.
- Rational users would prefer to access common pool resources without any own commitment, since every commitment has its price: music uploads consume bandwidth, CPU time and data storage; multi-hop routing reduces bandwidth and battery power available to each user since users

have to forward data packets that do not belong to them; peer-to-peer computing applications consume CPU cycles and energy; virtual antenna arrays may drain a battery of a terminal.

- They are vulnerable to various kinds of malicious behaviour, including denial of service attacks and distribution of malware.

We should mention that the necessity to understand cooperation and trust increases with the size of the networks involved, since in the larger networks both the probability of antisocial behaviour and the possible gains from misbehaviour (or illegal actions) are higher. In the recent years much of research has been focused on networking aspects of the cooperative communication systems. However, more effort is required to understand the fundamental issues related to dynamics of cooperative networks and development of trust relationships in these networks. Without understanding these issues, the work invested in development of different protocols and transmission networks may bring only suboptimal results.

2. Sustaining Cooperation

In this chapter we examine three possible approaches to enforcing cooperation and accountability in cooperative communication systems:

1 *Reputation and trust model*, based on reciprocity between recognisable nodes.
2 *Tax and reward model*, based on the strong authentication of the nodes.
3 *Payment model*, relying on (micro-)payments between anonymous nodes.

The approach chosen for the particular application may differ depending on the environment, *e.g.*, in a conventional business environment a payment or strong authentication might be the best solution, whereas in the scientific community a reciprocity-based mechanism may be more useful. Another important feature is that these accountability models require different levels of identification of nodes. For example, solutions based on the reputation mechanism and taxation require every node to have a unique and persistent identity, whereas micropayments can be used in a fully anonymous environment.

We focus in particular on distributed accountability mechanisms based on indirect reciprocity and micropayments. Despite that accountability in a fully distributed environment is a difficult challenge, cooperation in such systems can be also fostered by employing so-called *supernodes*. Such supernodes could have an additional utility from sustaining a cooperative communication system (*e.g.*, they are altruist or enjoy to be popular) or be paid to provide such service by a third party (*e.g.*, local authorities). Also, several hybrid solutions involving a mix of different accountability mechanisms are possible. Local communities might want to build confederations, interconnecting into wide area networks

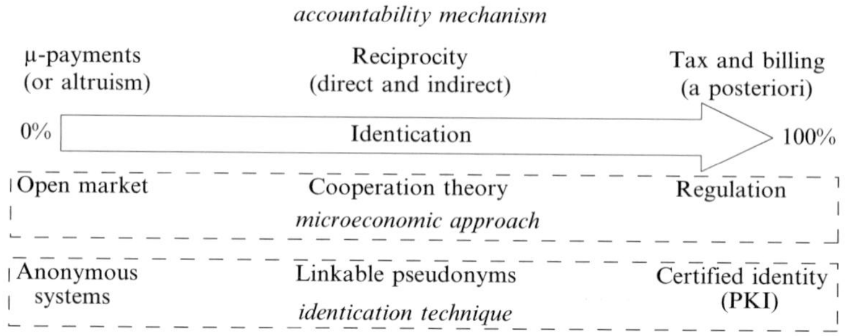

Figure 10.1. Microeconomic models and relationship between identification and incentives in cooperative systems.

and providing free services to their members on mutual terms. This might be achieved by means of third party authentication brokers belonging to a backbone network.

Distributed accountability mechanisms introduce many challenges. For example, in reputation-based systems nodes can falsely report other nodes as misbehaving. Such false reports can be generated by a malicious node in order to perform various attacks on the system, *e.g.*, in order to partition an ad hoc network by claiming that some nodes included in the path are misbehaving. Also, several technology-specific cheating strategies are possible, *e.g.*, in wireless ad hoc networks a node could limit its transmission power so that the signal is strong enough to be overheard by the previous node but too weak to be received by the true recipient.

In this chapter we show how to model cooperative communication systems using game-theoretic framework, based on models developed within theory of cooperation and microeconomics. These models include several variants of well-known Prisoner's Dilemma (PD) game, such as iterated PD, bilateral PD, and n-player PD games (also known as public goods or common-pool resources). However, it is worth to briefly discuss the more general criteria, which can be used in order to judge if a system consisting of autonomous and rational players is successful. The criteria that can be used for judging the effectiveness of a designed cooperation mechanism include, e.g, compatibility with individual rationality; dynamic stability; computational efficiency; distribution and communication efficiency; and Pareto efficiency. The most frequently used normative criterion of modern economics is the *Pareto optimality* (or *efficiency*) [Kreps, 1990]. A state of the system is said to be a *Pareto optimum* if it would be impossible to improve the well-being of one individual without harming at least one other individual. A Pareto optimum is not necessarily unique. Deciding which of the Pareto optima is preferred requires a social welfare function that weights the utility levels of individuals according to some normative rule. Under ideal conditions, market economies automatically end in a position of Pareto optimality, provided that all individuals maximise their utility. However, these ideal conditions rule out the existence of externalities, public goods, etc., and therefore they are not fulfilled in the case of cooperative communication systems. Economies that do not achieve Pareto optimality are said to result in *market failure*. However, it is important to keep in mind that the cost of correction of a market failure may, on occasion, outweigh the benefits associated with correction.

Finally, we should mention that the idea of treating a communication network as an economic system is not a new one. The microeconomic principles have been applied in communications systems to solve such problems as selling excess CPU time and other resources [Chavez et al., 1997], fair cost-sharing for multicasting [Jain and Vazirani, 2001; Feigenbaum et al., 2001], sharing of computing resources [Ng et al., 2003], control of transmitting power in wireless systems [Heikkinen, 1999; Heikkinen, 2000; Saraydar et al., 2002], allocating work flow and tasks [Nisan and Ronen, 1999; Ronen, 2000], resource sharing and allocation in the grid systems [Buyya et al., 2001], dissemination of private information [Kleinberg et al., 2001], intrusion detection [Kodialam and Lakshman, 2003], data replication and mirroring, as well as searching and selling of information [Rasmusson and Janson, 1999].

Below we discuss in more detail the challenges and opportunities introduced by various accountability mechanisms. We also discuss some proposed implementations of accountability mechanisms in distributed communication systems.

Payment-Based Accountability

One of the basic ways of accountability is payment for provided services. There are two main approaches to electronic payments. One is based on transfer of the value between accounts belonging to the sides of the transaction, *e.g.*, credit card systems. Another relies on a transfer of value tokens, or electronic coins, between participants (*e.g.*, e-cash, or Mobile Cash protocol in [Tracz and Wrona, 2001]). The particular advantage of systems based on *e-cash* payments is that they are able to provide accountability, and sustain cooperation, even in the case of lack of any identification mechanisms. However, such systems still require some *structural trust*, *i.e.*, trust in the value of money. Also, digital environment introduces a need for much smaller payment amounts, possibly in fractions of cents, which can be used to pay for such fine grained services like bandwidth, computational power and storage.

In addition to the obvious performance penalty, using payments (or in particular micropayments) for accountability in cooperative systems may also cause a structural inefficiency. The structural challenges are that excessive charging can dissuade nodes from using the system, creating a *social inefficiency*, and that micropayment systems tend to favour wealthy users. An important issue, which should also be taken into consideration, when designing any payment mechanisms for cooperative communication systems is a growing concern that complicated pricing mechanisms may not be accepted by the end users (see an interesting discussion in [Odlyzko, 2001]). From the buyer's point of view, every buying transaction requires comparison of the purchase price of the good to its personal value. This introduces a significant mental cost, which sets the basic lower bounds on transaction costs. Hence, comparing the personal value of large set of low-priced goods might require a mental expenditure greater than the prices of those goods [Szabo, 1997]. This is confirmed by empirical results. For example users regularly choose flat rate plans for fixed telephony and the Internet access, even when their usage history shows that a metered plan would cost less. However, in the context of cooperative communications it is important to note that introducing a flat rate membership does not have impact on the equilibrium that arises in the system, although it can affect decisions made by agents about participation. Therefore, some other conveniency solutions may have to be applied in order to overcome the usability problems. For example, a broker system, where brokers absorb price fluctuations and offer users a simple charging scheme may be an answer to end-user complexity of pricing solutions.

In addition to enforcing cooperation, micropayments can also be used to achieve optimal resource allocation among participants, especially in the case of congestion. In fact, in the case of burst-congestion, *e.g.*, flooding-attacks, dynamic introduction of micropayments may be sufficient. On the other hand, in the case of resources displaying a cumulative congestion patterns, *e.g.*, data storage, micropayments have to be used continuously. However, by changing the charging values and other mechanism-specific payment parameters, the level of accountability and security offered by a system can be changed.

Various economic models for selling resources and services have been proposed to be employed in distributed systems [Buyya et al., 2002; Cole et al., 2003; Crowcroft et al., 2003]. Unfortunately the basic assumptions of many classical pricing schemes are quite limiting and unrealistic. For example, many contributions assume the strong homogeneity of the users (*i.e.*, existence of system-wide parameters), existence of efficient algorithms for pricing, and that the taxes incurred on the nodes cause no disutility to network users (*i.e.*, they are refunded equally to all users, which may be practically infeasible either because of the complexity of the process or because taxes are represented by non-fungible good, such as time delay). The priority should be given to the technical efficiency and conceptual simplicity of the chosen scheme. In addition, in cooperative communication systems typically endpoint nodes are not the only ones involved in a transaction, and there might be several intermediate nodes involved, too. This adds an additional level of complexity to the pricing and payment mechanism. There are several possible strategies for handling micropayments in systems with intermediaries. In the case of *end-to-end payments* end nodes do not worry about what happens to intermediaries. Such schemes do not provide any congestion control or accountability to intermediaries, and therefore do not solve any of the problems being a cause for use of the micropayments in the first place. In the case of *pairwise payments* micropayments are involved into every transaction between every pair of peers. An interesting example of pairwise payments is the Packet Trade model explained in [Buttyán and Hubaux, 2001]. The potential disadvantages of pairwise payments are possibly high overhead and delay introduced by the payments and that if a single malicious operator is in charge of both end nodes, these two peers may be able to extract work from the intermediate nodes without paying. The *amortised pairwise payments* iteratively decrease cost of transaction, when moving through the system. This approach solves some of the problems of pairwise payments, especially in the case of payments, which involve transfer of real monetary value. However, it leaks information regarding the route length, which might be a danger in some systems (*e.g.*, FreeHaven). Furthermore, sender may often not know the route length in advance. In some cases, *e.g.*, Gnutella search, branching out through the network may make payments prohibitively expensive. In the *all points payments* sender pays each peer engaged in the protocol, intermediate and endpoint alike. An example of such approach in the Packet Pursue model in [Buttyán and Hubaux, 2001]. Here, the same problems of a path length and branching as in amortised pairwise-payments arise. Also, such solution in order to be practical requires a non-interactive payment mechanism. All points model might be implemented *e.g.*, using a limited number of blinded tickets per time period per user, issued by each node. However, most of blind signatures use interactive protocols, which are unsuitable for these types of applications.

An important question is how to provide optimal initial conditions for a system relying on payment-based accountability. In order to provide bootstrapping, an initial credit might be provided to the members of the network. The system may also rely on an assumption that nodes will provide on the average the same amount of services to each other. Thus payment could be implemented by issuing of promissory notes, which could be either cancelled as a result of further cooperation between nodes or required to be bought back by the node in the case of the lack of cooperation or suspicion of free-riding. Some of the services might be also provided based on tipping: the intermediary nodes may pass tipping suggestions to the served nodes.

The micropayment mechanisms, which are the most interesting in the context of accountability in cooperative networks, can be generally divided into payments involving transfer of redeemable (*e.g.*, monetary) value and payments consisting of proof-of-work.

The payment mechanisms involving transfer of redeemable or monetary value are called *real-value payments* or *fungible payments*. A payment system may be implemented either as an *open payment system*, using payment values commonly accepted also outside of the system, or as a *closed payment system* using payment tokens, which have a value only inside of the system. Examples of commonly used closed payment systems are various loyalty programs, such as frequent flyer miles. In either case, there must be some kind of control of how this value is generated and it can be redeemed. The most obvious solution is involvement of some existing financial institution. Another possibility is that electronic cash needed for transactions between system participants might be issued by *meta-operators*. Meta-operators can be entities, which are to some extent trusted by all market participants (*i.e.*, their identity is known and there is a legal framework for controlling their behaviour). The last two requirements allow for *bootstrapping* of the electronic trust and economical system by means of classical trust and legal framework. In addition to issuing, redeeming and tackling counterfeiting (*e.g.*, by providing verification and penalisation mechanisms) the third party authority might also have to take care of other monetary issues. One of such issues is controlling amount of money within the system in order to prevent possible inflation and deflation. In particular, joining or leaving of system by disproportionally *rich* parties may cause a large fluctuation in the amount of available *money* and lead to possible disruption of services.

Despite of requiring involvement of some third party, the real-value payments have several interesting and practical features. For example, fungible payment in addition to encouraging cooperation and efficient use of resources can also generate revenue for the system operator. Also, fungible payments, which can be actually redeemed for real-world money, provide additional, *i.e.* financial, defence against resource misuse. However, a problem with some of the payment schemes, requiring complex coin verification, may be an attacker flooding the system with counterfeit coins and eating up computational resources through a verification process alone.

As an alternative to fungible payments, and in order to avoid additional infrastructure and dependence on external parties, non-fungible payments based on *proof-of-work (POW)* can be used [Dwork and Naor, 1992]. Micropayments with a proof of work are based on showing that the user has performed some computationally difficult, and thus expensive, task. In most of such systems the client is asked to solve a crypto-puzzle. The difficulty of puzzle may be increased for subsequent transactions in order to exponentially decrease the utility a node gets from the increased consumption of the resources. In the case of a proof-of-work with a trap door, proof-of-work is hard to compute without knowledge of some secret, but easy to compute given this secret. This knowledge-dependent complexity can be used, *e.g.*, by third party providers or local authorities to easily generate POWs for sale. Currently, mechanisms based on POW are mainly used in order to prevent denial of service attacks by making it expensive, in terms of CPU time, to carry out such an attack [Oram, 2001]. In more sophisticated cases, POW charging may be started only if a possible DoS attack has been detected. Nevertheless, proofs-of-work might not be successful against an attacker with a large computational capacity.

An example of POW based payment system is a *hash cash* [Back, 2003]. In the hash cash a user is asked to guess the input to a hash function through a brute-force calculation. The hardness of the problem may be controlled by specifying the amount of bits to be guessed. This kind of problem is known as a k-bit partial hash collision. A variant of the hash cash is a client puzzle. A puzzle can be seen logically as a hash value for which a client needs to find the input value. In order to decrease the chance that a client might just guess the solution to the puzzle, each puzzle could optionally be made of multiple sub-puzzles that the client must solve individually. The hash collision schemes may be solved more easily by parallelisation. It is a vulnerability if the goal of non-fungible payments is to guard against DoS attacks. In proof of time schemes, client is requested to spend some amount of time, instead of work, in order to solve a problem. An example of a time-lock puzzle is the LCS35 time capsule [Oram, 2001]. These types of puzzles are designed to be intrinsically sequential in nature. The interesting idea is to construct POW schemes which enable reuse of the POWs computations for a separate and useful value computation. One of such schemes is *bread pudding protocol*, which defines a POW that can be reused for minting MicroMint coins [Oram, 2001].

Electronic Payments

As it was already mentioned, payment is one of the most robust mechanisms for inducing cooperation among selfish agents. Mobile payment is a fundamental enabler not only for the cooperative communication systems, but also for a huge number of mobile commerce applications, including banking, shopping, betting, trading, ticketing, entertainment, and gaming. However, mobile payments face a number of difficult challenges, both from the technological and the business point of view [Wrona and Schuba, 2001a; Wrona and Schuba, 2001b]. First of all, the cryptographic and security functionality offered by mobile devices at least at the application layer is typically much more limited than in fixed networks. Additionally, the performance of the wireless devices is naturally limited because the systems have to operate using battery energy and radio communication interface. Thus, often it might not be possible to access the existing payment infrastructure from a mobile device in the same way as from a regular PC. On the other hand, designing a dedicated mobile payment solution requires a know-how exchange and cooperation between different stakeholders in the mobile network and electronic payment area.

The mobile-specific challenges are commonly overcome by using either proxy solutions and relying on an existing payment infrastructure, or by defining a new payment protocol specifically designed for mobile terminals. There is also a chance for operators to be successful with their own payment solutions, as the existing roaming and billing systems might offer a good platform for payment processing.

Two types of payment solutions are commonly used in mobile environment in order to address the challenges described in the previous section. The *proxy solutions* allow access of existing payment protocol through the use of a proxy server. The *dedicated solutions* involve payment protocols designed specifically for mobile devices. The decision whether to use a proxy or a non-proxy solution for a particular mobile systems is often determined by a business model, and not technical considerations.

An example of a practical and secure mobile electronic cash system that combines macro and micropayment functionality and offers a strong protection of users' privacy is Mobile Cash protocol [Tracz and Wrona, 2001]. Nevertheless, our evaluation suggest that some of the proposed fine-grained accountability mechanism for ad hoc networks and peer-to-peer systems, which are based on pay-per-packet approach may be impractical. We believe that a corse-grained accountability is a more reasonable approach for cooperative communication systems, and wireless ad hoc networks in particular. Such coarse-grained accountability could be performed either on a per-session basis, *i.e.*, a user prepaying or promising to pay for a some amount of transmitted packets, or on a per-relationship basis, *i.e.*, involving a payment for a pseudonym or deposit of a *bond* that commits a user to a cooperative behaviour. In the first case there is inevitably some amount of risk involved, in the case when one of the transaction partners defect. However, such misbehaviour on part of either the payee or the payer could be further mitigated by incorporating appropriate reputation mechanisms. In any case, the amount of money involved in the transactions would be probably low enough to justify some amount of risk involved.

Tax/Reward Mechanisms

Another successful mechanism for dealing with economic externalities is *taxing*. There are several possibilities for implementing tax/reward accountability scenarios in cooperative communication systems. For example, a supervising institution can know the total number of packets sent in the network and the amount of packets forwarded by every user. Periodically it could compare an average amount of sent packets per user with the amount forwarded by the user. If the result is negative, user receives a payback of a part of his monthly fee. Another, rather unscalable, solution would be to check the balance of every user after forwarding every packet. Forwarded packets would have to be *time-stamped* in order to guard against some sophisticated reply attacks, *e.g.*, based on the fact that multiple forwarding of very short packets might be cheaper than sending longer packets. A variation of the above approaches would be to count the exact amount of sent data, and not packets or connections. In order to provide stronger and more fair accountability node might be issued with signed receipts acknowledging forwarding of data. Some more sophisticated scenarios involving rewards are possible, too. For example, every receipt could be a part of a puzzle, which might be used to construct *e.g.*, ecash coins. Another scenario could be the use of lottery tickets as receipts, with the rewards founded from the monthly fees.

An example of tax/reward accountability in communication systems is *cumulus pricing*, which combines a flat-rate and a pay-per-use scheme [Stiller et al., 2001]. Users pay a subscription fee, which includes some amount of free usage. Users who exceed this amount receive penalty points, while those who use less receive rewards points. This approach is analogical to the tax/reward mechanism discussed before.

All tax/reward mechanisms ,in order to function properly, require the central authority to be well informed about the behaviour and traffic generated by every node. The assumption that all observations are made directly by the central authority may be infeasible in most cases. One solution is, of course, to implement the observing party in a distributed manner, with multiple well-located observers. Another solution would be to rely on the behaviour and usage information reported by nodes participating in the network. In this case, tax/reward become conceptually very similar to a reputation-based system, where nodes are periodically charged and rewarded, depending on the value of their reputation.

Both fungible payment schemes and tax/reward accountability require involvement of external parties and/or some infrastructure. In contrast, direct and indirect reciprocity schemes enable, at least in theory, building of fully self-organising and distributed accountability mechanisms.

Direct reciprocity. The fundamental requirement on a direct reciprocity system is the ability to recognise a partner which was involved in a past interaction and connect him to an individual history of experiences. Direct reciprocity requires an existence of only relationship pseudonyms, *i.e.* persistent identity is meaningful only for a particular relationship. This offers several advantages and disadvantages. For example, it is possible to construct pseudonyms based on computationally efficient symmetric cryptography techniques (*i.e.* involving a shared secret). Also, only possible reputation fraud is a pseudonym secret theft, and the damage is limited to the reputation in the relationships for which the secrets have been stolen. Finally, there is no need for a trusted third party. However, there are also some disadvantages of using relationship pseudonyms. First of all, it might be very difficult to transfer reputation built up in one relationship to another. Secondly, similar to symmetric cryptographic systems, relationship pseudonyms introduce larger demands on memory, which grow exponentially with the number of pseudonyms (or nodes) in the network. Thus, such a solution is better suited for systems which are small or have relatively static interaction patterns. Interesting option may be a hybrid system, where one starts with several direct reciprocity relationships, using the same public-secret key pair as a pseudonym, and then moves to indirect reciprocity.

As it was already mentioned, mutual authentication, or recognisability, of partners in a direct reciprocity can be based on computationally efficient shared secret solution, thus eliminating the need for computationally expensive public key operations in every transaction. Nevertheless, in order to establish a shared secret, some kind public key mechanism will be required, *e.g.*, use of Diffie-Hellman key agreement protocol [Menezes et al., 1997]. However, this interaction will take place once per a lifetime of the relationship.

Pseudonyms used for direct reciprocity can be unidirectional (*i.e.* personal pseudonyms, different for every party) or bidirectional (*i.e.* a single pseudonym, identifying a particular relationship). They can also be persistent, lasting for a whole life-time of a relationship, or changed periodically, even after every transaction, using *e.g.*, pseudonym chaining mechanism. In a pseudonym chaining mechanism, node uses the last pseudonym to authenticate itself (can be sent in plaintext), and to distribute a new pseudonym (in ciphertext) to be used for a subsequent identification. In such a scenario, there might be a need for a separate, persistent or chained, encryption key as well.

Indirect reciprocity and reputation systems. Reputation has been a fundament of fair trade and exchange of goods and services throughout the history. Buyers and merchants have known each others' identities and were thus able to asses the past performance of their trading partners. The persistent identity and possibility of recognition of humans were providing means for implementing an effective punishment either by using a legal framework or (more often) by relying on collective action. For example, in order to prevent misbehaviour, a trigger strategy might be agreed upon among the honest agents, so that the deviation of one single player might be met by the reaction of all the others. However, in some cases the cost of coordination for such a collective punishment might be prohibitive.

In the digital world reputation systems offer a possibility of protecting the true identity of agents by relying on pseudonyms instead of true identities.

Nevertheless, reputation systems using pseudonymous identities require particularly careful development. It should not be possible to gain a good reputation too easily. Also, it should not be possible to loose it too quickly, due to short-lived operational problems. An example of a critical flow in a reputation system is indiscriminately accepting an input from just any single user. A malicious user can create a large number of accounts supporting each other. This problem, so-called *pseudo-spoofing* is one of the major problems in pseudonymous systems. There are several possible solutions to pseudo-spoofing. The most obvious one is using certified identities, depending on a PKI. But identity does not automatically imply accountability. Also, system can aim to ensure that every pseudonym is controlled by a distinct person. Another simple approach is to just monitor users behaviour for evidence of pseudo-spoofing. Some reputation systems might be also susceptible to so-called *creeping death attack*, in which bad nodes can move upward on the reputation spectrum by eroding reputation of the nodes above them (see [Dingledine and Syverson, 2003]). The timing attacks on reputation systems may lead to figuring out what kind of rating influenced a published score by a certain amount. This problem can be reduced by pooling approach, in which some number of ratings are kept in a pool, and, with every new arrival, a randomly chosen rating from the pool is aggregated into the score. However, this method may be susceptible to flooding attacks.

Another problem is transferring pseudonymity credentials between the participants of the system. Reputations and identities do not bind as tightly to people online as they do in the physical world. Reputations can be sold or stolen with a single password. The most common example of this phenomenon can be found in online multi-player games. It leads to an emergence of a whole market in trading game identities for cash online, *e.g.*, using eBay auctions. Several solutions have been proposed to the credential transfer problem. A simple method is to embed an important piece of information, such as a credit card number, into the credential (*e.g.*, into the password for an account). More sophisticated approach may make use of secret key certificates concept [Brands, 1999]. The drawback of this approach is that it makes an accidental losing of private credential information (*i.e.* password or private key) more expensive for the user, too. Some other sophisticated solutions to anonymous or pseudonymous credentials have been proposed in [Chaum, 1982; Chaum, 1985; Damgard, 1990; Chen, 1995; Camenisch and Lysyanskaya, 2001; Lysyanskaya et al., 1999]. One more solution is to make each pseudonym valid only for a certain number of logins. In order to prolong its validity, the user must prove that he is the same person. Also, an additional identification based on the behavioural pattern of the user can be used in order to discover a possible transfer of credentials. However, all such methods have important privacy implications, and in the case of cooperative communication systems preventing credential transfer does not seem to be a critical issue.

In many cases additional security mechanisms can be put in place if we allow involvement of an external party or emergence of some semi-persistent infrastructure. For example, pseudonyms might be sold only by a trusted third party, *e.g.*, as blindly signed certificates, making it expensive to change the identity often. In more organised systems certain users can be picked to become moderators, who can assign ratings to other users. In order to prevent fictious ratings, interacting nodes could provide blinded (anonymous) receipts for transactions. Without such a receipt, the reputation system would not accept the feedback. Also, running periodical statistical tests, independent of the actual reputation aggregation, can help to detect suspicious behaviour in the system.

The difficult question in reputation systems is how to introduce the newcomers. Some systems, *e.g.*, FreeHaven, assume that some participants will be altruistic enough to try out the new nodes to test their trustworthiness. Some other systems provide some initial value based on the exogenous references from another interaction arena.

An individual's trustworthiness is orthogonal to his ability to give reliable feedback on performance of others. Thus, separate scores should be used for nodes' performance and the credibility of their recommendations. Of course, providing reputation ratings can be also interpreted as a valuable service provided by an agent to the community. Therefore the rater himself may be a subject of reputation ratings concerning the reliability of his ratings as perceived by other members of the system. The question of meta-reputation is how to determine the reputation of the recommenders. In any case, since a claim cannot be taken as an absolute true, nodes are left with the task of determining the credibility of the claimer. The problem of claimer's credibility can be avoided by using reputation derived only from direct observations. However, this approach can slow down considerably the process of building up (or loosing) the reputation and lead to decrease of its accuracy, too. In general, decentralised systems require a broadcast mechanism in order for all nodes to keep their reputation standings synchronised. However, there are also some more efficient reputation dissemination mechanisms (cf. [Wrona, 2005]).

The simplest way of building a reputation system is treating reputations as probabilities. More complex semantics of reputation are discussed, *e.g.*, in [Wrona, 2005]. For example, context-aware ratings can give stronger weights to the ratings obtained in the similar situations. The difficult question here is how to pick up, and find, related categories. A higher reputation can be also a reward in itself, and thus can be used as a motivation mechanism. Just as the statistics pages for SETI@home encourage participation, publicly ranking generosity creates an incentive to participate. The incentives of public recognition explain most actual current peer-to-peer node operators.

In addition to fostering cooperation, reputation systems can also be used in order to provide security against malicious attacks. A malicious node would participate in the system only to compromise its security or reliability. In doing so, however, a dishonest agent will have to consider the costs of reaching and maintaining a position from which those attacks are effective, which will probably involve gaining reputation and acting as an honest node for an extended period of time.

There is also a whole family of systems based on the observation that there is a direct relation between the trust and monetary value. For example, reputation can be treated as a capital, which can be spent, earned or even sold [Seigneur et al., 2002]. Also, there is a connection between required level of trust and the perceived level of risk. This relation can be encompassed by introducing insurance-like systems [Reagle, 1996; Reiter and Stubblebine, 1999]. In a related vein, [Zacharia et al., 2000] have investigated agent systems employing pricing schemes explicitly dependent on the reputation of service providers.

Identification and Accountability

As it has been already mentioned earlier, the choice of a particular accountability mechanism for a cooperative communication system depends on the level of identification of participants, which is provided by the system. In general, three different levels of identification are possible: *certified identity*, *pseudonymity*, and *anonymity*, see also Figure 10.1. In real-life systems both complete assurance of identity and total anonymity are very difficult to achieve. At the same time, one of the widely spread misconception is that strong accountability requires personal identification. In fact, the usage of identity for accountability introduces new security problems. Identity theft was the leading type of consumer fraud in the U.S.A. in 2003: 42% of complaints compiled by the Federal Trade Commission involved identity theft as reported by [Associated Press, 2004].

There are several reasons why in cooperative communication systems we have to treat identity in a special way. First of all, unique and persistent identification of peers and their operators is difficult. Second, there might be no good way to asses the history and predict the performance of other peers. Further, the legal procedures are generally too costly and too slow to be applicable in the case of cooperative communication systems. And finally, peers' privacy may require anonymity or pseudonymity. However, it is worth to keep in mind that in a truly anonymous system all decisions must be based on immediately and temporarily available information. This leaves electronic cash (micro-)payments as the only reasonable way to establish accountability in fully anonymous systems.

Anonymity may be defined as unlinkability of action and an identifier of the subject. In order to enable anonymity, a set of subjects with potentially the same attributes must always exist. In [Pfitzmann and Köhntopp, 2000], *anonymity* is defined as the state of being not identifiable within a set of subjects, the *anonymity set*. The strength of anonymity increases with the size of the respective anonymity set. Alternatively, [Levine, 2000] defined anonymity of an entity in the network with respect to some other single entity. The rational behind is that anonymous protocols may not provide the same degree of anonymity with respect to every other entity in the network. Levine defined also several degrees of anonymity, which are defined with respect to probability that some other entity from the same set has initiated a connection.

Unlike other security services, anonymity does not depend only on sender and receiver – it also requires trust in the infrastructure to provide protection and that sufficient number of other nodes use the infrastructure. For example, as anonymous communication systems use cumulated traffic to hide single messages, the users are always better off in a crowded system. High traffic is necessary not only for strong anonymity, but it also enables better performance, since a system that processes only light traffic must delay messages significantly to achieve the adequate anonymity.

A *pseudonym* is an identifier of a subject. The subject that may be identified by the pseudonym is the *holder* of the pseudonym. Whereas anonymity and certified identity are the extremes with respect to linkability to subjects, pseudonymity includes all degrees of linkability to a subject. By using the same pseudonym more than once, the holder may establish a reputation. On the other hand, the stronger anonymity requires the pseudonyms to be changed more often.

Pseudonyms may be classified according to two basic criteria. The first one is the *initial knowledge* of the linking between the pseudonym and its holder. Here we can differentiate between three classes of pseudonyms. An example of a *public pseudonym* is a phone number and its owner listed in a phone book. An *initially non-public pseudonym* can be exemplified by a bank account number. *Initially unlinkable pseudonyms* can consist of, *e.g.*, biometrics data. The second criterion is *linkability* due to use of pseudonyms in different contexts. Here, we can differentiate between five classes of pseudonyms. *Person pseudonyms* are substituted for the holders name. *Role pseudonyms* are limited to use for a specific role. In the case of *relationship pseudonyms*, a different pseudonym is used for each interaction partner. Accordingly, in the case of *role-relationship pseudonyms* a different pseudonym is used for each role and for each interaction partner. Finally, *transaction pseudonyms* provide a different pseudonym for each transaction. The other important properties of pseudonyms include limitation of a number of pseudonyms per subject, guaranteed uniqueness, transferability to other subjects, transferability of attributes of one pseudonym to another, possibility and frequency of pseudonyms

Table 10.1. Comparison of the signed content formats. The *performance* analysis has been divided into two separate parts: *message size* and *overhead* caused by the additional data and *processing time* for creation, transmission and verification of signed content. The ratings range from [−−] (*not satisfactory*) to [++] (*very satisfactory*).

Encoding scheme	Security	Interoperability	Complexity	Performance	
				Size	Time
S/MIME implicit	++	++	++	−−	+
S/MIME explicit	++	++	++	+	++
XML Digital Sig.	++	+	−	+	−

changeover, limitation in number of uses, validity, possibility of revocation, participation of users or other parties in forming of the pseudonyms.

A digital pseudonym can be implemented as a public cryptographic key, with the holder proving his knowledge of the corresponding private key. Similarly, a digital identity can be implemented as a public key certificate, a digital signature of a certification authority (CA), binding a public key to a subject. A *certified identity* is the use of unique and permanent identity as an identifier of the subject. The identity is certified by a trusted third party. Certified identity is also sometimes called a *verinym* or a *true name* [Goldberg, 2000]. Anonymity services should ideally provide a guarantee of *forward secrecy*, *i.e.* an adversary should not be able to recover security-critical information, such as the true name of the entity, after the transaction has taken place [Goldberg, 2000].

Signed content plays an important role in most of the accountability mechanisms discussed within this chapter. In particular, signed content can be used in order to provide authentication and linkability of nodes, using both pseudonyms and certified identity. Also, signed content is critical for ensuring integrity and authentication of recommendation and reputation information, as well as for enabling payment mechanisms.

Signed content formats. A brief evaluation of some common *signed content encoding schemes* is presented in Table 10.1, taken from [Perlines-Horman et al., 2001]. We have compared the *explicit S/MIME* format [Galvin et al., 1995], together with the *implicit S/MIME* [Ramsdell, 1999], and the *XML Digital Signature* format [Eastlake et al., 2001]. We have chosen to focus on formats with the highest probability of being used for mobile applications. However, there are some other popular signed content formats, such as *Open Pretty Good Privacy (Open PGP)* [Callas et al., 1998], signed *Java Archive (JAR)* files [JAR, 1999], and *Microsoft Authenticode* [Authenticode, 1996].

Table 10.2. Comparison of certificate validation mechanisms.

Cert Validation Mechanism	Security	Interop.	Off-line	Complexity	Performance Size	Performance Scal.
Short-lived Certs	−	++	+	−	+	+
CRL's	+	++	++	++	−−	−−
OCSP	++	++	−	+	++	++
XKMS	−−	+	−	+	−	−

Certificate validation mechanisms. Another important aspect related to certified identity is mechanism enabling validation of public-key certificates. Despite of being a necessary component of public-key system, validation mechanism can often introduce by itself new vulnerabilities in the system. For example, the fact that CRL-based validation systems can collapse in the case of sudden surge of requests for CRL updates, has been demonstrated on Jan. 7, 2004, when VeriSign has experienced a sudden and and dramatic increase in the number of requests by Windows-based clients [VeriSign, 2004]. These requests were caused by an expiration of a CRL, which was included in some widely deployed security patches from 3rd party providers. In Table 10.2 we present a summary of evaluation of different *certificate validation mechanisms*, [Perlines-Horman et al., 2006]. We have compared *short-lived certificates* [X.509, 1997; Housley et al., 1999], *Certificate Revocation Lists (CRL's)* [Berkovits et al., 1994], *Online Certificate Status Protocol (OCSP)* [Myers et al., 1999], and *XML Key Management Specification (XKMS)* [XKMS, 2001].

Our analysis does by no means cover all possible validation techniques. In particular, several optimisations have been proposed to the original CRL model, focusing either 1) on improving mechanisms for generation and distribution of CRLs or 2) on efficient implementation of the revocation data structure. The goals of these optimisations are to reduce required communications bandwidth, *e.g.*, by using delta-CRLs [Housley et al., 1999] and segmented CRLs [Cooper, 1999], as well as the storage overhead at the client, and the peak request load experienced by the revocation information repository [Wohlmacher, 2000]. The peak request rate can be reduced, *e.g.*, by using over-issued CRLs [Cooper, 1999], windowed revocation [McDaniel and Jamin, 2000] and sliding window delta-CRLs [Cooper, 2000].

The advanced data revocation structures include Certificate Revocation System [Micali, 1996; Micali, 1997; Micali, 2002] that requires CA to periodically publish a message for every issued certificate which states if it was revoked or not. This approach reduces significantly the query communications costs, as the end entities do not have to download possibly large CRLs, however it also increases the CA-to-directory communication overhead. Use of the Certificate Revocation Tree (CRT) [Kocher, 1998] is another way of enabling the verifier to get a short proof that the certificate was not revoked. [Kikuchi et al., 1999] evaluated performance of balanced CRT implementation with S-expressions [Rivest, 1997]. [Naor and Nissim, 2000] proposed use of authenticated dictionaries, based on optimised search trees such as 2-3 trees or Randomised Search Trees, in order to provide an efficient structure for the revocation data. They have also shown how the authenticated dictionaries can be combined with the incremental cryptography mechanism in order to provide efficient updates. [Muñoz et al., 2003] have implemented an authenticated dictionary for certificate revocation using Merkle Hash Tree and 2-3 trees. [Goodrich et al., 2001] presented an implementation of authenticated dictionary using skip lists and commutative hashing.

3. Dynamics of Cooperative Communication Systems

As discussed in the introduction, in the case of the cooperative communications system it is important to understand how the network is behaving dynamically in the case of situation that different users ("players") are choosing to use different cooperation strategies. We want to avoid any cooperative network solution that is highly unstable, in a sense that a small number of misbehaving users could cause either collapse of the network, or large fluctuation in offered quality of service (more properly quality of experience).

Below we introduce a dynamic game-theoretic model of cooperation in ad hoc networks, based on *evolutionary game theory*. The use of game theory in general has gained recently popularity in the communications engineering field. We are here using evolutionary game theory as an extensive tool for dynamicity analysis. We use a limited, but illustrative, example to show the main promise it as an analytical method. We do not claim to give an exhaustive analysis, we are instead of trying to advocate the use of efficient tool. Moreover, we show how the evolutionary game based models can be used to find and study stability criteria for cooperative networks under some relatively simplistic conditions.

We are mostly interested in understanding the dynamic behaviour and stability of ad hoc networks from the cooperation point of view. The model includes uncooperative *bad nodes* and different cooperative nodes, *i.e.*, *good nodes*, which employ either unconditional or conditional cooperation strategies. Our model enables us to make predictions about possible equilibrium points of the network composed of selfish and learning nodes, which can dynamically adjust their strategies in order to maximise their own payoffs. In particular, we show that if an ad hoc network implements a reputation mechanism, all long term equilibrium points of the system will include cooperating nodes. In fact, in most of the equilibrium points, the cooperators will constitute a majority of the nodes. We believe that this new modelling approach, borrowing from biology and physics, can have broader applications for studying dynamics of distributed communication systems.

The presented evolutionary game-theoretic model uses some simplifying assumptions, since we wanted to obtain analytically tractable solutions. However, in our experience analytical models are useful in order to understand some of the underlying fundamentals, and quite often even simple game models are able to predict the behaviour of real systems with a good enough level of confidence. The large simulation studies can naturally cope with more complex models, and in our other work we have also used extensive simulations. We also present our multi-agent simulation model of cooperative communication systems, *i.e.* wireless ad hoc networks, and we investigate the simulation results for different accountability mechanisms: tax/reward mechanism, direct reciprocity and indirect reciprocity.

Both our analytical and simulation models are by design generic cooperation models and their applications are quite manifold. We focus on study of incentives to support packet forwarding and routing in wireless ad hoc and mesh networks. However, our models could be also easily extend to cover *e.g.* cooperation in peer-to-peer file sharing applications, distributed backups over mesh networks, peer-to-peer computing, and other cooperative networks.

Analytical Model of Cooperation in Ad Hoc Networks

One of the frequently asked questions in the context of ad hoc networks is if a networking concept solely based on cooperation among selfish nodes can be sustainable. Since an ad hoc network can be interpreted as a *common-pool resource*, it is susceptible to the classical *free-riding* problem, well known in microeconomics and in social sciences. The source of this social dilemma is that it is advantageous for every node to refrain from forwarding other nodes' packets, and thus save its own energy. But if a substantial number of nodes would follow this selfish strategy, the network would break down completely, depriving all nodes of communication services. Finding possible ways to overcome the free-riding problem in ad hoc networks has been an area of active research lately. The most of existing contributions focus on design and simulation of two main mechanisms: micropayments [Buttyán and Hubaux, 2001; Buttyan and Hubaux, 2003], and reputation-based accountability, *e.g.*, [Marti et al., 2000; Buchegger and Boudec, 2002]. Considerably less work has been put into more fundamental understanding of underlying mechanisms. [Michiardi and Molva, 2003] have examined effectiveness of collaborative monitoring techniques and reputation mechanism, using both cooperative and non-cooperative game theory. [Urpi et al., 2003] have proposed a model of cooperation based on Bayesian games. However, the existing models deploy static game theory and focus on the classical Nash equilibrium concept as a predictor of the nodes' behaviour. Thus these models are not well suited to answer questions concerning the dynamic behaviour of the system. Such questions include predicting the state of the ad hoc network consisting of evolving nodes, which can be reached in the long run, or identifying the theoretical limits on the values of the protocol parameters (*e.g.*, minimal required reliability of a reputation information) needed for the effective accountability in ad hoc networks.

In this section we propose a dynamic model of cooperation in ad hoc networks, based on *evolutionary game theory*. This model enables us to make predictions about possible states of the network composed of the selfish learning nodes, which can dynamically adjust their strategy in order to maximise its own payoff. The dynamic model is also required to better understand the evolution of cooperative networks, and most importantly, probe stability of algorithms and networks themselves. As far as we are aware, the issue of stability has not been emphasised in the previous works. In the following sections we present rationales behind our game-theoretic model of ad hoc network, and show how we can use our model to study dynamics of such networks.

General assumptions. In order to make the problem of cooperation in ad hoc networks analytically tractable we make the following simplifying assumptions.

First, we assume that the *bad nodes* use a static strategy, *i.e.*, they always defect (ALLD). This simplifies the problem, since we do not have to care about sophisticated cheating strategies, *e.g.*, occasional *cashing* of high reputation value or defecting against low reputation nodes and newcomers. In our simulation-based study we have also investigated performance of some more sophisticated Machiavellian strategy.

Second, we assume that there are two kinds of good nodes: *unconditional cooperators* (ALLC), and *conditional cooperators*, so-called *reciprocitors* (REPC). The unconditional cooperators can be interpreted in one of the following ways: (a) legacy nodes, without support for trust/reputation mechanisms; (b) low-end devices, with too constrained resources for implementing the trust and reputation logic; (c) altruistic nodes; (d) *seed nodes* installed by the service providers or communities to establish some (minimum) level of support towards cooperation. Conditional cooperators are nodes, which base their cooperation on the reputation of the other node(s). We assume that conditional cooperating nodes implement only two actions, they either cooperate or do not engage in the interaction.

Stages of the game. The process of participation in a wireless ad hoc network can be modelled from the simple trust/reputation point of view as a game consisting of the following steps: *inspecting reputation*, *cooperating/defecting*, *monitoring*, and *punishing/rewarding*.

The first two stages can be modelled as a single stage, which we call *conditional cooperation game*. The conditional cooperation game can be modelled either as a *simultaneous game* or as a *sequential game* with three strategies: *cooperate, defect,* and *conditionally cooperate*. The simultaneous game, which is analysed in this paper, can be seen as a *packet trading* game, where both players decide simultaneously if they should forward communications originating from the opponent or not. These kinds of games are well-studied in game theory, with Prisoner's Dilemma being one of the best known examples. Nevertheless, the applicability of this scenario to the wireless environment requires that both nodes engage in a constant, or synchronised communications (*i.e.*, either both nodes transmit continuously or they wait with sending data for the trading opportunity). In the sequential game, one player has some need to send data, and the other player makes a decision whether to help or not. The third possibility is a hybrid game, where every player, either player, or none of the players might need to send data in the particular round of the game. In this case, the game can take a form of a mutual or unilateral cooperation dilemma, or a game might not be played at all.

The constant cooperation and defection strategies can be used to model behaviour of unconditional cooperators and defectors respectively. The conditional cooperation strategy involves a process of inspecting reputation of the opponent, which might involve sending reputation queries, if the adequate reputation information is not available at node or is outdated.

The process of conditional cooperation and monitoring can be interpreted as a punishment/reward strategy in the game-theoretic sense. The punishment (or reward) in our case consists of two stages: reporting of the observed misbehaviour to the reputation network, and an ostracism (or cooperation) by conditional cooperators, *i.e.*, these are conditional cooperators that are executing punishment of the known defectors or give reward to known cooperators.

If the network includes conditional cooperating nodes, some sort of reputation reporting is required, and there is a lot of previous work in this field [Resnick et al., 2000]. Reputation system can be seen as a separate logical reputation network, with its own inevitable challenges to cooperation. The process of monitoring and reporting behaviour of other nodes can be described as a single game, which we call *cooperation monitoring game*. Such cooperation monitoring game is an *asymmetric game* with actors and observers. The actors can implement any of three strategies described in the conditional cooperation game. The observers can implement one of three actions: *ignore*, *monitor and keep the result for itself*, and *monitor and report result to the reputation network*. In our analysis, in order to simplify the scenario, we assume that the monitoring stage involves only two possible actions: *ignore* or *monitor and report*. In our analysis, we assume that an observer is a role played in the second stage (*i.e.*, the monitoring stage) of the conditional cooperation game with monitoring by a node which was an actor in the first, conditional cooperation, stage. Respectively, the actor of the monitoring game is observer's opponent from the first stage of the game. This game represents a case of a packet trading between two nodes, who can report their interaction experience to the reputation network in order to punish or reward the other player. However, another possibility is a game where the role of an observer is played by another player, who was not involved in the conditional cooperation stage. This is a situation, were nodes can observe interactions between other nodes and report facts of proper and improper behaviour. In a general case there can be a large, and possibly random, number of observers for every interaction. Thus such a game scenario can be extremely difficult to analyse, and we limit our analytic model to the scenarios involving two players playing both roles of actors and observers in a two stage two player game.

Connection between a monitoring stage and a conditional cooperation stage could be captured via reliability of reputation information, described by the parameter $p \in (0.5, 1]$. If the monitoring and reporting behaviour is getting less frequent, then the available reputation information is less reliable and the value of parameter p is decreasing. The less reliable reputation data causes a decrease in the expected payoff of conditional (and unconditional) cooperators, at the same time increasing the payoff of unconditional defectors.

Simultaneous conditional cooperation game with monitoring. As described earlier, the *simultaneous conditional cooperation game with monitoring* consists of two stages: cooperation stage and monitoring stage. Below we discuss the simultaneous games played by the nodes in both of these stages and describe how these stages can be combined to form a two-stage conditional cooperation with monitoring game.

Conditional cooperation game. A conditional cooperation game is a symmetric simultaneous game, where every player can choose one of the three possible strategies: defect (D), cooperate (C), or verify (V). The verify strategy involves checking of reputation of the opponent and results in taking either a *verified cooperate* (C_v) or a *verified exit* (E_v) action. We assume that the trust decision (or the reputation value) correctly identifies a cooperator C with a probability $p_1 > \frac{1}{2}$ and correctly identifies a defector with a probability $p_2 > \frac{1}{2}$. For simplicity, we assume that $p_1 = p_2 = p$. We also assume that the verification process incurs a cost δ on a verifier. In the general case, the verification process can also incur a cost η on the verified peer, *e.g.* in the case when it is up to the verified peer to prove its good reputation. Here we assume that $\eta = 0$. We also assume that in the case of the negative result of the verification, the verifier does not engage in interaction, and announces his decision to the opponent (and the outside world). This announcement is required since peers have to be able

Player 1 \ Player 2	D (defect)	C (cooperate)
D (defect)	P, P	T, S
C (cooperate)	S, T	R, R

Figure 10.2. The payoff matrix of Prisoner's Dilemma game.

to differentiate between an exit action and defection. The requirement to perform announcements leads to some power consumption issues, which should be studied in the future. The *exit* announcement results in a cost e to the exiting party (*i.e.* the verifier) and a zero payoff for the opponent. We can assume that the cost of sending an exit notification is similar to the cost c_f of forwarding the message. This assumption has two advantages: (1) the physical cost of sending exit notification is similar or equal to the cost of forwarding a message (this is true if there is no big difference between the size of notification and the message to be forwarded. This is the case if the game is played for every packet or for not so large chunks of data), and (2) it guards against misbehaviour relying on an *always exit* strategy.

In our analysis we assume that the payoff of the game without inspection is as in the standard Prisoner's Dilemma game (see *e.g.* [Fudenberg and Tirole, 1991]), depicted in Figure 10.2. In order to give a physical meaning to the values R, S, T, and P, we assume that:

1 The reward R to a node for mutual collaboration is equal to value m_f of having its own message forwarded minus cost c_o of sending own message to the forwarding node, and minus a cost c_f of forwarding a message belonging to the other node:

$$R = (m_f - c_o - c_f).$$

2 The value T of temptation to defect is equal to the value m_f of having own message forwarded minus cost of sending this message:

$$T = (m_f - c_o).$$

3 The sucker payoff S is equal the cost of sending own message c_o and cost c_f of forwarding the other node's message and then lost utility m_d from the delaying the delivery of the own message m:

$$S = -(c_o + c_f + m_d).$$

4 The punishment P for mutual defection is equal to the cost c_o of sending own message (which is dropped by the other node) and cost of delaying the delivery of the own message m_d:

$$P = -(c_o + m_d).$$

In order to simplify the analysis, we assume that:

$$c_o = c_f = c \qquad \text{and} \qquad m_f = m_d = m.$$

Since we have to fulfil the requirement for the Prisoner's Dilemma, $T > R > P > S$, we have the following constraints:

$$\begin{aligned} m - c &> m - 2c, \\ m - 2c &> -c - m, \\ -c - m &> -2c - m. \end{aligned}$$

Since $c, m \in \mathbb{R}^+$, these constraints can be reduced to $m > \frac{c}{2}$. If we also want to fulfil additional PD constraint, $R > \frac{T+P}{2}$, then $m > c$. These are quite reasonable assumptions, since the utility m gained by a node from sending a message should be always higher than the cost c incurred by the node, or a rational node would never try to send a message.

If we add a third strategy to the game (V) resulting in two additional actions (C_v and E_v), the game structure changes to the matrix presented in Figure 10.3. In order to simplify the notation we define

$$\pi_C(V) = Rp + S(1-p) \qquad \pi_V(C) = Rp - e(1-p) - \delta$$
$$\pi_D(V) = T(1-p) + Pp \qquad \pi_V(D) = S(1-p) - ep - \delta$$
$$\pi_V(V) = p^2R - (1-p)pS - e(1-p) - \delta$$

Adding the monitoring stage. The cooperation monitoring game is an asymmetric game with actors and observers. The actors are involved in a conditional cooperation game and can implement any of three strategies described for the conditional cooperation game. The observers can implement one of two strategies: ignore (I) or monitor and report (R). The monitoring gives correct results with the probability q, $q > \frac{1}{2}$, and incurs cost ρ to the node. We can assume $\rho = 0$ since any node has to monitor the radio channel anyhow. Strictly speaking $\rho \neq 0$ in the case of radio systems that employ clever sleeping strategies. The reporting incurs an additional cost ϵ to the node. We also assume that the fact of observing behaviour of other nodes increases the knowledge of the observer, giving him the payoff v_k, which satisfies $v_k \geq \rho + \epsilon > 0$. The payoff

		Player 2		
		C	D	V
	C	R R	S T	$\pi_C(V)$ $\pi_V(C)$
Player 1	D	T S	P P	$\pi_D(V)$ $\pi_V(D)$
	V	$\pi_V(C)$ $\pi_C(V)$	$\pi_V(D)$ $\pi_D(V)$	$\pi_V(V)$ $\pi_V(V)$

Figure 10.3. The payoff matrix of conditional cooperation game.

matrix of the conditional cooperation game with monitoring and reporting is derived according to the following rules:

1 Reliability of observation is q, *i.e.* an observer correctly identifies cooperation/misbehaviour with probability $q > \frac{1}{2}$.

2 If the observers ignore behaviour of their neighbours, or the node has no neighbours, the final payoff is the payoff after the first move, and ignoring nodes receive the payoff i (value of ignorance). We assume $i = 0$.

3 If the nodes monitor and report their observations to the network-wide reputation system, the payoff of the observed nodes is increased by h ($h > 0$) in the case of proper behaviour or decreased by l ($l > h$) in the case of misbehaviour. Thus, for the observed node quantities h and l represent an increase and decrease in the likelihood of cooperation respectively, when interacting with conditional cooperating nodes.

4 We assume that exit behaviour is recorded by the reputation system. This is done in order to guard against *always exit* strategy, which might be profitable if $e < C$, as well as against denial-of-service attacks. However, it does not have any direct positive or negative influence on the node's reputation.

In the following text we denote the payoffs of the actor from the first stage of the game by π_C, π_D, and π_V, where the subscript denotes the strategy played by the actor in the cooperation stage. We also assume that the cost π_R of reporting is given by

$$\pi_R = q(v_k - \epsilon) - \rho.$$

The resulting normal form of the game is depicted in Figure 10.4.

The problem with the game presented in Figure 10.4 is that from the point of view of the Player 3 (the observer) it can be seen as a static decision problem, since the observer's payoff depends only on his own choice and not on

Observer

		I	R
	C	π_C, i	$\pi_C + qh$, π_R
Actor	D	π_D, i	$\pi_D - ql$, π_R
	V	π_V, i	$\pi_V + qh$, π_R

Figure 10.4. The normal form of simultaneous monitoring and reporting game.

the actions of the other. Thus this game does not capture the interconnection between being an observer and an actor, which is at the heart of the real life ad hoc network cooperation dilemma. However, we can modify the game from Figure 10.4 in order to accommodate for the case when the nodes involved in the interactions are also observers, *i.e.*, they can influence their own and their opponents payoffs by observing and reporting the opponents behaviour. We assume that a node cannot report its own behaviour in order to increase its own reputation. The normal form of such simultaneous conditional cooperation game with monitoring is presented in Figure 10.5. We have simplified the notation used in the payoff matrices by introducing the following symbols:

$$\pi_C^R(C) = R + qh \qquad \pi_D^R(V) = \pi_D(V) - ql$$
$$\pi_C^R(D) = S + qh \qquad \pi_V^R(C) = \pi_V(C) + qh$$
$$\pi_C^R(V) = \pi_C(V) + qh \qquad \pi_V^R(D) = \pi_V(D) + qh$$
$$\pi_D^R(C) = T - ql \qquad \pi_V^R(V) = \pi_V(V) + qh$$
$$\pi_D^R(D) = P - ql$$

The final game. As the number of strategies accessible to every player in this game amounts to 24 ($3 \cdot 2^3$), the corresponding normal form of the game would be a 24 by 24 matrix with 576 payoff pairs. The possible strategies in a two stage simultaneity game with three possible actions in the first stage can be described as tuples (a, bcd), where a is an action taken by the player in the first stage, and b, c, and d denote actions taken by the player in the case when his opponent has played in the stage one action 1, 2, or 3 respectively. Thus, the possible strategies in our case of simultaneous conditional cooperation game with reporting would include: (C, IRI), (D, III), (V, RRR), and so on. We focus on analysing only *open-loop strategies, i.e.* (C, III), (C, RRR), (V, III), (V, RRR), (D, III), and (D, RRR). These strategies represent strategies where nodes decide in

	C	D	V
C	R, R	S, T	$\pi_C(V)$, $\pi_V(C)$
D	T, S	P, P	$\pi_D(V)$, $\pi_V(D)$
V	$\pi_V(C)$, $\pi_C(V)$	$\pi_V(D)$, $\pi_D(V)$	$\pi_V(V)$, $\pi_V(V)$

(I,I)

	C	D	V
C	$\pi^R_C(C)$, $R + \pi_R$	$\pi^R_C(D)$, $T + \pi_R$	$\pi^R_C(V)$, $\pi_V(C) + \pi_R$
D	$\pi^R_D(C)$, $S + \pi_R$	$\pi^R_D(D)$, $P + \pi_R$	$\pi^R_D(V)$, $\pi_V(D) + \pi_R$
V	$\pi^R_V(C)$, $\pi_C(V) + \pi_R$	$\pi^R_V(D)$, $\pi_D(V) + \pi_R$	$\pi^R_V(V)$, $\pi_V(V) + \pi_R$

(I,R)

	C	D	V
C	$\pi^R_C(C) + \pi_R$, $\pi^R_C(C) + \pi_R$	$\pi^R_C(D) + \pi_R$, $\pi^R_D(C) + \pi_R$	$\pi^R_C(V) + \pi_R$, $\pi^R_V(C) + \pi_R$
D	$\pi^R_D(C) + \pi_R$, $\pi^R_C(D) + \pi_R$	$\pi^R_D(D) + \pi_R$, $\pi^R_D(D) + \pi_R$	$\pi^R_D(V) + \pi_R$, $\pi^R_V(D) + \pi_R$
V	$\pi^R_V(C) + \pi_R$, $\pi^R_C(V) + \pi_R$	$\pi^R_V(D) + \pi_R$, $\pi^R_D(V) + \pi_R$	$\pi^R_V(V) + \pi_R$, $\pi^R_V(V) + \pi_R$

(R,R)

Figure 10.5. The payoff matrix of cooperation monitoring game.

Table 10.3. Parameters used in the game definition.

	Meaning	Value
c	Cost of sending a message	$c > 0$
m	Value of having own message sent	$m > c$
p	Probability of a correct identification of a cooperator and defector during verification	$1 \geq p > \frac{1}{2}$
δ	Cost of verification	$\delta > 0$
q	Probability of successful monitoring	$1 \geq q > \frac{1}{2}$
ρ	Cost of monitoring	$\rho \geq 0$
ϵ	Cost of reporting	$\epsilon \geq 0$
k	Value of increased knowledge	$k \geq 0$
h	Value of increased reputation	$h \geq 0$
l	Value of decreased reputation	$l \geq h$

advance what strategy to play in both stages of the game, *i.e.*, the knowledge about the information set in which every player has finished up after the first stage of the game has no influence on the strategy chosen in the second stage. Thus the open-loop strategies are functions of time alone as opposed to the *closed-loop strategies*, which condition players actions on the history of the play so far. It is typically much easier to characterise the open-loop equilibria of a given game than the closed-loop ones, since the closed-loop strategy space is much larger. It is also known that the open-loop equilibria may be a good approximation to the closed-loop ones [Fudenberg and Tirole, 1991], if there is a high number of *small* (or so-called *infinitesimal*) players, in which case an unexpected deviation by an opponent might have little influence on a player's optimal play. In the case of infinitesimal players the outcome of an open-loop equilibrium is subgame-perfect.

After limiting our analysis to open-loop strategies, as well as taking into account that the resulting game is symmetric, we can simplify the payoff matrix to the form presented below:

$$\mathbf{A} = \begin{bmatrix} R & \pi_C^R(C) & S & \pi_C^R(D) & \pi_C(V) & \pi_C^R(V) \\ R + \pi_R & \pi_C^R(C) + \pi_R & S + \pi_R & \pi_C^R(D) + \pi_R & \pi_C(V) + \pi_R & \pi_C^R(V) + \pi_R \\ T & \pi_D^R(C) & P & \pi_D^R(D) & \pi_D(V) & \pi_D^R(V) \\ T + \pi_R & \pi_D^R(C) + \pi_R & P + \pi_R & \pi_D^R(D) + \pi_R & \pi_D(V) + \pi_R & \pi_D^R(V) + \pi_R \\ \pi_V(C) & \pi_V^R(C) & \pi_V(D) & \pi_V^R(D) & \pi_V(V) & \pi_V^R(V) \\ \pi_V(C) + \pi_R & \pi_V^R(C) + \pi_R & \pi_V(D) + \pi_R & \pi_V^R(D) + \pi_R & \pi_V(V) + \pi_R & \pi_V^R(V) + \pi_R \end{bmatrix}$$

All parameters, which are used in the model are summarised in Table 10.3.

Modelling the dynamic behaviour. In order to model the dynamics of the system, we have to find a way to describe the changes in the behaviour of rational network nodes with time. Several dynamics have been proposed for use in evolutionary games [Hofbauer and Sigmund, 1998]. From analytic point of view, especially attractive is the replicator dynamics, which leads to a system of deterministic difference or differential equations, commonly used in biological models. Despite its originally biological motivation, several authors have recently noted that the replicator dynamics can emerge from simple learning models [Samuelson, 1998]. The process by which agents choose their strategies is most likely to resemble the replicator dynamics, if imitation is an important component of the learning process. This is due to the fact that a process of imitation shares many of the features of biological reproduction, where the most successful agents give a rise to additional agents playing the same strategy. Since the assumption of learning by imitation in the context of ad hoc networking seems to be a reasonable one, we assume that the ad hoc system adheres to the replicator dynamics, *i.e*

$$\dot{p_i} = p(\pi_i - \bar{\pi}).$$

Another assumption made by replicator dynamics, and which might not be true in the ad hoc networks, is that interaction pattern between agents is completely random. In fact, the completely random interaction pattern can not be universally true and trivial counter examples can be found *e.g.* in the case of small networks. However, in many cases it seems likely to be a reasonable approximation.

In game theory an important concept is that of the *Nash equilibrium* [Fudenberg and Tirole, 1991]. It can be shown [Gintis, 2000; Hofbauer and Sigmund, 1998] that if an evolutionary game satisfies the replicator dynamics, then:

1 If $\pi*$ is a Nash equilibrium of the evolutionary game, $\pi*$ is a fixed point of the replicator dynamics.

2 If $\pi*$ is an evolutionary equilibrium or a focal point of the replicator dynamics, then it is a Nash equilibrium.

In our investigation we focus on asymptotically stable equilibria of the system, which can be seen as a refinement of the Nash equilibrium concept, as some of the Nash equilibria of the game may not be asymptotically stable [Weibull, 1997; Samuelson, 1998]. In particular, it can be shown that already the weaker criterion of Lyapunov stability results in Nash equilibrium solutions. It can be also shown that if x is an asymptotically stable equilibrium point of an evolutionary game with replicator dynamics, than it is a *perfect Nash equilibrium* of the game. Informally speaking, a Nash equilibrium is perfect if it is resistant to small errors committed by players (*i.e.*, players not playing Nash profile strategies).

In our case we consider an evolutionary game with 6 pure strategies. The payoff of player i who interacts with player j is denoted by π_{ij}. If $\mathbf{x} = (x_1, \ldots, x_6)$ is the frequency of each type in the population, the expected payoff to player i

is then

$$\pi_i(\mathbf{x}) = \sum_{j=1}^{6} x_j \pi_{ij},$$

and the average payoff in the game is

$$\bar{\pi}(\mathbf{x}) = \sum_{i=1}^{6} x_i \pi_i(\mathbf{x}).$$

The replicator dynamics for this game is then given by

$$\dot{\mathbf{x}}_i = \mathbf{x}_i(\pi_i(\mathbf{x}) - \bar{\pi}(\mathbf{x})). \tag{10.1}$$

Dynamics of simultaneous conditional cooperation game with monitoring. In our analysis, we assume the following mapping between the strategies and values of the index i:

$$((C, III), (C, RRR), (D, III), (D, RRR), (V, III), (V, RRR)) \\ \mapsto (1, \ldots, 6).$$

We assume that the frequencies of unconditional cooperators, unconditional defectors, and conditional cooperators in the system at the time t are $x_i(t)$, $i = 1, \ldots, 6$.

We shall consider first-order nonlinear differential equations of the form

$$\dot{\mathbf{x}} = \mathbf{f}(t, \mathbf{x}), \tag{10.2}$$

where t is a time variable ($t \in \mathbb{R}$), $\mathbf{x}$ is a vector, $\mathbf{x} \in \mathbb{R}^n$, and $\mathbf{f}$ is a vector-valued function $\mathbb{R}^{n+1} \rightarrow \mathbb{R}^n$. In the following we will consider only *autonomous differential equations*, in which the independent variable t does not occur explicitly, *i.e.*,

$$\dot{\mathbf{x}} = \mathbf{f}(\mathbf{x}), \tag{10.3}$$

where $\mathbf{x} = (x_1, \ldots, x_6)$ and $\sum_{i=1}^{6} x_i = 1$. Since the payoff matrix $\mathbf{A}$ of the game has a form presented previously, the replicator dynamics equation can be written as

$$\dot{\mathbf{x}}_i = \mathbf{x}_i((\mathbf{Ax})_i - \mathbf{x} \cdot \mathbf{Ax}). \tag{10.4}$$

Thus, we obtain a system of six first-order nonlinear autonomous differential equations.

The stability of a nonlinear dynamical system can be analysed by investigating the behaviour of the system in the neighbourhoods of its *equilibrium points* (also called *rest points* or *critical points*).

Analysing stability of open-loop strategies. Due to the fact that $x_1, \ldots, x_6$ are the frequencies of strategies, and the following constraint for the replicator dynamics

$$\dot{x}_1 + \dot{x}_2 + \dot{x}_3 + \dot{x}_4 + \dot{x}_5 + \dot{x}_6 = 0,$$

we only need the first five equations in order to study the dynamics of the system. Hence we can substantially reduce computational complexity of the analysis. It is worth to mention that the above constraint does not imply that the system should be always closed, *i.e.* it allows for a study of systems where the number of nodes leaving and joining the system are independent of the strategy type. This is a realistic assumption for many of the systems. In some other cases, such as *e.g.* an ad hoc network formed at in a train or at a bus, the assumption of a conservative system can be true at least in the meta-stable conditions.

After solving the right sides of the system equations defined by eq. (10.4) for equality to zero, we have found a set of 25 rest points. Then, we have linearised the system at the rest points by calculating the Jacobian matrices and their eigenvalues. As already stated, asymptotic stability of the equilibrium points require all eigenvalues to be negative. Additional constraints are introduced by the fact that x_i are the frequencies of the strategies in the system, *i.e.*

$$0 \leq x_i \leq 1 \quad \text{and} \quad 0 \leq \sum_{i=1}^{5} x_i \leq 1.$$

The final constraints are introduced by possible values of parameters, see also Table 10.3. Together they lead us to the set of only six equilibrium points, which can be asymptotically stable.

An equilibrium point of particular interest is the situation when all players use strategy (C, RRR), as it is an asymptotically stable equilibrium for any values of the parameters of our model. This is a very optimistic result, showing that a fully cooperative behaviour is always an equilibrium.

Another equilibrium point, e_2, corresponds to the situation when all player use strategy (V, RRR). The necessary and sufficient stability constraint on values of p and $\frac{m}{c}$ is depicted in Figure 10.6. There are no constraints on values of other parameters of the model.

Equilibrium point e_3 corresponds to a situation when all players use either strategy (D, RRR) or (V, RRR). It can also be shown that in this equilibrium point the frequency of defectors, *i.e.* players using (D, RRR), is always higher than the frequency of cooperators (*i.e.* players using strategy (V, RRR)). The necessary constraints for asymptotical stability of this equilibrium are depicted in Figure 10.7. The necessary stability constraints on values of p and $\frac{l}{h}$ are depicted in Figure 10.8. There are no constraints on values of other parameters of the model.

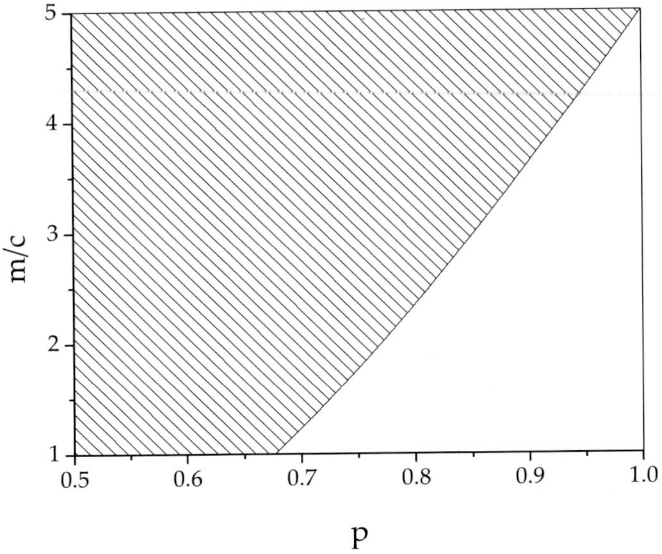

Figure 10.6. Necessary and sufficient relations between values of p and $\frac{m}{c}$ for which e_2 is asymptotically stable. The curve is representing $\frac{m}{c} = \frac{-3+2p+6p^2}{2-2p+p^2}$.

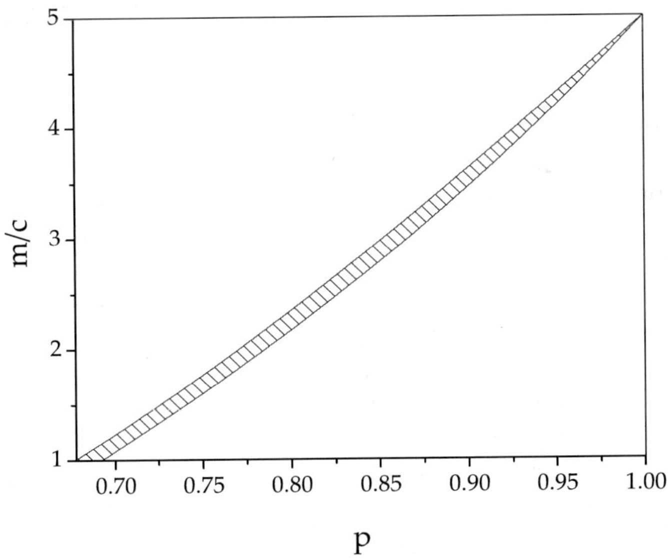

Figure 10.7. Necessary relation between values of p and $\frac{m}{c}$ for which e_3 and e_4 are asymptotically stable. The upper curve is representing $\frac{m}{c} = \frac{-3+2p+6p^2}{2-2p+p^2}$. The lower curve is representing $\frac{m}{c} = \frac{7-12p}{p-2}$.

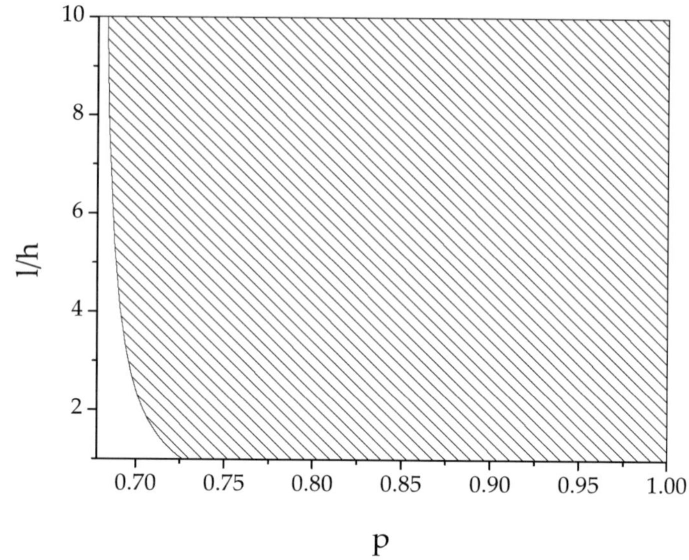

Figure 10.8. Necessary constraints on values of p and $\frac{l}{h}$ for which e_3 is asymptotically stable. The curve is representing $\frac{l}{h} = \frac{-2(p-1)}{-5+4p+5p^2}$.

A situation when all players use either strategy (D, III) or (V, III) corresponds to the equilibrium point e_4. It can be also shown that in this equilibrium point, similar to e_3, the frequency of defectors (*i.e.* players using (D, III)) is always higher than the frequency of cooperators (*i.e.* players using strategy (V, III)). The necessary constraints are depicted in Figure 10.7. The necessary stability constraints on values of $\frac{l}{h}$ can not be represented graphically in an easy way. There are no constraints on values of other parameters of the model.

Equilibrium point e_5 corresponds to a situation, where every player use one of three strategies: (C, RRR), (D, RRR) or (V, RRR). In this equilibrium point, frequency of defectors (*i.e.* players using (D, RRR)) is always lower than the frequency of cooperators (*i.e.*, players using strategy (C, RRR) or (V, RRR)). The necessary constraints are depicted in Figures 10.9 and 10.10. There are no constraints on the values of other parameters of the model.

Equilibrium point e_6 corresponds to a situation where every player use one of three strategies: (C, III), (D, III) or (V, III). In this equilibrium point, frequency of defectors (*i.e.* players using (D, III)) is always lower than the frequency of cooperators (*i.e.* players using strategy (C, III) or (V, III)). The necessary constraints are depicted in Figure 10.11. The constraints on values of l and h cannot be represented graphically in a simple form. There are no constraints on values of other parameters of the model.

From communications systems engineering point of view it is interesting to compare what are the requirements on the reliability of reputation information

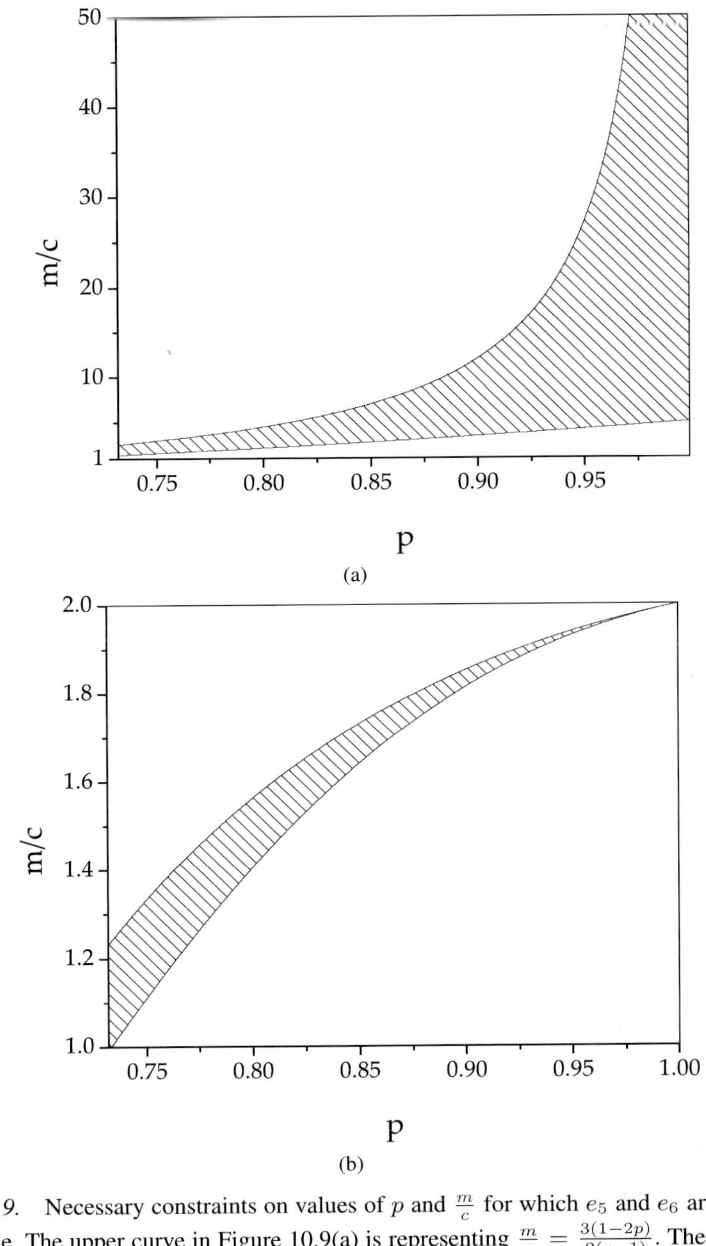

Figure 10.9. Necessary constraints on values of p and $\frac{m}{c}$ for which e_5 and e_6 are asymptotically stable. The upper curve in Figure 10.9(a) is representing $\frac{m}{c} = \frac{3(1-2p)}{2(p-1)}$. The lower curve in Figure 10.9(a) is representing $\frac{m}{c} = U_1$. The upper curve in Figure 10.9(b) is representing $\frac{m}{c} = U_2$. The lower curve in Figure 10.9(b) is representing $\frac{m}{c} = \frac{5p-3}{p^2}$.

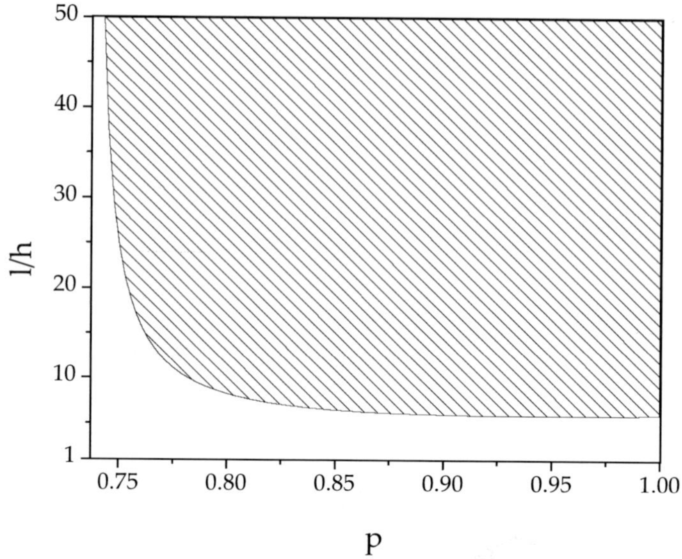

Figure 10.10. Necessary constraints on values of p and $\frac{l}{h}$ for which e_5 is asymptotically stable. The curve is representing $\frac{l}{h} = \frac{2+5p-13p^2}{5-9p+3p^2}$.

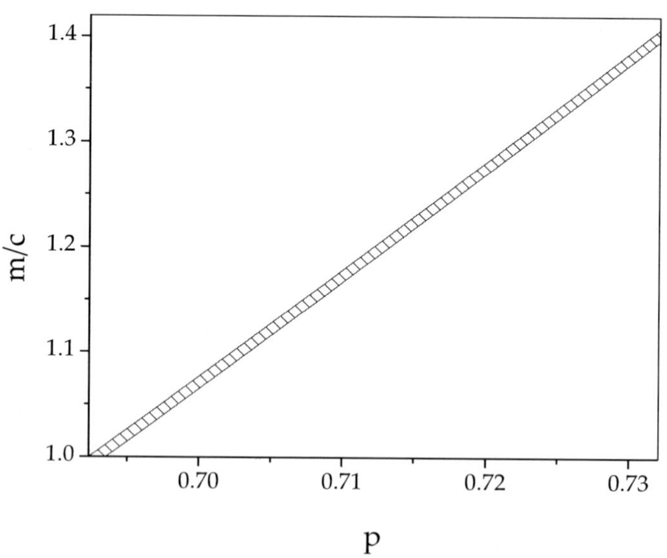

Figure 10.11. Additional necessary asymptotic stability constraints on p, m, and c for e_6. The upper curve is representing $\frac{m}{c} = \frac{7-12p}{p-2}$. The lower curve is representing $\frac{m}{c} = U_1$.

Table 10.4. Necessary asymptotic stability constraints on p.

Equilibrium point	Necessary constraints
e_1	$p \in \left(\frac{1}{2}, 1\right]$
e_2	$p \in \left(\frac{1}{2}, 1\right]$
e_3	$p \in \left(\frac{\sqrt{29}-2}{5}, 1\right]$
e_4	$p \in \left(\frac{\sqrt{29}-2}{5}, 1\right]$
e_5	$p \in \left(\frac{9-\sqrt{21}}{6}, 1\right]$
e_6	$p \in \left(\frac{9}{13}, 1\right]$

p, which are necessary for asymptotic stability of different equilibrium points (Table 10.4). The stability of the system is obviously something that is desirable, or even required in some cases. If we concentrate our discussion on a single design parameter p (*i.e.* reliability of reputation information available in the system), it is interesting to note that in the case of e_1 and e_2 (*i.e.* all nodes using either strategy (C, RRR) or (V, RRR)) the only constraint is $p > \frac{1}{2}$. Hence a practical engineering of systems with possibly full cooperative equilibriums does not put any extra hard constraints on the reliability of the employed reputation system. For example, in order to enable a partially cooperative equilibrium e_6 a much higher reliability of reputation information is required (*i.e.* $p > \frac{9}{13}$).

Simulating Cooperation in Ad Hoc Systems

An alternative to analytical modelling of cooperative communication systems is offered by simulations. Simulation models enable us to investigate more realistic models and take into account more variables and environmental features, which would make an analytical model intractable. In order to simulate the efficiency of these several strategic and structural solutions for inducing cooperation between selfish nodes, which have been discussed earlier, we have built a model using *RePast (REcursive Porous Agent Simulation Toolkit)*. RePast is a Java multi-agent simulation platform developed at University of Chicago [RePast, 2003]. We have chosen RePast after evaluating several possible simulation environments, due to its relative ease of implementation of complex behaviour and its extendibility. The other options included *cellular automata* [Wolfram, 2002], *CORMAS* (Common-Pool Resources and Multi Agent Systems) [Thébaud and Locatelli, 2001], and *Swarm* [Luna and Perrone, 2002].

Our simulation model covers two strategic, *i.e.* direct and indirect reciprocity, and two structural, *i.e.* tax/reward and micropayments, accountability mechanisms. In addition to the individual agent preferences, our simulation model

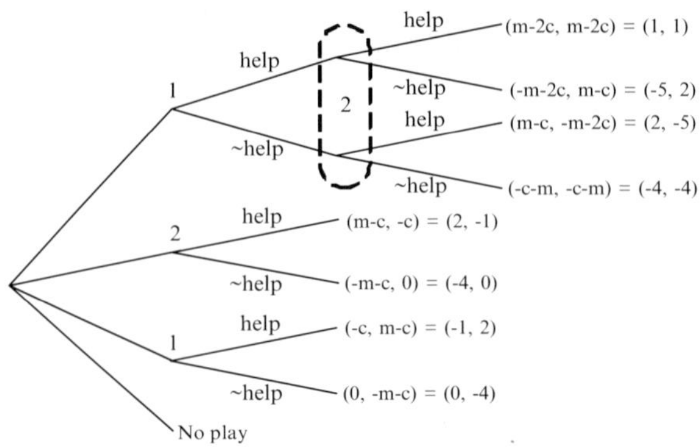

Figure 10.12. Bilateral IPD game used in simulation.

includes the three other main elements that affect emergence and maintenance of cooperation: *interaction processes*, *strategy space*, and *adaptive process*.

Interaction process. Our simulation model consists of a population of individuals with different strategies, placed on a 20 by 20 torus grid and playing bilateral IPD games with each other. We have decided to use a torus grid, as it makes it possible to simulate periodic boundary conditions. Also, implementation of nodes' mobility and of re-occurring individuals is easier when using a torus. The size of the torus has been chosen after some experimentation and we believe that it is large enough to model many realistic ad hoc networks. Very large scale ad hoc networks, such as ones used in the military context, are not considered explicitly. Nevertheless, we believe that it is unlikely that increasing the size of the population beyond some limit would introduce any new qualitative phenomena.

The initial population is generated by evenly distributing the agents throughout the strategy space. In each subsequent period each individual plays a small sample of other agents from the population - the exact number and selection procedure is determined by the interaction process being used. The number of iterations of the IPD two agents play is determined by a random variable. After all agents have played, agents are changed by the adaptive process being used for the run.

Cooperation is modelled as a bilateral cooperation game, using an Iterated Prisoner's Dilemma payoff matrix. The extended form of a single iteration of a game is depicted in Figure 10.12. The game start with randomly determining the need of every player for cooperation. In the case of ad hoc networks, this stage can be interpreted as a number of packets every of the player would like to send at the particular moment. In order to deliver the packets, node requires help (*i.e.* forwarding) by the other player. The number of packets for every player is a random variable with a Poisson distribution. Three cases are possible:

1 None of players have any packets to send, and there is no need for interaction at all.
2 The players have same amount of packets, and game turns into an n-stage simultaneous IPD game.
3 The players have different amount of packets, n and m, $n > m$ to be sent, and game consists of m-stage simultaneous IPD game plus $n - m$ stages of unilateral help by one of the nodes.

Context preservation and a *social structure* may play an important role in maintaining cooperation within a population of adaptive agents. We have decided to model two different aspects of context preservation: (1) various mobility patterns of agents, and (2) social network enabling the spreading of reputation information among nodes. The three simple cases of nodes mobility include:

No mobility: the fixed nodes interacting only with their Moore neighbours of range one. The *Moore neighbourhood* of range one of a grid cell is the set of 8 cells which share a vertex or edge with that cell. The Moore neighbourhood is also known as the *8-neighbours*.

Random walking: the nodes can after every step of simulation randomly decide to move n steps in any direction or stay in place.

Random jumping: the nodes can move to a randomly chosen place in the simulation space.

Strategy space. In our simulations we have modelled nodes implementing five different strategies: ALLC, ALLD, TFT, ATFT and SREP (Sporas reputation-based cooperators). In addition to typical defect and cooperate choices, all players have also a choice of temporarily or ultimately leave the system. After every round, each agent makes a decision if he should play or pause in the next round. The decision is based on the average utility $\bar{u}_i^j$ achieved by the node i during the round j. It also depends on the individual cooperation threshold t_i of the node i, *i.e.* the node makes a decision $d(j+1)$ about pausing or not in the round $j+1$ according to the following rule:

$$d(j+1) = \begin{cases} \text{play} & : \quad \bar{u}_i^j \geq t_i \\ \text{pause} & : \quad \text{otherwise} \end{cases}$$

Every decision to pause increases (by 1) the value of $V_i^e(j)$, denoting node's dissatisfaction from the repeating pausing. If $V_i^e(j)$ reaches a critical value V_{max}, the node decides to leave the system definitely. However, a forgetting process ϕ is modelled by decreasing the V_i^e after every round, *i.e.*,

$$V_i^e(j+1) = \delta V_i^e(j),$$

where $\delta = 0.75$, $V_i^e(0) = 0$, and $V_{\text{max}} = 2$.

In the case of SREP strategy, an action of a node is determined not by the result of the last interaction with other node, but by the reputation enjoyed by the other node in the system. Currently, in our simulations, we use *Sporas*. In its original version, Sporas provides a global reputation value for each member of an online community. However, we have modified the Sporas algorithm in order to accommodate also more distributed scenarios, *e.g.*, involving *supernodes* acting as *reputation hubs*. In Sporas, once a user has received at least one rating, his reputation will be higher than that of a newcomer and user is always worse off if he switches identities. However, the major limitation of the original Sporas is that it treats all the new users very unfavourably, assigning to them the lowest possible reputation. In our simulations we have experimented with different initial values of reputation in order to remove this economic inefficiency.

Adaptive process. Several different adaptive, or learning, processes have been proposed for use in game theoretic settings. The most commonly used are: *fitness-based imitation* and *conformism* with the strategy, which is most popular among the nodes met by the learning agent. We have modelled both of these adaptive processes in our simulation environment. In the case of utility-based adaptation, after every round r_j, every node n_i changes, with probability P_{adapt}, his strategy to the strategy of the most successful node n^*, belonging to the set of all nodes which were involved in an interaction with node n_i during the round r_j. The probability P_{adapt} has the same value for all nodes, and is constant during the whole simulation run. The success of every nodes is measured by an average individual utility $\bar{u}_i^j$ achieved by the node n_i during the round r_j. When copying with errors, two classes of errors are possible: *comparison errors*, which result in copying from the wrong agent, and *copy errors*, which result in making copying errors from the right agent. These copying errors can be included as parameter in the simulation model.

Modelling of individual nodes. The model of individual nodes consists of several parameters describing player's nature and his preferences, *i.e.*, player's fitness, initial endowment, cooperation threshold, and valuation of both shared and consumed resources.

Player's fitness, f_i, describes how *attractive* is the particular node for the other nodes. The more attractive the node the more satisfied are other nodes when it is present in their cluster/network. The exact definition of attractiveness depends on a particular peer-to-peer application. In general, it may mean popularity of a node as a possible receiver of the communication, *i.e.* some people act as *social hubs* receiving a disproportional large portion of network traffic. This might be caused by several reasons. For example, in file sharing networks, nodes may differ in regard to the variety and quality of the offered content.

In wireless ad hoc networks it might be bandwidth, routing paths composition, as well as transmission range and power that define attractiveness of the node to others. On the other hand, in distributed computing systems, the service quality, including speed and computations reliability, might be the deciding factor.

Individual player's fitness is a random variable with a Pareto distribution. Pareto distribution has been found to be a good approximation of both *wealth* and *social position* in human societies and therefore has been used in our model to describe players' attractiveness to each other, as well as their initial endowment.

Initial player's endowment describes the starting *capital* of a player when entering the collaborative communications system. This parameter, represented by a random variable with a Pareto distribution, can have multiple interpretations:

- initial amount of money in the case of micropayment solutions.
- initial resource level (*e.g.* battery power, disk space, etc). In this case it might be better to represent it as a vector.

Every nodes has his *individual valuation* of resources *shared* with other nodes and *consumed* from other nodes. This valuations are random variables with a Beta distribution. As it is not completely clear what distribution the valuation of shared and consumed resources has among the agents, the use of Beta distribution has enabled us to experiment with different probability distributions by just changing the appropriate parameters. The random variables V_i^s and V_i^c, describing shared and consumed values respectively as perceived by the node i, are used in the case of interaction based on micropayments. In this case interaction between nodes i and j takes place only if:

- node i requires help from node j and $V_i^c \geq V_j^s$ (*i.e.* the value, which node i is ready to pay for help is greater or equal to a value which node j demands for his help, or

- node j requires help from node i and $V_j^c \geq V_i^s$ (*i.e.* the value, which node j is ready to pay for help is greater or equal to a value which node i demands for his help.

The price to be paid for the service is set to be V^s of servicing node.

Cooperation threshold describes the minimal utility from all the games played in a single time step, which is expected by a player. Value of cooperation threshold in the agents' population has a Beta distribution. If player's utility in the period is lower than cooperation threshold, player decides to exit system for one round.

Player's *utility function* describes overall satisfaction of the node from the fact of participating in the network. The average utility of the node n_i received during the round r_j is denoted as $\bar{u}_i^j$ and is calculated as

$$\bar{u}_i^j = \bar{\pi}_i^j + s(C_i^j) + f(C_i^j) + \bar{e}_i^j,$$

where $\bar{\pi}_i^j$ is an *average payoff* of all games played by the node n_i during the round r_j, $s(C_i^j)$ is a *cluster size utility* of the node n_i, $f(C_i^j)$ is a *cluster fitness utility* of the node n_i, and $\bar{e}_i^j$ is an *average earning* of the nodes in the case of using micropayments strategy. *Clusters* are defined as groups of agents, where any two agents can be connected by a multi-hop connection. In the initial phase the world consists of a single cluster, but as agents decide to leave the system, it might be divided in several independent clusters. The cluster to which nodes n_i belongs during the round r_j is denoted as C_i^j. The *cluster size utility* function, $s(C_i^j)$, is defined as a relative size of the node's n_i cluster C_i^j during the round r_j when compared to the initial size of the world $|W^0|$. Since the initial world is a torus/square grid with a side length d, value of $|W^0|$ can be calculated as d^2:

$$s(C_i^j) = \frac{|C_i^j|}{|W^0|} = \frac{|C_i^j|}{d^2}.$$

The *cluster fitness utility*, $f(C_i^j)$, describes the relative fitness of the nodes participating in the cluster C_i^j, when compared to the accumulative fitness of the whole world, W^j:

$$f(C_i^j) = \frac{\sum_{n_k \in C_i^j} f_k}{\sum_{n_l \in W} f_l}.$$

The *average payoff*, $\bar{\pi}_i^j$ is calculated as an average payoff from the k games played by node n_i during the round r_j:

$$\bar{\pi}_i^j = \begin{cases} \frac{1}{k}\sum_{t=1}^{k} \pi_i^t & : \quad k > 0 \\ 0 & : \quad k = 0. \end{cases}$$

The *average earning*, $\bar{e}_i^j$ is used only when simulating micropayments - in all other cases it is set to zero. Average earning is calculated as an average change in the capital of node in the k games played by the node n_i during the round r_j:

$$\bar{e}_i^j = \begin{cases} \frac{\Delta c_i}{k} = \frac{c_i^{n+k} - c_i^n}{k} & : \quad k > 0 \\ 0 & : \quad k = 0. \end{cases}$$

Modelling dissemination of reputation and network topology. In order to model accurately the process of spreading reputation information in ad hoc systems, we should take into account the topology and communication patterns among nodes. The study of the topology and dynamics of various complex networks existing in real world has been recently a very active research area [Albert and Barabási, 2002; Barabási, 2002]. A large spectrum of naturally emerging networks has been studied in order to discover the common underlying principles in their topologies and evolution. It was discovered that almost all known complex network share three common features:

1 A short average path length, scaling logarithmically with the size of the network (*small-world* characteristic).

2 A large clustering coefficient (existence of well interconnected *cliques*), which is largely independent of the network size.

3 A power-law distribution of degree of the nodes (existence of very well connected *hubs*).

Various theoretical models have been proposed in order to capture the main features of existing complex networks. [Watts, 1999] has proposed a scheme for constructing small world graphs from regular lattice, by exposing every link in the lattice to a *rewiring* process, *i.e.* replacing, with some small probability p, the existing link with a new link to another, randomly chosen, node in the graph. This process can be interpreted, *e.g.* by observing that most people are friends with their immediate neighbours, but some people have also a few friends who are far away. The probability p is also called a *rewiring factor*. The networks build according to the Watts-Strogatz model are characterised by a short average path length and large clustering coefficient, however they have a relatively homogeneous topology, with all nodes having approximately the same degree.

Another approach to modelling complex real-life networks, so-called *scale free* model, has been proposed in [Albert and Barabási, 2002; Barabási, 2002]. The algorithm is based on *growth* and *preferential attachment*. Starting with a small number n of nodes, at every time step a new node with $m \leq n$ edges that link the new node to m different nodes already present in the network. When choosing the nodes to which the new node connects, the probability p that new node will be connected to node i depends on the degree d_i of node i. The scale-free model is characterised by a power-law distribution of nodes' degrees and a short average path length, but its clustering coefficient, although much larger then in the case of random graphs, is decreasing, according to a power-law, with the size of the network. Thus it is not fully consistent with experimental results and measurements. It is worth to mention that use of small-world and scale-free topologies in communication networks can have important security implications ([Albert et al., 2000; Barabási et al., 2003]), which should be taken into account, when designing a system.

In our simulation environment we have modelled two different cases of dissemination of reputation information:

1 *Global reputation.* This is an ideal case, when the reputation information is disseminated (almost) immediately and available to all the interested parties. This case can be interpreted as an existence of a central reputation server or of a global memory space where all nodes can store and access reputation ratings of other nodes.

2 *Local reputation.* This is a case where reputation is stored and processed by *supernodes*, *i.e.* a small amount of nodes acting as reputation hubs, storing and replying queries concerning a reputation of other nodes. Every node is connected to only one supernode. In our model there is no global synchronisation and supernodes do not share their reputation databases. Thus, if two nodes, reporting to two different reputation supernodes meet, they can not rely on reputation information concerning the opponent.

In the case of local reputation, we have assumed that reputation information is disseminated over a *social network* overlying the ad hoc communication network. The social network, over which dissemination of reputation information takes place, is fully independent of the physical topology of the ad hoc communications system. It can be seen as a *virtual network* modelling a subjective view of world, and trust relationships, as perceived by a particular node. We have modelled the exchange of reputation information over both a scale-free and a Watts-Strogatz small-world networks with various values of the *rewiring factor*. The small-world connectivity is obtained by applying a two step process [Watts, 1999]. In the first step we place all nodes in a regular lattice on the torus, with all nodes being connected to all their neighbours within *connection radius*, which is a parameter. In the second step some of the edges of the graph obtained in the step one are *rewired* to other, randomly chosen, nodes, thus introducing *shortcuts* in the network. The probability of rewiring is a model parameter, so-called *rewiring factor*. The scale-free network is built using a growth process [Albert and Barabási, 2002]. The best connected nodes in the neighbourhood are chosen to be reputation hubs. These are given the role of storing and calculating reputation values for the nodes connected to them. Thus, we avoid using a centralised server solution, at the same time reducing the computational and storage overhead put on individual nodes. Using hubs, forming a small-world network, as reputation repositories enable also a better reputation spreading in the case of highly mobile environments. This is a good example of how the research on network topologies and growth can have direct applications even to the systems security design.

4. Conclusions and Discussion

Stability analysis of cooperative systems is important and, as we have shown in this chapter, it is feasible in both analytical and simulation form. So far not much work has been done in this field and most of existing results have been accomplished in the last five years. Investigation of stability of cooperative systems is a promising and fertile ground for the further research. Seeing the behaviour of nodes as strategic games is an important insight, and one should not that this analysis can be done at the several layers of OSI stack.

We have proved that in our evolutionary model of cooperation in ad hoc communications systems, a cooperative behaviour is an asymptotically stable equilibrium in the case of dynamic systems with agents learning by imitation. We have also shown that in our model, a situation when almost all nodes cooperate in forwarding of packets and report other nodes' misbehaviour, is always an asymptotically stable equilibrium. The other cooperative equilibrium, where (almost) all nodes base their cooperation on reputation of other nodes and report experienced misbehaviour, is asymptotically stable for a wide value range of the system parameters (as depicted in Figure 10.6). On the contrary, the non-cooperative equilibria, where most of nodes fail to forward packets, are asymptotically stable only for a much more limited value range of system parameters (as depicted in Figures 10.7 and 10.8). These results have very optimistic character and suggest that, if well designed, ad hoc systems employing reputation-based and trust-based policies can lead to effective and self-organising communications networks. However, the required infrastructure may still include reputation servers, or super-nodes, which will perform aggregation and storing of reputation information. The high level of optimism is further justified by empirical results, which show that free-riding is not the first impulse of most people. In fact, already an existence of a punishment mechanism (*i.e.* conditional cooperators) can often induce a high level of cooperation between humans, without need for the mechanism to be used in practice.

References

Albert, R., Jeong, H., and Barabási, A.-L. (2000). Error and attack tolerance of complex networks. *Nature*, 406(6794):378–382.

Albert, Réka and Barabási, Albert-László (2002). Statistical mechanics of complex networks. *Reviews of Modern Physics*, 47(74).

Associated Press (2004). Altered egos: Identity theft is number-one consumer complaint. *Communications of the ACM*, 47(4).

Authenticode (1996). Microsoft authenticode technology. Available online at: http://msdn.microsoft.com/workshop/security/authcode/authwp.asp.

Back, Adam (2003). *The Hashcash Proof-of-Work Function*. The Internet Engineering Task Force (IETF). draft-hashcash-back-00.txt.

Barabási, Albert-László (2002). *Linked: The New Science of Networks*. Perseus Books.

Barabási, Albert-László, Dezső, Zoltán, Ravasz, Erzsébet, Yook, Soon-Hyung, and Oltvai, Zoltán (2003). Scale-free and hierarchical structures in complex networks. In *Conference Proceedings of The American Institute of Physics*, volume 661, pages 1–16.

Berkovits, S., Chokhani, S., Furlong, J., Geiter, J., and Guild, J. (1994). *Public Key Infrastructure Study: Final Report*. The MITRE Corporation.

Brands, Stefan (1999). *Rethinking public-key infrastructures and digital certificates–building in privacy*. MIT Press, Cambridge, Massachusetts, London, England.

Buchegger, Sonja and Boudec, Jean-Yves Le (2002). Performance analysis of the CONFIDANT protocol (Cooperation Of Nodes: Fairness In Dynamic Ad-hoc NeTworks). In *Proceedings of MOBIHOC'02*, EPFL Lausanne, Switzerland. ACM.

Buttyán, Levente and Hubaux, Jean-Pierre (2001). Nuglets: A virtual currency to stimulate cooperation in self-organized mobile ad hoc networks. Technical Report DSC/2001/001, Institute for Computer Communications and Applications, Department of Communication Systems, Swiss Federal Institute of Technology.

Buttyan, Levente and Hubaux, Jean-Pierre (2003). Stimulating cooperation in self-organizing mobile ad hoc networks. *ACM/Kluwer Mobile Networks and Applications*, 8(5).

Buyya, R., Abramson, D., and Giddy, J. (2001). A case for economy Grid architecture for service-oriented Grid computing. In *Proceedings of the 15th International Parallel and Distributed Processing Symposium (IPDPS'01)*, Los Alamitos, CA. IEEE Computer Society.

Buyya, Rajkumar, Abramson, David, Giddy, Jonathan, and Stockinger, Heinz (2002). Economic models for resource management and scheduling in Grid computing. *Concurrency and Computation: Practice and Experience*, 14(13-15):1507–1542.

Callas, J., Donnerhacke, L., Finney, H., and Thayer, R. (1998). *OpenPGP Message Format (RFC2440)*. The Internet Engineering Task Force (IETF).

Camenisch, J. and Lysyanskaya, A. (2001). An efficient system for non-transferable anonymous credentials with optional anonymity revocation. In *Advances in Cryptology – Eurocrypt'01*, volume 2045 of *Lecture Notes in Computer Science*, pages 93–118. Springer.

Chaum, D. (1985). Security without identification: Transaction systems to make big brother obsolete. *Communications of the ACM*, 28(10):1030–1044.

Chaum, David (1982). Blind signatures for untraceable payments. In Rivest, R. L., Sherman, A., and Chaum, D., editors, *Advances in Cryptology – Crypto'82*, pages 199–204, New York, USA. Plenum Publishing.

Chavez, Anthony, Moukas, Alexandros, and Maes, Pattie (1997). Challenger: A multi-agent system for distributed resource allocation. In Johnson, W. Lewis and Hayes-Roth, Barbara, editors, *Proceedings of the 1st International Conference on Autonomous Agents (Agents'97)*, pages 323–331, New York. ACM Press.

Chen, L. (1995). Access with pseudonyms. In Dawson, E. and Golic, J., editors, *Cryptography: Policy and Algorithms*, volume 1029 of *Lecture Notes in Computer Science*, pages 232–243. Springer.

Cole, Richard, Dodis, Yevgeniy, and Roughgarden, Tim (2003). Pricing networks with selfish routing. In *Proceedings of the 1st Workshop on Economics of Peer-to-Peer Systems*, Berkeley, California.

Cooper, D. A. (1999). A model of certification revocation. In *Proceedings of the 15th Annual Computer Security Applications Conference*, pages 256–264.

Cooper, D. A. (2000). A more efficient use of delta-CRLs. In *Proceedings of the IEEE Symposium on Security and Privacy*, pages 190–202.

Crowcroft, Jon, Gibbenst, Richard, Kelly, Frank, and Östring, Sven (2003). Modelling incentives for collaboration in mobile ad hoc networks. In *Proceedings of the Workshop on Modeling and Optimization in Mobile, Ad Hoc and Wireless Networks*, INRIA, Sophia-Antipolis, France.

Damgard, I. B. (1990). Payment systems and credential mechanism with provable security against abuse by individuals. In *Advances in Cryptology – Crypto'88*, volume 403 of *Lecture Notes in Computer Science*, pages 328–335. Springer.

Dingledine, R. and Syverson, P. (2003). Reliable MIX cascade networks through reputation. In Blaze, Matt, editor, *Financial Cryptography (FC'02)*, volume 2357 of *Lecture Notes in Computer Science*.

Dwork, Cynthia and Naor, Moni (1992). Pricing via processing or combating junk mail. In *Advances in Cryptology – Crypto'92*, pages 139–147.

Eastlake, D., Reagle, J., and Solo, D. (2001). *XML-Signature Syntax and Processing (RFC3075)*. The Internet Engineering Task Force (IETF).

Feigenbaum, Joan, Krishnamurthy, Arvind, Sami, Rahul, and Shenker, Scott (2001). Approximation and collusion in multicast cost sharing. In *Proceedings of the 3rd ACM Conference on Electronic Commerce, Tampa FL*. ACM.

Fudenberg, Drew and Tirole, Jean (1991). *Game Theory*. MIT Press.

Galvin, J., Murphy, S., Crocker, S., and Freed, N. (1995). *Security Multiparts for MIME: Multipart/Signed and Multipart/Encrypted (RFC1847)*. The Internet Engineering Task Force (IETF).

Gintis, Herbert (2000). *Game Theory Evolving*. Princeton University Press, Princeton, New Jersey.

Goldberg, Ian (2000). *A Pseudonomous Comunications Infrastructure for the Internet*. PhD thesis, University of California at Berkeley.

Goodrich, M., Tamassia, R., and Schwerin, A. (2001). Implementation of an authenticated dictionary with skip lists and commutative hashing. In *Proceedings of the DARPA Information Survivablity Converence and Exposition (DISCEX'01)*, volume 2, pages 68–82. IEE Press.

Heikkinen, Tiina (1999). A minimax game of power control in a wireless network under incomplete information. Technical Report 99-43, DIMACS, Piscataway, 08854 New Jersey.

Heikkinen, Tiina (2000). Resource allocation in a distributed network. Technical Report 2000-40, DIMACS, Piscataway, 08854 New Jersey.

Hofbauer, Josef and Sigmund, Karl (1998). *Evolutionary Games and Population Dynamics*. Cambridge University Press.

Housley, R., Ford, W., Polk, W., and Solo, D. (1999). *Internet X.509 Public Key Infrastructure Certificate and CRL Profile (RFC2459)*. The Internet Engineering Task Force (IETF).

Jain, Kamal and Vazirani, Vijay (2001). Applications of approximation algorithms to cooperative games. In *Proceedings of STOC'01, July 6-8, 2001, Hersonissos, Crete, Greece*. ACM.

JAR (1999). Jar file specification. Available online at: http://java.sun.com/j2se/1.3/docs/guide/jar/jar.html.

Kikuchi, Hiroaki, Abe, Kensuke, and Nakanishi, Shohachiro (1999). Performance evaluation of public-key certificate revocation system with balanced hash tree. In *Proceedings of the International Workshops on Security (IWSEC)*, pages 204–212.

Kleinberg, Jon, Papadimitriou, Christos, and Raghavan, Prabhakar (2001). On the value of private information. In *Proceedings of the 8th Conference on Theoretical Aspects of Rationality and Knowledge*.

Kocher, Paul (1998). On certificate revocation and validation. In *Financial Cryptography (FC'98)*, volume 1465 of *Lecture Notes in Computer Science*, pages 172–177. Springer.

Kodialam, Murali and Lakshman, T. V. (2003). Detecting network intrusion via sampling: A game theoretic approach. In *Proceedings of the IEEE INFOCOM'03*.

Kreps, David (1990). *A Course in Microeconomic Theory*. Prentice Hall.

Levine, John (2000). Why the Internet won't be metered – point-to-point and flat rates are king. IBM Developer Works: Startup Resources.

Luna, Francesco and Perrone, Alessandro, editors (2002). *Agent-Based Methods in Economics and Finance: Simulations in Swarm*, volume 17 of *Advances in Computational Economics*. Kluwer.

Lysyanskaya, A., Rivest, R., Sahai, A., and Wolf, S. (1999). Pseudonym systems. In *Selected Areas in Cryptography*, volume 1758 of *Lecture Notes in Computer Science*. Springer.

Marti, Sergio, Giuli, T. J., Lai, Kevin, and Baker, Mary (2000). Mitigating routing misbehavior in mobile ad hoc networks. In *Mobile Computing and Networking*, pages 255–265.

McDaniel, Patrick and Jamin, Sugih (2000). Windowed certificate revocation. In *Proceedings of the IEEE INFOCOM'00*, pages 1406–1414.

Menezes, Alfred, van Oorschot, Paul, and Vanstone, Scott (1997). *Handbook of Applied Cryptography*. CRC Press.

Micali, Silvio (1996). Efficient certificate revocation. Technical Report TM-542b, Massachusetts Institute of Technology, Laboratory for Computer Science.

Micali, Silvio (1997). Efficient certificate revocation. In *Proceedings of the RSA Data Security Conference*.

Micali, Silvio (2002). NOVOMODO: Scalable certificate validation and simplified PKI management. In *Proceedings of the 1st Annual PKI Research Workshop*.

Michiardi, Pietro and Molva, Refik (2003). A game theoretical approach to evaluate cooperation enforcement mechanisms in mobile ad hoc networks. In *Proceedings of the Workshop on Modeling and Optimization in Mobile, Ad Hoc and Wireless Networks*, INRIA, Sophia-Antipolis, France.

Muñoz, J., Forné, J., Esparza, O., and Soriano, M. (2003). Implementation of an efficient authenticated dictionary for certificate revocation. In *Proceedings of the 8th IEEE International Symposium on Computers and Communications (ISCC'03)*, pages 238–243.

Myers, M., Ankney, R., Malpani, A., Galperin, S., and Adams, C. (1999). *X.509 Internet Public Key Infrastructure, Online Certificate Status Protocol – OCSP (RFC2560)*. The Internet Engineering Task Force (IETF).

Naor, Moni and Nissim, Kobbi (2000). Certificate revocation and certificate update. *IEEE Journal on Selected Areas in Communications*, 18(4):561–566.

Ng, Chaki, Parkes, David, and Seltzer, Margo (2003). Strategyproof computing: Systems infrastructures for self-interested parties. In *Proceedings of the 1st Workshop on Economics of Peer-to-Peer Systems*, Berkeley, California.

Nisan, Noam and Ronen, Amir (1999). Algorithmic mechanism design. In *31 Annual ACM Symposium on Theory of Computing (STOC99)*.

Odlyzko, Andrew (2001). Internet pricing and the history of communications. *Computer Networks*, 36:493–517.

Oram, Andy (2001). *Peer-to-Peer: Harnessing the Power of Disruptive Technologies*. O'Reilly.

Perlines-Horman, Tomás, Wrona, Konrad, and Holtmanns, Silke (2001). Performance evaluation of signed content formats. Technical report, WAP Forum.

Perlines-Horman, Tomás, Wrona, Konrad, and Holtmanns, Silke (2006). Evaluation of certificate validation mechanisms. *Computer Communications*, 29(3).

Pfitzmann, Andreas and Köhntopp, Marit (2000). Anonymity, unobservability, and pseudonymity – a proposal for terminology. In Federrath, Hannes, editor, *Designing Privacy Enhancing Technologies: Proceedings of International Workshop on Design Issues in Anonymity and Unobservability*, volume 2009 of *Lecture Notes in Computer Science*, pages 1–9, Berkeley, CA, USA. Springer.

Ramsdell, B. (1999). *S/MIME Version 3 Message Specification (RFC2633)*. The Internet Engineering Task Force (IETF).

Rasmusson, Lars and Janson, Sverker (1999). Agents, self-interest, and electronic markets. *Knowledge Engineering Review*, 14(2):143–150.

Reagle, Joseph (1996). Trust in electronic commerce: The convergence of cryptographers and economists. *First Monday*, 1(2).

Reiter, Michael K. and Stubblebine, Stuart G. (1999). Authentication metric analysis and design. *ACM Transactions on Information and System Security*, 2(2):138–158.

RePast (2003). RePast Web site: http://repast.sourceforge.net/.

Resnick, Paul, Zeckhauser, Richard, Friedman, Eric, and Kuwabara, Ko (2000). Reputation systems: Facilitating trust on the Internet. *Communications of the ACM*, 43(12):45–48.

Rivest, Ronald (1997). *S-Expressions*. The Internet Engineering Task Force (IETF). draft-rivest-sexp-00.txt.

Ronen, Amir (2000). Algorithms for rational agents. In *Proceedings of the Conference on Current Trends in Theory and Practice of Informatics*, pages 56–70.

Samuelson, Larry (1998). *Evolutionary Games and Equilibrium Selection*, volume 1 of *MIT Press Series on Economic Learning and Social Evolution*. MIT Press.

Saraydar, Cem, Mandayam, Narayan, and Goodman, David (2002). Efficient power control via pricing in wireless data networks. *IEEE Transactions on Communications*, 50(2):291–303.

Seigneur, J.-M., Abendroth, J., and Jensen, C. (2002). Bank accounting and ubiquitous brokering of trustos. 7th CaberNet Radicals Workshop.

Stiller, Burkhard, Gerke, Jan, Reichl, Peter, and Flury, Placi (2001). A generic and modular Internet charging system for differentiated services and a seamless integration of the cumulus pricing schemes. *Journal of Network and Systems Management*, 9(3).

Szabo, Nick (1997). Formalizing and securing relationships on public networks. *First Monday*, 2(9).

Thébaud, O. and Locatelli, B. (2001). Modelling the emergence of resource-sharing conventions: An agent-based approach. *Journal of Artificial Societies and Social Simulation*, 4(2).

Tracz, Robert and Wrona, Konrad (2001). Fair electronic cash withdrawal and change return for wireless networks. In *Proceedings of the ACM Workshop on Mobile Commerce (MobiCom'01)*, Rome, Italy.

Urpi, A., Bonuccelli, M., and Giordano, S. (2003). Modelling cooperation in mobile ad hoc networks: A formal description of selfishness. In *Proceedings of the Workshop on Modeling and Optimization in Mobile, Ad Hoc and Wireless Networks*, INRIA, Sophia-Antipolis, France.

VeriSign (2004). Verisign update on certificate revocation list expiration. Press release.

Watts, Duncan J. (1999). *Small Worlds: The Dynamics of Networks between Order and Randomness*. Princeton Studies in Complexity. Princeton University Press, Princeton, New Jersey.

Weibull, Jörgen (1997). *Evolutionary Game Theory*. MIT Press.

Wohlmacher, Petra (2000). Digital certificates: A survey of revocation methods. In *Proceedings of the ACM Multimedia Conference*.

Wolfram, Stephen (2002). *A New Kind of Science*. Wolfram Media.

Wrona, Konrad (2005). *Cooperative Communication Systems*, volume 1 of *Aachener Beiträge zu Netzen und Systemen der Funkkommunikation*. Mainz Verlag.

Wrona, Konrad and Schuba, Marko (2001a). Mobile payments – state of the art and open problems. In *Proceedings of the 2nd ACM International Workshop on Electronic Commerce (Middleware'01)*, Heidelberg, Germany.

Wrona, Konrad and Schuba, Marko (2001b). Security for mobile commerce applications. In *Proceedings of the IEEE/WSES International Conference on Multimedia, Internet, and Video Technologies (MIV'01)*, Malta.

X.509 (1997). *Recommendation X.509: Technology – Open Systems Interconnection – The Directory: Authentication Framework*. International Telecommunication Union - Telecommunication Standardization Sector (ITU-T).

XKMS (2001). *XML Key Management Specification (DRAFT Version)*. VeriSign and Microsoft and webMethods.

Zacharia, Giorgos, Moukas, Alexandro, Boufounos, Petros, and Maes, Pattie (2000). Dynamic pricing in a reputation brokered agent mediated knowledge marketplace. In *Proceedings of the 33rd HICSS*. IEEE.

Chapter 11

POWER CONSUMPTION AND SPECTRUM USAGE PARADIGMS IN COOPERATIVE WIRELESS NETWORKS

The way out of the energy trap!

Frank H. P. Fitzek
Aalborg University
ff@kom.aau.dk

Persefoni Kyritsi
Aalborg University
persa@kom.aau.dk

Marcos D. Katz
Samsung Electronics
marcos.katz@ieee.org

Abstract: One of the most important and challenging topics for future wireless communications is the power consumption of wireless and mobile handheld devices. In Chapter 14 we underlined the importance of this issue and predicted that future terminals will consume even more power. The dramatic increase of power consumption can ultimately be attributed to the emergence of advanced services and it may result in an energy trap following the linear extension approach for future wireless communication systems. Therefore we address the problem of increased power consumption and introduce some solutions. In Chapter 18 the power consumption is also considered and solutions are given by means of task splitting in cooperative networks. Here we focus on the power that is consumed in the wireless transmission and reception process. Once more we advocate the use of cooperation, but this time the target is to reduce the power consumption for the receiving process of multicast services. The potential power saving using

Frank H. P. Fitzek and Marcos D. Katz (eds.), Cooperation in Wireless Networks: Principles and Applications, 365–386.

cooperation is not limited to multicast services, but they are used in this chapter for illustrative purposes.

Keywords: power consumption, energy, OFDM, TDMA

1. Motivation

As shown in Chapter14, the ever-increasing power consumption of future wireless terminals may be one of the most limiting factors for future wireless communication systems. New services as well as new transmission techniques will use more and more power. Moreover, future wireless communications are likely to take place at higher and less congested frequency bands, where path loss is larger, and higher transmission powers are needed to keep the same coverage area. In this chapter we introduce one possible solution to reduce the power consumption at the transmission level using cooperative strategies. As we will see in the following, the proposed solution is fully in line with the designing rule of 4G ($S_{4G} \sim 1/P_{4g}$) suggested in Chapter 14. For purely illustrative purposes, we present our approach in the context of multicast services supported by multiple description coding (MDC), but we stress that the concept is not limited to that particular application and it can be applied in any field of cooperative communication.

2. System under Investigation

The system under investigation follows the architecture of omnipresent cellular systems adding cooperation among wireless terminals. We assume the setup of Figure 11.1: A number of wireless terminals (WTs) are distributed over a given coverage area of an access point (AP). All terminals in a given group (A, B, or C) are interested in the same multicast service. The multicast service is provided using multiple description coding as described in Chapter 16. The MDC example is not a necessary condition, but it helps us to illustrate the idea. Furthermore MDC introduces robustness to the cooperative group allowing cooperative entities to leave and join the group whenever they want. Terminals may join or leave seamlessly the group without stopping the service for the other group members. The MDC sub–streams are transmitted to the terminals from the access point. We focus on J wireless terminals that are in close proximity and form a particular cooperative group. In general there are multiple cooperating groups with different numbers of cooperating terminals in each. For clarity, but without loss of generality, we assume that the number of transmitted sub–streams is also set to J. We investigate now different scenarios in terms of service quality and power consumption. We distinguish two possible operating strategies for the terminals:

- Autonomous Operation (No Cooperation)

  The J terminals do not cooperate and try to receive all J sub–streams in a stand alone fashion over the multicast downlink communication. This

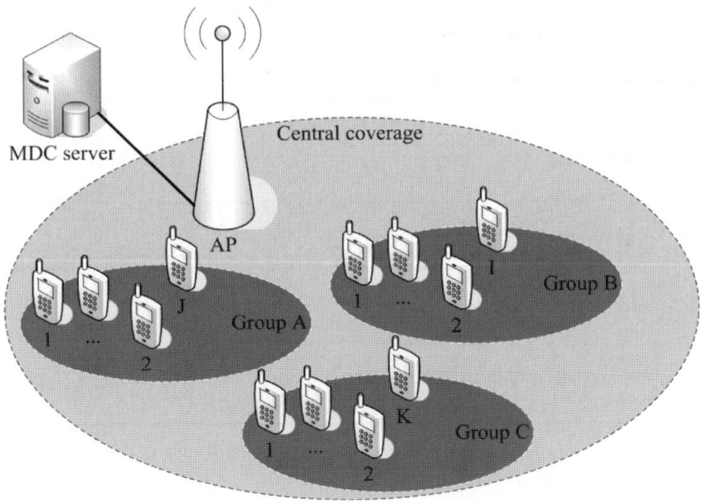

Figure 11.1. Example of cooperative groups with one central access point.

strategy is related to a certain service quality referred to as S_{NoCoop} and the power required to provide such a service P_{NoCoop}. P_{NoCoop} is the overall power consumed by a given terminal including cellular and short range communication.

- Cooperative Operation (Terminals Cooperate With Each Other)

  Each terminal receives one out of J MDC sub–streams, and the terminals within a cooperating group exchange these among each other. For this purpose the terminals need an additional connection to communicate with each other, and the connections will take place over *short–range communication* links. The target is to provide the same service quality as in the non–cooperative case ($S_{NoCoop} = S_{Coop}$). The power consumed in this case P_{Coop} may be different from P_{NoCoop}. We are only interested in the power consumption of the terminals without considering any power issues at the access point, which is assumed to be powered by a fixed line.

3. Time Division Multiple Access Cooperation

Our reference scenario is multicast downlink transmission, referred hereafter as Scenario 1. The MDC sub–streams are transmitted from the access point over a given radio interface at a given transmission rate R_c (where c stands for central or cellular) using the time division multiple access principle. The terminals receive the service at the same rate investing the power $P_{c,rx}$ (this power is consumed in the circuitry in order to perform all the operations necessary for the signal down–conversion and amplification, as well as the signal processing). In the following we indicate transmitted packets with solid lines and received

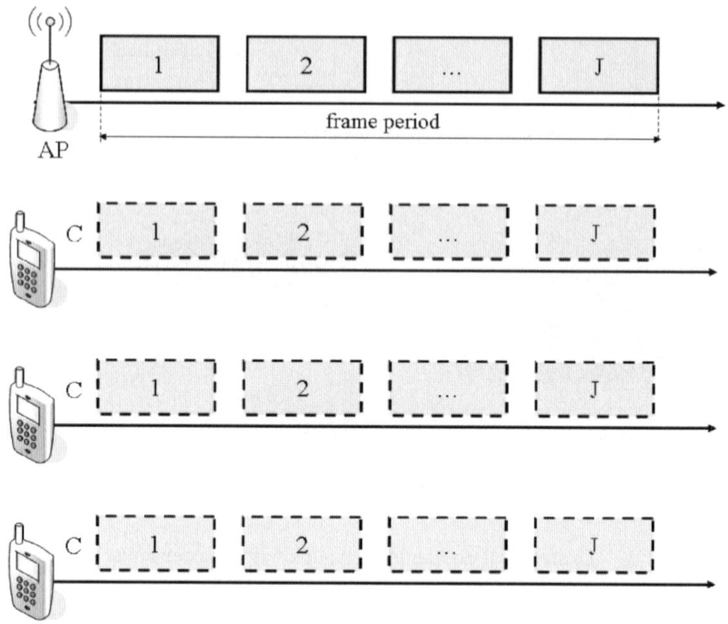

Figure 11.2. Scenario 1: Non cooperative reception of the J sub–streams.

packets with dashed lines. Figure 11.2 shows the non–cooperative transmission: the sub–streams are transmitted in a packetized form using time division. Within one *frame period* each sub–stream sends one packet as given in Figure 11.2. To receive the best service quality, each terminal needs to receive all packets of the sub–streams transmitted by the central air interface.

In the case of J cooperating terminals, only one packet per terminal per frame is received over the central air interface. The terminals have to agree on the disjoint reception of the J packets. For the exchange within the cooperating group we will consider two possible mechanisms. The first possible mechanism, referred to as Scenario 2, is shown in Figure 11.3 and assumes that the terminals are capable of using the central and the short–range communication interface at the same time. The second option (Scenario 3) assumes that only one radio link can be active for transmission or reception at any given time, and therefore activity alternates between the central and the short– range communication link, as shown in Figure 11.4.

In Figure 11.3 the first terminal receives one packet from the access point and forwards this packet within its cooperating group. The exchange of the cooperative packets need to be done in the next frame (after all J packets have been received). This has to be taken into account for the resulting delay, but this aspect is beyond the scope of this chapter. As all the other terminals do the same, the missing $J-1$ packets are received over the short–range communication link.

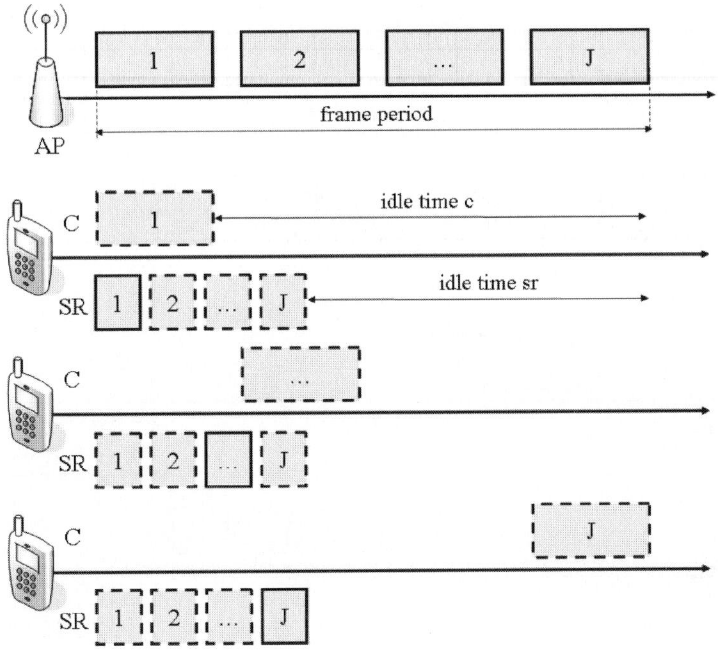

Figure 11.3. Scenario 2: Cooperative reception of the J sub–streams.

In order to compute the power levels, we assume that for the packet reception from the central access point a power level of $P_{c,rx}$ is needed. Furthermore power levels $P_{sr,tx}$ and $P_{sr,rx}$ are needed to send and receive a packet over the short range (sr), respectively. As given in the figures, the time to send or receive on the short–range is assumed to be shorter than the time needed to receive from the central AP. This is motivated by rate adapted systems such as IEEE802.11a/g. We inherently assume that higher rates can be achieved on the short–range link relative to the central link, because we expect that user proximity implies lower loss links among the members of a cooperative group. This assumption is essential for the success of the proposed cooperation scheme. In contrast to the example given in Figure 11.4, the central entity (AP) does not need to be aware of this kind of cooperation.

The cooperation scheme given in Figure 11.4 distinguishes two phases for the data transmission/reception. The first phase is dedicated to communication between the terminals and the AP. During this phase, the terminals receive the disjoint packets. The second phase is dedicated to inter-terminal communication, and it is during this phase that the exchange takes place. During this phase, the central entity stops its transmission and waits for the exchange to be completed. It is therefore obvious that the central entity needs to be aware of this kind of cooperation.

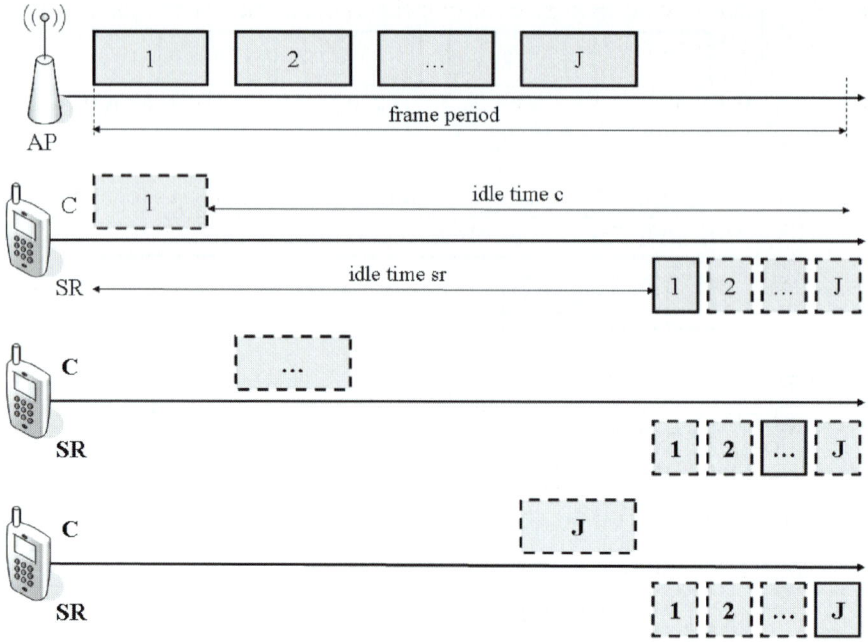

Figure 11.4. Scenario 3: Cooperative reception of the J sub–streams.

Homogeneous Cooperation Capabilities

In this subsection we focus on power consumption for cooperative terminals that have the same cooperative capabilities. We refer to this scenario as *homogeneous cooperation capabilities*. Let us first calculate the power consumption for autonomous (non–cooperating) terminals. The terminal receives always with the same power level P_{NoCoop}:

$$P_{NoCoop} = P_{c,rx} \tag{11.1}$$

To calculate the power consumption in the case of cooperating terminals, we assume that all possible rates among the cooperating terminals are the same. Later we will relax this assumption. The power P_{Coop} can be broken down into its contributions from the central communication (reception of one sub–stream and the related idle time) and the short–range communication (transmission of one sub–stream, reception of $(J-1)$ sub–streams, and a potential idle time). We define:

- $P_{c,rx}$ as the power (energy over unit time) consumed by the terminal for the reception on the centralized radio link, that is from the AP.

- $P_{c,i}$ as the power (energy over unit time) consumed when the radio link to the AP is idle.

- $P_{sr,rx}$ as the power (energy over unit time) consumed by the terminal for the reception on the short–range radio link.

- $P_{sr,tx}$ as the power (energy over unit time) consumed by the terminal for the transmission on the short–range radio link.

- $P_{sr,i}$ as the power (energy over unit time) consumed by the terminal when the short–range radio link is idle.

The total power consumed is

$$\begin{aligned} P_{Coop} = & \underbrace{c_{c,rx} \cdot P_{c,rx} + c_{c,i} \cdot P_{c,i}}_{cellular\ contribution} + \\ & \underbrace{c_{sr,tx} \cdot P_{sr,tx} + c_{sr,rx} \cdot P_{sr,rx} + c_{sr,i} \cdot P_{sr,i}}_{short\ range\ contribution} \end{aligned} \tag{11.2}$$

where

- $c_{c,rx}$ is the proportion of time spent on reception on the radio link to the AP.

- $c_{c,i}$ is the proportion of time the link to the AP is idle.

- $c_{sr,rx}$ is the proportion of time spent on reception on the short–range radio link.

- $c_{sr,tx}$ is the proportion of time spent on transmission on the short–range radio link.

- $c_{sr,i}$ is the proportion of time the short–range link is idle.

For Scenario 2, when the short–range link and the link to the access point are simultaneously active, the power consumed by the cooperative terminal is given as:

$$\begin{aligned} P_{Coop}^{Sc2} = & \underbrace{\frac{1}{J}}_{c_{c,rx}} P_{c,rx} + \underbrace{(1 - \frac{1}{J})}_{c_{c,i}} P_{c,i} + \\ & \underbrace{\frac{1}{J \cdot Z}}_{c_{sr,rx}} P_{sr,tx} + \underbrace{\frac{J-1}{J \cdot Z}}_{c_{sr,tx}} P_{sr,rx} + \underbrace{(1 - \frac{1}{Z})}_{c_{sr,i}} P_{sr,i} \end{aligned} \tag{11.3}$$

For Scenario 3, when the short–range link and the link to the access point are sequentially active, the power consumed by the cooperative terminal is:

$$P_{Coop}^{Sc3} = \underbrace{\frac{\frac{1}{J}}{1+\frac{1}{Z}}}_{c_{c,rx}} P_{c,rx} + \underbrace{\frac{1+\frac{1}{Z}-\frac{1}{J}}{1+\frac{1}{Z}}}_{c_{i,rx}} P_{c,i} + \underbrace{\frac{\frac{1}{J\cdot Z}}{1+\frac{1}{Z}}}_{c_{sr,rx}} P_{sr,tx} + \underbrace{\frac{\frac{J-1}{J\cdot Z}}{1+\frac{1}{Z}}}_{c_{sr,tx}} P_{sr,rx} + \underbrace{\frac{1}{1+\frac{1}{Z}}}_{c_{sr,rx}} P_{sr,i}. \tag{11.4}$$

We observe that the time required to send the same amount of bits is larger in Scenario 3 than in Scenario 2 by a factor $(1+\frac{1}{Z})$. This has implications with respect to delay, which is beyond the scope of this chapter. However, it might be useful to show the relationship between the total energy requirement for transmission in the various scenarios. For simplicity we normalize the energies E_{Coop}^{Sc2} and E_{Coop}^{Sc3} with respect to the energy consumed in Scenario 1, and define the efficiency ratios η_{Sc2} and η_{Sc3}.

$$\eta_{Sc2} = \frac{E_{Coop}^{Sc2}}{E_{NoCoop}} = \frac{P_{Coop}^{Sc2} T_{Coop}^{Sc2}}{P_{NoCoop} T_{NoCoop}} = \frac{\frac{1}{J}P_{c,rx}+(1-\frac{1}{J})P_{c,i}+\frac{1}{JZ}P_{sr,tx}+\frac{J-1}{JZ}P_{sr,rx}+(1-\frac{1}{Z})P_{sr,i}}{P_{c,rx}} \tag{11.5}$$

$$\eta_{Sc3} = \frac{E_{Coop}^{Sc3}}{E_{NoCoop}} = \frac{P_{Coop}^{Sc3} T_{Coop}^{Sc1}}{P_{NoCoop} T_{NoCoop}} = \frac{P_{Coop}^{Sc3}}{P_{NoCoop}}\left(1+\frac{1}{Z}\right) = \frac{\frac{1}{J}P_{c,rx}+(1+\frac{1}{Z}-\frac{1}{J})P_{c,i}+\frac{1}{JZ}P_{sr,tx}+\frac{J-1}{JZ}P_{sr,rx}+P_{sr,i}}{P_{NoCoop}} \tag{11.6}$$

Now we can apply different current technologies for the central and the short–range communication. In Table 11.1 we show the power levels and data rates motivated by measurements and report of [Atheros Communications, 2003]. Our investigations rely on the data rates provided by the physical layer of the IEEE802.11a or IEEE802.11g standard, as specified in [IEEE Std 802.11a, 1999] and [IEEE Std 802.11g, 2003], respectively. IEEE802.11a and 11g are based on Orthogonal Frequency Division Multiplex (OFDM), where multiple modulation schemes in combination with different coding rates are used. The combination of coding rates and modulation leads to multiple data rates starting at 6 Mbit/s up to 54 Mbit/s. The bit rate on a wireless link depends on the channel quality, which in turn depends heavily on the distance. Once again, let us underline our assumption that the data rate supported by the short–range needs to be larger than the data rate on the link to the AP in order for the proposed cooperative scheme to work.

Table 11.1. Parameters for the Analysis.

Description	Name	Value	Unit
Receiving power from central AP	$P_{c,rx}$	0.90	W
Power while idle	$P_{c,i}$	0.04	W
Receiving power over short–range	$P_{sr,rx}$	0.90	W
Transmitting power over short–range	$P_{sr,tx}$	2.00	W
Power for short–range while idle	$P_{sr,i}$	0.04	W
Rate for the central link	R_c	12.00	Mbit/s*
Rate for the short–range link	R_{sr}	54.00	Mbit/s

* the data rate was chosen to provide multicast transmission of the sub–streams; larger values would not allow the successful decoding of the data by terminals far away from the access point, if these terminals were to operate independently.

As a first result, Figure 11.5 shows the normalized energy consumed versus the number of cooperating entities for three different scenarios. In Scenario 2 each terminal has two WLAN network interface cards, while in Scenario 3 only one network interface card is needed. The energy consumed in the non cooperating case does not change and is normalized to unity, while the two cooperating strategies use less energy as the number of cooperating entities increases. A slightly larger power consumption is observed in Scenario 3 relative to Scenario 2. The power consumed in the cooperating case is approximately 50% of that in the non cooperating scenario for six cooperating terminals. It can also be noted from Figure 11.5 that the benefit from cooperation saturates with the number of cooperating terminals for both cooperating scenarios.

For a more detailed investigation we show how the total power can be broken into its components, *i.e.*, how much corresponds to each activity (transmission, reception or idleness) on each of the given air interfaces. Figure 11.6 illustrates the results for Scenario 2. For a large number of cooperating terminals the power spent on the receiving part from the central AP decreases. The power spent for the transmission in the short–range also decreases with the number of users. The power during the idle time on the central link and during reception of the sub–streams over the short–range communication increases as the number of cooperating users increases. The power for the idle time on the short–range link is constant since it only depends on the ratio of the achievable rates.

Generating Costs of Cooperating Sub–Streams

We note that, the separation of the original data stream into several sub–streams introduces an increase in the amount of data to be sent out. This kind of overhead is caused by the additional IP headers per sub–stream and

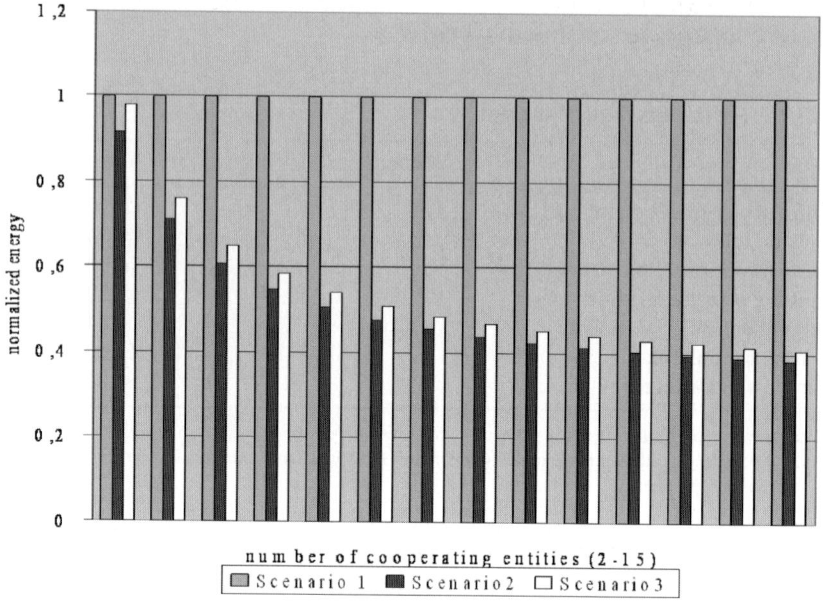

Figure 11.5. Normalized energy versus number of cooperating terminals for all three scenarios with two WLAN cards.

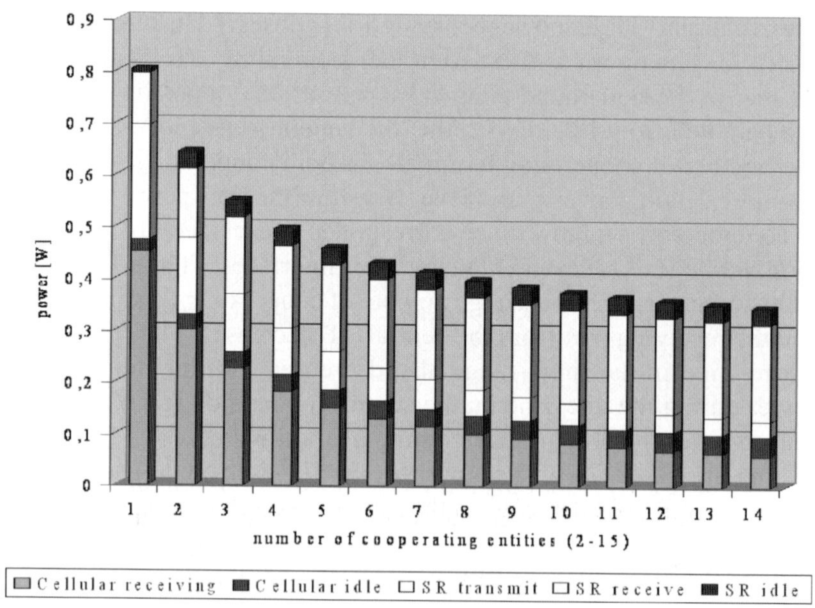

Figure 11.6. Detailed power consumption versus number of cooperating terminals for Scenarios 2.

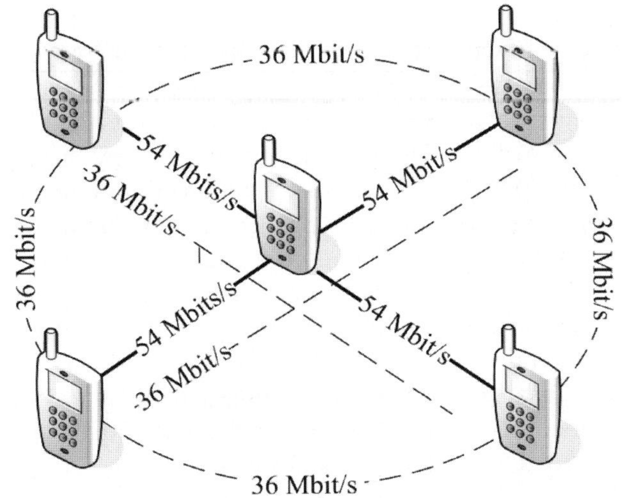

Figure 11.7. Example of an star configuration for the cooperating group with one terminal sending with 54 Mbit/s and others with 36 Mbit/s.

potentially by the encoding overhead. The IP overhead increases linearly with each additional sub–stream. The encoding overhead is more difficult to describe as it depends on the encoder used and the source which has to be encoded. In order to take this overhead into account, we should scale the above results by a function $f(J) \geq 1$, that depends on the number of data streams into which the original data are to be split (in our case this is also the number of cooperating users). Additional techniques such as header compression can help reduce the associated IP overhead (*i.e.*, reduce the value of the function $f(J)$) as given in Chapter 17.

Heterogeneous Cooperation Capabilities

So far we have assumed that the rate on the link between any two terminals within one cooperative group is the same for all of the members of the cooperating group. However, this might not always be the case. In the case of J terminals within one cooperative group, there are $J(J-1)/2$ possible links on the short–range communication with potentially different data rates. Using (11.3) or (11.4), we simply need to find out the minimum of all maximal available rates per link and set R_{sr} to that value. Doing so would still allow all cooperating terminals to successfully communicate, but this would also reduce the power savings dramatically. Therefore better strategies need to be found. To illustrate this, we consider next two possible examples.

Assuming a star configuration for the cooperative group as given in Figure 11.7, the terminal in the middle can communicate with the others at a rate of

54 Mbit/s, while all other communications have an achievable rate of 36 Mbit/s (clearly the achievable rate is the same in either direction of the communication link between any two terminals). In this first example we could derive a cooperation strategy whereby each terminal sends out its data over the short–range communication at a given rate. This rate is given by the minimum (over the links to the other $(J-1)$ terminals) of the maximum achievable rate on any link (dependent on the quality of the link). This may improve the situation compared to the initial scenario. The used power for cooperation differs now from that given in (11.3) as Z is not the same for all terminals and needs to be defined for each terminal individually. Furthermore the power saving gain also differs among the terminals.

If we assume a scenario where the short–range and the central link are allowed to operate simultaneously and the data rates on the link from the AP are all equal, then the power needed for terminal k is given by

$$\begin{aligned} P_{Coop,k} = \; & \tfrac{1}{J} \cdot P_{c,rx} + \left(1 - \tfrac{1}{J}\right) \cdot P_{c,i} + \\ & \tfrac{1}{J \cdot Z_k} \cdot P_{sr,tx} + \tfrac{1}{J} \cdot \textstyle\sum_{i=1, i \neq k}^{J} \tfrac{R_c}{R_{sr,i}} \cdot P_{sr,rx} + c_{sr,i} \cdot P_{sr,i} \end{aligned} \tag{11.7}$$

where

$$Z_k = \frac{R_{sr,k}}{R_c} \tag{11.8}$$

and the new value of c_{sr} (c_c remains the same) for Scenario 2 is

$$c_{sr,i} = 1 - \frac{1}{J} \cdot \sum_{i=1}^{J} \frac{R_c}{R_{sr,i}}. \tag{11.9}$$

The engineering cost of these more advanced schemes lays in the increased synchronization requirements among terminals.

The second example, given in Figure 11.8, is characterized by one outlying terminal. While four terminals could send to each other with a rate of 54 Mbit/s, the link to the exposed terminal has a maximum rate of 36 Mbit/s. In this example we have the aforementioned dilemma. The clustered terminals may agree to not cooperate with the exposed terminal to achieve the largest energy saving gain, which equals 0.608 for the normalized energy regarding Table 11.2 (we assumed a data rate of 12 Mbit/s from the access point and the energy levels given in Table 11.1). The exposed terminal would then be forced to receive the full information from the access point with the normalized energy level of 1.000. If all terminals decide to cooperate, they have to send at the common rate of 36 Mbit/s. In this case the related normalized energy level is 0.680. But in this example there are also intermediate levels of cooperation possible such that the clustered terminals agree to cooperate with the exposed one by sending two (they

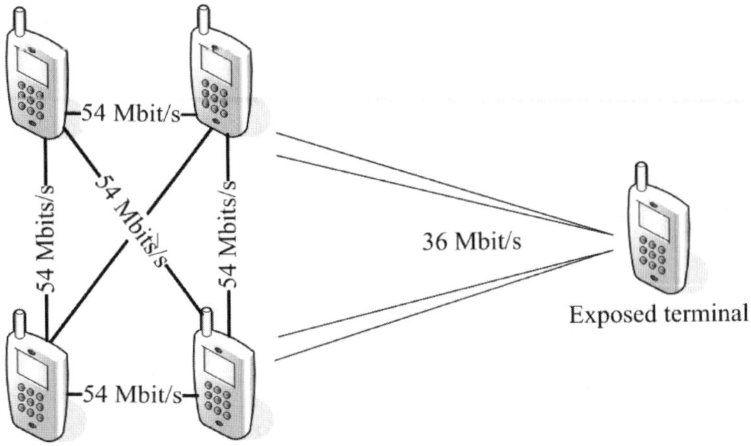

Figure 11.8. Cooperative group with some clustered and one exposed terminal.

want to be nice and generous, but any number may be motivated here instead of two) descriptors at a rate of 36 Mbit/s in exchange of the single descriptor from the exposed terminal. This solution has some charm as the exposed terminal will not get the full video quality and may change its position towards the clustered terminals. This behavior is well known for voice communications where users tend to move to achieve a better receiving position (such as the window in an indoor environment).

In Table 11.2 the scenarios with two and four exposed terminals are also given. In this scenario the exposed terminals communicate with a data rate of 36 Mbit/s with the clustered groups and among each other. For two exposed terminals an overall cooperation (six terminals with data rate of 36 Mbit/s) will require a normalized energy level of 0.634. Still the dilemma exists as the energy consumption level of the cluster group would be smaller (0.608). If the clustered group decide against cooperation with the exposed group, the exposed group would also not cooperate within each other as this cooperation would need more normalized energy than the self-sufficient (autarchic) central reception with 1.088 and 1.000 respectively. In the case of four exposed terminals the dilemma vanishes as the normalized energy level by overall cooperation (eight terminals at 36 Mbit/s) with 0.577 is lower than that of a cooperating group (0.608) with four members, aiming 54 Mbit /s among each other.

So far we have investigated scenarios with omnipresent technologies such as the well known wireless local area networks. Those technologies are not designed specifically to support cooperation, but we show that potential benefits of cooperation exist even with those techniques. Future technology such as ultra–wideband for short–range with high data rate will increase the potential of cooperative communication even more. Providing higher data rates in the

Table 11.2. Example1: Cooperation Matrix for the Clustered and Exposed Terminal.

Scenario		**Normalized Energy**		
		partial cooperation		full cooperation
cluster	exposed	cluster	exposed	
4	1	0.608	1.000	0.680
4	2	0.608	1.088*	0.634
4	4	0.608	0.748	0.577

*as this is larger than the stand alone power, the terminals may dismiss cooperation and receive directly from the access point.

cellular systems between base station and terminal always come along with increased costs in terms of power consumption and complexity. The exploitation of the short–range combined with cooperative techniques seems to be a promising way to support *virtual* high data rate. Instead of having two air interfaces for the short–range and cellular links, we highlight the potential of a unified air interface for short and cellular communication in the next section.

4. Orthogonal Frequency Division Multiple Access Cooperation

In the previous sections we assumed that the communication from the AP to the users and the communication among users happen over two different air interfaces. In this section we assume a common interface and that the two types of links are free to partition the bandwidth that is available to the system, and are allowed to access their respective parts of the spectrum at the same time. The transmission is based on a frequency division scheme, such as Discrete Multi Tone (DMT) or Orthogonal Frequency Division Multiplexing (OFDM). Let us assume that the total system bandwidth BW can be considered as a set of N_{sub} sub-carriers, each with a bandwidth $BW_{sub} = \frac{BW}{N_{sub}}$.

We assume that the access point (AP) allocates equal power on each subcarrier, so that the total transmitted power is P_t. Therefore the power allocated to each subcarrier is $\frac{P_t}{N_{sub}}$.

Let $g^{A \to B}(n)$ be the channel gain for the n-th subcarrier on the link from A to B. Due to channel reciprocity, we expect $g^{A \to B}(n) = g^{B \to A}(n)$. We distinguish the following types of gains:

- $g^{AP \to U_i}(n)$, which correspond to the links from the access point to the i-th user, and

- $g^{U_j \to U_i}(n)$, which correspond to the links between the j-th and the i-th users.

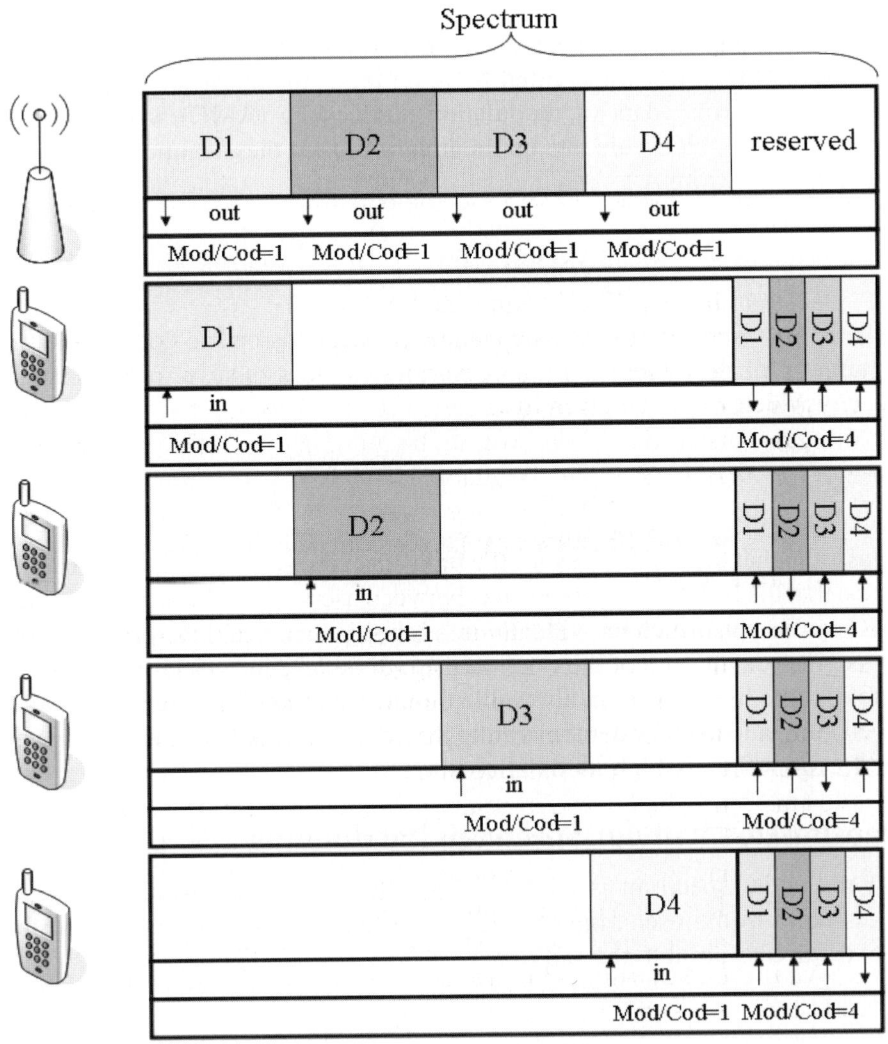

Figure 11.9. Spectrum partitioning: The base station sends out four data streams (D1, D2, D3, D4) on the downlink frequencies on the left. Each one is received by a different terminal, which in turn transmits the data stream to its neighbors over the short–range frequencies on the right, and receives the others on the rest of the short–range frequencies.

We expect that the difference in the length of the links between the access point and the users and the links between users will bias the channel gains such that

$$< g^{U_j \to U_i}(n) > \quad \geq \quad < g^{AP \to U_i}(n) >, \tag{11.10}$$

where $< \cdot >$ denotes the expectation of the argument $(\cdot)$.

Let P be the power transmitted by A on the n-th subcarrier. It is assumed that if A performs adaptive modulation and coding (AMC) with a view to maximizing the achievable rate on the link to B, then the maximum achievable rate on the link from A to B is

$$R^{A \to B} = log_2 \left(1 + \frac{1}{\gamma} \frac{P}{\sigma^2} g^{A \to B}(n) \right) \tag{11.11}$$

σ^2 is the receiver noise at B. The constant γ depends on the coding loss and the target probability of error, and describes the loss relative to the Shannon capacity which would be given by

$$C^{A \to B} = log_2 \left(1 + \frac{P}{\sigma^2} g^{A \to B} \right). \tag{11.12}$$

We assume that the coding loss for the links from the access point to the users is γ_c, and that the coding loss for the links between users is γ_{sr}. We also assume that AMC can be performed individually on each subcarrier, and therefore the total rate is given as the sum of the rates on the individual subcarriers. In a practical system, there is a maximum allowable modulation rate. This constraint would affect our results, but for demonstrating the concept, we assume that our system can decode infinitely high modulation sizes.

Transmission without Spectrum Partitioning

If the entire spectrum is used for the downlink communication from the access point to the users, then the total rate that can be achieved on the link from the access point to the i-th user is:

$$R^{AP \to U_i}_{cell\ only} = \sum_{n=1}^{N_{sub}} log_2 \left(1 + \frac{1}{\gamma} \frac{P_t}{N_{sub} \sigma^2} g^{AP \to U_i}(n) \right) \tag{11.13}$$

The total transmitted power and rates on each subcarrier are selected so that service can be provided to the users in the cell area with a certain outage probability.

In our case we investigate a group of users that can all access the broadcast channel independently, and we assume that the coding and modulation rates are selected so that all the users receive the same quality of service. They are therefore limited by the achievable rates on the link to the weakest user. The achievable rate on the n-th subcarrier is:

$$R^{dl}_{cell\ only}(n) = log_2\left(1 + \frac{1}{\gamma_{cell}}\frac{P_t}{\sigma^2}min_i\left(g^{AP\rightarrow U_i}(n)\right)\right) \tag{11.14}$$

and the total achievable rate is

$$R^{dl}_{cell\ only,total} = \sum_{n=1}^{N_{sub}} R^{dl}_{cell\ only}(n) \tag{11.15}$$

Transmission with Spectrum Partitioning

Let us now assume that the spectrum is partitioned into two blocks as given in Figure 11.10:

- The first block contains N_{cell} subcarriers and is used by the access point for cellular downlink transmission. Let $V^{U_i}_{cell}$ be the set of subcarriers allocated to downlink communication to the i-th user, and let $N^{U_i}_{cell} = |V^{U_i}_{cell}|$ be the number of these subcarriers ($N_{sr} = |V_{sr}|$). Then

$$\sum_{i=1}^{N_U} N^{U_i}_{cell} = N_{cell} \tag{11.16}$$

  where N_U is the number of users.

- The second block contains N_{sr} subcarriers and is used by the users for the short range communication among them. Let V_{sr} be the set of these subcarriers. This is further partitioned into N_U sets of the form $V^{U_i}_{sr}$, where N_U is the number of users. $V^{U_i}_{sr}$ is the set of subcarriers used by the i-th user for transmission on the short–range link, and let $N^{U_i}_{sr} = |V^{U_i}_{sr}|$ be the number of these subcarriers. The i-th user receives data from all the other $j \neq i$ users in the remaining $N_{sr} - N^{U_i}_{sr}$ subcarriers. Therefore

$$\sum_{i=1}^{N_U} N^{U_i}_{sr} = N_{sr}. \tag{11.17}$$

The system bandwidth is the same as before and therefore

$$N_{sub} = N_{cell} + N_{sr}. \tag{11.18}$$

We first concentrate on the links from the access point to the users. Following this scheme, the AP uses fewer subcarriers, and therefore the question becomes what happens to the transmit power, relative to the case of no spectral partitioning. Let $P_{t,cell}$ denote the transmit power from the AP. One option is that the total transmit power is kept constant ($P_{t,cell} = P_t$), and another is that the total transmit power scales according to the number of subcarriers used for downlink transmission, while keeping the power per subcarrier constant ($\frac{P_{t,cell}}{N_{cell}} = \frac{P_t}{N_{sub}}$).

Moreover, the AP has to decide how to allocate the available frequencies to the users, *i.e.*, how to partition the set of N_{cell} frequency subcarriers into N_U sets of various sizes. If that has been determined, the downlink rate to the i-th user on the n-th subcarrier ($n \in V_{cell}^{U_i}$) is

$$R_{dl}^{U_i}(n) = log_2\left(1 + \frac{1}{\gamma_{cell}}\frac{P_{t,cell}}{N_{cell}\sigma^2}g^{AP\to U_i}(n)\right), \tag{11.19}$$

and the total rate to the n-th user is

$$R_{dl,tot}^{U_i} = \sum_{n\in V_{cell}^{U_i}} R_{dl}^{U_i}(n). \tag{11.20}$$

Clearly the total downlink rate of transmission is

$$R_{coop,tot}^{dl} = \sum_i R_{dl}^{U_i}. \tag{11.21}$$

We observe that

$$R_{dl}^{U_i}(n) \geq R_{cell\ only}^{dl}(n) \tag{11.22}$$

because:

- The transmission is not limited by the minimum user gain:

$$g^{AP\to U_i}(n) \geq g^{AP\to U_i}(n), \tag{11.23}$$

- In the case where the total transmitted power from the AP is kept constant, we have more power available per subcarrier($\frac{P_{t,cell}}{N_{cell}} \geq \frac{P_{t,cell}}{N_{sub}}$).

Let us assume that the i-th user uses total power $P_{t,sr}^{U_i}$ for the cooperative transmission, and that this power is divided equally on all subcarriers in $V_{sr}^{U_i}$. This power might be the same for all users within the cooperating group for reasons of fairness, however we allow for the available transmit powers to vary among users. The AMC on each subcarrier in $V_{sr}^{U_i}$ is determined so that all the users in the cooperative group achieve the same rate. Therefore the maximum total transmission rate on the short–range communication link from the i-th user is:

$$R_{sr}^{U_i} = \sum_{n\in V_{sr}^{U_i}} log_2\left(1 + \frac{1}{\gamma_{sr}}\frac{P_{t,sr}^{U_i}}{N_{sr}^{U_i}\sigma^2}\min_{j\neq i} g^{U_i\to U_j}(n)\right). \tag{11.24}$$

The full problem involves the partitioning of the entire set of subcarriers into two sets (one for the downlink communication between the access point and the users, and one for the communication among the users), and further partitioning of each set into N_U sets, so that the conditions above are satisfied. The criterion for the optimal frequency partitioning can be the maximization of the achievable rate or the minimization of the total power. Clearly the complexity is prohibitive.

In this chapter we make some simplifying assumptions on the spectrum partitioning.

- A fixed percentage α of the total number of subcarriers is allocated to the communication on the short–range links ($N_{sr} = \alpha N_{sub}, N_{cell} = (1-\alpha)N_{sub}$). Figure 11.10 shows the cases, where α equals 0,0.5, and 0.75.

- The specific subcarriers that are allocated to each type of link are pre-determined. Figure 11.10 shows for example two different ways to allocate 75% of all subcarriers to the short range link.

- The specific subcarriers that are allocated to each user for each type of link are pre-determined (*i.e.*, the sets $V_{sr}^{U_i}$ and $V_{cell}^{U_i}$ are predetermined).

- All users are assigned the same number of subcarriers on both the link to the AP and the short–range link (the algorithm that determines what user is allocated which subcarrier is shown later).

The motivation for the use of cooperation in a scenario like this would be power saving. Clearly, if there is no penalty associated with data reception from the AP or on the short–range link, then the terminals have no motivation to cooperate and expend their power to transmit a data stream to their neighbors. Therefore we define the following costs:

- $P_{cell,Rx}(N)$ is the power consumed per bit received on the link to the AP. It is a function of the number of subcarriers used for the transmission. For example, one would expect the processing cost per bit to be reduced as the size of the Fast Fourier Transform (FFT) as in an OFDM system decreases. For simplicity, we assume it does so linearly.

- $P_{sr,Rx}(N)$ is the power consumed per bit received on the short–range link. It is also a function of the number of subcarriers used for the transmission.

- $P_{t,sr}^{U_i}$ is the power consumed for the transmission on the short–range link. It is determined by the battery level at each user terminal.

The users would be willing to cooperate if the following constraints are satisfied:

- The total achievable rate is the same in the cases with and without spectrum partitioning (otherwise they would just connect to the AP directly):

$$R^{dl}_{coop,tot} \geq R^{dl}_{cell\ only,tot}. \tag{11.25}$$

- The rate of transmission on the short–range link cannot be larger than the rate of reception on the link from the AP (the rates should be supported on both types of links).

$$R^{U_i}_{sr} \geq R^{U_i}_{dl}. \tag{11.26}$$

- There is a power benefit from the cooperation:

$$P_{cell,Rx}(N_{cell}) + (N_U - 1)P_{sr,Rx}(\frac{N_{sr}}{N_U}) + P^{U_i}_{t,sr} \leq P_{cell,Rx}(N_{sub}). \tag{11.27}$$

  Under the assumption that the power consumed for the reception of a data stream scales proportionately to the number of subcarriers used and that a fixed fraction α of the available subcarriers is allocated to the short–range communication (Figure 11.10 shows the cases, where α equals 0,0.5, and 0.75), this equation becomes

$$(1-\alpha) + \alpha\frac{N_U - 1}{N_U} + \frac{P^{U_i}_{t,sr}}{P_{cell,Rx}N_{sub}} \leq 1. \tag{11.28}$$

Given the definitions above, the achievability of all the constraints is a function of the available powers, the coding complexity, the noise level, and the channel gains from the AP to the users and among users.

As mentioned earlier, the optimal solution would allow the adaptive allocation of subcarriers to the various types of link.

Assuming that the number of subcarriers to be allocated to each user is known, then the optimal algorithm for the subcarrier allocation to the users for the downlink transmission is as follows:

Let us assume that we want to allocate N_{sub} subcarriers to N_U users, so that each one of them gets N_{sub}/N_U of them.

- Step 1: Initialization

  Define the set of users $S_U = \{U_1, U_2, \ldots, U_{N_U}\}$ and a set of subcarrier indices $B = \{1, \ldots, N_{sub}\}$, and construct a matrix G of dimensions $N_U \times N_{sub}$ such that $G_{i,j} = g^{AP\rightarrow U_i}(j)$. Also define N_U sets of the form $S_i = \{\}, i = 1, \ldots, N_U$.

  Set n = 1.

- Step 2: Find maximum.

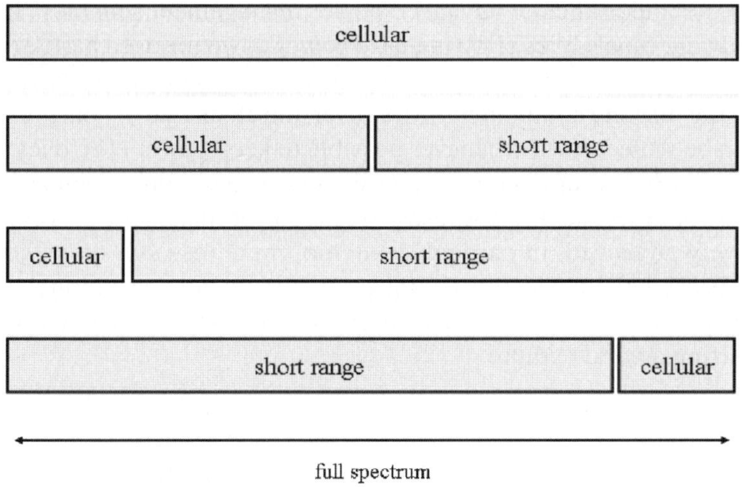

Figure 11.10. Examples where α equals 0,0.5, and 0.75.

Find $i \in S_U, j \in B$ such that $g^{AP \to U_i}(j)$ is maximum.

- Step 3: Allocate to user.

  $S_i \leftarrow S_i \cup \{j\}$

- Step 3: Exclusion

  If $|S_i| = N_{sub}/N_U$, then $S_U \leftarrow S_U - \{U_i\}$. Remove i–th row of the matrix G.

- Step 4: Advance

  $B \leftarrow B - \{j\}$. If $B \neq \{\}$, go to step 2.

A similar algorithm can be applied for the allocation of the subcarriers on the short–range communication link.

5. Conclusion

This chapter explored power consumption paradigms in cooperative networks. The purpose of the present research was twofold, first, to show the potential of power savings using cooperative information reception in two different widely used wireless technologies, and secondly, to introduce a OFDM–based common air interface, as we expect to be used in the 4G wireless communication systems. This kind of common air interface allows us to dynamically set the ratio between cellular and short link capacity. This is an important feature as the capacity on the cellular as well as on the short–range links depends significantly on the number of cooperating terminals. One of the main conclusions of this chapter is the importance of advanced power management schemes. These include hardware capabilities allowing us to power down unsused parts or turn-off not required functionalities. Sleep modes can be implemented on chip as well as in parts using discrete components. Moreover, the clock rate of some processing blocks can be scaled down whenever possible to keep power efficiency high. In addition to hardware, protocols need to be designed to allow dedicated switch off periods for power saving purposes. While power management technologies are relatively advanced, in particular on-chip, more research on power-aware protocols is needed. Indeed, most of the protocols are not considering dedicated switch off periods. Protocol design for cooperative networking is a important and promising area to explore.

References

Atheros Communications (2003). *Power Consumption and Energy Efficiency Comparison of WLAN Products*, white paper edition.

IEEE Std 802.11a (1999). *Wireless LAN Medium Access Control (MAC) and Physical Layer (PHY) specifications – High-speed Physical Layer in the 5 GHz Band*. IEEE Standard for Information technology.

IEEE Std 802.11g (2003). *IEEE Std 802.11g-2003*. Amendment to IEEE Std 802.11, 1999 Edn. (Reaff 2003) as amended by IEEE Stds 802.11a-1999, 802.11b-1999, 802.11b-1999/Cor 1-2001, and 802.11d-2001.

Chapter 12

COOPERATIVE ANTENNA SYSTEMS

From a Practical Channel Perspective

Patrick C. F. Eggers
Antennas and Propagation Division, Aalborg University
pe@cpk.auc.dk

Persefoni Kyritsi
Antennas and Propagation Division, Aalborg University
persa@kom.aau.dk

István Z. Kovács
Antennas and Propagation Division, Aalborg University
istvan.kovacs@ieee.org

Abstract: The practical performance of cooperative communication links is presented with respect to channel behavior and capacity potential. Measurements have been performed for different application scenarios, as well as idealized simulations of controlled synthetic generated channels. The evaluation is given with respect to capacity and data rate. We believe this chapter contains the first simultaneous common band (both transmit and receive) channel measurements.

The relative power level of the user-to-user cross link has paramount importance in the successful cooperation operation and consequently system performance heavily relies on sufficient power control ability. This poses particular difficulties for near field terminals (like hand-helds) and for short range channels where the base to user link budget still is comparable to the user-to-user link.

Conclusions are based on different system scenarios in narrow band, wideband and ultra wideband operation.

Keywords: antenna systems and algorithms, channel characterization, simultaneous full matrix channel sounding, capacity, practical cooperation.

Frank H. P. Fitzek and Marcos D. Katz (eds.), Cooperation in Wireless Networks: Principles and Applications, 387–419.

1. Introducing Antenna Cooperation

Cooperative operation of a wireless system is only feasible if cross-layer design is allowed. The efficiency of any cooperative scheme has strong interdependencies with all the layers of the OSI model. For example the link level cooperation involves aspects such as link control and dynamic resource allocation, while on the medium access layer cooperation introduces flexibility and poses strict requirements in terms of modem flexibility. The efficiency of cooperative schemes depends on hardware properties such as the transceiver dynamic range and the capacity and diversity potential of the antenna system. This chapter takes a realistic look at the performance potential of an antenna system if that uses cooperative transmission.

Definitions of Concepts and Parameters

The parameters most often used to describe the behavior of a radio link to a specific user equipment are:

- Channel Impulse Response (CIR). This describes the link state, and is often partially or fully available at either or both ends of the communications link in terms of channel state information (CSI).
- Noise and Interference levels. These are commonly determined in the context of the signal to noise ratio SNR, or the signal to noise plus interference ratio $SNIR$.
- Reliability/link robustness. This is closely related to the stability of the CSI.
- Terminal performance indicators. These are more hardware related parameters that reflect aspects such as power consumption, form factor etc.

The parameters above are related to the achievable/desirable bit error rate (BER) or frame error rate (FER) on that link. The most common performance measures used for the evaluation of an antenna system are:

- Capacity C. This is the maximum achievable data rate that can be transmitted over the link with an arbitrarily low probability of error. It depends on the CIR, the size of the system and the amount of CSI that is available at the transmitting and the receiving ends.
- Throughput R_b. This describes the actual achievable data rate over the communications link and differs from the capacity due to practical limitations, such as finite allowable modulation size for the data transmission or higher layer overhead requirements. This is the main parameter used for the evaluations in this chapter.

- Radio link quality (RLQ). This is evaluated based on the success rate of decoding a specified control, synchronization or signaling channel. This parameter can be directly used as a reliability measure of the analyzed radio link.

The motivation behind choosing cooperative transmission over single link operation, is that potentially one of more performance indicators can be improved. The focus here is on capacity and data rate.

In the following subsection we provide examples of existing radio system scenarios with enhancement potential if cooperative operation is used. The next sections 3 and 4 present some generic investigations into antenna system performance in a cooperative context.

Scenarios and Applications

The various communication environments can be classified according to different parameters such as bandwidth, range, morphology etc. These parameters tend to be strongly linked. We introduce the following classification of wireless systems where cooperative transmission can be used. We want to illustrate the fact that the motivation behind cooperative operation differs from one scenario to the other.

- Cellular systems with longer range coverage and low data rate.

  In this scenario cooperative transmission would probably be implemented as a piggy back short range cooperative links, in order to enhance the common throughput.

- Private Mobile Radio (PMR).

  In this case it is assumed that some terminals can have similar range to the cellular case, while others are only within a short range from each other, so as to operate in high stress safety/distress situations. In this scenario, the motivation behind cooperative transmission would be the improvement of the link reliability, whereas increasing the system throughput would be of a lesser concern.

  As a practical of a PMR system, the Terrestrial Trunked RAdio (TETRA) PMR system is chosen. The narrow band channel properties and channel measurement results are briefly discussed in Section 3.0, and the cooperative techniques are analyzed in Section 4.0.

- Wireless Local Area Network (WLAN).

  The inherent assumption for cooperative transmission in this scenario is that the terminals handle data and require internet access. In this case, both individual throughput as well as overall system throughput are of interest.

- Personal Area Network (PAN).

  This scenario is a heterogenous setup of varying application possibilities with varying data rate requirements. The main motivators for cooperation would be flexibility and low power consumption.

  For the characterization of the channel in a PAN scenario, multi-channel, multi-user, mobile-to-mobile ultrawideband (UWB) radio sounding experiments in several indoor scenarios were reported in [MAG, 2005]. These kinds of empirical data sets can be used for the analysis of the propagation channel in the more complex, ad-hoc communication scenarios typical to wireless PANs (WPAN) [MAG, 2005]. Furthermore, the use of dual antenna setup in the handset devices allows also the analysis of possible diversity and co-operative schemes. Due to the significant signal dynamics in the proximity of the human body, cooperative schemes could be potentially needed in order to enhance the reliability and provide higher overall system throughput.

  Using the UWB mobile-to-mobile radio channel properties presented in Section 3.0, the co-operative techniques which can potentially improve the reliability and capacity of the ad-hoc communication are discussed in Section 4.0.

  Moreover, sideband simultaneous cooperative channel sounding at 5.2 GHz, has been performed in the short range. This measurement scenario corresponds to a combination of the WLAN and WPAN situations described above. The details of the measurement campaign are given in Section 3.0. The user terminals were hand-helds associated to a PAN scenario, while the access point was placed on the ceiling as in a typical WLAN implementation.

- Sensor networks.

  In this case, the main motivation for cooperative transmission is the fact that sensor nodes are power limited. Cooperation is expected to contribute to power savings, as well as range extension over the monitored area.

An alternative classification approach would distinguish the cooperating networks with respect to the applications served by this system. For example, we can distinguish:

- Single user based applications such as personal voice/phone calls.

  In this case, the cooperation can be among users. Additional motivation should be provided to the ones that participate in the cooperation but do not benefit from its content. It can also be considered as cooperation among devices as in the BAN/ PAN scenario. Assuming that there is motivation for users to cooperate even if they do not get an immediate gain (*e.g.* better support for their individual application), the expected benefit from cooperation is an effective multi-user diversity operation. Therefore, we expect this technique to increase robustness against channel variations and CSI estimation.

- Multi user based applications such as broadcasting.

  In broadcast applications, a common content needs to be transmitted to a large number of users under strict quality requirements. This scenario provides an obvious environment where cooperation is expected to be beneficial, *i.e.*, it is expected to enhance the performance with a fractional link effort towards the broadcasting service. Therefore here the scope is higher total throughput/capacity at the user side (with the consumption of the same or lower resources such as power), when the direct non-cooperating link between a single user and the access point might or might not be insufficient. In the example of a WLAN scenario, an application that could be enhanced by cooperation is video streaming.

2. Antenna Systems and Algorithms : Foundations and Principles

Antenna Systems

Here we consider a terminal that can simultaneously support the cellular links and the short range support links. It is assumed that these have close proximity operation (possibly with the same transceiver). From the transceiver point of view, operation in neighboring bands with very high power level differences (much more than 20 dB) is very problematic. Consequently, it would also be beneficial if the antenna system itself could provide some sort, of duplexing action (possibly up to about 20dB). This would reduce the stress on the receiving part of the terminal in terms of return link energy feedback (clearly the problem is less severe if the major link and the short range link operate in a coordinated time-sharing fashion). For access point terminals, space separability is an obvious solution to this prolem. However, for more compact terminals where links are achieved through very closely spaced antennas (having different transceivers or capability of splitting the radio signal in two), or on the same antenna element (if we assume single transceiver operation), then the elements can help to provide some sort of separation between the ports. Here there are two obvious possibilities for this dual port discrimination:

- Polarization : for example with dual port patch antennas

- Mode excitation : by invoking different modes on different ports on the same physical patch [Vaughan and Andersen, 1984]

Both solutions are realizable for free space terminals (as in note-book terminals), but become much more difficult for near field loaded terminals such as hand-helds. With this in mind, new antenna designs must be sought with respect to particular casing and handling.

Another very important issue that affects hand-helds at 1800 MHz is a 7-10 dB drop in antenna efficiency with respect to free space, due to near field loading effects in normal handling [Pedersen et al., 2000]. At 5Ghz this might be worse. For the major link (cellular link), this loss appears at one end of the link, but appears at both ends when we look at links between mobile terminals. Thus, the short range links for this sort of terminals are particularly power handicapped, possibly loosing practical achievable gains (the power amplifier on the support links would need to be 'cranked up'). The link budget threshold for the short range link needs to be large enough to absorb this.

On top of the terminal antenna inefficiency, body blockage/shadowing can be very severe and abrupt, and it can completely dominate short range person to person links, when the users are using hand-helds. In this case even more link margin is required, further diminishing the gain potential of cooperation.

From an antenna point of view, more free space operated terminals such as note-book computers, may appear better suited for the first beneficial cooperative antenna deployments.

From the diversity and capacity point of view, the more spread the antenna elements are within the environment, the higher the expected gain potential. Here there is intuitively a large potential for cooperative operation. Most personal terminals have compact antenna systems that might provide micro diversity against short term fading. When it comes to long term/shadowing diversity and beneficial capacity gain through spatial multiplexing, both access point and user terminal antenna systems need to be in each other's 'near field' [Vaughan and Andersen, 2003]. However, when exploiting multiple different user terminals as one large antenna system, we get wide spatial spread antenna system to provoke 'near field' situations for shadowing diversity and capacity gain. This thoughs requires similar average power on all links. The difficulty is how to coordinate and share the power in the overall system operation. Particular difficulties appear when the objective is to increase the capacity, because that requires instant CSI from the complete system to provide decomposed eigen state information for all the links. Therefore it is very likely that heavy practical limitations will appear for such operation.

Rate Improvement with Relay Systems: Theoretical Analysis

In the following we assume that the purpose is to deliver an information content to a specific user. This can be done directly from the access point (AP) or through different users. We investigate the conditions under which relaying is beneficial and quantify the benefit in terms of the improvement in achievable data rate.

Theoretical results. We begin by making some fundamental definitions that we need for the theoretical analysis that follows. In a narrowband system where signal detection is impaired by additive white Gaussian noise (AWGN), the channel capacity (maximum data rate that can be transmitted with an arbitrarily low probability of error) is given by the well known Shannon formula:

$$C = log_2(1 + SNR) \tag{12.1}$$

where the signal to noise ration SNR is defined as

$$SNR = g\frac{P_t}{\sigma^2}. \tag{12.2}$$

P_t is the transmitted power, g is the instantaneous channel gain (amplification/attenuation) introduced by the physical channel, and σ^2 is the variance of the thermal noise at the receiver.

In a real life system, the achievable data rate is limited by the allowable amount of coding, the size of the transmitted data packets, and the allowable receiver complexity. We approximate the true achievable rate as

$$C = log_2(1 + \frac{1}{\gamma}SNR), \tag{12.3}$$

where the factor γ includes the effects of coding losses etc as in [Catreux et al., 2000]. The inherent assumption is that adaptive modulation and coding are performed. Clearly, the lower the target bit error rate performance, the more stringent the coding requirements are, and therefore the higher the loss factor γ.

We assume that we have an access point and two users, $UE1$ and $UE2$. Let g_1 and g_2 be the channel gains from the access points to the two users, and let g_R be the channel gain for the link between the two users (due to the channel reciprocity the channel gain is the same in either direction of the communication link). $UE1$ is the target user and downlink data needs to be communicated to it from the AP. Therefore the link between the AP and $UE1$ is referred to as the direct link.

We investigate the following possible communication scenarios:

- Direct communication with the AP.

  $UE1$ communicates directly with the AP. In this case, the maximum achievable rate for this cellular communication is

$$R_{cell} = log_2(1 + \frac{P_t g_1}{\gamma_{cell}\sigma^2}), \tag{12.4}$$

where γ_{cell} describes the coding loss for transmission from the AP to $UE1$, and P_t is the total transmitted power.

- Communication using the relay link.

  The data that is destined for $UE1$ is transmitted from the AP at a transmit power level $\lambda_c P_{cell}$, and $UE2$ receives and them. The maximum rate on the link from the AP to $UE2$ is

$$R_c = log_2(1 + \frac{\lambda_c P_{cell} g_2}{\gamma_c \sigma^2}), \tag{12.5}$$

  where γ_c is the coding loss for the transmission from the AP in this case and λ_c describes the percentage of the total transmit power P_{cell} that is expended on this step.

  We only discuss decode and forward scenarios, where the $UE2$ decodes the data and retransmits them to the target receiver $UE1$. A different mathematical approach would apply in 'amplify and forward' situations, where $UE2$ would only amplify the received data and retransmit them. The inherent limitation in that situation is that part of the amplified signal is noise, and its amplification and retransmission constitutes a waste of system resources.

  The $UE2$ regenerates the information and sends it to $UE1$ over the link between them, with a transmit power of $\lambda_r P_{cell}$. The maximum rate on this link is

$$R_r = log_2(1 + \frac{\lambda_r P_{cell} g_r}{\gamma_r \sigma^2}), \tag{12.6}$$

  where γ_r is the coding loss for this relay transmission, and λ_r describes the percentage of power that is expended on the relay transmission.

  We assume that the total network power consumed is the same as in both cases ($P_{cell} = P_t$), and therefore:

$$\lambda_c + \lambda_r = 1. \tag{12.7}$$

  Let η_c be the time percentage of time dedicated to the transmission from the AP to $UE2$. Then the rest of the time ($\eta_r = 1 - \eta_c$), the relay link is active. Clearly, the maximum achievable data rate that can be achieved in this cooperative mode of transmission is:

$$R_{coop} = \max_{\eta_c,\eta_r,\lambda_c,\lambda_r} \left[\min\left[\eta_c R_c, \eta_r R_r\right]\right], s.t.\lambda_c + \lambda_r = 1, \eta_c + \eta_r = 1. \tag{12.8}$$

  The minmax problem shown above achieves its solution for

$$\eta_c R_c = \eta_r R_r \tag{12.9}$$

The maximization of R_{coop} involves the optimization over the variables λ_c, η_c (λ_r, η_r can be uniquely determined from λ_c, η_c). The solution of this problem is not trivial, and we propose a simplified approach. We observe that for a known λ_c (and therefore λ_r), the optimal choice for η_c, η_r is

$$\eta_c = \frac{R_r}{R_r + R_c}, \eta_r = \frac{R_c}{R_r + R_c}, \tag{12.10}$$

and the corresponding achievable rate is

$$R_{coop}(\lambda_c) = \frac{R_r R_c}{R_r + R_c}. \tag{12.11}$$

The choice of fixed values for λ_c might be motivated by the amplifier capabilities of each terminal.

- Suboptimal solution through $UE2$.

  In a simplified approach, we can assume that the decision is made between the direct and the relay link for a given λ_c. We define this suboptimal solution as

$$R_{sel,\lambda_c} = \max\{R_{cell}, R_{coop}(\lambda_c)\}. \tag{12.12}$$

- Optimal routing.

  For any channel realization, the data to $UE1$ are transmitted over the best of the direct and the relay link, and a search is performed over a predetermined set of values for the power ratio λ_c. Clearly, the finer the search in λ_c, the closer the solution is to optimal. In that case the achievable rate is

$$R_{sel} = \max_{\lambda_c \in \Lambda}\{R_{cell}, R_{coop}\}, \tag{12.13}$$

  where Λ is the set of possible values for λ_c.

Clearly, the rate performance of any of the three schemes described above depends on various parameters such as the relative average channel gain of the direct and relay links, the difference in coding requirements on the two types of links and the channel statistics (distribution and correlations).

An interesting observation should be made with respect to the coding losses on each type of link. The probability of bit error at $UE1$ (if the communication happens through $UE2$) is

$$P_e \approx (1 - P_e(AP \to UE2))P_e(UE2 \to UE1) + P_e(AP \to UE2) \\ (\geq P_e(AP \to UE2)), \tag{12.14}$$

where $P_e((source) \to (destination)))$ is the probability of bit error on the link from the origin 'source' to the target 'destination'. In order for the relay link to provide sufficient protection, if $\gamma_c = \gamma_{cell}$, we expect $\gamma_r \geq \gamma_c$.

Simulation results. In the following we investigate the effects of the channel parameters on the achievable rate in systems with relaying using simulations of the wireless channel.

We assume that the transmit power is normalized so that the average signal to noise ratio on the links from the access point to the UEs is 10dB. Figure 12.1 shows the median capacity of the direct and the relay channels, assuming that these are perfectly independent. We show the median values for R_{cell}, R_{sel,λ_c} and R_{sel} if the relay node $UE2$ can select among the values $\lambda_c \in \Lambda_C = \{0.25, 0.5, 0.75\}$. Finally we also set $\gamma_c = \gamma_r = \gamma_{cell} = 5dB$. We can clearly observe from these results that relaying can offer a diversity benefit. When the average gain on the relay link increases relative to the average gain on the direct link, relaying becomes the preferred option. Moreover, as this difference increases, additional benefits are accrued by smaller power allocation on the relay link.

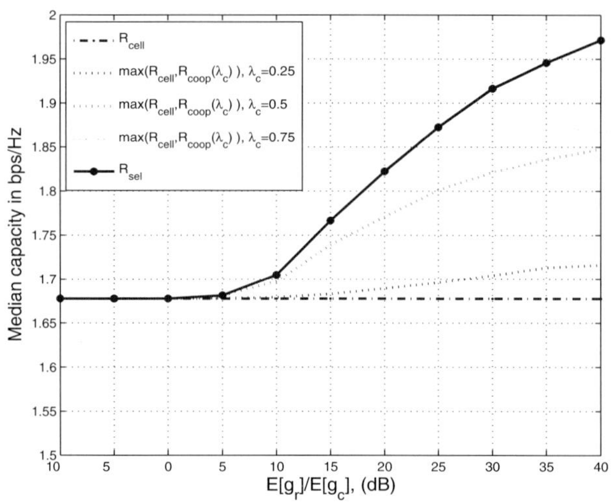

Figure 12.1. Median rate improvement with relaying.

Figure 12.2 shows the cumulative distribution functions of the achievable rate for independent direct and relay channels. For this figure too, we have assumed that the relay node $UE2$ can select among the values $\lambda_c \in \{0.25, 0.5, 0.75\}$, and that $\gamma_c = \gamma_r = \gamma_{cell} = 5dB$.

The steepness of the selection curves is due to the diversity benefit of relaying. Moreover, as the relative gain of the relay versus the direct link increases, the outage performance of the system improves.

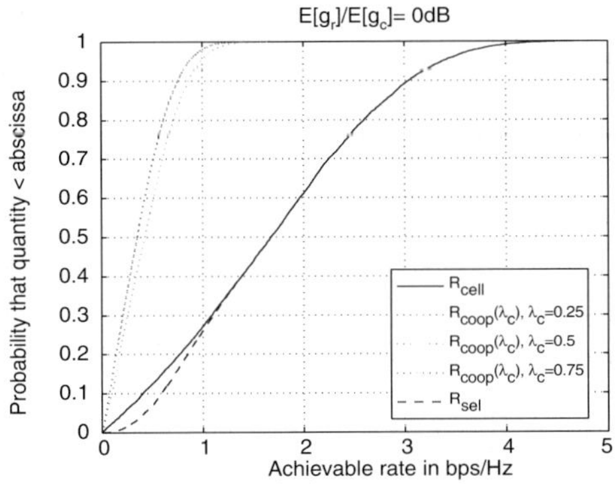

(a) $\frac{E[g_r]}{E[g_c]} = 0dB$

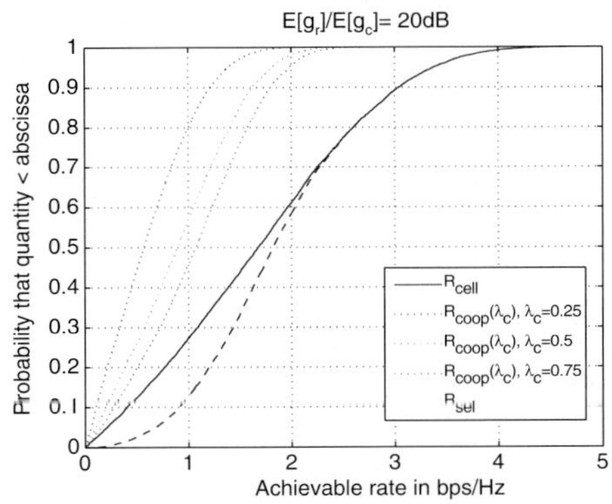

(b) $\frac{E[g_r]}{E[g_c]} = 20dB$

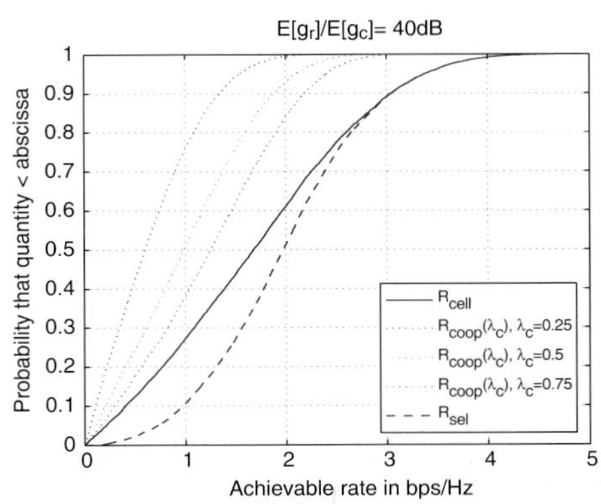

(c) $\frac{E[g_r]}{E[g_c]} = 40dB$

Figure 12.2. Cumulative distribution functions of the achievable rates for various relative channel gains.

Finally, we investigate the effect of varying the coding losses. Figure 12.3 shows the median value of the achievable rate for the direct and the relay channels for various values of the relative channel gains and relative coding losses. For this figure too, we have assumed that the relay node $UE2$ can select among the values $\lambda_c \in \{0.25, 0.5, 0.75\}$, and we have set $\gamma_c = \gamma_{cell}$.

We observe that as the coding requirements on the relay link become more stringent, a higher relative gain difference is required for the same rate performance.

In the context of investigating the effect of channel correlation, we have looked at situations where the links from the access point had power correlations up to 0.5. A similar range of values was investigated for the correlation between the relay link and the links from the access point. This range of values was motivated by the experimentally observed correlation values. However, our results have shown that for correlations up to 0.5 no significant degradation of the achievable rate is observed.

3. Channel Conditions, Measurements and Modeling: Practical Channels

Simultaneous Channel Sounding Principles

While radio link sounding can be performed in various traditional ways, the cooperative channel poses particular problems. Subchannel sharing among antennas on the same terminal (different links to the AP and among users) can cause huge differences in power levels on neighboring transmit and receive subbands. In order to perform simultaneous sounding for both transmission and reception on same the terminal, isolating measures need to be implemented.

Typical isolating elements (combiners/splitters, circulators, mixers, pin diode switches) give an isolation of around 20dB at 2-5 GHz. The very highest quality elements may have an isolation of up to 30-40 dB. Thus to achieve a minimum of 80 dB isolation between transmit and receive chains, cascading of isolating stages is needed.

Furthermore, leakage currents are a frequently encounteried problem with respect to achievable isolation. Thus, high standard RF implementation techniques need to be applied.

Characterization of Cooperative Channels in a Short Range Setup

The basic system for our measurements was the MIMO sounder of the Antennas & Propagation Division at Aalborg University [Kotterman et al., 2003]. It was rigged for cooperative link test measurements at 5.2 GHz, in a scenario resembling a small open office or internet café, see Figure 12.4. A common access point (AP) equipped with 4 $\lambda/2$ spaced monopoles and two user equipment

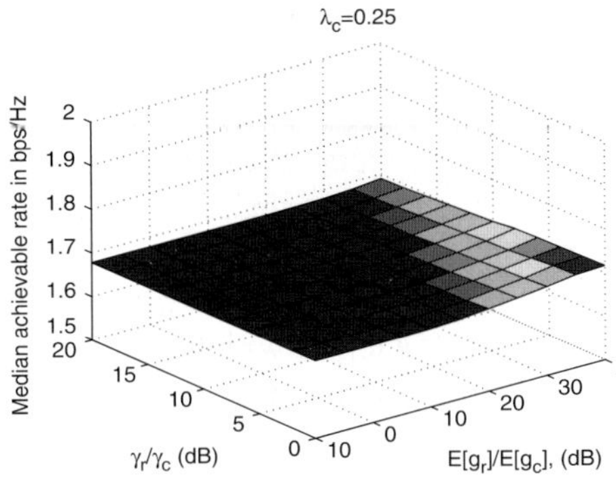

(a) $\lambda_c = 0.25$

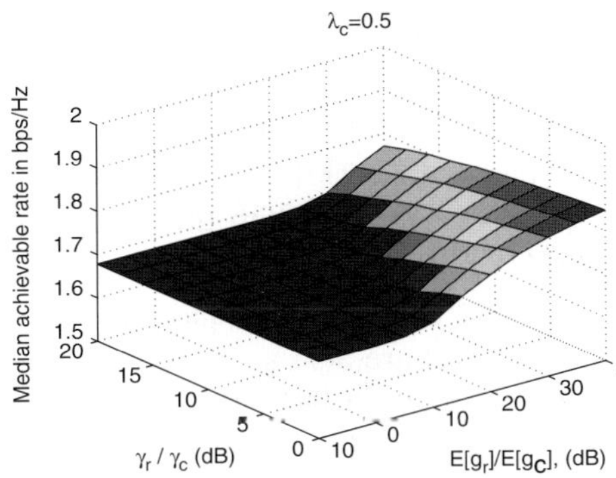

(b) $\lambda_c = 0.5$

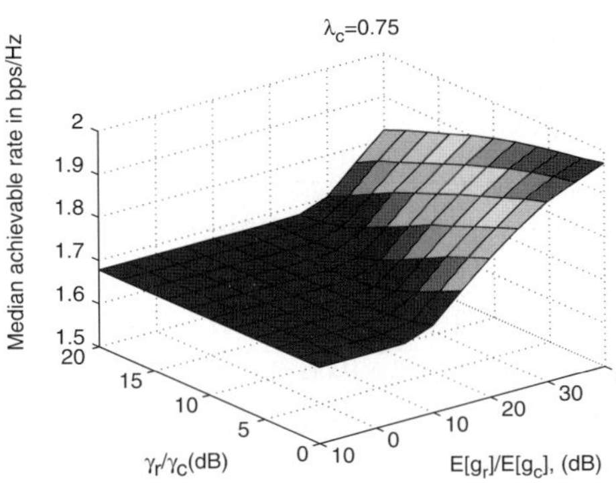

(c) $\lambda_c = 0.75$

Figure 12.3. Median achievable rate for various power allocations, relative channel gains and coding losses.

(UE) handsets with 4 patch elements around the rim on a $10 \times 5 \times 1$cm shell - formed a triangular connection. While the AP was transmitting (Tx) and UE1 was receiving (Rx), UE2 had both Tx and Rx capability with an RF isolation of 80 dB between TX and RX operation (with further sounding signal separation between TX and RX). As each terminal had 4 antennas, 4x4 multiple input-multiple output (MIMO) trunks appear between the terminals.

The scenario provides about equal range among all three terminals and is thus a particular difficult situation for cooperative operation, as normally cooperative benefits are associated with long range to the common point (AP) and very short ranges between user terminals.

The measurements were taken with two users moving along the tracks shown with dashed lines in Figure 12.4, each of which is approximately 8m in length. The persons were holding the terminals in a 'video mode' (terminal in front of the user bodies), while walking.

Figure 12.5 shows the mean power traces of the three 4x4 trunks along paths approximately 8m in length. What is immediately visible is that not only are the branch powers different between the links in the same 4x4 trunk, but the links between the UEs are allhave noticeably lower power (about 10dB) than the main trunks from the AP. The explanation most likely lies in the UE terminal operation on the user side: [Pedersen et al., 2000] found terminal attenuations of about 10dB relative to free space operation, when the terminal is used by a human user. In our case, the AP is placed on the ceiling and has free surroundings, while the handsets are subject to near field loading of the antennas due to the hands and torso of the users and are also subject to body shadowing. Thus the links between the UEs are subject to this disturbance at both ends while it only appears at one end for links to the AP.

This situation is very unfavorable for cooperative operation as the inter UE-UE links need a lot of extra power to become effective cooperative partners.

Table 12.1. Statistics of short term signal power correlations from the short range cooperative operation measurements.The mean traces of the actual channels are shown in Figure 12.5.

Links	$¡\rho¿$	σ_ρ	ρ_{min}	ρ_{max}
AP-UE1, AP-UE1	0.16	0.13	-0.12	0.54
AP-UE2, AP-UE2	0.13	0.15	-0.17	0.57
UE1-UE2, UE1-UE2	0.10	0.16	-0.20	0.55
UE1-AP, UE2-AP	0.00	0.07	-0.12	0.16
UE2-UE1, AP-UE1	0.01	0.16	-0.26	0.59
UE1-UE2, AP-UE2	0.11	0.11	-0.13	0.42

The short term fading distribution is shown in Figure 12.6. It is very similar to a Rayleigh distribution for the links in the three trunks. A slight tendency for the cross UE-UE links to have more dynamics is visible. This is expected as a mobile-to-mobile link in particular situations can exhibit double Rayleigh behavior ([Andersen and Kovács, 2002], [Kovács, 2002]). The short term link correlations are given in Table 12.4. It follows that all 4x4 trunks are practically decorrelated (three upper rows). What is interesting is that the three lower rows

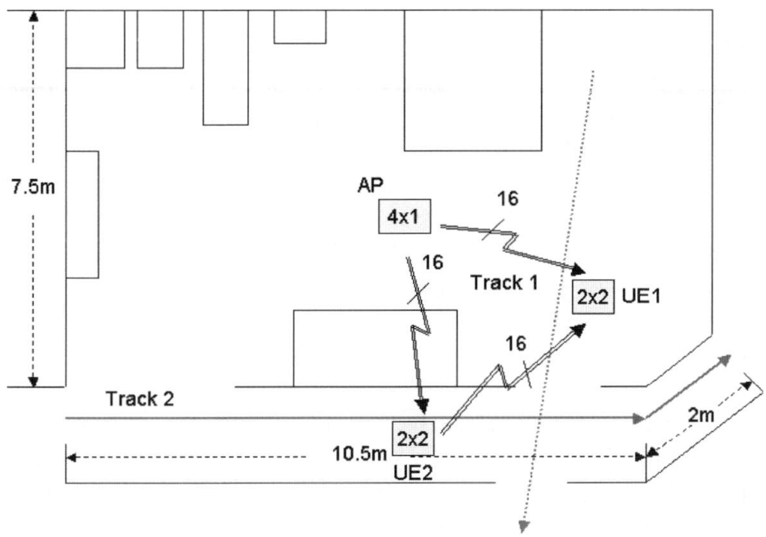

Figure 12.4. Sketch of indoor short range cooperative operation measurement scenario.

show more significant cross trunk correlations (one link from one trunk versus a link from another trunk). It appears that links having a UE as a common node have the tendency to exhibit slightly higher correlations than they would have if the AP were a common node. This can be explained by the near field loading on the UE side, which dominantly influences the channel state, no matter whether the link is associated with another UE or a AP at the other end.

Most of the theoretical work on relay and cooperative channels has been performed under very simplified assumptions for the channel properties. Namely all links have so far been assumed to be independent and of well known channel dynamics. However, the analysis above and the earlier discussion about small near field terminals shows that more details need to be included in the channel models, *e.g.*, joint versus disjoint shadowing, short term fluctuations, angular dispersion considered individually or for the whole group. Under this light, we present two more measurement campaigns that provide insight into the characterization of cooperative channels for different scenarios, and we use these measurement results to apply the theoretical algorithms.

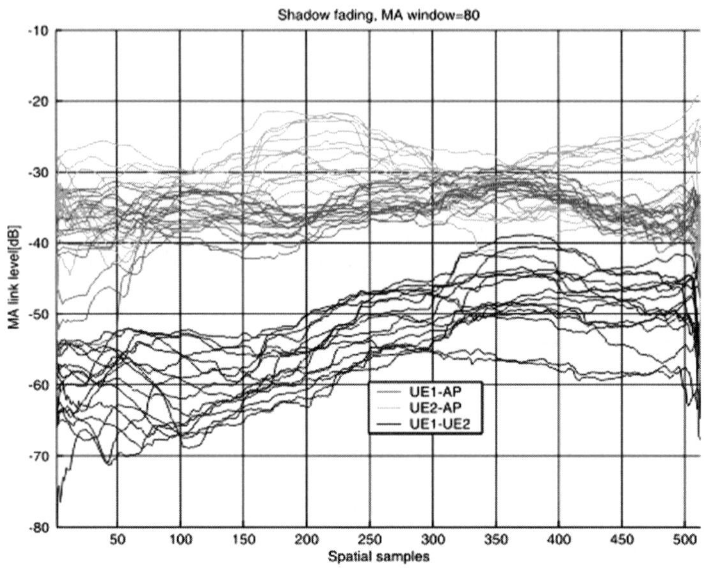

Figure 12.5. Shadow fading for all 4x4 trunks of the short range cooperative operation measurements.

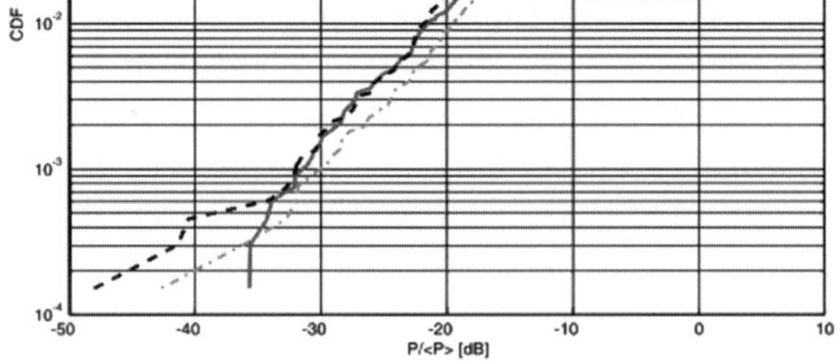

Figure 12.6. Short term signal envelope distributions of the short range cooperative operation measurements.

Public Mobile Radio - Mobile-to-Mobile Radio Channels

As the basis of our discussions for co-operative techniques in narrowband systems, we use the indoor-to-outdoor subset of the extensive radio channel investigations carried out in typical TETRA Direct Mode Operation (DMO) scenarios, which have been presented in [Kovács, 2002].

The narrowband radio links were measured simultaneously at a carrier frequency of 385MHz for the following modes of operation:

- *Mobile-to-Mobile Mode*: between two hand-held portables, one indoor (stationary/moving) UE1 and an outdoor (stationary/moving) UE2
- *Repeater Mode*: between an outdoor (stationary/moving) hand-held portable UE2 and a car mounted (outdoor), fixed terminal, CAR.

From the user/terminal dynamics (stationary/moving) point of view, four UE1-UE2 combinations and two UE1-CAR combinations have been investigated with this measurement setup. The UE1-UE2/CAR distances measured varied in the range of 15m to 50m, while the UE2-CAR distances were in the range between 5m to 10m.

All UE antennas used were linear polarized omni-directional, $\lambda/4$ antennas. The car mounted antenna was a vertically polarized low-profile antenna while the UE antennas were linear polarized but with random orientations due to the random user movement and handling of the equipment [Kovács, 2002].

Two main sets of outdoor-to-indoor measurements have been performed, corresponding to two different types of buildings: a 3 floor low-rise building and a 16 floor high-rise building. The indoor UE1 was moved to different floors of the buildings.

In our investigated DMO scenarios, the mobile terminals had low speeds, up to 3km/h. In combination with the low system bandwidth of 25kHz, this yields slow (relative to the TETRA symbol rate) and flat fading radio channels in all investigated scenarios. Furthermore, the measured DMO radio channels are characterized by a certain envelope distribution and auto/cross-correlation [Kovács, 2002]. The adaptive RF power control is available (optional) only in TETRA repeater/gateway operation and is not used in the TETRA UE-UE operation [ETS, 1999]. Thus, the local signal shadowing is another important factor at these slow mobile speeds, which contributes significantly to the link level performance degradation.

The average difference between the power levels measured with the car mounted antenna and the hand-held antenna, was found to be up to +10dB in the low-rise building scenarios and up to +7.5dB in the high-rise building scenarios.

The signal fading in the UE1/UE2-CAR radio channels can be well described with the classical Rayleigh/Rice models specified for TETRA DMO [ETS,

1999]. In contrast, in the dynamic UE1-UE2 scenarios when both mobile terminals were located in relatively dense scattering areas, the radio channel was shown to be significantly different and can be better described with a *multiple-Rayleigh* channel model [Andersen and Kovács, 2002], [Kovács, 2002].

Consequently, the link level performance analysis in [Kovács, 2002] also showed that, in scenarios where the classical channel models do not describe appropriately the radio propagation conditions, the predicted system performance is lower. Potential co-operative techniques could mitigate these channel effects and improve the overall system reliability.

Another important parameter for simple co-operative schemes, is the wideband power (RSSI) cross-correlation between the different radio links. The UE2-CAR radio channel has not been measured. Due to the scenario characteristics, a reasonable assumption for the UE2-CAR radio link is a Rician radio channel with a lower path loss and a very low signal power cross-correlation with the other two radio links, UE1-UE and UE1-CAR.

The obtained signal power cross-correlation for the measured UE1-UE2 and UE1-CAR channels showed values up to 0.6 for certain dynamic UE1-UE2 scenarios. Furthermore, in the high-rise building scenarios significant variations of the correlation coefficients (both envelope and wideband power) were determined for different building floors.

Indoor Short Range Ultra Wideband Channels

The dedicated measurement set-up used for the investigations presented in [MAG, 2005] allowed the full separation of all the simultaneously measured 16 radio links: 2 users (UE1 & UE2) x 2 TX antennas and 2 users (UE3 & UE4) x 2 RX antennas. The distances between the UE's varied in the range of 1m to 6m. The radio links between UE1-UE2 and UE3-UE4 have not been measured in this setup, thus certain assumptions have to be made when evaluating the co-operative schemes theoretically presented in Section 2.0.0.

On each user movement route, three terminal scenarios have been investigated emulating the *hand-held*, *PDA-held* and *belt-mounted* usage. These terminal usage cases differ in terms of radio propagation channel characteristics due to the different user hand/body proximity effects: almost free antennas in *hand-held*, partially covered antennas in *PDA-held* and antennas near a large dielectric body in *belt-mounted* scenarios.

The average differences between the power gain on the different radio links was in the order of 15dB, due mainly to different terminal scenarios and user movements. The distribution of the wideband power around the local average, corresponding to a measured bandwidth of 2.0GHz, was found to be well approximated with a log-normal distribution on all measured radio links, with an average standard deviation of 3.6dB.

The wideband power correlation results showed on average a low correlation level between the links belonging to different users, *i.e.* the links UE1/2-UE3/4, while high correlation levels (above 0.7) were measured between the radio links belonging to a pair of users, *i.e.* the links UE1-UE3, UE1-UE4, etc.. These results yield low diversity gain with the dual antenna terminals and relatively high multi-user diversity potential. The latter is beneficial for the performance of the co-operative techniques described theoretically in Section 2.0.0.

4. Radio Systems : Performance Investigation

Capacity of Short Range Cooperative Channels

The overall potential of the 4x4 MIMO trunks and the possible cooperative joint 4x4 MIMO (2 links from AP-UE1 and 2 links AP-UE2) have been investigated with respect to total capacity. Here three different power schemes have been used:

- No power control: the global level is adjusted so that the mean of the strongest trunk has power 1, see Figure 12.7,
- Moving average (MA) mean level power control compensation,: this would be typical in TDMA and FDMA systems, see Figure 12.8 and,
- Instant power control (limited to 40dB range in 1dB steps): this would be used in CDMA systems, see Figure 12.9.

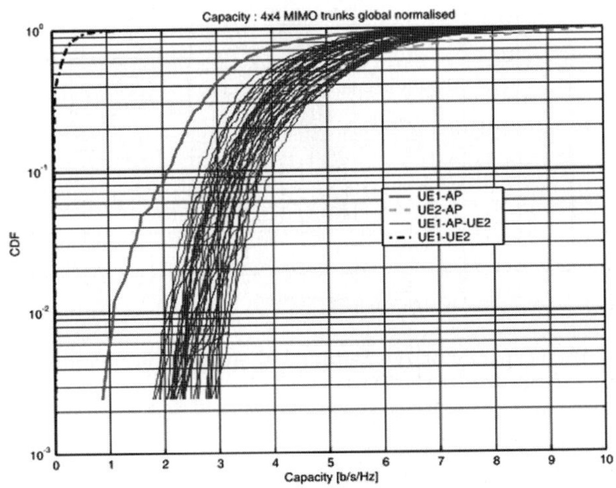

Figure 12.7. Total capacity of 4x4 MIMO with no power control.

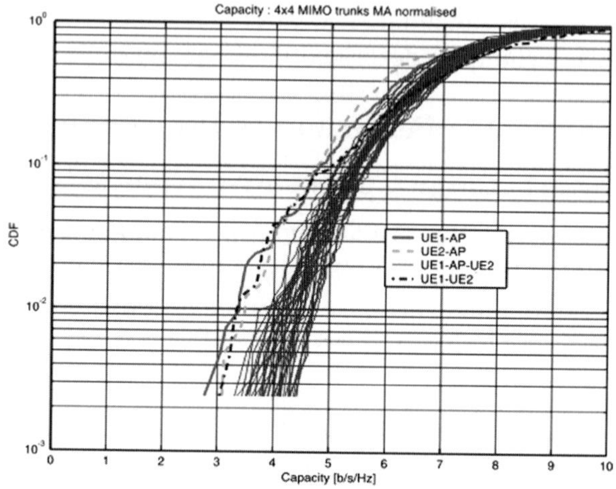

Figure 12.8. Total capacity of 4x4 MIMO with MA power control.

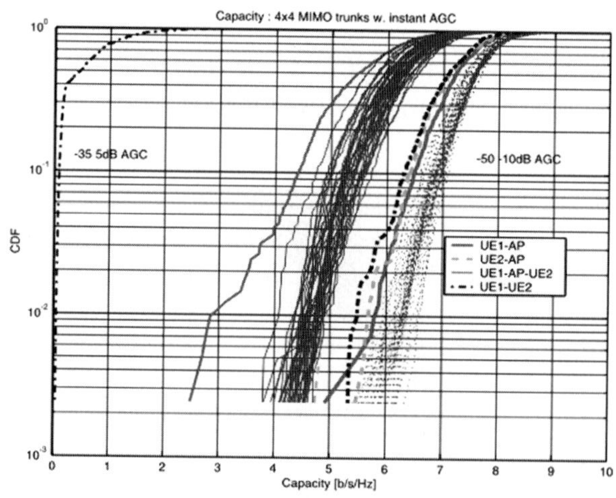

Figure 12.9. Total capacity of 4x4 MIMO with instant power control (40dB range).

Figure 12.7 shows that capacity is totally dominated by the strongest trunk/links and no benefits appear in any joint MIMO operation (dotted curves). No other trunk communication would be of any supporting use for overall capacity.

When all trunks and links are stabilized to the same mean power, as shown in Figure 12.8, there are clear benefits of using joint MIMO and the cross UE-UE link can support some cross communication to facilitate this.

For a CDMA like operation with instant power control with 40dB dynamic range in 1dB steps, the over-all dynamics are very critical. As seen in Figure 12.9 an 'offset' power control range (left curves) fails to equalize the cross UE-UE links and some of the weaker AP-UE links. If the power control range is adjusted to 'bring in' all the weaker links (right curves), higher capacity results can be achieved than those that would be achieved with just average power control.

Rate Improvement in Short Range Relaying Systems

In the following, we use the measurements described in the previous section and apply them to the relaying principles developed in Section 2.0.0.

The measurements involved multiple antennas at each transmitting and receiving end. The algorithms defined earlier can be generalized to multiple input-multiple output (MIMO) systems, but for our calculations we concentrate on the single input-single output (SISO) link performance. We therefore select one of the possible links as indicative of the overall link quality.

Moreover, we observed that the average measured gain to the two UE devices was different. This imbalance is reflected in our measurements. Namely for the transmit-receive combinations that we selected, we had $\frac{E[g_1]}{E[g_2]} = 3dB$ and $\frac{E[g_1]}{E[g_r]} = 15dB$, where $E[\cdot]$ stands for the expectation of the argument $(\cdot)$.

We analyzed three main cases, differentiated by the average power normalization applied. In all these cases, the average gain of the relay link is varied over a range of values relative to the link to the target user equipment.

- Rate improvement with $E[g_1] = E[g_2]$

  We look at the cases where all the links are normalized as in the theoretical analysis, and we isolate the effect of the simultaneous link fades.

  Figure 12.10 shows the median achievable rate improvement based on the measured data described in Section 3.0. We can see the similarity between the curves based on the measured data and the theoretical curves shown in Figure 12.1. We also observe that we need at least 10dB gain difference between the direct and the relay link for cooperation to be useful.

  It is very interesting to look at the cumulative distribution functions of the achievable rates, as shown in Figure 12.10. The comparison with the

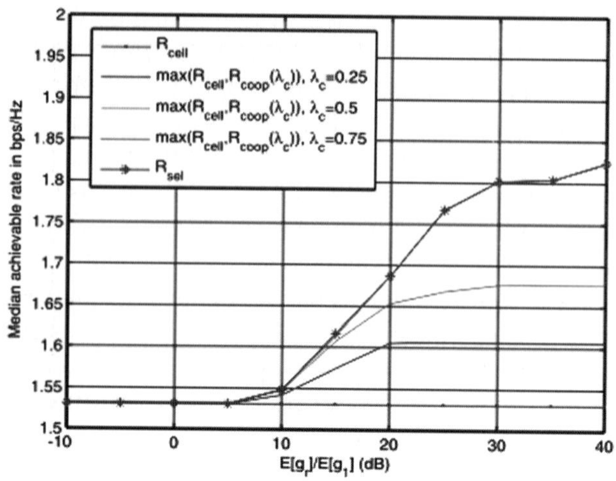

Figure 12.10. Median achievable rate improvement with relaying (normalized measured data such that $E[g_1] = E[g_2]$).

theoretical plots shows great similarity if the relay link has significantly higher average gain that the direct link. However when all the links have the same average gain, there is significant departure from the theoretical curves. This is due to the fact that the measured data display some slow fading, which is not taken into account in the normalization process. This is especially significant for the relay link which, as seen from Figure 12.5 has a 10dB fluctuation over the measured distances.

- Rate improvement with $E[g_1] = E[g_2] + 3dB$

  Here we look at the case where the link to the target user is normalized to unity average gain, and therefore the link to the relay has 3dB lower average gain.

  Figure 12.11 shows the median achievable rate, based on the normalized data. Clearly the fact that the direct link has a higher gain than the link to the relay station makes the use of relaying less likely, and a higher average gain is required on the relay link for it to be useful. Similar observations can be made from the cumulative distribution function plots shown in Figure 12.12.

- Rate improvement with $E[g_1] = E[g_2] - 3dB$

  We also look at the case where the roles of relay and target are interchanged. Then the link to the target user equipment is again normalized

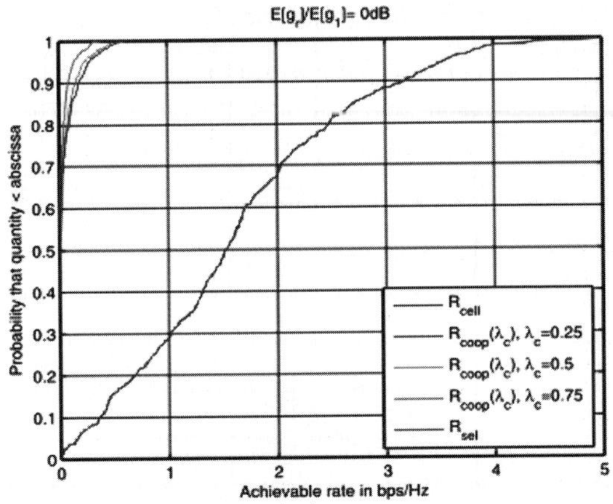

(a) $\frac{E[g_r]}{E[g_1]} = 0dB$

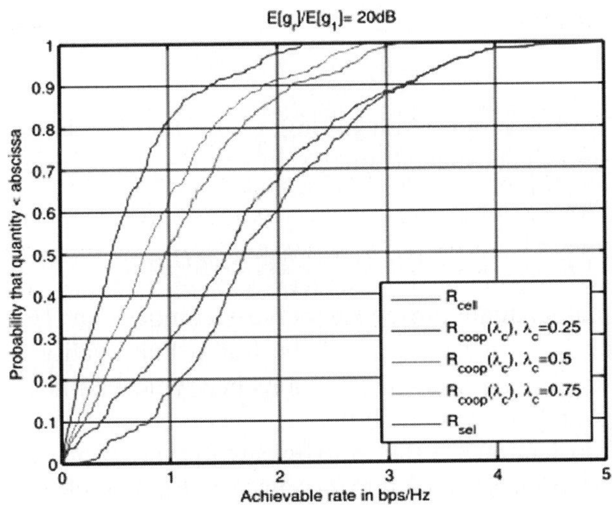

(b) $\frac{E[g_r]}{E[g_1]} = 20dB$

Figure 12.11. Rate cumulative distribution functions for two relative channel gains (normalized measured data such that $E[g_1] = E[g_2]$).

to unity average gain, and therefore the link to the relay has 3dB higher average gain.

Figure 12.13 shows the median achievable rate, based on the normalized data. Clearly the fact that the direct link has a lower gain than the link to the relay station makes the use of relaying more likely, and a lower

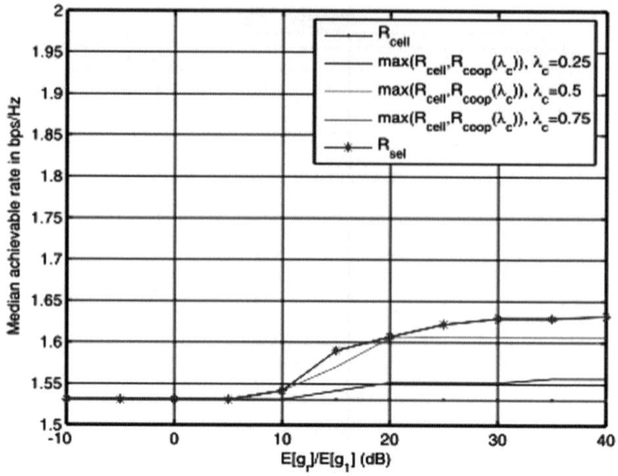

Figure 12.12. Median achievable rate improvement with relaying (normalized measured data such that $E[g_1] = E[g_2] + 3dB$).

average gain is required on the relay link for it to be useful. Similar observations can be made from the cumulative distribution function plots shown in 12.14.

Rate Improvement Results in PMR Systems

The main goal in applying co-operative techniques in TETRA mobile-to-mobile scenarios, is the increase of the Radio link quality for all considered UE-UE and UE-CAR radio connections in a typical ad-hoc scenario, thus increasing the overall system reliability.

As presented in Section 3.0, the radio links with the stationary car terminal in one of the ends, UE1/UE2-CAR, exhibit classical Rayleigh/Rice envelope distribution. The dynamic UE1-UE2 radio links have a double-Rayleigh envelope distribution and are subject to a more significant signal shadowing compared to the UE1/UE2-CAR radio links.

For the analysis of the co-operative schemes presented in Section 2.0.0 (TETRA DMO scenarios), we consider as transmitter the indoor UE1, while the outdoor CAR is used as relaying station for the communication link with the outdoor UE2. Given the average low correlation between the radio links in these scenarios, we have used for the UE2-CAR link a channel data set from another measurement run, from the same environment, which had strong Rician characteristics.

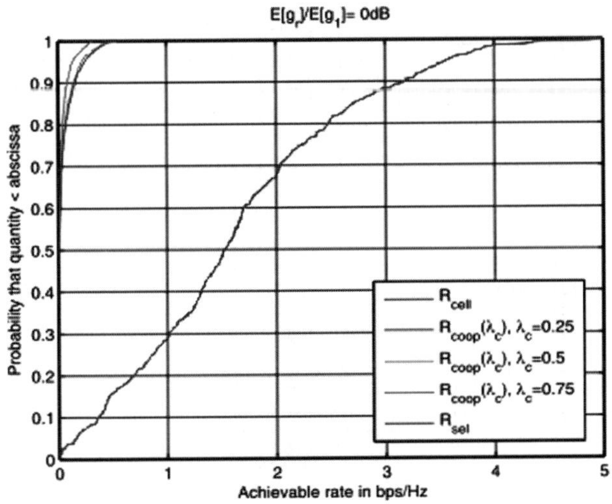

(a) $\frac{E[g_r]}{E[g_1]} = 0dB$

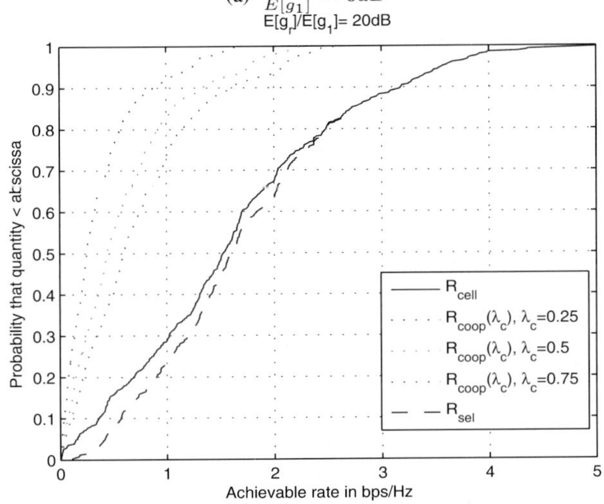

(b) $\frac{E[g_r]}{E[g_1]} = 20dB$

Figure 12.13. Rate cumulative distribution functions for two relative channel gains (normalized measured data such that $E[g_1] = E[g_2] + 3dB$).

In the following discussions, $E[g_c]$ is the average gain of the UE1-to-CAR channel and $E[g_r]$ is the average gain of the CAR-to-UE2 channel.

The median achievable rate improvement when using the sub-optimal scheme ($R_{sel}(\lambda_c)$) is given in Figure 12.16. The results for the co-operative schemes with fixed power allocation, *i.e.* a fixed λ_c, are presented in Figure 12.17.

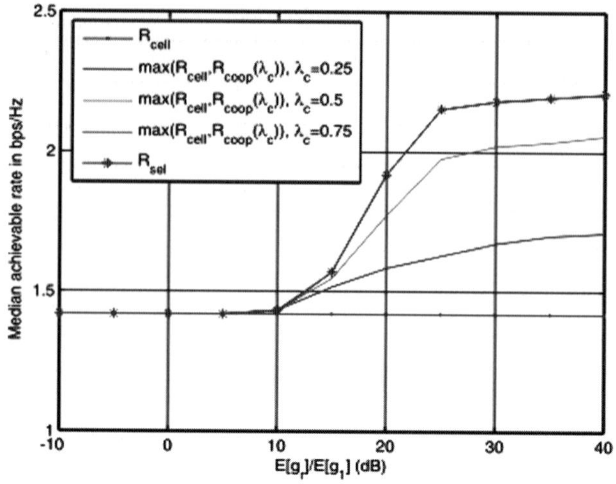

Figure 12.14. Median achievable rate improvement with relaying (normalized measured data such that $E[g_1] = E[g_2] - 3dB$).

Both results, with fixed power allocation (Figure 12.17) and with sub-optimal selection ($R_{sel}(\lambda_c)$)(Figure 12.16), show that for low channel gain ratios ($E[g_r]/E[g_c]$) there is no gain from using these co-operative techniques, even if the links exhibit very different (small and large-scale) fading characteristics. Actually, at low channel gain ratios, the large-scale channel fading has a strong influence on the achievable capacity and causes performance well below the theoretical results (see Figure 12.2).

For high channel gain ratios (above 10dB), by increasing the power used on the UE1-to-CAR link relative to the CAR-to-UE2 link, *i.e.* increasing λ_c, the fixed power allocation co-operative technique starts to be useful, and better system capacity is obtained. The optimal routing method further improves the median capacity compared to the single link transmission case.

Comparing the curves in Figure 12.17 with the theoretical curves in Figure 12.2 one can notice an increase in the probabilities for high data rates, above $\approx$1.0 bps/Hz, when a co-operative scheme is used. We believe this result can be explained by the combination of high gain and strongly Rician fading on the relay link, CAR-to-UE2, in the TETRA DMO scenarios analyzed.

Improvements in terms of interference control and optimal power terminal utilization can be obtained if all (or most of) the UE terminals in an ad-hoc TETRA scenario are configured to operate as repeater/gateway. This would enable the use of the adaptive power control mechanism, already provisioned in the TETRA DMO specification [ETS, 1999].

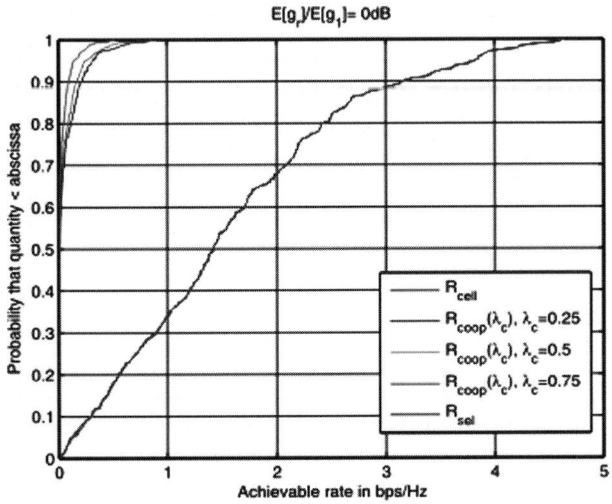

(a) $\frac{E[g_r]}{E[g_1]} = 0dB$

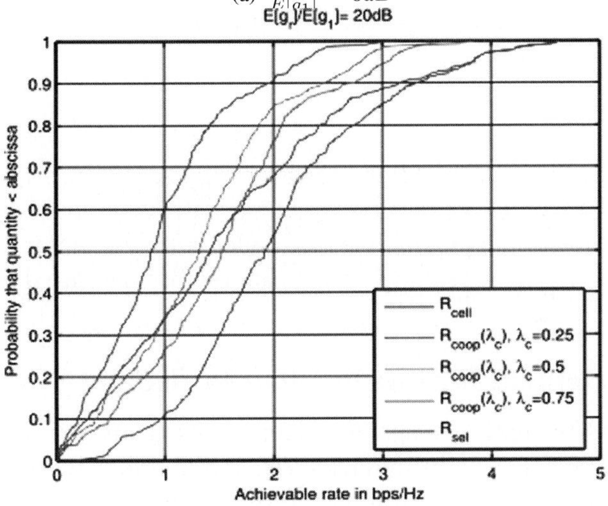

(b) $\frac{E[g_r]}{E[g_1]} = 20dB$

Figure 12.15. Rate cumulative distribution functions for two relative channel gains (normalized measured data such that $E[g_1] = E[g_2] - 3dB$).

More advanced co-operative schemes could also make use of the time-domain channel coherence time. The signal envelope auto-correlation lengths determined in our TETRA DMO investigations corresponded to ≈4.5 traffic time slots or 9 synchronization time slots in the low-rise building scenarios. The corresponding values were ≈8 traffic time slots or 16 synchronization

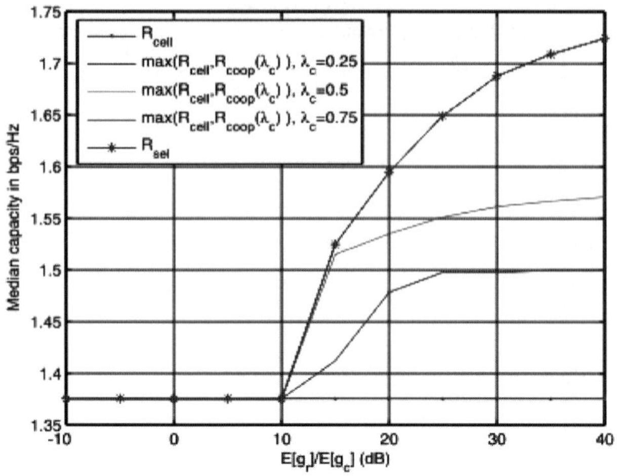

Figure 12.16. Median achievable rates in TETRA DMO scenarios versus relative channel gains (normalized measured data such that $E[g_1] = E[g_2]$).

time slots in the high-rise building scenarios [Kovács, 2002]. Therefore, a co-operative mechanism in these deployment scenarios can, for example, use an adaptive block interleaving length on the coded 4.8kbps and/or 2.4kbps traffic channels [ETS, 1999] depending on the link quality measured on the 'worst' link.

Rate Improvement Results in Ultra Wideband Systems

In UWB PAN systems, co-operative techniques, apart from the possible capacity improvements, can also reduce the required transmit power levels, thus reducing interference as well as power consumption.

The UWB channel characteristics in the investigated mobile-to-mobile PAN scenarios described in Section 3.0 have been used to evaluate the performance of the co-operative schemes introduced in Section 2.0.0. In this scenario, all terminals UE1,UE2,UE3 are mobile thus the signal fading characteristics are very similar, though de-correlated, on all analyzed radio links. For the discussion presented herein, only the SISO links have been used and two handset positions were considered for the UE1 and UE2/3, respectively.

The reader should note here, that in practice the required SNR levels (channel gains) have to be set to match a certain BER (PER) and these can vary depending on the technique used by the UWB system (multi-band (MBOA-UWB) [IEE, 2004b], direct sequence spread spectrum (DS-UWB) [IEE, 2004a], impulse radio (IR-UWB) [MAG, 2004], etc). For a low-number of users/terminals

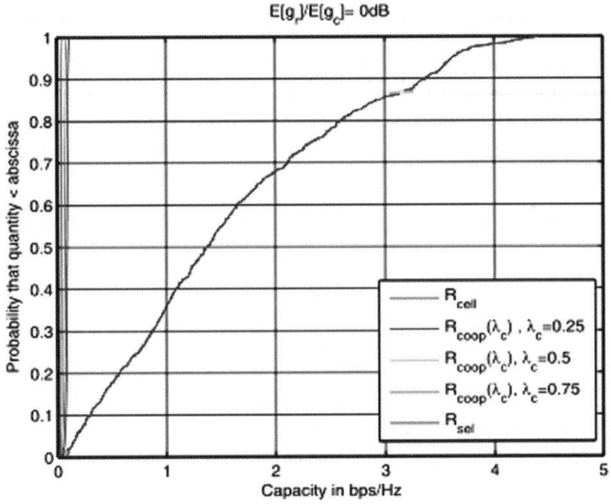

(a) $\frac{E[g_r]}{E[g_c]} = 0dB$

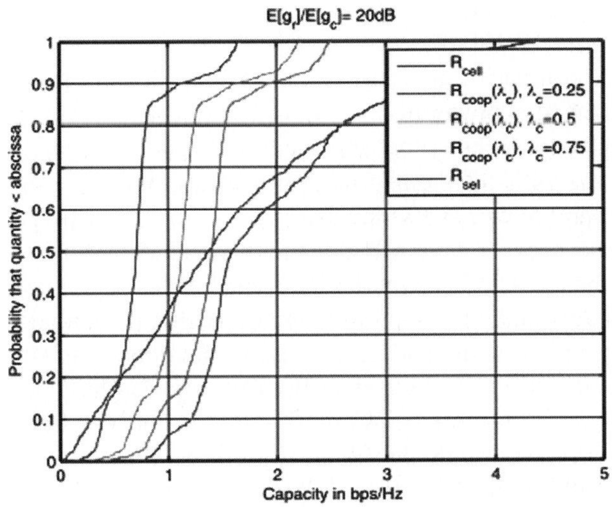

(b) $\frac{E[g_r]}{E[g_c]} = 20dB$

Figure 12.17. Rate cumulative distribution functions in TETRA DMO scenarios for two relative channel gains (normalized measured data such that $E[g_1] = E[g_2]$).

the multi-user interference can be considered to be fairly well mitigated by the appropriate multi-access scheme used in these UWB systems. These aspects are not considered here.

Due to the large communication bandwidths, above 500MHz, generally the wideband power parameter is characterised/ modelled in UWB in order to evaluate system performances. Furthermore, the narrowband envelope of an UWB signal has very low time/space correlation while the wideband power level can exhibit higher correlation. This leads to potentially different estimated system performance when using the two signal parameters.

The median achievable rate improvement when using the sub-optimal scheme ($R_{sel}(\lambda_c)$) is given in Figure 12.18. The results for the co-operative schemes with fixed power allocation are presented in Figure 12.19 when the narrowband power and wideband power (2.5GHz) levels are used respectively.

The median capacity CDF values in Figure 12.18, show similar results as presented above and highlight the observation that the relay link needs to have at least 15-20dB higher gain in order to gain from the use of this co-operative scheme. As a reminder, this threshold value was 10-15dB for the analyzed narrowband and wideband systems.

As in the case of the narrow and wideband systems analyzed in the previous sections, where the narrowband power signal has been used, capacity improvements can be expected only at high channel gain ratios between the relay and direct links (Figure 12.19b,d).

Furthermore, comparing the narrowband and wideband power results it is clear that using the total power of the received signal provides at least 20% better performance for capacities below the median 2bps/Hz (Figure 12.19a/b,c/d). The relative gain with different power allocation factors λ_c is higher when using the wideband power signal levels.

Similar to the case of narrowband and wideband systems, advanced co-operative schemes in UWB have also to consider the large-scale channel de-correlation time/distance. For example, the determined channel stationarity of 0.5s/0.25m in terms of wideband power and signal clustering statistics [MAG, 2005] suggests the need for long data interleaving depths, thus large buffering capabilities in the terminals. While this might not be a desirable solution for low-cost terminals, it can significantly improve the co-operative schemes implemented in the high end terminals. An additional dimension to be considered in co-operative schemes for UWB is frequency diversity, over the entire occupied frequency band or only in successive frequency sub-bands.

5. General Conclusions on Practical Antenna Cooperation

A general conclusion drawn from the different scenarios and simulated cooperative operation, is that branch signal correlations and short term statistics play a lesser role compared to the influence of the average power balance between the different inter-terminal links. This is due mainly to the path loss (im)balance but in some cases also due to the shadow fading.

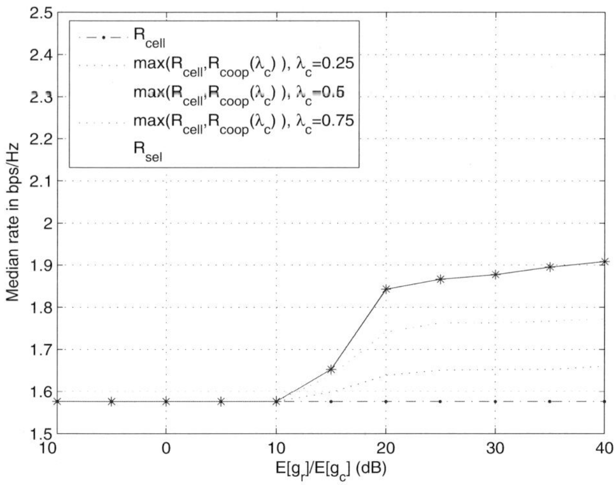

(a) Narrowband

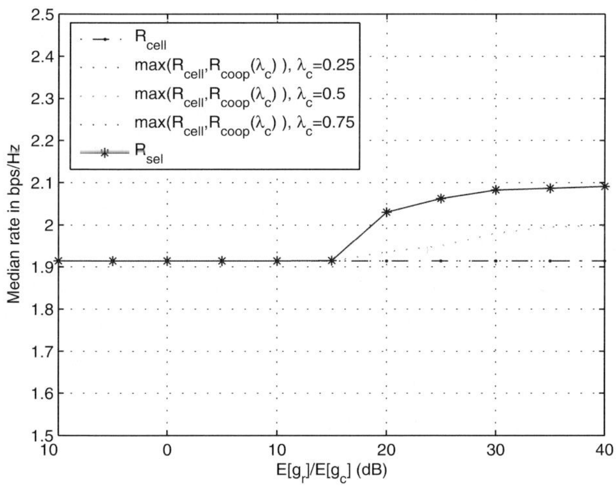

(b) Wideband

Figure 12.18. Median achievable rate improvement with relaying in UWB systems (normalized measured data such that $E[g_1] = E[g_2]$): a) narrowband; b) wideband.

The presented investigations have provided some very first experimental results with a very difficult co-band cooperative channel sounding setup. They revealed that for this very short range scenarios, hand-set terminals are not very suitable for cooperative operation. Rather longer range to common AP

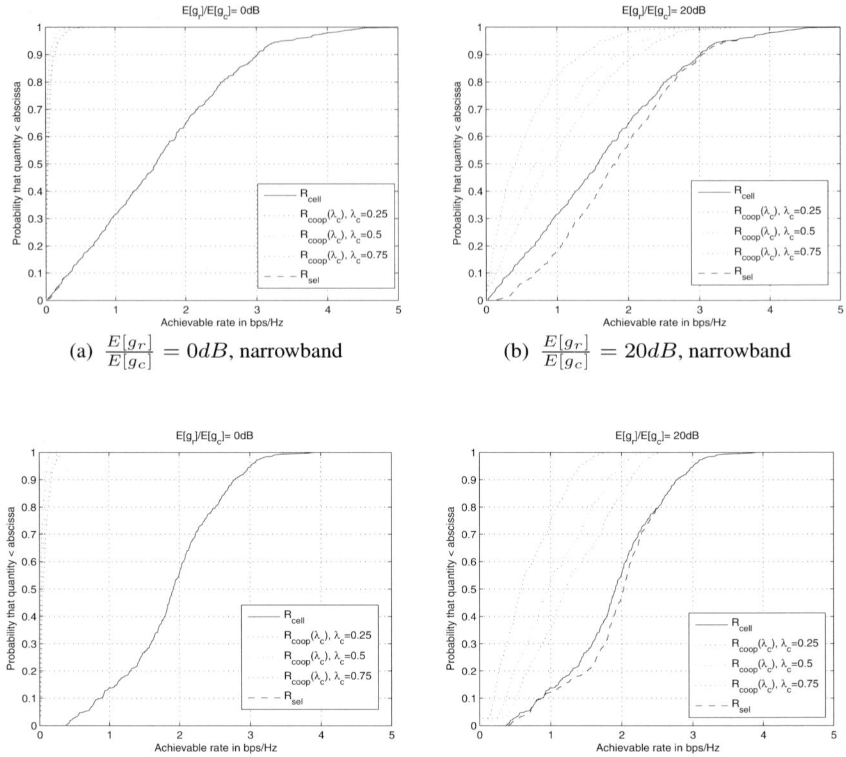

(a) $\frac{E[g_r]}{E[g_c]} = 0dB$, narrowband

(b) $\frac{E[g_r]}{E[g_c]} = 20dB$, narrowband

(c) $\frac{E[g_r]}{E[g_c]} = 0dB$, wideband

(d) $\frac{E[g_r]}{E[g_c]} = 20dB$, wideband

Figure 12.19. Rate cumulative distribution functions in UWB mobile-to-mobile scenarios for two relative channel gains (normalized measured data such that $E[g_1] = E[g_2]$): a-b) narrowband; c-d) wideband.

and/or more free space terminals (like note-books etc) should be considered for beneficial cooperative operation.

Furthermore the investigations reveal that sufficient power control capability is paramount for achieving beneficial effects of cooperative operation, regardless of the system bandwidth.

References

(1999). ETS 300 396-2: Trans-European Trunked Radio (TETRA); Technical requirements for Direct Mod Operation (DMO)- Part 2: Radio aspects. Technical report, European Telecommunication Standards.

(2004a). DS-UWB Physical Layer Submission to 802.15 Task Group 3a; IEEE P802.15-04/0137r3. Technical report, IEEE P802.15 Working Group for Wireless Personal Area Networks (WPANs).

(2004b). Multi-band OFDM Physical Layer Proposal for IEEE 802.15 Task Group 3a; IEEE P802.15-03/268r3. Technical report, IEEE P802.15 Working Group for Wireless Personal Area Networks (WPANs).

(2004). My personal Adaptive Global Net (MAGNET); Candidate Air Interfaces and Enhancements - Deliverable d.3.2.2a. Technical report, European IST-507102.

(2005). My personal Adaptive Global Net (MAGNET); PAN Channel Characterisation (Part I and II) - Deliverable D.3.1.2b. Technical report, European IST-507102.

Andersen, J. Bach and Kovács, I.Z. (2002). Power Distributions Revisited. *COST 273 Meeting*.

Catreux, S., Driessen, P. F., and Greenstein, L. J. (2000). Simulation results for an interference-limited multiple-input multiple-output cellular system. *IEEE Communications Letters*, Vol. 4(11):334–336.

Kotterman, W.A.T., Pedersen, G.F., Olesen, K., and Eggers, P.C.F. (2003). Cable-Less Measurement Setup for Wireless Handheld Terminals. *12th IEEE International Symposium on Personal, Indoor and Mobile Radio Communications (PIMRC)*, Vol. 1:B112–B116.

Kovács, István Z. (2002). *Mobile-to-Mobile Radio Communication Channels*. Aalborg University, Center For PersonKommunikation, Aalborg.

Pedersen, G. F., Tartiere, M., and Knudsen, M. B. (2000). Radiation efficiency of handheld phones. *Proc. IEEE 51st Vehicular Technology Conference (VTC)*, Vol. 2:1381–1385.

Vaughan, R. G. and Andersen, J. Bach (1984). A Multiport Patch Antenna for Moble Communications. *Proc. 14th European Microwave Conference*, pages 697–712.

Vaughan, R. G. and Andersen, J. Bach (2003). *Channels, Propagation and Antennas for Mobile Communications*. IEE Press, London, UK.

Chapter 13

DISTRIBUTED ANTENNAS: THE CONCEPT OF VIRTUAL ANTENNA ARRAYS

Mischa Dohler
France Télécom R&D, France
mischa.dohler@francetelecom.com

A. Hamid Aghvami
King's College London, UK
hamid.aghvami@kcl.ac.uk

Abstract: We introduce a communication paradigm where spatially adjacent mobile terminals or nodes cooperate and thereby form a virtual transceiver entity, which we refer to as virtual antenna array (VAA). It is the prime aim of this chapter to shed some historical background on the developments around the concept of VAAs, as well as to analyse and synthesise some specific topologies. As such, by means of prior derived closed form capacity expressions, we will derive cross-layer optimised communication protocols for distributed and cooperative relaying VAAs. These protocols are shown to be robust, of low complexity, and to perform near-optimum; they are hence easily applicable to cellular, ad hoc and wireless sensor networks. As of today, VAA-type communication topologies have gained significantly in research momentum, mainly due to their ability to boost capacity and their inherent attribute of scalability. Indeed, it is in hot-spots where an increasing amount of users competes for the same capacity; however, hot-spots bring along an increasing number of terminals which, if cooperating, counteract the decrease in available system capacity by further increasing it.

Keywords: distributed wireless systems, cooperative wireless systems, virtual antenna arrays, Shannon capacity, cross-layer optimisation, communication procotols

Frank H. P. Fitzek and Marcos D. Katz (eds.), Cooperation in Wireless Networks: Principles and Applications, 421–461.

1. Introduction

In December 1999, on the quest for a high capacity and inherently self-scaling wireless communication system, we discussed for the first time the possibility of applying space-time coding techniques to spatially adjacent mobile terminals, as depicted in Figure 13.1. After some considerations, we understood that this simple idea solved a couple of problems at once, *i.e.*,

- **Antenna Issues:** MIMO techniques were already then known to boost the system capacity but, with limited terminal size, the provision of multiple antennas was a considerable implementation problem. Our distributed approach hence facilitated the applicability of MIMO techniques to single antenna terminals.

- **Capacity Issues:** Capacity, particularly in hot spot cellular systems, was and still is a crucial issue. The application of space-time coding techniques hence promised to increase the down and up link capacities and/or coverage.

- **Scalability Issues:** An increasing number of users competing for the same over-the-air bandwidth drain the system capacity fast. An increase in users, however, comes along with an increase in the number of available antenna elements and thereby also in capacity, given that these antenna elements cooperate.

What we only guessed is that there were plenty of problems to be solved on the way to make such a system reality. What we certainly did not anticipate is that half a decade later a major bulk of available research literature would deal, in one way or another, with this idea applied to cellular, ad hoc and sensor networks.

We hence decided to present the idea shortly after to the UK Mobile Virtual Centre of Excellence (www.mobilevce.com), a consortium consisting of seven academic institutions and about 20 leading telecommunications companies. The proposed approach found considerable interest and became an integral part of the core 3 research phase, where a more general distributed and cooperative space-time coding approach was henceforth patented, see [Dohler et al., 2001]. Since single antenna terminals form a mutually communicating entity, the concept was initially termed Artificial Antenna Array (AAA), however, because of the already used triple-A acronym, renamed to *Virtual Antenna Array (VAA)*.

In its infancy, the concept of VAA evolved from the contributions by [Vodafone, 1999], and [Harrold and Nix, 2000], on relaying and by [Telatar, 1999], and [Alamouti, 1998], on MIMO communication aspects. Other excellent research has been performed in parallel, all of which led to the currently flourishing research area of distributed and cooperative wireless communication

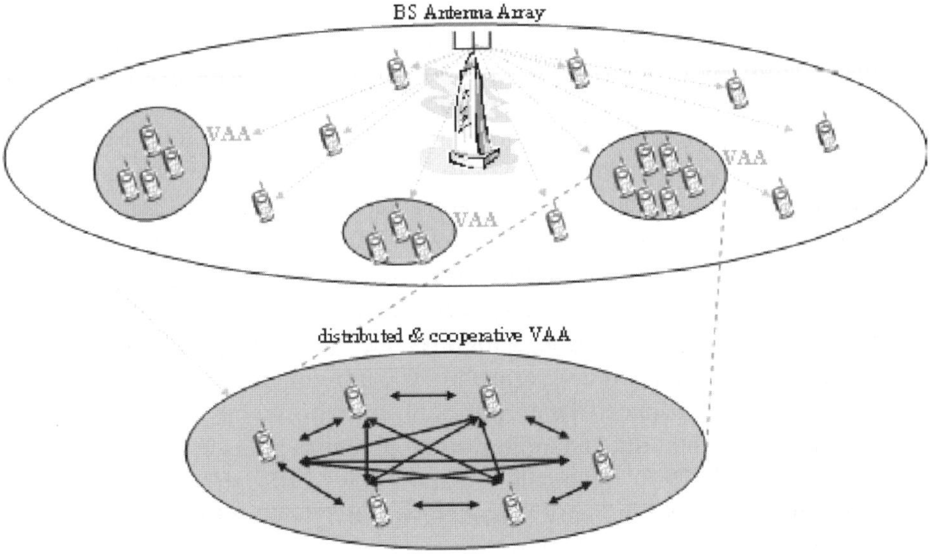

Figure 13.1. Virtual Antenna Arrays in a cellular deployment.

networks. That cooperation is beneficial is not obvious but perhaps best exemplified by means of this book, which is collaboratively written by us and our colleagues, the later all being pioneers and outstanding experts in this field.

In this chapter dedicated to VAA, we will briefly review the state-of-the-art back then and now. We will highlight the differences to other approaches exposed in the open literature and in this book. We will then dwell on a few basic principles related to the application and applicability of such cooperative communication structures. This is followed by some Shannonian description of a more general VAA topology, where we also propose low-complexity resource allocation strategies. The proposed protocol is then applied to some selected communication scenarios, from which also some design guidelines are derived.

2. Background & State-of-the-Art

The work exposed in [Vodafone, 1999], was undertaken from 1996 until approximately 2000 within the scope of the Universal Mobile Communications System Terrestrial Radio Access (UTRA) Concept Group Epsilon and mainly driven by Vodafone. The system was referred to as Opportunity Drive Multiple Access (ODMA), the main purpose of which was to increase the high data rate coverage within a cell. It is a relaying protocol and not a stand-alone PHY layer technology, which was the reason why it had been rejected as a potential candidate for UTRA. However, due to its capacity benefits, it is now an optional protocol for UTRA TDD/CDMA.

The methodologies suggested in [Vodafone, 1999], have been studied in [Harrold and Nix, 2000]. In this study, SISO relaying has enabled an extension of the serving area of a BS by utilising relaying mobile terminals (r-MTs) at the coverage edge to provide data services to MTs out of reach. It has been demonstrated that, although each r-MT consumes additional power to accommodate the relaying process, all MTs in the network gain on average in performance. Simple relaying protocols have also been suggested which are based on shortest distance relaying. The gains of the relaying process have been attributed to the non-linear pathloss equation, which reduces the aggregate pathloss when breaking a long distance into several shorter communication distances.

As for the MIMO aspects, the elegant analysis exposed in [Telatar, 1999], has enabled a fundamental understanding of the potential gains offered by ergodic and non-ergodic MIMO channels. Telatar showed that the capacity offered by the wireless channel increases drastically when the number of transmit and receive antennas is increased. With the exposed analysis on uncorrelated flat Rayleigh fading MIMO channels, Telatar has opened the door to evaluating the capacity offered by MIMO channels obeying more general conditions, *e.g.*, different fading statistics, different correlation, flat or frequency selective fading.

The work by [Alamouti, 1998], has triggered a revolution on the way pre-processing at the transmitting side is viewed. It is certainly far from the complex theories exposed in later works by [Tarokh et al., 1999], and [Tarokh et al., 1998]; however, it was the first transmission scheme which allowed the deployment of transmit diversity as opposed to the well established receive diversity.

Although these contributions were the primary trigger for the invention of Virtual Antenna Arrays and related distributed communication systems, they have not been the first and only contributions in the field of relaying and distributed MIMO systems. For this reason, a short summary on the state of the art related to the work in this chapter is given below. Note that, because each chapter can be read stand-alone, some overlap with the contributions of other chapters is unavoidable. Note further that, due to the sheer volume of contributions in this field, the summary is far from complete and we apologize beforehand for having potentially omitted some key contributions.

Relaying Communication Systems

The method of relaying has been introduced by [van der Meulen, 1971], and has also been studied by [Sato, 1976]. A first rigorous information theoretical analysis of the relay channel has been exposed by [Cover and el Gamal, 1979], with more details in his book [Cover and Thomas, 1991].

In these contributions, a source MT communicates with a target MT directly and via a relaying MT. In [Cover and el Gamal, 1979], the maximum achievable communication rate has been derived in dependency of various communication scenarios, which include the cases with and without feedback to either source MT or relaying MT, or both. The capacity of such a relaying configuration was shown to exceed the capacity of a simple direct link. It should be noted that the analysis was performed for Gaussian communication channels only; therefore, neither the wireless fading channel has been considered, nor have the power gains due to shorter relaying communication distances been explicitly incorporated into the analysis.

Only in the middle of the 90s, research in and around the Concept Group Epsilon revived the idea of utilising relaying to boost the capacity of wireless networks, thereby leading to the concept of ODMA [Vodafone, 1999]. As already mentioned, the power gains due to the shorter relaying links have been the main incentive to investigate such systems to reach MTs out of BS coverage. However, the emphasis of the study was its applicability to cellular systems, as well as a suitable protocol design. The research did not encompass more theoretical investigations into capacity bounds, transmission rates or outage probabilities.

Key milestones into the above-mentioned theoretical studies have been the contributions by [Sendonaris et al., 1998]. In their study, a very simple but effective user cooperation protocol has been suggested to boost the uplink capacity and lower the uplink outage probability for a given rate. The designed protocol stipulates a MT to broadcast its data frame to the BS and to a spatially adjacent MT, which then re-transmits the frame to the BS. Such a protocol certainly yields a higher degree of diversity because the channels from both MTs to the BS can be considered uncorrelated. The simple cooperative protocol has been extended by the same authors to more sophisticated schemes, which can be found in the excellent contributions [Sendonaris et al., 2003a], and [Sendonaris et al., 2003b]. Note that in its original formulation [Sendonaris et al., 1998], no distributed space-time coding has been considered.

The contributions by [Wornell, 2000], are a conceptual and mathematical extension to [Sendonaris et al., 1998], where energy-efficient multiple access protocols have been suggested based on decode-and-forward and amplify-and-forward relaying technologies. It has been shown that significant diversity and outage gains are achieved by deploying the relaying protocols when compared to the direct link. Again, in its original formulation, no distributed space-time coding has been considered.

The case of distributed space-time coding has been analysed by [Laneman, 2002], in his PhD dissertation. In his thesis, information theoretical results for distributed SISO channels with possible feedback have been utilised to design simple communication protocols taking into account systems with and without temporal diversity, as well as various forms of cooperation. He has demonstrated that cooperation yields full spatial diversity, which allows drastic transmit power savings at the same level of outage probability for a given communication rate. A vital asset of his thesis is also a discussion on the applicability of the suggested protocols to cellular and ad-hoc networks. However, in its original formulation, [Laneman, 2002], does not incorporate an analysis of distributed-MIMO multi-stage communication systems as exposed in this chapter. Nonetheless, the analysis exposed here can be used to design protocols similar to the ones in [Laneman, 2002].

[Gupta and Kumar, 2000], were the first to statistically analyse the information theoretically offered throughput for large scale relaying networks. They showed that under somewhat ideal situations of no interference, hop-by-hop transmission and pre-defined terminal locations, capacity per MT decreases by $1/\sqrt{M}$ with an increasing number of MTs M in a fixed geographic area. They also showed that if the terminal and traffic distributions are random, then the capacity per terminal decreases even in the order of $1/\sqrt{M \log M}$. The analysis in [Gupta and Kumar, 2000], has been extended by the same authors to more general communication topologies, where the interested reader is referred to the landmark paper [Gupta and Kumar, 2003].

Furthermore, [Grossglauser and Tse, 2002], have shown that mobility counteracts the decrease in throughput for an increasing number of users in a fixed area. The protocols suggested therein benefit from the decreased power for a hop-per-hop transmission for decreasing transmission distances. It also benefits from the location variability due to mobility, *i.e.*, a packet is picked up from the source MT by any passing by r-MT and only re-transmitted (and hence delivered) when passing by the target MT.

MIMO Communication Systems

Contributions on MIMO systems have flourished ever since the publication of the landmark papers by [Telatar, 1999], and [Foschini and Gans, 1998], on capacity and [Foschini, 1996], [Alamouti, 1998], and [Tarokh et al., 1999] & [Tarokh et al., 1998], on the construction of suitable space-time transceivers.

As for the BLAST system introduced by [Foschini, 1996], a transmitter spatially multiplexes signal streams onto different transmit antennas which are then iteratively extracted at the receiving side using the fact that the fades from any transmit to any receive antenna are uncorrelated and of different strength. The BLAST concept has ever since been extended to more sophisticated systems, a good summary of which can be found in [Vucetic and Yuan, 2003]. Note that these systems require a quasi-static (or slow-fading) channel as the iterative cancellation process requires a precise knowledge on the channel coefficients.

[Alamouti, 1998], introduced a very appealing transmit diversity scheme by orthogonally encoding two complex signal streams from two transmit antennas, thereby achieving a rate one space-time block code. His work was then mathematically enhanced by the landmark paper of [Tarokh et al., 1998], who essentially exposed various important properties of space-time block codes. [Tarokh et al., 1999], also showed how to construct suitable space-time trellis codes which were shown to yield diversity and coding gain. Many other contributions on coherent and differential space-time block and trellis code design followed, a summary of which is beyond the scope of this overview.

MIMO relaying systems. A landmark contribution on relaying systems deploying multiple antennas at transmitting and receiving side has been made by [Gupta and Kumar, 2003]. The network topology exposed therein is the most generic one can think of, *i.e.*, any MT may communicate with any other MT in the network such as to achieve a maximum system capacity. This is in contrast to the scheme depicted in Figure 13.2, which considers only stage-by-stage relaying. They have also derived an information theoretic scheme for obtaining an achievable communication rate region in a network of arbitrary size and topology. The analysis showed that sophisticated multi-user coding schemes are required to provide the derived capacity gains. Note also that the exposed theory is fairly intricate, which makes the design of realistic communication protocols a difficult task.

Specific distributed space-time coding schemes have also been suggested in recent years, *e.g.*, [Stefanov and Erkip, 2003]. In this publication, two spatially adjacent MTs cooperate to achieve a lower frame error rate to one or more destination(s), where a quasi-static fading channel has been assumed. Distributed space-time trellis codes have been designed which maximise the performance for the direct link from either of the MTs to the destination and the relaying link.

Position of our Research

The work during recent years on Virtual Antenna Arrays for the Mobile-VCE originally endeavoured to embed the VAA concept into existing and emerging communication systems. It was understood, however, that a deployment, capacity and performance analysis for generic communication topologies as depicted in Figure 13.2 would be more beneficial in understanding relaying systems.

To simplify analysis and understanding, investigations first concentrated on an end-to-end scenario where a given source MT communicates with target MT separated by various relaying hops. Analysis was then extended to the case where each relaying hop contained more than one relaying MT, henceforth referred to as the relaying stage. In due course it became apparent that enough problems were unsolved for such a communication scenario, some of which are exposed in this book chapter.

During the period of our research, the contribution by [Gupta and Kumar, 2003], emerged. Their topology can be seen as a generalisation of the one depicted in Figure 13.2, which is the reason why it is referred to as a 'fairly' generic communication scenario throughout this chapter. With hindsight to [Gupta and Kumar, 2003], the approach in this chapter can be seen as a bridge between the intricate information theoretical description of the maximum achievable sum rate of large scale networks and the information and performance theory needed to deploy comparably simple communication protocols as introduced by [Laneman, 2002].

With this in mind, it is the aim of the herein developed analysis to design communication protocols which yield optimum or near-optimum end-to-end data throughput for an information source communicating with an information sink via a given number of topologically imposed relaying stages. As will be demonstrated, such protocols have to guarantee an optimum assignment of resources to each relaying stage as a function of the average channel conditions. These protocols are henceforth referred to as *fractional resource allocation strategies*.

For a potential deployment, these strategies have to be as simple and robust as possible. Their role is to allocate resources in terms of power (PHY), bandwidth and frame duration (MAC) to each relaying stage dependent upon of the prevailing channel conditions. Such an optimum allocation hence relates to the

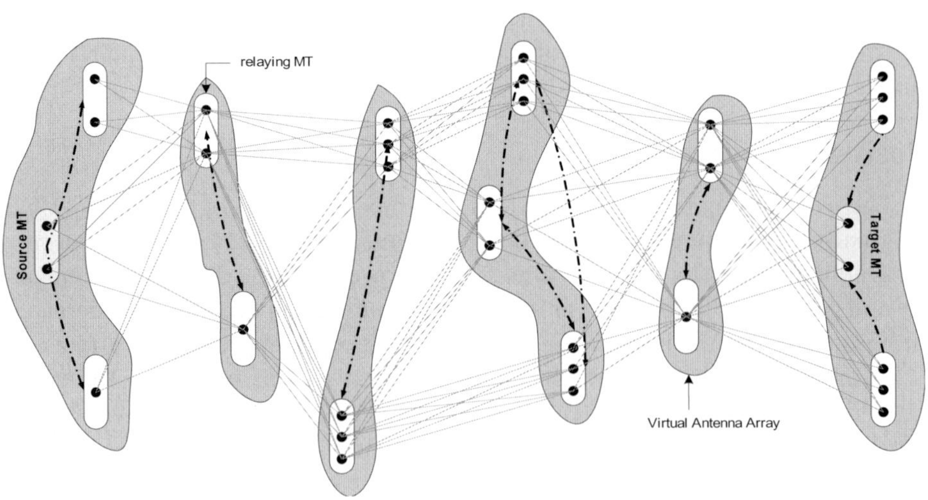

Figure 13.2. Distributed and cooperative MIMO multi-stage communication system.

area of cross-layer design, which, in our case, is applied to a reservation based communication protocols (as opposed to random access protocols).

To derive optimum allocation algorithms, a thorough understanding of the offered end-to-end capacity is needed first. Only after that, allocation strategies can be developed which achieve maximum end-to-end throughput. Since the notion of capacity assumes transceivers of infinite (or very high) complexity, their deployment is only justified in systems which deploy such transceivers. A system operating below the capacity limit, simply because it utilises less complex transceivers, may require different allocation strategies to achieve maximum throughput. A derivation of these fractional allocation strategies, albeit available in [Dohler, 2003], is beyond the scope of this introductory chapter.

Once the throughput maximising protocols are derived for an end-to-end communication link, they can then be enhanced to allow optimum throughput in multiple end-to-end communication links. This can be seen as a further step towards the realisation of generic systems as introduced by [Gupta and Kumar, 2003]; however, the major aim is only to equip the reader with some analytical tools which allow him to design cross-layer optimised protocols for such more general topologies. Before we start the analytical exposure, however, the subsequent section summarises some basic applications of VAAs to modern wireless communication systems.

3. Basic Application Principles

The VAA principle of distributed and cooperative terminals communicating among each other or a BS is now projected onto known communication paradigms, such as the cellular type network, the WLAN type network, the ad hoc network, and emerging sensor networks. The descriptions are by no means exhaustive but easily applied to other forms of network topologies. More details about potential deployment scenarios with standards, such as GSM, UMTS, HiperLAN2, IEEE802.x, Bluetooth, UWB, etc, can be found in [Dohler et al., 2001].

Cellular-Type Networks

Cellular 2G and 2.5G are well deployed, 3G and WMAN networks are being rolled out and, as of 2005, we dream of 4G - all of which struggle with the same problem of providing sufficient data rates to the data hungry end user applications. The problem is currently not serving a single user, but a large amount of users requiring large data rates at the same time. As already mentioned in the introduction, the cooperative VAA approach fortunately exhibits a self-scaling behaviour where an increase in competing users is compensated for by an increase in capacity.

The underlying principle for cellular-type deployment is depicted in Figure 13.1. A base station array consisting of several antenna elements transmits a space-time encoded data stream to the associated mobile terminals which can form several independent VAA groups. Each mobile terminal within a group receives the entire data stream, extracts its own information and concurrently relays further information to the other mobile terminals. It then receives more of its own information from the surrounding mobile terminals and, finally, processes the entire data stream. The wired links within a traditional receiving antenna array are thus replaced by wireless links. The same principle is applicable to the uplink, where a synchronised space-time encoded data stream is emitted from the VAA group.

In this situation, the VAA accomplishes a special type of network which bridges cellular and ad-hoc concepts to establish a heterogenous network with increased capacity. It calls for intelligent synchronisation, relaying and data scheduling algorithms, the exact realisation of which depends on the access scheme, choice of main link technology, choice of relaying technology, technological limits, number of antennas within a given geographical area and other factors, *e.g.*, the ability of the cellular system to synchronise users, etc.

A more specific example shall illustrate the previously mentioned deployment, where a VAA is embedded into a 3G communication system. Here, the direct link between BS and MTs is based on 3G W-CDMA as described by [Holma and Toskala, 2000]. For the relaying link a current standard with direct mode communication capabilities is required, which is chosen to be Bluetooth because it is deployed in virtually every current MT. Therefore, MTs which happen to be in communication range of the Bluetooth transceiver form a VAA in the sense that they start supporting each other via cooperative and spatially distributed communication links. They continue communicating with the BS using the W-CDMA link and, at the same time, relay further captured information to the other MTs within the VAA group utilising Bluetooth, thereby increasing the end-to-end link and system capacity; this has been quantified in [Zeng et al., 2002].

In the cellular context, the deployment of VAAs creates various problems which need to be addressed. For instance, the ability of the terminals to transmit and receive simultaneously and thus to operate in full duplex mode. The duplex communication problem can be solved by assuming that the frequency bands for the main link (BS to MT) and the relaying links (MT to MTs) differ. However, such duplex deployment still poses serious constraints on the MTs' RF chains. Particularly, if the receiving main link band and the transmitting relaying band are not separated sufficiently far apart in the frequency domain, the transmitter front-end duplex filter may not be able to protect the receiving branch sufficiently well. However, I believe that problems like these are either already solved (*e.g.*, MEMS) or will be solved in the near future with the ever increasing technological advances.

Of further importance is the actual relaying process when deployed with current standards. Similar to satellite transponders, the signal can be retransmitted using a transparent or regenerative relay. A transparent relay is generally easier to deploy since only a frequency translation is required. However, additions to the current standards are required. For a simpler adaptation of VAA to current standards, regenerative relays ought to be deployed. This generally requires more computational power, but is also known to increase the capacity of the network.

WLAN-Type Networks

WLANs – initially feared by most operators but then rightly embraced by them – are an important constituent of our daily lives. As of 2005, the majority of data hot spots at airports, train stations, cafes, etc, are connected to a WLAN access point. The great advantage of WLANs is that they do not require any license to be deployed and, if there is an internet backbone available, deployment does not take longer than 1min. This, of course, poses the problem of interference between different WLAN access points.

In any case, also WLAN networks suffer from capacity problems, particularly at coverage edges. To overcome this drawback, a VAA-type deployment could yield significant deployment gains. In contrast to cellular-type networks, WLAN terminals actually rarely suffer from serious power constraints, thereby easing the distributed cooperation among terminals. Indeed, most WLAN users have notebooks which are easily connected to some mains usually available in airports or cafes; in any case, compared to the power consumption of the screen, processor and memory, the wireless link consumes little power.

To exemplify deployment, one could imagine a notebook or PDA to be equipped with a WLAN system and a shorter range Bluetooth or UWB system; while the terminal communicates with the access point via the WLAN interface, the short range standard facilitates the required cooperation. One could also imagine to use the same WLAN standard for cooperation as had been envisaged by HiperLAN2 which accommodated direct communication between terminals.

The enhanced coverage achieved by the deployment of distributed and cooperative WLAN terminals may even one day lead to a complete coverage of city centres or other large scale geographic regions.

Ad Hoc Networks

Networks which form spontaneously without any prior imposed infrastructure are referred to as *ad hoc networks*. They have already been applied to battle-field equipment, but still need to find their way into more civilised applications. Theoretically, a node in an ad hoc network can communicate with any other node. Practically, this poses serious challenges in terms of stability, synchronisation, data flow, etc; for this reason, the concept of guaranteed QoS is foreign to ad hoc networks. An ad hoc system designer, however, will endeavour to find a suitable trade-off between maximum capacity and minimum latency, jitter and signalling overhead.

In the context of ad hoc networks, VAA-type distributed and cooperative communication systems have the great advantage of offering superior capacity and link stability. In ad hoc environments, links can easily break or be interrupted due to mobile transmitters, receivers and clutter in-between; a distributed link clearly suffers from such effect with less probability.

Sensor Networks

Sensor networks have been around already for decades, however, their wireless extension have become fashionable only in the past years. While the academic world loves them because of their sheer infinite degrees of freedom to do research and publish scientific papers, the commercial world hopes that one day this Globe will be monitored by large clusters of wireless nodes sensing motion, mood, temperature, humidity, light, pollution, etc, and thereby generating large revenues.

Be they academic or commercial, sensor nodes suffer from one common problem, *i.e.*, they need to consume the least possible power whilst still maintaining satisfactory sensing capabilities. Although a distributed approach yields significant transmit power savings, it is not clear to date whether such approach will also be beneficial in sensor networks. The main reason is because the nodes operate at such tight transmit power budget that the power of reception and (re)processing starts to matter. Our feeling is that with a suitable cooperative protocol and under very specific conditions, VAA-type communication as depicted in Figure 13.3 will be beneficial. It is hence an interesting research exercise for the future to find these conditions and build these cooperative protocols.

4. Closed-Form Capacity Expressions

Capacity is a concept related to the vast area of Information Theory, a branch of science which really commenced after the publication of Shannon's legendary monogram on "A Mathematical Theory of Communication"; see [Shannon, 1948]. More than half a century has passed since, during which major achievements in the field of theoretical and practical communications have been achieved. A brilliant overview on the milestones of Shannon's information theory from its very infancy until the year 1998 has been compiled by [Verdu, 1998].

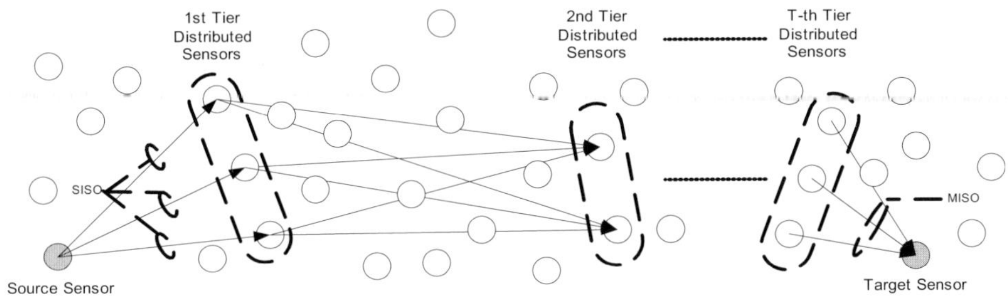

Figure 13.3. Distributed sensor network, where a source sensor communicates with a target sensor via a number of sensor tiers, each of which is formed of distributed relaying sensors.

The Shannon information theory predicts the achievable error-free communication rate for a communication system with given input distribution, transmission power, noise power, and bandwidth. The maximum achievable rate is referred to as the *channel capacity*. The theory also suggests design rules of how such an error-free transceiver can be constructed at the expense of infinite complexity; however, note that the theory does not suggest design criteria for transceivers operating at the capacity limits with finite complexity.

Information theorists thought that with the investigation of SISO systems the research area was almost closed, when [Telatar, 1999], and [Foschini and Gans, 1998], suggested utilising the additional spatial dimension to further boost capacity. Their contributions gave birth to the nowadays well established research branch of MIMO communication systems. The multiplexing scheme suggested by Foschini and Gans resulted in practical MIMO BLAST-like systems, whereas Telatar's contribution formed a profound mathematical foundation for further developments in MIMO information theory. Both contributions passed almost unnoticed, until Tarokh published minute code design rules which allowed the derived capacity bounds to be approached; see [Tarokh et al., 1999], and [Tarokh et al., 1998]. His work was inspired by the works of Telatar, Foschini and Gans, and Alamouti, and so is this work on distributed VAA systems.

Although communication at the MIMO capacity limit requires a transceiver of infinite complexity, this limit serves very well as a general characterisation and differentiation of communication systems. It is hence the aim of this chapter to deal with generic issues related to ergodic MIMO capacity, as well as specific issues applicable to distributed VAA systems. Although many theorems on the capacity behaviour of point-to-point and distributed MIMO channels have been proven to date, some novel and useful results are exposed here. In particular, we will derive a closed-form expression of the ergodic MIMO capacity assuming a generic transceiver and one with deployed orthogonal space-time block codes (O-STBCs). We will then introduce suitable approximations to these capacity expressions, because, as will be shown later, optimising a VAA-type system with the closed-form ergodic capacity expressions is impossible. Note that non-ergodic channels, which are characterised by the outage probability for a given communication rate, are beyond the scope of this chapter. Finally, also frequency selective fading channels have not been considered, so as not to clutter mathematical developments. The extension to non-ergodic and/or frequency selective channels is tedious but straightforward.

System Model

A distributed wireless MIMO transceiver model is depicted in Figure 13.4. It is assumed that spatially distributed information sources communicate with spatially distributed information sinks via a channel spanned by t inputs and r outputs. The channel is henceforth referred to as a distributed multiple-input-multiple-output (MIMO) channel.

Communication is achieved by properly encoding the information s at the transmitters across the temporal and spatial dimensions to produce a given space-time codeword. Transmitters are allowed to cooperate prior to such codeword construction, where we, in contrast to other works, assume that such cooperation is free of errors. This distributed codeword is then transmitted with average power S and received by each receiver with an additive average noise power N. The receiver cooperate and perform appropriate decoding to yield an estimate $\hat{s}$ of the originally transmitted information. It is the aim of this section to assess the achievable communication rate in dependency of the distributed communication scenario.

To produce a neat mathematical representation of the communication system, let $\mathbf{x} \in \mathbb{C}^{t\times 1}$ be the spatial codeword transmitted over the t transmitters at any time instant. With the required constraint on average transmission power, we ensure that $\mathrm{tr}(\mathrm{E}\{\mathbf{x}\mathbf{x}^{\mathrm{H}}\}) = \mathrm{tr}(\mathrm{E}\{\mathbf{S}\}) \leq S$, where $\mathrm{tr}(\cdot)$ denotes the trace operator and $\mathbf{x}^{\mathrm{H}}$ is the Hermitian to $\mathbf{x}$. If all transmission powers $S_{i\in(1,t)}$ are equal, then $S_i = S/t$. The generally complex channel realisation from transmitter

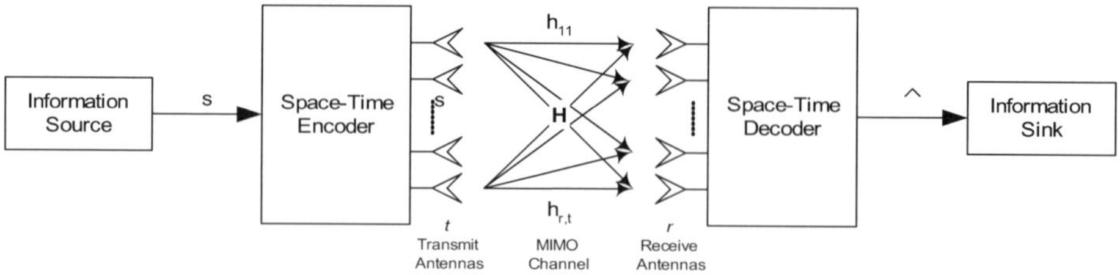

Figure 13.4. Multiple-Input-Multiple-Output Transceiver Model.

$i \in (1,t)$ to receiver $j \in (1,r)$ is denoted as h_{ij}. The channel realisations h_{ij} are henceforth referred to as sub-channels and may obey different statistics. They are conveniently grouped into a channel matrix $\mathbf{H} \in \mathbb{C}^{r\times t}$, where

$$\mathbf{H} = \begin{pmatrix} h_{11} & h_{12} & \cdots & h_{1,t} \\ h_{21} & h_{22} & \cdots & h_{2,t} \\ \vdots & \vdots & \ddots & \vdots \\ h_{r,1} & h_{r,2} & \cdots & h_{r,t} \end{pmatrix} \tag{13.1}$$

Due to the spatial separation, we assume that $\mathbf{H}$ is full-rank of rank $\min(t,r)$. That implies that at least $\min(t,r)$ sub-channels are mutually independent. Because of the flat-fading assumption, the received vector $\mathbf{y} \in \mathbb{C}^{r\times 1}$ can now be written as $\mathbf{y} = \mathbf{H}\mathbf{x} + \mathbf{n}$, where $\mathbf{n} \in \mathbb{C}^{r\times 1}$ is the noise vector containing the random noise samples from each receiver with average noise power N. The noise vector belongs to an r-dimensional complex zero-mean circular symmetric Gaussian distribution with variance N per dimension, *i.e.*, $\mathbf{n} \sim \mathcal{N}_c(\mathbf{0}_r, N\cdot\mathbf{I}_r)$, where $\mathbf{0}_r$ and $\mathbf{I}_r$ denote respectively an all-zero and identity matrix of dimensions $r \times r$.

Exact MIMO Capacity

The capacity of an ergodic channel is obtained by averaging over all realisations of $\mathbf{H}$ for a capacity maximising codebook covariance matrix $\mathbf{S}$. The dependency between capacity and codebook covariance $\mathbf{S}$ and channel statistics $\mathbf{H}$ is fairly complicated, which is the reason why capacity expressions for only very few special cases could be determined. The case where each entry in $\mathbf{H}$ obeys the same uncorrelated Rayleigh fading statistics, has been derived by [Telatar, 1999]; Other special cases for correlated channels with Ricean fading have been derived later. The case of different sub-channel statistics has not been solved today in satisfactory form and, if it will be solved, we expect the expressions to be of fairly intricate form. For that reason, we will simply assume that the average distance between the same cooperative VAA nodes is small compared to the distance between the VAA groups, which allows us to assume that each sub-channel obeys approximately the same statistics. We will further assume that this statistics is complex Gaussian, *i.e.*, Rayleigh, and its average power is normalised to unity.

Under these assumptions, we invoke Telatar's landmark MIMO capacity theorem, *i.e.*,

$$\begin{aligned} C &= \int_0^\infty m \log_2\left(1 + \frac{\lambda}{t}\frac{S}{N}\right) \cdot \\ &\quad \frac{1}{m}\sum_{k=0}^{m-1} \frac{k!}{(k+n-m)!}\left[L_k^{n-m}(\lambda)\right]^2 \lambda^{n-m} e^{-\lambda} \cdot d\lambda \end{aligned} \tag{13.2}$$

where λ is the unordered eigenvalue of the associated Wishart matrix with parameters $m \triangleq \min\{r,t\}$ and $n \triangleq \max\{r,t\}$, and $L_k^{n-m}(\lambda)$ is the associated Laguerre polynomial of order k. The capacity can also be expressed as

$$C = \mathrm{E}_\lambda \left\{ m \log_2 \left(1 + \frac{\lambda}{t} \frac{S}{N} \right) \right\} \tag{13.3}$$

with

$$pdf_\lambda(\lambda) = \frac{1}{m} \sum_{k=0}^{m-1} \frac{k!}{(k+n-m)!} \left[L_k^{n-m}(\lambda) \right]^2 \lambda^{n-m} e^{-\lambda} \tag{13.4}$$

where $\mathrm{E}_\lambda\{\cdot\}$ represents the statistical expectation with respect to λ and $pdf_\lambda(\lambda)$ is the probability density function (pdf) of λ.

The capacity in (13.2) is given in integral form, to which an iterative and explicit closed form expression are derived below. The advantage of such a development is that long Monte-Carlo simulations are avoided; further, it proves generally useful in a variety of problems relating to the computation of MIMO capacity. Note that other explicit solutions have been found in the meantime, *e.g.*, by [Kang and Alouini, 2002], or [Shin and Lee, 2003]; however, they are not straightforward applicable to the resource allocation problems we will deal with in the subsequent section.

The derivation of a closed form expression of (13.2) is performed in two stages: The pdf (13.4) is evaluated first, and then the expectation (13.3) is calculated. To this end, the associated Laguerre polynomial of order k is expressed through the Rodrigues representation; see [Gradshteyn and Ryshik, 2000], §8.970.1,

$$L_k^{n-m}(\lambda) = \sum_{l=0}^{k} (-1)^l \frac{(k+n-m)!}{(k-l)!(n-m+l)!l!} \lambda^l \tag{13.5}$$

Inserting (13.5) into (13.4) gives

$$\begin{aligned} pdf_\lambda(\lambda) = \frac{1}{m} \sum_{k=0}^{m-1} \frac{k!}{(k+d)!} \Bigg[& \sum_{l=0}^{k} A_l^2(k,d)\, \lambda^{2l} + \\ & \sum_{l_1=0}^{k} \sum_{\substack{l_2=0, \\ l_2 \neq l_1}}^{k} (-1)^{l_1+l_2} A_{l_1}(k,d)\, A_{l_2}(k,d)\, \lambda^{l_1+l_2} \Bigg] \lambda^d e^{-\lambda} \end{aligned} \tag{13.6}$$

where

$$d \triangleq n - m \tag{13.7}$$

$$A_l(k,d) \triangleq \frac{(k+d)!}{(k-l)!\,(d+l)!\,l!} \tag{13.8}$$

The derived pdf can now be inserted into (13.2), which yields for the capacity in [bits/s/Hz]

$$
\begin{aligned}
C = \sum_{k=0}^{m-1} \frac{k!}{(k+d)!} \Bigg[& \sum_{l=0}^{k} A_l^2(k,d)\, \hat{C}_{2l+d}(a) + \\
& \sum_{l_1=0}^{k} \sum_{\substack{l_2=0, \\ l_2 \neq l_1}}^{k} (-1)^{l_1+l_2} A_{l_1}(k,d)\, A_{l_2}(k,d)\, \hat{C}_{l_1+l_2+d}(a) \Bigg]
\end{aligned} \tag{13.9}
$$

where $a \triangleq \frac{1}{t}\frac{S}{N}$ and $\hat{C}_\zeta(a)$ is defined as

$$
\hat{C}_\zeta(a) \triangleq \int_0^\infty \log_2(1 + a\lambda)\, \lambda^\zeta\, e^{-\lambda}\, d\lambda \tag{13.10}
$$

Due to its frequent occurrence, $\hat{C}_\zeta(a)$ is henceforth referred to as the *Capacity Integral*. Note that according to [Gradshteyn and Ryshik, 2000], §4.337.2,

$$
\hat{C}_0(a) = -e^{1/a}\, \mathrm{Ei}\,(-1/a)\, /\, \log(2) \tag{13.11}
$$

where $\mathrm{Ei}(\zeta) \triangleq \int_{-\infty}^{\zeta} \frac{e^t}{t} dt$ is the exponential integral. $\mathrm{Ei}(\zeta)$ is related to $\mathrm{ExpInt}(\zeta)$ typically found in mathematical programmes via $\mathrm{ExpInt}(\zeta) = -\mathrm{Ei}(-\zeta)$. Further, for $\zeta > 0$

$$
\hat{C}_\zeta(a) = \zeta \cdot \hat{C}_{\zeta-1}(a) + \frac{1}{\log(2)} \int_0^\infty \frac{a\lambda^\zeta}{1 + a\lambda}\, e^{-\lambda}\, d\lambda \tag{13.12}
$$

The remaining integral can be expressed in closed form, where, from [Gradshteyn and Ryshik, 2000], §3.353.5,

$$
\begin{aligned}
\int_0^\infty \frac{a\lambda^\zeta}{1 + a\lambda}\, e^{-\lambda}\, d\lambda &= (-1)^{\zeta-1}\, (1/a)^\zeta\, e^{1/a}\, \mathrm{Ei}\,(-1/a) \\
&\quad + \sum_{k=1}^{\zeta} (k-1)!\, (-1/a)^{\zeta-k}
\end{aligned} \tag{13.13}
$$

$\hat{C}_\zeta(a)$ can thus be obtained through ζ iterations in (13.12). $\hat{C}_\zeta(a)$ can also be expressed in an explicit way by consecutively performing the ζ iterations, which finally yields

$$
\begin{aligned}
\hat{C}_\zeta(a) = \frac{1}{\log(2)} \sum_{\mu=0}^{\zeta} \frac{\zeta!}{(\zeta-\mu)!} \Bigg[& (-1)^{\zeta-\mu-1} (1/a)^{\zeta-\mu} e^{1/a} \mathrm{Ei}(-1/a) \\
& + \sum_{k=1}^{\zeta-\mu} (k-1)! (-1/a)^{\zeta-\mu-k} \Bigg]
\end{aligned} \tag{13.14}
$$

Hence, (13.9) constitutes a closed solution for the MIMO capacity C with $\hat{C}_\zeta(a)$ given either in iterative form (13.12) or in explicit form (13.14). The asymptotic expression for above capacity expressions at large SNRs is

$$C \to m \log_2 \left(\frac{1}{t} \frac{S}{N} \right) + \frac{1}{\log(2)} \left[\sum_{\mu=1}^{m-1} \frac{m-\mu}{d+\mu} + m \left(\sum_{\mu=1}^{d} \frac{1}{\mu} \right) - \mathcal{C} \right] \quad (13.15)$$

where $\mathcal{C} \approx 0.577$ is the Euler-Mascheroni constant. From the asymptotic expression, we conclude that deploying transmit diversity is a waste of resources if the number of transmit elements does not match the number of receive elements. The deployment of receive diversity yields notable gains due to the additional independent noise samples. A linear increase in capacity is achieved if the number of transmitters equates to the number of receivers. This is an important observation, which dictates the formation requirements of VAA-type systems.

Exact O-MIMO Capacity

Orthogonal space-time block codes inherently orthogonalise the MIMO channel. They are known to reduce the MIMO channel into parallel SISO channels, which drastically simplifies capacity analysis. The channel is henceforth referred to as the Orthogonal-MIMO (O-MIMO) channel. This section is dedicated to the capacity analysis of O-MIMO channels, where the cases of different statistics and attenuations are dealt with. This will be vital later for the allocation of optimum resources to MTs belonging to a VAA.

Note that, strictly speaking, Shannon capacity is understood to be the maximum mutual information a given channel can offer between source and sink, independent of the signal processing at either end. In subsequent analysis, however, the maximum mutual information a given channel with applied space-time block coding can accomplish is simply referred to as the capacity of the O-MIMO channel.

To briefly summarise the functioning, the space-time block encoder receives s encoded symbols $x_1, x_2, \ldots, x_s$ from the channel encoder, which are part of a longer codeword $\mathbf{x}$. These are encoded with an orthogonal space-time coding matrix $\mathcal{G}$ of size $d \times t$, where d is the number of symbol durations required to transmit the space-time code word, and t is the number of (distributed) transmit elements. At each time instant $1 \leq k \leq d$, the space-time encoded symbol $c_{k,i} \in \mathcal{G}$ is transmitted from the ith distributed transmit element, where $i = 1, \ldots, t$. Such encoding may come at a decrease in transmission rate R, defined as $R \triangleq s/d$. Note that space-time block coding does not provide any coding gain, which is accomplished by an outer channel code; however, the space-time block encoder provides a diversity gain which allows the outer channel code to yield better performance.

The use of orthogonal space-time block codes is known to reduce the MIMO channel into a single SISO channel with modified channel statistics; see *e.g.*, [Larsson and Stoica, 2003], or [Nabar et al., 2002]. For fixed channel realisations $\mathbf{H}$, the normalised capacity in [bits/s/Hz] over such an O-MIMO channel can be expressed as

$$C = R\log_2\left(1 + \frac{1}{R}\frac{\|\mathbf{H}\|^2}{t}\frac{S}{N}\right) \tag{13.16}$$

where $\|\mathbf{H}\|$ denotes the Frobenius norm of $\mathbf{H}$, the square of which is given as

$$\|\mathbf{H}\|^2 = \sum_{i=1}^{t}\sum_{j=1}^{r}|h_{ij}|^2 = \mathrm{tr}\left(\mathbf{H}\mathbf{H}^{\mathbf{H}}\right) \tag{13.17}$$

From (13.17), it is clear that $\|\mathbf{H}_{t\times r}\| = \|\mathbf{h}_{1\times t\cdot r}\|$, where $\mathbf{h} \triangleq \mathrm{vect}(\mathbf{H})$. Furthermore, the following is adopted to simplify notation: $u \triangleq t\cdot r$, $\lambda_i \triangleq h_i h_i^*$, $\lambda \triangleq \|\mathbf{h}\|^2 = \sum_{i=1}^{u} h_i h_i^* = \sum_{i=1}^{u}\lambda_i$, and $\gamma_i \triangleq \mathrm{E}\{h_i h_i^*\}$.

Generally, the sub-channel gains $\gamma_{i\in(1,u)}$ in the VAA setup can be different where some gains may be repeated. There shall hence be $g \leq u$ distinct sub-channel gains, which are henceforth referred to as $\hat{\gamma}_{i\in(1,g)}$ with each of them being repeated $\nu_{i\in(1,g)}$ times. In this case, [Dohler, 2003], demonstrated that the $pdf_\lambda(\lambda)$ of the only eigenvalue can be obtained in closed form as

$$pdf_\lambda(\lambda) = \sum_{i=1}^{g}\sum_{j=1}^{\nu_g} K_{i,j}\cdot\frac{\lambda^{j-1}}{\Gamma(j)\cdot(\hat{\gamma}_i)^j}e^{-\lambda/\hat{\gamma}_i} \tag{13.18}$$

where

$$K_{i,j} = \frac{1}{(\nu_i - j)!\,(-\hat{\gamma}_i)^{\nu_i - j}}\frac{\partial^{\nu_i - j}}{\partial s^{\nu_i - j}}\left[\prod_{\substack{i'=1,\\ i'\neq i}}^{g}\frac{1}{(1 - s\hat{\gamma}_{i'})^{\nu_{i'}}}\right]_{s=1/\hat{\gamma}_i} \tag{13.19}$$

This allows the capacity of the O-MIMO channel with unequal but possibly repeated channel coefficients to be expressed as

$$C = R\sum_{i=1}^{g}\sum_{j=1}^{\nu_g}\frac{K_{i,j}}{\Gamma(j)}\cdot\hat{C}_{j-1}\left(\frac{1}{R}\frac{\hat{\gamma}_i}{t}\frac{S}{N}\right) \tag{13.20}$$

where $\hat{C}_\varsigma(a)$ is the Capacity Integral defined in (13.10) and solved in (13.14).

The number of different scenarios obeying (13.20) is certainly infinite. To demonstrate its applicability, it is assumed that a terminal with two (uncorrelated) transmit elements communicates with two distributed but cooperating

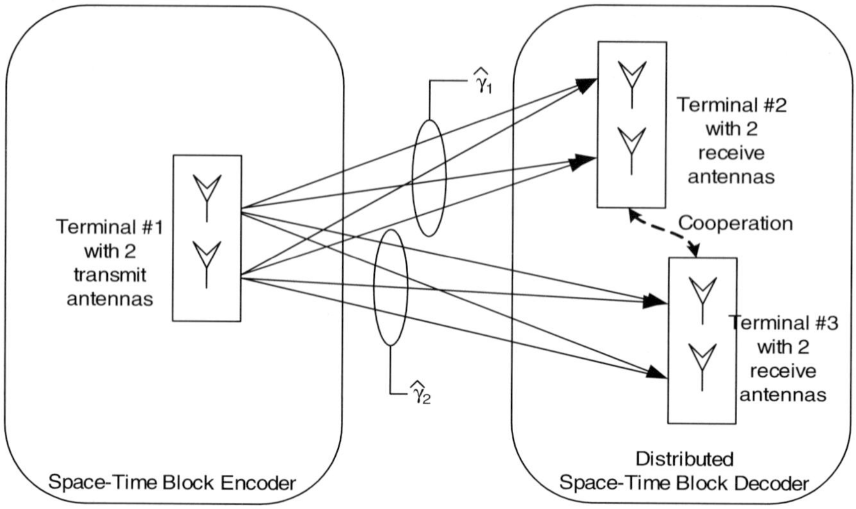

Figure 13.5. Distributed STBC communication scenario with one transmitter and two cooperating receivers, all of which possess two antenna elements.

target terminals, each possessing two (uncorrelated) receive antennas. This scenario is depicted in Figure 13.5; it could correspond to the case where an access point in an office communicates with a remote VAA group consisting of two terminals.

The spatial proximity between the elements of the same terminal results in the same channel attenuations from the first terminal to the second terminal, henceforth denoted as $\hat{\gamma}_1$ with repetition $\nu_1 = 4$, and from the first terminal to the third terminal, henceforth denoted as $\hat{\gamma}_2$ with repetition $\nu_2 = 4$. The coefficients $K_{i=\{1,2\},j\in(1,4)}$ can be obtained by simply performing the required differentiations to arrive at

$$K_{\{1,2\},j\in(1,4)} = \frac{1}{3!}\frac{(7-j)!}{(4-j)!}\frac{\left(-\hat{\gamma}_{\{2,1\}}/\hat{\gamma}_{\{1,2\}}\right)^{4-j}}{\left(1-\hat{\gamma}_{\{2,1\}}/\hat{\gamma}_{\{1,2\}}\right)^{8-j}} \tag{13.21}$$

which allows one to calculate (13.20) in closed form for the given scenario.

Figure 13.6 depicts the normalised Shannon capacity in [bits/s/Hz] versus the SNR in [dB] for the above-given scenario. In the case of equal channel coefficients, the expectation of the square of the Frobenius norm of the normalised channel coefficients would yield u; here $u = 2\cdot 4$. For this reason, the power of the two unequal channel coefficients is chosen such that $\sum_i^8 \gamma_i = \sum_i^2 \hat{\gamma}_i \equiv 8$. The particular case where $\hat{\gamma}_1 : \hat{\gamma}_2 = 2 : 1$ was chosen, *i.e.* $\hat{\gamma}_1 = 16/3$ and $\hat{\gamma}_2 = 8/3$. The capacities of each individual link are shown, as well as the capacity when both target terminals cooperate and hence realise a 2×4 O-MIMO system. The latter case yields an expected increase in capacity due to additional receive diversity.

Note that the additional resources in terms of relaying power and bandwidth required to maintain the cooperation are not incorporated into current analysis. It is a fair assumption, however, that the cooperating terminals are spatially sufficiently close as to neglect the relaying power compared to the transmission power at the access point. Furthermore, it can be assumed that with many such VAA groups re-using the relaying bandwidth, the additional bandwidth can also be asymptotically neglected.

Approximate MIMO & O-MIMO Capacities

Distributed and cooperative MIMO communication systems require optimum resource allocation algorithms which, in the case of FDMA-based relaying, assign each relaying terminal within a relaying stage a fractional bandwidth αW and a fractional power βS, such as to maximise the end-to-end capacity. It will later be shown that the normalised capacity of a MIMO link with given

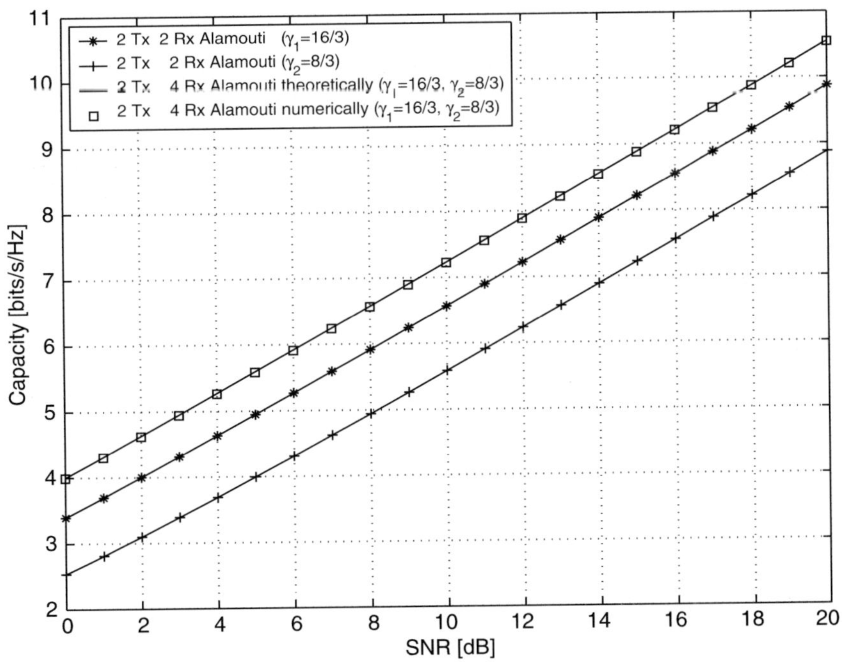

Figure 13.6. Capacity versus SNR for the scheme of Figure 13.5; $\hat{\gamma}_1+\hat{\gamma}_2 \equiv 8$, $\hat{\gamma}_1 : \hat{\gamma}_2 = 2:1$.

fractional bandwidth and power allocations can be expressed as

$$C = \alpha \cdot \mathrm{E}_\lambda \left\{ m \log_2 \left(1 + \lambda \frac{\beta}{\alpha} \frac{\gamma}{t} \frac{S}{N} \right) \right\} \tag{13.22}$$

where $m = \min\{t, r\}$, γ is associated with the pathloss and the expectation is generally calculated with the aid of (13.4) or (13.18). The optimisation process clearly involves some form of differentiation with respect to the fractional resources α and β; however, with reference to the intricate formulations of the capacities for MIMO and O-MIMO channels, an analytical approach remains intractable, which is mainly due to the logarithmic integration kernel in *e.g.*, (13.2). However, given (13.9) and (13.20), one can apply a functional approximation as suggested in [Dohler and Aghvami, 2005], *i.e.*, $\log_2(1+x) \approx \sqrt{x}$. The suggested approximation simplifies (13.22) to

$$C \approx \alpha \sqrt{\frac{\beta}{\alpha}} \sqrt{\gamma \frac{S}{N}} \cdot \Lambda(t, r) \tag{13.23}$$

This clearly decouples the fractional resources α and β from the MIMO capacity term $\Lambda(t, r) = \mathrm{E}_\lambda\{m\sqrt{\lambda/t}\}$.

The expectation with respect to the unordered eigenvalue λ is evaluated following exactly the same approach as exposed by (13.4)–(13.10), to arrive at

$$\begin{aligned}\Lambda(t, r) = \frac{1}{\sqrt{t}} \sum_{k=0}^{m-1} \frac{k!}{(k+d)!} \Bigg[& \sum_{l=0}^{k} A_l^2(k, d)\, \hat{L}_{2l+d} \\ & + \sum_{l_1=0}^{k} \sum_{\substack{l_2=0, \\ l_2 \neq l_1}}^{k} (-1)^{l_1+l_2} A_{l_1}(k, d)\, A_{l_2}(k, d)\, \hat{L}_{l_1+l_2+d} \Bigg]\end{aligned} \tag{13.24}$$

Furthermore, with reference to (13.9), the capacity integral $\hat{C}_\zeta(a)$ is replaced by the capacity approximation integral $\hat{L}_\zeta$, which is defined and solved with the aid of [Gradshteyn and Ryshik, 2000], §3.381.4, as $\hat{L}_\zeta \triangleq \int_0^\infty \sqrt{x}\, x^\zeta\, e^{-x}\, dx = \Gamma(\zeta + 3/2)$. Interestingly, the expectation in (13.23) with the pdf of (13.4) can be calculated in a more compact form with the aid of [Gradshteyn and Ryshik, 2000], § 7.414.4.1,

$$\begin{aligned}\Lambda(t, r) &= \frac{1}{\sqrt{t}} \sum_{k=0}^{m-1} \frac{k!}{(k+d)!} \int_0^\infty \sqrt{\lambda} \left[L_k^d(\lambda) \right]^2 \lambda^d e^{-\lambda} d\lambda \\ &= \frac{1}{\sqrt{t}} \sum_{k=0}^{m-1} \frac{k!}{(k+d)!} \frac{\Gamma^3(d+k+1)\Gamma(d+\frac{3}{2})\Gamma(k-\frac{1}{2})}{(k!)^2 \Gamma(d+1)\Gamma(-\frac{1}{2})} \times \\ &\quad {}_3F_2(-k, d+\frac{3}{2}, \frac{3}{2}; d+1, \frac{3}{2}-k; 1)\end{aligned} \tag{13.25}$$

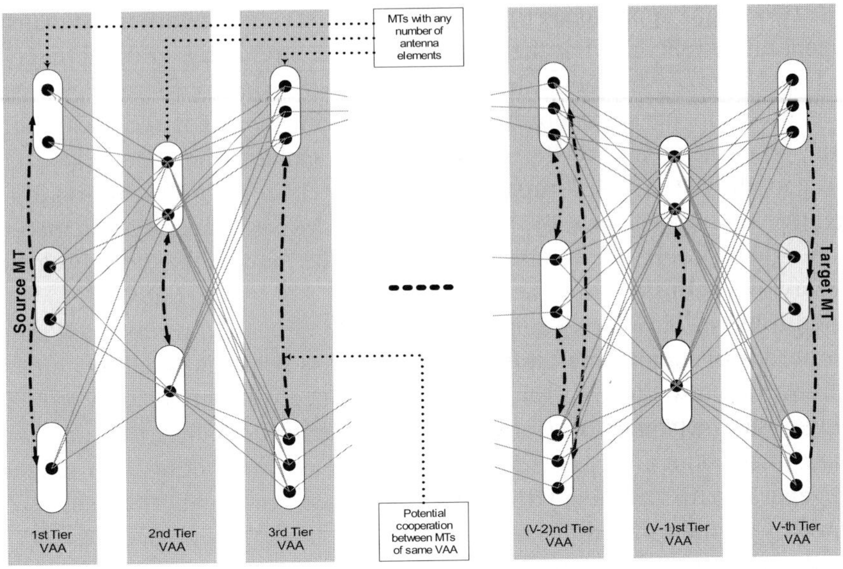

Figure 13.7. Distributed-MIMO multi-stage communication system.

where ${}_3F_2(\cdot)$ is the generalised hypergeometric function with three parameters of type 1 and two parameters of type 2. The same approach is taken to calculate the approximated capacity for O-MIMO channels, to arrive at

$$\Lambda(t,r) \quad = \quad \frac{R}{\sqrt{t}} \sum_{i=1}^{g} \sum_{j=1}^{\nu_g} K_{i,j} \cdot \sqrt{\hat{\gamma}_i} \cdot \frac{\Gamma(j+1/2)}{\Gamma(j)} \tag{13.26}$$

Finally note that a tighter approximation to the logarithmic function is possible, see *e.g.*, [Dohler and Aghvami, 2005]; however, we will utilise the simpler square root expression in the subsequent chapter.

5. Resource Allocation Protocols

It is the aim of this chapter to analyse and optimise the behaviour of cooperative distributed MIMO multi-stage communication networks with the aid of the theory developed in the previous chapter. An example realisation of the named communication network is depicted in Figure 13.7. Of major interest here is to maximise the end-to-end data throughput by optimally assigning resources in terms of frame duration, frequency band and transmission power to each of the terminals. These resources are usually constrained, thus calling for effective allocation strategies. Since resources have to be shared among all terminals involved, the allocation strategies to be developed are referred to as *fractional resource allocation strategies*.

The two major approaches to accomplish a relaying network are transparent or regenerative relaying. For the transparent case, a relaying terminal receives a signal stream on one frequency band and directly translates it for re-transmission onto another. In the regenerative case, a relaying terminal receives a signal stream, processes it and re-transmits it. From a capacity perspective it is clear that regenerative relaying outperforms transparent relaying. Additionally, regenerative relaying allows the deployment of MIMO capacity enhancement technologies at each relaying stage. These are the main reasons why regenerative relaying is considered in here.

The problem formulation to similar resource allocation problems with partial solutions to achieve maximum throughput has been analysed by numerous researchers up to today. All solutions to the respective optimisation problems, however, require some form of numerical optimisation. It is the aim of this chapter to introduce for the first time explicit resource allocation strategies for regenerative distributed MIMO multi-stage communication scenarios constrained by a total utilised power S, bandwidth W and frame duration T.

To this end, we will commence with some underlying system assumptions and a general description of how the cooperation and the end-to-end data transmission is accomplished. We will then derive the explicit allocation protocols for ergodic MIMO and O-MIMO channels.

System Model

The description here relates to the generalised VAA multi-stage communication system as depicted in Figure 13.7. Consider, a source mobile terminal (s-MT) communicating with a target mobile terminal (t-MT) via a given number of relaying mobile terminals (r-MTs). Spatially adjacent r-MTs are grouped into VAAs, thereby forming a relaying VAA (r-VAA) tier. The s-MT and t-MT themselves might be a member of a VAA, henceforth referred to as source VAA (s-VAA) and target VAA (t-VAA), respectively. The system of a s-VAA communicating with a t-VAA via several tiers of r-VAAs is referred to as a VAA multi-stage communication system. The optimum choice of r-MTs, as well as their optimum grouping into VAAs, is beyond the scope of this chapter. It is therefore assumed that the fractional resource allocation algorithms developed here are applied to the given topology.

The s-MT, t-MT and r-MTs may possess any number of antenna elements, depicted as large dots in Figure 13.7. Furthermore, the MTs may or may not cooperate among each other within the same VAA tier. The cooperative link is shown as a dash-dotted line. Each r-MT of the same r-VAA transmits the prior agreed spatial branch of a space-time code word, where the encoding bases on the received and detected symbol from the previous r-VAA tier. The resulting MIMO sub-channels are shown as grey lines.

Clearly, a cooperative deployment yields a higher capacity; however, at the expense of additional complexity, relaying power and bandwidth. The latter two are assumed to be negligible in the current analysis, as justified in the preceding chapter. The increase in capacity is thus due to more complex transceivers only, which have to accomplish intra (cooperation) and inter (multi-stage) VAA relaying. It is also assumed here that the intra (cooperation) VAA communication process is error-free. Note further that not all available antenna elements need to be active for the intra and/or inter VAA relaying process.

From a Shannon point of view, *i.e.*, $C = \lim_{T\to\infty}(WT\log_2(1+S/N)/T)$, there are two basic access methodologies available: frequency division multiple access (FDMA) and time division multiple access (TDMA). FDMA-based regenerative relaying implies that the totally available bandwidth W is orthogonally or non-orthogonally partitioned among the relaying VAA tiers; communication may occur continuously over the entire frame duration T. On the other hand, TDMA-based regenerative relaying implies that the total frame duration T is orthogonally or non-orthogonally partitioned into slots among the relaying VAA tiers; communication occurs over the entire bandwidth W.

For orthogonal relaying, available resources are divided such that no interference between the relaying stages occurs. Thus, bandwidth/frame has to be fractioned into non-overlapping frequency-bands/slots such that at any time they are used by only one relaying link. On the other hand, non-orthogonal relaying allows resources to be re-used among stages, which leads to interference between the relaying VAA tiers.

For either type of relaying, we have assumed that the average channel conditions of the full relaying chain is available to every node. This assumption clearly precludes very fast-changing topologies but is realistic enough for typically occurring communication scenarios with some form of feedback mechanisms.

The encoding, distributed relaying and decoding process is described for an FDMA-based relaying system as follows.

- **Source MT.** In an FDMA-based relaying system, the s-MT continuously broadcasts the data to the remaining r-MTs in the first relaying VAA tier, utilising negligible power and bandwidth, and possibly not deploying all of its available antenna elements.

- **First relaying VAA tier.** The first VAA relaying tier is formed by q_1 spatially adjacent MTs (including the s-MT). Each of the involved MTs possesses $n_{1,i}$ antenna elements for inter VAA relaying purposes, where the first subscript relates to the first VAA relaying tier and $1 \le i \le q_1$.

  After cooperation between the s-MT and the remaining r-MTs of the first relaying VAA tier, the data is space-time encoded according to a given code book with $t_1 = \sum_{i=1}^{q_1} n_{1,i}$ spatial dimensions. Each MT then transmits only $n_{1,i\in(1,q_1)}$ spatial dimensions such that no transmitted codeword is duplicated. Transmission from the first relaying VAA tier is accomplished at frequency band W_1 with total transmission power S_1.

- **Second relaying VAA tier.** The second VAA relaying tier is formed by q_2 spatially adjacent MTs such that their inclusion into the VAA yields capacity benefits to the communication system.

  Each of the q_2 MTs possesses $n_{2,i\in(1,q_2)}$ antenna elements. Some MTs may cooperate among each other, thereby forming Q_2 clusters, where $1 \leq Q_2 \leq q_2$. The case of $Q_2 = 1$ represents the scenario where all MTs cooperate, whereas $Q_2 = q_2$ means that none of the MTs cooperate. The former case clearly yields the best performance, however, at the expense of additional transceiver complexity to realise the cooperation; also, additional bandwidth and power are required to accomplished the relaying process. The later case yields less gains; however, it will be shown that the performance of such a system still outperforms a traditional SISO relaying system.

  The jth cluster is assumed to contain $r_{2,j}$ receive antennas, where $1 \leq j \leq Q_2$ and $\sum_{i=1}^{q_2} n_{2,i} = \sum_{j=1}^{Q_2} r_{2,j}$. Therefore, Q_2 MIMO channels are created, each with t_1 transmit antennas and $r_{2,j\in(1,Q_2)}$ receive antennas.

  After cooperation, the data is space-time decoded and re-encoded according to a given code book with $t_2 = \sum_{i=1}^{q_2} n_{2,i}$ spatial dimensions. Again, each MT then re-transmits only $n_{2,i\in(1,q_2)}$ spatial dimensions such that no re-transmitted code word is duplicated. Re-transmission from the second relaying VAA tier is accomplished at frequency band W_2 with total transmission power S_2.

- **vth relaying VAA tier.** The reception, cooperation, de-coding, re-encoding and re-transmission process is congruent to the proceedings described above. Again, Q_v MIMO channels are created. All of these MIMO channels will have t_{v-1} transmit antennas and $r_{v,j\in(1,Q_v)}$ receive antennas. After cooperation, the data is space-time decoded and re-encoded according to a given code book with $t_v = \sum_{i=1}^{q_v} n_{v,i}$ spatial dimensions. Re-transmission from the vth relaying VAA tier is accomplished at frequency band W_v with total power S_v.

- **Vth relaying VAA tier.** The final relaying tier contains the t-MT. Similar to the 1st tier, only cooperative MTs are considered here (no cooperation between the r-MTs and the t-MT would terminate the data flow in the respective r-MTs). Therefore, there will be one MIMO channel with t_{V-1} transmit antennas and $\sum_{i=1}^{q_V} n_{V,i}$ receive antennas.

- **Target MT.** After cooperation between the r-MTs and the t-MT, the data is space-time decoded and passed on to the information sink in the t-MT.

A TDMA-based system operates exactly like the above-described FDMA-based relaying system, with the only difference that all fractional bandwidths $W_{v\in(1,K)}$ need to be replaced by fractional frame durations $T_{v\in(1,K)}$. Here, K denotes the number of relaying stages and is related to the number of VAA relaying tiers via $K = V - 1$.

For any scenario, the total communication duration is normalised to T and the total bandwidth to W. For orthogonal TDMA and FDMA-based relaying systems, $T = \sum_{v=1}^{K} T_v$ and $W = \sum_{v=1}^{K} W_v$ respectively. For non-orthogonal (interfering) relaying systems, the sum of all utilised fractional resources need to add-up to T and W respectively.

End-to-End Throughput

Throughput is defined as the information delivered from source towards sink. This requires a certain duration of communication T and frequency band W. Subsequent analysis will therefore refer to the normalised (spectral) throughput Θ in [bits/s/Hz].

An ergodic channel offers a normalised capacity C in [bits/s/Hz] with 100% reliability, which allows relating capacity and throughput via $\Theta = C$. Therefore, maximising the throughput Θ is equivalent to maximising the capacity C. As defined by Shannon, capacity relates to error-free transmission. Hence, if a certain capacity was to be provided from source to sink, all channels involved must guarantee error-free transmission. From this it is clear that the end-to-end capacity C is dictated by the capacity of the weakest link.

For a cooperative VAA multi-stage relaying network with V relaying tiers as depicted in Figure 13.7, there will be $V - 1 = K$ stages each comprised of multiple MIMO channels. At each of the stages, partial cooperation may take place on the receiving side. As an example, the transmission stage from the vth VAA relaying tier to the $(v+1)^{st}$ is enlarged in Figure 13.8, where the three receiving terminals cooperate such as to yield two clusters. Generally, the clustering yields Q_{v+1} MIMO channels with t_v transmit antennas and $r_{v+1,j\in(1,Q_{v+1})}$ receive antennas. For the example below, $Q_{v+1} = 2$, $t_v = 5$, $r_{v+1,1} = 3$ and $r_{v+1,2} = 3$.

Prior to optimising the end-to-end capacity, the weakest of all Q_{v+1} MIMO channels has to be determined at each relaying stage. If the distances between the r-MTs of the same relaying VAA tier are negligible compared to the inter-VAA distances, then the strength of a MIMO channel can be measured by the number of receive antennas. It is generally desirable to guarantee mutual cooperation between terminals such that all created MIMO channels offer the same capacity, which can be achieved if they have the same number of receive

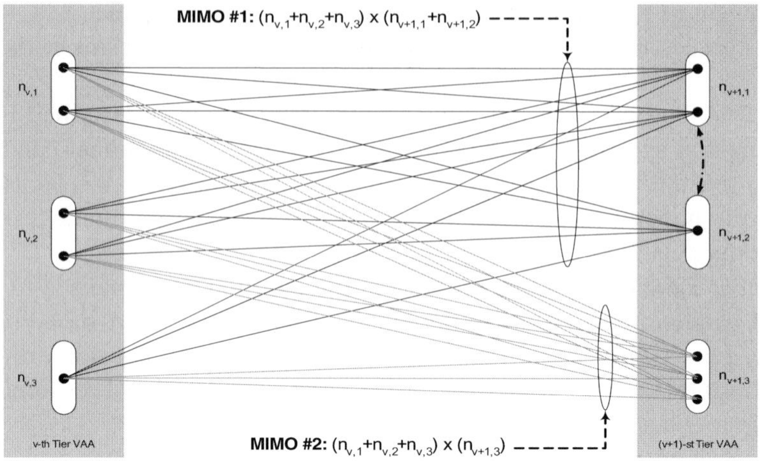

Figure 13.8. Established MIMO channels from the vth to the $(v+1)^{st}$ VAA relaying tier.

antennas. In Figure 13.8 for example, two r-MTs cooperate with a total of three receive antennas, which equates to the number of receive antennas of the non cooperating r-MT, hence achieving an optimal relaying solution.

Since optimisation has to be performed on the weakest link (or one of the equally strong links) at each of the K relaying stages, notation can be simplified further. To this end, it is assumed that the vth relaying stage has t_v antennas acting as transmitters and, to simplify subsequent notation, $r_v \triangleq \min_{j\in(1,Q_{v+1})}\{r_{v+1,j}\}$ antennas acting as receivers. This is henceforth denoted as $(t_1 \times r_1)/(t_2 \times r_2)/\ldots/(t_K \times r_K)$.

The capacity C_v of the vth relaying stage is hence determined by t_v and r_v, and the average occurring channel conditions. For an FDMA-based relaying VAA network, it is thus the aim to find for all $v = 1,\ldots,K$ stages the fractional allocations of bandwidth α_v and power β_v for given channel conditions λ_v and γ_v so as to maximise the minimum capacity C, *i.e.*,

$$C = \sup_{\boldsymbol{\alpha},\boldsymbol{\beta}}\Big\{\min\{C_1(\alpha_1,\beta_1,\lambda_1,\gamma_1),\ldots,C_K(\alpha_K,\beta_K,\lambda_K,\gamma_K)\}\Big\} \quad (13.27)$$

where $C_v(\alpha_v,\beta_v,\lambda_v,\gamma_v)$ denotes the dependency of the capacity in the vth link on the fractional resource allocations α_v and β_v, and on the channel conditions λ_v and γ_v. The optimisation is performed over the fractional sets

$\boldsymbol{\alpha} \triangleq (\alpha_1, \ldots, \alpha_K)$ and $\boldsymbol{\beta} \triangleq (\beta_1, \ldots, \beta_K)$, which are constrained by

$$\sum_{v=1}^{K} \alpha_v \equiv 1 \tag{13.28}$$

$$\sum_{v=1}^{K} \beta_v \equiv 1 \tag{13.29}$$

The normalised capacity of the vth stage is given as

$$C_v = \alpha_v \cdot \mathrm{E}_{\lambda_v} \left\{ m_v \log_2 \left(1 + \lambda_v \frac{\gamma_v}{t_v} \frac{\beta_v}{\alpha_v} \frac{S}{N} \right) \right\} \tag{13.30}$$

where the expectation is evaluated with any of the applicable pdfs given in a previous section. In [Dohler, 2003], it has been proven that (13.28)–(13.30) also holds for TDMA based VAA networks; however, a fractional bandwidth α_v translates into a frame duration α_v, whereas a fractional transmission power β_v for an FDMA-based system translates into a transmission power β_v/α_v for a TDMA-based system.

With above parameter constraints, increasing one capacity inevitably requires decreasing the other capacities. The minimum is maximised if all capacities are equated and then maximised. Hence, α_v is obtained by equating (13.30) for all $v = 1, \ldots, K$, which is derived to be

$$\alpha_v = \frac{\prod_{w \neq v} \mathrm{E}_{\lambda_w} \left\{ m_v \log_2 \left(1 + \lambda_w \frac{\gamma_w}{t_w} \frac{\beta_w}{\alpha_w} \frac{S}{N} \right) \right\}}{\sum_{k=1}^{K} \prod_{w \neq k} \mathrm{E}_{\lambda_w} \left\{ m_v \log_2 \left(1 + \lambda_w \frac{\gamma_w}{t_w} \frac{\beta_w}{\alpha_w} \frac{S}{N} \right) \right\}} \tag{13.31}$$

The end-to-end capacity $C = C_1 = \ldots = C_K$ is obtained by inserting (13.31) into (13.30), *i.e.*,

$$C = \frac{\prod_{w=1}^{K} \mathrm{E}_{\lambda_w} \left\{ m_v \log_2 \left(1 + \lambda_w \frac{\gamma_w}{t_w} \frac{\beta_w}{\alpha_w} \frac{S}{N} \right) \right\}}{\sum_{k=1}^{K} \prod_{w \neq k} \mathrm{E}_{\lambda_w} \left\{ m_v \log_2 \left(1 + \lambda_w \frac{\gamma_w}{t_w} \frac{\beta_w}{\alpha_w} \frac{S}{N} \right) \right\}} \tag{13.32}$$

Equation (13.32) is conveniently expressed as

$$C = \left[\sum_{k=1}^{K} \frac{1}{\mathrm{E}_{\lambda_k} \left\{ m_v \log_2 \left(1 + \lambda_k \frac{\gamma_k}{t_k} \frac{\beta_k}{\alpha_k} \frac{S}{N} \right) \right\}} \right]^{-1} \tag{13.33}$$

which constitutes a $2K$-dimensional optimisation problem with respect to (w.r.t.) the fractional bandwidth and power allocations α_k and β_k, respectively.

Resource Allocation Strategies

Using Lagrange's method, see *e.g.*, [Ben-Tal and Nemirovski, 2001], for maximising (13.33) under constraints (13.28) and (13.29), suggests the Lagrangian

$$\begin{aligned} \mathcal{L} &= \left[\sum_{k=1}^{K} \frac{1}{\mathrm{E}_{\lambda_k}\left\{m_v \log_2\left(1+\lambda_k \frac{\gamma_k}{t_k}\frac{\beta_k}{\alpha_k}\frac{S}{N}\right)\right\}}\right]^{-1} \\ &\quad +\iota\left[1-\sum_{k=1}^{K}\alpha_k\right]+\kappa\left[1-\sum_{k=1}^{K}\beta_k\right] \end{aligned} \tag{13.34}$$

which is differentiated w.r.t. α_k K times and then w.r.t β_k another K times. The resulting $2K$ equations are equated to zero and the system of equations is resolved in favour of any α_k and β_k, where ι and κ are chosen so as to satisfy (13.28) and (13.29).

Clearly, a pdf given in the form of (13.4) leads to $2K$ equations which are not explicitly resolvable in favour of the sought variables. This is the main reason why no explicit resource allocation strategy has been developed to date, where only numerical optimisation routines can be found in the literature. Three explicit resource allocation protocols are derived below.

(1) A near-optimum fractional bandwidth and fractional power allocation protocol can be derived by invoking the approximation introduced in the previous section. To this end, it is suggested to reduce the $2K$ dimensional optimisation problem (13.33) to a K-dimensional optimisation problem by optimising w.r.t. $\frac{\beta_k}{\alpha_k}$, the ratio between the fractional power and bandwidth allocation. In [Dohler, 2003], it has been shown that

$$\sum_{k=1}^{K} \frac{\beta_k}{\alpha_k} \approx K \tag{13.35}$$

Applying approximation (13.23) and taking (13.35) into account, (13.33) can be simplified to

$$C \approx \left[\frac{1}{\Lambda\left(t_1,r_1\right)\sqrt{\gamma_1 \frac{S}{N}}\sqrt{K-\sum_{k=2}^{K}\frac{\beta_k}{\alpha_k}}}+\sum_{k=2}^{K}\frac{1}{\Lambda\left(t_k,r_k\right)\sqrt{\gamma_k\frac{S}{N}}\sqrt{\frac{\beta_k}{\alpha_k}}}\right]^{-1}$$

which constitutes now a $(K-1)$-dimensional optimisation problem w.r.t. $\frac{\beta_k}{\alpha_k}$ with $k = 2,\ldots,K$. The approximate capacity gain $\Lambda(t,r)$ in dependency of the number of transmit antennas t and receive antennas r can be taken from (13.24)–(13.26) for the appropriate communication scenarios. The maximum end-to-end capacity is obtained by equating the first derivative of C

w.r.t. $\frac{\beta_2}{\alpha_2}, \ldots, \frac{\beta_K}{\alpha_K}$ to zero. Instead of maximising C it is more convenient to minimise $1/C$, *i.e.*,

$$\frac{\partial\left(\frac{1}{C}\right)}{\partial\left(\frac{\beta_2}{\alpha_2}\right)} = \ldots = \frac{\partial\left(\frac{1}{C}\right)}{\partial\left(\frac{\beta_K}{\alpha_K}\right)} \equiv 0 \tag{13.36}$$

After partial differentiation, one obtains

$$\begin{aligned}
\frac{1}{\Lambda(t_1,r_1)\sqrt{\gamma_1\frac{S}{N}}\left[K-\sum_{v=2}^{K}\frac{\beta_v}{\alpha_v}\right]^{\frac{3}{2}}} - \frac{1}{\Lambda(t_2,r_2)\sqrt{\gamma_2\frac{S}{N}}\left[\frac{\beta_2}{\alpha_2}\right]^{\frac{3}{2}}} &= 0 \\
&\vdots \\
\frac{1}{\Lambda(t_1,r_1)\sqrt{\gamma_1\frac{S}{N}}\left[K-\sum_{v=2}^{K}\frac{\beta_v}{\alpha_v}\right]^{\frac{3}{2}}} - \frac{1}{\Lambda(t_K,r_K)\sqrt{\gamma_K\frac{S}{N}}\left[\frac{\beta_K}{\alpha_K}\right]^{\frac{3}{2}}} &= 0
\end{aligned}$$

The $K-1$ equations can be resolved for any β_v/α_v which yields

$$\frac{\beta_v}{\alpha_v} \approx K \cdot \frac{\prod_{w\neq v}\sqrt[3]{\gamma_w \cdot \Lambda^2(t_w,r_w)}}{\sum_{k=1}^{K}\prod_{w\neq k}\sqrt[3]{\gamma_w \cdot \Lambda^2(t_w,r_w)}} \tag{13.37}$$

The fractional bandwidth allocations $\alpha_{v=(1,\ldots,K)}$ can now be obtained by substituting (13.37) into (13.31). The fractional power allocations $\beta_{v=(1,\ldots,K)}$ can finally be obtained by inserting the prior obtained α_v into (13.37) and solving for β_v. Since the derived fractional resource allocation rules are the result of various approximations, they have to be applied with care.

First, one has to make sure that the constraints (13.28) and (13.29) hold for the derived α_v and β_v. Therefore, it is suggested to derive $K-1$ coefficients α_v and β_v, and then obtain the remaining two α_v and β_v from (13.28) and (13.29). Second, it is suggested to obtain the $K-1$ coefficients α_v and β_v from the first $K-1$ strongest links, where the strength is determined by $\gamma_w \cdot \Lambda^2(t_w, r_w)$. Third, the obtained K capacities $C_{v\in(1,K)}$ are not entirely equal, which is again due to the approximations deployed. The end-to-end capacity utilising the above-given technique is hence obtained by choosing the minimum of all C_v. The flowchart in Figure 13.9 summarises the method in obtaining the fractional bandwidth and power allocations.

(2) An equal fractional bandwidth but optimised fractional power allocation protocol is derived here to facilitate a comparison with the above-developed

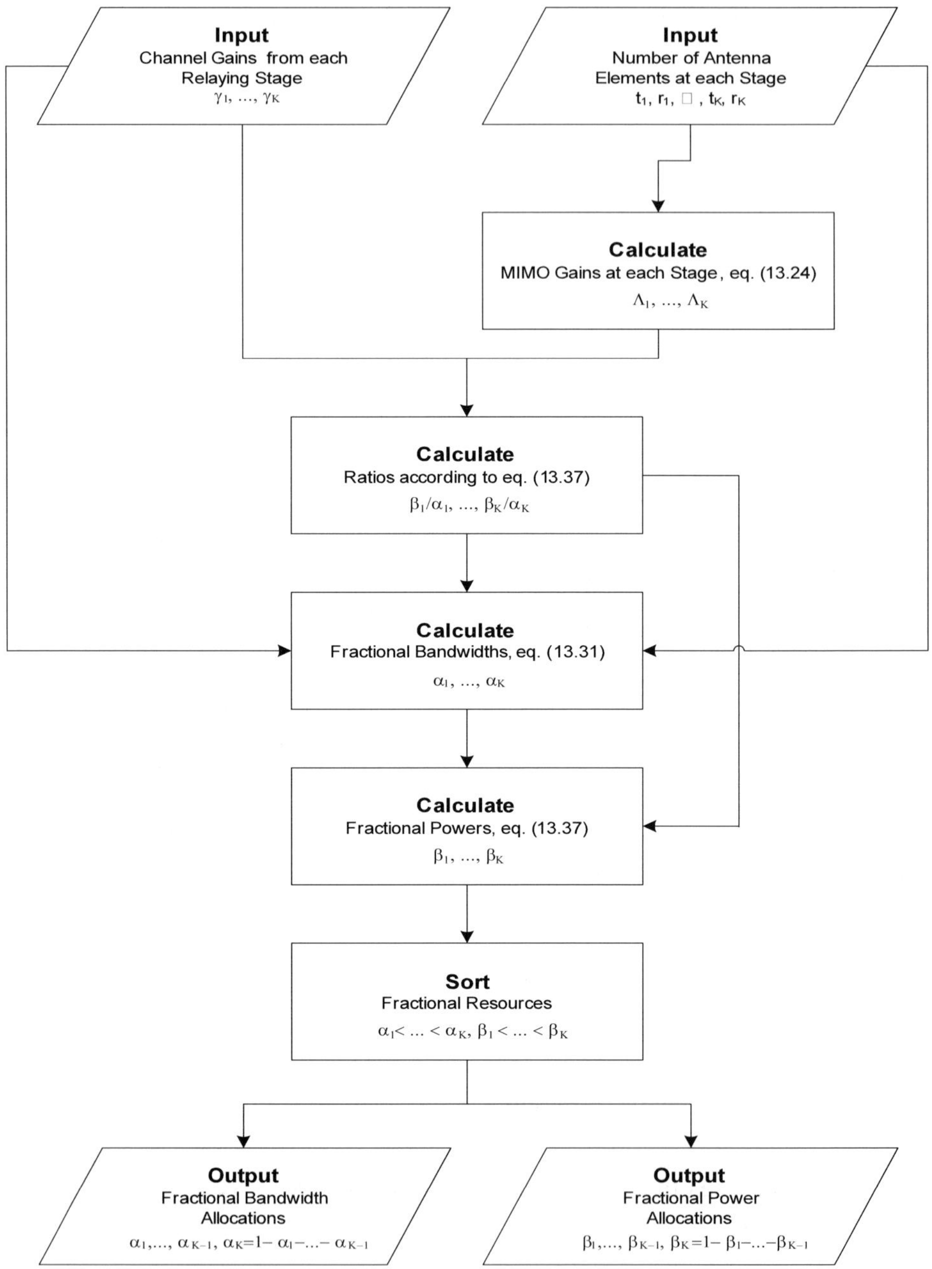

Figure 13.9. Flowchart specifying the algorithmic method for determining the fractional bandwidth and power.

fractional resource allocation algorithm. With the fractional power β_v to be optimised and equal bandwidth $\alpha_v = 1/K$, (13.30) turns into

$$C_v = \frac{1}{K} \cdot \mathrm{E}_{\lambda_v} \left\{ m_v \log_2 \left(1 + \lambda_v \beta_v K \frac{\gamma_v}{t_v} \frac{S}{N} \right) \right\} \tag{13.38}$$

$$\approx \sqrt{\beta_v} \sqrt{\frac{\gamma_v}{K}} \sqrt{\frac{S}{N}} \Lambda(t_v, r_v) \tag{13.39}$$

Equating all capacities C_v given by (13.39) and applying constraint (13.29) allows one to resolve the set of equations in favour of any β_v as

$$\beta_v \approx \frac{\prod_{w \neq v} \gamma_w \cdot \Lambda^2(t_w, r_w)}{\sum_{k=1}^{K} \prod_{w \neq k} \gamma_w \cdot \Lambda^2(t_w, r_w)} \tag{13.40}$$

which, when inserted into (13.38), yields the end-to-end capacity C. Again, the obtained capacities C_v do not entirely coincide because of the approximation chosen. The end-to-end capacity will therefore be dominated by the smallest of all. Note finally that the case of optimised bandwidth and equal power has not been considered here because of negligible applicability.

(3) An equal fractional bandwidth and power allocation protocol yields for the capacity at each stage simply $C_v = \frac{1}{K} \cdot \mathrm{E}_{\lambda_v} \left\{ m_v \log_2 \left(1 + \lambda_v \frac{\gamma_v}{t_v} \frac{S}{N} \right) \right\}$, and the end-to-end capacity C is obtained by choosing the minimum of all C_v.

6. Case Studies & Observations

The developed fractional resource allocation algorithms are assessed below for various VAA relaying scenarios. The simplest scenario is the 2-stage relaying scenario with only one relaying VAA tier. In addition to this, the 3-stage relaying configuration is assessed. More relaying stages have not been analysed here due to the lengthy numerical optimisation.

The obtained graphs are generally labelled on the parameter p defined as

$$p \triangleq \left[10 \log_{10} \left(\frac{\gamma_1}{\gamma_1} \right), 10 \log_{10} \left(\frac{\gamma_2}{\gamma_1} \right), \ldots, 10 \log_{10} \left(\frac{\gamma_K}{\gamma_1} \right) \right] \tag{13.41}$$

which characterises the relative strength in dB of the K relaying stages with respect to the first stage.

The 2-Stage VAA Relaying Scenario

The precision and applicability of the derived resource allocation strategies is assessed here for various antenna configurations of the 2-stage VAA relaying scenario.

(1) Single Antenna Element. The derived resource allocation strategies are obviously also applicable to traditional relaying networks with one antenna element per MT. The precision of the developed fractional resource allocation algorithm is assessed in Figure 13.10(a). It depicts the optimum end-to-end capacity obtained via numerical optimisation on (13.34) and the approximate end-to-end capacity obtained from (13.37), (13.31) and (13.30) versus the SNR in the first relaying stage. The graphs are labelled on the parameter p as defined in (13.41) with $K = 2$, where the second relaying channel is 10dB and 5dB stronger than the first one, equally strong than the first one, and 5dB and 10dB weaker than the first one.

It can be observed that the exact and developed end-to-end capacities almost coincide. The error was found not to exceed 3% for any of the depicted cases. The developed explicit resource allocations are hence a powerful tool in obtaining a near-to-optimum end-to-end capacity without the need for lengthy numerical optimisations. The algorithm is shown to be applicable for channels with attenuations differing by magnitudes.

Figure 13.10(b) compares the obtained end-to-end capacities of various allocation strategies, where the curves are labelled on $p = ([0, 10], [0, 0], [0, -10])$dB. The numerically obtained optimum allocation strategy is depicted together with the developed strategies of optimised bandwidth and power, equal power but optimised bandwidth, and equal bandwidth and equal power. When both links are equally strong, *i.e.*, $p = [0, 0]$dB, then all of the considered allocation strategies yield the same end-to-end capacity. This is obvious because for the given symmetric communication scenario, resources have to be shared equally between the relaying terminals.

When the second link is 10dB weaker, *i.e.*, $p = [0, -10]$dB, then optimising bandwidth and power or optimising power only yields close to optimum performance. This is because for low SNR, $\log(1 + x) \approx x$, which, with reference to (13.30), makes the optimisation problem independent of the bandwidth α_v. When no optimisation is performed then the end-to-end capacity is dictated by the weakest link, here the second link which is 10 times weaker than the first one. The capacity is considerably lower than for the optimised cases; at an SNR of 6dB a loss in rate of 40% can be observed, whereas at a rate of 0.4 bits/s/Hz approximately 40% more power is required.

When the second link is 10dB stronger, *i.e.*, $p = [0, 10]$dB, then optimising bandwidth and power yields close to optimum performance, whereas only optimising power does not. This is because for high SNR, the dependence of the end-to-end capacity on the bandwidth α_v increases. At an SNR of 6dB, a relative loss of approximately 10% occurs. As for the case where no optimisation is performed, the end-to-end capacity is dictated by the first link which yields the same end-to-end capacity as for $p = [0, 0]$dB. Here, a relative loss of 30% occurs at an SNR of 6dB, or 50% more power is required to maintain a rate of 1

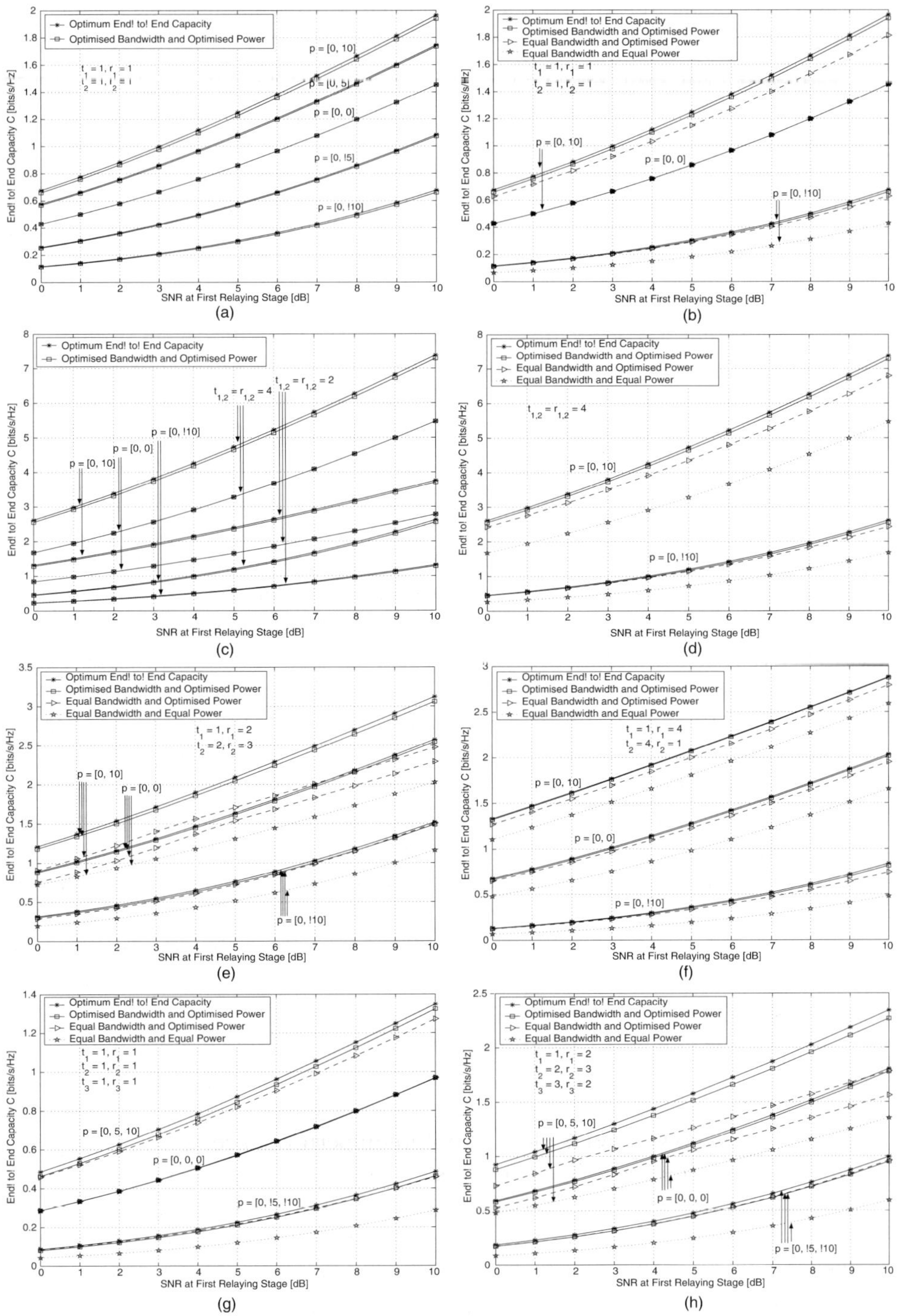

Figure 13.10. End-to-end throughput for optimum, near-optimum and sub-optimum resource allocation protocols for various two and three stage relaying networks.

bit/s/Hz. Note that the absolute loss in bits/s/Hz is much higher as for the case of $p = [0, -10]$dB.

For decreasing differences in the attenuations between the links, the relative and absolute errors in the maximum achievable end-to-end capacities of all developed allocation strategies decrease. The example of 10dB difference has been chosen to obtain some upper bounds on the occurring errors.

(2) Multiple Antenna Elements. The algorithms are now scrutinised for communication scenarios where the MTs possess multiple but equal number of antenna elements. Figure 13.10(c) is the equivalent to Figure 13.10(a), with the only difference that each MT possesses two or four transmit and receive antennas. Again, the occurring errors between derived allocation strategy and an optimum allocation is below 3%.

Figure 13.10(d) is the equivalent of Figure 13.10(b), with the only difference that each terminal possesses 4 antenna elements; note that the $p = [0, 0]$dB case has been omitted here due to the symmetric communication scenario. The same comments on the precision of the algorithms as above apply. For a second link being 10 times weaker than the first link, the allocation of optimised bandwidth/power and power only yield close to optimum performance, whereas no optimisation leads to a loss of 40% at an SNR of 6dB, or approximately 75% more power is required to maintain 1 bit/s/Hz. When the second link is 10 times stronger than the first one, then the losses from optimum to optimised power only is about 8%, whereas from optimum to no optimisation a loss of about 30% occurs at an SNR of 6dB or, alternatively, 85% more power is needed to accomplish 4 bits/s/Hz.

The demonstrated performance gains and power savings clearly underline the merit of the developed fractional resource allocation algorithms.

(3) Differing Antenna Elements. The importance of the developed strategy, however, becomes apparent when the 2-stage communication scenario is optimised for terminals with a different number of antenna elements. The precision of the fractional resource allocation algorithm, as well as its performance gains when compared to sub-optimal solutions, is exposed in Figures 13.10(e) and 13.10(f).

In particular, Figure 13.10(e) depicts the case where a s-MT with one antenna element communicates with a t-MT (or t-VAA) with three elements via a relaying stage, which effectively provides two relaying antennas. The asymmetry of the gains provided by the respective distributed-MIMO relaying stages causes the sub-optimum allocation strategies not to overlap with the optimum one for $p = [0, 0]$dB. Furthermore, non-linearities can be observed in the end-to-end capacity for the case of optimised power only, which is due to the approximation utilised in the derivation of the allocation strategy. In fact, one can observe a breakpoint which divides the zones where one or the other approximate capacity dominates the end-to-end capacity.

However, the optimised fractional bandwidth and power allocation strategy yields close to optimum performance, even if the link attenuations and the created MIMO configurations differ significantly. Additionally, the gains obtained from a bandwidth/power optimised system when compared to a power optimised or non-optimised system increase with the balance between both links decreasing. For instance, for $p = [0, 10]$dB, the first (1×2) MIMO link is much weaker than the second (2×3) link. At an SNR of 6dB the capacity losses of a power optimised system is then 20% and of a non-optimised system a considerable 40%. Alternatively, the power required to maintain 2 bits/s/Hz is about 55% higher for the power optimised system, whereas a non-optimised system requires 120% (!) more power.

Figure 13.10(f) depicts a $(1 \times 4)/(4 \times 1)$ scenario. The same tendencies as already described can be observed; additionally, the precision of the developed fractional bandwidth/power allocation strategy is once more corroborated.

The 3-Stage VAA Relaying Scenario

The 3-stage relaying scenario is dealt with in less detail as for the 2-stage case, which is due to the increased simulation times and the increased number of potentially different communication scenarios. To this end, Figures 13.10(g) and 13.10(h) depict the performance of the developed resource allocation algorithms for the 3-stage communication scenario.

Explicitly, Figure 13.10(g) deals with the case of only one antenna element per MT, *i.e.*, a $(1 \times 1)/(1 \times 1)/(1 \times 1)$ relaying scenario. The case of equally strong links when $p = [0, 0, 0]$dB yields an equal performance for any of the allocation strategies, which is again due to the scenario's symmetry. The allocation strategies, however, deviate from the numerically obtained optimum when the links are unbalanced; the error was found to be below 3%. Figure 13.10(h) shows the performance of the allocation strategies for the relaying scenario of $(1 \times 2)/(2 \times 3)/(3 \times 2)$. Here, the approximation error was found not to exceed 5%.

In summary, sufficiently precise fractional bandwidth and power allocation algorithms have been developed for a variety of distributed-MIMO multi-stage networks communicating over ergodic channels. The exposed algorithms are of very low complexity, yet they perform near-optimum. That renders a numerical optimisation within each mobile terminal superfluous.

Note that the complexity of numerical optimisation routines is prohibitively high and hence not applicable. For example, to find an optimum fractional bandwidth and power allocation for a simple example as depicted in Figure 13.10(g), the numerical optimisation required 5min per point on a Pentium III, 800MHz. That is in contrast to the developed algorithms, which take a fraction of a second to be calculated.

Conclusions

This chapter has introduced the concept of virtual antenna arrays, a communication paradigm where spatially adjacent terminals cooperate among each other and thereby form a MIMO-like communication system. We have briefly elaborated on the historical developments related to VAA, from its infancy in 1999 to analytical approaches today. It has been observed that the applicability of VAA-type systems has shifted from the cellular context to wireless ad hoc and sensor networks.

The reason for the gaining momentum of cooperative and distributed communication topologies, such as VAAs, is their ability to boost capacity and their inherent attribute of scalability. Indeed, it is in hot-spots where an increasing amount of users competes for the same capacity; however, they bring along an increasing number of terminals which, if cooperating, counteract the decrease in available system capacity by further increasing it.

It has then been the aim to develop applicable resource allocation protocols for a fairly generic relaying topology which encompasses a data flow from a data source towards a data sink via a given relaying topology. It has been assumed that Shannon transceivers are available, that the channel is ergodic, and that each terminal is aware of the average channel conditions in the network. We have also assumed that the terminals posses a physical layer (PHY) which is capable of adapting its transmission power; we have further assumed that a reservation-based medium access control (MAC) protocol is in place which regulates the access to the wireless medium by means of adaptive frame-durations (TDMA) or bandwidths (FDMA).

Under these assumptions, we derived near-optimum resource allocation protocols of low complexity and high reliability, where throughput maximising transmission power, bandwidth and frame duration have been exposed. In fact, the performed analysis constitutes a cross-layer optimisation, because PHY and MAC are dynamically adapted so as to guarantee optimum end-to-end performance.

The development of such cross-layer optimised communication protocols has been facilitated by the exact solutions to the Shannon capacity of general MIMO systems and orthogonalised MIMO systems with arbitrary sub-channel statistics. Some approximate expressions to these capacities have also been developed which were vital in deriving the allocation protocols in analytically closed form.

The exposed work is, of course, far from complete. One can think of many more distributed and cooperative communication topologies, some of which are addressed in this book by our colleagues; others still need to be discovered and analysed. Many research issues are still untouched, *e.g.* the case of interference-limited distributed relaying, wideband and/or non-ergodic fading channels, transceivers of finite complexity, etc. This, however, is left to the reader to be discovered.

References

Alamouti, S. M. (1998). A simple transmit diversity technique for wireless communicationss. *IEEE J-SAC*, 16(8).

Ben-Tal, A. and Nemirovski, A. (2001). Lectures on modern convex optimization. *SIAM, Philadelphia.*

Cover, T. and el Gamal, A. (1979). Capacity theorems for the relay channel. *IEEE Trans. Inform. Theory*, 25(5):572–584.

Cover, T. and Thomas, J. A. (1991). *Elements of Information Theory.* John Wiley & Sons, Inc.

Dohler, M. (2003). *Virtual Antenna Arrays.* PhD thesis, King's College London.

Dohler, M. and Aghvami, A. H. (2005). On the approximation of MIMO capacity. *IEEE Transactions on Wireless Communications, Letter*, 4(1): 30–34.

Dohler, M., Said, F., Ghorashi, A., and Aghvami, A. H. (June 2001). Improvements in or relating to electronic data communication systems. *Patent Publication No. WO 03/003672.*

Foschini, G. J. (1996). Layered space-time architecture for wireless communication in a fading environment when using multi-element antennas. *Bell Labs Technical Journal*, 1(2):41–59.

Foschini, G. J. and Gans, M. J. (1998). On limits of wireless communications in a fading environment when using multiple antennas. *Wireless Personal Communications*, 6:311–335.

Gradshteyn, I. S. and Ryshik, I. M. (2000). *Table of Integrals, Series, and Products.* Academia Press, 6th edition.

Grossglauser, M. and Tse, D. (2002). Mobility increases the capacity of ad hoc wireless networks. *IEEE ACM Trans. on Networking*, 10(4).

Gupta, P. and Kumar, P. R. (2000). The capacity of wireless networks. *IEEE Trans. Inform. Theory*, 46(2):388–404.

Gupta, P. and Kumar, P. R. (2003). Towards an information theory of large networks: An achievable rate region. *IEEE Trans. Inform. Theory*, 49: 1877–1894.

Harrold, T. J. and Nix, A. R. (2000). Capacity enhancement using intelligent relaying for future personal communications system. *Proceedings of VTC-2000 Fall*, pages 2115–2120.

Holma, H. and Toskala, A. (2000). *W-CDMA for UMTS: Radio Access for Third Generation Mobile Communications.* John Wiley & Sons, Inc.

Kang, M. and Alouini, M. S. (2002). On the capacity of mimo rician channels. *0th Allerton Conference on Communication, Control, and Computing*, pages 936–945.

Laneman, J. N. (2002). *Cooperative Diversity in Wireless Networks: Algorithms and Architectures*. PhD thesis, MIT.

Larsson, E. G. and Stoica, P. (2003). *Space-Time Block Coding for Wireless Communications*. Cambridge University Press.

Nabar, R. U., Bölcskei, H., and Paulraj, A. J. (2002). Outage performance of space-time block codes for generalized mimo channels. *submitted to IEEE Trans. on Inform. Theory*.

Sato, H. (1976). Information transmission through a channel with relay. *The Aloha System, University of Hawaii, Honolulu, Tech. Rep. B76-7*.

Sendonaris, A., Erkip, E., and Aazhang, B. (1998). Increasing uplink capacity via user cooperation diversity. *IEEE ISIT*, page 196.

Sendonaris, A., Erkip, E., and Aazhang, B. (2003a). User cooperation diversity - part i: System description. *IEEE Transactions on Communications*, 51(11):1927–1938.

Sendonaris, A., Erkip, E., and Aazhang, B. (2003b). User cooperation diversity - part ii: Implementation aspects and performance analysis. *IEEE Transactions on Communications*, 51(11):1939–1948.

Shannon, C. E. (1948). A mathematical theory of communication. *Bell Syst. Tech. J.*, 27(4):379–423, 623–656.

Shin, H. and Lee, J. H. (2003). Closed-form formulas for ergodic capacity of mimo rayleigh fading channels. *IEEE ICC*, pages 2996–3000.

Stefanov, A. and Erkip, E. (2003). Cooperative space-time coding for wireless networks. *Proc. IEEE ITW*.

Tarokh, V., Jafarkhani, H., and Calderbank, A. (1999). Space-time block codes from orthogonal design. *IEEE Trans. Inform. Theory*, 45(5):1456–1466.

Tarokh, V., Seshadri, N., and Calderbank, A. (1998). Space time codes for high data rate wireless communication: performance criterion and code construction. *IEEE Trans. Inform. Theory*, 44(2):744–765.

Telatar, E. (1999). Capacity of multi-antenna gaussian channels. *European Trans. on Telecomm.*, 10(6):585–595.

van der Meulen, E. (1971). Three-terminal communication channels. *Adv. Appl. Prob.*, 3:120–154.

Verdu, S. (1998). Fifty years of shannon theory. *IEEE Trans. on Inform. Theory*, 44(6):2057–2078.

Vodafone (1999). Opportunity driven multiple access. *3rd Generation Partnership Project, Technical Specification Group Radio Access Network*, (3G TR 25.924 V1.0.0).

Vucetic, B. and Yuan, J. (2003). *Space-Time Coding*. John Wiley & Sons, Inc.

Wornell, J. N. Laneman G. W. (2000). Energy-efficient antenna sharing and relaying for wireless networks. *IEEE WCNC*.

Zeng, Z., Dohler, M., and Aghvami, A. H. (2002). System performance of a w-cdma based network with deployed vaa. *ICT 2002*.

Chapter 14

COOPERATION IN 4G NETWORKS

Cooperating in a heterogeneous wireless world

Marcos D. Katz
Samsung Electronics
marcos.katz@ieee.org

Frank H. P. Fitzek
Aalborg University
ff@kom.aau·dk

Abstract: This chapter explores the forthcoming generation of mobile communication systems - the Fourth Generation or 4G - with particular emphasis on motivating the use cooperative techniques within its vast realm. The purpose of this chapter is two-fold, to approach and define 4G from different perspectives, and to identify opportunities for cooperation. A number of key 4G challenges are discussed and some solutions based on cooperative techniques are considered. 4G is usually seen as a convergence platform, where heterogeneous networks coexist. We consider that a tighter interaction among networks is beneficial, leading towards a wider and more ambitious approach to 4G, namely seeing 4G as a cooperating platform where resources are shared and traded by the constituent networks. 4G is in principle a fertile ground for developing and applying cooperative concepts and techniques, offering for the fist time opportunities to apply these ideas in a broad domain, at many different levels and within and across layers.

Keywords: future wireless networks, heterogeneous systems, research challenges

1. Introduction

The ever flourishing mobile and wireless communications scene entered this century with unprecedented momentum, as it is easily perceived by observing

Frank H. P. Fitzek and Marcos D. Katz (eds.), Cooperation in Wireless Networks: Principles and Applications, 463–496.

the unrelenting research and development activities, as well as the increasing levels of acceptance and penetration of such technology on a worldwide scale. Today, at the same time that the deployment of the third generation (3G) of mobile communications systems - also known as IMT-2000 networks - is taking place, the next generation systems are already being conceived and developed. Indeed, at this 3G introductory stage, efforts are being made at a global scale to envision, define and develop the successor mobile communication system. Systems beyond IMT-2000 are commonly referred to as the *Fourth Generation* or *4G* in short. Preliminary exploration tends to show that a great deal of useful and interesting services could be developed under the assumption that a ubiquitous, high-speed wireless access is available. The opposite is also true: future users will be attracted by rich-content based services that pervasively interact with the environment. Thus, it appears that one of the main driving forces for 4G development is the growing demand for higher data throughput in virtually every possible scenario. The key players in the 4G development process are terminal and infrastructure equipment manufacturers, academia, operators, service providers, regulatory bodies and governmental agencies. Considering the complex interaction among the aforementioned players and taking into account that these diverse parties do not necessarily share the same interests, goals and time plans, no one would be surprised to realize that finding a universal definition of 4G has been a very elusive task, even after several years of activities and numerous attempts in the literature. In this chapter we first endeavor to describe and define 4G, highlighting the common views shared by the research and development community. We consider as the main official guideline the ITU-R Framework Recommendation M.1645 [ITU, 2003], which delineates the research goals for system capabilities. The 4G arena is inherently fragmented, as involved parties represent various typically non-aligned sectors. We can see that the worldwide 4G development is following several paths, with target solutions that can be *complementary* (co-existing) as well as *competing* (mutually exclusive).

In this chapter we will initially look at 4G from different perspectives in an attempt to identify its most important characteristics and capabilities. As it will be discussed later, *heterogeneity* and *convergence* are two of the most distinctive features of 4G, and they apply to networks, terminals and services. 4G offers opportunities to the designer to widely adopt several recently developed technologies. One of the most important connotations of 4G is the departure from many conventional solutions used in previous generations. Multi-antenna techniques, justifiably identified as one of the key enabling technologies, mean the departure of relatively simple single-antenna transceivers to systems supporting several parallel receive and transmit branches. Network architectures are expected to be highly diversified in 4G, with a more balanced participation of centralized and distributed network approaches. Interaction among wireless entities is expected to be considerably strengthened in a mutual effort to better use resources and improve performance, leading to cooperation. Cooperating wireless entities include not only concrete or tangible devices like a wireless terminal in the visible domain of the user, but also, and most importantly for designers, layers (of the OSI stack), algorithms, networks, processors, etc. In this chapter we will focus on cooperative techniques, particularly those which appear to have potential to solve many of the technical challenges of 4G. Many of these technical solutions will be first introduced with 4G. Probably in most of the cases we cannot expect that technology will be fully exploited. Initial multi antenna design will favor simple configurations, in particular at the terminal side, where very few antennas will be used, expectedly not more than two in small form factor devices. As far as cooperative techniques are concerned, 4G will be the very first real communication system where these techniques will be implemented, mostly to solve some problems of particular networks, like range extension, enhancement of quality of service (QoS) and others. Rather simple solutions are expected to prevail initially, for instance low-complexity mechanisms implemented at the physical and MAC layers.

2. Defining 4G

During recent years we have witnessed innumerable attempts aiming to define 4G, examples of some recent approaches can be found in [Kim and Prasad, 2006a], [Frattasi et al., 2006], [Bria et al., 2001], [Kupetz and Brown, 2003], [Katz and Fitzek, 2005]. Despite huge efforts by industry and academia a well established and widely accepted definition of 4G has not yet emerged. Moreover, though the term 4G is widely used, it is not endorsed by all involved parties. Other denominations as Beyond 3G (B3G), In particular, the ITU refers to *beyond IMT-2000* in lieu of 4G. A vertical approach to 4G, the *linear vision* tends to see 4G as a linear extension of current 3G systems, basically aiming for higher data rates. This vision is limited to highlight the high speed capabilities of future communication systems. The horizontal approach, or *concurrent vision* of 4G is based on the integrative role of 4G as a convergence platform of several networks, and includes the linear vision as one of its constituent component networks. The latter approach is well in line with the visions of the ITU. Indeed, the ITU-R Recommendation M.1645 [ITU, 2003], states that future wireless communications systems could be realized by functional fusion of existing, enhanced and newly developed elements of current 3G systems, nomadic wireless access systems and other wireless systems with high commonality and seamless inter-working. The ITU approach is generous and flexible, truly allowing legacy systems (2G and 3G), the products of their evolutionary development, and new systems to coexist, each being a component part of a highly heterogeneous network, the 4G network. Backward compatibility and interoperability are key characteristics of 4G. We can expect that the 4G arena will be highly competitive as telecom (mobile) and IT (wireless) communication industry will contend to attain an important share of the business. Note that the words *mobile* and *wireless* are often used to emphasize and differentiate the conventional cellular (wide-area) and local-area approaches, respectively. In the sequel of this section, we will discuss 4G from different perspectives, in an attempt to get a comprehensive insight into what is understood by 4G.

A Multifaceted Approach to 4G

In this section we will approach and describe 4G systems form different perspectives in an attempt to provide a comprehensive overview of the future wireless communication systems.

Brief historical perspective. In a period spanning not more than a quarter of century three mobile communication generations were developed and deployed. The first generation of mobile communication systems, denoted by 1G, provided voice-only services. Users were separated in the frequency domain by implementing Frequency Division Multiple Access (FDMA) in the analog domain. The first generation systems already exploited the basic concepts of mobile communications, namely a centralized cellular architecture. The concept of handover to provide uninterrupted communications across wide area cells as well as roaming across regions or countries. The Second Generation (2G), introduced in the 1990's, made mobile telephony truly popular and widespread, being the reigning mobile technology of today. From the communications perspective, 2G meant the departure from an analog world to the digital one, with all the advantages that this implied. These Time Division Multiple Access (TDMA) based systems offered evident advantages to end users and operators, including high quality voice services, primitive though very popular data services (*i.e.*, Short Message Services), global mobility, increased network capacity, etc. In particular GSM (Global System for Mobile Communications), the most representative 2G system, and its immediate successors represent the most widespread mobile system today. The so called 2.5G extended 2G with data service and packet switching capabilities, bringing Internet into the mobile personal communications. 2G was designed from the beginning as an evolving platform, from which emerged the High Speed Circuit Switched Data (HSCSD), the General Packet Radio System (GPRS) and the Enhanced Data Rates for GSM Evolution (EDGE). 3G, exploiting Code Division Multiple Access (CDMA) techniques, was developed in the late 1990's and is now being deployed globally. Among the key characteristics of 3G, also known as Universal Mobile Telecommunication System (UMTS), we highlight the support of higher data rates (*e.g.*, a few hundred Kbit/s typically) and as well as the provisioning of intersystem handover, *e.g.* 3G-WLAN. Like its predecessor,

3G was conceived as an evolutionary platform, with an evolutionary phase targeting data rates into the 10 Mbit/s range. The enhanced data capability of 3G is the driving force behind mobile multimedia services. As users can now wirelessly and quickly access information databases we are starting to glimpse the birth of person-to-machine communications. The wireless personal communications history is very short but indeed extremely dynamic and rich in accomplishments. The most impressive fact is perhaps to realize that as today (2006) one third of the world population is a mobile phone subscriber, while at the beginning of the 1980's there were virtually none. By year 2010 half of the globe's population is expected to own a mobile handset. At that time, the most optimistic figures already foresee the launching of 4G systems. It is clear that the explosive growth and predicted subscriber base form a very fertile ground for the development of a new communication system. Mobile users are likely to expect a variety of new interactive and on-demand services exploiting high-speed data transfer and location-based capabilities, among others. The upcoming 4G system is projected to solve still-remaining problems of the previous generations and, moreover, to provide a convergence platform that will offer clear advantages in terms of services as well as coverage, bandwidth, spectrum usage, and devices. The terms *convergence platform* specifically refer to the fact that 4G is seen as a wireless ecosystem whose components are different wireless networks interworking in harmony. The world today is comprised of mutually exclusive wireless networks, which are now beginning to work with each other. The needs for better interconnected wireless systems are obvious: better, more reliable and continuous service, wider coverage, and other benefits that mobile users will certainly appreciate. We are witnessing today the beginning of this trend, and this is giving us a clue of what we can expect in the future, the convergence of wireless networks. Thus, we could interpret the future 4G network as a system where heterogeneous networks (3G, WLAN, new very-high speed wireless networks, etc.) would interoperate in a seamless manner. Moreover, from a broader perspective, the overall composite network, a true mosaic of networks, would appear as a single network. Future users will not see the underlying complexity, they will simply be always connected to *the network*.

A user-centric approach to 4G. During the development of previous and current mobile communication generations, industry mainly focused on the appropriate technology for providing voice and basic data communications. Services and applications were then developed based on technical capabilities supported by those systems. This approach worked well in 2G, because of the simplicity and novelty of the services offered. In 3G, services were developed in a later phase and despite the enhanced information transfer capabilities, we are finding it difficult to lure users. It is becoming more and more evident that 4G will need to be approached from a different perspective if we want to ensure its commercial success. Technology should be developed to match user's needs, and not the other way around. Indeed, putting the user in the center of the development aims to guarantee a long-lasting, sound and profitable future for 4G. Rather than being attracted by figures like high throughput numbers, users are drawn by useful, convenient and enjoyable services. These services could certainly exploit high data rate capabilities, but it is the services not the data rates that appeal to users. It should be also mentioned that the patterns in mobile user behavior today differ greatly from those prevalent during the development of 2G and 3G. Broadband wired internet is finding its way to every home and users are likely to expect comparable features on the move. The actual trend towards the diversification of terminal capabilities supporting high-quality audio as well as still and video imaging opens up a new world in terms of services and applications. These new terminal capabilities are expected to have an impact not only on the overall data traffic but also on the typically assumed uplink-downlink data traffic imbalance. Indeed, data requirements are expected to increase substantially in the uplink direction. Ultimately, mobile users will become content providers.

A user-centric approach tries to see and understand the user in different contexts aiming to extract some information that will lead to the development of possible scenarios and ultimately, their associated services. The user can be considered as *a*) an isolated individual with personal and (somewhat) unique needs, *b*) a member of a distinctive group with some common characteristic and *c*) an infinitesimal constituent part of the society. These setups will give us some hints on the user needs and expectations, which in turn will be starting point for identifying services likely to appeal the users. Some work has already been done in identifying scenarios and developing services, mostly at the Wireless World Research Forum (WWRF) [WWRF, 2006], in particular in the Book of Visions [BoV, 2001] as well as at the Mobile IT Forum (mITF) [mITF, 2006], in the "Flying Carpet" report [Carpet, 2004]. These fundamental issues are also discussed at the Samsung 4G Forum. We highlight here the following user trends:

- *Mobile user as an information sink/source*: Users are avid consumers of information. Accessing information and knowledge is always valued by users. Information comes in many forms, multimedia formats becoming the de facto presentation style. Mobile users are also becoming avid producers of information, sharing pictures, video, and commentary from wherever they are.

- *Multi-access services*: There are different approaches describing this trend. In general it refers to the delivery of services to multiple devices over multiple networks. It can be also understood as the delivery of service to a terminal equipped with multiple air interfaces in a fashion that more than one air interface are simultaneously exploited. Regardless of the interpretation, multi-access services describes very important emerging areas of research with high potential for developing countless applications. *Pervasive connectivity*: Ubiquity of services drastically increases the value of the information. The wider the coverage and the quicker the retrieval, the better.

- *Personalization of devices and services*: Personal preferences and the uniqueness of each user should be taken into account to allow differentiation of users.

- *Simplicity*: User friendliness is highly valued. Natural, transparent, intuitive and minimal interaction between man and machine will help reduce the gap between people and technology.

- *Predisposition to interact*: Interaction with other mobile users will become more tangible, leading to positive attitude towards cooperation and ultimately paving the way for cooperative communications among users. Users would consent to have their terminals processing signals other than their own provided there is a clear benefit from cooperation (*e.g.*, increased QoS, reduced communication cost, etc.) and provided the interaction is secure (*e.g.*, no possibility for the signal to be tapped by intermediate nodes).

- *User-driven mobile technology*: High data rates alone do not appeal users; useful and attractive services/applications exploiting high data throughputs are likely to please the users.

- *Core life values*: Technological innovations supporting the well-being movement are expected to be widely accepted. Vital user values related to mobile technology include health, closer circles (family, friends), security, environmental values, etc.

Various promising 4G scenarios have been identified. They include typical everyday life situations with the potential to unlock user needs for connectivity and bandwidth. Typical mobile scenarios are: E-commerce, business/work, private life (home/free-time), vehicular, public places, entertainment, education, health-care, travel, etc. Numerous associated services to these scenarios have been identified, trying to match user values and expectations. Examples are personal manager/assistant (finance, health, security, information, etc.), home manager/assistant (control, comfort, security, maintenance, etc.), news/weather report delivery, travel agent/mobile tourist guide, mobile gaming, mobile shopping, positioning-related services (tightly complementing the above mentioned services), and many others. Additional information on 4G scenarios and services can be found in [WWRF, 2006], [mITF, 2006], [Kim and Prasad, 2006b] and [Karlson et al., 2004].

An integrative approach: 4G as a convergence platform. In this section we explore the integrative approach of 4G in particular from the perspective of the aforementioned ITU-R Recommendation M.1645 [ITU, 2003]. We can see that a paradigm shift may be needed to define future wireless communications systems. Indeed, previous and current generations (1G-3G) refer mainly to cellular systems while future systems (4G) are seen to encompass several access approaches, with cellular (wide-access) and nomadic (local-access) being the main component networks. 4G attempts to combine several complementary communications networks into a single network.

Convergence of Heterogeneous Networks in 4G: One of the tenets mostly associated with 4G that is *Being Always Connected, Everywhere, Anytime*. Fulfilling such a simple principle demands unprecedented efforts on the part of the designers of future wireless communication systems. The apparent simplicity and transparency enjoyed by users of future 4G systems has an enormous price to be paid by manufacturers, research community and standardization bodies. Their goal is to make a network of highly eclectic networks appear as a single, simple and everywhere reaching network. In Figure 14.1 the 4G domain is depicted as a conjunction of two well-known mobile (wide-area coverage, cellular) and nomadic (local-area coverage, short-range) developments, corresponding to the upper and lower portions of the figure, respectively. Cellular mobile communications have enjoyed a steady growth in terms of achievable throughput. Every generation of cellular systems (1G, 2G, 3G) offered marked improvements over the preceding one. The same applies with the nomadic (local-area) access, characterized by low-mobility and moderate-to-high data rates. The transitional period between 3G and 4G is sometimes known as Beyond 3G (B3G), where further enhancements are expected. As time goes by, the mobile approach acquires more characteristics of its nomadic counterpart, *e.g.*, supporting higher data throughput, while the same applies for the nomadic systems, which gradually inherit attributes of mobile systems, like higher mobility, seamless coverage, better use and reuse of radio resources and voice/data/multimedia capabilities. Telecommunication manufacturers (the mobile sector) tend to focus on the cellular components of 4G, including both an evolutionary path aiming to further enhance the current mobile systems and an innovative path concentrating on developing new technical solutions. These approaches correspond to the upper right corner of Figure 14.1 (path 1). Equally, IT companies (the wireless sector), with background business in local access systems are more inclined to see 4G as enhanced extensions of current short-range communication systems (lower right corner in Figure 14.1, path 2). In addition to the cellular and WLAN paths there is a third and somewhat more recent development also likely to be incorporated into 4G, namely through enhancements based on wireless Metropolitan Area Network (WMAN), see path 3 in Figure 14.1. In terms of data throughput, mobility and coverage WMANs represent a mid-way point between local- and wide-area approaches. Academia, regulatory bodies and other parties with less economic ties to the wireless business tend to favor a more unified and balanced vision of 4G and its constituent technologies. Summarizing, worldwide 4G development is following several paths with target solutions that can be complementary (co-existing) as well as mutually exclusive (competing). Evolution is an important component of this development, in particular taking into account the integration and further enhancement of legacy systems. In addition, development of novel approaches might be required to cope with some of the stringent requirements.

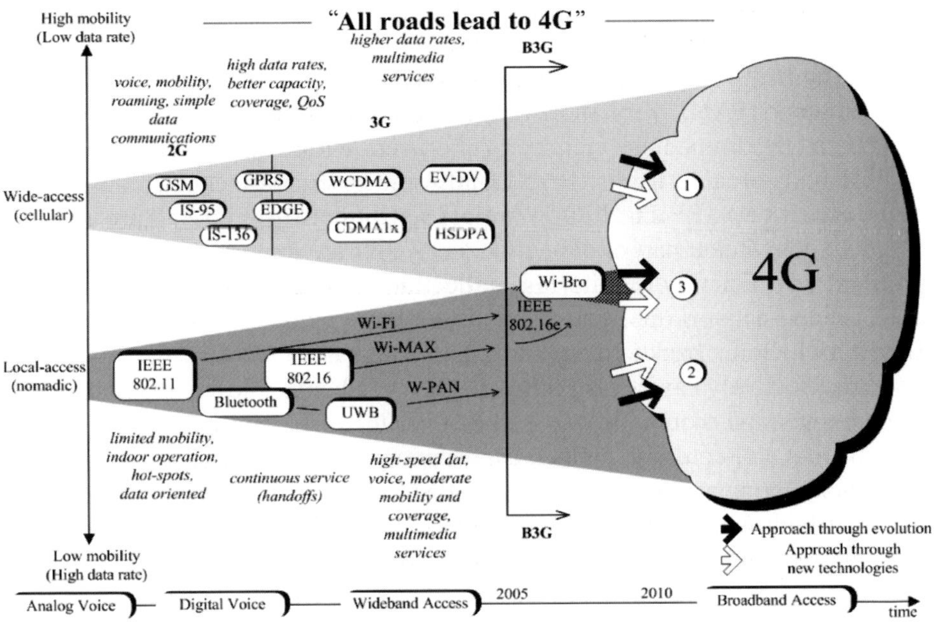

Figure 14.1. Approaching 4G through convergence of cellular, nomadic (WLAN) and metropolitan (WMAN).

4G can be approached from the network coverage standpoint, by looking at how different wireless services are provided at different geographical scales. Figure 14.2 shows the network hierarchy, starting with a distribution layer at the largest scale. This layer provides large geographical coverage with full mobility, though links may convey a chunk of composite information rather than signals from individual subscribers, for instance broadcast services such as DAB and DVB. Next in the hierarchy is the cellular layer, with typical macro-cells of up to a few tens of kilometers. This network also provides full coverage, full mobility but now connections are intended to cater to individual users directly. Global roaming is an essential component of 2G cellular systems, *e.g.*, GSM. Note that the cellular layer encompasses both macro and micro cells. The metropolitan layer or network, of which IEEE 802.16 [802.16, 2004], HyperMAN [HiperMAN, 2005] and Korean WiBro [WiBro, 2004] are typical examples, provides urban coverage with a range of a few kilometers at the most, with moderate mobility and moderate data speed capabilities. In a further smaller scale and moving to the local-area layer, *e.g.*, indoor networks or short-range communications, the network provides here access in a pico-cell, typically not larger than a few hundred meters, to fulfill the high capacity needs of hot-spots. Nomadic (local) mobility is supported as well as global roaming. 3G makes use of the cellular layer (typically micro-cells) in combination with hot-spots (WLAN), through vertical handovers, to provide coverage in dedicated areas. The next in the wireless network hierarchy is reserved for the personal area network (PAN), very-short range communication links (typically 10 m or less) in the immediate vicinity of the user. Within this layer we can also enclose body area networks (BAN), and some other sub-meter wireless short-range access (*e.g.*, RFID, NFC). Wireless sensor networks (WSN) are also one essential constituent part of 4G networks. WSNs are important solutions to the problem of efficiently monitoring, collecting and distributing information in a distributed network made of (typically) a large number of nodes [Cook and Das, 2004]. Going back to the paradigm of a pervasive 4G wireless network, in order to effectively have an unlimited reach while being able to support a variety of data rates, 4G would have to embrace all the described network layers. In other words, 4G could be defined as a convergence platform taking in as well as working along and across WBAN, WPAN, WLAN, WMAN, WSN, cellular and distribution networks.

Convergence of Heterogeneous Terminals in 4G: As 4G is a network of heterogeneous networks, it follows that it needs to support heterogeneous terminals. There is no archetype of a 4G terminal, they will come in different shapes and sizes, and they will have different communication capabilities and additional functions. 4G terminals will fall into a broad range of devices from pen-like to conventionally shaped portable mobile communication devices to PDAs, laptops and other devices, including cars. As 4G involves not only man-to-man but also man-to-machine and machine-to-machine communication, 4G transceivers are expected to be integrated in wide range of devices, *e.g.*, office equipment, home appliances, etc. As far as the user terminal is concerned, the current trend is to have either single-mode or multi-mode terminals. Even though both approaches could easily find considerable market share, the latter

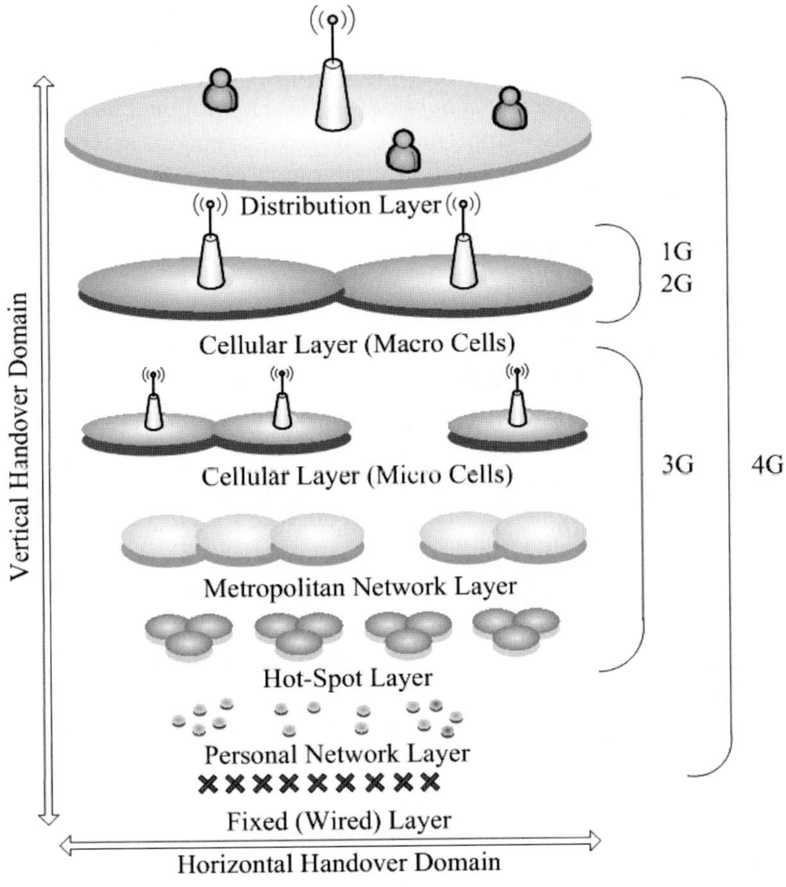

Figure 14.2. Network layers coverage.

will inherently better match the capabilities of 4G networks, namely handling multimedia information of various types supporting advanced services. Multi-functional devices represent the convergence of several technologies. Multi-functionality means several air interfaces on board (*e.g.*, wide-area, local-area, very short range), audio, imaging, positioning and other features. Terminal convergence is possible because there is a 4G network ecosystem supporting pervasive connectivity. This convergence will allow users to have simultaneous or independent access to different networks with a single terminal. 4G will ultimately facilitate and expedite the *three-screen* convergence, bringing together the TV, PC and mobile phone screens into a single portable device.

Convergence of Heterogeneous Services in 4G: Heterogeneous networks and terminals need to be finally complemented by heterogeneous services. In other words, heterogeneous services imply a wide range of services able to operate across different networks and in various types of terminals. In addition, convergence is essential in this context as the concept of multi-access services is becoming a reality.

Technology perspective. We see 4G as a continuum of wireless technologies providing ubiquitously a wide range of seamless connectivity options. The most critical technical features of 4G are the pervasive provision of seamless connectivity as well as the support of high data rates in moderate-to-high mobility environments. The integration of eclectic wireless networks needs to be done at the IP networking layer, where the cohesive role of IP is paramount to enable wide seamless connectivity across heterogeneous networks. An all-IP network, embracing the access and core networks, is the most straightforward and effective way to integrate all the possible different networks constituting the 4G network. Horizontal and vertical handovers will assure seamless intra- and inter-network connectivity, respectively. As far as modulation and multiple access schemes for 4G are concerned, it is widely agreed that multicarrier techniques have full potential to attain high data throughput at high mobility. OFDM (Orthogonal Frequency Division Multiplexing) and its multiuser extension OFDMA (Orthogonal Frequency Division Multiple Access) are the main component techniques for 4G. Usually OFDM is combined with other access techniques, typically CDMA and TDMA, to allow, among others, more flexibility in multi-user scenarios. Multicarrier CDMA (MC-CDMA), another technique with great potential, can be seen as a special case of OFDM. OFDM and CDMA are robust against multipath fading, which is a primary requirement for high data rate wireless access techniques. Overlapping orthogonal carriers OFDM results in a spectrally efficient technique. Each carrier conveys lower-data rate bits of a high-rate information stream, hence it can cope better with the intersymbol interference (ISI) problem encountered in multipath channels. The delay-spread tolerance and good utilization of the spectrum has put OFDM techniques in a rather dominant position within future communications. From a technology standpoint, 4G should overcome limitations and solve the problems of the previous generations. The difficulty for CDMA to achieve very high data rates in interference limited multi-user, multi-rate environments puts the mentioned multicarrier techniques in a unique strategic position in 4G. Also, another problem in current wireless systems is the difficulty of providing a full range of multi-rate services with different QoS requirements due to the constraints imposed on the core network by the air interface standard (it is not a fully integrated system). 4G needs to tackle also the lack of end-to-end seamless transport mechanism. Other important constraints of current mobile systems are the limited availability of spectrum and its particular allocation as well as the difficulty of roaming across distinct service environments in different frequency bands.

Number-wise, 4G will favor short–range links, and air interfaces supporting local access are expected to be omnipresent. In addition to conventional narrow and wide band transmission techniques, ultra wide band (UWB) techniques have lately received considerable attention, in particular as a nonintrusive, low power and low cost alternative to other short–range communications methods [Porcino and Hirt, 2003]. In addition, optical wireless communications is also a viable alternative for short–range links. Optical wireless systems can be used not only for point-to-point links, like those standardized through the Infrared Data Association (IrDA) [IrDA, 2006], but also for full-mobility indoor applications based on either infrared or visible light [O'Brien and Katz, 2005a], [O'Brien and Katz, 2005b] and [Tanaka et al., 2003]. Among the main advantages of optical wireless systems, we mention their almost unlimited bandwidth, and inherent security, as the optical signal is confined within the operational scenario. Moreover, in optical systems no RF radiation is generated, consequently no interference pollution nor possible health hazards are produced, thus they are well suited to sensitive environments.

Even though there is not a widespread consensus on the main technical characteristics of 4G systems, several important features are commonly underlined. The main 4G features are shown in Table 14.1.

Geographical perspective. Even though enormous efforts on 4G research and development are global, it is worth noticing that 4G visions, interpretations and emphasis are not identical. Following the paradigm of generational changes, it was originally expected that 4G would follow sequentially after 3G and emerge as an ultra-high speed broadband wireless network [Bohlin et al., 2004]. As Asia pioneered 3G development, it rapidly became involved in 4G developments based on a linear extension of the cellular 3G and focusing chiefly on the high data rate aspects. This linear vision is still the prevalent approach to 4G in Asia, where notably Korea, Japan, China and India are the major players. In North America, emphasis on the high-data rate side of 4G has prevailed,

Table 14.1. Key characteristics of future 4G systems.

Data transfer capability	100 Mbps (wide coverage) 1 Gbps (local area) Design targets representing overall cell throughput.
Networking	All-IP network (access and core networks) Plug & Access network architecture An equal-opportunity network of networks
Connectivity	Ubiquitous Mobile Seamless Continuous
Network capacity	10-fold that of 3G.
Latency	Connection delay ¡ 500 ms Transmission delay ¡ 50 ms
Cost	Cost per bit: 1/10-1/100 that of 3G Infrastructure cost: 1/10 that of 3G
Connected entities	Person-to-person Person-to-machine Machine-to-machine
4G Keywords	Heterogeneity of networks, terminals and services Convergence of networks, terminals and services Harmonious wireless ecosystem Perceptible simplicity, hidden complexity Cooperation as one of its underlying principles.

though mainly through the development of wireless local area networks. More recently Asia and America have concentrated on the development and enhancement of metropolitan area networks. On the other hand, the aforementioned concurrent approach is often identified as the European vision of 4G. Indeed, the European Commission (EC) envisions that 4G will ensure seamless service provisioning across a multitude of wireless systems and networks, from private to public, from indoor to wide area, and provide an optimum delivery via the most appropriate (*i.e.*, efficient) network available. This view emphasizes the heterogeneity and integration of networks and new service infrastructures, rather than increased bandwidth *per se*. European research and development activities reflect quite closely such an integrative approach.

3. Cooperation Opportunities in 4G

In this section we will explore different approaches to cooperation in 4G networks, with especial emphasis on solving inherent practical problems of such a communication system. After identifying some technical challenges, we will discuss how cooperative techniques have the potential to tackle many of the identified problems. We mainly underline cooperation taking place among heterogenous networks.

Challenges in 4G

The discussed concept behind 4G is certainly fascinating. However, the research and development community needs to solve many problems to make the 4G vision a reality. Instead of concentrating extensively on the numerous 4G challenges, in this section we would rather focus on some key technical challenges that the designers will be confronting. Then, we will identify cooperative techniques with potential to solve these problems. Many 4G link-network- and system-level setups can be described by models representing cooperative actions between interacting entities. These entities can be in principle either abstract or concrete objects of the wireless communication network, from conceptual OSI layers to actual signal processing or components in the physical world. Cooperation is not only confined to model such a complex wireless system, but most importantly, cooperative techniques can solve or ease many technical problems, as those hinted below. Intra-layer cooperation has probably received most of the attention in the literature, mostly on the MAC and PHY layers. Inter-layer cooperation has recently been the research focus of several studies that consider cross-layer design and optimization in 4G.

Some important technical challenges are a direct consequence of the fundamental nature of 4G: the pervasive provision of a wide range of wireless connectivity over different fixed, nomadic and mobile scenarios and supporting an array of diverse terminal devices. Heterogeneity of networks, terminals and ultimately services poses one of the key challenges as they need to inter-work seamlessly. Seamless connectivity in 4G means temporal and spatial continuity in the service provision within and across networks, as Figure 14.2 suggested. Of the several 4G access components, the most challenging is, without doubt, the design of a new air interface supporting high speed connectivity, with a per user data throughput of one or two orders of magnitude higher than those found in in 2G and 3G, particularly in environments with moderate to high mobility. Spectrum is scarce and this trend will become pronounced as one considers the continued explosive growth of mobile subscribers as well as the emergence of new services exploiting broadband capabilities.

One of the key challenges that we identify is the difficulty to fulfill the foreseeable increase in power demand of future 4G terminals. We particularly consider this challenge as essential, as it has not only crucial technical implications but also affects user acceptance and eventually the success of 4G as a whole. The capability of being wireless of any terminal is ultimately dictated by the battery powering the device. Frequent recharging or replacing of the battery makes terminals and associated services unappealing [TNS, 2005]. Unfortunately we are witnessing already now rather reduced operational times in current devices (*e.g.*, mobile phones, wireless-enabled PDAs). Another illustrative and discouraging example speaks by itself: 3G terminals are typically shipped with two batteries. Figure 14.3 exemplifies the evolution of power demand in past, present and future mobile generations. One of the fundamental challenges in 4G is thus to break conventional design rules targeting advanced services without the need to increase considerably the power requirements. Services are considered as one of the most important factors for the success of 4G systems. In terms of service provision the following paradigm shift is taking place in different generations.

$$Service_{2G} = constant \tag{14.1}$$
$$Service_{3G} \sim f(place) \tag{14.2}$$
$$Service_{4G} \sim f(place, time, terminal, user) \tag{14.3}$$

Breaking the design rules can be interpreted as departing from many conventional technical approaches aiming to a weaker dependence between services and power drain. Cooperative techniques have the potential to break the design rules by sharing or distributing tasks among cooperating entities, for instance by exploiting particular arrangements of these entities, by considering cooperative capabilities already at the design stage of the communication layers and including intra- and inter-layer aspects of cooperation, etc. One of the goals here would be to drastically reduce the dependence between service and power requirements. In 3G systems, as we move to higher data rates (*i.e.*, advanced services) we need to increase considerably the transmitted power, as suggested by Equation 14.1. Under the same cellular network architecture and assuming that multiple antennas are used on the terminal it is difficult to think that things would change in 4G. This trend would pose unacceptable practical constrains in 4G as it will make terminals even more power hungry. New techniques should be sought aiming to make the dependence between service quality and power consumption less dominant, ideally as hinted by Equation 14.4. We believe that cooperative techniques can help us to loosen this dependence, as is shown for instance in Chapter 11.

$$Power_{3G} \sim f(Service_{3G}) \tag{14.4}$$
$$Power_{4G} \sim f(1/Service_{4G}) \tag{14.5}$$

Power consumption in future terminals is undoubtly going to increase considerably due to the following facts:

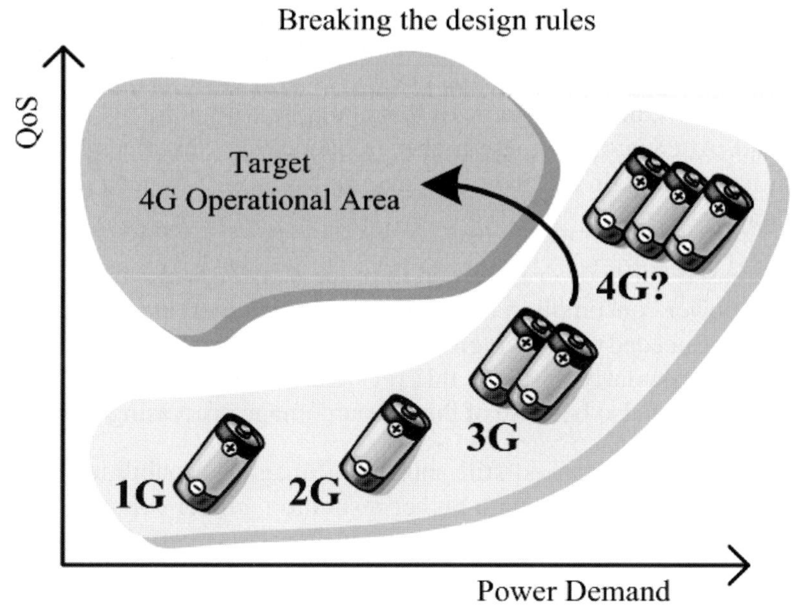

Figure 14.3. Power consumption of past, present and future mobile communication generations.

- Higher data rates: As energy per bit decreases, in order to operate with acceptable signal-to-noise ratios, the transmitted power needs to be increased.

- One of the most mentioned characteristics of 4G is the ability to provide users with a continuous connection, or as it is typically referred to *"always being connected"*. From the battery standpoint, this can also be interpreted as *"always being drained"*.

- 4G favors the emergence of multi-standard, multi-function terminals. Multi-standard particularly means the support of multiple air interfaces which can, in principle, be used simultaneously.

- Multi-antenna techniques are acknowledged as one of the key enabling technologies for enhancing link and network performance in 4G. In particular MIMO-based spatial multiplexing allows increasing data throughput by several folds. Multiple antennas on the terminals mean multiple transceivers on board, consequently boosting power consumption. MIMO techniques can help us to solve many 4G problems (*e.g.*, high throughput, QoS, coverage, etc.) but form the portable terminal implementation point of view, it brings also several challenges.

- Due to current spectrum allocation and high demand of additional bands, it is expected that frequency for future wireless systems will be allocated to less congested higher frequency bands. This means a shift from the 1 to the 2 GHz band to frequencies that could lie within the 3.4 to the 5 GHz band. Attenuation in these higher frequencies in significantly higher. In order to maintain the power budget transmitting ends need to use higher power.

- Increased DSP power is needed to process faster, wider bandwidth data. The power consumption of the processing unit increases as higher clock rates are needed to support greater processing power. Video signal processing is particularly costly in this regards, not only for the high data rate required but also because of the onboard image processing requirements.

- The inherent advanced still and video imaging capabilities of 4G terminals mean displays with higher resolution, higher contrast and higher frame displaying rates. These all having an adverse effect on power consumption.

- Audio capabilities, particularly high-fidelity applications, sometimes incorporating stereo loudspeakers on the handheld device.

- Terminals contain large amounts of mass memory, and one is witnessing that this trend is on the rise. The use of semiconductor memory and particularly the recently introduced hard-disk equipped terminals increase power consumption significantly.

- New services may also unfavorably impact power consumption. For instance, they may exploit user location information obtained by using either a onboard satellite receiver (GPS/Galileo positioning system), or based on processing specific system signaling for that purpose (*e.g.*, time-of arrival, triangularization, etc.)

From the 4G terminal manufacturer perspective the power consumption problem is critical, not only technically but also taking into account the market expectations from a newly introduced technology. The long operational time capability of terminals is both satisfying and vital for users; it gives them a truly wireless experience. This feature has been put at the top of the wish list by consumers as shown recently in [TNS, 2005], and therefore it must be taken seriously by the industry, and indirectly, by the research community. As a concrete example, Figure 14.4 shows the power requirements of different wireless terminal generations, including an approximate power consumption breakdown, [source: Nokia Corp.]. In terms of power consumption we have moved from a relatively low 1-2 Watt range in the first generations to around twice of that in 3G. The prospective for the future does not look encouraging in this aspect, as one could easily expect again another doubling in the power consumption figure. Many of the above listed factors having a direct impact on power consumption are directly related to the basic communications and signal processing capabilities of the terminal and, as shown in Figure 14.4, they account for roughly 50 percent of the power budget. Therefore, any reduction in the power consumption in these functionalities will have a substantial impact of the overall power demand.

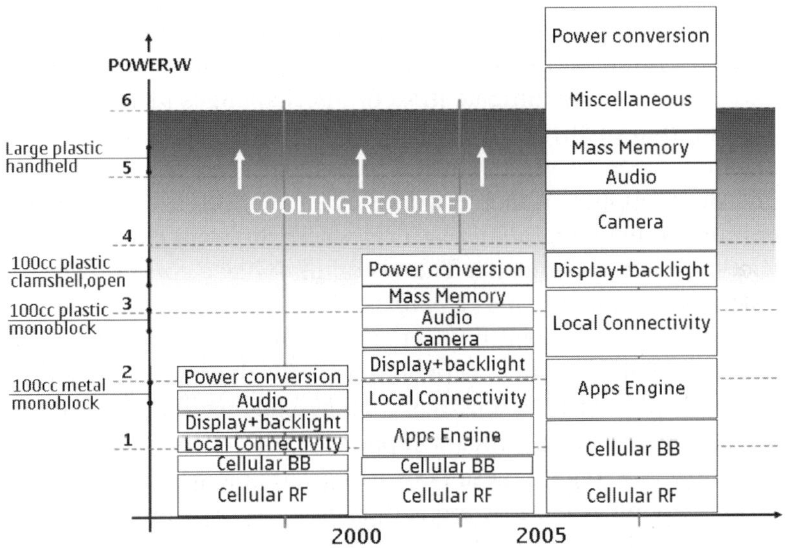

Figure 14.4. Power consumption breakdown of wireless terminals, Figure courtesy of Nokia.

One could wishfully expect that the above mentioned problems will be solved by the use of batteries with high energy density. Fuel cells will provide at the best a ten-fold increase in energy density, as compared to the best batteries today. Fuel cell technology is not yet fully mature, though optimistic predictions foresee the mass introduction of portable cells by the time 4G will be launched. New battery technologies will eventually solve the high power consumption problem by compensating for the power drain. However, high power consumption means also considerable heat dissipation and hence careful terminal design, followed by usually expensive technical solutions, in particular in small form factors. Heat management becomes a need when power levels already exceed a few watts. High power dissipation in terminals means that the temperature of the small handheld devices would rise to unpleasant values for the user, regardless of the effectiveness of the cooling system used. Thus, even though powerful batteries would eventually solve the problem of the high power demand, from a practical point of view it is highly desirable to the reduce effectively the overall power consumption. As Figure 14.3 suggests, we need to find a new operational regime characterized by low power consumption, without sacrificing system capabilities. In the sequel of this chapter we would discuss how cooperative technique could be used to reduce the power consumption as well as solve to some of the mentioned problems.

Cooperative Techniques in 4G: Identifying Research Opportunities

In this section we motivate the use of cooperative techniques in 4G wireless networks. We mostly focus on concepts helping to alleviate the increased power consumption of future terminals, though we strongly believe that cooperative techniques can be used to mitigate most of the challenges identified in the previous section. Two of the most distinctive characteristics of the 4G wireless communication systems were referred to as *heterogeneity* and *convergence*. The former term applies in the network, terminal and service domains, while the latter refers to the integrating platform where different networks, terminals and services operate and coexist. The vision of 4G as a *convergence platform* can be extended to consider it as a *cooperating platform* where heterogeneous networks operate, coexist and interact. As mentioned before, the main components are wide-area cellular networks as well as short-range local access networks. Typically, these two network approaches were considered as competing, though, given the significant support of both concepts by the industry, the current view is to regard these networks as complementing each other. We shall see that moving a step forward and creating synergy through cooperation between wide and local area networks can actually pay off. This will be one of the main 4G cooperative strategies advocated in this chapter. Another major attribute of the fourth generation systems is the availability of a *fine resource granularity*, unseen in previous generations, that can be exploited to design the system. Indeed multi-carrier, multi-antenna techniques, some of the *de facto* 4G assumptions for the physical layer, give to the designer unprecedent degrees of control on the use of time-, frequency- and space-domains. At the same time these extended and more finely partitioned domains make the problem of optimizing the use of resources in a multidimensional space a truly challenging design task.

From a network topology standpoint, 4G encompasses centralized (a.k.a. infrastructure, cellular) and decentralized (a.k.a distributed, infrastructureless, ad hoc) networks. Cooperation within each particular network is certainly possible. In most of the cases, due to the existence of either a centralized or distributed control strategy, coordinating the cooperative efforts leads to different approaches. Chapters 7 and 8 deal specifically with cooperative techniques for distributed (non-centralized) and centralized networks, respectively. Later in this section we particularly consider cooperative approaches across the two type of networks. At the physical layer, among the most representative cooperative approaches are the basic relaying (or multi-hop) techniques based on amplify-and-forward (AF) [Laneman and Wornell, 2000] and decode-and-forward (DF) [Sendonaris et al., 1998] as well as the concept of coded cooperation [Hunter and Nosratinia, 2002] [A. Nosratinia, 2004]. A unified framework for cooperation, comparing various schemes is proposed in [Herhold et al., 2005]. In general, a two-hop solution appears as a good engineering compromise in cellular networks, while in distributed networks a generic multi-hop approach is usually considered. In the cellular context the second hop need not necessarily be over a wireless link. Several approaches targeting 4G systems consider that the hop between the relaying node and the base station takes place on a fixed link using either a wireline or optical fiber, as in the concepts of Distributed Base Stations (DBS) [Adachi, 2001], [Clark et al., 2001] or Radio Over Fiber (ROF) [Al-Raweshidy and Komaki, 2002], [Way, 1993]. Unlike with the full wireless multi-hop techniques, these hybrid concepts do not exploit diversity provided by the multiple received signals.

A pragmatic approach to cooperation in 4G. Cooperation in 4G can be approached from several angles. We highlight here chiefly those cooperative techniques specifically exploiting some of the basic characteristics of 4G, and/or tackling some of its inherent practical problems. For our purposes we model the interactions taking place in the system with a number of variables. Service (Q) is the ultimate deliverable that the system needs to provide to the user. Such a provision has a cost, measured in resources needed to transfer the required signal to the end user with a given quality. For our purposes we consider power (P), complexity (C) and spectral efficiency (S) as the main contributors in the cost equation. As identified previously, P, C and S represent fundamental challenges in 4G. Figure 14.5 illustrates the interaction of these key practical factors. By cooperating at a given layer (of the OSI protocol stack) or across different layers, we can in principle have some degree of control on how the aforementioned resources are used. Even though cooperation can take place just between two entities (*e.g.*, typically terminals, base stations or functional parts of them), in general the setup for cooperation is understood to include more than the original source and destination nodes. These additional nodes share their resources to help the source node to convey reliably its message to the destination, as in multihop (relaying) networks. One may next ask, what are the incentives for these nodes to cooperate? When a fixed relay node is used, as done typically in cellular networks to increase coverage, the question is irrelevant as the only function of such a node is to help others. However, the situation is different for an autonomous node (*e.g.*, wireless terminal), where sharing resources means in practice letting others use its battery, giving others priority to exchange information, etc. Being selfish has its price and hence *reciprocity* could be seen as the main driving force inducing cooperative behavior. Of course, cooperativeness can be also sparked by other factors, like benefits to cooperative users granted by the network (*e.g.*, access priority), operator (*e.g.*, reduced service price), among others. In some scenarios, like in personal area networks, several wireless nodes in immediate closeness to the user serve him or her, thus, the incentives for cooperation are inherently embedded in the relationship to the user and associated nodes.

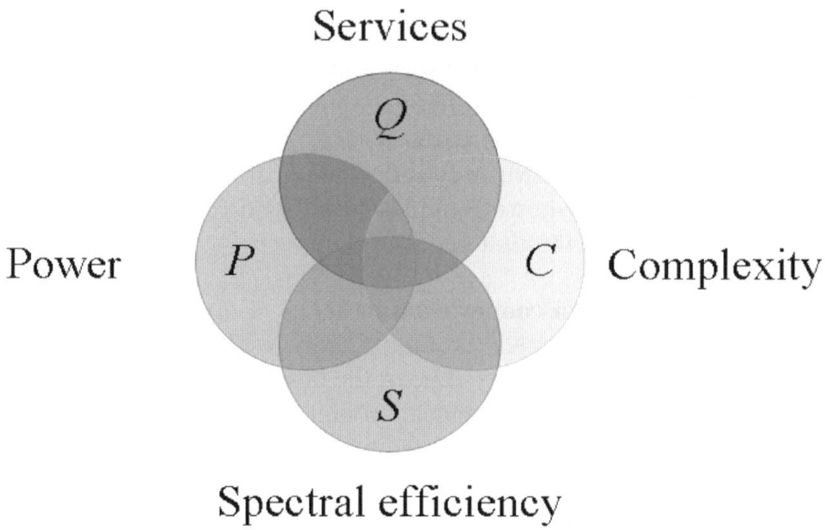

Figure 14.5. Key practical challenges in 4G.

In the following subsections we will explore how cooperative techniques could be used to solve or at least alleviate many important practical problems likely to emerge in 4G.

Cooperation and power efficiency. Let's approach first cooperation form the possible power saving benefits, leading ultimately to an extended operational time of the terminal. Having a number of nodes collaborating by relaying the signal from the source can be seen indirectly as reducing the equivalent average distance (end-to-end) between source and destination. This, together with the diversity gain obtained by the use of multiple signal paths to reach the destination, results in a less stringent power budget for the source, as compared with the direct (non-relaying) case. Of course, the gain comes from the fact that a comparable QoS is attained at destination with reduced power expenditure P at the source. However, one should look carefully to the overall cost of cooperation. The coordinating efforts to bring and keep cooperation among entities consumes valuable resources. The mechanisms for identifying potential cooperating nodes as well as announcing and maintaining the cooperation require exchange of information which in turn consumes power (for transmission, reception and processing), reduces spectral efficiency S (due to the required overhead), and adds in principle complexity C to the system. In some cases the additional power consumption resulting from cooperation may not be an issue, but this cannot be straightforwardly generalized for all types of networks. In typical terminals of cellular systems (*e.g.*, 2G, 3G) up to half of the power consumption comes from communications-related functions like baseband processing, RF and connectivity functions, as depicted in Figure 14.4. In terms of additional power consumption, the cost of cooperation are by no means negligible in such a cellular scenario, though the answer depends on the number of involved nodes, type of protocols used, etc. In wireless sensor networks the energy needed for establishing and maintaining a link could be considerably lower than that required for other onboard functions. Some cooperative techniques exploit the availability at the source transmitter of all channel state information (CSI), an assumption that could result in prohibitive practical implementations in some cases. Certainly, increased latency is another typical consequence of cooperation. Multi-hop techniques are not always energy-efficient, in general their efficiency depends on the type of scenario [Min and Chandrakasan, 2003]. Chapter 18 considers a different approach to achieve better power efficiency, namely task computing by distributing the efforts among a number of cooperative nodes. Power awareness in distributed (ad hoc) wireless networks is a fundamental design issue [Goldsmith and Wicker, 2002], and the problem of attainting energy efficiency is particularly exacerbated in wireless sensors networks, where it is vital for a very large number of nodes to exhibit extremely low power consumption [Min et al., 2002], [Rhee et al., 2004]. Even though an extension of the operative time of the terminal has been recognized as one of the main driving forces for using cooperative strategies, smaller transmitted powers result in less generated heat and reduced radio emissions, which in turn generate less interference to other nodes of the network and less potentially hazardous electromagnetic radiation to the user.

Cooperation and complexity. One of the major challenges of 4G is developing advanced equipment and services at relatively low cost. These targets are sometimes conflicting, as in the case of infrastructure. Providing wide coverage with high-speed connectivity at the high frequency bands likely to be allocated to 4G means that a dense network of base and relaying stations will need to be deployed. In order to guarantee the success of 4G from its launching, terminal prices should be attractively low. Terminal complexity and its associated cost can be traded at the expense of more complex (though expensive) infrastructure. Cooperation can be in principle exploited to bring terminal complexity down by distributing the tasks among several cooperating units. Terminals form a wirelessly connected grid of nodes, each contributing to a shared resource pool with some particular (local) resources, like computational capability, etc. Terminals could then be simple, with plain communicational and processing capabilities, but with enabled cooperative capacity. Under these assumptions, terminals could improve their communicational and processing capabilities proportionally to the number of collaborating units. From a different perspective, an ideal design rule could target relatively simple terminals assuming that a large number of these will cooperate. Figure 14.6 illustrates some of the discussed concepts to reduce terminal complexity (or enhance QoS) in 4G. An example of such a scenario is the use of Multiple Description Coding (MDC) by several wireless terminals, where every intervening terminal acts as the source of one descriptor. The more cooperating units the better the QoS [Fitzek et al., 2005]. In future 4G systems the possible gains (in complexity reduction) would be large if terminals were, by design, enabled with cooperative functions. The cost to support the simplicity of terminals, namely additional power P and overhead, increases with the number of interacting terminals, but still, trading complexity (cost) with power and spectrum is a viable engineering alternative to be considered.

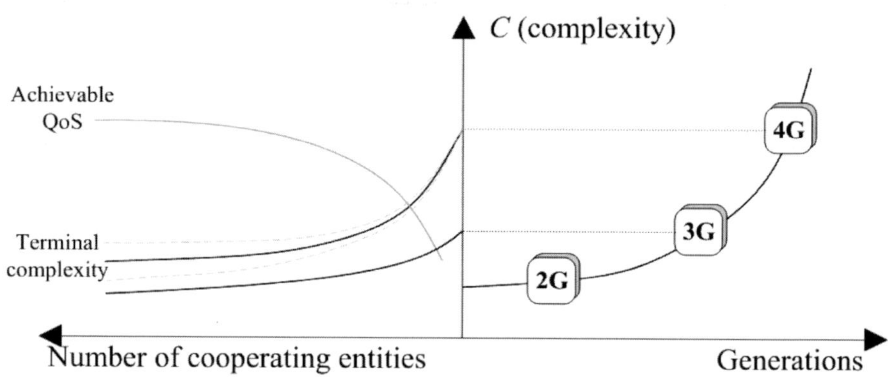

Figure 14.6. Can cooperation help us to reduce terminal complexity in 4G?.

Cooperation and spectral efficiency. Spectrum has always been a very limited radio resource and it is expected that 4G will dramatically accentuate this trend. We can also resort to cooperation to efficiently use the spectrum. This is not necessarily a straightforward task, as overhead needed to support cooperative techniques also tend to consume that valuable resource. *Cognitive Radio* techniques aim to better usage of the spectrum by sensing the environment over a wide bandwidth and dynamically allocating users to temporarily unused bands, thus boosting spectral efficiency [Weiss and Jondral, 2004], [Cabric et al.,

2005]. To be effective, the cognitive behavior needs to be complemented with flexible and cooperative systems. Knowledge of the spectrum is not a necessary condition to increase spectral efficiency. In the next section we will consider some examples (though not based on cognitive principles) where cooperation between networks is exploited to increase spectral efficiency. Cooperation between heterogeneous networks has the potential to improve spectral efficiency, particularly when the interacting networks make use of licensed and unlicensed spectra.

Exploring cooperation in heterogeneous networks. The fact that 4G is fundamentally a platform embracing different networks makes certainly 4G the ideal setting for exploring inter-network cooperation. It is interesting first to consider the relationship between the two prevailing networks in the 4G context, namely cellular networks for wide area access and ad hoc networks for local access. Already a legacy feature from 3G networks, we highlight the importance of network convergence leading to *coexisting networks*, an approach that can be also interpreted to mean *competing networks*. More recent visions tend to see these networks in a slightly amicable manner, namely as *complementary networks*, though interactions between networks take place mostly for (vertical) handover purposes. We advocate a closer and synergetic interaction, leading to *cooperating networks*. These evolving visions are illustrated in Figure 14.7, where the typical contending network solutions are seen in a different light, namely within the framework of *cooperating heterogeneous networks*. Cooperation between networks is a rapidly emerging research area. A summary of the research activities being carried our at the Cooperative Network Working Group of the WWRF can be found in [Politis et al., 2004], where network cooperation is mostly approached from the transport and network layers. The Ambient Networks project [Network, 2006] addresses the problem of cooperation in heterogeneous networks, particularly where networks belong to different providers or exploit different access technologies [Niebert et al., 2004], [Ahlgren et al., 2005]. The ultimate goal of these projects is to ensure seamless operation between heterogeneous networks, a fundamental requirement in 4G. Cooperation in this context refers to the mechanisms and architecture required to support the automatic and simple (to the user) provision of end-to-end connectivity regardless of the interacting networks and access technology involved. Many fundamental challenges related to interworking between different networks are being tackled by these projects.

It is interesting to investigate the potential of cooperation between heterogeneous networks beyond the goal of achieving inter-operativity, aiming to a more synergetic and dynamic interaction. To get a better insight on these potentials, an example of a possible 4G specific cooperation approach is discussed next, aiming, among others, to enhance spectral efficiency. Let's assume a multicast

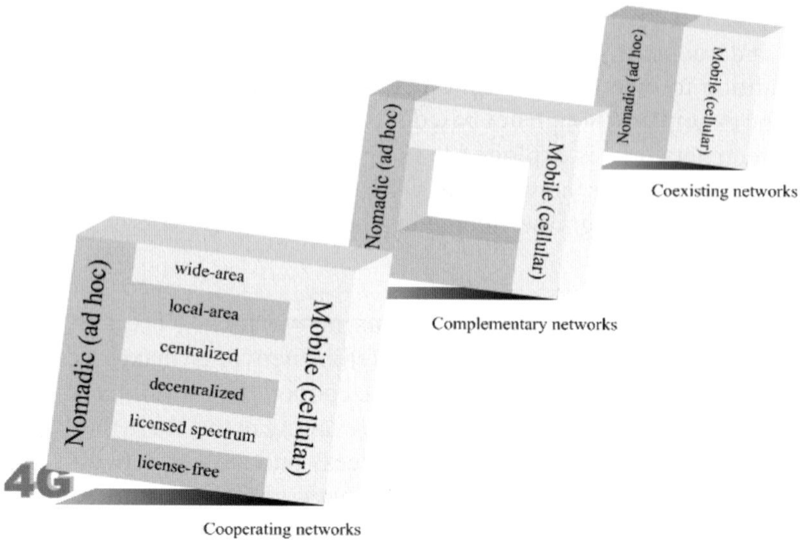

Figure 14.7. Recent views on the role of heterogeneous networks in 4G: Getting mobile and nomadic networks to cooperate.

service being provided by a base station. A number of terminals using the same multicast service and being in relatively close proximity form a given *local group*. Figure 14.8 depicts the considered scenario. Due to the channel fluctuations, there will be situations in which the signal cannot be successfully decoded by particular terminals. In such cases, the cellular network would need to resend a given packet, consuming additional resources, namely, spectrum and power. However, a *local retransmission* by any of the group members would do the same, with no additional spectral cost as license-exempt frequency bands are used for the ad hoc communication. In terms of energy, the local retransmission is carried our at low-level power using a short-range link, instead of a global retransmission from the base station, involving much higher power levels, and generating more interference. In a small-size group, cellular retransmissions would be needed occasionally, but spectral efficiency increases with group size.

We briefly discussed the concept of cooperation in heterogeneous networks, and highlighted the possible benefits of the interaction between a wide-access (cellular) and a short-range network. As exemplified in Figure 14.2, the constituent 4G elements range from distribution (*e.g.*, broadcast) networks down to personal networks, and thus, the possibilities for inter-network cooperation are in principle numerous. Note that fruitful cooperation strategies between more than one network can be devised around a single user, assuming that his terminal is equipped with multiple air interfaces. In principle cooperation between

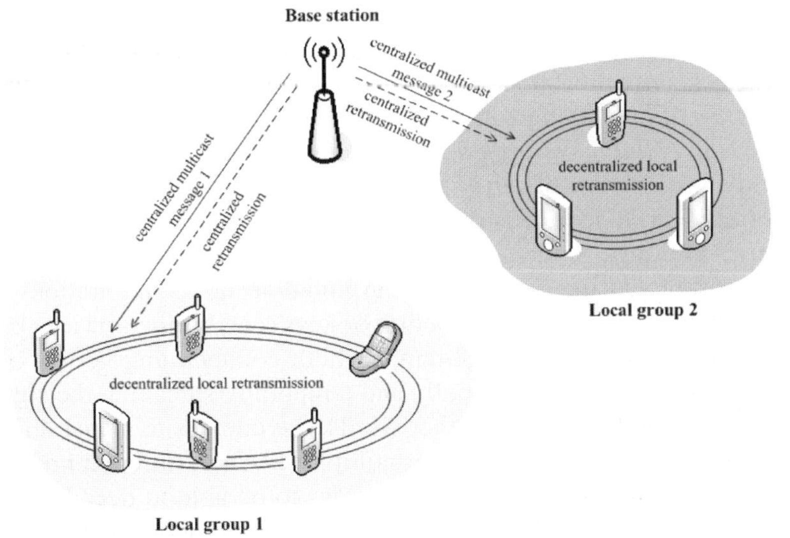

Figure 14.8. Network cooperation: Bringing together mobile and nomadic networks.

networks can take place at any of the OSI layers. Next, and to motivate further exploration, we mention a few promising concepts and scenarios:

- **Exploiting *air interface diversity* at the terminal**: Cellular (wide-area) links support typically less data throughput than their counterpart links for local-access. At the destination the data rate or QoS can be improved by combining multiple signals jointly provided by the cellular and ad hoc networks. Multiple description coding could be used for this purpose, for instance for video streaming, as suggested in [Frattasi et al., 2005a]. Multiuser diversity exploits the fact that in a multiuser environment a base station is likely to find at least one user with good channel conditions, making possible high-data throughput transmissions to such destinations. However, at a given time, multiuser diversity may not necessarily favor a particular high-data rate user, a fact that would become an issue in low mobility scenarios. Extending the centralized network with a locally distributed one, formed by an ad hoc group surrounding the target user, will help to convey more efficiently the high-data rate information to the destination. The larger the ad hoc group, the better are the chances to find a good overall multihop path to the destination. Thus, through cooperation between a cellular and ad hoc network we can in principle fulfil very well the request of the destination while the source is efficiently used.

- **Security**: Peer-to-peer communications over a short-range link is certainly a very feasible approach to exchange information. However, such a direct communication could be seen sometimes as risky, from the point of view of interacting with an untrusted (or unknown) counterpart. Thus, through *cellular-controlled short-range communication* the base station could take the role of verifying, authenticating and making secure a given transaction. If service is requested over a short-range link to a machine (*e.g.*, printing, vending machines, content retrieval) the infrastructure network could intervene providing initial secure configurations for the transaction (including distribution of keys) and billing services [Frattasi et al., 2005b]. In another example, both collaborating networks could also boost security by spatially and temporally spreading the signal targeting the destination. In other words, the composite signal arriving to the destination will come, time-multiplexed, from different nodes, some of them in different networks. In order to be able to decode the target signal the destination will need all the signal components to be present. Assuming that some of the components are provided by a wireless local-area network, reconstructing the signal is impossible for a node not in close proximity to the destination.

- **Local retransmission**: As discussed before, the interaction between centralized and decentralized networks can be exploited to improve, among others, spectral and power efficiency. Typically, centralized approaches (*e.g.*, cellular networks) consume spectrum and require more power, while decentralized approaches (*e.g.*, WLAN) operate in unlicensed frequency bands and require lower power levels. Cooperation between these two networks will aim to use as much as possible short-range links, bringing advantages to users (in the ad hoc networks) as well as to the operators.

- **Synchronization**: For some purposes local synchronization may be required, leading to a common reference time among a number of nodes. This common timing could be defined at different layers, *e.g.*, physical and application layers. In the former, a distributed process may need a precise common temporal reference, which may not be straightforwardly distributed by a central entity. By combining master-slave and mutual synchronization approaches provided by the cellular and ad hoc networks respectively, local synchronization can be obtained. At the application layer, some services shared by a group may need to have a common timing reference, *e.g.* aligning video and audio signals on the group users, as suggested in [Frattasi et al., 2005b].

We have mostly considered the interaction between centralized and decentralized networks, represented in this section mostly by cellular and WLAN and WPAN systems. 4G will consist of several other network technologies and associated air interfaces with high potential for cooperation, also comprising WMAN, sensor networks, near-field communications, RFID and other wireless networks.

4. Discussions and Conclusions

In this chapter, we have explored 4G from various perspectives, aiming to identify opportunities for applying cooperative techniques, and taking into account some practical issues. Global efforts to find a universal and well accepted definition of 4G are slowly paying off. A consensus on the 4G vision is hard to come by, because the perceptible departure from a cellular-oriented mobile communications generation paradigm towards an extended "all-encompassing network" approach. Even though a great deal of players in the 4G R&D arena tend to approach 4G from different and seemingly conflicting business interests, a number of common visions have recently emerged. The role of WWRF as a global consensus making organization on future wireless communications is significantly helping to create common understanding on 4G. We highlight here the integrative role of 4G, serving as a converging platform where a heterogeneous wireless networks will operate, coexist and cooperate. Virtually every imaginable wireless network will be integrated to the 4G network, from broadcasting networks down to wireless personal area networks, including wireless wide and local networks, sensor networks, etc. As a whole, 4G would be seen as a single, monolithic and simple network where a myriad of different terminals with different capabilities can be connected. Heterogeneity and convergence are the words that best describe 4G in terms of networks, terminals and services. It seems convenient to extend the rather established notion of 4G as a convergence platform to consider cooperative aspects, in particular those pertaining to the beneficial interaction between heterogeneous networks. In short, we extend the concept of coexistence and complementarity of networks in 4G to also embrace cooperativity.

We briefly discussed several technical challenges of 4G, mostly taking into consideration implementation aspects of terminals. In addition to improving basic performance measures like data throughput, coverage, QoS, etc., a number of practical but critical issues needs to be properly addressed by the research community to guarantee a successful 4G, including effective solutions to reduce power consumption and complexity in the terminal while boosting spectral efficiency. Power, complexity and spectral efficiency are key resources that can be traded in different ways to achieve a desired level of performance, and *cooperation is a promising resource-trading framework* for future wireless networks. We emphasized that cooperation has the potential to solve many of the challenges of 4G. There is already a vast and rapidly growing body of literature showing numerous advantages of cooperation in wireless networks. However, additional efforts are needed to better understand limitations and practical aspects of cooperative techniques in wireless systems. Undeniably, cooperation in wireless networks brings advantages, but its overall impact on system design, performance and practical implementation needs to be studied in more detail. Initially cooperation was confined to particular layers, typically physical and MAC layers, though, through cross-layer design, 4G is extending it to the inter-layer domain. Moreover, 4G opens up new possibilities, specifically cooperation among different component networks. Figure 14.9 illustrates, layer- and network-wise, the realm for cooperation in different communication generations. We note that the possibilities for cooperation are open but we cannot take for granted that this will happen. More research efforts are needed to better position these techniques within the developing course of 4G. A few motivating examples of inter-network cooperation were briefly discussed in this chapter, to unearth the potential of cooperation. Cooperative techniques should be considered as one of the fundamental enabling technologies for 4G.

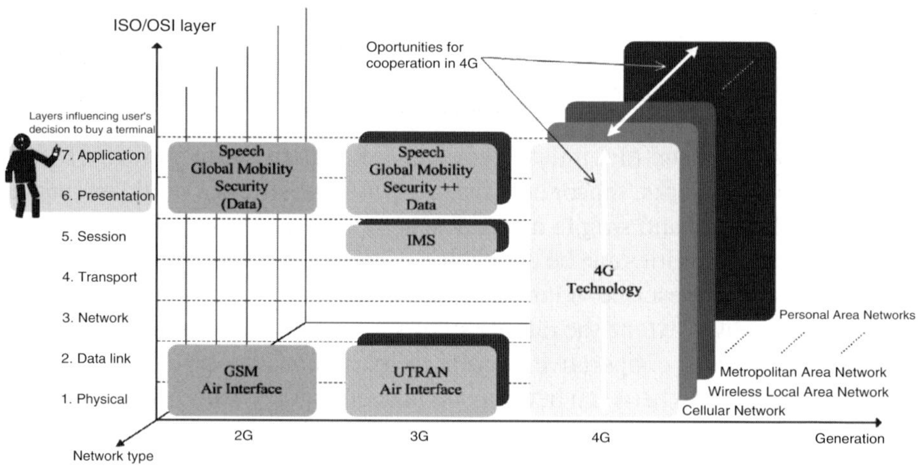

Figure 14.9. Domains for cooperation in 4G mobile and wireless networks.

Where will the user stand in a cooperative wireless world? Cooperation decisions can be inside or outside the realm of user choice. Many techniques can inherently exploit cooperation, the user is not part of the process. However, in some cases a given user can take an important role by deciding to share or not share with others the resources of his or her terminal. If clear and appealing incentives for cooperation are provided, like improvement and/or cost reduction of services, cooperation-enabled terminals costing less than non-cooperating ones, etc., users will be motivated to cooperate by sharing their resources. Thus, in a cooperation-enabled wireless world incentives would encourage users to cooperate. In such a scenario noncooperative behavior will be discouraged by the implicit punishment of missing the benefits. Supporting cooperation, security issues need to be solved to ensure that users of cooperating nodes cannot obtain other users' signals being processed by their terminals. In Figure 14.9 we show that the decisions made by an user when buying a terminal are influenced basically by features defined in upper layers, typically presentation and application layers. In the future the "cooperation-capability" feature could be a strong selling point for terminals, where ultimately users, manufacturers and operators would benefit.

In this chapter, inter-network cooperation was identified as a potentially promising area of research that can help us to solve several important practical challenges of 4G. We need to exploit the synergetic effects of combining centralized and distributed networks, taking advantages of the interaction between licensed and unlicensed frequency bands, wide and local area coverage, public and private, as well as high and low mobility networks. In order to *enhance* performance by cooperation, we need to both *enable* cooperation by design and *encourage* cooperation by clear incentives.

References

802.16, IEEE (2004). Ieee 802.16-2004 wireless standard. http://www.ieee802.org/16/.

Nosratinia, A., Hunter, T. E., Hedayat, A. (2004). Cooperative communication in wireless networks. *IEEE Communications Magazine*, pages 74–80.

Adachi, F. (2001). Wireless past and future - evolving mobile communication systems. *IEICE Trans. on Funamentals of Elec. and Comm.*, E84- A(1): 55–60.

Ahlgren, Bengt, Eggert, Lars, Ohlman, Börje, and Schieder, Andreas (2005). Ambient networks: Bridging heterogeneous network domains. In *16th IEEE International Symposium on Personal Indoor and Mobile Radio Communications (PIMRC 2005)*, Berlin, Germany.

Al-Raweshidy, H. and Komaki, S., editors (2002). *Radio over Fiber Technologies for Mobile Communications Networks*. Artech House.

Bohlin, E., Lindmark, S., Bjrkdahl, J., Weber, A., Wingert, B., and Ballon, P. (2004). *Future of Mobile Communications in the EU: "Assessing the Potential of 4G",*. ESTO Publications.

BoV (2001). Book of visions. http://www.wireless-world-research.org/general_information/Bookofvisions/BoV1.0/BoV/BoV2001v1.1B.pdf.

Bria, A., Gessler, F., Queseth, O., Stridh, R., M.Unbehaun, Wu, J., and Zander, J. (2001). 4th-generation wireless infrastructure: Scenarios and research challenges. *IEEE Personal Communications*.

Cabric, D., Mishra, S. M., Willkomm, D., Brodersen, R., and Wolisz, A. (2005). "A Cognitive Radio Approach for Usage of Virtual Unlicensed Spectrum". In *14th IST Mobile and Wireless Communications Summit*, Dresden, Germany.

Carpet, Flying (2004). Flying carpet report. http://www.mitf.org/public_e/archives/Flying_Carpet_Ver200.pdf.

Clark, M. V., Willis, T. M., Greenstein, L. J., and J., A. (2001). Distributed versus centralized antenna arrays in broadband wireless networks. In *Proc., IEEE Veh. Technology Conf.*, pages 33–37.

Cook, D. J. and Das, S. K., editors (2004). *Wireless Sensor Networks – Smart Environments:Technologies, Protocols, and Applications*. John Wiley.

Fitzek, F. H. P., Yomo, H., Popovski, P., Prasad, R., and Katz, M. (2005). Descriptor Selection Schemes for Multiple Description Coded Services in 4G Wireless Communication Systems. In *The First IEEE International Workshop on Multimedia Systems and Networking (WMSN05) in conjunction with The 24th IEEE International Performance Computing and Communications Conference (IPCCC 2005)*, Phoenix, Arizona, USA.

Frattasi, S., Fathi, H., Fitzek, F. H. P., Katz, M., and Prasad, R. (2006). Defining 4G Technology from the User Perspective. *IEEE Network Magazine*.

Frattasi, S., Fitzek, F. H. P., Mitseva, A., and Prasad, R. (2005a). A Vision on Services and Architectures for 4G. In *1st CTIF B3G/4G Workshop*, Aalborg, Denmark.

Frattasi, S., Olsen, R. L., de Sanctis, M., Fitzek, F. H. P., and Prasad, R. (2005b). Innovative Services and Architectures for 4G Wireless Mobile Communication Systems. In *IEEE ISWCS*, Siena, Italy.

Goldsmith, A. and Wicker, S. (2002). Design challenges for energy-constrained ad hoc wireless networks. *IEEE Wireless Communications Magazine*, 9(4):8–27.

Herhold, P., Zimmermann, E. and Fettweis, G. (2005). Co-operative multi-hop transmission in wireless networks. *Computer Networks Journal*, 49(3).

HiperMAN (2005). Hiperman wireles standard. http://portal.etsi.org/radio/hiperman/hiperman.asp.

Hunter, T. E. and Nosratinia, A. (2002). Coded cooperation under slow fading, fast fading, and power control,. In *Proc. Asilomar Conference on Signals, Systems, and Computers,Pacific Grove, CA*.

IrDA (2006). Infrared data association (irda). http://www.irda.org.

ITU (2003). Recommendation itu-r m.1645,"framework and overall objectives of the future development of imt-2000 and systems beyond imt-2000". http://www.itu.int/ITU-R/publications/index.html.

Karlson, B., Bria, A., Lind, J., Lönnqvist, P., and Norlin, C. (2004). *Wireless Foresight: Scenarios of the Mobile World in 2015*. Wiley.

Katz, M. and Fitzek, F. H. P. (2005). On the Definition of the Fourth Generation Wireless Communications Networks: The Challenges Ahead. In *International Workshop on Convergent Technology (IWCT) 2005*, Oulu, Finland.

Kim, Y. K. and Prasad, R. (2006a). *4G Roadmap and Emerging Communication Technologies*. Artech House.

Kim, Y. K. and Prasad, R. (2006b). *4G Roadmap and Emerging Communication Technologies*. Artech House.

Kupetz, A. H. and Brown, K. T. (2003). 4G - A Look Into the Future of Wireless Communications. *Rollings Business Journal*.

Laneman, J. N. and Wornell, G. W. (2000). Energy-efficient antenna sharing and relaying for wireless networks. In *IEEE WCNC*, pages 7–12, Chicago, IL.

Min, R., Bhardwaj, M., Cho, S.-H., Ickes, N., Shih, E., Sinha, A., Wang, A., and Chandrakasan, A. (2002). Energy-centric enabling technologies for wireless sensor networks. *IEEE Wireless Communications Magazine*, 9(4):28–39.

Min, R. and Chandrakasan, A. (2003). Top five myths about the energy consumption of wireless communication. *Mobile Computing and Communications Review*, 7(1):65–67.

mITF (2006). Mobile it forum. http://www.mitf.org/index_e.html.

Network, Ambient (2006). Ambient network project. http://www.ambient-networks.org.

Niebert, N., Schieder, A., Abramowicz, H., Malmgren, G., Sachs, J., Horn, U., Prehofer, C., and Karl, H. (2004). Ambient networks: An architecture for communication networks beyond 3G. *IEEE Wireless Communications*, 11(2):14–22.

O'Brien, D. C. and Katz, M. (2005a). Optical wireless communications within fourth-generation wireless systems. *Journal of Optical Networking*, 4(6): 312–322.

O'Brien, D. C. and Katz, M. (2005b). White paper: Short-range optical wireless communications. In *WWRF 14th Meeting*, pages 1–22, Chicago, IL.

Politis, C., Oda, T., Dixit, S., Schieder, A., Lach, K.-Y., Smirnov, M., Uskela, S., and Tafazolli, R. (2004). Cooperative networks for the future wireless world. *IEEE Communications Magazine*, 42(9):70–79.

Porcino, D. and Hirt, W. (2003). Ultra-wideband radio technology: Potential and challenges ahead. *IEEE Commun. Mag.*, 7(41):66–74.

Rhee, S., Seetharam, D., and Liu, S. (2004). Techniques for minimizing power consumption in low data-rate wireless sensor networks. In *IEEE Wireless Communications and Networking Conference (WCNC)*, pages 1727–1731.

Sendonaris, A., Erkip, E., and Aazhang, B. (1998). Increasing uplink capacity via user cooperation diversity. In *IEEE ISIT*, page 156, Cambridge, MA.

Tanaka, Y., Komine, T., Haruyama, S., and Nakagawa, M. (2003). Indoor visible light data transmission system utilizing white led lights. *IEICE Trans. on Commun.*, 86(8):2440–2454.

TNS (2005). Two-day batter life tops wish list for future all-in-one phone device. Technical report, Taylor Nelson Sofres.

Way, W. I. (1993). Optical fiber based microcellular systems: An overview. *IEICE Trans. Commun.*, E76-B(9):1091–1102.

Weiss, T. A. and Jondral, F. K. (2004). Spectrum pooling: an innovative strategy for the enhancement of spectrum efficiency. *IEEE Communications Magazine*, 42(3):8–14.

WiBro (2004). Wibro wireless standard. http://www.itu.int/ITU-D/imt-2000/documents/Busan/Session3_TTA.pdf.

WWRF (2006). Wireless world research forum. http://www.wireless-world-research.org.

Chapter 15

COOPERATIVE TECHNIQUES IN THE IEEE 802 WIRELESS STANDARDS: OPPORTUNITIES AND CHALLENGES

Explicit Macro Cooperation in Practice

Kathiravetpillai Sivanesan
Samsung Electronics
k.sivanesan@ieee.org

David Mazzarese
Samsung Electronics
david.mazzarese@ieee.org

Abstract: Recently, cooperative techniques have drawn much attention in industry and academia for throughput enhancement, coverage extension and spectral efficiency improvement. They have long been used to improve the reliability and scalability of mesh communication networks. The IEEE 802 standards are concerned with the personal area network, the local area network, and the regional area network, among others. Each network has its limitation to deliver the required throughput and quality-of-service to the end users. Recently, there have been several attempts to adopt cooperative techniques into IEEE 802 standards to overcome those limitations. In the sequel, we address the opportunities and impending challenges in adopting emerging cooperative techniques in IEEE 802 standards.

Keywords: cognitive radio, cooperative techniques, mesh network, multihop techniques, relay.

Frank H. P. Fitzek and Marcos D. Katz (eds.), Cooperation in Wireless Networks: Principles and Applications, 497–514.

1. Introduction

IEEE wireless standards are addressed by the IEEE 802 LAN/MAN Standards Committee. Wireless Local Area Networks (WLAN), Wireless Personal Area Networks (WPAN) and Wireless Metropolitan Area Networks (WMAN) have been gaining much attention as they can offer easily deployable networks with high throughputs that fit the needs of bandwidth-demanding applications. These applications, such as multimedia, real-time video, VoIP (voice-over-IP), are foreseen to be driving the commercial embrace of next generation wireless networks. The LAN/PAN/MAN will be an integral part of the global network that will support ubiquitous wireless access. It is not surprising that the most advanced communication techniques have found their ways into the IEEE standards. In particular cooperative techniques are now being seriously considered in many Working Groups of the IEEE 802 standards committee.

Cooperative techniques have only recently received considerable attention. Theoretic as well as practical approaches have been taken. Theoretic problems, such as the capacity of networks using cooperative techniques, remain largely

Table 15.1. The IEEE 802 standardization activities that address cooperative techniques across the different Working Groups.

IEEE group	Scope	Operation scenario	Type of cooperation
802.15 TG 3, 4, 5	High rate wireless personal area network (WPAN)	Mesh networking	Cooperative retransmission
802.11s WG	Local Area Network (LAN) MAC enhancement for reliable and easily scalable network	Mesh Networking	Peer-to-peer cooperation
802.16-2004	Metropolitan Area Network (MAN) MAC enhancement for reliable and easily scalable network	Mesh Networking	Peer-to-peer cooperation
802.16 MMR-SG	Coverage extension, Throughput enhancement, Spectral efficiency improvement in MAN	Relay	Multihop relay, cooperative transmission
802.22	Wireless Regional Area Network (WRAN)	Fixed centralized point-to-multipoint for unlicensed operation in TV bands	Cognitive radios

unsolved. Yet, IEEE 802 standards have already started working on incorporating cooperative techniques into current standards development. Although it is clearly recognized that cooperative techniques offer great opportunities, this early adoption also poses a lot of challenges.

Cooperative techniques appear at several levels of the network:

- Cooperative transmission among mobile stations (in centralized or non-centralized networks)
- Cooperation among networks (*e.g.* for traffic load balancing, handover, spectrum sharing)
- Cooperation among mobiles and networks in unlicensed operation
- Cooperation between licensed and unlicensed spectrum users.

An interesting outcome of this challenge is where theory meets practice. Although a lot of fundamental theoretical results are not available (*e.g.* for mesh networks or cognitive radio networks even with simple channel models) practical approaches provide solutions for real-life systems. It is likely that theoreticians will also benefit from this approach.

Cooperation among networks that utilize different radio access technologies embodies one of the fundamental characteristics of foreseen 4G networks. In that sense, IEEE standards present a lot of opportunities to approach the necessary cooperative techniques that will need to be implemented in the more complex 4G networks.

In this chapter, we briefly summarize the main IEEE standard activities addressing cooperative techniques, namely mesh networks, cooperative or multihop relay, and spectrum sharing or cognitive radio techniques. Our main goal is to identify the opportunities and challenges in incorporating the cooperative techniques into the standards. The rest of the chapter is organized as follows: The mesh mode MAC layer enhancement for IEEE 802.11 is considered in section 2; The mesh techniques for 802.15 PAN networks are addressed in Section 3; The mesh operation and cooperative or multihop relay techniques for 802.16-2004 and 802.16e standards are considered in Section 4 and Section 5, respectively; Finally, spectrum sharing or cognitive radio techniques are addressed in Section 6 for the 802.22 standard.

2. Mesh MAC Enhancement in IEEE 802.11s

The IEEE 802.11 standard is concerned with wireless local area networks in unlicensed (ISM) bands in indoor environments such as office, home, etc. It evolved through 802.11a, 802.11b and 802.11g with maximum throughputs up to 54 Mbps. As laying wires in homes, offices and public areas is cumbersome and expensive, WLAN have become very popular recently. They also have the flexibility of allowing the terminals to move within the coverage area. The demand for higher throughput increases with the variety of services such as video, gaming, etc. The WLAN 802.11 standard has also evolved to deliver higher throughput to the terminals. The 802.11 WLANs can be deployed in either infrastructure mode or ad-hoc mode. In infrastructure mode, terminals are connected to an access point wirelessly and the access point is connected to either the wired or wireless backhaul network such as DSL, cable, IEEE 802.16, etc. The access point functions are similar to those of a base station in a cellular network. In the ad-hoc or mesh mode, the terminals communicate wirelessly with each other in the coverage range in a peer-to-peer fashion. There is no access point present in this type of network. As the mesh network is easily scalable and has the ability to reconfigure and self-heal around a blocked path, this architecture is reliable and preferred over the infrastructure network.

The new evolution of 802.11 standards, named 802.11n, has been discussed in the Task group (TG) 'n' to deliver about 10-fold higher throughputs than the current 54 Mbps [IEEE11web, 2006]. Recently, another evolution of 802.11 using mesh networking, named 802.11s, has been discussed in TGs. This Task Group is concerned with upgrading the 802.11 MAC layer operation to self-configuring and multihop topologies. It may support broadcast, multicast and unicast traffic in the network. There are a few network element functionalities defined in the TGs such as mesh point, mesh access point, and mesh portal. The mesh point is the basic element. It collects information about the neighboring mesh points, communicating with them and forwarding the traffic. The mesh access point is a mesh point that has the capability to function as the 802.11 access point. The mesh portal is a mesh point, which connects the mesh network and a non-802.11 network. Figure 15.1 depicts the network element functionalities.

The TGs received around 15 proposals for the initial call for proposals in June 2005. As of November 2005 only two main proposals remain on the table. They are the SEE mesh and the Wi-Mesh Alliance proposals. The SEE mesh proposal was put together by a consortium of major companies, included which Intel, Nokia, Motorola, NTT DoCoMo, Texas Instruments and Samsung. It introduced the concept of mesh portal for interoperability in mesh networks and to accommodate other 802.11 WLAN (old or new) services in the network. The Wi-Mesh alliance companies include Nortel Networks, Thomson, InterDigital Communications, NextHop Technologies, and Philips, among others. Their proposal was claimed to be equipment vendor independent and operable in indoor and outdoor situations. The usage models for 802.11s are categorized into four main items depending on the deployment, propagation characteristics and required service. The basic residential model contains a small number of nodes and its main characteristic is to provide a low-cost,

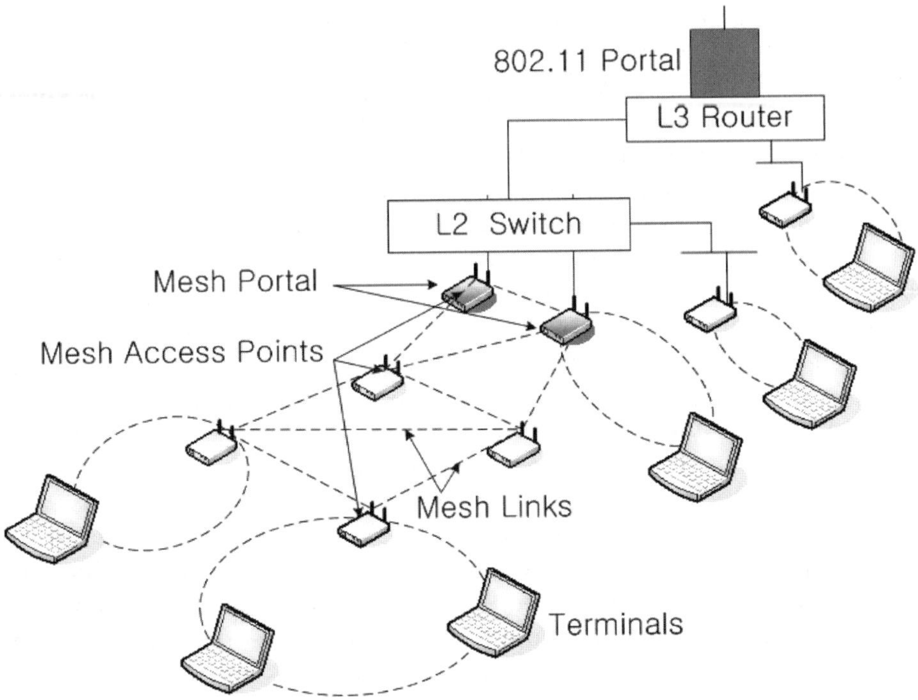

Figure 15.1. The IEEE 802.11s mesh operation from IEEE P802.11-04/0730d3 [802wireless-world, 2006].

high performance and easily deployable mesh network to remove the radio frequency dead-spots. Other usage models include the office, campus/public access network and public safety networks. The office and campus/public access models contain a relatively large number of nodes and a wider coverage area. The public safety model is to form a relatively smaller easily deployable network during emergency situations.

In summary, mesh networking is a suitable solution for LANs to provide easily scalable, reliable, flexible and cost effective networks.

3. Mesh Mode Operation in IEEE 802.15

The IEEE 802.15 Standards defines the physical and medium access control layers for short-range communications Wireless Personal Area Networks using the ultrawideband (UWB) technology. Data rates from 250 kbps (802.15.4) to 55 Mpbs (802.15.3), with communication distances from 1 to 75 meters, are expected. The IEEE 802.15.5 standard is the mesh extension to 802.15 [IEEE15web, 2006].

In comparison with the mesh operation in 802.11 networks described in the previous section, the 802.15 differs in the way terminals act as nodes in the mesh network. In 802.11, which is an infrastructure mesh, only Access Points are nodes of the mesh network, whereas in 802.15, which is a client mesh, user terminals are the nodes of the mesh network. As a result, the mesh control layer must now also address network performance and control, in addition to coverage and range extension. This feature requires collective behaviors to be implemented. Thus cooperation is required at the network level. In particular in large mesh networks, local routing decisions result in sub-optimal global routing, leading to unacceptable QoS performance. In order to guarantee QoS to critical applications, local network information must be shared globally. The challenges of propagating network information to every node lie in the overhead required for transmitting that information, and the delay between the time the information is sent by a node and received by all other nodes, which could render the information obsolete due to the time-varying nature of the Mesh WPAN. Nodes in the mesh network should therefore cooperate to propagate control and data streams of other nodes, hopefully resulting in a benefit for every single node in the network. Moreover, cooperative re-transmission mechanisms using nodes as relays, built on ARQ protocols and cooperative coding, also offer further enhancement to the physical and medium access control layers.

An important characteristic of WPAN is the low transmission powers due to energy-limited battery-operated devices. Another distinguishing feature of WPAN networks is proactive power management. It is well known that relaying, multihop and cooperative transmission techniques can help save energy. MAC protocols can also be designed to allow nodes to participate in cooperative routing for power savings, and to go into energy-saving modes as often as possible.

Cooperation is also often required for the coexistence or sharing of resources by collocated networks. In addition to contention-based access to the channel for delay-insensitive applications (with a carrier sense multiple access (CSMA) approach for collision avoidance), delay sensitive applications rely on beacons to ensure isochronous transmissions in IEEE 802.15.3. In the scenario of simultaneous operating mobile piconets, collisions of such beacons would prevent the successful transmission of delay-sensitive data. Cooperation between the piconets is thus necessary to avoid this undesirable situation. The beason mode of operation specifies a superframe structure with a subframe for the transmission of beacons, and a PAN coordinator to address coexistence. However, the beacon mode of operation is currently not allowed in the mesh mode.

To conclude on PAN, due to the short communication ranges, a mesh architecture is natural, but it requires advanced cooperative techniques in order to be scalable and reliable. The power-limited nature of the devices is also addressed by cooperative transmission and routing techniques.

4. Mesh Mode Operation in IEEE 802.16

The IEEE 802.16-2004 is an OFDM, OFDMA and single carrier based fixed wireless LAN/MAN standard in licensed bands of 10-66 GHz approved in June 2004. It improved and consolidated the previous standards such as 802.16-2001, 802.16a-2003, and 802.16c-2002. The MAC layer supports the point-to-multipoint and optional mesh network topology [IEEE, 2004]. The optional mesh mode operation was initially defined in the the 802.16a-2003 standard with basic signaling, message formats, etc. Subsequently, the mesh mode specifications were integrated and improved in the IEEE 802.16-2004 revised standard. Unlike the point-to-multipoint mode, there are no clearly separate downlink and uplink subframes in the Mesh mode. Each terminal communicates with a number of neighboring stations instead of communicating with a base station. There are a few terminals, which function as gateway to the backhaul network and provide some of the base station functions.

In the IEEE 802.16-2004 standard, centralized scheduling, distributed scheduling, and a combination of both scheduling schemes are used. If centralized scheduling is employed, the mesh base station nodes functions are similar to the base station in the point-to-multipoint mode. The mesh base station provides the control and scheduling decisions. When distributed scheduling is employed, all terminals, including the mesh base station, transmit their data after coordinating with the two-hop neighborhood and broadcast their scheduling information, such as available resources, requests and grants [IEEE, 2004]. It is assumed that no interference occurs between nodes that are two hops away. Thus, the mesh with two-hop neighborhood suffers from the hidden terminal problem [I.F. Akyildiz, 2005]. The inter node interference is one of the major factors affecting the network capacity and the scalability in mesh networks. If the inter node interference is taken into account in the radio resource allocation, better spectral efficiency may be obtained. In centralized scheduling, resources are allocated in a more centralized manner. The mesh base station gathers requests for resources in uplink and downlink from the terminals within a range of a few hops. It makes the decision and transmits the scheduling message which is not the actual schedule to the terminals. The terminals use a predetermined method to calculate the actual scheduling information depending on the system parameters [IEEE, 2004]. The mesh network with centralized scheduling has limited scalability. It can only support around 100 subscribers due to the structure of centralized scheduling messages.

5. Mobile Multihop Relay PHY/MAC Enhancement for IEEE 802.16e

The modification to PHY and MAC layers in the 802.16-2004 standard was considered in the IEEE 802.16e task group to include mobile and nomadic applications [IEEE16e, 2006]. The task group was approved in December 2002. It incorporated advanced techniques such as MIMO, LDPC codes, scalable OFDMA, adaptive modulation and coding (AMC), into the IEEE 802.16e to deliver broadband access to mobile and nomadic subscribers. This task group completed its activities and submitted the draft standard for approval by the IEEE standards committee in September 2005.

The demand for higher data rate keeps on increasing with new wireless services. Adaptive modulation and coding may cause non-uniform coverage of the IEEE 802.16e systems at the boundary of the cells. To overcome these impediments, modifications to the PHY and MAC of IEEE 802.16e was proposed and a study group was formed in July 2005 to develop methods for using multihop relay and cooperative techniques.

The study group was named the mobile multihop relay study group and defined its goals as coverage extension and throughput enhancement. These goals will be achieved through the modification of the frame structure and the addition of new protocols for relay operation, while keeping the backward compatibility for the point-to-multipoint mode in IEEE 802.16e. As the mesh type of operation is already incorporated in the IEEE 802.16-2004, it will not be considered in this study group. The other major requirements are that one end of the relayed path should be the a base station or a mobile station, and to efficiently provide a multihop or relay path to a mobile station or to a base station with a small number of hops. The operating scenarios under consideration in the mobile multihop relay study group are summarized in Table 15.2 [IEEE16, 2006].

Table 15.2. Table The topology and operating scenarios considered in IEEE 802.16 MMR-SG.

Topology	Scenario	
	Infrastructure	Client
Mesh operation	No	No
Fixed	Yes	Yes
Nomadic	Yes	Yes
Mobile	Yes	No

As described in the table the mobile client relay will not be considered by this Study Group due the complexity, battery life of the client relay, and security.

In recent years there has been a lot of interest in the industry and academia in multihop relays and cooperative diversity systems. At this point we need to make the distinction between relays and repeaters. The repeaters are the networks elements which receive, amplify and transmit without any baseband processing [of Visions 2003 (WWRF), 2006]. They are basically bidirectional amplifiers. They are normally used to extend the coverage in shadowed areas within a cell or extend the cell coverage. The use of repeaters is already addressed in the 3GPP cellular standard [3GPP, 2006a], [3GPP, 2006b].

Since the relay stations are not connected to the network backhaul, they are low cost and low power elements. The relays can be placed in such a way to reduce the propagation losses between the relay and the mobile users in order to improve the coverage and throughput. The relaying operation can be carried out in either the time or the frequency domain. With time domain relaying, the same frequency is used by the base station and the relay station, and they share the channel temporally. Different frequencies are used by the base station and the relay station with frequency domain relaying, and they transmit during the same time slot. The relays can employ forwarding schemes such as

- Amplify and forward
- Decode and forward
- Estimate and forward
- Store and forward

The detailed description of these schemes is out of the scope of this chapter, however we briefly discuss the impending challenges in adopting them. In the amplify and forward scheme, the received signal is just amplified by a fixed gain. It is essentially similar to an analog repeater and simpler to implement. However, interference enhancement may occur and instability may result in the system. In the decode and forward scheme the received signal is fully decoded and reencoded, and then transmitted by the relay station. It poses the danger of error propagation and higher latency. In the estimate and forward scheme, the data is estimated and transmitted by the relay station. It is similar to the decode and forward scheme but relatively simpler at the expense of the performance. It also has the drawback of error propagation and higher latency. In the store and forward scheme, the relay node in the relay chain receives the data, stores it and transmits it as required by the particular protocol. This scheme also has the drawback of higher latency and requires large buffer sizes. The relay or cooperative techniques can also be used in conjunction with advanced techniques such as MIMO, space time coding, adaptive modulation and coding, and advanced channel coding. We briefly summarize the candidate techniques proposed in the literature and their inherent challenges in their adoption for a standard.

1 Virtual Antenna array

A source multicasts the desired data to number of relays, which in turn retransmit the processed data to the destination. The destination may intelligently combine and process the received data to obtain higher throughput and spatial diversity [Dohler, 2003]. The challenges for this scheme are to obtain the channel state information at the relays, synchronization, cluster information for relays, etc.

2 Distributed MIMO and space time coding

A source transmits the desired data to a number of closely spaced relays. They fully decode the data and then using space-time coding or spatial multiplexing or any other advanced MIMO technique in distributed manner, they transmit the data to the destination [Laneman and Wornell, 2003]. As in virtual antenna arrays the channel state information may be needed at the relays and the relays may need to exchange their channel state information and other control information among them.

3 Coded cooperation

The channel coding and cooperative relaying are integrated in coded cooperation [Hunter and Nosratinia, 2005], [A. Nosratinia and Hedayat, 2004]. Different error correction schemes are used in the direct and the relayed paths depending on the channel conditions. In general, various channel coding methods can be used in this framework such as block codes including LDPC codes, convolutional codes, Turbo codes. The major impediments of this scheme are the decoding complexity and the large overhead in transmission.

In conclusion, there are many cooperative or multihop relay techniques proposed in the literature for coverage extension and throughput enhancement. However, the 802.16 mobile multihop relay study group is the first attempt to induct them into a standard. It is early to say whether cooperative techniques or other already known mature techniques serve the purpose effectively.

6. Cognitive Radio/Spectrum Sharing Techniques in IEEE 802.22

The Working Group 802.22 on Wireless Regional Area Networks (WRAN) was created in November 2004 to address the use of cognitive radios in unlicensed spectrum operation in TV bands [WG, 2006]. This approach was prompted by a Notice of Proposed Rule Making from the FCC, which was released in December 2003 [Commission, 2004]. This Notice of Proposed Rule Making proposes to open the licensed TV band in the United States to unlicensed operation by spectrum-agile devices, provided they do not interfere with incumbent license users. This is a new approach to spectrum management, prompted by the observation that licensed spectrum is mostly unused in certain locations and at certain times. Thus the current approach of allocating the spectrum has been recognized to be inefficient in the light of the shortage of spectrum unanimously observed.

The scope of cognitive radios capabilities in IEEE 802.22 is more limited than the original definition of Mitola [Mitola, 2000], for which a comprehensive review is available in [Haykin, 2005]. However, even in its limited approach to the problem, the FCC opened the door to many challenges in the definition of a standard adopting these principles. The IEEE 802.22 WRAN Working Group aims for a fixed wireless broadband access in regional areas where TV bands will be largely unoccupied most of the time. The scope of receivers with cognitive capabilities is thus limited to switching spectrum bands and controlling their emitted power to avoid creating interference to nearby TV receivers. TV receivers located inside the noise-protected contour of a TV station are entitled to protection. The noise-protected contour is defined by the quality of TV reception in terms of the value of the Desired-to-Undesired ratio at the TV receiver. For NTSC TV, it is referred to as the Grade B contour [O'Connor and A., 1968]. However even such a limited approach in a very particular scenario poses the challenges of cooperative sensing, cooperative decision-making, cooperative power control, coexistence among such unlicensed networks operating in the same band, and coexistence with other types of unlicensed devices, and the design of efficient physical and medium access control layers to support these requirements.

In October 2005, the Working Group has approved the functional requirements upon which proposals have been submitted in November 2005. The main lines of the functional requirements [IEEE802.22, 2006] in term of interference management can be summarized as follows. The WRAN is a large area network operating in rural or sub-urban areas, where a base station will cover a cell of radius from 33 km up to 100 km where propagation conditions permit. Broadband Internet services will be delivered to the Consumer Premise Equipment (CPE), which is fixed and possibly professionally installed at the user's home or office.

The CPEs and the WRAN base stations have an obligation to protect all TV receivers in the TV bands within the noise-protected contour of a licensed TV operation. Apart from TV stations, other incumbent users include Part 74 devices in the United States (wireless microphones), Public Land Mobile Radio System (PLMRS) services, and emergency services, which must also be protected whenever they appear within the interference range of the WRAN. While TV stations mostly change on a monthly, or possibly on a daily basis if they are turned off during the night, the behavior of wireless microphone users are more unpredictable in space and time. Sensing periods and durations, as well as the range of frequencies sensed by one CPE, will determine how well an incumbent service can be protected. Cooperative sensing should be performed from all sensing measurements in order to provide an accurate and updated state of the radio scene, which can be used to control transmission parameters of the CPEs to meet the interference requirements.

The operation of the WRAN is point-to-multipoint, with the base station controlling the CPEs in a Master/Slave relationship. The restrictions on CPEs capabilities can be listed as follows:

- CPEs can only transmit when being told by the base station.
- CPEs can only sense as told by the base station.
- CPEs can only change their parameters (transmit power, modulation, error control code, antenna beam) when ordered by the base station.

However, even though the base station controls the CPEs, cooperation is required in the way that sensing measurements are collectively used at the base station to achieve a better detection and a better radio resource utilization, and cooperative power control is required to meet the interference requirements at the boundaries of the noise-protected contours.

The fewer degrees of freedom of the CPEs, compared to the more general cognitive radios of Mitola, are meant to provide more assurance that incumbent license users will be protected. For example, an ad-hoc network is not allowed to be formed by co-located CPEs, even locally. This requirement avoids bursts of signaling for network set-up, and collisions that necessarily occur in CSMA/CA types of systems. In fact, even very short bursts, on the order of a few milliseconds, can dramatically disrupt the operation of a wireless microphone that is transmitting live action from a sports game. However, the centralized operation also has the advantage to allow for advanced cooperative sensing techniques, and to simplify coexistence between overlapping WRAN cells operated by the same or different operators. When all information is collected and analyzed by the base station, the centralized decision can make better use of radio resources. Given that TV stations are also fixed in space and change rarely with time, the radio resource allocation can be optimized given that the base station benefits from a sufficient amount of time to perform the radio-scene analysis and the optimization. The trade-off will be an increase in signaling for reporting all sensing information, and increased computational requirements at the base station, whereas some coexistence problems could have been be solved locally by the CPEs if they were allowed to.

The usual problems encountered in detecting the presence of licensed users and adapting one's transmitter functions to limit the amount of interference to the licensed users have been addressed in the academic literature, but many problems remain unsolved. In particular we can list the following issues that are directly related to the IEEE 802.22 standard:

- Hidden node problem: a CPE estimates its distance to the noise-protected contour of a TV station by measuring the TV signal it receives. If the CPE is in a region affected by shadow fading, it might determine that there is no TV signal or make a wrong estimate of its distance to the nearest TV receiver, resulting in its decision to transmit with a larger power than would be allowed to meet the interference requirements.

- Even though it has been proven that licensed users can still operate in the presence of licensed spectrum users, it is still not clear whether a cognitive radio network can achieve any useful throughput [Hoven and Sahai, 2005]. In this sense, the IEEE 802.22 WRAN standard should provide such a proof.

- Cooperative sensing is needed to improve the detection threshold of incumbent signals. However the probability of false alarm must be tightly controlled and efficient algorithms for decision fusion and data fusion are needed for that purpose. A related problem arises from the lack of accurate models of high order statistics for shadow fading, which are required to determine the probabilities of detection and false alarms accurately.

- Cooperative power control must be provided by the base station in order to ensure the interference levels created by simultaneously transmitting CPEs do not exceed the incumbents' thresholds. It is known that a sea of unlicensed users acts as an equivalent single unlicensed user experiencing a path loss exponent decreased by two [Tandra and Sahai, 2005] in the propagation channel between its transmitter and a nearby TV receiver.

- Sensing range vs. interference range for heterogeneous devices: coexistence with incumbent users that transmit at low power is a problem, given that the interference range of the CPE can be larger than the detection range of the incumbent user (*e.g.* a wireless microphone).

- Coexistence between unlicensed spectrum users: game theoretic approaches will most likely provide the required solution. However without centralized control they might result in trial and errors or transient states leading to contentions that create unwanted interference to incumbent users. Even though contention-based principles could be used to access control channels it is not clear whether these solutions could be used to access traffic channels. Game theoretic approaches have been considered in the literature for spectrum sharing of a few devices, or between at most two networks. Yet it has been recognized that they could sometimes lead to solutions that are far from optimal even in these simple cases. For instance, a Nash equilibrium could result in a very inefficient use of the spectrum with a very low throughput compared to easily found heuristic solutions [Clemens and Rose, 2005]. Game theory could provide sets of rules to use in deterministic algorithms. Yet it must be demonstrated that these rules are scalable to large networks.

- Coexistence between wide-area incumbent and license-exempt networks. The coverage area of the WRAN is of the same order as the coverage area of a TV station. In general, the cognitive networks that have been considered in the literature have a much smaller coverage than the incumbent service, which allows the unlicensed user to use low powers and still achieve acceptable throughput. However in the case of the WRAN, large powers will need to be radiated by the WRAN base station and CPEs, resulting in more stringent spatial constraints for operation.

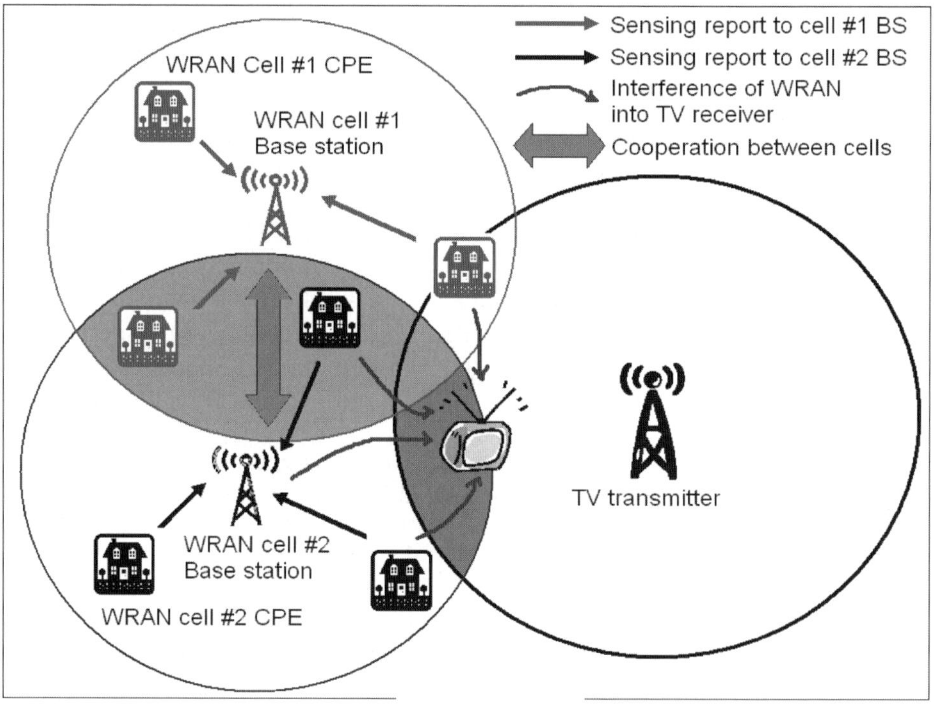

Figure 15.2. Operating scenario for two overlapping WRAN cells coexisting with a television broadcasting station.

Figure 15.2 illustrates some of the above mentioned issues. In this figure, two partially overlapping WRAN cells operate near an operating TV station. Both WRAN cells could belong to the same operator, or to different operators. In both cases, cooperation is required between the two cells to ensure that:

- The interference created by all CPEs into nearby TV receivers within the noise protected contour of the TV operation does not exceed the interference threshold.
- The CPEs in the overlapping (blue) area can coexist and achieve useful data rates.

Cooperation is also seen as the collection of sensing reports from all CPEs belonging to the same cell. The base station of that cell collects this information, and performs cooperative sensing. Sensing reports are coordinated among CPEs by the base station, which controls the frequency bands sensed by each CPE, the sensing integration time, the sensing period, and the type of interference being targeted by specific sensing techniques, like a simple energy detector or a cyclostationary feature detector. Using database location information, the base station can make sure that sensing is performed accurately while providing small sensing periods to guarantee useful operation of the WRAN for high data rate and delay-sensitive applications. Cooperation between base stations operated by the same or different operators could consist of several options. It is unlikely that different providers would share location information of their own CPEs to protect their market interests. However, wide-area sensing measurements results and the density of CPEs could be shared for the coexistence (blue) and interference into incumbent (green) regions. Negotiations need to take place between the base stations to dedicate operating and backup channels to each cell in the overlapping regions, while leaving as many degrees of freedom as possible in other areas to get the maximum benefit of opportunistic spectrum access.

Another issue that has not been broadly addressed is the amount of control information needed to operate a cognitive radio network. The 802.22 approach is that of a point-to-multipoint network with centralized decision-making to decide whether cognitive radio devices are allowed to operate in a certain frequency band with a maximum allowable transmit power. This approach, even though it should provide more stability and security for the incumbent license users, puts a burden on control channels, which need to convey a large amount of information between the users' terminals and the base station.

Another important issue is the protection of devices licensed under Part 74 of the FCC. These devices, such as wireless microphones, do not occupy spectrum for a long time, but are very sensitive to interference. They also operate on a short-range, much shorter than the range of the WRAN. Therefore, WRAN devices would have difficulty detecting the presence of Part 74 devices, but they would create unacceptable levels of interference to the Part 74 devices. In that context, cooperative sensing is absolutely necessary, and appropriate and novel protocols for spectrum occupancy must be designed. A Study Group within the 802.22 has been created in September 2005 to enhance the detection and protection or Part 74 devices. One of the principles that this Study Group will look at is the possibility of requiring the use of a beacon by Part 74 devices, such that the beacon's detection range will match the interference range of CPEs into Part 74 devices. The necessity of pilots has been recognized in the academic world [Tandra and Sahai, 2005], also because the detection threshold of incumbent users signals must be much lower than the decoding threshold of these signals in order for cognitive radios to offer the appropriate protection.

Finally, with respect to the emerging technologies related to cognitive radios that are expected to find their way into IEEE standards, an effort is ongoing for defining these technologies more accurately than is actually available. The IEEE P1900 Standard Series on Next Generation Radio and Spectrum Management sponsored by the IEEE Electromagnetic Compatibility Society and the IEEE Communications Society, have been approved early 2005. The IEEE P1900.1 Working Group will develop standard terms, definitions and concepts for spectrum management, policy defined radio, adaptive radio and software defined radio. The IEEE P1900.2 Working Group will develop a recommended practice for interference and coexistence analysis, while the IEEE P1900.3 Working Group will develop a recommended practice for conformance evaluation of software defined radio (SDR) software modules. A brief introduction to this new standard can be found in [Siller and Boutaba, 2005].

To conclude on the IEEE 802.22 Working Group activities, even a limited-scope cognitive radio network poses tremendous challenges, for which no definite solution exits neither in the academic nor in the industry world. This standard provides a unique opportunity for industries and universities to create a cognitive radio network with solid foundations, and to address some of the most fundamental problems of cognitive radios.

7. Conclusions

The IEEE standards provides a forum where industry and academia can jointly promote advanced technologies into emerging standards. These standards address communication networks ranging from Personal Area Network to Regional Area Networks. Each standard presents its own challenges, and many of the proposed solutions rely on cooperative techniques. Cooperation appears at the terminal level, at the network level, and between networks. The main applications of cooperative techniques have emerged from the introduction of mesh networks (11, 15 and 16) with the use of relays and cooperative transmission, as well as to address the coexistence between 802 standard networks themselves and with licensed spectrum users. It is expected that as these networks grow, and as their numbers grow, cooperative behaviors will be the key to scalability and reliability of these networks. Cooperative techniques have just found their way into the IEEE 802 standards, and still a lot remains to be accomplished to ensure the goals will be reached successfully.

References

3GPP (2006a). 3GPP 25.106, UTRA repeater radio transmission and reception.

3GPP (2006b). 3GPP 25.956, UTRA repeater planning guidelines and system analysis.

802wirelessworld (2006). 11-04-0730-03-000s-draft-core-terms-and-definitions-802-11s.doc. http://www.802wirelessworld.com.

Nosratinia, A. Hunter T. E. and Hedayat, A. (2004). Cooperative communication in wireless networks. *IEEE Communications Magazine*.

Clemens, N. and Rose, C. (2005). Intelligent power allocation strategies in an unlicensed spectrum. In *Proceedings of the IEEE Dynamic Spectrum Access Networks Conference (DySPAN 2005)*, Baltimore, MD.

Commission, Federal Communications (2004). Notice of Proposed Rule Making. ET Docket no. 04-113.

Dohler, M. (2003). *Virtual Antenna Arrays*. PhD thesis, King's College London, University of London, Strand, London.

Haykin, S. (2005). Cognitive Radio: Brain-Empowered Wireless Communications. *IEEE Journal on Selected Areas in Communications*.

Hoven, N. and Sahai, A. (2005). Power Scaling for Cognitive Radio. In *Proceedings of IEEE WirelessCom 05 Symposium on Emerging Networks, Technologies and Standards*, Hawaii, USA.

Hunter, T. E. and Nosratinia, A. (2005). Diversity through coded cooperation. *IEEE Transactions on Wireless Communications*.

IEEE (2004). IEEE Std 802.16-2004. IEEE Standard for Local and metropolitan area networks Part 16: Air Interface for Fixed Broadband Wireless Access Systems.

IEEE11web (2006). IEEE 802.11 Standard Web Site. http://grouper.ieee.org/groups/802/11/.

IEEE15web (2006). IEEE 802.15 Standard Web Site. http://www.ieee802.org/15/.

IEEE16 (2006). C80216mmr-05/d040. http://www.ieee802.org/16.

IEEE16e (2006). IEEE P802.16e/D12. IEEE Standard for Local and metropolitan area networks Part 16: Air Interface for Fixed and mobile Broadband Wireless Access Systems, Amendment for Physical and Medium Access Control Layers for Combined Fixed and Mobile Operation in Licensed Bands.

IEEE802.22 (2006). Functional Requirements for the 802.22 WRAN Standard. doc.: IEEE 802.22-05/0007r46, http://www.ieee802.org/22/.

Akyildiz, I. F. Wang, X. and Wang, W. (2005). Wireless mesh networks: a survey. *to be published in Elsevier, Computer Networks*.

Laneman, J. N. and Wornell, G. W. (2003). Distributed space-time-coded protocols for exploiting cooperative diversity in wireless networks. *IEEE Transactions on Information Theory*, pages 2415–2425.

Mitola, J. (2000). *Cognitive radio: An integrated agent architecture for software defined radio*. PhD thesis, Royal Inst. Technology (KTH), Stockholm, Sweden.

O'Connor, See and A., Robert (1968). Understanding Television's Grade A and Grade B Service Contours. *IEEE Transactions on Broadcasting*.

of Visions 2003 (WWRF), Book (2006). Relay-based deployment concepts for wireless and mobile broadband cellular radio. http://www.wireless-world-research.org/.

Siller, C. and Boutaba, R. (2005). The President's Page, Standards - A New Challenge for COMSOC. *IEEE Communications Magazine*.

Tandra, R. and Sahai, A. (2005). Fundamental limits on detection in low SNR under noise uncertainty. In *Proceedings of IEEE WirelessCom 05 Symposium on Emerging Networks, Technologies and Standards*, Hawaii, USA.

WG, IEEE 802.22 (2006). Ieee 802.22 working group on wireless regional area networks. http://www.ieee802.org/22/.

Chapter 16

COOPERATIVE COMMUNICATION WITH MULTIPLE DESCRIPTION CODING

Expanding the Cooperation to the Realm of Source Encoding

Morten Holm Larsen
Aalborg University - Department of Telecommunication Technology
mhl@kom.aau.dk

Petar Popovski
Aalborg University - Department of Telecommunication Technology
petarp@kom.aau.dk

Søren Vang Andersen
Aalborg University - Department of Telecommunication Technology
sva@kom.aau.dk

Abstract: Multiple Description Coding (MDC) is a source coding technique where the source is encoded into two or more descriptions. The descriptions are self–sufficient in the sense that each description can provide a distorted version of the source information, while the distortion is decreased as more descriptions are utilized at the decoder. MDC was proposed as a coding scheme to gain robustness to packet loss over a communication network in a scenario with single source and single destination. In this chapter, we study how the MDC can be applied to support cooperative communications. In particular, we focus on Multiple Description Lattice Vector Quantizer (MDLVQ) and suggest optimal design methods for MDLVQ in a cooperative network. Next, we propose a novel scheme, termed MDC with Conditional Compression (MDC–CC). The basic observation behind MDC–CC is that the availability of timely feedback presents the need for robustness, originally implied by the MDC scheme. The source encoding with

Frank H. P. Fitzek and Marcos D. Katz (eds.), Cooperation in Wireless Networks: Principles and Applications, 515–545.

MDC–CC is done in such a way that upon having a feedback from the destination, the encoding overhead can be removed at any time by a cooperative node or an active network. We show an implementation where MDC–CC utilizes the highly structured design of MDLVQ and thus produces a very elegant solution, where the computational complexity becomes almost negligible. Finally, we introduce three generic scenarios for cooperative communication with MDC and, in particular, MDC–CC: data delivery with cooperative sources, data delivery with cooperative destinations and data delivery with meshed cooperation.

Keywords: higher layer cooperative networking, source coding, lattice vector quantizers, multiple description coding (MDC), MDC with conditional compression (MDC–CC), cooperative communication, cooperative sources, cooperative destinations, meshed cooperation.

1. Introduction

The paradigm of cooperative communication has recently gained significant attention in relation to the wireless communications. The initiating observation is that the broadcast nature of the wireless medium offers a possibility for a group of terminals to cooperate by sharing their antennas and thus creating a distributed multi–antenna entity, see *e.g.* [Nosratnia et al., 2004]. In an exemplifying scenario, such entity communicates with the Base Station (BS), this provides the terminals with a better service as compared to the case when each terminal communicates with the BS independently. Such method of communication creates a diversity effect, termed *cooperative diversity* and is concerned mainly with the physical and link layer of the protocol stack. At the network layer cooperation also occurs as explained *e.g.* by [Gupta and Kumar, 2000]. As an example, in multi–hop wireless networks, a communication node can act as a router that forwards packets on behalf of other nodes.

The instances of cooperative communication mentioned above are exploiting the benefits of cooperative communication, regardless of the information source that produces the communicated data. The solution space for cooperative communication can be further expanded by bringing the cooperation further up in the protocol stack and considering the source coding aspect. In this chapter we propose and analyze the suitability of source coding schemes based on Multiple Description Coding (MDC) for the scenarios with cooperative communication. We give examples of how to design the source coding schemes within the multiple description (MD) paradigm to account for the fact that the descriptions are transmitted through a cooperative communication network.

Source Coding and Cooperative Communication

To get started with the cooperative aspects of the source encoding, in this section we introduce several motivating examples. First, let us consider an example of cooperative networking where two terminals cooperate on a realtime application download, such as a video–on–demand (VoD). A straightforward cooperation scheme can be one in which the source sends the whole information to one of the terminals and that terminal forwards the information to the other terminal. With a more sophisticated scheme, a coarse information about the source is sent to both terminals and a different refinement of information is sent to each terminal. If a terminal receives only the information sent directly from the source to it, then the video is shown with a low quality. To obtain the full video quality, the terminal should receive also the refinement information sent to the other terminal. In this simple example, the cooperation among the terminals, *i.e.* the exchange of refinement information, increases the video quality.

The theoretical framework for splitting information is known in the information theory as Multiple Description (MD) coding. MD was first introduced in the 1970s. A thorough review of the history of MD with applications and algorithms can be found in [Goyal, 2001]. The encoding of the source information is done in a way that multiple descriptions are produced to describe each chunk of source information. Each description contains coarse information about the source, but by using all generated descriptions this can be completely recovered. The fundamental MD encoding–decoding concept is typically described and analyzed in a framework with two channels and three receivers, as shown in Figure 16.1.

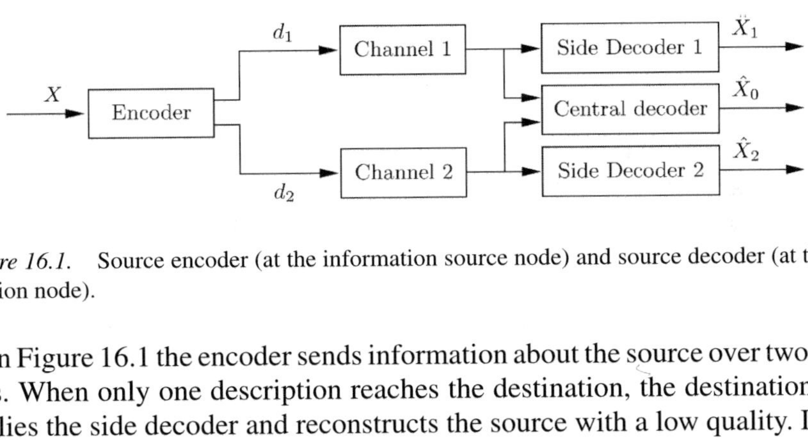

Figure 16.1. Source encoder (at the information source node) and source decoder (at the destination node).

In Figure 16.1 the encoder sends information about the source over two channels. When only one description reaches the destination, the destination node applies the side decoder and reconstructs the source with a low quality. If both descriptions reach the destination, then the central decoder is used and source information is reconstructed with a high quality. The system on Figure 16.1 can be generalized to M channels and $2^M - 1$ receivers.

Now, Figure 16.2 depicts a simple cooperative network, where BS is transmitting the two descriptions d_1 and d_2 to terminal 1 and terminal 2, respectively. Subsequently, the two terminals exchange descriptions through the cooperative link.

In this system, if terminal 1 receives only d_1 from BS, it reconstructs the source with a low quality. If, in addition to d_1, terminal 1 receives d_2 from

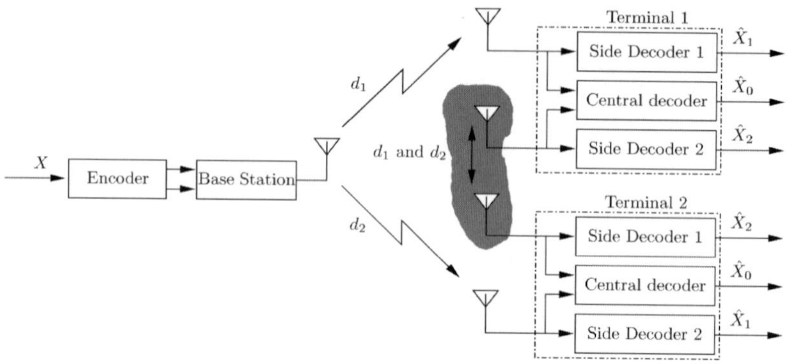

Figure 16.2. A simple cooperative network with two terminals and a Base Station (BS). The gray cloud denotes the cooperative link.

terminal 2, it applies the central decoder and hence reconstructs the source with a high quality. In the case where d_1 does not reach terminal 1, but only d_2 is received through the cooperative link, the second side decoder is used and the source is reconstructed with a low quality. This mechanism constitutes a first example on how Multiple Descriptions can be used in Cooperative Communication.

The incorporated robustness in the MD introduces a rate overhead for similar distortion level when compared with encoding the source for a single channel and decoder. Hence, when designing the encoder and decoder we can choose the amount of overhead. Conversely, this means that when the quality provided by the central decoder is fixed, the higher overhead we allow, the higher quality is provided by each side decoder. In the classical MD framework shown on Figure 16.1, the overhead and the design of encoder and decoders are determined from the rates and loss probabilities on the two channels by [Østergaard et al., 2004]. In practice, the loss probabilities are normally not known explicitly, but only estimates of the loss probabilities are known with an uncertainty. In [Larsen et al., 2005] it is shown how this uncertainty can be taken into account when designing the encoder and decoders.

Broadcast is frequently used in streaming real–time data to many users in a wireless environment. The existing broadcast schemes use a special case of MD commonly referred to as *layered coding*. In layered coding, the first description contains a coarse information and the following descriptions are only containing refinement information. Thus, in the case where the first description is lost, the following descriptions are approximately useless. Under a certain assumption, the combination of cooperative networking and multiple descriptions is also advantageously applied in the broadcast scenario. The critical assumption is the existence of a fast feedback between the cooperating terminals. If the terminals cooperate with a fast feedback mechanism, then in case of information loss over the broadcast channel, a terminal can request the information from the other terminals via the fast feedback. An exemplifying scheme for this method is shown in Figure 16.3.

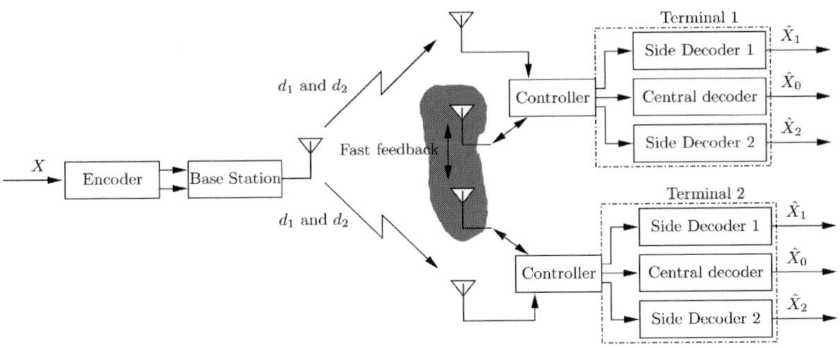

Figure 16.3. Scenario for cooperative reception of broadcast information with base station and two terminals.

Organization of this Chapter

After briefly introducing the motivating examples above, the next section provides sufficient mathematical detail on Multiple Description Coding (MDC) to do the analysis of its use for Cooperative Networking. In particular, we focus on the Multiple Description Lattice Vector Quantizer (MDLVQ) as we will use the lattice structure in the design examples throughout this chapter. Subsequently, Section 3 describes how to optimize the MD quantizer design for the cooperative scenario from Figure 16.2, and Section 3 further describes how the design result from [Østergaard et al., 2004] and [Larsen et al., 2005] can be used to design MD quantizers for the cooperative scenario from Figure 16.2. In Section 4 we introduce a novel MDC scheme, termed *MDC with Conditional Compression (MDC–CC)*. With MDC–CC, the compression of the source information can be done by any node in the network after such node gets information about what has already been received at the destination. We show that by using a lattice vector quantizer, the MDC–CC scheme becomes very elegant and the computational complexity becomes almost negligible. Section 5 discusses the presented methods and casts them into a set of more generally defined scenarios in which the combination of MDC and cooperative networking should be considered. Finally, the last section concludes the chapter.

2. Multiple Description Coding (MDC) Basics

In an MD system for two channels, the encoder sends information about the source over the two channels, with rate R_i bits per source symbol [bpss] for each channel, $i \in \{1, 2\}$. Each channel may either be in working or non–working state and this is not known at the encoder. The destination node uses side decoder 1 to reconstruct the source when channel 1 is in the working state and channel 2 is in the non–working state. Similarly, side decoder 2 is used to reconstruct the source when channel 2 is in the working state and channel 1 is in the non–working state. When both channels are in the working state the central decoder is used to reconstruct the source. Although this system can be generalized to M channels and $2^M - 1$ receivers, all fundamental concepts presented in this chapter can be explained in the simpler two channel case.

The principles in this chapter applies with multiple description designs in general, such as the scalar and vector designs proposed in [Vaishampayan, 1993; Vaishampayan and Domaszewicz, 1994]. However, in the special case of lattice structured quantizers, the principles have elegant implementations with very low computational complexity. The design framework of lattice structured multiple description quantizers is extensively described and analyzed in the literature. See *e.g.* [Sergio et al., 1999; Vaishampayan et al., 2001; Goyal et al., 2002] for a thorough introduction to this field. In the following, we give an outline of the main issues from this framework that we will need to convey central concepts related to design for cooperative networks.

Lattice Vector Quantizer

Let the real lattice $\Lambda \subset \mathbb{R}^L$ be a collection of lattice points λ (reconstruction points), where Λ is generated by a generating matrix $\mathbf{G} \in \mathbb{R}^{L \times L}$:

$$\Lambda = \{\lambda : \lambda = \mathbf{G}\xi, \xi \in \mathbb{Z}^L\}. \tag{16.1}$$

The generating matrix $\mathbf{G}$ is a set of linearly independent basis vectors $\mathbf{v}$, which span the lattice, $\mathbf{G} = [\mathbf{v}_1 \ \mathbf{v}_2 \ \cdots \ \mathbf{v}_L]$. The region of source vectors $\mathbf{x}$ that quantize to a lattice point λ is called a Voronoi region $V(\lambda)$, and given by

$$V(\lambda) \triangleq \{\mathbf{x} \in \mathbb{R}^L : \|\mathbf{x} - \lambda\|^2 \leq \|\mathbf{x} - \lambda^*\|^2, \forall \lambda^* \in \Lambda\}, \tag{16.2}$$

where the distortion measure $\|\cdot\|^2 \triangleq \frac{1}{L}\mathbf{x}^T\mathbf{x}$ is chosen to be the normalized 2–norm. The normalized 2–norm is typically used, even in the cases for which the map between 2–norm and the perceptual distortion measure is complex. In such cases, *e.g.* voice, audio and video coding, such map is typically provided by use of compandors and weighting filters. The L dimensional volume of a Voronoi region is the determinant of the generator matrix

$$\nu = \det[G]. \tag{16.3}$$

Figure 16.4 shows two very common lattices in two-dimensional space, Z_2 and A_2, where the generating matrices are

$$Z_2 : \mathbf{G} = \begin{bmatrix} 1 & 0 \\ 0 & 1 \end{bmatrix} \text{ and } A_2 : \mathbf{G} = \begin{bmatrix} 1 & -\frac{1}{2} \\ 0 & \frac{\sqrt{3}}{2} \end{bmatrix}. \tag{16.4}$$

The encoding algorithm is straightforward for Z_L lattice, but for A_L the encoding algorithm is nontrivial. A fast encoding algorithm for A_L is given in [Conway and Sloane, 1982a].

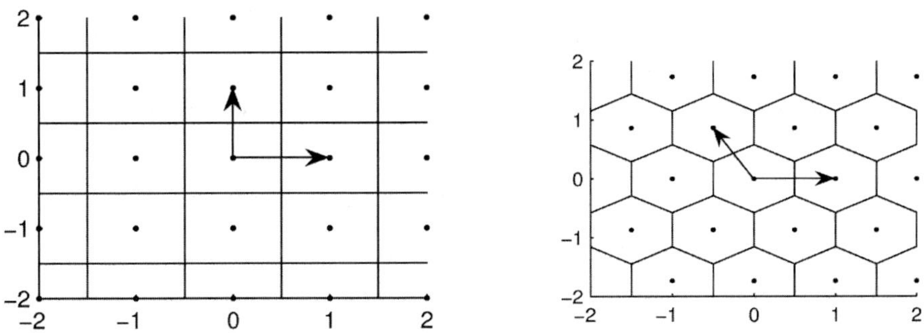

Figure 16.4. An example of a Z_2 lattice (left) and an A_2 lattice (right). The dots are lattice points λ, the lines bounds the voronoi regions and arrows are the basis vectors.

When evaluating the average distortion for a given lattice quantizer, let $X \in \mathbb{R}^L$ be an arbitrary i.i.d source and $f_{\mathbf{x}}(\mathbf{x})$ be the probability density function (pdf).

$$D = \sum_{\lambda \in \mathbf{\Lambda}} \int_{V(\lambda)} f_{\mathbf{x}}(\mathbf{x}) \|\mathbf{x} - \lambda\|^2 d\mathbf{x}. \tag{16.5}$$

Often, it is useful to analyze the distortion for high resolution of the quantizer. In high resolution, the Voronoi region is small and we can assume a locally constant probability density, $f_{\mathbf{x}}(\mathbf{x}) \approx f_\lambda$ for $\mathbf{x} \in V(\lambda)$. Thus, we can find the probability for a given λ by

$$P_\lambda \triangleq Pr\left(X \in V(\lambda)\right) \approx f_\lambda \nu_\lambda, \tag{16.6}$$

where ν_λ denotes the volume of the Voronoi region, $V(\lambda)$. From the structure of the lattice, we see that the volume is constant and Eq. (16.5) can be written as

$$D \approx \sum_{\lambda \in \mathbf{\Lambda}} \frac{P_\lambda}{\nu} \int_{V(\lambda)} \|\mathbf{x} - \lambda\|^2 d\mathbf{x}. \tag{16.7}$$

Quantization error will, due to the geometrical structure, yield

$$\int_{V(\lambda)} \|\mathbf{x} - \lambda\|^2 d\mathbf{x} = \int_{V(0)} \|\mathbf{x}\|^2 d\mathbf{x}, \tag{16.8}$$

where $V(0) = V(\lambda_{\mathbf{0}})$ and $\lambda_{\mathbf{0}} = [00 \cdots 0]^T$, [Gray, 1990]. Traditionally, the normalized second–order moment $G(\mathbf{\Lambda})$ is defined as

$$G(\mathbf{\Lambda}) \triangleq \frac{\int_{V(0)} \|\mathbf{x}\|^2 d\mathbf{x}}{\nu^{1+2/L}}, \tag{16.9}$$

and has been tabulated for many lattice structures, see *e.g* [Conway and Sloane, 1999; Conway and Sloane, 1982b]. Few of these are summarized in Table 16.1. The average distortion can hence be found from the volume of the Voronoi region and this table as

$$D \approx G(\mathbf{\Lambda})\nu^{2/L}. \tag{16.10}$$

For an arbitrary volume, we can observe from Eq. (16.9) that the smallest second–order moment is obtained by a sphere lattice for the 2–norm. Spheres can not be packed to fill the space and therefore cannot be used to constitute a quantizer, but the second moment of a sphere gives an analytical lower bound. However, from the table we see that A_2 is very close to the second moment of the sphere, and it is well known that A_2 is the optimal lattice for two dimensions, see *e.g.* [Conway and Sloane, 1999]. Unfortunately, the problem of constructing an optimal lattice for higher dimensions is still unsolved, but there exist lattices that perform relatively close to the sphere lower bound in higher dimensions, *e.g.* the Leech lattice in 24 dimensions, Λ_{24}. When analyzing the lattice vector

$\mathbf{\Lambda}$	$G(\mathbf{\Lambda})$	$n \rightarrow \infty$	$n = 24$	$n = 2$
Z_n	$\frac{1}{12}$	$\frac{1}{12}$	-	$\frac{1}{12}$
A_n	$\frac{1}{(n+1)^{1/n}}\left(\frac{1}{12}+\frac{1}{6(n+1)}\right)$	$\frac{1}{12}$	-	0.0802
Λ_{24}	Monte Carlo Simulation	-	0.06561	-
Sphere	$\frac{\Gamma(n/2+1)^{2/n}}{(n+2)\pi}$	0.0585	0.0647	0.0796

Table 16.1. Second moment for the most popolar lattice, ([Conway and Sloane, 1999]).

quantizer, we assume an entropy coding that maps a source symbol ξ to a variable bit rate. The mapping is made such that the length of the binary sequence to which ξ is mapped is inversely proportional to the probability of occurrence of ξ. This mapping is exploited in several data compression algorithms such as Huffman coding, see *e.g* [Cover and Thomas, 1991]. It can be shown that a Huffman coder can encode to an average bit rate of the entropy plus 1 bit. From this point we will assume that the entropy coder can encode arbitrarily close to the entropy. The entropy in bit per dimension for a lattice vector quantizer, when assuming high resolution is given by [Gray, 1990]:

$$R = -\frac{1}{L}\sum_{\lambda\in\mathbf{\Lambda}}\int_{V(\lambda)} f_{\mathbf{x}}(\mathbf{x})d\mathbf{x}\log_2\int_{V(\lambda)} f_{\mathbf{x}}(\mathbf{x})d\mathbf{x} \tag{16.11}$$

$$\approx h(X) - \frac{1}{L}\log_2(\nu). \tag{16.12}$$

To summarize, we list our observations about the benefits that emerge from the highly structured nature of the lattice:

- As any lattice point λ can be regenerated from ξ and the generating matrix $\mathbf{G}$, there is no need to store the reconstruction point.

- A fast encoding algorithm exists for Z_n and A_n. This is described in [Conway and Sloane, 1982a].

- There is a closed–form expression for the entropy, and hence an expression of the achievable average bit rate.

- There is a simple expression for the average distortion for the lattice quantizer. The second moment of this distortion is typically given in the form of a table.

We'll make beneficent use of each of these properties when we return to cooperative communication later in this chapter.

EXAMPLE 16.1 *Let us design a two dimensional lattice vector quantizer for a system with a unit variance Gaussian source, a A_2 lattice and with a rate constraints $R = 3$ bit pr. dimension. The differential entropy of a unit variance Gaussian source is $h(X) = \frac{1}{2}\log_2(2\pi e)$, [Cover and Thomas, 1991]. First we find the volume of the Voronoi region by Eq. (16.11),*

$$\nu = 2^{L(h(X)-R)} = 0.1334. \tag{16.13}$$

Next, we determine the generator matrix for this system by scaling the matrix for A_2 given by Eq. (16.4) as,

$$\mathbf{G}' = \mathbf{G}c. \tag{16.14}$$

The scaling constant "c" can be determined from the volume constrain in Eq. (16.13),

$$\begin{aligned} \nu &= \mid \mathbf{G}c \mid & (16.15)\\ &= \prod_{i}^{L} \lambda_i c & (16.16)\\ &= c^L \mid \mathbf{G} \mid, & (16.17) \end{aligned}$$

where λ_i is the i'th eigenvector of $\mathbf{G}$. The generator matrix can now be determined as,

$$\mathbf{G}' = \mathbf{G}\sqrt[L]{\frac{\nu}{\mid \mathbf{G} \mid}} = \begin{bmatrix} 0.3925 & -0.1963 \\ 0 & 0.3399 \end{bmatrix}. \tag{16.18}$$

Now, for example, the source input $x = [-0.1\ 0.6]^T$ is quantized to $\lambda = [0\ 0.6799]^T$ or $\xi = [1\ 2]^T$. This can be realized from Figure 16.4 by dividing the source input x by c.

Review of the Geometrical Relationship

Before explaining the method for constructing the unbalanced multiple description vector quantizer (MDLVQ) proposed in [Diggavi et al., 2002], we now review some important geometrical relationships between two lattices. We define a sublattice $\mathbf{\Lambda}'$ to be geometrically similar to the lattice $\mathbf{\Lambda}$, if $\mathbf{\Lambda}' \subseteq \mathbf{\Lambda}$ and $\mathbf{\Lambda}'$ can be obtained by scaling and rotating $\mathbf{\Lambda}$. The generator matrix for the $\mathbf{\Lambda}'$ is given by

$$\mathbf{\Lambda}' = \mathbf{UG}. \tag{16.19}$$

The number of lattice points λ that are included in a Voronoi region of the sublattice $V'(\lambda')$ is denoted N. The requirements for similar sublattices of Z_2 and A_2 are derived in [Conway et al., 1999], where it is found that for Z_2 there exists a similar sublattice when N has the form $N = a^2 + b^2$, where $a, b \in \mathbb{Z}$. The possible combinations of a and b yields

$$N = 1, 2, 4, 5, 8, 9, 10, 13, 16, 17, 18, 20, \cdots \tag{16.20}$$

This result can be found in [Sloane, 2005] as sequence A1481. The rotating and scaling matrix $\mathbf{U}$ for the similar sublattice of Z_2 is found by

$$\mathbf{U} = \begin{bmatrix} a & -b \\ b & a \end{bmatrix}. \tag{16.21}$$

For A_2 there exists a similar sublattice when N has the form $N = a^2 - ab + b^2$, where $a, b \in \mathbb{Z}$. The possible combinations of a and b yields the sequence,

$$N = 1, 3, 4, 7, 9, 12, 13, 16, 19, 21, 25, 27, \cdots \tag{16.22}$$

which again can be found in [Sloane, 2005], sequence A3136. The rotation and scaling matrix for A_2 is

$$\mathbf{U} = \begin{bmatrix} a + b\cos(2\pi/3) & -b\sin(2\pi/3) \\ b\sin(2\pi/3) & a + b\cos(2\pi/3) \end{bmatrix}. \tag{16.23}$$

Figure 16.5 shows an example of a geometrically similar sublattice of A_2, where $a = 4$ and $b = 3$. It can be seen from the figure that the sublattice includes exactly $N = 13$ lattice points, as expected. We define a sublattice $\mathbf{\Lambda}'$ to be clean if $\mathbf{\Lambda}$ does not intersect with the boundary of the Voronoi region of $\mathbf{\Lambda}'$. The geometrical relationship for a clean respectively non–clean sublattice of Z can be illustrated as in Figure 16.6, where $\mathbf{\Lambda}$ and two sublattices $\mathbf{\Lambda}'$ are generated by an even and an odd scaling, respectively. From this figure we observe how an even N will result in intersection between lattice points and the boundary of the $\mathbf{\Lambda}'$ Voronoi region. Conversely, when N is odd, the boundary will not intersect with lattice points and hence the sublattice is clean.

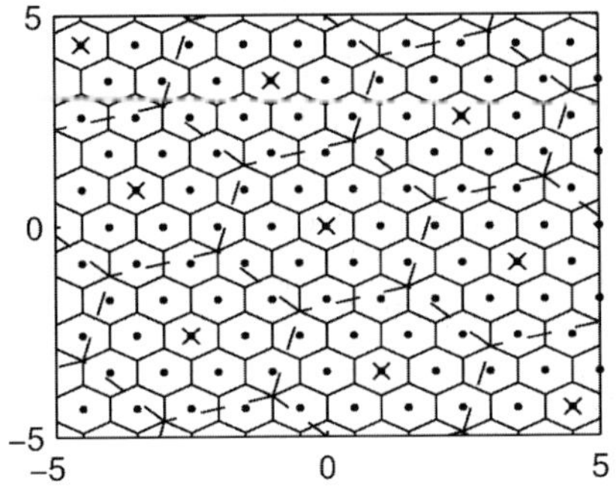

Figure 16.5. An example of an A_2 lattice and a similar sublattice (dashed) with $N = 13$.

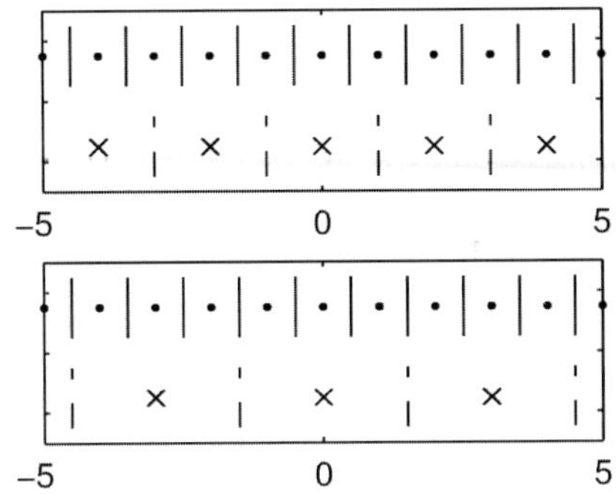

Figure 16.6. In the top an example of a Z_1 lattice and a clean sublattice where $N = 3$. In the bottom a sublattice where $N = 2$ and not clean.

The condition for a clean sublattice in two dimensions is investigated in [Conway et al., 1999] and clean sublattice of Z_n for higher dimensions are solved in [Diggavi et al., 2002]. Condition for a clean sublattice of Z_2 is that N must be odd and have the form $N = a^2 + b^2$, where $a, b \in \mathbb{Z}$. The possible combinations of a and b yields

$$N = 1, 5, 9, 13, 17, 25, 29, 37, 41, \cdots \tag{16.24}$$

as can be found in [Sloane, 2005], sequence A57653. For making a clean sublattice of A_2, N must have the form $N = a^2 - ab + b^2$, where a and b are relatively prime. This condition yields the possible combinations,

$$N = 1, 7, 13, 19, 31, 37, 43, 49, 61, \cdots \tag{16.25}$$

as can be found in [Sloane, 2005], sequence A57654. The rotating and scaling matrix $\mathbf{U}$ for clean sublattice of Z_2 and A_2 are expressed in Eq. (16.21) and Eq. (16.23), respectively. In the example on Figure 16.5 with $N = 13$, it can be seen that the sublattice is both similar and clean.

Multiple Description Lattice Vector Quantizers

The task of designing a Multiple Description Lattice Vector Quantizer (MDLVQ) has been an active area over an extended period of time and thereby addressed by many authors, *e.g.* [Vaishampayan et al., 2001; Diggavi et al., 2002; Goyal et al., 2002; Zhao, 2004; Diggavi et al., 2002; Østergaard et al., 2005]. The general unbalanced and asymmetric MDLVQ design was proposed by Diggavi, Sloane and Vaishampayan in [Diggavi et al., 2002]. Our scheme of MDC with conditional compression (MDC–CC), described in Section 4 employs this MDLVQ, described below.

The structure of the MDLVQ is shown in Figure 16.7. It operates as follows: a source vector is quantized to a lattice point $\lambda \in \mathbf{\Lambda}$. To send the information about λ over the two channels, a label function α is applied. We will from this point denote sublattice points with a subscript, *e.g.* $\lambda_\mathbf{1} \in \mathbf{\Lambda_1}$. The label function maps λ to a pair of sublattice points $(\lambda_\mathbf{1}, \lambda_\mathbf{2})$. We assume that the label function is one–to–one, so when both channels work, the inverse mapping $\alpha^{(-1)}$ will reconstruct the lattice point λ. Conversely, when only channel i works the reconstruction point of the source is the sublattice point $\lambda_\mathbf{i}$.

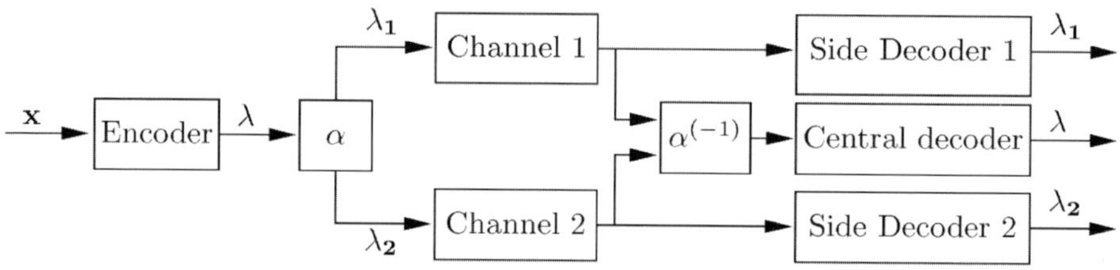

Figure 16.7. Block diagram of a two channel lattice vector quantizer.

To reduce the complexity of the label function α, the following three constraints on the MDLVQ are imposed:

- Constraint 1: The two sublattices $\mathbf{\Lambda_i}$ are geometrically similar and clean to $\mathbf{\Lambda}$, hence the reused index number N_i is given for $i \in 1, 2$, respectively.
- Constraint 2: There is a product sublattice $\mathbf{\Lambda_s}$ of $\mathbf{\Lambda_1} \cap \mathbf{\Lambda_2}$ that is geometrically similar to $\mathbf{\Lambda}$ and has reused index, $N_s = N_1 N_2$.
- Constraint 3: The label function must satisfy the shift-property, which means that $\alpha(\lambda + \lambda_\mathbf{s}) = \alpha(\lambda) + \lambda_\mathbf{s}, \forall \lambda \in \mathbf{\Lambda}, \lambda_\mathbf{s} \in \mathbf{\Lambda_s}$.

Encoding and decoding procedures for MDLVQ. The encoding procedure for an MDLVQ is a two step procedure, as illustrated in Figure 16.7. First the input vector $\mathbf{x}$ is quantized to the closed lattice point in $\mathbf{\Lambda}$,

$$\lambda = Q(\mathbf{x}). \tag{16.26}$$

Second, the label function maps the lattice point to the two sublattice points $\lambda_\mathbf{i}$, that are transmitted over the two channels:

$$(\lambda_\mathbf{1}, \lambda_\mathbf{2}) = \alpha(\lambda). \tag{16.27}$$

The two side decoding procedures are to find the best reconstruction point for a given λ_i. It is not guaranteed that the optimal reconstruction point is the sublattice point λ_i, as shown on Figure 16.7. This suboptimality induce a relatively small additional distortion, which may be neglected. Alternatively, one possible solution is to store a codebook containing reconstruction points, obtained by a Lloyd algorithm as investigated by [Zhao, 2004]. When both channels are working, the central decoding procedure is used as shown on Figure 16.7. The lattice point λ is found by the inverse label function $\alpha^{(-1)}$, hence $\lambda = \alpha^{(-1)}(\lambda_1, \lambda_2)$.

Average distortion measurement. When both channels are working the inverse label function will reconstruct the lattice Λ, and we can therefore obtain the average central distortion from Eq. (16.10) as:

$$D_0 \approx G(\mathbf{\Lambda})\nu^{2/L}, \tag{16.28}$$

where ν is the volume of a Voronoi region. The average side distortion can be found by the distortion between input $\mathbf{x}$ and the sublattice point λ_i. By assuming that λ is the centroid of its Voronoi region and high resolution, it is shown in [Vaishampayan et al., 2001] that the side distortion can be expressed as:

$$D_i = D_0 + \sum_{\lambda \in \Lambda} \|\lambda - \alpha_i(\lambda)\|^2 P_\lambda, \tag{16.29}$$

where P_λ is the probability of the lattice point λ.

Rate for a MDLVQ. The entropy in bit per dimension for a MDLVQ assuming high resolution is given in [Diggavi et al., 2002],

$$R_0 \approx h(p) - \frac{1}{L}\log_2(\nu). \tag{16.30}$$

The entropy for a sublattice is derived in [Vaishampayan et al., 2001] and the entropy on each channel is found in [Diggavi et al., 2002] to be,

$$R_i = R_0 - \frac{1}{L}\log_2(N_i). \tag{16.31}$$

Before explaining how the label function is constructed, we will give a design example of an MDLVQ and illustrate the label function.

EXAMPLE 16.2 (AN EXAMPLE OF MDLVQ) *Let us make the simples and possible example of an asymmetric MDLVQ by using the A_2 or Z_2 lattices. To obey the similar and clean constraints for the MDLVQ design, we must determine two small numbers of N that are different to make the MDLVQ asymmetric ($N = 1$ is trivial). The smallest combination is obtained by Z_2 with $N_1 = 5$ and $N_2 = 9$. And for simplicity we choose the simplest generator matrix for Z_2,*

$$\mathbf{G} = \begin{bmatrix} 1 & 0 \\ 0 & 1 \end{bmatrix}. \tag{16.32}$$

As described in Section 2.0 the two generator matrices for the sublattices can be found:

$$\mathbf{G_1} = \begin{bmatrix} 2 & -1 \\ 1 & 2 \end{bmatrix}, \mathbf{G_2} = \begin{bmatrix} 3 & 0 \\ 0 & 3 \end{bmatrix}. \tag{16.33}$$

The product sublattice $\mathbf{\Lambda_s}$ can for this Z_2 example be found by:

$$\mathbf{G_S} = \mathbf{U_1 U_2 G} = \begin{bmatrix} 6 & -3 \\ 3 & 6 \end{bmatrix}. \tag{16.34}$$

The lattice and the three sublattices are shown on Figure 16.8, where we can verify that $N_1 = 5$, $N_2 = 9$ and $N_s = 45$. Furthermore, it can be realized when $\mathbf{\Lambda_s}$ is clean and we apply the shift–property that a label function describing the 45 lattice points in $V_s(0)$ can cover $\mathbf{\Lambda}$. The label function describing the 45

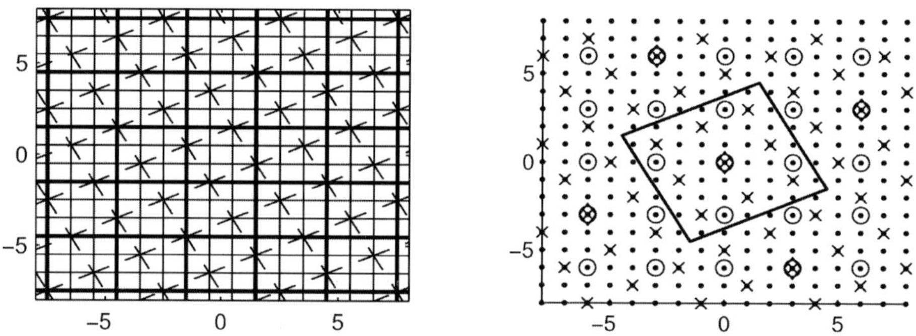

Figure 16.8. A MDLVQ example with $N_1 = 5$ and $N_2 = 9$, where the Voronoi regions are shown on the left and the centroids on the right. On the left, the solid line is the lattice $\mathbf{\Lambda}$, dashed line is the sublattice $\mathbf{\Lambda_1}$, wide line is the sublattice $\mathbf{\Lambda_2}$. On the right, the solid line is the Voronoi region $V_s(0)$.

lattice points are shown in Table 16.8. How to generate this label function will be explained in Section 2.0.0.

From the label function in Table 16.8 we can calculate the average distortions in the high resolution case from equations (16.28) and (16.29),

$$D_0 = \frac{1}{12}, \quad D_1 = 0.4833, \quad D_2 = 2.3500. \tag{16.35}$$

The rates for the example are

$$R0 = 2.0471, \quad R1 = 0.8861, \quad R2 = 0.4621. \tag{16.36}$$

Construction of the label function. Constructing the label function, a Lagrangian cost function is formulated which can be interpreted as the average

	(−3,1)	(−2,−1)	(−1,2)	(−1,−3)	(0,0)	(1,3)	(1,−2)	(2,1)	(3,−1)
(−6,3)	(−3,1)								
(−6,0)	(−4,1)	(−3,−1)							
(−3,3)	(−3,2)		(−2,2)						
(−3,0)	(−3,0)	(−2,−1)	(−1,1)		(−1,0)				
(−3,−3)		(−2,−2)		(−1,−3)					
(−3,−6)				(−2,−3)					
(0,6)			(−1,3)			(0,3)			
(0,3)			(−1,2)		(0,1)	(1,3)		(2,2)	
(0,0)	(−2,1)	(−2,0)	(0,2)	(−1,−2)	(0,0)	(1,2)	(1,−1)	(1,1)	(2,−1)
(0,−3)		(−1,−1)		(0,−3)	(0,−1)		(0,−2)		
(0,−6)				(−1,−4)			(1,−3)		
(3,6)						(1,4)			
(3,3)						(2,3)		(2,1)	
(3,0)					(1,0)		(1,−2)	(2,0)	(3,0)
(3,−3)							(2,−2)		(3,−2)
(6,0)								(3,1)	(4,−1)
(6,−3)									(3,−1)

Table 16.2. Label function α: The lattice points λ for a given λ_1 (rows) and λ_2 (column).

distortion, where γ_0 is the probability for using the central decoder, γ_i is the probability for using the i'th side decoder and γ_3 is the probability that both channels are not working. Using (16.29) the Lagrangian cost is,

$$J = \gamma_0 D_0 + \sum_{i=1}^{2} \gamma_i D_i + \gamma_3 \tag{16.37}$$

$$= (\gamma_0 + \gamma_1 + \gamma_2) D_0 + \sum_{i=1}^{2} \gamma_i \sum_{\lambda \in \Lambda} \|\lambda - \alpha_i(\lambda)\| P_\lambda + \gamma_3. \tag{16.38}$$

Constructing a good label function is to minimize the Lagrangian cost function. The central distortion and γ_3 can be neglected, since they are independent of the label function. In [Diggavi et al., 2002], the complexity is reduced by using constraint number 3, such that only the lattice points in $V_0 = \{\lambda \in \mathbf{\Lambda} : \lambda \in V_s(0)\}$ need to be assigned. Furthermore, assume the P_λ is constant over the V_0, such that the design objective is to minimize:

$$\sum_{\lambda \in V_0} (\gamma_1 \|\lambda - \alpha_1(\lambda)\| + \gamma_2 \|\lambda - \alpha_2(\lambda)\|). \tag{16.39}$$

The procedure for generating the label function as proposed in [Diggavi et al., 2002] is as follows:

1 Determine the indexes N_1,N_2, the lattice Λ, the sublattices Λ_1,Λ_2 and the Lagrangian multipliers γ_1 and γ_2. We will later give specific

examples and guidelines for how to determine these factors in the cooperative context.

2 Determine the two sets, $\eta_1 = V_0 \cap \mathbf{\Lambda_1}$ and $\eta_2 = V_0 \cap \mathbf{\Lambda_2}$, and the sets,

$$\zeta_i(\lambda_\mathbf{i}) \quad = \quad \{\lambda_\mathbf{j} \in \mathbf{\Lambda_j} : \lambda_\mathbf{j} \in V_0 + \lambda_\mathbf{i}\}, \quad \forall \lambda_\mathbf{i} \in \eta_i. \tag{16.40}$$

Determine all possible combinations of sublattice points by:

$$\epsilon_0 = \{(\lambda_\mathbf{i}, \lambda_\mathbf{j}) : \lambda_\mathbf{i} \in \eta_i, \lambda_\mathbf{j} \in \zeta_i(\lambda_\mathbf{j})\}, \tag{16.41}$$

where $(i, j) = (1, 2)$ or $(2, 1)$.

3 Matching the combinations in ϵ_0 to the lattice points in V_0, such that Eq. (16.39) is minimized, can be considered as a Mixed Integer linear Programming (MIP) problem. The optimization will result in two equivalent label functions for ϵ_0 when $i = 1$ and $i = 2$.

In [Diggavi et al., 2002], the authors first construct a label function that covers V_0 and subsequently extends the label function to the entire lattice using the shift–property. Hence, they can avoid quantization of the sublattice $\mathbf{\Lambda_s}$ in the encoding and decoding procedures. This will certainly require storage of a large label function or a mathematical description of the label function. In this chapter, for notational convenience, we will use the reduced label function, but the usage of the extended label function is straightforward.

EXAMPLE 16.3 (AN EXAMPLE OF MDLVQ (CONTINUED)) *Let us, finally, return to the MDLVQ example from Section 16.2 to show how the label function can now be generated by the procedure described in Section 2.0.0.*

In Step 1 all variables are known except $\gamma_1 = 0.0533$ *and* $\gamma_2 = 0.0438$. *We will later explain how* γ_i *should be for a given loss probability. In step 2 we determine all the sublattice points in* V_0 *and*

$$\begin{aligned} \eta_1 &= \{(-1,-3),(-2,-1),(-3,1),(1,-2),(0,0), \\ &\quad (-1,2),(3,-1),(2,1),(1,3)\} \end{aligned} \tag{16.42}$$

$$\eta_2 = \{(-3,0),(0,-3),(0,0),(0,3),(3,0)\}, \tag{16.43}$$

where η_1 *and* η_2 *are shown as x-marks and circles on Figure 16.8(right), respectively. In step 3, we determine the sets* ζ_1 *and* ζ_2. *In step 4, with the optimization over all the elements in* ϵ_0, *we find the label function as shown in Table 16.8.*

3. Optimizing Multiple Description Coding for losses in the Cooperative Context

The described MDLVQ can be optimized for the cooperative scheme shown on Figure 16.2. With an egoistic behavior, each participant in the cooperative scheme will demand that MDLVQ is optimized such that it gets the most out of the cooperation. This egoistic behavior maps into single terminal optimization of the MDLVQ. Subsequently, we will show that this optimization is not optimal for the MDLVQ cooperation scheme. A compromise among the participants in the cooperation can be to minimize the overall average distortion. We explain this in Section 3.0 .

The Single Terminal Optimization of the MDLVQ

Optimizing the MDLVQ in the cooperative scheme as shown on Figure 16.2 subject to either terminal 1 or terminal 2 will yield two different MDLVQ designs. In this section we first optimize to one of the terminals and then illustrate that this is not optimal for both of the terminals. Optimizing the MDLVQ design for terminal 1 is equivalent when optimizing to MDLVQ scheme when only one terminal is considered. The single terminal problem has been thoroughly analyzed in [Østergaard et al., 2004] for the high resolution case. The main results are outlined in the following. First, the side distortion for the MDLVQ is found in [Diggavi et al., 2002] to be,

$$D_i \approx \frac{\gamma_j^2}{(\gamma_i + \gamma_j)^2} G(\mathbf{\Lambda_s}) 2^{2h(p)} 2^{-2(R_1+R_2-R_0)}, \tag{16.44}$$

where $(i, j) = (1, 2)$ or $(2, 1)$ and γ_i is probability for receiving description i at terminal 1. The central distortion is given in Eq. (16.28),

$$D_0 \approx G(\mathbf{\Lambda})\nu^{2/L}. \tag{16.45}$$

Then, assuming an entropy constrain on the three channels, we note that $N_i\nu$ become a constant when combining Eq. (16.30) and Eq. (16.31),

$$N_i\nu = 2^{L(h(p)-R_i)} = c_i. \tag{16.46}$$

Combining Eq. (16.46) and Eq. (16.31) we get:

$$\frac{c_i}{\nu} = 2^{L(R_0-R_i)}. \tag{16.47}$$

Now, using Eq. (16.46) and Eq. (16.47), we can write the average distortion as:

$$\overline{D} = \gamma_0 G(\mathbf{\Lambda})\nu^{2/L} + \frac{\gamma_1\gamma_2^2 + \gamma_2\gamma_1^2}{(\gamma_1 + \gamma_2)^2} G(\mathbf{\Lambda_s}) \left(c_1 c_2\right)^{2/L} \nu^{-2/L} + \gamma_3. \tag{16.48}$$

Finally, we can determine the optimal volume ν by putting its derivative with respect to ν equal to zero. The solution leads to the optimal volume:

$$\nu = \left(\frac{\gamma_1\gamma_2}{\gamma_0(\gamma_1 + \gamma_2)}\right)^{L/4} \left(\frac{G(\mathbf{\Lambda_s})}{G(\mathbf{\Lambda})}\right)^{L/4} \sqrt{c_1 c_2}. \tag{16.49}$$

From Eq. (16.46) and Eq. (16.49) we determine the optimal N_i as follows.

$$N_i = \left(\frac{\gamma_0(\gamma_1 + \gamma_2)}{\gamma_1 \gamma_2} \right)^{L/4} \left(\frac{G(\mathbf{\Lambda_s})}{G(\mathbf{\Lambda})} \right)^{-L/4} \sqrt{\frac{c_i}{c_j}}. \tag{16.50}$$

This specifies an MDLVQ design optimized for one of the terminals. As we'll see in the following example, this design is however not always optimum for all involved terminals.

EXAMPLE 16.4 *Let us design an MDLVQ for a system with a unit variance Gaussian source, a Z_2 lattice and with a rate constraints $R_1 = 6$, $R_2 = 5.5$ and no rate constrain on the cooperative link. The loss probability on the three channels are $p_1 = 0.01$, $p_2 = 0.08$ and $p_c = 0.01$, and the three γ's for terminal 1 can be determined,*

$$\gamma_0 = 0.9017, \quad \gamma_1 = 0.0883 \quad \textit{and} \quad \gamma_2 = 0.0091. \tag{16.51}$$

Now, we determine the two constants c_i by Eq. (16.46), where the differential entropy of a unit variance Gaussian source is, $h(X) = \frac{1}{2}\log_2(2\pi e)$, such that the two constants are:

$$c_1 = 0.0042, \quad c_2 = 0.0083. \tag{16.52}$$

It becomes straightforward to find the optimal ν and N_i by Eq. (16.49) and Eq. (16.50):

$$\nu = 5.51 \cdot 10^{-4}, \quad N_1 = 7.6, \quad N_2 = 15. \tag{16.53}$$

In a similar manner, we can determine the loss probability for terminal 2:

$$\gamma_0 = 0.9017, \quad \gamma_1 = 0.0183 \quad \textit{and} \quad \gamma_2 = 0.0784, \tag{16.54}$$

and then determine the optimal ν and N_i with respect to terminal 2,

$$\nu = 7.39 \cdot 10^{-4}, \quad N_1 = 5.6, \quad N_2 = 11.3. \tag{16.55}$$

Note how the different loss probabilities, seen by the two terminals, lead to different quantizer designs as optimum for each of the two terminals.

From the example, we can conclude that sometimes optimization for each of the terminals is not feasible as they have to share the same encoder. In some cases with similar loss probabilities and because of cleanness and similarity constraints on the two sublatices the two MDLQ designs may indeed turn into the same MDLQ design, but in general egoistic behavior is suboptimal for some involved terminals.

Minimization of the Mean Distortion

A fair method to share the network resources can be to minimize the mean distortion over all terminals that cooperate. The mean distortion over all terminals is:

$$\overline{D} = \frac{1}{2}\Big((\gamma_0^{(1)} + \gamma_0^{(2)})D_0 + (\gamma_1^{(1)} + \gamma_1^{(2)})D_1 + (\gamma_2^{(1)} + \gamma_2^{(2)})D_2 + (\gamma_3^{(1)} + \gamma_3^{(2)})\Big), \qquad (16.56)$$

where $\gamma_i^{(k)}$ is the γ_i for the k–th terminal. Therefore, we can determine the mean optimum MDLVQ design by adopting the interpretation that the system has only one terminal, with parameters:

$$\gamma_k = \frac{\gamma_k^{(1)} + \gamma_k^{(2)}}{2} \qquad (16.57)$$

for $k \in \{0, 1, 2, 3\}$, and then use the procedure described in Section 3.0. This approach generalize in a straight forward manner to any number of terminals. We illustrate this in the following continuation of the example.

EXAMPLE 16.5 ***[Continued]*** *Minimization of the mean distortion for the above example yields three new parameters:*

$$\gamma_0 = 0.9017, \quad \gamma_1 = 0.0533 \quad \textit{and} \quad \gamma_2 = 0.0438. \qquad (16.58)$$

The optimal volume and reused index for the MDLVQ is,

$$\nu = 9.40 \cdot 10^{-4}, \quad N_1 = 4.4, \quad N_2 = 8.9. \qquad (16.59)$$

From this, we can conclude that a good MDLVQ design can be $N_1 = 5$ and $N_2 = 9$. Note that this is the acutal design carried out in Section 2.0 when neglecting a scaling factor.

The rounding–off for the reused index N applied in the example is not described in the theory. However, the optimal N can be determined from the design of the possible N by evaluation of performance of each. Another method is to use an unstructured Multiple Description Vector Quantizer as described in [Koulgi et al., 2003] rather than the structured lattice, thus avoiding the lattice conditions on N.

Networks with Time–Varying Loss Probabilities

In practical cooperative networks, the exact loss probabilities for the channels in the cooperative network are not known at design time. Rather, these probabilities are known only with stochastic uncertainty, or they are known to be time–varying quantities during application of the coding system. In both cases, we can see the loss probability p_i and p_c as stochastic variables described by the pdf $f(p_i)$ and $f(p_c)$. The mean distortion over all terminals in a network with time–varying loss probabilities is given as follows.

$$\begin{aligned} \bar{d} &= \frac{1}{2}\iiint \Big((\gamma_0^1+\gamma_0^2)d_0+(\gamma_1^1+\gamma_1^2)d_1+(\gamma_2^1+\gamma_2^2)d_2+ \\ &\quad (\gamma_3^1+\gamma_3^2)\Big)\, f(p_1)f(p_2)f(p_c)dp_1dp_2dp_c. \end{aligned} \tag{16.60}$$

To minimize the mean distortion in the stochastic formulation we again determine the γ's,

$$\gamma_k = \frac{1}{2}\iiint \left(\Gamma_c^1+\Gamma_c^2\right) f(p_1)f(p_2)f(p_c)dp_1dp_2dp_c, \tag{16.61}$$

for $k \in \{0,1,2,4\}$, and then use the initial procedure described in Section 3.0. As an example of a cooperative network where stochastic and time–varying design is called for, we mention real–time media transmission using the real–time protocol (RTP). In this setting, the packet loss probabilities can be estimated at the receiver and fed back to the transmitter via the real–time control protocol (RTCP) in [Schulzrinne et al., 1996]. For this type of applications, we can design a bank of MDLVQ's, such that the encoder can select the most suitable MDLVQ design for the given loss probability. This bank construction is investigated in [Larsen et al., 2005] where a significant gain was found with increasing the number of designs in the bank.

4. MDC with Conditional Compression (MDC–CC)

As explained in the previous sections, a MD scheme introduces coding overhead in order to provide self–sufficient descriptions. Consider the case with two descriptions, d_1 and d_2. Each description is carrying information that is sufficient to obtain a low–quality replica of the original source information. This means that there is some portion in each of d_1 and d_2 that is carrying identical information about the source. Assume that d_1 has been received at the destination. Then it is not necessary that the full d_2 is sent to the destination, but only the information from d_2 that is not contained in d_1. We denote this information as $(d_2|d_1)$ and call it a conditional description provided that d_1 is available.

An important consequence of this concept is that the compression of the source information is not performed at the source, but in a node that lies on the path between the source and destination. This concept is shown on Figure 16.9. The source node S produces two descriptions and sends d_1 and d_2 through two disjoint paths, passing through X and Y, respectively. Assume that the feedback from destination D is not timely available at S, but it is available at X or Y. Now let d_1 arrive through X at the destination D, then D can inform Y about this through the fast feedback channel. With such an information, Y can re–code d_2 and send only the conditional information $(d_2|d_1)$. Hence, the information that traverses the path from Y to D is compressed, as one example of this solution D co–insides with X or Y and X and Y are cooperating terminals. In the sequel we describe the proposed realization of the MDC–CC in the MDLVQ framework.

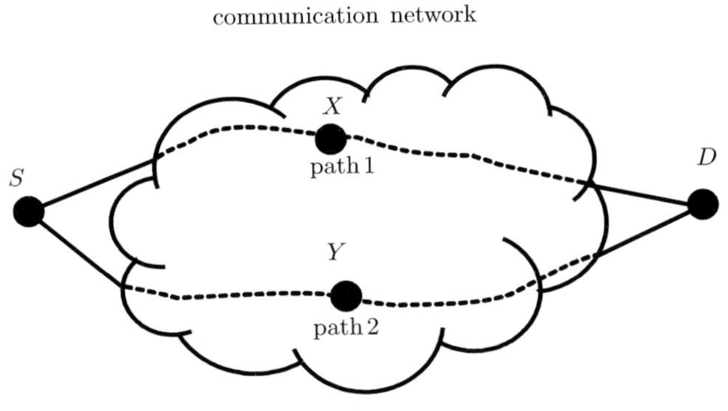

Figure 16.9. A communication network that provides two paths from S to D. X and Y are intermediate nodes along the paths The feedback from D is not available timely at S, but can be available at X or Y.

MDLVQ for MDC–CC

We start by making an interpretation of the MDLVQ, inspired by the design algorithm described in Section 2. In the design algorithm of the MDLVQ, the lattice points in V_0 are assigned by the label function and, subsequently, the label function is expanded to R^L. In this new scheme, we keep the reduced label function for V_0 and reintroduce the pre–encoder $\lambda_s = Q_s(X)$, equivalent to the shift–property. We transmit λ_s (neglecting the small offset in λ_s^+) over both channels and the relative refinement information λ_1^* and λ_2^* over each channel, instead of transmitting the λ_1 and λ_2 over each channel, as shown on Figure 16.10. Regarding the decoding, when only description i is received, then the reconstruction point is based on information from λ_s and λ_i^*. Conversely, when both descriptions are received the reconstruction is based on λ_1^*, λ_2^* and λ_s or λ_s^+. Clearly, when applying the central decoder, λ_1^* is a conditional description provided that λ_s^+ and λ_2^* are received, as denoted $(d_1|d_2)$ in last section. The interpretation for λ_2^* is analogous. In the next two subsections we will describe how to construct the new relative label function β, and furthermore it will be shown that the entropies in the MDLVQ interpretation are maintained.

Clearly the distortions are maintained since the interpretation has no impact on the λ and λ_1 and λ_2, at the decoders.

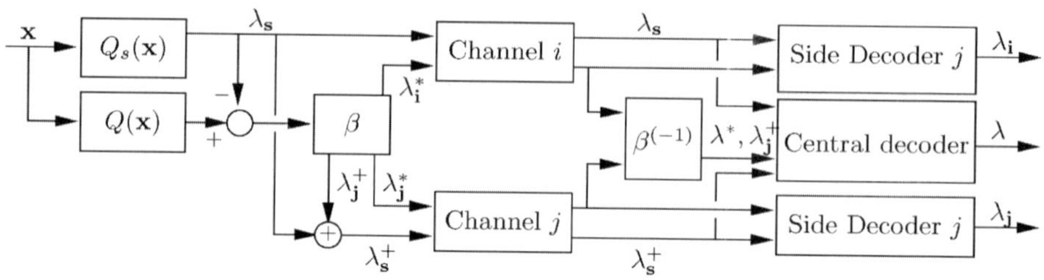

Figure 16.10. An interpretation of a MDLVQ.

Construction of the relative label function. In Section 2.0.0 where the label function was constructed, all combinations of the sublattice points η_i and the sublattice points in ζ_i were considered. By definition, all sublattice points in η_i are included in V_0, but the sublattice points in ζ_i belong to a larger set. Each sublattice point $\lambda_j \in \zeta_i$ can be described by a relative sublattice point, $\lambda_j^* \in \eta_j$ and a corresponding offset sublattice point λ_j^+. The relationship between the $\lambda_j \in \zeta_i$ and the λ_j^* is:

$$\lambda_j^* = \lambda_j + \lambda_j^+, \tag{16.62}$$

where $\lambda_j^+ \in \Lambda_s$. It is built–in the optimization in Step 3, that a combination of a sublattice point in η_i and a sublattice point in η_j is only used once, to obey the shift–property. Thus, we can after designing the label function compress the label function to only include combinations of sublattice points from η_i and η_j, without any conflict. For simplicity, we also denote the sublattice points $\lambda_i^* \in \eta_i$. Thus, we can construct a relative label function, β, that maps a lattice point $\lambda \in V_0$ to λ_i^*, λ_j^* and λ_j^+, as shown on Figure 16.10.

EXAMPLE 16.6 (AN EXAMPLE OF MDLVQ (CONTINUED)) *Let us return to the MDLVQ example from Example 16.2 to show how the label function α can be reformulated to a relative label function β, where $i = 1$ and $j = 2$. For each lattice point λ in Table 16.8 the corresponding sublattice point λ_j is reduced to λ_j^*, will result in Table 16.3. Furthermore, for each lattice point λ, the difference between λ_j and λ_j^* will result in λ_j^+, as shown on Table 16.4. The label function β is completely described by Table 16.3 and 16.4. From Table 16.3 we note that all the combinations of the sublattice points in η_1 and η_2 are used, which will be used in next section.*

	(−3, 0)	(0, 3)	(0, 0)	(0, −3)	(3, 0)
(−3, 1)	(−3, 0)	(−4, 1)	(−2, 1)	(−3, 2)	(−3, 1)
(−2, −1)	(−2, 1)	(−3, −1)	(−2, 0)	(−1, −1)	(−2, −2)
(−1, 2)	(−1, 1)	(−1, 2)	(0, 2)	(−2, 2)	(−1, 3)
(−1, −3)	(−1, −4)	(−2, −3)	(−1, −2)	(0, −3)	(−1, −3)
(0, 0)	(−1, 0)	(0, 1)	(0, 0)	(0, −1)	(1, 0)
(1, 3)	(2, 3)	(1, 3)	(1, 2)	(1, 4)	(0, 3)
(1, −2)	(1, −3)	(2, −2)	(1, −1)	(0, −2)	(1, −2)
(2, 1)	(2, 1)	(2, 2)	(1, 1)	(3, 1)	(2, 0)
(3, −1)	(3, −1)	(3, −2)	(2, −1)	(4, −1)	(3, 0)

Table 16.3. Label function β: The relative lattice points λ^* for a given λ_1^* (column) and λ_2^* (rows).

	(−3, 0)	(0, 3)	(0, 0)	(0, −3)	(3, 0)
(−3, 1)		(−6, −3)		(−3, 6)	(−9, 3)
(−2, −1)		(−6, −3)			(−6, −3)
(−1, 2)				(−3, 6)	(−3, 6)
(−1, −3)	(3, −6)	(−3, −9)			(−6, −3)
(0, 0)					
(1, 3)	(6, 3)			(3, 9)	(−3, 6)
(1, −2)	(3, −6)	(3, −6)			
(2, 1)	(6, 3)			(6, 3)	
(3, −1)	(9, −3)	(3, −6)		(6, 3)	

Table 16.4. Label function β: The offset lattice points λ_2^+ for a given λ_1^* (column) and λ_2^* (rows). On the empty places the offset lattice point is zero.

Rate computation for a MDC–CC in the MDLVQ framework. To determine the entropies in the MDC-CC framework we will assume high resolution, as in Section 2.0. The entropy of a sublattice is derived in [Vaishampayan et al., 2001] for a given reused index and assuming high resolution. Thus, we can realize that the entropy for the productive sublattice is,

$$R_s \quad = \quad R_0 - \frac{1}{L}\log_2(N_1 N_2). \tag{16.63}$$

To determine the entropy of the offset lattice point λ_s^+, we first realize the following: That λ_s and λ_s^+ is adjacent (in space) and therefore the $Pr(\lambda_s) \approx Pr(\lambda_s^+)$, when assuming high resolution. Thus the entropy of λ_s^+ is equal to the entropy of λ_s. A similar approximation for λ_i is taken in [Vaishampayan et al., 2001], when calculating the side distortion. Another way to see this is, when λ^* is close to the centroid of V_0 then λ_j^+ is 0. Conversely, when λ^* is close to the boundary of the Voronoi region of V_0 then λ_j^+ is not-zero. Which was also the case in Table 16.4. So, when assuming high resolution, then the cardinality of η_j is large and the probability for λ^* is close to the boundary goes towards 0. Thus, the probability for $\lambda_j^+ \neq 0$ goes towards 0 and thereby the entropy of λ_s^+ goes towards the entropy of λ_s.

Applying the label function β it is guaranteed that all the combination of $\lambda_i \in \eta_i$ and $\lambda_j \in \eta_j$ is used and only once. Furthermore, when assuming high resolution, this implies equal probability for all $\lambda_j \in V_0$, enables us to determine the entropy of the refinement information and thereby the entropy of the conditional compression,

$$R_j^* = \frac{1}{L}\log_2(N_i). \tag{16.64}$$

We can then verify that the entropy is maintained, by $R_s + R_j^*$ and compared with R_j from Section 2.0.0. On the other hand, for low resolution it can also be argued that the entropy on the channel is maintained. Because, the information in λ_j from the classical MDLVQ scheme is exactly the same information in λ_s^+ combined with λ_j^* in the MDC–CC scheme. Where conversion between them can always be performed by ether a quantization $\lambda_s^+ = Q_s(\lambda_j)$ or with a careful design of the entropy coder that keeps the information about λ_j^* and λ_s^+ separated. The interpretation for the entropy of the channel i is analogous.

Encoding and decoding procedure MDC–CC in the MDLVQ framework. The encoding procedure for MDC–CC with MDLVQ is a three step procedure, as illustrated in Figure 16.10. First the input vector $\mathbf{x}$ is quantized to the closest sublattice point in $\mathbf{\Lambda}_\mathrm{s}$,

$$\lambda_\mathrm{s} = Q_s(\mathbf{x}). \tag{16.65}$$

The second step, is to quantize the input vector to the closest lattice point λ in $\boldsymbol{\Lambda}$,

$$\lambda = Q(\mathbf{x}). \tag{16.66}$$

In the third step, the sublattice point λ_s is subtracted from the lattice point λ. This ensures that $\lambda' = \lambda - \lambda_s$ is included in V_0 and the label function β can be applied, similar to the shift–property in the design algorithm. Applying the label mapping, we get the two relative sublattice points and the offset sublattice point,

$$(\lambda_i^*, \lambda_j^*, \lambda_j^+) = \beta(\lambda'). \tag{16.67}$$

Finally, the two multi–points are transmitted over the two channels:

$$\text{Channel } i : (\lambda_s, \lambda_i^*), \quad \text{Channel } j : (\lambda_s^+, \lambda_j^*). \tag{16.68}$$

The side decoding procedures are simply the sum of the multi-points, $\lambda_i = \lambda_s + \lambda_i^*$ for side decoder i and $\lambda_j = \lambda_s^+ + \lambda_j^*$ for side decoder j. When both channels are working, the decoding procedure is a two–step procedure, as shown on Figure 16.10. The first step is to find the relative lattice point λ^* and the offset lattice point λ_j^+ by applying the inverse label function, $(\lambda^*, \lambda_j^+) = \beta^{(-1)}(\lambda_1^*, \lambda_2^*)$. Subsequently the relative lattice point is added to the sublattice λ_s in order to reconstruct the lattice point, when λ_s is available. Otherwise, when λ_s^+ is available the lattice point is reconstructed by: $\lambda = \lambda_s^+ - \lambda_j^+ + \lambda^*$.

EXAMPLE 16.7 *Let us now revise the the cooperative scheme from Figure 16.2, where two terminals exchange information over the cooperative link. Let terminal 1 receive* d_1*, which contains* $\lambda_s + \lambda_1^*$. *In order to obtain the high quality only the relative refinement information* λ_2^* *needs to be transmitted over the cooperative link, since terminal 1 already knows* λ_s. *For this scheme with conditional compression to work, the terminal 2, before compressing the description* d_2*, must be sure that terminal 1 has received* d_1. *Hence, a full cooperative protocol operates with adaptation to the feedback: if terminal 2 receives positive feedback that terminal 1 received* d_1*, then compressed description is forwarded. Otherwise, with negative feedback or no feedback at all, terminal 2 forwards the full description* d_2.

It can be observed that, regarding the source coding, the network is an *active actor* when MDC–CC is used. Compared to this, in case of conventional MDC or LC, the network is a passive actor and only the source node does the source coding and compression. We can say that such operation of MDC–CC is a representative case of a cross–layer optimization in the protocol design.

5. Discussion

Having presented the MDC basics as well as specific MDC optimizations and schemes suitable for cooperative communications, in this section we will discuss three generic scenarios in which source coding based on MDC appears as a suitable solution within the paradigm of cooperative networking.

- Data delivery with Cooperative Sources (CS–scenario)
- Data delivery with Cooperative Destinations (CD–scenario)
- Data delivery with Meshed Cooperation (MC–scenario)

We will see that the overall efficiency of those cooperative scenarios naturally increases when MDC–CC is applied.

Data delivery with Cooperative Sources (CS–scenario)

In this scenario the whole network can be considered as a distributed source of information for the destination node D. Therefore, the data transmission can take advantage of the diversity provided by the network, such as path diversity. We illustrate this scenario through two examples.

Example CS–1. This scenario has been depicted on Figure 16.9. The nodes X and Y can be considered as distributed sources of a correlated information, since the descriptors forwarded by them are correlated. If X and Y are not co-operating, then each of them is "blindly" forwarding the appropriate description generated at S. If X and Y are mutually coordinated in transmitting the data to D, then they can be considered as cooperative sources of information. As we have explained in Section 4, this cooperation is realized through the use of MDC–CC. The cooperation between X and Y can be initiated by the destination through the feedback paths or via a link between X and Y. This scenario sets a stage for building protocols for cooperative streaming. As a special case, the nodes X and Y can be two wireless access points (APs) and the destination node D can be a terminal which lies in the radio range of both APs.

Example CS–2. The distributed storage has been outlined as an appropriate application for MDC in [Goyal, 2001]. For example, such is the case where a multimedia content is stored on several locations with multiple description (MD) encoding, such that each location (content server) contains a single descriptor pertained to the video stream. Consider the example on Figure 16.11. With a conventional MDC, the user would require a whole description from each content server. If MDF–CC is utilized, then the user needs to get the full description d_2 from the server S_2, while it retrieves the conditionally compressed descriptions $(d_1|d_2)$, $(d_3|d_2)$ from the servers S_1, S_3. This decreases the overall traffic in the network. Again, in this case the cooperation among the servers can be initiated by feedback from the user or through the usage of direct coordination among S_1, S_2, S_3. Note that layered coding cannot provide such a forwarding mechanism in this scenario because the server that should provide

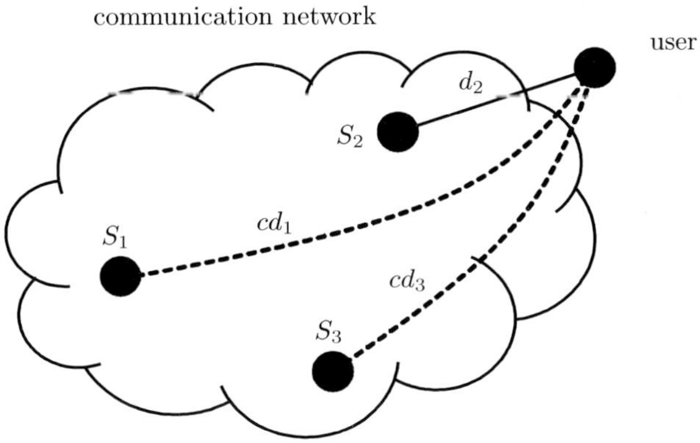

Figure 16.11. Access to multimedia content that is stored by MD encoding in three content servers S_1, S_2 and S_3. The user gets a full description from the closest server (d_2 from S_2), while it retrieves the compressed descriptions cd_1, cd_3 from the other two servers.

the full description is not predefined. That is, if another user is close to S_3, then the compressed descriptions should come from S_1 and S_2.

Data delivery with Cooperative Destinations (CD–scenario)

In these scenarios the source data is broadcasted to several destination terminals, while the terminals use the communication links among them to cooperate and thus enhance each other's reception of the broadcasted data.

Example CD–1. Figure 16.12 illustrates a broadcast scenario in which the feedback link from the terminals MS_1, MS_2 to the source base station BS is not available. The BS encodes the information with two descriptions d_1, d_2 and transmits them over the air. Depending on what has been received at each terminal, the cooperative link between the terminals is used for MS_1 to transmit a whole or compressed description to the terminal MS_2 and vice versa. In this case the MDC–CC scheme reduces the traffic on the cooperative link. Furthermore, MDC–CC appears to be an essential ingredient of this scenario. If the capacity of the cooperative link is less than the capacity of the broadcast transmission.

Example CD–2. To illustrate this scenario we can again use Figure 16.12, but in this case we assume that the links from each terminal to the BS are bi directional. The usage of multiple descriptions enables dynamic compression and routing of the broadcasted information. This can mean, for example, that only the description that is lost is forwarded through the cooperative link. The BS should initially transmit both full descriptions, but upon request for

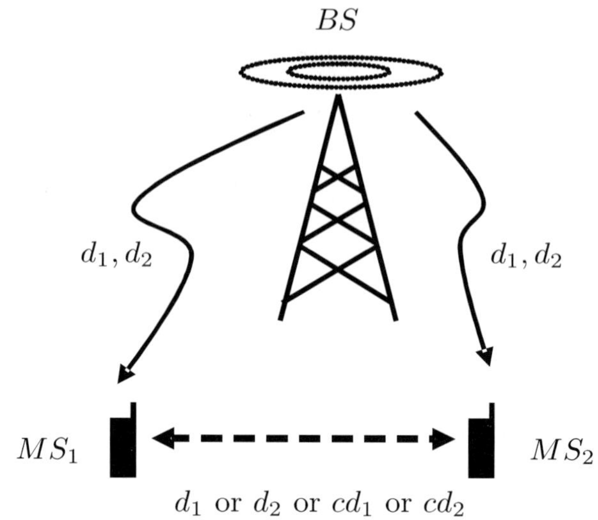

Figure 16.12. Broadcast scenario with cooperative destinations (MS_1 and MS_2) when the link from the source (base station BS) is unidirectional such that feedback to the source is not available.

retransmission from MS_1 and/or MS_2 the compressed descriptions can be provided by either BS or the other MS.

Data delivery with Meshed Cooperation (MC–scenario)

In this scenario each node involved in the communication can be a source of information and a destination. Such can be the case of video–conferencing or gaming. The source information of $node_1$ is encoded by two descriptions d_{12} and d_{13} and they are sent through the links l_{12} and l_{13}, respectively. Similarly, $node_2$ sends d_{21}, d_{23} through l_{21}, l_{23}, and $node_3$ sends d_{31}, d_{32} through l_{31}, l_{32}. Depending on the link conditions, each node can forward compressed or full description on behalf of another node. For example, if there are no errors, $node_1$ can forward the compressed descriptor cd_{31} to $node_2$, such that $node_2$ is able to completely reconstruct the source information of $node_3$, since it receives d_{32} through l_{32}. However, if the link conditions on l_{32} become bad, then $node_1$ forwards the full description d_{31}, while the quality at $node_2$ degrades gracefully.

6. Conclusion

In this chapter, we have presented ideas that relate the paradigm of cooperative communications to the problems of source encoding and compression.

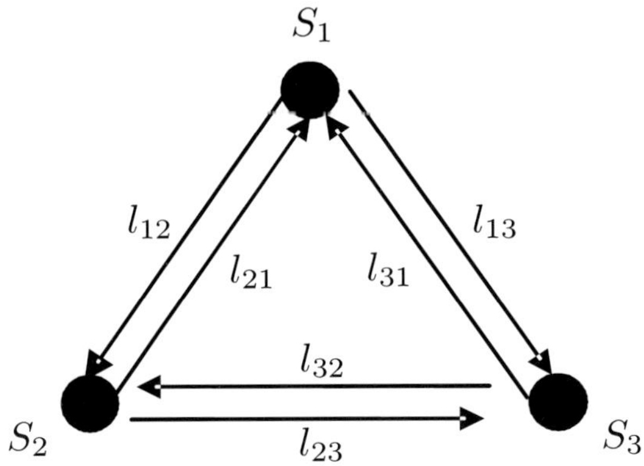

Figure 16.13. Meshed cooperation with three nodes, where each node is a source and a destination of information. l_{ij} denotes unidirectional link between nodes i and j.

The possibilities for conducting cooperative communication can be vastly expanded when the communication protocols use the features of the transmitted data in terms of source encoding. Our starting point is represented by the class of Multiple Description Coding (MDC) methods, in which multiple descriptions are produced information. For balanced MDC, each description is sufficient to restore the source information with certain rate distortion and the rate distortion is decreased as more descriptions of the same source symbol are received and used in the decoding. This chapter brings three distinctive contributions. The *first contribution* is related to the multiple description lattice vector quantizer (MDLVQ), which is a practical procedure for MDC where the quantizer is highly geometrically structured. We have shown how to optimize the design of MDLVQ in the cooperative context. The *second contribution* is a proposal of a novel MDC scheme, termed MDC with Conditional Compression (MDC–CC). This scheme emerges from the joint consideration of the source encoding and the networking and, although general, we elaborate MDC–CC in case of two descriptions, d_1 and d_2 per source information. The basic observation is that, once a node X in the network that contains description d_2 has the information that the destination already has received d_1, then it can compress the description d_2 before forwarding it to the destination. It can be observed that the concept of MDC–CC can move the compression task at any node in the network instead of solely the source node. We also introduce implementation of MDC–CC based on MDLVQ. Finally, the *third contribution* introduces several scenarios for cooperative communication in which the features of MDC–CC can boost the performance of the cooperative scheme. We also provide a taxonomy for the cooperative scenarios based on MDC. This taxonorny exposes a field of open questions on the MDC–CC concept. Answering a few of these questions are topics of our currect research.

References

Conway, J. and Sloane, N. (1982a). Fast quantizing and decoding and algorithms for lattice quantizers and codes. *Information Theory, IEEE Transactions on*, 28(2):227–232.

Conway, J. and Sloane, N. (1982b). Voronoi regions of lattices, second moments of polytopes, and quantization. *Information Theory, IEEE Transactions on*, 28:211–226.

Conway, J. H., Rains, E. M., and Sloane, N. J. A. (1999). On the existence of similar sublattice. *Canadian J. Math.*, 51:1300–1306.

Conway, J. H. and Sloane, N. J. A. (1999). *Sphere Packings, Lattices and Groups*. Springer-Verlag.

Cover, Thomas M. and Thomas, Joy A. (1991). *Elements of Information Theory*. John Wiley.

Diggavi, S. N., Sloane, N. J. A., and Vaishampayan, V. A (2002). Asymmetric multiple description lattice vector quantizers. *Information Theory, IEEE Transactions on*, 48(1):174–191.

Goyal, V. K (2001). Multiple description coding: compression meets the network. *Signal Processing Magazine, IEEE*, 18:74–93.

Goyal, V. K., Kelner, J. A., and Kovacevic, J. (2002). Multiple description vector quantization with a coarse lattice. *Information Theory, IEEE Transactions on*, 48:781–788.

Gray, Robert M. (1990). *Source Coding Theory*. Kluwer Academic Publishers.

Gupta, P. and Kumar, P. R. (2000). The capacity of wireless networks. *Information Theory, IEEE Transactions on*, pages 388–404.

Koulgi, P., Regunathan, S. L., and Rose, K. (2003). Multiple description quantization by deterministic annealing. *Information Theory, IEEE Transactions on*, 49:2067–2075.

Larsen, M. H., Arnbak, K. D., and Andersen, S. V. (2005). Optimization of multiple description quantizers for stochastic and time-varying loss probabilities. *IST 2005*.

Nosratnia, A., Hunter, T. E., and Hedayat, A. (2004). Cooperative communication in wireless networks. *IEEE Communications Magazine*, pages 74–80.

Østergaard, J., Jensen, J., and Heusdens, R. (2004). Entropy constrained multiple description lattice vector quantization. *IEEE International Conference on Acoustics, Speech, and Signal Processing (ICASSP)*, 4.

Østergaard, J., Jensen, J., and Heusdens, R. (2005). n-channel symmetric multiple-description lattice vector quantization. In *IEEE Proc. Data Compr. Conf.*, pages 378–387.

Schulzrinne, H., Casner, S., Frederick, R., and Jacobson, V. (1996). RFC 1889: Rtp: A transport protocol for real-time applications.

Sergio, S. D., Vaishampayan, V. A., and Sloane, N. J. A. (1999). Multiple description lattice vector quantization. *Data Compression Conference (DCC)*.

Sloane, N. J. A. (2005). The on-line encyclopedia of integer sequences.

Vaishampayan, V. A. (1993). Design of multiple description scalar quantizers. *Information Theory, IEEE Transactions on*, 39(3):821–834.

Vaishampayan, V. A. and Domaszewicz, J. (1994). Design of entropy-constrained multiple-description scalar quantizers. *Information Theory, IEEE Transactions on*, 40:245–250.

Vaishampayan, V. A., Sloane, N. J. A., and Servetto, S.D. (2001). Multiple-description vector quantization with lattice codebooks: design and analysis. *Information Theory, IEEE Transactions on*, 47:1718–1734.

Zhao, D. Y. and Kleijn, W. B. (2004). Multiple-description vector quantization using translated lattices with local optimization. *Global Telecommunications Conference (GLOBECOM)*, 1:41–45.

Chapter 17

COOPERATIVE HEADER COMPRESSION

Exploiting cooperation for overhead reduction in wireless networks

Tatiana K. Madsen
Dept. of Communication Technology, Aalborg University, Denmark
tatiana@kom.aau.dk

Abstract: Header compression is gaining more and more attention as an efficient method for transport overhead reduction in IP-based networks. This is especially actual for bandwidth-limited links as in cellular systems for example. One of the main challenges in designing of header compression algorithms is context maintenance at the decompressor side. We present a compression approach based on the cooperative behavior of multiple streams. Cooperative Header Compression (COHC) is characterized by high robustness towards the transmission errors, low complexity and no need for the feedback channel from the receiver. Different application fields and scenarios for COHC are presented, including cellular, local and multi-hop networks. Applying cooperation principles in header compression allows higher bandwidth savings compared with the conventional algorithms.

Keywords: IP header compression, cooperation, transport overhead reduction, multiple channels

Using multiple channels, performance gain can be achieved by exploiting their different propagation characteristics and, as a result, different error patterns. To benefit from multi-channel and multiple-flow communication, cooperation techniques are required. In this chapter we advocate the approach when a number of IP packet flows transmitted over multiple channels exhibit the cooperative behavior in order to reduce the transport overhead. We present a

Frank H. P. Fitzek and Marcos D. Katz (eds.), Cooperation in Wireless Networks: Principles and Applications, 547–566.

header compression scheme where packet streams help each other to survive transmission errors without ruining the decompression procedure. Cooperative compression applied for communication between a wireless terminal and a base station or between wireless terminals leads to the efficient usage of the limited resources. In particular, significant bandwidth savings can be achieved increasing operators' revenue and decreasing cost for the users.

1. Header Compression Principles

Header Compression is a method for reduction in transport overhead. By transporting data over an IP-based network, the related overhead in terms of additional header information can constitute a large portion of the packet. For many services and applications *e.g.* Voice over IP, interactive games, messaging etc. the payload of an IP packet has almost the same size or even a smaller size than the header. Using RTP/UDP/IP suit results in 40 bytes of header for IPv4; with IPv6 there is a total of 60 bytes of overhead. Compressing IP headers provides bandwidth savings and facilitates the efficient usage of the limited and expensive resources. Header compression techniques allow header reduction to 4 bytes [RFC2508, 1999] or even smaller [RFC3095, 2001]. An additional benefit that the header compression potentially provides, is the decrease of packet errors, since smaller packets are less error–prone.

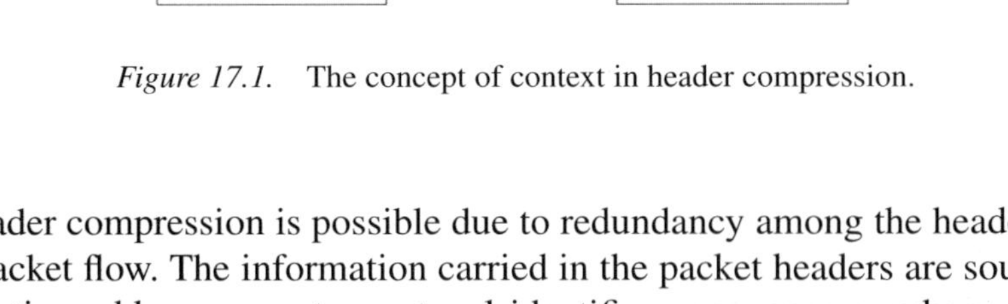

Figure 17.1. The concept of context in header compression.

Header compression is possible due to redundancy among the header fields of a packet flow. The information carried in the packet headers are source and destination addresses, ports, protocol identifiers, sequence numbers etc. The majority of the header fields (except so-called random fields, as *e.g.* TCP/UDP checksum) remains the same or only slightly changes for a single flow. Changes in the headers of two successive packets can be efficiently compressed using *differential (delta) coding*: the compressor removes redundancy from the incoming packet using information from the past packets, called the *context*. The decompressor maintains the context and uses it to reconstruct the header of the incoming packet (Figure 17.1). After the context between the compressor and decompressor is established, the packets with compressed headers are transmitted on the link. Figure 17.2 illustrates delta coding approach, as it is used in Compressed TCP Header Compression [RFC1144, 1990] for example.

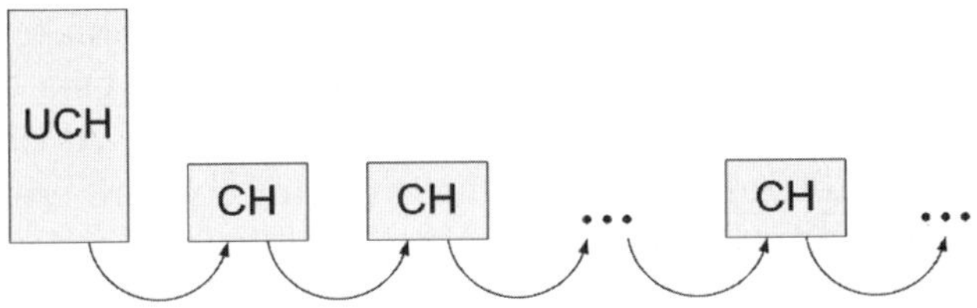

Figure 17.2. Delta coding approach.

Unfortunately, the bandwidth savings achieved by header compression can be easily jeopardized due to vulnerability of the compression procedure towards packet losses. Loss of one packet can lead to the inconsistences in the context of the decompressor and result in the failure of the decompression procedure for the subsequent packets. We refer to this situation as *error propagation.*

To prevent the context re–synchronization, a context repair mechanism is required. If a return channel is available between a receiver and a sender, the sender can use feedback information from the receiver to know when to transmit a packet with an uncompressed header. Whenever a compressed header suffers from a transmission error, a full context update has to be sent. Using this approach the repair mechanism is triggered exactly when it is needed. At the same time, due to propagation and processing delays, several packets with compressed headers will be discarded at the decompressor until the context update is received. One should also note that this method requires additional signalling that increases complexity of the scheme and can potentially degrade efficiency. This algorithm closely resembles header compression schemes such as IP header compression [RFC2507, 1999] or Robust Header Compression (ROHC), Optimistic and Reliable modes [RFC3095, 2001].

When the feedback is unavailable, the synchronization of the context is achieved by periodic refreshments of the states. The context update is performed at the beginning of each packet frame. How often the updates should be done depends a lot on the channel error rate and the propagation environment. Since no feedback information is required, this approach is characterized by low complexity and it supports multicast/ broadcast scenarios. We refer to this method as *framed delta coding* (FDC). In ROHC [RFC3095, 2001], Unidirectional-mode, designed especially for links without a return channel, can serve as an example of framed delta coding: in U-mode transitions between compression states are done based on periodic timeouts and irregularities in the header field patterns.

The above-described strategies for header compression can be characterized as *reactive*: the action (*i.e.* the context update) is taken when we learn or predict that synchronization is lost or is about to be lost. Conceptually a different approach is *proactive*: rather than waiting to respond to the re–synchronization after it happens, we control the situation using preventive measures. To implement the proactive approach in header compression will mean including some extra information in the compressed packet header in order to make the decompressor context more robust towards transmission errors. One obvious way is to use stronger coding. In [Suryavanshi and Nosratinia, 2005a; Suryavanshi and Nosratinia, 2005b] it is shown that by using forward error correction (FEC) techniques based on Reed-Solomon and convolutional codes, many of the otherwise lost compressed packets can be recovered. The main problem in this approach is long delays introduced at both the compressor and decompressor sides. Efficient error-recovery requires long code-words that will be distributed over many compressed packets. But if some symbols are lost, they can not be recovered until a prescribed number of bits have been received correctly. Thus, even though the throughput can be improved significantly, due to introduced large delays this approach is unusable for many applications.

It turns out that applying cooperative principles to header compression leads to a robust proactive strategy for context synchronization. First, *Cooperative Header Compression* has been proposed in [Fitzek et al., 2005a] as a compression scheme for multichannel communication. Additional information is appended to the compressed packets transmitted over different channels. Multiple channel diversity allows robustness against packet losses without the need of the feedback from the receiver. To update the context for each new incoming packet, the decompressor might use the information provided by the neighboring channels. The cooperative behavior of packet streams gives the name to the proposed scheme, Cooperative. Next section explains the details of COHC.

2. Cooperative Header Compression

Header compression is usually considered in a single-channel, single-flow configuration. We start with expanding and applying compression into multichannel communication systems. Compared with a single channel, usage of multiple channels allows more flexibility in the system design, robustness and higher capacity; it can also provide support for reliability and Quality of Service. For example, increase in system performance can be achieved by exploiting path diversity effects or QoS can be provided in a wireless network domain by using multichannel MAC protocols. In the following we define "*channel*" as a resource that is used for transmission of a certain IP packet flow. Typically, multiple channels are considered as physically different entities (*e.g.* OFDM sub carriers), each characterized by a particular delay, bandwidth and error rate. We expand the notion of multiple channels by including also the case of physically identical but logically different channels. As an example one can consider streaming of multimedia data that usually consists of several individual RTP-based streams. Clients that receive multimedia, *e.g.* video clip, typically receive two IP streams, one for audio and one for video. We propose this generalized approach to the multiple channels from the stand point of header compression. Link layer protocols do not differentiate the packets transmitted over one physical channel; header compression entities treat each IP stream (between the same source-destination pair) separately regardless of whether they are sent over different physical channels or not.

When multiple channels are available between a sender and a receiver, any of the header compression schemes can be applied for each channel individually to reduce transport overhead. However, in this case the multi-channel header compression will inherit all the properties of a corresponding scheme for a single-channel communication. For example, framed delta coding will suffer from the error propagation problem. To overcome the limitations of delta coding but still enjoy other attractive properties on this scheme, Cooperative Header Compression has been proposed in [Fitzek et al., 2005a]. Exploiting the availability of the multiple channels, joint header compression of parallel streams is performed.

COHC is based on the concept of an additional information container (AIC). To each packet with a compressed header some extra information, an AIC, destined for the neighboring channels is appended. The AIC is used to repair a corrupted context of neighboring compression entities. It is not the purpose of the AIC to repair the whole packet (including the payload) but only to retrieve the current context at the decompressor.

By introducing AICs the problem of error propagation can be reduced. An AIC should carry information sufficient to update a context of a particular neighboring stream in case of transmission errors. The AIC transmitted over a given channel can refer to the context of the same channel or to the context of any neighboring channel. At the same time, there are no limitations in the channel or time domain: an AIC can be used to repair a compressed header of a packet received simultaneously, or to rebuild previous or upcoming packets. These choices are up to the designing process. Depending on the information carried by an AIC and the way it is constructed, a different number of errors can be sustained. What options for AICs' design should be preferred depends on the particular system's characteristics and requirements. The general approach for AICs' transmission over multiple channels is given on Figure 17.3.

We further illustrate the ideas of Cooperative Header Compression by describing one possible COHC implementation. In addition to the compressed header for each packet, the header compression entity includes one AIC for each neighboring channel in the same time domain. Compressing a packet header, each compressor generates the related AIC as well: in the simplest case the AIC can be just a copy of the compressed header. The neighboring compression

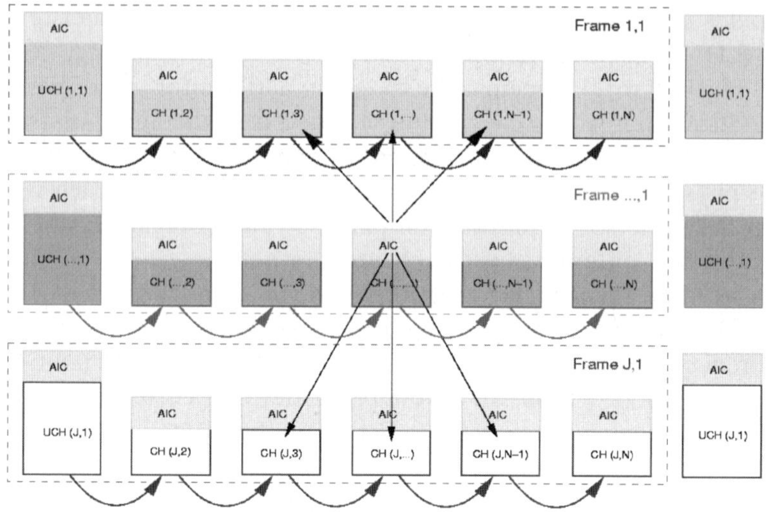

Figure 17.3. The concept of context in header compression.

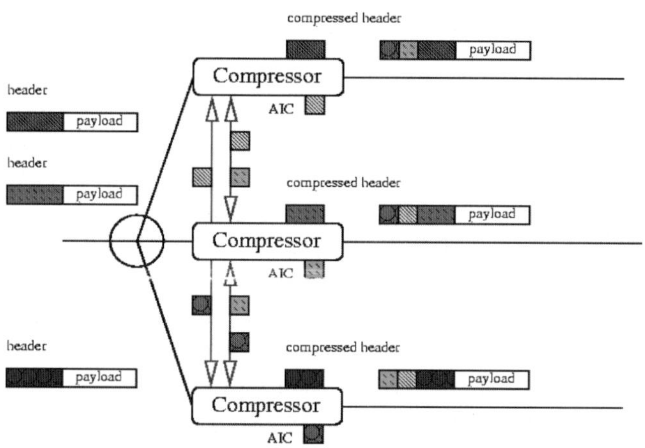

Figure 17.4. AICs construction for three cooperative channels.

entities exchange the generated AICs (Figure 17.4). After the AICs are passed to other compressors, the compressed headers, AICs and payload are composed in packets and are ready to be transmitted over multiple channels.

In this implementation only AICs with the same time instants can be used to update the context in case of packet errors. Figure 17.5 illustrates how using

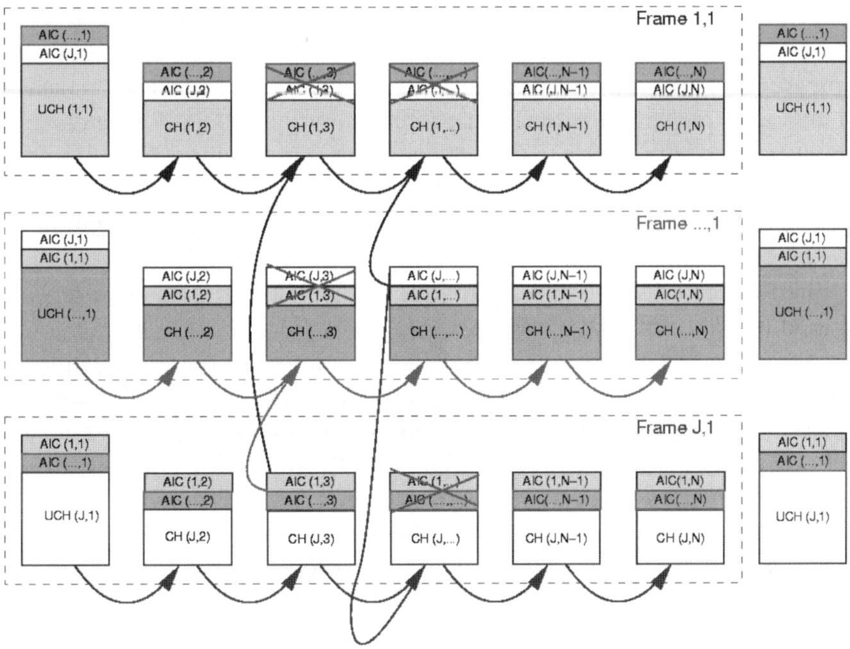

Figure 17.5. Context healing in COHC by using AICs.

AICs loss propagation can be reduced or completely avoided. The example shows that the two packet errors on the first channel caused by transmission errors has no impact on the following compressed headers. Even if the whole packet (1, 3) on channel 1 is lost, we do not apply a cross to the compressed header $CH(1, 3)$ as it can be healed by $AIC(1, 3)$. In this example we have five packet errors due to propagation errors, but no packet lost due to error propagation.

This approach has proven to be efficient when the parallel flows are synchronized, that is, all channels between a sender and a receiver have the same delay characteristics within the granularity of IP packets. For many application scenarios the synchronous channels is a feasible assumption. If IP datagrams are segmented into smaller data link packets, jitter can be expected between IP packets. In case there is a limited number of retransmissions on the link layer or no retransmission at all (*e.g.* broadcasting), the jitter is bounded and will not have a significant impact on the performance of the scheme. Asynchronous channels can be brought 'in tact' by introducing a buffer at the receiver side. If the buffer solution is not desirable due to the delay, design using time-domain separated AICs should be considered.

In the presented implementation the compressed packet can be healed if at least one channel delivers a correct packet. That is, if the channels' behavior is uncorrelated or, in other words, the errors are independent in the channel domain, COHC efficiently maintains the context synchronization between the compressor and decompressor. If the context can be successfully retrieved with each incoming packet, then the full update is not required and the frame length can be increased. This leads to the further bandwidth savings. One should note that here we are speaking about the correlation and error pattern on the IP packet level. If a loss of an IP packet on one channel implies with high probability that packets on other channels are also corrupted, then AICs interleaving can be applied to break the channel correlation. COHC with AICs interleaving is described in [Madsen et al., 2005b].

As we can see from Figure 17.5, the bigger number of parallel channels are, the smaller is the probability that propagation loss will occur. At the same time, it means that with the increase in the number of channels L, the size of the compressed header grows as $H_{CH} + (L-1)H_{AIC}$ where H_{CH} and H_{AIC} are sizes of a compressed header and an AIC respectively. To keep high compression gain, the size of additional appended information should be made as small as possible. It turns out that for the independent packet errors, the number of three cooperative channels is enough to achieve high robustness of the scheme under a large range of channel error rates [Fitzek et al., 2005a; Madsen et al., 2005c]. To minimize further the amount of appended bytes, no information belonging to a packet itself should be included in an AIC, but only the information necessary to recover the context. Thus, it is not necessary to include UDP checksum in the AIC, since it corresponds to the whole packet, but is not required to update the context. In this way, using CRTP header compression [RFC2508, 1999] as underlying framed delta coding scheme, the size of an AIC can be reduced from 4 bytes to 2 bytes. For some applications the size of AICs can be reduced even more. For example, considering Multiple Description Coding (MDC) streaming the size of an AIC can be made equal to zero, that is no extra information will be carried by channels [Madsen et al., 2005d]. Due to the high correlation of the header fields of MDC sub-streams produced from the same audio or video source, cooperative decompression of MDC descriptors reduces the error propagation problem.

Cooperative Header Compression can be considered as a general compression strategy for parallel IP streams between the same source-destination pair. One of its advantage is low complexity and its suitability for unidirectional links. We have presented the possible modes of operation of COHC when framed delta coding is considered as an underlying compression scheme. In principle, nothing prevents us to apply COHC on top of any other known header compression algorithm. Using AICs in the schemes with feedback can help to save bandwidth due to reduction in signalling.

3. Application Fields of the Cooperative Header Compression

In general, COHC presents an efficient solution for overhead reduction over bandwidth-limited links, both wired and wireless. Examples of links and networks where COHC can be applied are:

- low-speed serial links;
- cellular networks (3G and beyond);
- short-range and local networks, *e.g.* IEEE WLAN 802.11;
- meshed and ad hoc networks (multihop networks).

The development of header compression schemes started in a wired domain for low bandwidth links such as PSTN. Wired links are usually characterized by low error rates and the error propagation problem does not occur frequently. Therefore, applying COHC in wired networks can give a limited advantage compared with other compression schemes. It is not the case for wireless networks that experience the lossy behavior with high bit error rate (BER) coupled with the limited bandwidth. Cooperative schemes are, first of all, targeted to wireless environments, to start with, cellular networks.

Cellular networks have provided the users with the possibility of always being reachable no matter where they are. Cellular networks of the second generation have been mostly developed for voice transmission. 3G systems and beyond are expected to support a broad range of IP-based data services, including audio, video, email, gaming etc. But the case of cellular links still remains technically demanding. Header compression is vital for data services support in cellular networks, *e.g.* ROHC [RFC3095, 2001] became an integral part of the Third-Generation Partnership Project (3GPP-UMTS) specification. To keep the complexity of compression algorithms low, thus providing support for heterogeneous terminals, COHC can serve as one of the tools helping to overcome the bandwidth limitations. Figure 17.6 illustrates possible applications of COHC in next generation cellular systems when a mobile terminal establishes parallel connections with multiple base stations or with one base station or when a relay is used for coverage extension. In the three scenarios indicated on Figure 17.6, introducing cooperative behavior of multiple channels can significantly increase system capacity, *i.e.* more users will be served or larger data rate per user will be experienced.

The wireless local area networks (WLAN) support different data applications, including Voice over IP, and present convenient data infrastructure for

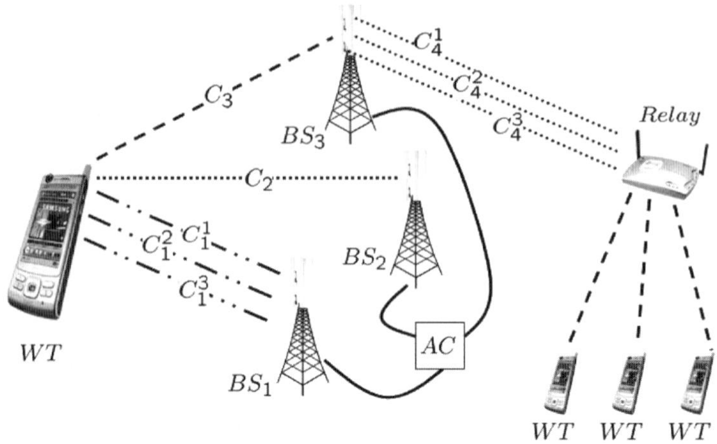

Figure 17.6. Possible Application Fields of the Cooperative Header Compression Mechanism in next generation cellular networks.

homes and offices. Most of these networks use unlicensed spectrum, thus suffering from high bit error rates due to the interference. The recent development indicates that header compression schemes can be successfully applied in WLAN [Bormann, 2005], making it possible to save bandwidth and reduce delays by inherently sending smaller packets. One of the scenarios we are envisioning for COHC application in WLAN might be where multiple base stations (BS) serve one or several wireless terminals (WT) in a cooperative manner. BSs are connected to an access controller and the streaming data intended for one or several recipients is split into several substreams by a controller (Figure 17.7). Cooperative compression of the headers of the streams is also performed at the controller. Each substream is forwarded to a BS for further wireless transmission to WT. Due to propagation characteristics of wireless medium, the receipt quality of links between a wireless terminal and BSs can be different. COHC can help to survive a temporal increase in BER in one or several channels without loosing context synchronization [Madsen et al., 2005c].

In multi-hop wireless networks, communication between two nodes is carried out through a number of intermediate nodes. In contrast to meshed networks, consisted of static nodes, the relaying nodes in ad hoc networks are in general mobile. The mesh topology of multi hop networks implies that it is possible to establish multiple paths between a source and a destination. Figure 17.8 shows a multi-hop network example with three channels. Each channel is composed of several hops between a sender and a receiver. Since a multi-hop network can consist of heterogeneous nodes, the intermediate nodes can vary a great amount with respect to their processing capabilities and thus their

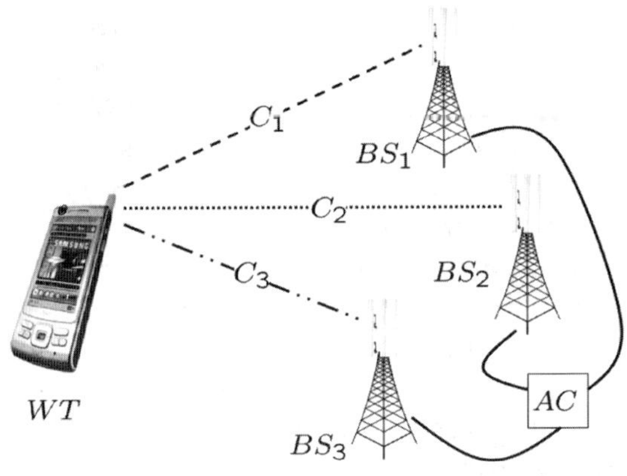

Figure 17.7. Application of COHC in WLAN.

offered services. Some of the nodes might not be able to perform header compression. One solution is not to apply header compression over links where such nodes are either a sender or a receiver. Alternatively, header compression and decompression should be performed only at the end nodes. We advocate this later approach, since it leads to greater bandwidth savings and gives us the possibility of applying COHC. The COHC implementation in meshed networks is discussed in [Fitzek et al., 2005a]. Speaking about ad hoc networks, a special routing protocol is required that is able to provide synchronous multiple paths and does not use IP header information for packet forwarding. What is more, COHC should be adapted to deal with the varying number of channels between a source and a destination, since in ad hoc networks the number of available routes can change in time due to the node mobility. More details about COHC in ad hoc networks can be found in [Madsen et al., 2005a].

The above scenarios illustrate cooperation on the system level when the entities belonging to the same communication chain cooperate in order to use their resources in the most efficient way. The truly cooperative behavior using header compression is given in the next scenario. As it is shown in Figure 17.9, we assume that each WT has the capability of communicating with the BS and simultaneously with other terminals (by using either the same or different air interfaces). Each terminal is receiving a data stream with compressed headers in order to increase data-rates. At the same time, users receive AICs intended for other users. AICs exchange is performed using short-range connections. The question is why in the presented scenario egoistic users should cooperate? Would it not be better to receive from a BS some extra information that can help to recover your own context instead of getting information destined to the others and afterwards spend resources on forwarding it? The benefit of

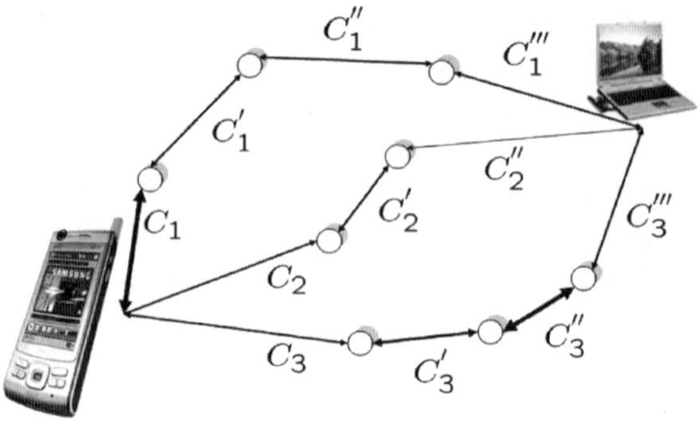

Figure 17.8. Presence of multiple channels in a multi-hop network.

cooperation comes from the following observations. The extra information is needed to update the context only when a packet is corrupted. If a radio path between a BS and a user is greatly deteriorated by the instantaneous channel conditions, the whole IP packet, including a compressed header, an AIC and a payload, is lost. Thus, a user is not able to help him/ herself. A neighboring user might be experiencing good channel conditions and might be able to deliver a correct AIC. The cooperative users are rewarded by keeping their decompressors operational.

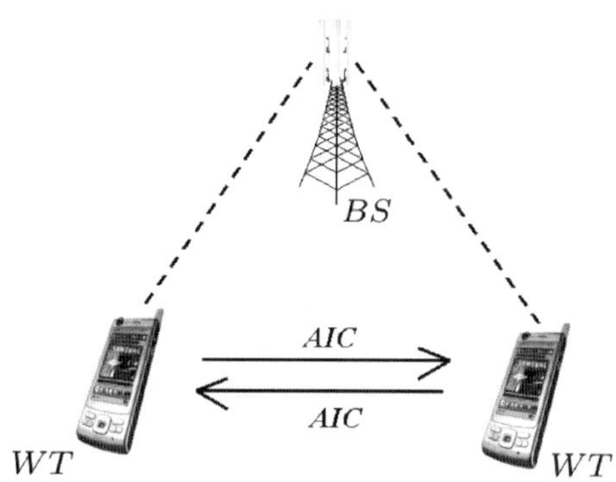

Figure 17.9. Cooperation among terminals by AICs exchange through a short-range link.

The need of multi channel communication can be also dictated by applications. As examples, one can consider Multiple Description Coding (MDC) or Multi Layered Coding (MLC) that split a single audio or video source into multiple descriptors. The transport overhead for MDC or MLC grows linearly with the number of descriptors: for each descriptor an overhead of 40 bytes for IPv4 or 60 bytes for IPv6 should be taken into account. To make MDC and MLC attractive for bandwidth-limited links, the total overhead can be reduced by means of header compression schemes. It is shown that COHC is particularly suitable for combination with MDC [Madsen et al., 2005d; Fitzek et al., 2005b]. In Figure 17.10 the encoder and network overhead for the foreman video sequence using MDC is given. Applying COHC the overhead can be kept close to the encoding overhead (encoding overhead gives us the lowest achievable bound). For example, if three descriptors are used, the total overhead can be reduced from 150% to approx. 50%.

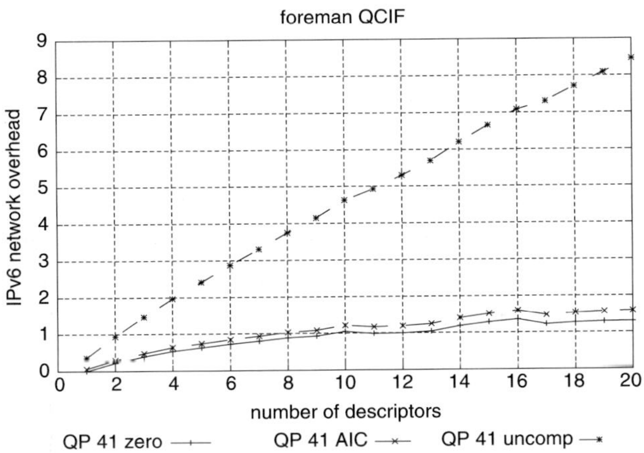

Figure 17.10. Network overhead (RTP/UDP/IPv6) for the `foreman` video sequence and the quantization parameter 41.

4. Tradeoff Between Compression Gain, Robustness and Bandwidth Savings

To estimate how efficient one or another header compression scheme is, we are interested to what extend it is able to compress the original header. The parameter called *compression gain* is defined as a difference between the sizes of the uncompressed and the compressed headers normalized to the uncompressed header:

$$CG = \frac{U - C}{U} = 1 - \frac{C}{U}, \tag{17.1}$$

where C and U are the sizes of compressed and uncompressed headers, respectively. The compression gain can be interpreted as the potential *bandwidth savings* in case of absolutely reliable links. Indeed, for the framed delta coding the bandwidth savings can be calculated as:

$$S = \frac{N \cdot U - \sum_{i=1}^{N} C_i}{N \cdot U}, \tag{17.2}$$

where N denotes the frame length. If the communication link is reliable, *i.e.* no transmission errors are experienced, then no framing is required. In the long run the initial uncompressed header becomes negligible and the bandwidth savings can be estimated by using formula 17.1.

Table 17.1 shows compression gain for different schemes. COHC does not achieve the highest compression gain compared with other schemes, since the size of the compressed headers is increased due to the AICs. One should note, however, that in case of error-prone links compression gain can not be directly related to the achievable bandwidth savings or the increase in the capacity of the system. And what we are interested in at the end of the day is the achievable savings. Not always the highest compression gain will correspond to the highest savings. Small compressed headers make the system more vulnerable towards the re-synchronization of the compressor and decompressor. In some cases it can be preferable not to compress some of the header fields at all. There is a clear trade–off between the compression gain and the robustness of the scheme. Both these parameters can be traded for the complexity of the system.

Table 17.1. Compression gain for framed delta coding (FDC) and cooperative scheme (COHC).

Header Compression scheme	Compression Gain
FDC (RFC 2508)	90%
FDC, UDP checksum disabled (RFC 2508)	95%
COHC for three cooperative channels	80%
COHC for MDC streaming	90%

In order to introduce a parameter that realistically reflects the behavior of a compression scheme, the impact of the transmission errors and the error propagation problem on the savings should be taken into account. The *average expected bandwidth savings* can be evaluated as:

$$E[S] = \frac{1}{N} \sum_{k=1}^{N} P_{\Delta}(k) \cdot S_{\Delta}(k), \tag{17.3}$$

where $P_{\Delta}(k)$ is the probability to receive and decompress the k-th packet correctly and $S_{\Delta}(k)$ represents savings due to compression of the k-th packet. $S_{\Delta}(k)$ is given by

$$S_{\Delta}(k) = \begin{cases} 1 - \frac{C}{U} & k \geq 2, \\ 0 & k = 1. \end{cases} \tag{17.4}$$

For a general evaluation, we model a communication channel using uncorrelated bit errors. This is a useful approximation if a heavy interleaving is applied. We denote the probability of a bit sent and received successfully (unsuccessfully) as P^g_{bit} (P^e_{bit}). The probability of a packet consisting of $x = H + X$ bytes of a header and a payload being correctly transmitted is

$$P^e(x) = 1 - (P^g_{bit})^{8 \cdot x} = 1 - P^g(x). \tag{17.5}$$

For the framed delta coding, a packet is decompressed correctly only if all the previous packets in the frame are received correctly. Therefore,

$$P_{F\Delta}(k) = P^g(U_{F\Delta} + X) \cdot P^g(C_{F\Delta} + X)^{k-1} \tag{17.6}$$

where $U_{F\Delta}$, $C_{F\Delta}$ and X are the sizes of the uncompressed and compressed headers for FDC and payload, respectively. Substituting formula 17.6 in formula 17.3, we obtain:

$$\begin{aligned} E[S_{F\Delta}] &= \left(1 - \tfrac{C_{F\Delta}}{U_{F\Delta}}\right) \cdot \tfrac{1}{N} \sum_{k=2}^{N} P^g(U_{F\Delta} + X) \cdot P^g(C_{F\Delta} + X)^{k-1} \\ &= \left(1 - \tfrac{C_{F\Delta}}{U_{F\Delta}}\right) \cdot \tfrac{P^g(U_{F\Delta}+X) \cdot P^g(C_{F\Delta}+X) \cdot \left(1 - P^g(C_{F\Delta}+X)^{N-1}\right)}{N \cdot (1 - P^g(C_{F\Delta}+X))} \end{aligned} \tag{17.7}$$

For cooperative compression with L cooperating channels, in order to decompress a packet correctly it should be received without errors and the synchronization between the compressor and decompressor should not be lost. The context is maintained if at least one packet on L channels is received correctly at every previous step. Thus, the probability to keep the synchronization at m-th packet in the frame is given by[1]

$$P_S(C_{C\Delta} + X) = 1 - \prod_{i=1}^{L} P^e_i(C_{C\Delta} + X). \tag{17.8}$$

For FDC this probability is $P^g(C_{F\Delta} + X)$, that is, it is much smaller. We can conclude that the robustness of COHC is higher; in other words, COHC can sustain a larger amount of errors without loose of the synchronization.

Probability of decompressing the k-th packet correctly for COHC can be calculated as:

$$P_{C\Delta}(k) = P^g(C_{C\Delta}+X) \cdot \left(1 - \prod_{i=1}^{L} P^e_i(U_{C\Delta}+X)\right) \cdot \left(1 - \prod_{i=1}^{L} P^e_i(C_{C\Delta}+X)\right)^{k-2}, k \geq 2 \tag{17.9}$$

Using formulas 17.3 and 17.9 we obtain the final expression for the expected bandwidth savings for COHC:

$$E[S_{F\Delta}] = \left(1 - \frac{C_{C\Delta}}{U_{C\Delta}}\right) \cdot \frac{P_S(U_{C\Delta} + X) \cdot P^g(C_{C\Delta} + X) \cdot \left(1 - P_S(C_{C\Delta} + X)^{N-1}\right)}{N \cdot (1 - P_S(C_{C\Delta} + X))}. \tag{17.10}$$

One can notice a clear similarity between formulas 17.7 and 17.10: substituting in 17.10 the probability to keep the synchronization with the corresponding probability for FDC, we come to formula 17.7.

In case there are L channels with different BER, the averaging of the obtained L expected bandwidth savings should be additionally made:

$$E_L[S_{F\Delta}] = \frac{1}{L}\sum_{i=1}^{L} E[S_{F\Delta}](P_i^g). \tag{17.11}$$

We illustrate the corresponding expected bandwidth savings for the framed delta coding and cooperative scheme in Figure 17.11. For COHC the curves are given for the three cooperative channels. The same BERs are assumed for all channels. We observe that for the $BER = 10^{-3}$ COHC shows savings that are three times higher than for the FDC. This is due to the ability of the cooperative scheme to cope with the transmission errors. If BER is lower, the difference in the performance of COHC and FDC becomes smaller, and when $BER < 10^{-5}$, FDM achieves higher savings. Indeed, under good channel conditions the error propagation problem will rarely occur, and there is no need to send extra information AICs.

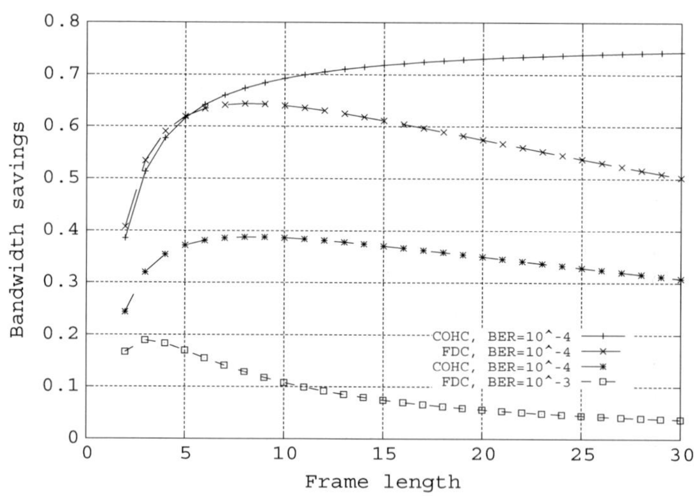

Figure 17.11. Expected bandwidth savings for the framed delta coding and cooperative compression (with three cooperative channels) with different frame length N and different BER. Payload X is 40 bytes.

In Figure 17.12 we take a closer look at the COHC performance under different channel conditions. As BER is increased from 10^{-2}, we observe the rapid gain in the expected savings; but for $BER > 10^{-4}$ the gain is marginal. The curve for $BER = 10^{-6}$ almost coincides with the one for $BER = 10^{-5}$, and therefore, it is not given on Figure 17.12. We can conclude that cooperative behavior can significantly improve the performance of the system, especially under bad channel conditions where the other compression approaches fails.

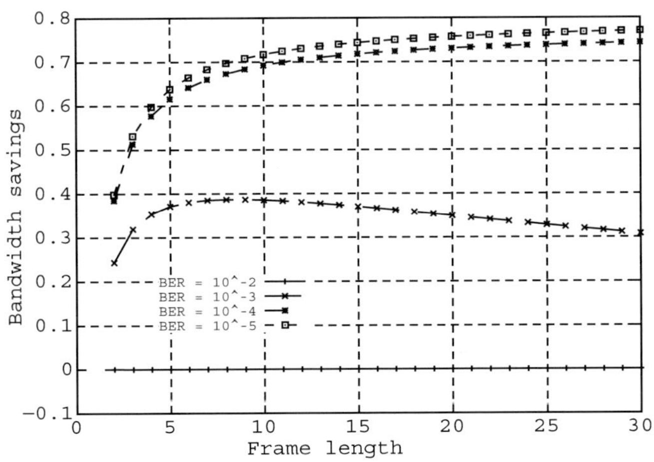

Figure 17.12. Expected bandwidth savings for COHC with different BER. $L = 3$; $X = 40$ bytes.

Figure 17.13 illustrates the influence of the number of cooperative channels on the performance of COHC. First, with the increase in the number of channels, the achievable savings go up, but if we choose a large number of channels, *e.g.* $L = 8$, the savings drop. The case of $L = 8$ corresponds to the compressed header of the size 18 bytes. At the same time the robustness for $L = 4$ is almost as good as for $L = 8$. Therefore, three or four cooperative channels is a good designing choice for this particular COHC implementation. We furthermore note that the expected savings rise with the frame length and decline after a maximum is reached. It is desirable to choose N as a function of both L and BER such as the expected savings are maximized.

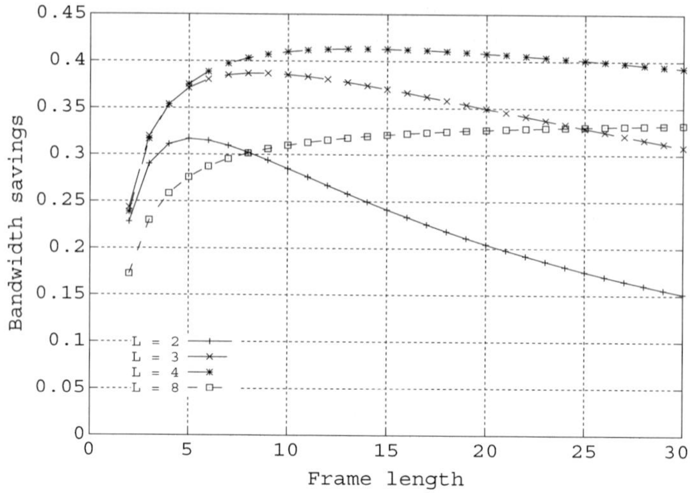

Figure 17.13. Expected bandwidth savings for COHC with different number of cooperative channels. $BER = 10^{-3}$; $X = 40$ bytes.

5. Conclusion

A novel approach towards header compression for wireless IP-based networks has been presented in this chapter. If multiple channels, physical or logical, are available between a sender and a receiver, cooperative behavior of parallel streams can boost the performance of compression algorithms in terms of bandwidth savings and robustness. AICs' exchange between the cooperating channels/users allows survival from the instantaneous deteriorated channel conditions without loosing the synchronization between the compressor and decompressor.

One possible implementation of COHC has been described in detail. We are using framed delta coding as an underlying compression principle. In case of a packet loss, AICs from neighboring channels in the same time domain help to update the context without the need of sending a packet with an uncompressed header. By estimating the expected bandwidth savings for the case of independent channel errors, we show that using three cooperative channels is a good designing choice for this particular implementation.

The design of AICs has a big impact on the overall performance of the scheme. Depending on the information carried by an AIC and the way it is constructed, a different number of errors can be sustained. We note that in case of MDC or MLC the size of an AIC can be reduced to zero. This makes COHC particularly suitable for combination with MDC and MLC for audio or video streaming. Other possible application fields for COHC are outlined.

Notes

1. indexing $C\Delta(F\Delta)$ shows that a value corresponds to COHC (FDC) scheme; indexing i refers to the values for the i-th channel.

References

Bormann, C. (2005). ROCH over 802, Internet Draft (work in progress).

Fitzek, F. H. P., Madsen, T. K., Popovski, P., Prasad, R., and Katz, M. (2005a). Cooperative ip header compression for parallel channels in wireless meshed networks. In *Proceedings of IEEE International Conference on Communication.*

Fitzek, F. H. P., Sheahan, R., Madsen, T. K., and Prasad, Ramjee (2005b). Fixed/mobile convergence from the user perspective for new generation of broadband communication systems. In *Proceedings of 2nd International CICT Conference on Next Generation Boroadband.*

Madsen, T. K., Fitzek, F. H. P., and Nethi, S. (2005a). Cooperative header compression for ad hoc networks, technical report, aalborg university.

Madsen, T. K., Fitzek, F. H. P., Prasad, R., and Katz, M. (2005b). Ip header compression for media streaming in wireless networks. In *Proceedings of VTC Fall 2005.*

Madsen, T. K., Fitzek, F. H. P., Takatori, Y., Prasad, R., and Katz, M. (2005c). Cooperative ip header compression using multiple access points in 4g wireless networks. In *Proceedings of IST Mobile Summit.*

Madsen, T. K., Takatori, Y., Fitzek, F. H. P., Prasad, R., and Katz, M. (2005d). Zero-aic header compression with multiple description coding for 4g wireless networks. In *International Workshop on Convergent Technology (IWCT) 2005.*

RFC1144 (1990). V. Jacobson, Compressing TCP/IP Headers for Low-Speed Serial Links, Request for Comments 1144.

RFC2507 (1999). M. Degermark, B. Nordgren and S. Pink, IP Header Compression, Request for Comments 2507.

RFC2508 (1999). S. Casner and V. Jacobson, Compressing IP/UDP/RTP Headers for Low-Speed Serial Links, Request for Comments 2508.

RFC3095 (2001). C. Bormann, C. Burmeister, M. Degermark, H. Fukushima, H. Hannu, L-E. Jonsson, R. Hakenberg, T. Koren, K. Le, Z. Liu, A. Martensson, A. Miyazaki, K. Svanbro, T. Wiebke, T. Yoshimura, and H. Zheng, RObust

Header Compression: ROHC: Framework and four profiles: RTP, UDP, ESP, and uncompressed, Request for Comments 3095.

Suryavanshi, V. and Nosratinia, A. (2005a). Convolutional coding for resilient packet header compression. In *Proceedings of IEEE GLOBECOM*.

Suryavanshi, V. and Nosratinia, A. (2005b). A hybrid arq scheme for resilient packet header compression. In *Proceedings of Asilomar Conference on Signals, Systems and Computers*.

Chapter 18

ENERGY AWARE TASK ALLOCATION IN COOPERATIVE WIRELESS NETWORKS

Bestow and Down-throttle, Unburdening the Unhappy Batteries

Anders Brodlos Olsen

Center for Indlejrede Software Systemer (CISS), Aalborg University, Fredrik Bajers Vej 7B, 9220 Aalborg Øst, Denmark

abo@kom.aau.dk

Peter Koch

Center for TeleInFrastruktur (CTIF), Aalborg University, Niels Jernes Vej 12, 9220 Aalborg Øst, Denmark

pk@kom.aau.dk

Abstract: This chapter aims to describe the framework of energy aware task allocation in a cooperative network. It is motivated by the general acceptance that energy consumption is a significant design limitation for battery powered systems and also that cooperative terminals have the willingness to make unselfish assignments. Therefore, utilizing the cooperative aspect and the fact that workload computation has a square energy relation, potentials of cooperative task computing are evaluated. By an assumption of traditionally distributed or parallel systems, energy models are proposed for computational and communicating components, where system terminals are delimited to contain processing and communication components only. The methodology of dynamic voltage scaling is used to make low energy workload execution on the computational units and the energy overhead for distributing workload is accounted for by a parameterized energy model. Simulation results show a realistic energy gain of as much as three times compared to terminals selfishly executing identical workloads. On the contrary, overheads related to workload distribution will also result in overall increased energy consumption, making a careful and wise decision on task allocation evident.

Frank H. P. Fitzek and Marcos D. Katz (eds.), Cooperation in Wireless Networks: Principles and Applications, 567–591.

Keywords: cooperative networks, task allocation, energy optimization, dynamic voltage scaling, multi-processors

1. Introduction

Standing on the doorstep to the next generation of mobile wireless systems (also referred to as fourth generation or simply 4G) issues related to resource optimization, especially energy consumption, are of great importance. This need is manifold; integration of multiple functional devices, support of multiple communication standards, multimedia applications, adaptive transmission schemes, new envisioned services, and so on. 4G developments are also driven by concepts like anywhere and anytime, seamless access, adaptive air interfaces, adaptive quality of services, flexibility, efficiency, etc., therefore calling for new flexible implementation strategies. In order to facilitate these initiatives, concepts like software defined radio (SDR) and reconfigurable radios are introduced, composed of programmable devices being less energy efficient. It is generally accepted that energy consumption is one of the most significant limiting design constraints and issues related to energy optimization in embedded systems have received increased attention over the last decade. The main driving force is the lacking improvement in battery technology, together with the before mentioned trends, making the gap between required and obtainable battery capacity constantly wider.

Recently, concepts of cooperative wireless networks are introduced, with the overall purpose to join forces in order to reach a better joint QoS. By nature, wireless communication links are prone to error and great efforts are spend to overcome these, such that high capacity data links can be provided. Cooperative methods have mainly been suggested as means for alleviating some of these errors, where the basic principle is to share resources for diversity. In [Nosratinia et al., 2004] a survey of a number of physical layer cooperative issues is given and in broad terms relaying concepts are the main philosophy. Examples are; detect and forward, amplify and forward, and coded cooperation. In [Politis et al., 2004] cooperative networks at higher layers are discussed, outlining the cooperative network architecture. The key principles are that systems should be layered on demand, reuse of module blocks, multiple services, and end-to-end connectivity across access technologies. The majority of cooperative concepts are covering transmission and reception issues all with the goal of enhancing the individual and/or group wireless link capacity for the cooperative terminals. Nevertheless, cooperation is not only limited to these issues; in this work principles of cooperative task computing is motivated for cooperative willing terminals connected by a short range wireless technology. As illustrated in Figure 18.1, the overall principle is to distribute tasks among cooperative terminals, such that workload is wisely balanced making it possible to

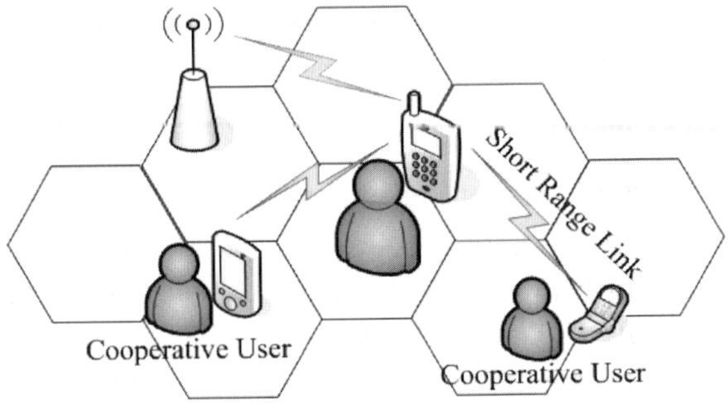

Figure 18.1. The principle cooperation scenario between terminals within a cellular network. A user receives input over a centralized link, which initiates workload on the terminal and by using a short range wireless links the workload is offloaded to other cooperative users.

operate the individual terminals at a lower performance level. At the individual terminals, energy conserving methods are utilized, such that the workload is executed using minimum energy. This is obtained by Dynamic Voltage Scaling (DVS) methods, which utilizes knowledge of workload and its timing constraints. It is also proven that DVS are able to generate energy optimal workload executions, as shown by [Aydin et al., 2001]. Later on the principles of DVS will be explained in more detail. Therefore, using the motivation of reducing energy consumption and cooperative terminals, the scheme of energy aware task allocation in cooperative wireless networks is proposed.

Distributing workload to other network elements is not a new concept distributed and parallel computing, together with GRID concepts are well known principles in wired systems, but wireless replicas are emerging. However, the goals for these principles are to enhance the computational capabilities or throughput of the systems, where the goal for this work is to optimize energy consumption. Workload throughput and energy reduction is to some extend contradicting parameters, as workload is typically defined by a number of computational operations together with timing constraint and energy consumption is a function on operations executed in a time period. Therefore is energy reduction and workload throughput a fine balance where both must be considered.

In the following short and not excursive surveys of related distributed computing are given: Wireless GRID is a relative resent proposed scheme and [McKnight et al., 2004] presents a conceptual overview. The used observation is that today's mobile devices contain an advanced computational capability, together with a wide range of sensors, *e.g.* digital cameras, GPS receivers, etc. The overall philosophy is to share resources among devices in the network, using similar concepts as seen in wired GRID systems. Obviously, concern regarding sharing resources on portable devices (battery powered) is not trivial, as a fair consumption of individual resources must be respected. Nevertheless, the author's sees Wireless GRID's as a mean of providing new and advanced applications in mobile environments and even also as a mean for more efficient use of mobile resources.

Distributed and parallel computing is a well established research area with numerous literature references; however, for the energy aware counterpart the literature pool is significantly less. [Shirazi et al., 1995] provides an overview of various traditional static and dynamic scheduling approaches for parallel and distributed systems. As for single processor environments the methodology of DVS is used for achieving energy conservation and also divided into static and dynamic scheduling approaches. The general approach is to distribute the workload according to workload estimates and timing constraints, such that each individual processing element is able to run at its minimum speed while fulfilling the application timing specification. Overall, identical approaches are used for energy aware scheduling as known from traditional multiple processing, with significant difference that both energy and time must be optimized jointly in order to get the least energy consumption but also to fulfill timing constraints (when considering real time systems). Without making an excursive literature survey a few energy aware multiple processor scheduling schemes is worth mentioning: [Zhu et al., 2003] proposed a slack reclamation approach for a global scheduling (GSSR) and for a fixed order scheduling (FLSSR), showing that timing constraints are always meet and considerable energy saving is obtainable. The work of [Yu and Prasanna, 2002] formulates energy aware static allocation of independent tasks in a heterogeneous system, using an integer linear programming (ILP) approach and also a linearization heuristic.

In line with our work two groups have proposed work with related concepts: In [Yu and Prasanna, 2005] they study energy aware task allocation on a set of homogeneous processing units, where the applications contain a communicating periodic task-set, and each processing unit is supported by DVS technology. The optimizing metric is to balance the energy consumption of each processing unit with respect to its remaining energy level, such that the lifetime of the sensor network is maximized. An ILP-based approach, together with a three-phase heuristic model is proposed. Their simulation results show, for the best case, improvements of lifetime in the order of 2.5 times compared to when no DVS is used. Conceptually, our work is similar, although it deviates in a number of essential ways: 1) In [Yu and Prasanna, 2003] tasks are described using Directed Acyclic Graphs to describe the workload, where we instead use the definition of real-time task-sets. Using DAG and multiple ILP heuristic models seem as a static approach of allocating tasks among nodes, where we eventually are targeting a more dynamic environment and scheduling mechanism. 2) From DVS literature, issues of task slack time reclaiming are shown to be important, and by static approaches this is not possible, where dynamic DVS methods are able to utilize runtime accumulation of task slack times for further energy reduction. 3) Finally, we believe that task allocation on other terminals are not always beneficial, which thereby call for dynamic decision mechanisms for task distribution. Also related is the work of [Alsalih et al., 2005] and [Lu et al., 2005]. Overall the idea is that wireless devices in an area can cooperate on task execution, with the assumption that device have some diversity in battery size, energy consumption and etc. The devices in the cooperative group are categorized using various cost parameters, and by choosing the cheapest device, taking communication overheads etc into account, tasks are mitigated among the terminals. In [Alsalih et al., 2005] a static approach is proposed, and in [Lu et al., 2005] a dynamic approach is proposed. In contrast to our work, only traditional power management schemes for energy reduction on the individual terminals are considered, making the need for devices with different energy consuming profiles essential if energy gains are to be accomplished.

In previous publications we have introduced the concept of energy aware computing in wireless cooperative networks [Olsen et al., 2005b], [Olsen et al., 2005a], and [Olsen et al., 2005c]. The overall idea is: 1) an energy aware scheduler decides where tasks are going to be executed, 2) a short range wireless network protocol exchanges task data among terminals, and 3) efficient energy conserving methods perform the energy saving on the individual terminals. The scheduling mechanism uses cost functions for making the task allocation decision and the individual terminals are energy reduced by the concept of dynamic voltage scaling (DVS) which has been proven to make near optimal energy schedules of task-sets, [Aydin et al., 2001]. The overall operation criteria is formulated as: *Mobile terminals cooperating on energy aware task execution must individually gain from the cooperation, and are therefore willing to exchange task-sets. If terminals do not gain form the cooperation, they act selfishly and execute their tasks by them self.* However, when terminals have decided to cooperate the question of how many members in the cooperative group arises. In the following sections a discussion on energy aware task allocation among cooperative terminals will be made. A model taking an outset in traditional distributed/parallel systems is made, applying simplified but realistic models of the wireless inter-communication among processing units. Finally, simulations showing the potentials of the proposed method are presented, outlining the energy advantages but also the energy overhead hazards introduced by the wireless links.

2. Motivating Scenarios

Proposing energy aware task allocation in cooperative networks, the question of what applications are suitable for such a scheme arises. This section will discuss possible application areas – though in the light of that this scheme is targeted for next generation terminals or even other types of networks like sensor or PAN networks, therefore a broad general viewpoint is taken. Such discussion swiftly becomes diffused and colored by authors, but having in mind that these applications are well ahead into the future, only imagination is setting the limit to these. This discussion hopes to indicate the conditions for such distributable applications.

First of all, let's discuss unsuitable applications: Considering a modern mobile cell-phone, the trend is towards an extensive growth in applications that has nothing to do with voice communication, which still must be regarded as the main functionality. These arising applications (MP3, camera, low resolution multi-media, personal organizers, etc.) are not feasible for distribution of a simple reason – such applications are from an implementation viewpoint relatively simple, making it possible to make them energy efficient, and also, there already exists small low budgeted devices such as portable MP3 players, or PDA's. Hence, the target applications have a need for extensive computational power, beyond or on the edge of what is found in applications in modern cell phones or PDA's.

Coming to plausible applications: Such could be any application containing a sufficient degree of parallelism, or in scenarios where multiple applications are executed on a single platform, as a very general outset. All in all it has a computational complexity that calls for devices with fast and energy hungry processing capabilities. In gaming applications huge processing hunger is typically found, traditionally limiting such applications to stationary devices powered by wired power supplies. It is evident that gaming, as it does today, will occupy future generations of mobile devices, introducing *e.g.* 3D graphics and other computational heavy visual features. For such, it can be imagined that cooperative terminals can have an energy benefit, if the workload of the game is distributed among devices.

Going to more wireless sensor like applications; target recognition could also be a feasible application, where a number of photo-sensors can cooperate on finding a given object or person in a field of interest. Such application is already found in various military applications, but could easily be imagined as useful for civilian purposes. An already proposed distributed application is proposed by [Delaney et al., 2005], where energy aware distributed speech recognition for small embedded systems is proposed, with the overall functionality of transforming a speech sequence into a text sequence. Such an application is ideal for distributed computing, using some front-end signal processing the speech sequence is reduced and by comparing these reduced sequences to lookup-table like databases, text-strings can be generated. Obviously there is a bit more to it than described, but overall the reduced speech streams make communication less heavy and finding information in databases are in general a computational heavy task. Thereby, suitable tasks for distribution are generally defining as tasks with relative easy communication overhead and heavy computational demands.

Finally, an example using Multiple Description Coding (MDC) and distributed computing, where three coded streams are assumed and these have to be received, decoded and forwarded to a sink, connected by a short range wireless link. In this example, terminals include a centralized wireless link, a processing unit, and a short range wireless link, where the short range link is proportionally cheaper in terms of energy compared to the centralized link. Furthermore, the decoding has to be made under real-time constraints. Figure 18.2 illustrates the scenario for there different system setups. 1) A single terminal is receiving all three streams, decoding and forwarding them to the source. For the wireless links this implies three active periods and the processor must be able to decode all three streams within the period. 2) Three terminals are used to receive one each of the three MDC streams. The wireless links handle only a single stream and the processor also decodes only on stream. Energy wise, two additional sets of wireless links are active, but the processor only decodes one stream, implying the possibility of reducing the computational capability. 3) One central link is active and one terminal forward streams to two other terminals over the short range link. Each terminal still decodes and forwards one stream. The advantage being, that two central links can be idle, though introducing additional traffic on the short range link. Which of the three cases is most energy efficient depends on the ratio between the energy overheads on the wireless links. Under the assumption that the short range is significant less energy consuming, the third case will be likely to have an energy gain compared to the second case.

3. Energy Aware Computing in Cooperative Networks

Initially, lets consider an illustrative example of the proposed energy aware task allocation scheme, as indicated in Figure 18.3. Two scenarios are considered, one without and one with cooperation on task execution. Now assuming that a source initiates an application, receiving information or input events from a user, imposing a workload on the terminal and also loading the terminal to its maximal capabilities (the workload measure (W) is one). In the non-cooperative scenario the processor will execute the task-set by itself, using what corresponds to one energy unit. In the cooperative scenario terminals have negotiated a cooperation agreement. It is assumed that terminals being idle (or at least are very easy loaded) are willing to execute others workload, implying that a workload diversity within the cooperative group is present. Using adaptive performance scaling on the terminals (DVS), the energy consumption only becomes a quarter for executing half the workload, assuming an ideal performance scalable technology and also disregarding communication overhead. Hence, energy

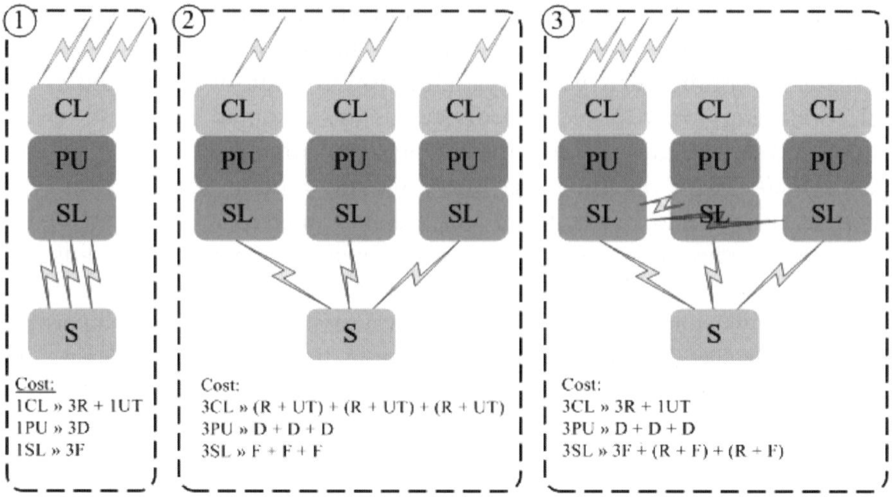

Figure 18.2. Multiple description coding example of task distribution among terminals for cooperative execution using centralized links to receive the task and short range links to forward the decode result.

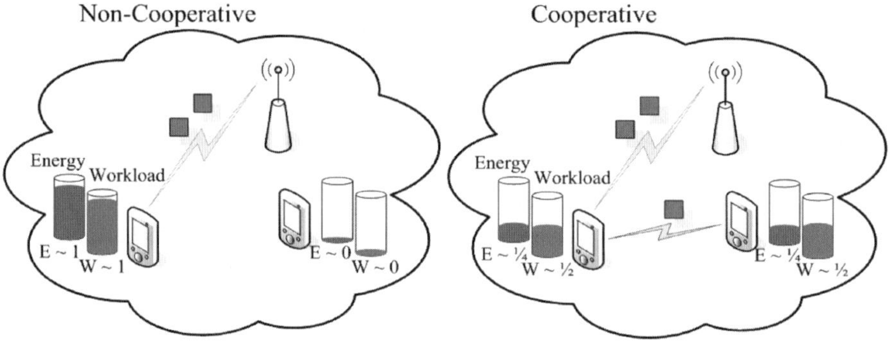

Figure 18.3. A non- and cooperative scenario of energy aware task allocation. Energy and workload levels are illustrated for both scenarios.

consumption is a quarter compared to the non-cooperative scenario, but as two terminals are cooperating the joint energy consumption becomes a half compared to the non-cooperative case.

The above example hopefully exemplifies the overall operation of the proposed scheme. Obviously, a number of issues are not targeted in this work, but are of great importance before the scheme is practically feasible. As the concept of cooperation is targeted, issues like willingness and fairness among the cooperative terminals are challenging subjects. First of all, presumed cooperative willingness of terminals: Next, how is fairness between terminals guarantied, where fairness could be on acceptation of workload, energy consumption, and joint energy conservation? Also, how do terminals identify that they are capable of executing other terminals tasks and most importantly what is the energy cost. Such issues could be enclosed under a service discovery aspect, which is somewhat related for all cooperative concepts. This will not be further investigated, assuming that cost and fairness aspects can be solved in a sort of *e.g.* accounting scheme. Security will obviously also raise huge questions, trust aspects among terminals and also data communications between the terminals. Reliability, since what if a cooperative fails to return received workload?

Assuming that cooperative willingness, fairness, security, and reliability are issues that are solvable, the following four subsections make a framework partitioning of the system parameters. They describe abstractions and definitions used for modeling energy aware cooperative task computing.

Workload. As no specific application is targeted within this work a high abstraction for workload description is needed. As this eventually becomes a scheduling problem, a traditionally scheduling workload definition is used. As mentioned, workload is initiated by an event, either received information or user input, loading the terminal with an application. Systems are traditionally divided into real-time or non real-time systems, where this work considers real-time applications, but not necessarily in its hard sense. Real-time workload is typically defined by task-sets, having tasks been defined by; arrival time (a), deadline time (d), workload (w), and perhaps also precedence. Utilizing an earliest deadline first (EDF) scheduling approach, also proven to be able to utilize a single processor system optimally, as shown by [Liu and Layland, 1973], a utilization factor is defined to describe the density of the workload. Utilization factor (U) is also known as a scheduling criterion, where EDF is able to load a system to its maximum capacity. EDF utilization is defined as:

$$U = \sum \frac{w_i}{d_i - a_i} \leq 1$$

Eventually, this is a system composed of multiple units processing the workload, although it is a multi-processing architecture where the individual units can be seen as self-contained units and therefore the above workload assumption. Hence the system analogy of cooperative terminals into a traditional parallel or distributed system as illustrated in Figure 18.4. Therefore, parallelism within the task-set is crucial, in order to be able to distribute the workload. There mainly exist two levels of Parallelism; 1) within the task where subparts are executed on different processing units, and 2) inter task parallelism where branches or threads can be executed in parallel.

Computational units. Portable battery powered systems normally consist of various system components, *e.g.*, displays, processing units, memories,

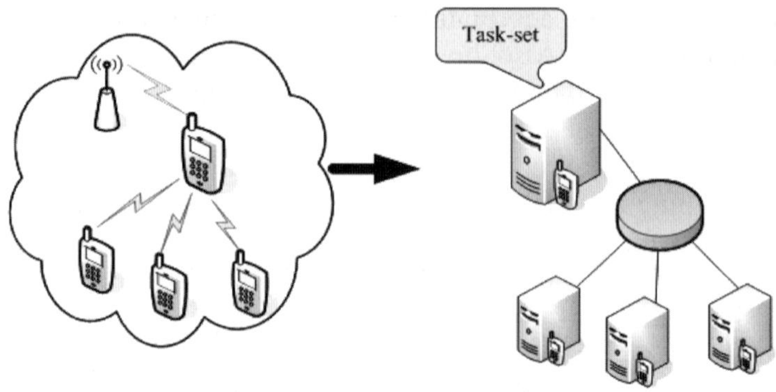

Figure 18.4. Analogy of cooperative terminals into a parallel/distributed system containing a number of computational elements connected by a given network architecture.

communication units, etc. Since no particular type of terminals is targeted for this framework, terminals are from an energy point limited to contain processing and communication units, where the later will be described in the following subsection.

For battery powered systems, methods for energy reduction have been an intensive research topic within the last decade. Dynamic Power Management (DPM) strategies in particular are widely used to overcome the poor advances in battery technology, which continuously is a significant design bottleneck. In [Benini and de Micheli, 1998] various DPM methods at different system levels are presented. On system level, the common method is to place system components in power-down modes, whenever they are inactive. The recent advantages in embedded programmable processors are the support for dynamic changes to the clock frequency and supply voltage. The DPM technology is also referred to as speed scaling. The technology is adopted by the majority of the large manufactures on the processor marked, *e.g.*, Intel with their SpeedStep® and AMD's PowerNow!™. Speed scaling uses information from software in order to decide the speed level of the system, a methodology also known as Dynamic Voltage Scaling (DVS).

The reason why DVS is efficient is because it utilizes the physics of the CMOS power relation, which today is the most common implementation technology for embedded systems. CMOS power dissipation is mainly determined by what is known as dynamic power dissipation:

$$P_{dynamic} = K \cdot C_L \cdot V_{dd}^2 \cdot f, \tag{18.1}$$

where K is the activity factor of the circuitry, C_L the equivalent load capacitance, V_{dd} the supply voltage, and f is the clock frequency. Another important physical parameter is the circuit delay, which determines the obtainable frequency of the unit:

$$T_{delay} = k\frac{V_{dd}}{(V_{dd} - V_{th})^{\alpha}} \propto \frac{1}{V_{dd}} \tag{18.2}$$

where k and α are technology constants, and V_{th} is the threshold voltage. The time delay is typically considered proportional to the inverse of the supply voltage, meaning that the lower V_{dd}, the higher delay and thereby the lower obtainable clock frequency. Thereby, becomes power dissipation a function on frequency and eventually the required performance demand. Utilizing the CMOS physics and information about the required performance needs from the task-set executed on the system, is what makes DVS a very energy effective way of executing workload. Traditionally when scheduling task-sets idle times will occur, because of various timing patterns for the individual tasks and for a traditional system, energy controlled by DPM, this idle time is used to place the system in low power modes. DVS on the other hand utilizes this idle time in order to prolong task execution, if allowed, and thereby reduce the over all energy consumption for executing the task-set. The golden rule in DVS methodologies is to run the system as slow as possible, without violating task-set timing constraints. DVS methods can be divided into two categories, as done by [Kim et al., 2002b], Inter-task and Intra-task methods. The Intra-task methods use static analysis and typical graph analysis methods, providing voltage frequency scaling points for the task. Inter-task methods are a more dynamic approach, also described as a task-by-task method, typically an extension of traditional scheduling algorithms like; Fixed Priority (FP) and Earliest Deadline First (EDF), where representative methods can be found in [Shin et al., 2000], [Pillai and Shin, 2002], [Aydin et al., 2001], and [Kim et al., 2002a]. Of the two DVS categories Inter-task DVS are the most investigated, and also most practical. Also, DVS methods for single processor systems are mostly investigated, but multi-processor variations are receiving more interest, also as multi-processor systems are the current trend in processor development.

To get an understanding of task scaling, an example is illustrated in Figure 18.5 (the bobble). The task is specified by 1) arrival time, 2) deadline time, and 3) workload. The overall idea is to utilize the task period by adjusting the speed of the processor and thereby gaining energy saving. Also in Figure 18.5 a simple example of a task–set is illustrated, containing two tasks with periodic arrival frequency. Without DVS, slack time is introduced and often handled as an idle task. The idea of the Inter–task DVS methods is to convert the slack time into energy saving by lowering the processor speed. Stretching or prolonging the execution time of some of the task, while maintaining the overall task-set timing constraints. This is illustrated in Figure 18.5 by the DVS scheduling policy proposed by [Shin et al., 2000].

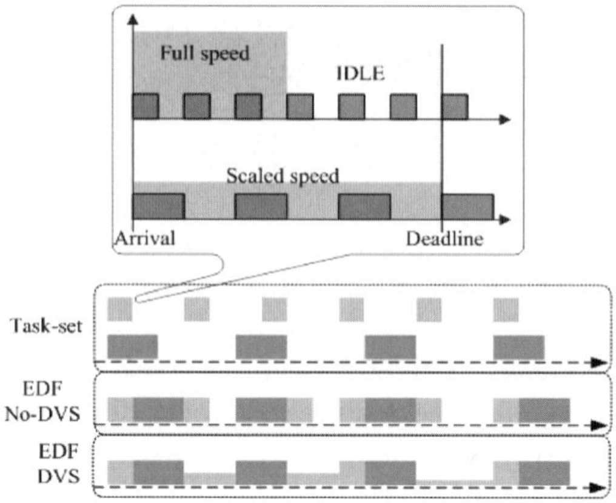

Figure 18.5. A task-set of two tasks with a periodic arrival. A traditional schedule indicating that slack time is introduced by the tasks-set, implying idle time for the processing unit. A DVS schedule stretches the execution time of the task by speed scaling, utilizing available time. In the bubble: a single task defined by its parameters, showing that task is utilized by scaling the speed.

Short range network. Wireless communication is often considered to account for a large amount of the power budget in a mobile device. Therefore, as wireless communication facilities are used to distribute workload, careful considerations of this overhead must be made. From known technologies like Bluetooth there is a support of adjustable power emission on the air interface according to distance. A technology like Wireless LAN is typically not energy aware and therefore a poor candidate. Hence, a definition of short range technology, where due to limited coverage range tolerable energy consumption is assumed.

For wireless communication systems various energy aware protocols are proposed, as surveyed by [Jones et al., 2001]. Energy conservation is typically considered at the physical layer, where a considerable amount of work is made. However, work at higher protocol layers is getting more focus, with the principle of controlling behavior patterns for the network. Also in [Jones et al., 2001] methods for data link, network, transport, OS/middleware, and application protocol layers are surveyed and discussed. From a power management point the method of placing the system in power modes is used, whereas dynamic methods or changes to transmission schemes are less seen.

When distributing tasks both parties in the reception are suffering from energy consumption. In Figure 18.6 an illustration of workload transmission is made. It is assumed that the terminal distributing workload, noted the local,

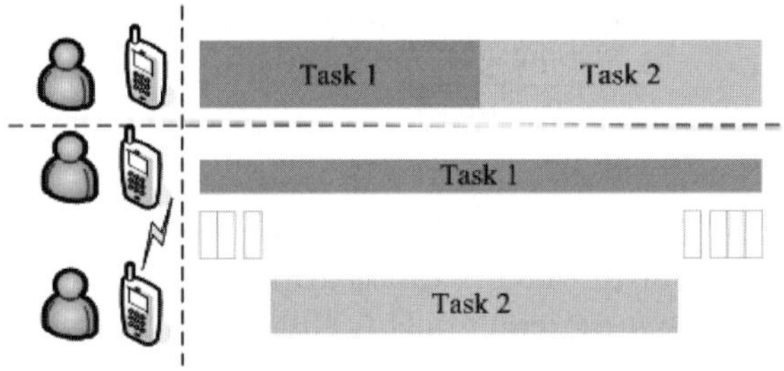

Figure 18.6. Workload distribution, showing overheads introduced by the communication and also the potential decreasing scaling potential on the remote terminal.

also is capable to execute workload while transmitting. On the network the overhead is mainly determined by the time it takes to transmit (active time), the function of the network link capacity and also on the level of power used to establish the link. Also from a time perspective the active time overhead will affect the scaling factor on the receiving terminal, noted the remote, decreasing the potential energy savings as illustrated in Figure 18.6.

Energy aware task allocation. The allocation of tasks is based on cost models for executing task 1) local, 2) remote, and 3) transmitting the task. As the energy cost of distributing tasks can exceed a selfish execution, it is important that the schedule is based on energy cost models. The cost model parameters are the task-set specification (T), the tasks of T (τ_i) and the energy operators of both the terminals and network devices, $E_x(\cdot)$; where $E_x(\cdot)$ for the terminals is convex functions related to the workload and for the network device a linear relation proportional to the communication level.

The overall distribution mechanism is divided into a number of functionalities. First, the energy cost of executing the entire task-set at the local terminal without cooperation:

$$E_{NoCoop} = E\left(T\right) \tag{18.3}$$

In the cooperative scenario it must be evaluated which and how many of the tasks that have to distributed:

$$E_{Coop} = E_L\left(\underset{i:1\rightarrow g}{\tau_i}\right) + E_{\mathrm{R}}\left(\underset{i:g+1\rightarrow n}{\tau_i}\right) + E_N\left(\underset{i:g+1\rightarrow n}{\tau_i}\right) \tag{18.4}$$

where the energy operators E_L, E_R, E_N are the energy at 1) the local terminal, 2) the remote terminals, and 3) on the wireless network device, respectively. The energy consumption for executing some of the tasks locally and some remotely, together with the overhead of transmitting the tasks to and from the remote terminal must be considered.

Distributing tasks among the local and remote terminals become a traditional optimization problem, with the challenge of finding the minimum energy consumption by placing tasks on the available terminals. The overall decision of executing all tasks locally or distributing them is a matter of choosing the cheapest energy consumption:

$$E = \min(E_{NoCoop}, E_{Coop}) \tag{18.5}$$

The runtime mechanism making the distribution decision is not within the scope of this chapter, but the potentials will be investigated based on the cost models described in the following section.

4. Modeling and Simulating Cooperative Energy Aware Computing

To be able to evaluate the potentials of energy aware task allocation for cooperative wireless networks a model using simplified but realistic assumptions are made. Summarizing system limitations, where energy is consumed by two sources, 1) the terminals and 2) the short range wireless network link. Also, workload is randomly distributable between the n terminals in the cooperative group, connected by $n-1$ communications links, as only local to remote terminal communication is considered and that workload cannot be further distributed from a remote terminal. The overall energy consumption can be described by:

$$E_{Tot} = \sum_{i=0}^{n-1} E_{PU}(i) + \sum_{i=0}^{n-2} E_{Net}(i) \tag{18.6}$$

where E_{PU} is the energy of the individual terminals, also indexed by PU as only the energy contribution from the processing units that are considered. E_{Net} is the energy cost for distributing tasks among the terminals in the cooperative group.

Terminal energy model. Using the assumption that the task-set is randomly distributable, meaning that it can evenly be distributed among n terminals, the energy at each PU can be expressed by Equation 18.7;

$$E_{PU} = E\left(\frac{U_{Task-set}}{n}\right) \tag{18.7}$$

where $U_{Task-set}$ is the utilization of the task-set, and $E(\cdot)$ is the energy operator for executing the task-set using an energy optimal execution speed. Two different models are used, 1) an idealized utilization squared (US energy model) and 2) a model based on measurements of a commercial available DSP processor from Analog Devices (AD energy model):

$$E\left(\frac{U_{Task-set}}{n}\right) = U_{PU}^2 \vee f(S|U_{PU}) \tag{18.8}$$

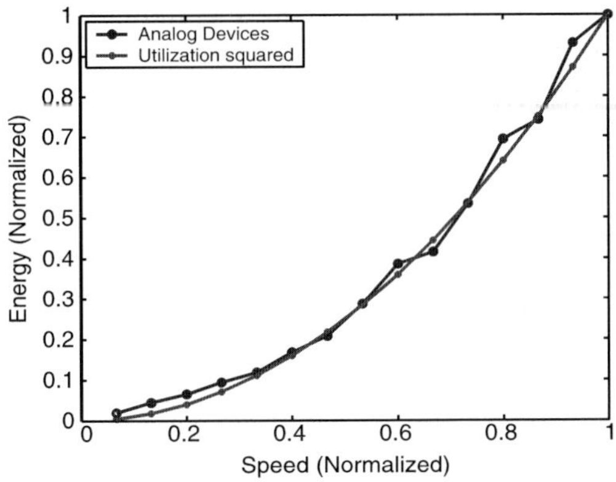

Figure 18.7. The utilization squared energy model, compared to measurements on an Analog Devices Blackfin 535 EZ-Kit Lite evaluation board. Normalized for speed and energy.

where U_{PU}^2 is the US model, a model commonly used in DVS literature and are shown to have realistic hardware relation. Secondly, the $f(S|U_{PU})$ model (AD model) expresses energy as a function of the speed required for a given task-set utilization. The speed value refers to the optimal pair of clock frequency and supply voltage for a given hardware architecture, the function is a look-up table where a given utilization refers to an energy level. In Figure 18.7 the two models are plotted, showing correlated shapes.

Network energy model. This model utilizes a high abstraction not targeting any particular networking technology, but includes parameters that overall will affect the energy consumption. Taking an outset in a simple communication relation, it is well known that communication is a relation of the network capacity, the amount of data that has to be transmitted, and the condition of the wireless link. By the assumption that a network is available for transmission at any time instance, issues like packet collision and other network interference will be neglected, as it is assumed that the network devices are able to handle the traffic on the network. Therefore, the most momentous parameters are 1) the time it takes to transmit a given task information (activity time), 2) the power consumption of the active network device, and 3) the number of tasks that have to be transmitted between the terminals. The energy consumption of transmitting information can be expressed as in Equation 18.9.

$$E_{Net} = f(t_{\tau_i} \cdot P_{Net}) \cdot \sum_{i=1:m} n_{\tau_i} , \tag{18.9}$$

where t_{τ_i} is the time it takes to transmit the τ_i'th task, P_{Net} the power consumption on the network devices, and n_{τ_i} is the number of tasks transmitted on the network (a maximum of m tasks is assumed).

The time expression of Equation 18.9 is a composition of the time it takes to establish the connection and the time it takes to transmit the information, as expressed in Equation 18.10;

$$t_{\tau_i} = t_{setup} + t_{bit} \cdot m_{\tau_i}, \tag{18.10}$$

where t_{setup} is the time for setting up the communication link, t_{bit} the time it takes to transmit a bit, and m_{τ_i} the number of bit for the data of the τ_i'th task. In this work, τ_i is used for describing the time load of the uniform task parts distributed among terminals.

Transmitting and receiving information normally comes with different cost, as expressed in Equation 18.11;

$$P_{Net} = P_T + P_R, \tag{18.11}$$

where P_T is the power for transmitting information and P_R is the power of receiving information on the network device. In this work, a superposition of the two is used. As joint energy overhead of the cooperative network is the objective, it is therefore not distinguished what the individual energy consumption of the terminals are.

Finally, a constraint on the network active time is needed, leading to the inequality in Equation 18.12;

$$\sum_{i=1}^{n-1} t_{\tau_i} \leq 1, \tag{18.12}$$

where the summation of task time overheads is less than or equal to 1, which denotes full activity on the network and has to be or less than fully utilized.

5. Effects of System Parameters

Based on the models described in the previous section various system parameters are now evaluated. The results will show the potentials but also some of the hazards of using energy aware task allocation in cooperative networks.

Ideal and cost free communication. Assuming ideal and cost free communication between terminals, optimal energy perspectives are considered. In the simulations both proposed energy models are used, providing an ideal scalable architecture, but also an example of a commercially available architecture. Where, due to hardware limitations the AD energy model will follow the speed and energy profile shown in Table 18.1.

In Figure 18.8 the results of cost free communication for n cooperative terminals are shown (where n is in the interval of one to ten terminals). As expected, the US model is a decaying relation with respect to the number of terminals. However, for the AD model it is clearly seen that the limited scaling

Table 18.1. Normalized speed values and energy consumptions for AD model, showing quantization due to hardware limitations.

Speed/# PU	1	1/2	1/3	1/4	1/5	1/6	1/7	1/8	1/9	1/10
Actual speed	1	.533	.333	.266	.2	.2	.2	.133	.133	.133
Energy	1	.287	.118	.094	.065	.065	.065	.044	.044	.044

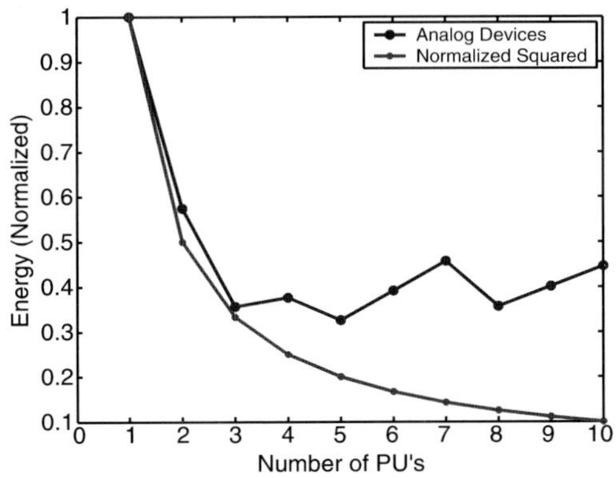

Figure 18.8. Energy consumption as a function of PU's, both illustrated for the US and the AD energy models. Showing the effect of hardware quantization.

capabilities are determining a saving floor. In Figure 18.8 it is observed that the two models correlate until 3 terminals, where above this number the effects of the hardware limitations become evident. The reason for this is the energy floor caused by the implementation technology captured in the AD model, which is not present in the more idealized US energy model. Actually, from Figure 18.7 this is clear for the lower speed settings, where the proportional deviations between the models are considerable.

For the AD model the minimum energy level is observed for a group of five members and compared to a non-cooperative energy consumption the energy gain is in the order of three times less energy used. Obviously, as ideal and cost free communication is assumed this sets the limit of achievable energy gains for the AD energy model. However, as this model is generated according to measurements on an actual available processor, it only sets the limit for this specific technology.

Task distribution. Finding the optimum ratio of workload distribution among the members in the cooperative group is likely to be determined by various system overheads. In these simulations this ratio is investigated, such that either all the workload is executed at the local or at the remote terminal(s). The simulations are conducted in a two and three terminal setup. For the three

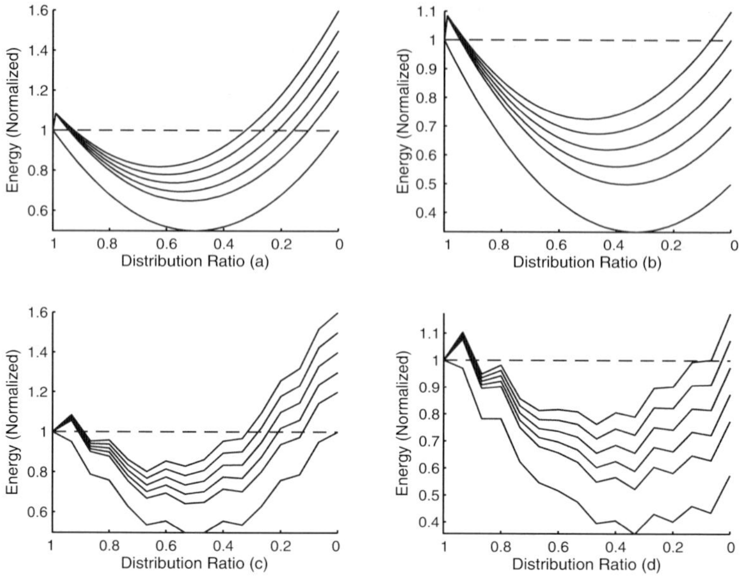

Figure 18.9. Effect of task distribution, where at the two left hand side plots are task distribution ratios on two terminals. The right hand side is task distribution ratios on three terminals. Shown at different network activity time levels.

terminal cases, the two remote terminals evenly divide the distributed workload. Therefore, if half the workload is distributed, the two remote terminals execute correspond to a quarter of the combined workload. The simulations are repeated for the two terminal energy models. Figure 18.9 shows the results, where the left hand side plots are the two terminal case, the right hand side plots are the three terminal case, and the two top plots are the US energy model. The network power level is fixed at one unit and the activity time is plotted for $\{0.0, 0.1, 0.2, 0.3, 0.4, 0.5\}$. Seen in sequence from bottom to top in each subplot. Uniquely, a small static energy contribution is introduced on the network devices in these simulations. Activating the network device is likely to introduce some energy overhead, and therefore when only a slight amount of the workload is distribution such an overhead must be noticeable. The level of this will depend on the network technology, and is randomly fixed to a 0.1 energy unit in these simulations. In each subplot a vertical dotted line indicates the energy consumption of a single terminal.

First, considering the two terminal case: When the network activity time is 0.0, meaning that communication is cost free and ideal, energy consumption is minimum when workload is evenly balanced among the terminals. Contributing to an energy saving that is half of the energy consumption compared to a selfish execution. Increasing the activity time, it becomes clear that load balancing is not resulting in minimum energy consumption. The tendency is towards a distribution ratio where more and more of the workload is executed on the local terminal. Identical trends are seen for the two energy models, whereas the AD model has a more distinct minimum point for higher activity times, due to its energy pattern.

Secondly, considering the three terminal case: Logical, when communication is cost free load balancing contributes with the minimum energy consumption, which is in the order of a factor three energy saving compared to a selfish execution. Coincident with the previous case, as activity time increases the minimum moves towards a higher ratio executed on the local terminal.

Effects of network model parameters. These experiments evaluate the effects of the network parameters' power level and activity time, where the two values are; $\{2, 1, 0.75, 0.5\}$, and $\{0.5, 0.4, 0.3, 0.25, 0.2, 0.15, 0.1, 0.05\}$, respectively. The workload is fixed to one utility unit, meaning that a single terminal will be fully utilized. In Figures 18.10 and 18.11 simulation results of the US and AD terminal energy model is shown, the four power levels are sequenced in the four subplots, and the activity times are from top to bottom also in each subplot. In particular it should be noted that only curves are shown such that Equation 18.12 is satisfied, *e.g.*, an activity factor of 0.5 only supports transmission to additional two terminals. In each plot the vertical line indicates the energy consumption of the workload executed on a single terminal.

Figures 18.10 and 18.11 make it obvious that the two network parameters have significant effects on the performance. For example, the top left subplot, where the power level is two, double the energy level for a terminal executing its maximum workload capacity, energy gains are possible only when network activity time is insignificant. Whereas, in the bottom right subplot, energy gains are significantly improved, becoming close to the energy minimum, as discussed in the ideal and cost free communication section. Interesting, even when the activity time on the network is simulated at a low level, the plots indicate that the number of terminals relatively quickly reach a minimum. Therefore, obviously, the optimal number of terminals is determined by these two parameters, but it is also clear that the number of terminals is limited to only a few. Specific terminals are limited to 6 for the setup in the lower right sub-plot of Figure 18.10. Contrary to this, the AD model indicate that the optimal number of terminals never exceed three, as indicated in Figure 18.11. This is again limited by the scaling capabilities of the specific hardware.

The effect of task-set utilization. At last the workload utilization factor is investigated, where the utilization factor is a measure of how many instructions per time unit a terminal is loaded with. In Figure 18.12, similar plots as for the previous experiment are shown, using identical activity time values and

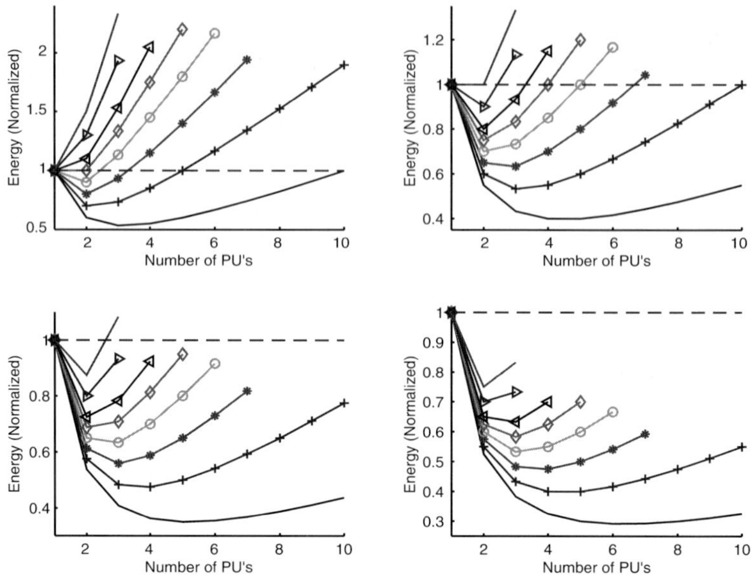

Figure 18.10. Network power and active time shown as a function of terminals and using the US energy model. Power shown in values {2, 1, 0.75, 0.5} seen in sequence from top left to bottom right. Task time load in values {0.5, 0.4, 0.3, 0.25, 0.2, 0.15, 0.1, 0.05} seen from top to bottom in each plot.

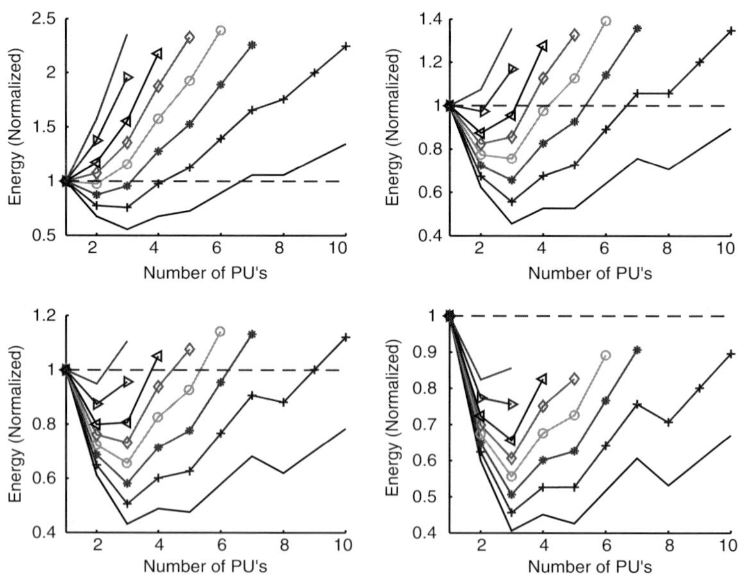

Figure 18.11. Network power and activity time parameters shown as a function of terminals and using the AD energy model. With similar parameter set-up as Figure 18.10.

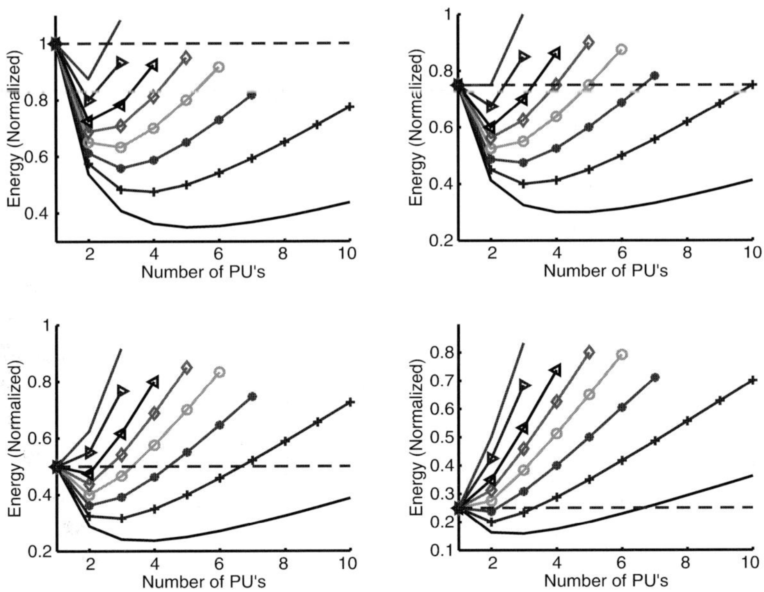

Figure 18.12. Task-set utilization effect on similar task network load time as showed in Figure 18.10 and showed for network power 0.75.

a network power level of 0.75. Workload utilizations of {1, 0.75, 0.5, 0,25} are illustrated, with a utilization of 1 in the left top sub-plot succeeded to a utilization of 0.25 in the bottom right sub-plot. In this experiment only the US terminal energy model is shown, as the AD model is having relative similar patterns.

Comparing the energy level from a single terminal to the four different utilization levels, it becomes clear that task-set utilization is a significant factor. For example, at the lowest utilization factor only limited activity times on the network enables workload profits, whereas increased utilization level allow for increased overheads on the network device. Intuitively it is also obvious, since the proportional speed scaling gain at low workloads are much less than for high workloads. Hence, from an energy perspective it is not worth distributing a small amount of workload, if as a starting point the workload is at a low level.

6. Summary

Motivated by the fact that energy is a limited resource for portable battery powered devices has brought about the concept of ener gy aware task allocation in cooperative wireless networks. The concept of cooperation is utilized as the methodology for motivating wireless network terminals to accept workload for execution from other members of the cooperative group. Terminals joining a cooperative group have to benefit with respect to a given metric, which for this work has been energy consumption. For the individual terminals, energy gains are achieved as an average over a given time period, whereas the cooperative group gain an energy saving instantaneous. Hence, terminals having limited workload are providing a task execution service to other members of the cooperative network, anticipating that similar service is returned. Technical challenges of making the cooperation fair are not targeted in this work, some non-trivial accounting and also service discovery methodologies have to be introduced.

The proposed scheme uses an abstraction of multiple processor architecture, which is connected by an arbitrary short range wireless technology. Terminals are abstracted into processing units, connected by a single hop or single tree-branch like communication network. The processing units are energy controlled by energy management, where the superior method of dynamic voltage scaling is utilized for energy conservation. Dynamic voltage scaling make benefit of the inherent convex shaped energy consumption, which is related to joint scaling of supply voltage and clock frequency. By this, system performance is scaled according to current workload, making energy efficient schedules of the workload. Modeling energy consumption of the processing units and also the interlinking wireless network, potentials of cooperative energy aware task execution is simulated.

From these simulations the following lessons are learned:

- When communication cost is disregarded, the optimal number of cooperative members is depended on the energy profile of the system. In the simulations two different models are utilized, where one is an ideal model following the energy function for CMOS technology, whereas the other is a model based on real measurements from a commercial DSP processor. Observations indicate that for the ideal model the more members the better, where the measured model show that five cooperative members are resulting in the minimum energy consumption. For the later case, factor three times the energy is saved in the optimum case, which obviously only related for this specific hardware architecture.

- Considerations of the ratio of workload executed at the various terminals, related to overheads on the network are simulated. From traditional multi-processor scheduling it is generally accepted that load balancing is the superior way of distributing workload, when communication overheads are not considered. The simulations show that load balancing is not optimal for network overheads, whereas a comparative ratio must be executed on the terminal distributing the workload.

- Not surprisingly, overheads introduced by distributing the task over the network are a factor that has to be taken into account. If ratios between the energy consumed on the processing units and the network are not proportionally feasible no energy savings are introduced by the proposed scheme, increased additionally energy consumption can be introduced. However, when these network overheads are proportionally beneficial, energy savings of a factor three is possible.

- Finally, the task-set utilization is a significant factor, together with the network overheads. Obviously, for workloads with a small utilization of the processor, the ratios of the network overheads also become increased and therefore quickly result in increased energy consumption. Whereas, workloads that highly utilize the processor also have more potential for energy savings and therefore also support higher overheads on the network.

Obviously, a scheme of this nature will have a number of advantages and disadvantages. Mainly, overheads due to distribution of tasks must be wisely considered in relation to the energy advantages introduced by off-loading workload. Therefore, eventually, the mechanism making the runtime distribution decision must be able to evaluate based on cost functions, similar to those proposed here. Deciding if workload should be distributed, the number of terminals within the cooperative group, and finally also the amount of workload targeted to the individual terminals.

References

Alsalih, W., Akl, S. G., and Hassanein, H. S. (2005). Energy-aware task scheduling: Towards enabling mobile computing over manets. In *19th International Parallel and Distributed Processing Symposium (IPDPS'05)*, Denver, USA.

Aydin, H., Melhem, R., Mosse, D., and Alvarez, P. M. (2001). Dynamic and aggressive scheduling techniques for power-aware real-time systems. In *Proceedings of the 22nd IEEE Real-Time Systems Symposium (RTSS'01)*, pages 95–105, Austin, TX, USA.

Benini, L. and de Micheli, G. (1998). *Dynamic Power Management: Design Techniques and CAD Tools*. Kluwer Academic Publishers.

Delaney, B., Jayant, N., and Simunic, T. (2005). Energy-aware distributed speech recognition for wireless mobile devices. *Design & Test of Computers*, 22(1):39–49.

Jones, C. E., Sivalingam, K. M., Agrawal, P., and Chen, J. (2001). A survey of energy efficient network protocols for wireless networks. *Wireless Networks*, 7(4):343–358.

Kim, W., Kim, J., and Min, S. L. (2002a). A dynamic voltage scaling algorithm for dynamic-priority hard real-time systems using slack time analysis. In *Proceedings of the Design Automation and Test in Europe (DATE'02)*, pages 788–794, Paris, France.

Kim, W., Shin, D., Yun, H. S., Kim, J., and Min, S. L. (2002b). Performance comparison of dynamic voltage scaling algorithms for hard real-time systems.

In *Proceedings of the Eighth IEEE Real-Time and Embedded Technology and Applications Symposium (RTAS'02)*, pages 219–228, San Jose, USA.

Liu, C. L. and Layland, J. W. (1973). Scheduling algorithms for multiprogramming in a hard-real-time environment. *J. ACM*, 20(1):46–61.

Lu, X., Hassanein, H., and Akl, S. (2005). Energy aware dynamic task allocation in mobile ad hoc networks. In *International Conference on Wireless Networks, Communications, and Mobile Computing (WirelessCOM'05)*, Maui, Hawaii, USA.

McKnight, Lee W., Howison, James, and Bradner, Scott (2004). Guest editors' introduction: Wireless grids–distributed resource sharing by mobile, nomadic, and fixed devices. *IEEE Internet Computing*, 8(4):24–31.

Nosratinia, A., Hunter, T. E., and Hedayat, A. (2004). Cooperative communication in wireless networks. *Communications Magazine*, 42(10):74–80.

Olsen, A. B., Fitzek, F. H. P., and Koch, P. (2005a). Energy aware computing in cooperative wireless networks. In *International Conference on Wireless Networks, Communications, and Mobile Computing (WirelessCOM'05)*, Maui, Hawaii, USA.

Olsen, A. B., Fitzek, F. H. P., and Koch, P. (2005b). Evaluation of cooperative task computing for energy aware wireless networks. In *International Workshop on Wireless Ad-hoc Networks (IWWAN'05)*, London, England.

Olsen, A. B., Fitzek, F. H. P., and Koch, P. (2005c). Optimizing the number of cooperating terminals for energy aware task computing in wireless networks. In *Wireless Personal Multimedia Communications (WPMC'05) Symposia*, Aalborg, Denmark.

Pillai, P. and Shin, K. G. (2002). Real-time dynamic voltage scaling for low-power embedded operating systems. In *Operating Systems Review (ACM), 18th ACM Symposium on Operating Systems Principles (SOSP'01)*, pages 89–102, Banff, Alberta, Canada.

Politis, C., Oda, T., Dixit, S., Schieder, A., Lach, K.-Y., Smirnov, M. I., Uskela, S., and Tafazolli, R. (2004). Cooperative networks for the future wireless world. *Communications Magazine*, 42(9):70–79.

Shin, Y., Choi, K., and Sakurai, T. (2000). Power optimization of real-time embedded systems on variable speed processors. In *Proceedings of International Conference on Computer-Aided Design (ICCAD)'00*, pages 365–368, San Jose, USA.

Shirazi, B. A., Kavi, K. M., and Hurson, Ali R., editors (1995). *Scheduling and Load Balancing in Parallel and Distributed Systems*. IEEE Computer Society Press, Los Alamitos, CA, USA.

Yu, Y. and Prasanna, V. K. (2002). Power-aware resource allocation for independent tasks in heterogeneous real-time systems. In *ICPADS '02: Proceedings of the 9th International Conference on Parallel and Distributed Systems*, page 341, Washington, DC, USA. IEEE Computer Society.

Yu, Y. and Prasanna, V. K. (2003). Energy-balanced task allocation for collaborative processing in networked embedded systems. In *LCTES '03: Proceedings of the 2003 ACM SIGPLAN conference on Language, compiler, and tool for embedded systems*, pages 265–274, San Diego, California, USA.

Yu, Y. and Prasanna, V. K. (2005). Energy-balanced task allocation for collaborative processing in wireless sensor networks. *Mob. Netw. Appl.*, 10(1-2):115–131.

Zhu, D., Melhem, R., and Childers, B. R. (2003). Scheduling with dynamic voltage/speed adjustment using slack reclamation in multiprocessor real-time systems. *IEEE Trans. Parallel Distrib. Syst.*, 14(7):686–700.

Chapter 19

COOPERATIVE CODING AND ITS APPLICATION TO OFDM SYSTEMS

Jerry C. H. Lin
Polytechnic University
clin17@poly.edu

Andrej Stefanov
Polytechnic University
stefanov@poly.edu

Abstract: We study OFDM systems with cooperative coding over frequency selective Rayleigh fading channels. We derive the pairwise error probability for the block-fading OFDM channel model. We use the derived pairwise error probability to get an upper bound on the frame error probability for the coded cooperative OFDM system. This bound is then utilized in the study of the diversity and coding gains achievable through cooperative coding in OFDM systems for various inter-user channel qualities. We consider the design of cooperative convolutional codes based on the principle of overlays and provide simulation results for different cooperation scenarios. We observe significant gains over conventional non-cooperative OFDM systems.

Keywords: diversity methods, error–correction coding, fading channels.

1. Introduction

Information transfer through wireless networks involves simultaneous communication among multiple source–destination pairs. Wireless local area networks may operate in infrastructure mode or as ad-hoc networks. In the infrastructure mode the coordination of these multiple communications is done

Frank H. P. Fitzek and Marcos D. Katz (eds.), Cooperation in Wireless Networks: Principles and Applications, 593–605.

via the access point. The access point processes all the signals transmitted from the sources (uplink) and forwards them to their respective destinations (downlink). In the ad–hoc mode on the other hand there is no fixed infrastructure and the terminals utilize other terminals as relays to transfer information from the source to its destination. Motivated by the diversity effects and power efficiency of communicating via relaying, recent research efforts have focused on cooperation among the terminals in the network (user-cooperation), demonstrating the advantages of user-cooperation regardless of the mode of the network operation.

In a cooperative network, two or more terminals share their information and transmit jointly as a virtual antenna array. This enables them to obtain higher data rates and it leads to decreased sensitivity to channel variations [Sendonaris et al., 2003]. The terminals share information by tuning into each other's transmitted signals and by processing the information they overhear through the inter-user channel. The cooperation still leads to performance improvements over single user transmission, even though the inter-user channel may be faded and noisy. The fact, that in practice, the relaying terminal cannot receive and transmit at the same time was incorporated in [Laneman et al., 2004], where the authors considered different protocols to achieve diversity gains such as amplify and forward or decode and forward. From a coding perspective these protocols resemble repetition coding, and there are more effective ways of designing channel codes.

In [Stefanov and Erkip, 2004] we demonstrated that an overall block fading channel model is appropriate in the case of user-cooperation, since the cooperating terminals observe independently faded channels towards the destination. This resulted in a framework for the design of cooperative channel codes optimized for user-cooperation. As the next generation of wireless local area networks (WLAN's) and cellular systems will utilize Orthogonal Frequency Division Multiplexing (OFDM) [Nee and Prasad, 2000], it is necessary to consider the analysis and design of cooperative codes in the context of OFDM systems.

2. System Model

We consider an OFDM system with K subcarriers. Each code word spans P adjacent OFDM words, and each OFDM word consists of K symbols, transmitted simultaneously during one time slot. Each symbol is transmitted at a particular OFDM subcarrier. We assume that the fading is quasi-static during each OFDM word, but varies independently from one OFDM word to another.

At the receiver, the received signal can be expressed in the frequency domain as follows

$$y[p,k] = H[p,k]c[p,k] + z[p,k] \tag{19.1}$$

where $k = 0, \ldots, K-1, p = 1, \ldots, P$, and $H[p,k]$ is the complex channel frequency response at the kth subcarrier and at the pth time slot. $c[p,k]$ and $y[p,k]$ are the transmitted signal and the received signal, respectively, at the kth

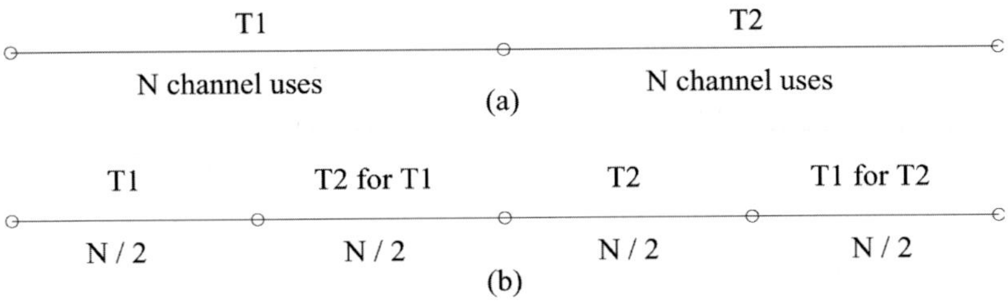

Figure 19.1. Time-division channel allocations: (a) orthogonal direct transmission and (b) orthogonal cooperative diversity transmission.

subcarrier and at the pth time slot. $z[p, k]$ is the circularly symmetric complex Gaussian noise with variance of $N_0/2$. The time domain channel response can be expressed as

$$h(\tau) = \sum_{l=1}^{L} \alpha(l)\delta(\tau - \frac{n_l}{K\triangle_f}) \tag{19.2}$$

where $\delta(\cdot)$ is the Dirac delta function, L denotes the number of non-zero taps and $\alpha(l)$ is the complex gain of the lth non-zero tap, whose delay is $\frac{n_l}{K\triangle_f}$, where n_l is an integer and $\triangle_f$ is the tone spacing of the OFDM system.

The channel frequency response between the transmit antenna and the receive antenna at the pth time slot and at the kth subcarrier is given by [Lu et al., 2002]

$$H[p, k] = H(pT, k\triangle_f) = \sum_{l=1}^{L} \alpha(l; pT)e^{-j2\pi k n_l/K} = h^H(p)w(k) \tag{19.3}$$

where $h(p) = [\alpha(1), \ldots, \alpha(L)]^H$ is the L-sized vector containing the time responses of all the non-zero taps and $w(k) = [e^{-j2\pi k n_1/K}, \ldots, e^{-j2\pi k n_L/K}]^T$ contains the corresponding DFT coefficients.

We adopt a time-sharing cooperative scheme similar to that of [Laneman et al., 2004; Stefanov and Erkip, 2004; Hunter and Nosratinia, 2002], as illustrated in Figure 19.1. Terminal T_1 transmits the first half of its codeword to the destination and T_2. If T_2 is able to decode it correctly, it then transmits the second half of T_1's codeword to the destination. If T_2 fails to decode it correctly, it notifies T_1 and T_1 then transmits the rest of the codeword itself. In the next transmission, the role of T_1 and T_2 are interchanged.

3. Performance Analysis of Coded Cooperative OFDM Systems

In this section, we analyze the performance of coded cooperative OFDM systems. We first derive the Chernoff bound on the pairwise error probability of the block fading OFDM channel, resulting from cooperation. We then utilize the pairwise error probability in the analysis of the frame error probability of coded cooperative systems. In particular, we study the achievable diversity order for various inter-user channel qualities.

Pairwise Error Probability for Block Fading OFDM Systems

In the block fading OFDM channel model resulting from cooperation, each block may have a different received signal-to-noise ratio and different number of non-zero channel taps. The user T_i-destination channels have L_i non-zero taps, $i = 1, 2$, respectively. Assuming that perfect channel state information is available at the receiver and by applying the Chernoff bound, the pairwise error probability (PEP) of transmitting codeword $\boldsymbol{c}$, while another codeword $\boldsymbol{e}$ is decoded at the receiver, is upper bounded by

$$P(\boldsymbol{c} \to \boldsymbol{e}|H) \leq \exp\{-(d_1^2(\boldsymbol{c}, \boldsymbol{e})\gamma_1 + d_2^2(\boldsymbol{c}, \boldsymbol{e})\gamma_2)\} \tag{19.4}$$

where $\gamma_i = \frac{E_{s_i}}{N_0}$, $i = 1, 2$, denotes the signal-to-noise ratio for the first half and second half of the channel codeword, respectively. $d_i^2(\boldsymbol{c}, \boldsymbol{e})$, $i = 1, 2$, can be expressed as

$$d_i^2(\boldsymbol{c}, \boldsymbol{e}) = \sum_{k=0}^{K-1} \sum_{p=(i-1)P/2+1}^{iP/2} |H_i[p, k]\epsilon[p, k]|^2 = h_i^H D_i h_i \tag{19.5}$$

where, $\epsilon[p, k]_{1\times1} = c[p, k] - e[p, k]$, and

$$D_{i_{L_i \times L_i}} = \sum_{k=0}^{K-1} \sum_{p=(i-1)P/2+1}^{iP/2} w_i(k)\epsilon[p, k]\epsilon^*[p, k]w_i^H(k) \quad i = 1, 2. \tag{19.6}$$

Note that $\epsilon[p, k]\epsilon^*[p, k]$ equals to 0 if the entries of codeword $\boldsymbol{c}$ and $\boldsymbol{e}$ corresponding to the kth subcarrier and the pth time slot are the same. Let $D_{(1)}$ denote the number of instances when $\epsilon[p, k]\epsilon^*[p, k] \neq 0$, $p = 1, \ldots, P/2$, $\forall k$; and let $D_{1_{eff}}$ denote the minimum $D_{(1)}$ over every possible pair of codewords [Lu et al., 2002; Schlegel and Costello, 1989]. Denoting $r_1 = rank(D_1)$, it follows that $\min_{c,e} r_1 \leq \min\{D_{1_{eff}}, L_1\}$. Similarly, $\min_{c,e} r_2 \leq \min\{D_{2_{eff}}, L_2\}$. We observe that D_1 and D_2 are non-negative definite Hermitian matrices. Hence, by an eigen-decomposition, we obtain

$$D_1 = V_1 \Lambda V_1^H \quad D_2 = V_2 \Phi V_2^H \tag{19.7}$$

where V_1 and V_2 are unitary matrices, while Λ and Φ are diagonal matrices with $\{\lambda_j\}_{j=1}^{r_1}$ and $\{\phi_j\}_{j=1}^{r_2}$ being positive eigenvalues of D_1 and D_2, respectively. All the L_1 elements, $\alpha_1(1), \ldots, \alpha_1(L_1)$, of $\{h_1\}$, and the L_2 elements,

$\alpha_2(1), \ldots, \alpha_2(L_2)$, of $\{h_2\}$, are assumed to be i.i.d. circularly symmetric complex Gaussian with zero means. Eq. (19.4) can be written as

$$P(\mathbf{c} \to \mathbf{e}|H) \leq \exp\left\{-\left(\gamma_1 \sum_{j=1}^{r_1} \lambda_j |\beta(j)|^2 + \gamma_2 \sum_{j=1}^{r_2} \phi_j |\kappa(j)|^2\right)\right\} \quad (19.8)$$

where $\beta(j) = [V_1^H h_1]_j$ and $\kappa(j) = [V_2^H h_2]_j$. Since V_1 and V_2 are unitary, $\beta(j)$ and $\kappa(j)$ are also i.i.d. circularly symmetric complex Gaussian with zero mean and their magnitudes $|\beta(j)|$ and $|\kappa(j)|$ are i.i.d. Rayleigh distributed. By averaging the conditional PEP over the Rayleigh distribution, the pairwise error probability is found to be

$$P(\mathbf{c} \to \mathbf{e}) \leq \left(\prod_{j=1}^{r_1} \lambda_j \prod_{j=1}^{r_2} \phi_j\right)^{-1} \gamma_1^{-r_1} \gamma_2^{-r_2} \quad (19.9)$$

where r_1 and r_2 are the diversity levels with maximum of L_1 and L_2, respectively. We observe that in the block fading model resulting from cooperation in an OFDM system, each block may have a different received signal-to-noise ratio and different number of nonzero channel taps. For the case when $\gamma_1 = \gamma_2 = \gamma$, the pairwise error probability expression simplifies to

$$P(\mathbf{c} \to \mathbf{e}) \leq \left(\prod_{j=1}^{r_1} \lambda_j \prod_{j=1}^{r_2} \phi_j\right)^{-1} \gamma^{-(r_1+r_2)}. \quad (19.10)$$

Note that the quasi-static fading case [Lu et al., 2002], can be readily obtained as a special case of the block fading OFDM model.

Frame Error Probability Analysis

Without loss of generality, we study the cooperative coding performance gains from the perspective of node T_1. Similar results would also be obtained for node T_2. The frame error probability (FEP) can be obtained as

$$P_f^{coop} = (1 - P_f^{in})P_f^{BF} + P_f^{in} P_f^{QS} \leq P_f^{BF} + P_f^{in} P_f^{QS} \quad (19.11)$$

where P_f^{in} denotes the FEP of the first half codeword over the inter-user channel, P_f^{BF} denotes the FEP over the block fading channel when the cooperation takes place, and P_f^{QS} denotes the frame error probability over the quasi–static fading T_1–destination channel which the destination observes if T_2 cannot decode T_1. Let γ_1 denote the received signal-to-noise ratio at the destination corresponding to the transmission from user T_1. Similarly, let γ_2 denote the received signal-to-noise ratio at the destination corresponding to the transmission from user T_2

and γ_{in} denote the received signal-to-noise ratio at user T_2 corresponding to the transmission from user T_1.

Utilizing the pairwise error probability for the block Rayleigh fading OFDM channel derived in the previous section and the union upper bound on the frame error probability, when node T_1 transmits in cooperation with node T_2, the upper bound on the frame error probability, P_f^{coop}, is

$$\begin{aligned} P_f^{coop} &\leq \left(\sum_{\mathbf{c}} \sum_{\mathbf{e} \neq \mathbf{c}} \frac{1}{\left(\prod_{b=1}^{2} \mu_b\right) \gamma_1^{r_1} \gamma_2^{r_2}} \right) \\ &+ \left(\sum_{\mathbf{c}} \sum_{\mathbf{e} \neq \mathbf{c}} \frac{1}{\left(\prod_{j=1}^{r_{in}} \delta_j\right) \gamma_{in}^{r_{in}}} \right) \\ &\quad \left(\sum_{\mathbf{c}} \sum_{\mathbf{e} \neq \mathbf{c}} \frac{1}{\left(\prod_{j=1}^{r} \xi_j\right) \gamma_1^{r}} \right) \end{aligned}$$

where r_b denotes the rank of the codeword difference matrices in the OFDM fading block b, $b = 1, 2$, and r denotes the rank of the codeword difference matrix between the two entire codewords of T_1. The μ_b's, $b = 1, 2$ are given by $\mu_1 = \prod_{i=1}^{r_1} \lambda_i$ and $\mu_2 = \prod_{i=1}^{r_2} \phi_i$, where the λ_i's and the ϕ_i's denote the nonzero eigenvalues of the product of the codeword difference matrix and its respective conjugate transpose for the OFDM fading block $b = 1$ and $b = 2$, respectively. The ξ_i's denote the nonzero eigenvalues of the product of the codeword difference matrix between the two entire codewords and its conjugate transpose. Similarly, the δ_i's denote the r_{in} nonzero eigenvalues of the product of the codeword difference matrix for the first half of the codeword and its conjugate transpose, utilized over the inter-user OFDM channel.

We consider the case when, $\gamma_1 \approx \gamma_2 \approx \gamma_{in} = \gamma$, that is all channels, including the inter–user channel, have similar quality. This assumption simplifies the diversity analysis and is quite reasonable at high signal-to-noise ratios in all channels. In this case, P_f^{coop}, can be approximately upper bounded by

$$\begin{aligned} P_f^{coop} &\leq \gamma^{-(L_1+L_2)} \left(\sum_{\mathbf{c}} \sum_{\mathbf{e} \neq \mathbf{c}} \frac{1}{\prod_{b=1}^{2} \mu_b} \right) \\ &+ \gamma^{-(L_1+L_{in})} \left(\sum_{\mathbf{c}} \sum_{\mathbf{e} \neq \mathbf{c}} \frac{1}{\prod_{j=1}^{L_{in}} \delta_j} \right) \left(\sum_{\mathbf{c}} \sum_{\mathbf{e} \neq \mathbf{e}} \frac{1}{\prod_{j=1}^{L_1} \xi_j} \right) \end{aligned}$$

Here we have assumed that all codeword difference matrices of interest are of full rank. Let $k = \min\{L_{in}, L_2\}$. At high signal-to-noise ratios, we have the following approximation

$$P_f^{coop} \approx \mathcal{K}_1 \, \gamma^{-(L_1+k)} \tag{19.12}$$

where the term $\mathcal{K}_1$ represents the coding parameters. This means that when all links have the same average quality, the diversity order achieved through cooperative coding depends on $k = \min\{L_2, L_{in}\}$, as indicated by the exponent of the signal-to-noise ratio. Note, that this diversity level is also indicated by the information theoretic cut-set bound [Cover and Thomas, 1991].

Good inter-user channel. Next, we focus on the case when the inter–user channel is very good, *i.e.*, it has a very high signal-to-noise ratio. This could represent the scenario when the two partners are located very close to each other. This means that P_f^{in} is small and we simply have $P_f^{coop} \approx P_f^{BF}$. Hence,

$$P_f^{coop} \approx \mathcal{K}_2 \, \gamma^{-(L_1+L_2)} \tag{19.13}$$

where $\mathcal{K}_2$ represents the coding parameters. We observe that when the inter-user channel quality is very good, the inter-user channel does not represent a bottleneck and a diversity level of $(L_1 + L_2)$ is achieved.

Poor inter-user channel. Finally, when the inter-user channel quality is poor, the inter-user channel signal-to-noise ratio, γ_{in}, will be lower than the signal-to-noise ratio of the user-destination channel. We can assume that $\gamma_{in}^{r_{in}} \leq C_{in}$, for all signal-to-noise ratios of interest. Hence, P_f^{coop} is upper bounded by the term $P_f^{in} P_f^{QS}$, yielding

$$P_f^{coop} \approx \frac{1}{C_{in}} \cdot \frac{\gamma^{-L_1}}{\min_{\mathbf{c},\mathbf{e}}\{(\prod_{j=1}^{L_{in}} \delta_j \prod_{j=1}^{L_1} \xi_j)\}} \tag{19.14}$$

where $\min_{\mathbf{c},\mathbf{e}}\{(\prod_{j=1}^{L_{in}} \delta_j \prod_{j=1}^{L_1} \xi_j)\}$ represents the dominant term in the union bound at high signal-to-noise ratios. In this case, the diversity level is only L_1. This is the same diversity level achieved by T_1 when there is no cooperation. However, there is still some coding gain, as indicated by the eigenvalue product, compared to the conventional OFDM system.

4. Simulation Results

In this section we provide numerical examples illustrating the performance of cooperative convolutional codes in OFDM systems. We assume that the OFDM system has $K = 128$ subcarriers. We consider the constraint length 7 convolutional code (133,171,117,165). This convolutional code belongs to the family of convolutional codes designed on the principle of overlays [Stefanov

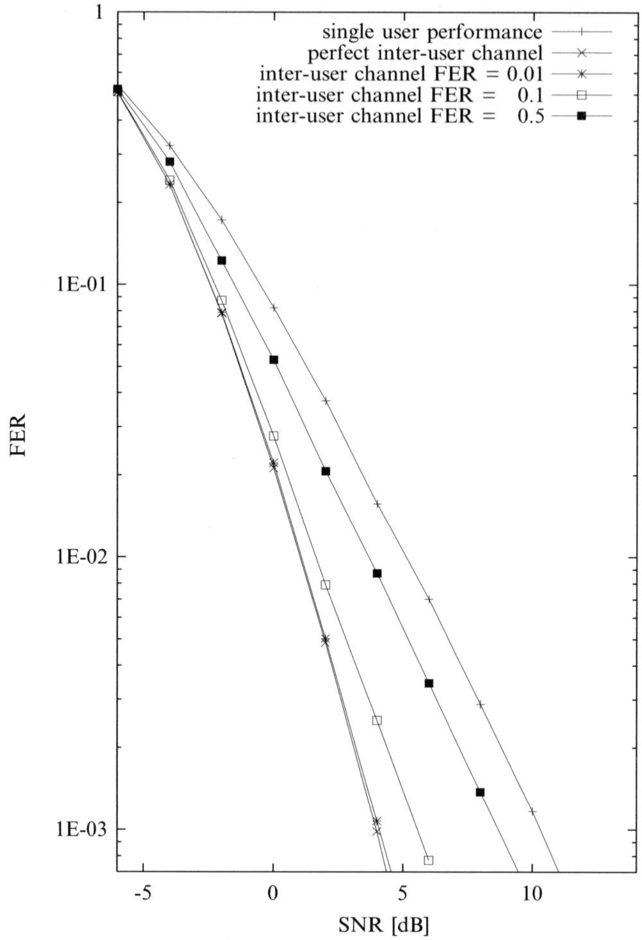

Figure 19.2. Single user performance vs. two user cooperation, for different inter-user channel qualities.

and Erkip, 2004; Gamal et al., 2003]. The coded bits are suitably multiplexed in order to achieve the maximum degree of diversity available in the cooperative communication scenario. We consider BPSK modulation. We assume maximum likelihood detection [Lin and Costello, 1983] and perfect channel state information at all the respective receivers. The frame size is 256 bits and each codeword spans $P = 2$ OFDM words. Both user-destination channels are assumed to have two taps, namely, $L_1 = L_2 = 2$.

Figure 19.2 illustrates the frame error rate (FER) performance comparison between the non-cooperative case and the cooperative case for different inter-user channel qualities. Both user-destination channels have similar quality. We observe that when the inter-user channel quality is very good, we achieve full diversity. The gain over the single user performance is about 6.5 dB at a FER of 10^{-3}. We note that even in the case when the inter-user channel FER is 0.5, we still obtain about 2 dB improvement at a FER of 10^{-3} as compared to the non-cooperative case.

Next we consider the scenario when one of the users has much better channel to the destination than the other partner. Figure 19.3 illustrates the FER performance for both users in this asymmetric scenario. We assume that user T_1 has better channel quality to the destination, *i.e.*, its signal-to-noise ratio is fixed at 10.3 dB, resulting in a FER of 10^{-3}. We observe the performance of both users as we vary the signal-to-noise ratio of user T_2. The inter-user channel FER is 10^{-1}. From Figure 19.3 it can be observed that both users benefit from cooperation. User T_1 achieves the FER of 10^{-3} when the signal-to-noise ratio of user T_2 is

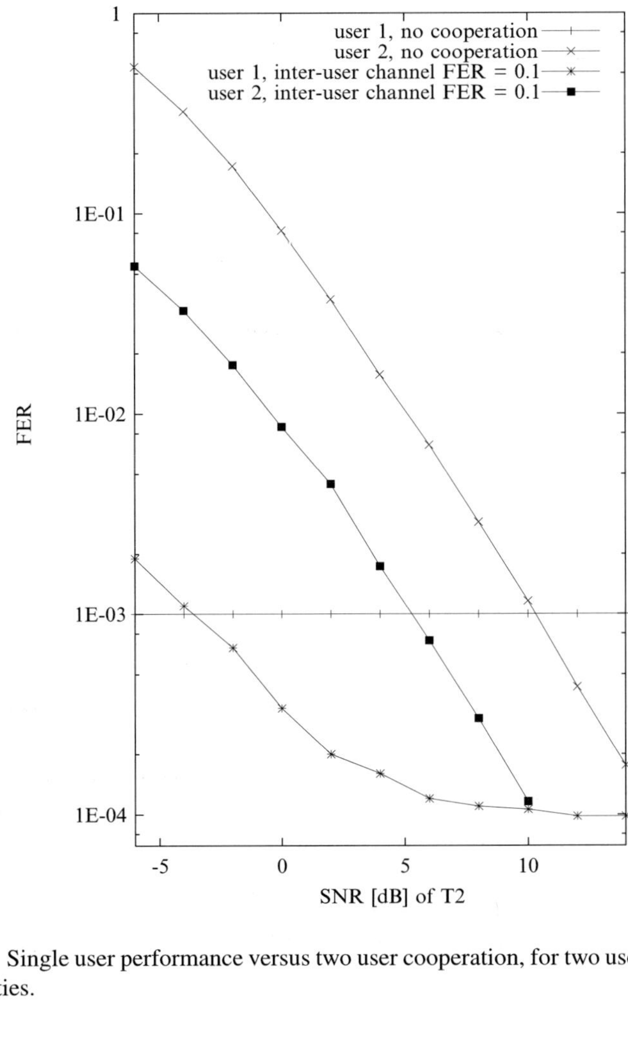

Figure 19.3. Single user performance versus two user cooperation, for two users with different channel qualities.

about -4 dB. At higher signal-to-noise ratios, its performance is even better than in the non-cooperative case. User T_2 also has significant gains, as it improves its performance by about 5 dB with respect to the non-cooperative case.

In the previous discussion, we considered the performance of the coded cooperative OFDM system for various inter-user channel qualities. Next, we will consider the case where the signal-to-noise ratio in the inter-user channel varies in proportion with the signal-to-noise ratio in the user-destination channel. Figure 19.4 represents the scenario when the inter-user channel has $L_{in} = 2$ taps. We consider two cases for the signal-to-noise ratio in the inter-user channel. In the first case the signal-to-noise ratio in the inter-user channel is approximately the same with the signal-to-noise ratios in the user-destination channels, *i.e.*, $\gamma_{in} \approx \gamma$. This could represent the scenario when all three nodes are at a similar distance from one another. We observe that the in this scenario

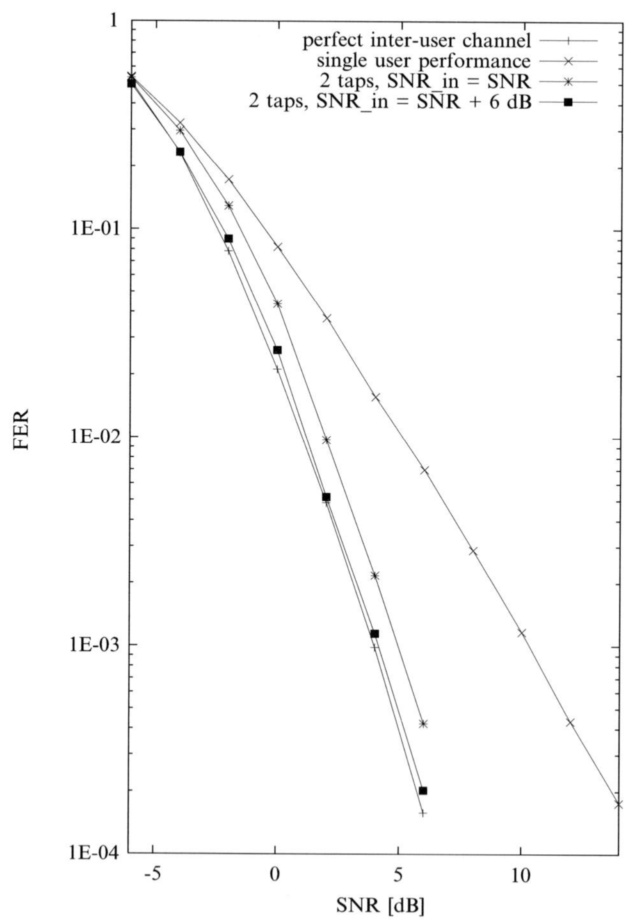

Figure 19.4. Cooperative coding example, the inter-user channel has two taps, $L_{in} = 2$.

the coded cooperative OFDM system achieves a FER of 10^{-3} at about 5.5 dB. This is only about 1.5 dB away from the performance with a perfect inter-user channel. It also results in a gain of almost 5 dB compared to the non cooperative case. In the second scenario, we consider the case when the signal-to-noise ratio in the inter-user channel is approximately 6 dB better than the signal-to-noise ratio in the user–destination channels, *i.e.*, $\gamma_{in} \approx \gamma + 6$ dB. This could represent the scenario when the two cooperating nodes are closer to each-other than to the destination. We observe that in this case the coded cooperative OFDM system essentially achieves the same performance that it would have with a perfect inter-user channel.

In Figure 19.5, we consider the case when the signal-to-noise ratio in the inter-user channel is approximately the same with the signal-to-noise ratio in the

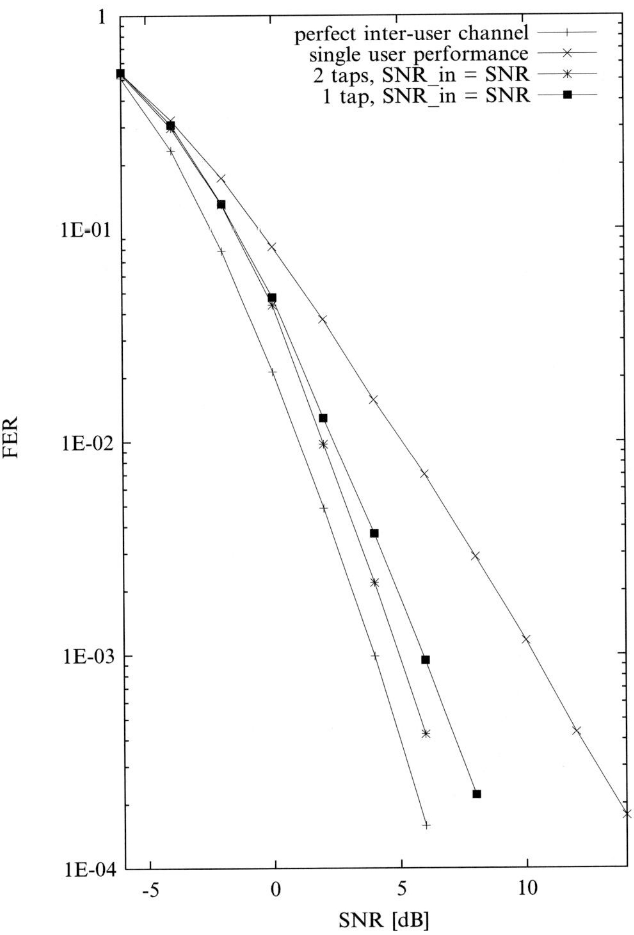

Figure 19.5. Cooperative coding example, $(\gamma_{in} \approx \gamma)$, the inter-user channel has either one or two taps, $L_{in} = 1$ or $L_{in} = 2$.

user-destination channels, *i.e.*, $\gamma_{in} \approx \gamma$. Again, this could represent the scenario when all three nodes are at a similar distance from one another. We focus on the case when the inter-user channel may or may not be frequency selective, *i.e.*, $L_{in} = 2$ or $L_{in} = 1$. In either case, the (133,171) convolutional code used in the inter-user channel achieves the best performance over that channel regardless whether it has frequency selectivity or not. As expected, we observe that the overall coded cooperative OFDM system achieves a better performance when there is frequency selectivity in the inter-user channel, as this leads to better performance of the inter-user channel code and allows cooperation to take place more often. Nonetheless, even in the case when there is no frequency selectivity in the inter-user channel, which represents the worst case, we observe that the coded cooperative system achieves a FER of 10^{-3} at about 6 dB, which is less than 2 dB away from the performance with a perfect inter-user channel. It also results in a gain of over 4 dB compared to the non-cooperative case.

5. Conclusions

We considered cooperative coding and its application to OFDM systems. We derived the Chernoff bound on the pairwise error probability for the block fading OFDM model and subsequently used it in the analysis of the frame error probability of the coded cooperative OFDM system. The performance analysis indicated that cooperative coding can provide increased diversity and coding gains over conventional OFDM systems. We also provided examples of convolutional codes based on the principle of overlays that could exploit the cooperative gains. We illustrated that the codes perform well for a variety of cooperation scenarios and inter-user channel qualities.

References

Cover, T. M. and Thomas, J. A. (1991). *Elements of Information Theory*. John Wiley and Sons, Inc.

Gamal, H. E., Hammons, A. R., and Stefanov, A. (2003). Space–time overlays for convolutionally coded systems. *IEEE Trans. Commun.*, 51:1601–1612.

Hunter, T. and Nosratinia, A. (2002). Cooperation diversity through coding. In *Proc. of IEEE ISIT*, Lausanne, Switzerland.

Laneman, J. N., Tse, D. N. C., and Wornell, G. W. (2004). Cooperative diversity in wireless networks: Efficient protocols and outage behavior. *IEEE Trans. Inform. Theory*, 50(12):3062–3080.

Lin, S. and Costello, D. J. (1983). *Error Control Coding: Fundamentals and Applications*. Prentice–Hall.

Lu, Ben, Wang, X., and Narayanan, K. R. (2002). LDPC–based space–time coded OFDM systems over correlated fading channels: Performance analysis and receiver design. *IEEE Trans. Commun.*, 50:74–88.

Nee, R. V. and Prasad, R. (2000). *OFDM for wireless multimedia communications*. Artech House Publishers.

Schlegel, C. and Costello, D. J. (1989). Bandwidth efficient coding for fading channels: Code construction and performance analysis. *IEEE J. Select. Areas Commun.*, pages 1356–1368.

Sendonaris, A., Erkip, E., and Aazhang, B. (2003). User cooperation diversity–part I: System description. *IEEE Trans. Commun.*, pages 1927–1938.

Stefanov, A. and Erkip, E. (2004). Cooperative coding for wireless networks. *IEEE Trans. Commun.*, 52(9):1470–1476.

Chapter 20

COOPERATIVE METHODS FOR SPATIAL CHANNEL CONTROL

From indoor to outdoor scenarios

Yasushi Takatori
NTT network innovation laboratories
takatori.yasushi@lab.ntt.co.jp

1. Introduction

As described in the preceding chapters, the demand for advanced wireless access systems is still growing. However, frequency resources are limited and most frequencies in the microwave band, which are suited to mobile wireless access, have already been assigned to various radio systems. Therefore, higher spectrum efficiency is indispensable. Spatial channel control (SCC) methods are attracting attention as one approach to achieve higher spectrum efficiency. There are four main expected effects of SCC, transmission diversity gain, array gain, spatial multiplexing gain and interference suppression. Since these effects cannot be expected to happen simultaneously, the SCC must be carefully designed to balance the effects to adapt to the target systems. Those effects are enhanced by cooperative control with multiple access points (APs)/base stations (BSs) rather than autonomous control, while the total system complexity becomes high. This chapter describes the cooperative SCC methods with multiple APs/BSs for both the high density hot-spots and the outdoor scenarios.

In the following, the basic SCC is briefly overviewed first. Next, multiple APs cooperation method is introduced for a high density hot-spots scenario. After that, the multiple BSs partial cooperation method to mitigate the co-channel interferences (CCI) in outdoor scenarios is described.

Frank H. P. Fitzek and Marcos D. Katz (eds.), Cooperation in Wireless Networks: Principles and Applications, 607–629.

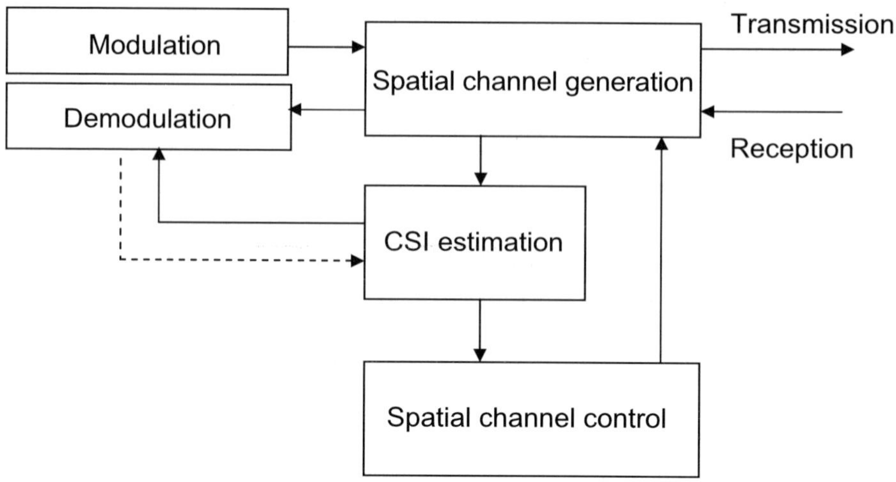

Figure 20.1. Block diagram of a spatial channel control technique.

2. Overview of SCC Methods

The basic idea of SCC was introduced almost four decades ago as a method to improve the performance of military radar. In this technique, array antennas are employed and both the main beam and the null points are adaptively controlled. By directing the main beam toward the desired signals and controlling nulls toward interfering signals, SCC has the potential to achieve very high values of signal to interference plus noise ratio (SINR).

Figure 20.1 shows a block diagram of SCC. First, channel state information (CSI) is estimated from the received signals in the CSI estimation block. In case of the feedback CSI scenario, CSI is estimated at the other node and then fed back. The fed back CSI is extracted from the demodulated signal streams and transferred to the CSI estimation block. CSI includes various channel parameters and the required CSI depends on the SCC method. For instance, the received power on each antenna branch is used as CSI in the selection diversity method, which is one of the simplest SCC methods. In more sophisticated SCC methods, the complex amplitude of the channel responses are often used as CSI. After CSI estimation, spatial channels are generated in the spatial channel generation block using analogue circuits or digital signal processing. Signal transmission/reception is then performed.

Figure 20.2 shows the general configuration of SCC in the reception mode with digital beamforming (DBF). In this configuration, RF signals are converted into IF signals or base band signals to be sampled at the analogue to digital converter (ADC) with a moderate sampling clock in each antenna branch. The

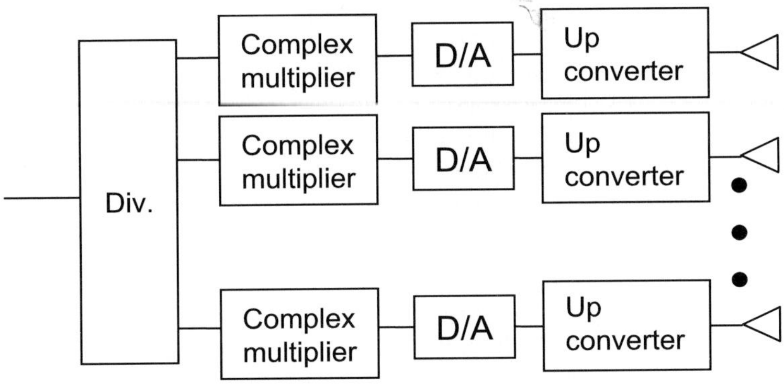

Figure 20.2. Configuration of digital reception beamforming.

sampled data is then stored in memory temporarily. Next, the data is multiplied by complex weights in a digital signal process in each branch and the weighted data is summed up. This procedure allows the reception beam to be generated by digital signal processing. CSI estimation involves the use of the temporarily stored data, so this configuration can support bursty data transmission while analogue beamforming cannot respond to the bursty signals within a received frame.

In SCC, multiple spatial channels can be controlled by multiple beams. The data rate can be increased by transmitting the different signal streams over multiple beams while it increases the complexity of the beamforming part especially with analogue beamforming approach. In DBF, common analogue devices at each antenna branch, *i.e.* amplifier, filter, down converter and ADC, are used for all spatial channels. This drastically reduces the hardware complexity and makes DBF very suitable for the multiple spatial channel generation scenario. Transmission using the configuration shown in Figure 20.3 can be expected to yield almost the same performance if accurate CSI is available at the transmitter.

In the following, we derive the optimum reception beamforming weight matrix which minimizes the mean squared error (MSE) in a multiple spatial channel scenario. For M_r antennas at the receiver with M_s spatial channels, the received signal vector, $\mathbf{r}$, can be written as follows,

$$\mathbf{r} = \mathbf{H}\mathbf{s}_d + \sum_{i=1}^{N_I} \mathbf{h}_{I,i} s_{I,i} + \mathbf{n}, \tag{20.1}$$

where $\mathbf{H}$ of an $M_r \times M_s$ matrix is the channel response matrix between M_s beams at the transmitter site and M_r antennas at the receiver site. The receiver site, $\mathbf{s}_d$ of $M_s \times 1$ vector as the desired signal vector, N_I is the number of interference signals, $\mathbf{h}_{I,i}$ as an $M_r \times 1$ vector is the channel response of the i-th

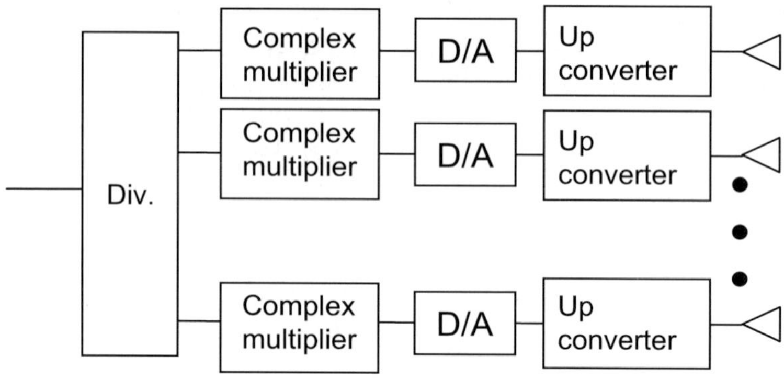

Figure 20.3. Configuration of digital transmit beamforming.

interference signal; $s_{I,i}$ is the i-th interference signal and **n** as an $M_r \times 1$ vector is the noise vector. Since the Wiener filter maximizes the SINR performance by using the minimum-mean-squared-error (MMSE) criteria with known reference signals, the optimum weight matrix at the beamforming circuit, **W** of $M_s \times M_r$, is expressed as follows,

$$\mathbf{W} = \mathbf{H}^H \mathbf{R}^{-1}, \tag{20.2}$$

where superscript ^{H} denotes conjugate transpose and **R** is the correlation matrix among reception antennas and defined as follows,

$$\mathbf{R} = \mathbf{H}\mathbf{H}^H + \sum_{i=1}^{N_I} \mathbf{h}_{I,i}\mathbf{h}_{I,i}^H + \sigma^2 \mathbf{I}_{M_R}, \tag{20.3}$$

where the noise power is assumed to be equal in each antenna branch. σ^2 is the noise power at each antenna branch and $\mathbf{I}_{M_R}$ is the $M_R \times M_R$ unit matrix. No correlations are assumed between desired and interference signals as well as the noise signals. To obtain the CSI, each data packet has a training signal part, which may be also used to establish synchronization. For the MMSE criteria, various weight updating algorithms have been developed, *e.g.* steepest decent algorithm, recursive least square (RLS) and direct matrix inversion (DMI). Gradient based algorithms such as the steepest decent algorithm have very low calculation complexity, but long training signals are required to converge the beam patterns used to generate the spatial channels. On the contrary, RLS or DMI achieves fast convergence but with high calculation complexity. If the accurate CSI is available at the receiver, DMI, which calculates the 20.2, can be used to obtain the optimum weight matrix.

By using the optimum weight matrix, SCC offers inter–symbol and co–channel interference suppression as well as desired signal enhancement. For the simplest WT scenario, a single carrier is used and SCC is employed for inter–symbol interference suppression. If the WT supports a moderate level of complexity, other techniques, *e.g.* OFDM or RAKE receiver with CDMA, can be used to mitigate the influence of long delayed waves. Thus SCC can only focus on the other function, *i.e.* co–channel interference suppression. In multiple input multiple output (MIMO) systems where array antennas are employed at both ends, this effect is used to generate multiple spatial channels. By transmitting different signal streams over multiple spatial channels, it enables the spatial division multiplexing (SDM). Therefore, it increases the transmission data rate without demanding new frequency resources.

In SDM systems, beamforming at both ends maximizes the mutual information if the CSI is available at the transmitter. The optimum transmit weight matrix is derived from the singular value decomposition (SVD) of the channel matrix $\mathbf{H}$, $\mathbf{H} = \mathbf{UDV}^H$, where $\mathbf{U}$ and $\mathbf{V}$ are unitary matrices, $\mathbf{D}$ is a diagonal matrix. To maximize the mutual information, $\mathbf{V}$ is used at a transmitter as a transmit weight matrix while $\mathbf{U}^H$ is uses as a reception weight matrix. In the practical scenario, the estimated CSI contains the error and it causes the interference among spatial channels. 20.2 is often employed to calculate the reception weight matrix to suppress the interferences. Note that 20.2 is identical to $\mathbf{U}^H$ in case of the no CSI estimation error.

In MIMO systems, it is well known that the channel capacity is increased almost linearly as the number of antenna branches increases. By introducing the cooperation techniques among multiple APs/BSs, large virtual array antennas are generated so that further system capacity improvement can be expected. In the following, two SCC methods are introduced for high density hot spots and outdoor multiple cell scenarios.

3. SCC with Multiple APs for High Density Hot Spots Scenario

As introduced in the preceding section, the MIMO technique is one of the most attractive candidates with respect to increasing spectrum efficiency. As can be seen in Figure 20.4, it allows the channel capacity to linearly increase with the number of antenna branches in an independent and identically distributed (i.i.d.) fading channel. It achieves even higher channel capacity if the CSI is known at the receiver as well as the transmitter. However, in general, MIMO is effective only in multipath-rich environments and its effectiveness is decreased in correlated fading environments such as the line of sight (LOS). Since there is only one strong spatial channel in a LOS scenario, beamforming at both a

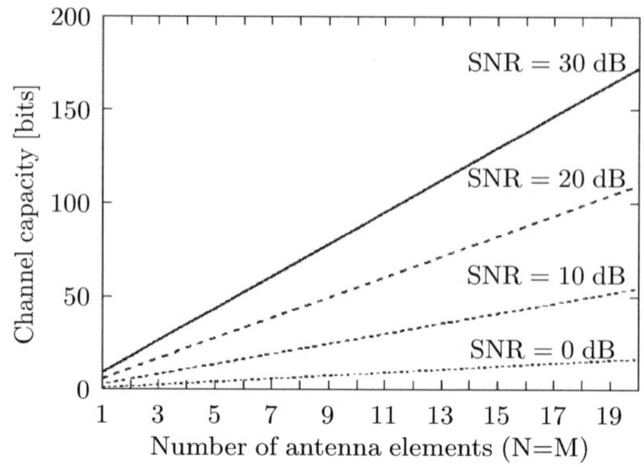

Figure 20.4. MIMO channel capacity in i.i.d. fading environments.

transmitter and a receiver is often used. This approach enhances the desired signal power and improve the SNR. However, it can not realize the spatial channel multiplexing effect and the channel capacity improvement is limited. Figure 20.5 shows the channel capacity normalized by that of SISO for various SNRs in the direct path environment. As this figure shows, the MIMO effect doesn't linearly increase in LOS environments and it is also found that only the slight channel capacity enhancement is expected with MIMO techniques in higher SNR region.

In the following, a SCC with multiple APs (MAP-SCC) is described, which adds single frequency network (SFN) technique to TDD-OFDM-MIMO and ensures the MIMO effect even in the LOS scenario. SFN with OFDM has

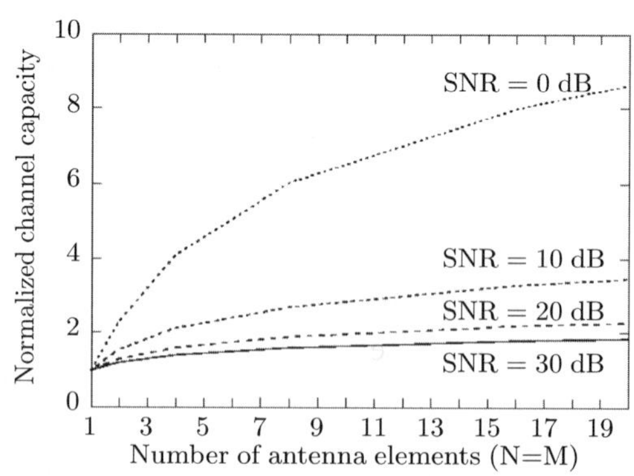

Figure 20.5. MIMO channel capacity in LOS environment.

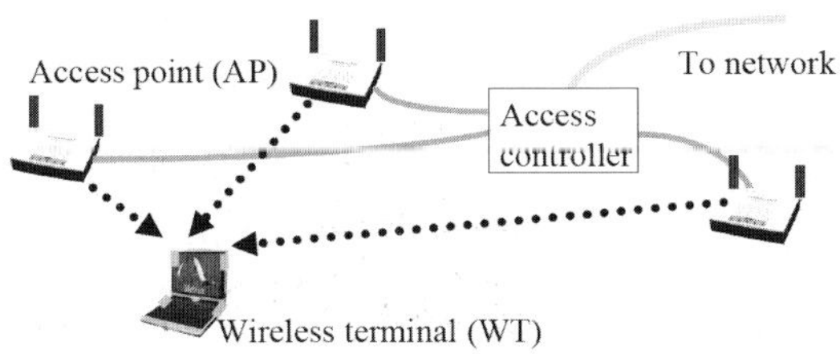

Figure 20.6. System configuration for a MAP-SCC.

been developed for broadcasting systems to improve the transmission quality in overlapping cells. MAP-SCC use SFN technique for MIMO data transmission with multiple APs so that multiple APs can cooperate with each other. This enables the increase of the spectrum efficiency with the number of APs while the autonomous data transmission scheme permits a significant drop in performance due to the strong interference imposed by the adjacent APs.

Figure 20.6 shows the configuration for a MAP-SCC in wireless LANs with OFDM. As this figure shows, multiple APs are connected to one access controller. Since OFDM can mitigate the influence of the delayed waves, this system can compensate the delay caused by the different distances between each AP-WT pair. Thus, it artificially generates a rich multipath environment and enhances the MIMO effect in both the uplink and the downlink. The data transmission scheme in each link with CSI at both the APs and WT is as follows.

In the uplink, OFDM data frames consist of a training period and data period. In the training period, known signals are transmitted from each WT antenna branch. Next, multiple OFDM data symbols are transmitted with multiple beams in each sub-carrier. All received signals at all APs are delivered to the access controller. At the access controller, all channel responses are estimated at the same time in each sub-carrier in the training period, and multiple spatial channels are optimized by the MMSE criterion. After that receiving data are separated using the multiple beam-forming network and are demodulated. Note that this approach supports TDD systems; the channel responses in the downlink can be considered to be equal to that in the uplink. By using the accurate calibration method at each AP, the generated beam patterns can be also employed in the downlink; the transmission power of each beam can be optimized by the water pouring theory. In the downlink, multiple OFDM data frames with training symbols are conveyed to the multiple beam-forming network and transmission signals for each antenna branch are generated. Those signals are then delivered to the APs and transmitted from the array antennas simultaneously. At each WT, channel responses are estimated in the training period and multiple beams are optimized by the MMSE criterion. The spatially multiplexed data are separated and demodulated.

In the following, we focus on the operation of MAP–SCC in downlink in direct path environments to clarify the spatial separation characteristic in highly correlated fading environments. Note that the operation can be easily translated to that in the uplink simply by exchanging the transmitter and the receiver.

In the downlink, the received signal in sub-carrier k at a WT can be expressed as the following equation.

$$\mathbf{r}_k = \mathbf{H}_k \mathbf{s}_k + \mathbf{n}_k, \tag{20.4}$$

where, $\mathbf{r}_k$ denotes the received signal vector, $\mathbf{H}_k$ is the channel matrix, $\mathbf{s}_k$ is the transmission signal vector and $\mathbf{n}_k$ is the noise vector. In the MAP-SCC scheme in direct path environments, the channel matrix can be rewritten as follows

$$\mathbf{H}_k = \mathbf{ADB}^H, \tag{20.5}$$

where superscript ^{H} denotes the transpose conjugation of a matrix, the l-th row vector of $\mathbf{A}$, $\mathbf{a}_l$, is the steering vector at the WT for the l-th AP, $\mathbf{D}$ is the diagonal matrix and $\mathbf{B}$ is defined by the following equation.

$$\mathbf{B} = \begin{bmatrix} \mathbf{b}_1 & 0 & \cdots & 0 \\ 0 & \mathbf{b}_2 & \ddots & \vdots \\ \vdots & \ddots & \ddots & 0 \\ 0 & \cdots & 0 & \mathbf{b}_L \end{bmatrix}, \tag{20.6}$$

where $\mathbf{b}_l$ is the steering vector at the l-th AP and L is the number of APs. Steering vectors $\mathbf{a}_l$ and $\mathbf{b}_l$ satisfy the following equation.

$$\begin{aligned} \mathbf{a}_l^H \mathbf{a}_l &= 1 \\ \mathbf{b}_l^H \mathbf{b}_l &= 1. \end{aligned} \tag{20.7}$$

The diagonal element of **D** is the magnitude of the channel response between each AP and WT. If identical array antennas are used at the APs and all antenna branches are assumed to be omni-directional, the l-th element of **D** can be expressed as follows.

$$d_{l,l} = \alpha_l \sqrt{MN}, \tag{20.8}$$

where α_l is the amplitude of the direct path between the l-th AP and the WT, M is the number of antenna branches at each AP, and N is the number of antenna branches at a WT. The correlation matrix in sub-carrier k can be written as follows.

$$\mathbf{R}_k = \mathbf{H}_k \mathbf{H}_k^H = \mathbf{A}\mathbf{D}^2\mathbf{A}^H. \tag{20.9}$$

The above equation shows that the correlation matrix does not depend on matrix **B**; it also indicates that the channel capacity can be determined for any AP array configuration.

If the CSI is completely unknown at the transmitter, the channel capacity is expressed as follows.

$$C_{unknown} = \log_2 \left[\det(\mathbf{I}_N + \frac{1}{\sigma^2 ML} \mathbf{H}\mathbf{H}^H) \right], \tag{20.10}$$

where $\mathbf{I}_N$ is the $N \times N$ identity matrix. If the propagation loss between each AP and WT is the same, α, and the thermal noise on each antenna branch is independent, the channel capacity can be simplified as follows.

$$\begin{aligned} C_{unknown} &= \log_2 \left[\det(\mathbf{I}_N + \frac{N\alpha^2}{\sigma^2 L} \mathbf{A}\mathbf{A}^H) \right] \\ &= \sum_{l=1}^{L} \log_2 \left[1 + \frac{N\alpha^2}{L\sigma^2} \lambda_l \right], \end{aligned} \tag{20.11}$$

where λ_l is the l-th eigenvalue of $\mathbf{A}\mathbf{A}^H$, σ^2 is the power of thermal noise. Thus, the channel capacity does not depend on M. This is because each AP does not know the WT location and it can not generate a beam directed toward any particular WT. If all APs are located in the same position, the rank of $\mathbf{A}\mathbf{A}^H$

becomes 1 and the channel capacity is minimized. If each AP has a different position and L eigenvalues become equal, the channel capacity is maximized. Thus, the channel capacity satisfies the following inequality.

$$\log_2\left[1+\frac{N\alpha^2}{\sigma^2}\right] \leq C_{unknown} \leq L\log_2\left[1+\frac{N\alpha^2}{L\sigma^2}\right]. \tag{20.12}$$

If perfect CSI can be assumed at the transmitter, the channel capacity can be written as follows,

$$C_{known} = \sum_{l=1}^{L}\log_2\left[1+\frac{N\alpha^2}{L\sigma^2}\lambda_l\gamma_l\right], \tag{20.13}$$

where γ_l indicates the transmission power as determined by the water pouring theory.

$$\gamma_l^{opt} = \left(\mu - \frac{L\sigma^2}{\lambda_l N\alpha^2}\right), \tag{20.14}$$

μ is a constant that satisfies the following equation to constrain the transmission power.

$$\sum_{l=1}^{L}\gamma_l = ML, \tag{20.15}$$

where $(x)_+$ implies

$$(x)_+ = \begin{cases} x & if \quad x \geq 0 \\ 0 & if \quad x < 0. \end{cases}$$

If all APs are occupy the same position, the rank of $\mathbf{AA}^H$ becomes 1 and the channel capacity is expressed as

$$C_{known,1} = \log_2\left[1+\frac{MNL\alpha^2}{\sigma^2}\right]. \tag{20.16}$$

If each AP has a different position and L eigenvalues become equal, the channel capacity can be expressed as

$$C_{known,L} = L\log_2\left[1+\frac{MN\alpha^2}{L\sigma^2}\right]. \tag{20.17}$$

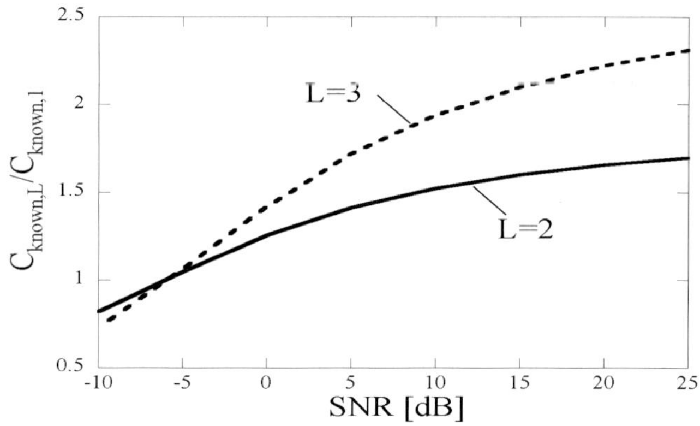

Figure 20.7. Influence of SNR on the channel capacity of MAP–MIMO in the direct path environment.

Locating all antenna branches in the same area yields the single AP MIMO system. Thus $C_{known,1}$ is equal to the channel capacity of the MIMO system with one AP. In Equation (20.16), it is found that the power of the direct path, α^2, is multiplied by the number of APs, L. Thus, the single AP MIMO improves the SNR performance by increasing the array gain. On the other hand, Equation (20.17) shows that the MAP-SCC enhances the spatial multiplexing effect by multiplying channel capacity in each spatial channel by L; the SNR of each spatial channel decreases as L increases. Therefore, the MAP-SCC improves the channel capacity in the high SNR scenario since the spatial multiplexing effect becomes larger than the array gain effect. In the following section, the channel capacity of the MAP-SCC is compared to that of the autonomous SCC and the environments that suit the MAP-SCC are clarified. Figure 20.7 shows the influence of SNR on the channel capacity of the MAP-SCC scheme in the direct path environment. In case of the practical SNR scenario, the SNR can exceed -6dB, $C_{known,L}/C_{known,1}$ becomes larger than 1.0 while it becomes less than 1.0 in the very low SNR region. This is because the channel capacity is sensitive to SNR in the low SNR environments and the array gain effect overwhelms the spatial multiplex effect. Figure 20.8 shows the relationship between channel capacity and AP location. In this calculation, the same circular array antenna is assumed at each AP and WT. The number of antenna branches at each AP and WT is four, the antenna branch space is 0.7 λ, and the number of APs are two and three. The distance between each AP and WT is assumed to be equal. In this figure, the horizontal axis indicates the angle between the first AP and the l-th AP. Each channel capacity is normalized by the channel capacity at $\theta = 0$ to clarify the spatial separation performance. The l-th AP is located at $(l-1)\theta/(L-1)$ and θ is varied. The SNR for the SISO channel

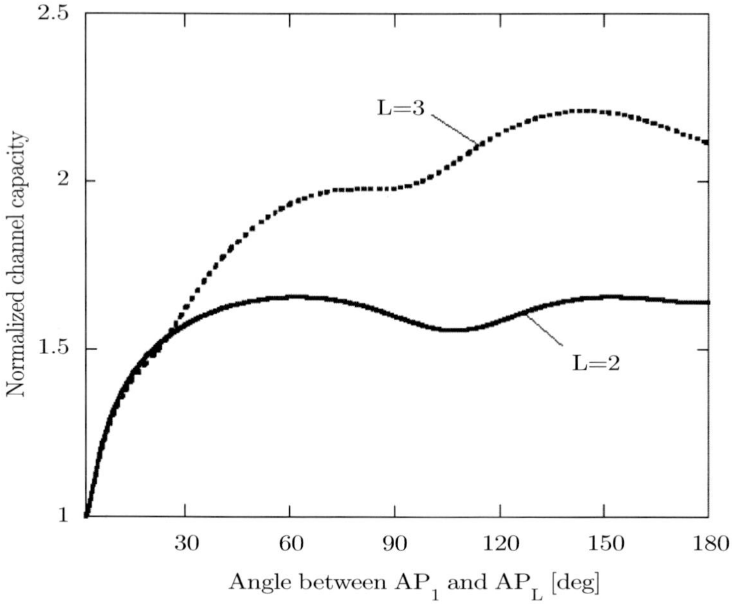

Figure 20.8. Channel capacity of MAP–MIMO in the direct path environment.

is set to 20 dB. As this figure shows, the channel capacity increases with the AP spread. In the case of three APs, the channel capacity becomes more than twice that with $\theta = 0$. These basic operations indicate that the MAP-SCC has the potential to achieve the higher channel capacity than the single AP MIMO scheme in highly correlated fading environments.

Figure 20.9 shows the cumulative probability of the average achievable total throughput of the MAP-SCC scheme comparing with that of the autonomous MIMO systems where each AP works independently. In this figure, d indicates the distance between the AP and WT. As this figure shows, higher throughput was achieved in the MAP-SCC scheme regardless of the distance between the AP and WT, while the improvement reduces as the distance increases. In case of $d = 50$ m, the highest throughput of the autonomous scheme is almost the same as that of the MAP-SCC scheme, because such high throughput can be achieved if each WT is close to the AP and the influence of the interference becomes negligible. However, the performance of the autonomous method degrades as the AP-WT distance decreases. This is because the interference increases as the distance decreases, while the MAP-SCC scheme improves the channel capacity through the cooperation of the multiple APs. These results suggest that the MAP-SCC scheme is stuitable for high density hot-spots scenario.

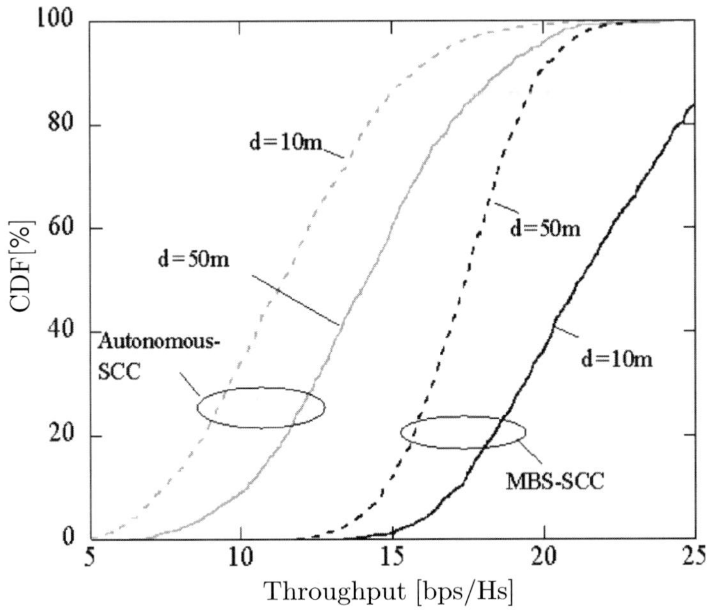

Figure 20.9. Cumulative probability of the achievable total throughput.

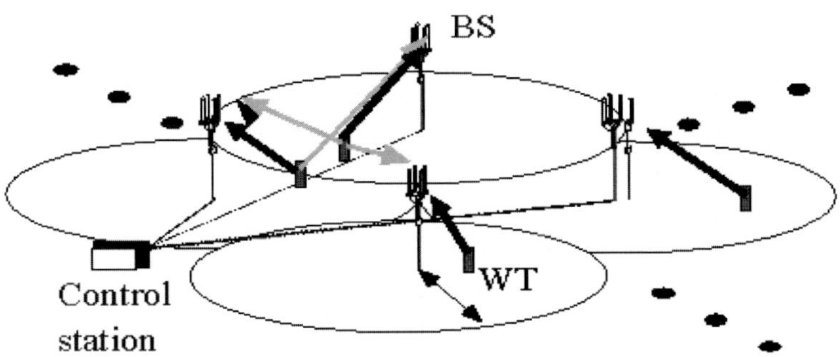

Figure 20.10. System configuration.

4. SCC with Multiple BSs for Multi-Cell Outdoor Systems

Figure 20.10 shows the multiple BSs (MBS) system model used in this section. In this model, only BSs that use the same frequency channel are considered; other BSs are ignored. BSs are connected by the wired network and the spatial channel is controlled by the control station in the cooperative SCC method

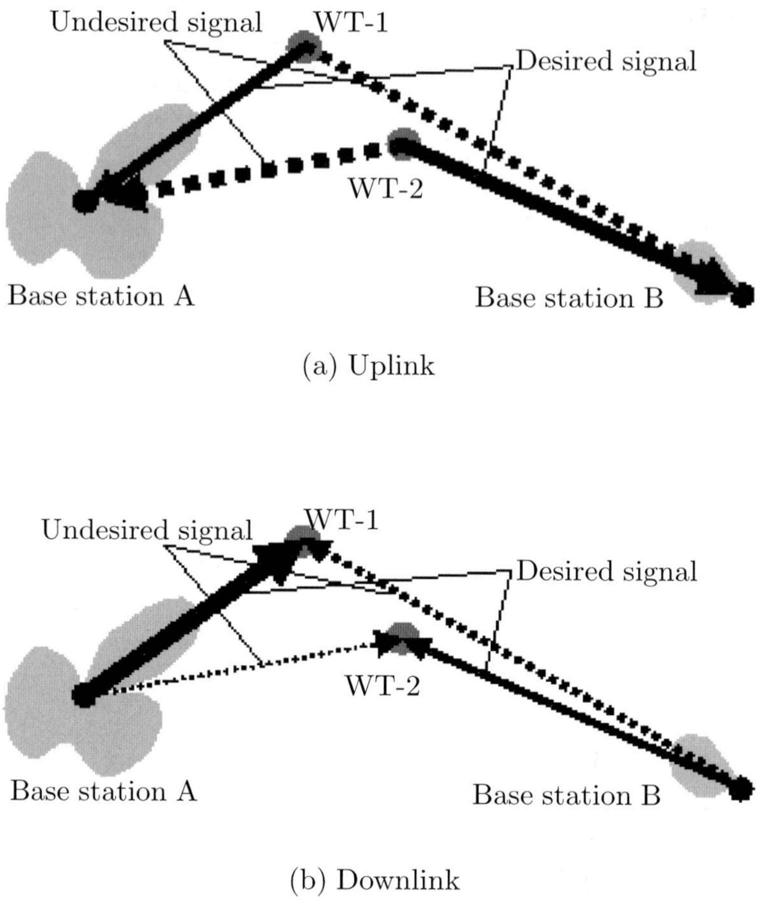

Figure 20.11. Asymmetrical environments in the uplink and the downlink.

which hereafter we denote MBS-SCC. On the other hand, each BS controls its spatial channel independently in the autonomous SCC method. In the following, the performances of the MBS-SCC method are shown comparing with those of the autonomous SCC method.

Figure 20.11 shows an example of the WT locations at which the signal to interference ratio (SIR) performance is degraded with autonomous control. The solid line indicates the path of desired signals and the dotted line indicates the path of undesired signals. As this figure shows, both WTs are located near the edge of each cell. In the uplink, the signal of WT-2 is degraded by the signal of WT-1 because the signals from WT-1 and WT-2 are coming from almost the same direction at BS-B while the signal of WT-1 is successfully received by generating a deep null toward WT-2. In the downlink, the signal for WT-1 is not received at WT-2 because BS-A generates a deep null toward WT-2 while the signal for WT-2 runs over the cell edge to WT-1 and the SIR of WT-1 is degraded. Thus, in the uplink, the SIR of WT-2 is degraded while the signal of WT-1 is degraded in the downlink. Therefore, the transmission quality becomes asymmetrical in the uplink and downlink. Moreover, the performance in the uplink is degraded compared to that in the downlink. In the following, the difference in transmission quality in the uplink and the downlink is described.

In general, the magnitude of MMSE weight vector decreases in case of the low output SINR. Thus, the magnitude of MMSE weight vector at BS-B becomes small because the transmission quality of WT-2 at BS-B is degraded. This weight vector is also used in the downlink, so the interference toward WT-1 decreases as Figure 20.11 shows. The SIR of WT-1 in the downlink becomes higher than that of WT-2 in the uplink. The SIR of WT-2 in the downlink becomes high because BS-A generates a deep null toward WT-2 although the desired signal power is decreased.

Taking the above discussion into consideration, it is necessary to consider the joint optimization of transmission power control and deep null generation to improve the transmission quality. In the following, a MBS-SCC method that optimizes both the transmission power control and beam patterns at the BSs is explained.

As Figure 20.11 shows, SIR performance of WT-1 can be improved by reducing the transmission power of BS-B. Moreover, if BS-A generates a deep null toward WT-2 at the same time, the required SIR of WT-2 can be obtained; a response not possible with the MMSE algorithm.

To realize the operation described above, the algorithm here simultaneously controls the beam pattern of multiple BSs. This can be actualized by adding an SCC block that connects the BSs via a network as depicted in Figure 20.10. Hereafter, we assume flat fading to simplify the explanation. For frequency selective fading channels, this method can be easily extended by adding terms of the channel responses for delayed signals. In the following in this chapter, we only investigate the performances of SCC in the ideal CSI estimation scenario because it is beyond this book to go into the imperfect CSI scenario. First of all, the SCC method in the downlink is introduced followed by that for the uplink.

In the downlink, the signal received at WT-m can be expressed as

$$r_m = \sum_{m_B=1}^{M_B} \mathbf{w}_{down,m_B}^H \mathbf{h}_{m_B,m} s_{down,m_B} + z_m \tag{20.18}$$

where $\mathbf{w}_{down,m_B}$ is the weight vector for beamforming at BS-m_B, $\mathbf{h}_{m_B,m}$ is the channel response between WT-m and BS-m_B, $s_{down,m}$ is the transmission signal for WT-m, z_m is the thermal noise at WT-m, and M_B is the number of BSs. BS-m is assumed to be linked to WT-m and the number of WTs, M_T, is equal to M_B. Using the autonomous SCC control method, the WTs experience unequal transmission quality which reduces the spectrum efficiency. To overcome this problem, the algorithm employs the sum of the squared error of all WTs as a quality measure. It can be expressed as

$$Q_{down} = \sum_{m=1}^{M_T} |g_{r,m} r_m - s_{down,m}|^2 + \epsilon \mathbf{w}_{down,m}^H \mathbf{w}_{down,m}, \qquad (20.19)$$

where $g_{r,m}$ is the reception gain from the antenna port to the input of the signal processor including automatic gain controller (AGC). The second term of the above equation is included to avoid excessively large weight values. The above equation also assumes that the thermal noise is generated at the low noise amplifier (LNA) so that no additional thermal noise at the AGC is seen as the AGC gain changes. Therefore, the SNR fluctuation caused by the AGC gain variation is neglected. By calculating the partial differential of the expected value of the above equation for the weight vector, $\mathbf{w}_{down,m_T}$, the following equation is obtained.

$$\mathbf{w}_{down,m_T} = g_{r,m_T} \left[\left(\sum_{m=1}^{M_T} g_{r,m}^2 \mathbf{h}_{m_T,m} \mathbf{h}_{m_T,m}^H \right) + \epsilon \mathbf{I} \right]^{-1} \mathbf{h}_{m_T,m_T} \qquad (20.20)$$

Furthermore, by calculating the partial differential with respect to the WT amplifier gain, the following equation is obtained.

$$g_{r,m_T} = \frac{Re\left[\mathbf{w}_{down,m_T}^H \mathbf{h}_{m_T,m_T}\right]}{\sum_{m=1}^{M_B} \mathbf{w}_{down,m}^H \mathbf{h}_{m,m_T} \mathbf{h}_{m,m_T}^H \mathbf{w}_{down,m} + \sigma_{m_T}^2} \qquad (20.21)$$

Where $\sigma_{m_T}^2$ is the thermal noise power at WT-m_T. These two equations minimize the total amount of error. Therefore, in the MBS-SCC method, weight vectors are determined by using the two equations above alternately in iterative fashion.

In the uplink, the signal received at BS-m can be expressed as

$$\mathbf{x}_m = \sum_{m_T=1}^{M_T} g_{t,m_T} \mathbf{h}_{m,m_T} s_{up,m_T} + \mathbf{n}_m \tag{20.22}$$

where g_{t,m_T} is TPC gain at WT-m_T, $\mathbf{n}_m$ is the thermal noise vector at BS-m. The following value is used as the uplink quality measure.

$$Q_{up} = \sum_{m=1}^{M_B} \left| \mathbf{w}_{up,m}^H \mathbf{x}_m - s_{up,m} \right|^2 + \epsilon g_{t,m}^2 \tag{20.23}$$

This measure includes pseudo noise to avoid excessively large transmission powers at each WT. By calculating the partial differentiation of the expected value of the above equation for weight vector, $\mathbf{w}_{up,m_B}$, the following equation is obtained.

$$\mathbf{w}_{up,m_B} = g_{t,m_B} \left[\left(\sum_{m=1}^{M_T} g_{t,m}^2 \mathbf{h}_{m_B,m} \mathbf{h}_{m_B,m}^H \right) + \sigma_{m_B}^2 \mathbf{I} \right]^{-1} \mathbf{h}_{m_B,m_B} \tag{20.24}$$

Similar to the downlink case, by calculating the partial differential with respect to the WT amplifier gain, the following equation is obtained.

$$g_{t,m_B} = \frac{Re\left[\mathbf{w}_{up,m_B}^H \mathbf{h}_{m_B,m_B}\right]}{\sum_{m=1}^{M_B} \mathbf{w}_{up,m}^H \mathbf{h}_{m,m_B} \mathbf{h}_{m,m_B}^H \mathbf{w}_{up,m} + \epsilon} \tag{20.25}$$

The above two equations are similar to those in the downlink. The difference is ϵ and σ_m^2. Since ϵ is set to avoid excessive transmission power, it should be as small as possible while not exceeding the transmission power limit at each station. If ϵ is set equal to σ_m^2, the weight vectors at BSs and the gain values at WTs in the downlink become equal to those in the uplink.

Figure 20.12 shows the convergence performance of the MBS in the downlink. Similar performance can be obtained in the uplink. In this simulation, the analysis model depicted in fig. 20.11 was used. The number of BSs was two and that of WT stations was two. In the initial state, all receiver gains at the WTs were assumed to be equal. When the iteration number is one, it indicates the performance of the autonomous SCC system.

As this figure shows, this algorithm converged within 100 iterations and achieved the SIR of 10 dB. It also shows that SIR of BS-B becomes less than 0 dB if there is only one iteration. It indicates that the MBS improves the

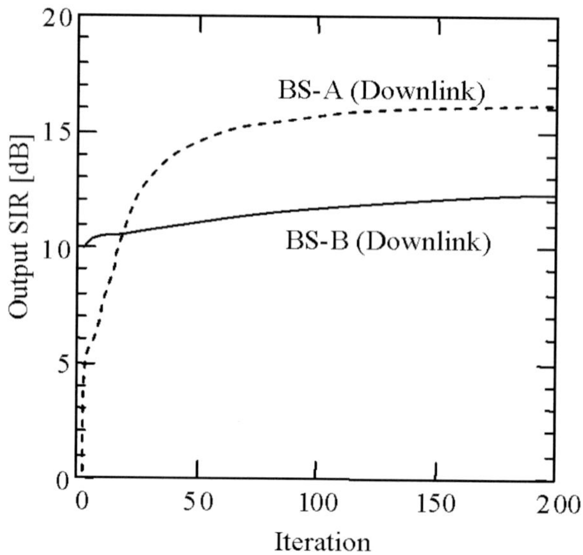

Figure 20.12. Convergence performance of the MBS system.

transmission quality by employing cooperative SCC control for multiple BSs. Figure 20.13 shows the operation of the MBS systems with examples. There are 36 BSs, 36 WTs and each BS communicates with just one WT. Omni-directional antennas are used for WTs and SCC is applied to all BSs. A circular array is used for SCC, the center frequency is 2.0 GHz, the number of branches is four, and the antenna branch spacing is 2.0 λ. The cell size of each BS was assumed to be 250 m. It was assumed that the delay profile was expressed as an exponential distribution. The delay spread was 0.1 T_s. The number of incoming waves was 100 for each link between BS-m_b and WT-m_t. In the autonomous SCC, TPC of WTs is employed to ensure that all received levels at the communicating BSs are identical. The autonomous MMSE algorithm is applied to SCC in the uplink at each BS and the weight vector obtained is also used for the downlink.

Figure 20.13 (a) shows, for the uplink, the interference WT locations and powers for the BS located at (1500 m, 500 m). The center of each circle indicates the location of an interference WT and the radius indicates the received interference power normalized by the desired signal power. In this case, the SIR was 0.2 dB. This figure shows that the number of interference signals was more

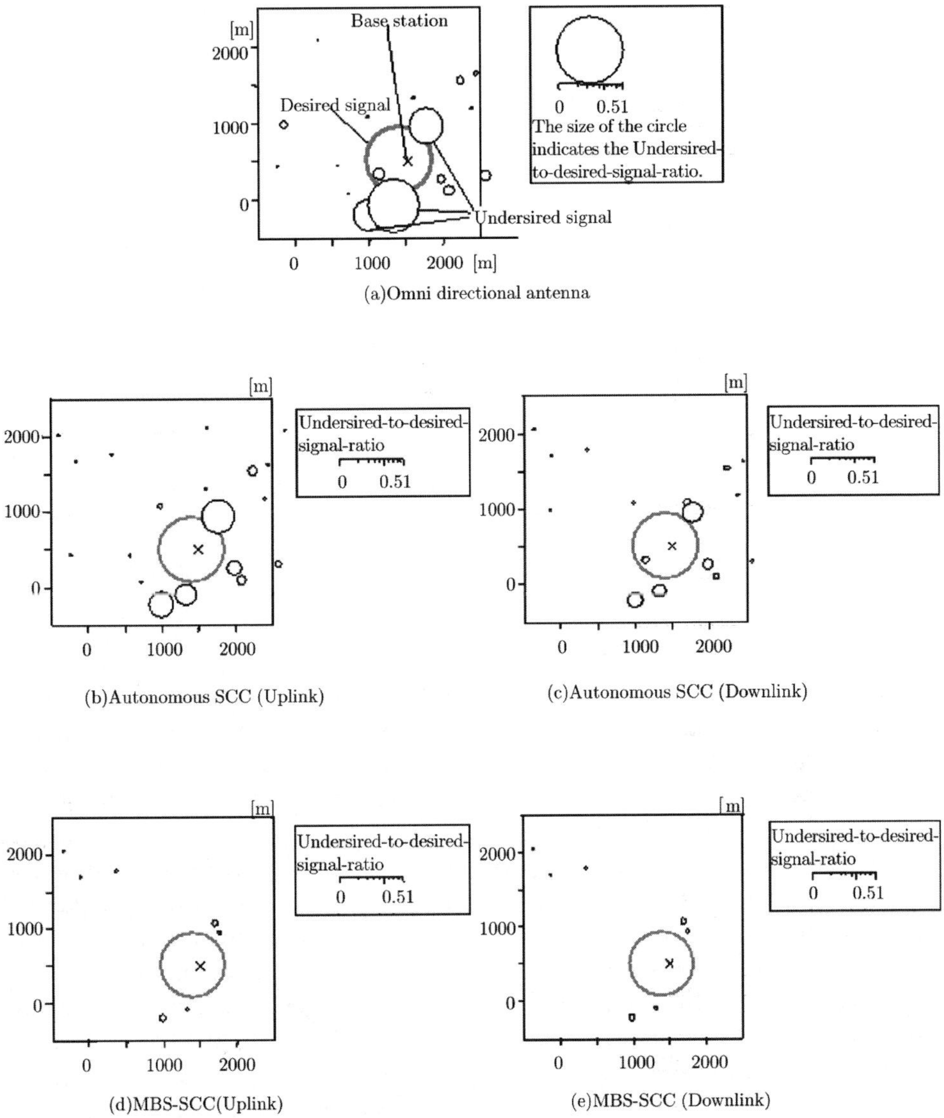

Figure 20.13. An example of locations and powers of interference signals.

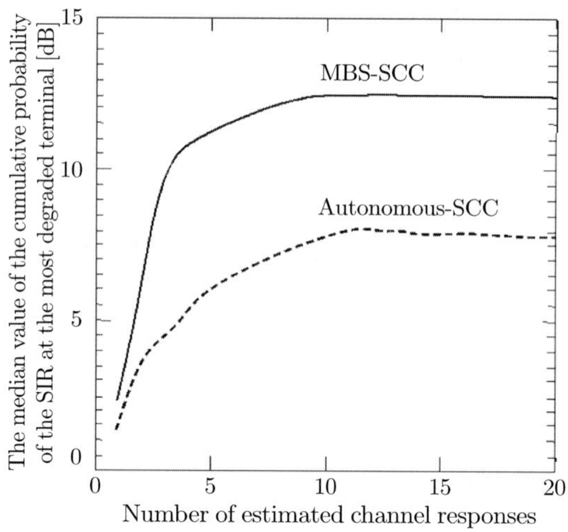

Figure 20.14. Influence of the distance between BSs.

than that of the antenna branches and so the performance was degraded significantly. Figure 20.13 (b) shows the results in the uplink with an autonomous SCC. It suppressed strong interference from (1314 m, -79 m). However, some interference signals remained and the transmission quality was still degraded. On the other hand, the transmission quality was improved in the downlink and the SIR of 7.8 dB was achieved as shown in Figure 20.13 (c). This is because the interference power was enhanced by the TPC of interference WTs in the uplink. For instance, when the desired WT came close to the BS, the transmission power of the desired WT became less than that of the other WTs and transmission quality was degraded. Finally, Figure 20.13 and Figure 20.13(e) show results gained with MBS-SCC. All interference signals were strongly suppressed and the SIR of 14 dB was achieved.

In deriving the MBS-SCC method, all channel responses were assumed to be used to determine the weight vectors. However, it is difficult to estimate the CSI for all WTs at each BS in actual systems. Figure 20.14 shows the interference suppression performance versus the number of estimated channels. The y axis indicates the median value of SIR of the worst downlink. The channel responses are estimated in order of the magnitude of the channel response. The remaining channel responses, those not estimated, are assumed to be zero vectors. As this figure shows, the performance of MBS-SCC converges with three channel responses and performance superior to that of autonomous systems is achieved. It is also found that as the number of estimates increases, the performance is slightly degraded. This is because the convergence performance is degraded with a fixed number of iterations as the number of interference sources increases. This result indicates that the MBS-SCC method outperforms the autonomous SCC if just three channel responses are estimated at each BS.

5. Summary

This chapter introduced cooperative SCC methods for both outdoor scenario and high density hot-spots scenario. Both methods use a control station/access controller to inter–connect several APs/BSs. In cellular systems, since BS spacing is larger than that of APs, the partial cooperation approach is introduced. By optimizing the transmission power and the beam patterns at the BSs, the transmission quality is improved while the system complexity also increases. On the other hand, the full cooperation approach is used for the high density hotspots scenario. It also has the potential to improve the achievable data rate with the cooperation of multiple APs. Those results indicate that the approach of multiple node cooperation has significant potential to improve the system capacity further in future wireless systems.

References

Andersen, J. Bach (2000). Array gain and capacity for known random channels with multiple element arrays at both ends. *IEEE J. Select. Areas Commun.*, 18:2172–2178.

Andersson, S., Millnert, M., Viberg, M., and Wahlberg, B. (1991). An adaptive array for mobile communication systems. *IEEE Trans. Veh. Tech.*, 40: 230–236.

Cho, K. and Hori, T. (2000). Smart antenna systems actualizeing sdma for future wireless communication. In *International Symposium on Antenna and Propagation*, volume 4, pages 1477–1480.

Cioffi, J. M. and Kailath, T. (1984). Fast, rls, transversal filters for adaptive filtering. *IEEE Trans. on ASSP*, 32(2):304–337.

Gerlach, D. and Paulraj, A. (1994). Spectrum reuse using transmitting antenna arrays with feedback. In *Acoustics, Speech, and Signal Processing, ICASSP*, volume 4, pages 97–100.

German, G., Spencer, Q., Swindlehurst, L., and Valenzuela, R. (2001). Wireless indoor channel modeling: statistical agreement of ray tracing simulations and channel sounding measurements. In *International Conference on Acoustics Speech and Signal Processing (ICASSP '01)*, volume 4, pages 2501–2504.

Ichitsubo, S., Furuno, T., Nagato, T., Taga, T., and Kawasaki, R. (1996). 2ghz-band propagation loss prediction in urban area; antenna heights ranging from ground to building roof. In *Technical Report of IEICE, AP96–15*, volume 1, pages 73–78.

Lo, T. K. Y. (1999). Maximum ratio transmission. *IEEE Trans. on Commun.*, 47:1458–1461.

Medbo, J., Harryson, F., Asplund, H., and Berger, J. E. (1996). Measurements and analysis of a mimo macrocell outdoor-indoor scenario at 1947mhz. In *IEEE VTC 2004 Spring*, volume 1, pages 73–78.

Miyashita, K., Nishimura, T., Ohgane, T., Ogawa, Y., Takatori, Y., and Cho, K. (2002). High data-rate transmission with eigenbeam-space division multiplexing (e-sdm) in a mimo channel. In *IEEE VTC*, volume 3, pages 302–1306.

Monzingo, R. A. and Miller, T. W. (1980). *Introduction to Adaptive Arrays*. John Wiley & Sons, NY.

Nishimori, K., Cho, K., Takatori, Y., and Hori, T. (2001). Automatic calibration method using transmitting signals of an adaptive array for tdd systems. *IEEE Trans. Veh. Tech.*, 50(6):1636–1640.

Norklit, O., Eggers, P., Zetterberg, P., and Andersen, J. B. (1996). The angular aspect of wideband modelling and measurements. In *IEEE International Symposium on Spread Spectrum Techniques and Applications*, volume 1, pages 73–78.

Paulraj, A. and et al. (2003). *Introduction to space–time wireless communications*. Cambridge university press.

PRASAD, R. (2004). *OFDM for Wireless Communications Systems*. Artech House.

Rashid-Farrokhi, F., Liu, K. J. R., and Tassiulas, L. (1998a). Transmit beamforming and power control for cellular wireless systems. *IEEE Trans. J. Select. Areas Commun.*, 16:1437–1450.

Rashid-Farrokhi, F., Tassiulas, L., and Liu, K. J. R. (1998b). Joint optimal power control and beamforming in wireless networks using antenna arrays. *IEEE Trans. on Comm.*, 46(10):1313–1324.

Rebhan, R. and et al. (1993). On the outage probability in single frequency networks for digital broadcasting. *IEEE Trans. on Broadcasting*, 39: 395–401.

Reed, I. S., Mallett, J. D., and Brennan, L. E. (1974). Rapid convergence rate in adaptive arrays. *IEEE Trans. Aerosp. Electron. Syste.*, AES-10(6):853–863.

Saleh, A. A. M. and Valenzuela, R. A. (1987). A statistical model for indoor multipath propagation. *IEEE J. Select. Areas Commun.*, 5:128–137.

Shiu, D. S. and et al. (2000). Fading correlation and its effect on the capacity of multi–element antenna systems. *IEEE Trans. on Commun.*, 48:502–513.

Takatori, Y., Cho, K., Nishimori, K., and Hori, T. (2000). Adaptive array employing eigenvector beam of maximum eigenvalue and fractionally-spaced tdl with real tap. *IEICE Trans. Commun.*, E83–B(8):1678–1687.

Wiener, Nobert (1949). *Extrapolation, Interpolation, and Smoothing of stationary time series, with engineering applications*. Cambridge Technology Press of the Massachusetts Institute of Technology.

GLOSSARY

1G First generation of mobile communication systems

2G Second generation of mobile communication systems

3GPP 3rd Generation Partnership Project

3G Third generation of mobile communication systems

4G Fourth generation of mobile communication systems

AF amplify-and-forward

AIC Additional Information Container

ALLC Always cooperate (strategy used by unconditional cooperators)

ALLD Always defect (strategy used by unconditional defectors)

AP Access point

ATFT Anti-tit-for-tat strategy

AWGN Additive White Gaussian Noise

Amplify-and-forward A relay protocol where the relay retransmits a scaled version of its received analog signal.

B3G beyond IMT-2000

BER Bit Error Rate

BLAST Bell-Labs Layered Space-Time (BLAST) coding is a technique applied to MIMO transceivers, which multiplexes different data streams onto different transmit antennas. This requires signal processing at the receiving side to extract the various streams, which is facilitated by the spatial signatures of the MIMO channel; this technique allows the realisation of high transmission rates.

BLUE Best Linear Unbiased Estimator

BS Base Station

Broadcast Channel A communication system where a single transmitter sends potentially different information to multiple users.

CA Certification Authority

CDMA Code Division Multiple Access. A technology where each user modulates its transmission symbols with a spreading code before transmission. The spreading codes of different users are often orthogonal.

CMOS Complementary Metal-Oxide Semiconductor technology, both N-type and P-type transistors are used to realize logic functions. Today, CMOS technology is the dominant semiconductor technology for microprocessors, memories and application specific integrated circuits.

COHC Cooperative Header Compression. Header compression approach that is based on the cooperative behavior of multiple IP streams and is characterized by high robustness and bandwidth savings.

CORMAS Common-Pool Resources and Multi Agent Systems

CPU Central Processing Unit

CRL Certificate Revocation List

CRTP Compressed Real Time Protocol. Header compression for the RTP/UDP/IP suit presented in RFC 2508.

CRT Certificate Revocation Tree

CSI Channel State Information

CSI Channel state information

CSMA Carrier Sensing Multiple Access

Capacity The capacity of a channel is the maximum achievable error-free communication rate for a communication system with given input distribution, transmission power, noise power, and bandwidth.

Coded network A network where nodes are capable of performing network coding.

Cooperative Destinations The source data is broadcasted to several destination terminals, while the terminals use the communication links among them to cooperate and thus enhance each other's reception of the broadcasted data.

Cooperative Sources A communicaton scenario in which two or more nodes in a network are cooperating to deliver the data to the destination.

Cooperative System A system where distributed terminals cooperate in a coherent manner so as to improve the system performance, is referred to as cooperative system.

Cross-Layer Optimisation The process of jointly optimising various OSI layers of a wireless communication system, is referred to as cross-layer optimisation.

Cut-set theorems A class of theorems that give upper bounds, and sometimes achievable rates, for flow in networks. This flow may correspond to information flow, or flow of a fluid through a network of pipes, or any other physical quantity.

DBS Distributed Base Stations

DF decode-and-forward

DMO Direct mode operation

DPM Dynamic Power Management

DPM Dynamic Power Management, a class of methods or policies placing various system components in less power consuming modes, or compleat power down modes.

DVS Dynamic Voltage Scaling

DVS Dynamic Voltage Scaling, scheduling methodology utilizing knowledge of task-set specification, task timing and workload, to adjust the performance of a programmable processor. This is done by dynamical changes to the processor supply voltage and clock frequency.

Decode-and-forward A relay protocol where the relay first decodes its received signal, then transmits a signal that is derived from the decoded information.

Degraded relay channel A channel where the signal received at the destination is a corrupted version of the signal received at the relay.

Distributed System A system where mobile terminals or nodes are spatially separated, however, communicate, is referred to as a distributed system.

Diversity Order The slope of the bit error rate (BER) vs. the signal-to-noise ratio (SNR) at high SNR. Diversity order is a measure of the number of independent data-paths from the source to the destination in a communication system. Common forms of communication diversity include temporal, spatial, spectral and multiuser diversity.

Diversity Order The slope of the bit error rate (BER) vs. the signal-to-noise ratio (SNR) at high SNR. Diversity order is a measure of the number of independent datapaths from the source to the destination in a communication system.

EC European Commission

EDF Earliest Deadline First

EDGE Enhanced Data Rates for GSM Evolution

ESS Evolutionary Stable Strategy

Ergodic Channel If the wireless channel varies over the transmitted codeword so that all its moments are the same from codeword to codeword, then the channel is referred to as an ergodic channel; fast fading channels approximately realise an ergodic channel.

Estimate-and-forward A relay protocol where the relay transmits an estimate of its received analog signal without decoding but potentially compressed.

FDC Framed Delta Coding

FDMA The multiple access scheme where terminals use prior assigned, generally non-overlapping, frequency bands, is referred to as frequency division multiple access

FEC Forward Error Correction

FER Frame Error Rate

FLSSR Fixed Order Scheduling

Full-duplex A communication node is said to operate in full-duplex mode when it can simultaneously transmit and receive in the same frequency band.

GPRS General Packet Radio System

GPS Global Positioning System

GSM Global System for Mobile Communications

GSSR Global Scheduling

HSCSD High Speed Circuit Switched Data

Half-duplex A communication node is said to operate in half-duplex mode when it can simultaneously transmit and receive only if the transmitted and received signals are orthogonal in time.

Hyperarc A generalized arc that starts at a single node and ends at one or more nodes.

Hypergraph A collection of nodes and hyperarcs.

ILP Integer Linear Programming

IMT-2000

IPD Iterated Prisoner's Dilemma game

IPv4 Internet Protocol version 4

IPv6 Internet Protocol version 6

ISI Intersymbol Interference

ITU International Telecommunication Union

IrDA Infrared Data Association

JAR Java Archive

LCS35 MIT LCS' 35th aniversary *Time Capsule of Innovations*

LDPC (Low-density Parity-check) Code A class of block codes characterized by sparse parity-check matrices.

LVQ A Lattice Vector Quantizer (LVQ) is a quantizer which utilizes the design of a highly structured lattice.

Layered coding In layered coding, the first description contains a coarse information and the following descriptions are only containing refinement information.

MAC The medium access control layer of a wireless communication system is referred to as MAC. It controls the way the mobile nodes access the wireless medium and hence compete for wireless resources.

MC-CDMA Multicarrier CDMA

MDC–CC MDC with Conditional Compression (MDC–CC) is a method where the MDC encoding overhead can be removed by any node in the network, provided that this node has already a feedback information from the destination that the overhead is unnecessary.

MDC Multiple Description Coding (MDC) is a source coding technique where the source is encoded into two or more descriptions. The descriptions are self–sufficient in the sense that each description can provide a distorted version of the source information, while the distortion is decreased as more descriptions are utilized at the decoder.

MDLVQ Multiple Description Lattice Vector Quantizer (MDLVQ) is a lattice quantizer realization of the MDC paradigm.

MIMO A transceiver or channel with multiple inputs and outputs is referred to as multiple-input multiple-output (MIMO) transceiver or channel.

MQAM M-ary Quadrature Amplitude Modulation

Meshed Cooperation Each node involved in the communication can be a source of information and a destination. Such can be the case of video-conferencing or gaming.

Mobile VCE The Mobile Virtual Centre of Excellence (Mobile VCE, MVCE) is a consortium where researchers from about seven academic institutions are subsidised by about 20 international telecommunications companies to perform research into future communication paradigms. For more details, consult www.mobilevce.com.

Multi-hop Communication Communication from a source to a destination through a chain of intermediate nodes, where each intermediate node communicates only with the node immediately preceding it and immediately following it in the chain.

Multiple-access Channel A shared channel where multiple sources transmit to a single destination in the same frequency band.

NFC Near Field Communication

Network-coding A network information processing paradigm where intermediate nodes combine and encode received information before forwarding it, as contrasted to traditional communication where nodes are restricted to the passive role of forwarding information without processing.

Non-Ergodic Channel If the wireless channel varies over the transmitted codeword so that its moments are not necessarily the same from codeword to codeword, then the channel is referred to as an non-ergodic channel; slow fading channels approximately realise a non-ergodic channel.

OCSP Online Certificate Status Protocol

OFDMA Orthogonal Frequency Division Multiple Access

OSI Open Systems Interconnection

PAN Personal Area Network

PD Prisoner's Dilemma game

PGP Pretty Good Privacy

PHY The physical layer of a wireless communication system is referred to as PHY. It is responsible for the encoding/decoding of data, execution of power control signals, etc.

PMR Private Mobile Radio

POW Proof-of-work

Power Control Adjusting transmission power based on channel condition to achieve a given target. For example, transmission at high power when the channel is good, and low power when the channel is bad, achieves the target of increasing overall rate for a given average power constraint.

QPSK Quadrature Phase Shift Key

REPC Reputation-based cooperation

RFC Request For Commands

RFID Radio Frequency Identification

RLQ Radio Link Quality

ROF Radio Over Fiber

ROHC Robust Header Compression. Header compression scheme designed especially for the operation in wireless cellular networks with highly error-prone links and long round-trip times. Described in RFC 3095.

RTP Real Time Protocol

RePast Recursive Porous Agent Simulation Toolkit

Relay Channel A three-terminal communication channel where communication from a source to a destination is aided by a third terminal called the relay.

Relay Protocol The information-processing strategy employed at the relay for retransmitting received information in a relay channel. For example, the relay may choose between retransmitting the received analog signal without decoding, or it may decode the received signal and re-encode it before retransmission.

Routed network A network where nodes are not capable of performing network coding and can only forward or replicate packets. This is the case in the conventional approach to networking. Cf. CODED NETWORK.

S/MIME Secure/Multipurpose Internet Mail Extensions

SISO A transceiver or channel with a single input and output is referred to as single-input single-output (SISO) transceiver or channel.

SNR Signal to Noise Ratio

SREP Sporas reputation-based cooperation strategy

STBC Space-Time Block Codes (STBC) is a signal processing technique applied to MIMO transceivers, which facilitates the exploitation of the diversity provided by the wireless channel; this effect is visible as an increase in the steepness of the outage and error rate probability curves.

STTC Space Time Trellis Codes (STTC) is a signal processing technique applied to MIMO transceivers, which facilitates the exploitation of the diversity provided by the wireless channel and in addition provides a coding gain; this effect is visible as an increase in the steepness of the outage and error rate probability curves, as well as a shift towards lower signal-to-noise ratios.

Space-time Code A channel coding technique in multi-antenna systems, where the antennas as well as time are treated as dimensions of the channel. Space-time codes can be used to yield higher diversity or to achieve higher communication rates than in possible with single-antenna communication.

Spreading Code A noise-like sequence with which a transmission symbol is modulated (multiplied) with the goal of spreading the information in the symbol among different dimensions in the time-frequency plane.

TCP Transport Control Protocol

TDMA The multiple access scheme where terminals use prior assigned, generally non-overlapping, time slots, is referred to as time division multiple access

TFT Tit-for-tat strategy

UDP User Datagram Protocol

UE User equipment

UMTS Universal Mobile Telecommunication System

UWB Ultra wideband, greater than 25% relative bandwidth

VAA A communication paradigm where spatially adjacent mobile terminals or nodes cooperate and thereby form a virtual transceiver entity, is referred to as virtual antenna array (VAA).

WLAN Wireless Local Area Network

WMAN Wireless Metropolitan Area Network

WSN Wireless sensor networks

WT Wireless Terminal

WWRF Wireless World Research Forum

XKMS XML Key Management Specification

XML eXtensible Markup Language

mITF Mobile IT Forum

Index